GEOMETRY, TOPOLOGY AND PHYSICS

Graduate Student Series in Physics

Other books in the series

GRADUATE STUDENT SERIES IN PHYSICS

Series Editor:
Professor Douglas F Brewer, MA, DPhil
Emeritus Professor of Experimental Physics, University of Sussex

GEOMETRY, TOPOLOGY AND PHYSICS

SECOND EDITION

MIKIO NAKAHARA

Department of Physics
Kinki University, Osaka, Japan

Taylor & Francis
Taylor & Francis Group
New York   London

Published in 2003 by
Taylor & Francis Group
6000 Broken Sound Parkway NW, Suite 300
Boca Raton, FL 33487-2742

Published in Great Britain by
Taylor & Francis Group
2 Park Square
Milton Park, Abingdon
Oxon OX14 4RN

© 2003 by Taylor & Francis Group, LLC

No claim to original U.S. Government works
Printed in the United States of America on acid-free paper
10 9 8 7 6 5

International Standard Book Number-10: 0-7503-0606-8 (Softcover)
International Standard Book Number-13: 978-0-7503-0606-5 (Softcover)

Library of Congress Cataloging-in-Publication Data

Catalog record is available from the Library of Congress

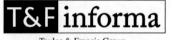

Taylor & Francis Group
is the Academic Division of T&F Informa plc.

Visit the Taylor & Francis Web site at
http://www.taylorandfrancis.com

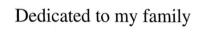

CONTENTS

PREFACE TO THE FIRST EDITION

This book is a considerable expansion of lectures I gave at the School of Mathematical and Physical Sciences, University of Sussex during the winter term of 1986. The audience included postgraduate students and faculty members working in particle physics, condensed matter physics and general relativity. The lectures were quite informal and I have tried to keep this informality as much as possible in this book. The proof of a theorem is given only when it is instructive and not very technical; otherwise examples will make the theorem plausible. Many figures will help the reader to obtain concrete images of the subjects.

In spite of the extensive use of the concepts of topology, differential geometry and other areas of contemporary mathematics in recent developments in theoretical physics, it is rather difficult to find a self-contained book that is easily accessible to postgraduate students in physics. This book is meant to fill the gap between highly advanced books or research papers and the many excellent introductory books. As a reader, I imagined a first-year postgraduate student in theoretical physics who has some familiarity with quantum field theory and relativity. In this book, the reader will find many examples from physics, in which topological and geometrical notions are very important. These examples are eclectic collections from particle physics, general relativity and condensed matter physics. Readers should feel free to skip examples that are out of their direct concern. However, I believe these examples should be the *theoretical minima* to students in theoretical physics. Mathematicians who are interested in the application of their discipline to theoretical physics will also find this book interesting.

The book is largely divided into four parts. Chapters 1 and 2 deal with the preliminary concepts in physics and mathematics, respectively. In chapter 1, a brief summary of the physics treated in this book is given. The subjects covered are path integrals, gauge theories (including monopoles and instantons), defects in condensed matter physics, general relativity, Berry's phase in quantum mechanics and strings. Most of the subjects are subsequently explained in detail from the topological and geometrical viewpoints. Chapter 2 supplements the undergraduate mathematics that the average physicist has studied. If readers are quite familiar with sets, maps and general topology, they may skip this chapter and proceed to the next.

Chapters 3 to 8 are devoted to the basics of algebraic topology and differential geometry. In chapters 3 and 4, the idea of the classification of spaces with homology groups and homotopy groups is introduced. In chapter 5, we define a manifold, which is one of the central concepts in modern theoretical physics. Differential forms defined there play very important roles throughout this book. Differential forms allow us to define the dual of the homology group called the de Rham cohomology group in chapter 6. Chapter 7 deals with a manifold endowed with a metric. With the metric, we may define such geometrical concepts as connection, covariant derivative, curvature, torsion and many more. In chapter 8, a complex manifold is defined as a special manifold on which there exists a natural complex structure.

Chapters 9 to 12 are devoted to the unification of topology and geometry. In chapter 9, we define a fibre bundle and show that this is a natural setting for many physical phenomena. The connection defined in chapter 7 is naturally generalized to that on fibre bundles in chapter 10. Characteristic classes defined in chapter 11 enable us to classify fibre bundles using various cohomology classes. Characteristic classes are particularly important in the Atiyah–Singer index theorem in chapter 12. We do not prove this, one of the most important theorems in contemporary mathematics, but simply write down the special forms of the theorem so that we may use them in practical applications in physics.

Chapters 13 and 14 are devoted to the most fascinating applications of topology and geometry in contemporary physics. In chapter 13, we apply the theory of fibre bundles, characteristic classes and index theorems to the study of anomalies in gauge theories. In chapter 14, Polyakov's bosonic string theory is analysed from the geometrical point of view. We give an explicit computation of the one-loop amplitude.

I would like to express deep gratitude to my teachers, friends and students. Special thanks are due to Tetsuya Asai, David Bailin, Hiroshi Khono, David Lancaster, Shigeki Matsutani, Hiroyuki Nagashima, David Pattarini, Felix E A Pirani, Kenichi Tamano, David Waxman and David Wong. The basic concepts in chapter 5 owe very much to the lectures by F E A Pirani at King's College, University of London. The evaluation of the string Laplacian in chapter 14 using the Eisenstein series and the Kronecker limiting formula was suggested by T Asai. I would like to thank Euan Squires, David Bailin and Hiroshi Khono for useful comments and suggestions. David Bailin suggested that I should write this book. He also advised Professor Douglas F Brewer to include this book in his series. I would like to thank the Science and Engineering Research Council of the United Kingdom, which made my stay at Sussex possible. It is a pity that I have no secretary to thank for the beautiful typing. Word processing has been carried out by myself on two NEC PC9801 computers. Jim A Revill of Adam Hilger helped me in many ways while preparing the manuscript. His indulgence over my failure to meet deadlines is also acknowledged. Many musicians have filled my office with beautiful music during the preparation of the manuscript: I am grateful to J S Bach, Ryuichi Sakamoto, Ravi Shankar and Erik Satie.

Mikio Nakahara
Shizuoka, February 1989

PREFACE TO THE SECOND EDITION

The first edition of the present book was published in 1990. There has been incredible progress in geometry and topology applied to theoretical physics and *vice versa* since then. The boundaries among these disciplines are quite obscure these days.

I found it impossible to take all the progress into these fields in this second edition and decided to make the revision minimal. Besides correcting typos, errors and miscellaneous small additions, I added the proof of the index theorem in terms of supersymmetric quantum mechanics. There are also some rearrangements of material in many places. I have learned from publications and internet homepages that the first edition of the book has been read by students and researchers from a wide variety of fields, not only in physics and mathematics but also in philosophy, chemistry, geodesy and oceanology among others. This is one of the reasons why I did not specialize this book to the forefront of recent developments. I hope to publish a separate book on the recent fascinating application of quantum field theory to low dimensional topology and number theory, possibly with a mathematician or two, in the near future.

The first edition of the book has been used in many classes all over the world. Some of the lecturers gave me valuable comments and suggestions. I would like to thank, in particular, Jouko Mikkelsson for constructive suggestions. Kazuhiro Sakuma, my fellow mathematician, joined me to translate the first edition of the book into Japanese. He gave me valuable comments and suggestions from a mathematician's viewpoint. I also want to thank him for frequent discussions and for clarifying many of my questions. I had a chance to lecture on the material of the book while I was a visiting professor at Helsinki University of Technology during fall 2001 through spring 2002. I would like to thank Martti Salomaa for warm hospitality at his materials physics laboratory. Sami Virtanen was the course assisitant whom I would like to thank for his excellent work. I would also like to thank Juha Vartiainen, Antti Laiho, Teemu Ojanen, Teemu Keski-Kuha, Markku Stenberg, Juha Heiskala, Tuomas Hytönen, Antti Niskanen and Ville Bergholm for helping me to find typos and errors in the manuscript and also for giving me valuable comments and questions.

Jim Revill and Tom Spicer of IOP Publishing have always been generous in forgiving me for slow revision. I would like to thank them for their generosity and patience. I also want to thank Simon Laurenson for arranging the copyediting, typesetting and proofreading and Sarah Plenty for arranging the printing, binding and scheduling. The first edition of the book was prepared using an old NEC computer whose operating system no longer exists. I hesitated to revise the book mainly because I was not so courageous as to type a more-than-500-page book again. Thanks to the progress of information technology, IOP Publishing scanned all the pages of the book and supplied me with the files, from which I could extract the text files with the help of optical character recognition (OCR) software. I would like to thank the technical staff of IOP Publishing for this painstaking work. The OCR is not good enough to produce the L^{A}T$_E$X codes for equations. Mariko Kamada edited the equations from the first version of the book. I would like to thank Yukitoshi Fujimura of Peason Education Japan for frequent T$_E$X-nical assistance. He edited the Japanese translation of the first edition of the present book and produced an excellent L^{A}T$_E$X file, from which I borrowed many L^{A}T$_E$X definitions, styles, diagrams and so on. Without the Japanese edition, the publication of this second edition would have been much more difficult.

Last but not least, I would thank my family to whom this book is dedicated. I had to spend an awful lot of weekends on this revision. I wish to thank my wife, Fumiko, and daughters, Lisa and Yuri, for their patience. I hope my little daughters will someday pick up this book in a library or a bookshop and understand what their dad was doing at weekends and late after midnight.

Mikio Nakahara
Nara, December 2002

HOW TO READ THIS BOOK

As the author of this book, I strongly wish that this book is read in order. However, I admit that the book is thick and the materials contained in it are diverse. Here I want to suggest some possibilities when this book is used for a course in mathematics or mathematical physics.

(1) A one year course on mathematical physics: chapters 1 through 10. Chapters 11 and 12 are optional.

(2) A one-year course on geometry and topology for mathematics students: chapters 2 through 12. Chapter 2 may be omitted if students are familiar with elementary topology. Topics from physics may be omitted without causing serious problems.

(3) A single-semester course on geometry and topology: chapters 2 through 7. Chapter 2 may be omitted if the students are familiar with elementary topology. Chapter 8 is optional.

(4) A single-semester course on differential geometry for general relativity: chapters 2, 5 and 7.

(5) A single-semester course on advanced mathematical physics: sections 1.1–1.7 and sections 12.9 and 12.10, assuming that students are familiar with Riemannian geometry and fibre bundles. This makes a self-contained course on the path integral and its application to index theorem.

Some repetition of the material or a summary of the subjects introduced in the previous part are made to make these choices possible.

NOTATION AND CONVENTIONS

The symbols $\mathbb{N}, \mathbb{Z}, \mathbb{Q}, \mathbb{R}$ and $\mathbb{C}$ denote the sets of natural numbers, integers, rational numbers, real numbers and complex numbers, respectively. The set of quaternions is defined by

$$\mathbb{H} = \{a + b\boldsymbol{i} + c\boldsymbol{j} + d\boldsymbol{k} \mid a, b, c, d \in \mathbb{R}\}$$

where $(1, \boldsymbol{i}, \boldsymbol{j}, \boldsymbol{k})$ is a basis such that $\boldsymbol{i} \cdot \boldsymbol{j} = -\boldsymbol{j} \cdot \boldsymbol{i} = \boldsymbol{k}$, $\boldsymbol{j} \cdot \boldsymbol{k} = -\boldsymbol{k} \cdot \boldsymbol{j} = \boldsymbol{i}$, $\boldsymbol{k} \cdot \boldsymbol{i} = -\boldsymbol{i} \cdot \boldsymbol{k} = \boldsymbol{j}$, $\boldsymbol{i}^2 = \boldsymbol{j}^2 = \boldsymbol{k}^2 = -1$. Note that $\boldsymbol{i}$, $\boldsymbol{j}$ and $\boldsymbol{k}$ have the 2×2 matrix representations $\boldsymbol{i} = \mathrm{i}\sigma_3$, $\boldsymbol{j} = \mathrm{i}\sigma_2$, $\boldsymbol{k} = \mathrm{i}\sigma_1$ where σ_i are the Pauli spin matrices

$$\sigma_1 = \begin{pmatrix} 0 & 1 \\ 1 & 0 \end{pmatrix} \qquad \sigma_2 = \begin{pmatrix} 0 & -\mathrm{i} \\ \mathrm{i} & 0 \end{pmatrix} \qquad \sigma_3 = \begin{pmatrix} 1 & 0 \\ 0 & -1 \end{pmatrix}.$$

The imaginary part of a complex number z is denoted by $\mathrm{Im}\, z$ while the real part is $\mathrm{Re}\, z$.

We put c (speed of light) $= \hbar$ (Planck's constant$/2\pi$) $= k_{\mathrm{B}}$ (Boltzmann's constant) $= 1$, unless otherwise stated explicitly. We employ the Einstein summation convention: if the same index appears twice, once as a superscript and once as a subscript, then the index is summed over all possible values. For example, if μ runs from 1 to m, one has

$$A^\mu B_\mu = \sum_{\mu=1}^{m} A^\mu B_\mu.$$

The Euclid metric is $g_{\mu\nu} = \delta_{\mu\nu} = \mathrm{diag}(+1, \ldots, +1)$ while the Minkowski metric is $g_{\mu\nu} = \eta_{\mu\nu} = \mathrm{diag}(-1, +1, \ldots, +1)$.

The symbol $\square$ denotes 'the end of a proof'.

The first term in the middle expression vanishes since q is a solution to the Euler–Lagrange equation. Accordingly, we obtain

$$\sum_k \delta q_k(t_i) p_k(t_i) = \sum_k \delta q_k(t_f) p_k(t_f) \qquad (1.14)$$

where use has been made of the definition $p_k = \partial L/\partial \dot{q}_k$. Since t_i and t_f are arbitrary, this equation shows that the quantity $\sum_k \delta q_k(t) p_k(t)$ is, in fact, independent of t and hence conserved.

Example 1.2. Let us consider a particle m moving under a force produced by a spherically symmetric potential $V(r)$, where r, θ, ϕ are three-dimensional polar coordinates. The Lagrangian is given by

$$L = \tfrac{1}{2} m[\dot{r}^2 + r^2(\dot{\theta}^2 + \sin^2 \theta \dot{\phi}^2)] - V(r).$$

Note that $q_k = \phi$ is cyclic, which leads to the conservation law

$$\delta\phi \frac{\partial L}{\partial \dot{\phi}} \propto mr^2 \sin^2 \theta \dot{\phi} = \text{constant}.$$

This is nothing but the angular momentum around the z axis. Similar arguments can be employed to show that the angular momenta around the x and y axes are also conserved.

A few remarks are in order:

- Let $Q(q)$ be an arbitrary function of q. Then the Lagrangians L and $L + \mathrm{d}Q/\mathrm{d}t$ yield the same Euler–Lagrange equation. In fact,

$$\frac{\partial}{\partial q_k}\left(L + \frac{\mathrm{d}Q}{\mathrm{d}t}\right) - \frac{\mathrm{d}}{\mathrm{d}t}\left[\frac{\partial}{\partial \dot{q}_k}\left(L + \frac{\mathrm{d}Q}{\mathrm{d}t}\right)\right]$$
$$= \frac{\partial L}{\partial q_k} + \frac{\partial}{\partial q_k}\frac{\mathrm{d}Q}{\mathrm{d}t} - \frac{\mathrm{d}}{\mathrm{d}t}\frac{\partial L}{\partial \dot{q}_k} - \frac{\mathrm{d}}{\mathrm{d}t}\frac{\partial}{\partial \dot{q}_k}\left(\sum_j \frac{\partial Q}{\partial q_j}\dot{q}_j\right)$$
$$= \frac{\partial}{\partial q_k}\frac{\mathrm{d}Q}{\mathrm{d}t} - \frac{\mathrm{d}}{\mathrm{d}t}\frac{\partial Q}{\partial q_k} = 0.$$

- An interesting observation is that Newtonian mechanics is realized as an extremum of the action but the action itself is defined for *any* trajectory. This fact plays an important role in path integral formation of quantum theory.

1.1.3 Hamiltonian formalism

The Lagrangian formalism yields a second-order ordinary differential equation (ODE). In contrast, the Hamiltonian formalism gives equations of motion which are first order in the time derivative and, hence, we may introduce flows in the phase space defined later. What is more important, however, is that we can make the symplectic structure manifest in the Hamiltonian formalism, which will be shown in example 5.12 later.

Suppose a Lagrangian L is given. Then the corresponding **Hamiltonian** is introduced via Legendre transformation of variables as

$$H(q, p) \equiv \sum_k p_k \dot{q}_k - L(q, \dot{q}), \tag{1.15}$$

where $\dot{q}$ is eliminated in the left-hand side (LHS) in favour of p by making use of the definition of the momentum $p_k = \partial L(q, \dot{q})/\partial \dot{q}_k$. For this transformation to be defined, the Jacobian must satisfy

$$\det\left(\frac{\partial p_i}{\partial \dot{q}_j}\right) = \det\left(\frac{\partial^2 L}{\partial \dot{q}_i \dot{q}_j}\right) \neq 0.$$

The space with coordinates (q_k, p_k) is called the **phase space**.

Let us consider an infinitesimal change in the Hamiltonian induced by δq_k and δp_k,

$$\delta H = \sum_k \left[\delta p_k \dot{q}_k + p_k \delta \dot{q}_k - \frac{\partial L}{\partial q_k}\delta q_k - \frac{\partial L}{\partial \dot{q}_k}\delta \dot{q}_k\right]$$

$$= \sum_k \left[\delta p_k \dot{q}_k - \frac{\partial L}{\partial q_k}\delta q_k\right].$$

It follows from this relation that

$$\frac{\partial H}{\partial p_k} = \dot{q}_k, \qquad \frac{\partial H}{\partial q_k} = -\frac{\partial L}{\partial q_k} \tag{1.16}$$

which are nothing more than the replacements of independent variables. **Hamilton's equations of motion** are obtained from these equations if the Euler–Lagrange equation is employed to replace the LHS of the second equation,

$$\dot{q}_k = \frac{\partial H}{\partial p_k} \qquad \dot{p}_k = -\frac{\partial H}{\partial q_k}. \tag{1.17}$$

Example 1.3. Let us consider a one-dimensional harmonic oscillator with the Lagrangian $L = \frac{1}{2}m\dot{q}^2 - \frac{1}{2}m\omega^2 q^2$, where $\omega^2 = k/m$. The momentum conjugate to q is $p = \partial L/\partial \dot{q} = m\dot{q}$, which can be solved for $\dot{q}$ to yield $\dot{q} = p/m$. The Hamiltonian is

$$H(q, p) = p\dot{q} - L(q, \dot{q}) = \frac{p^2}{2m} + \frac{1}{2}m\omega^2 q^2. \tag{1.18}$$

Hamilton's equations of motion are:

$$\frac{dp}{dt} = -m\omega^2 q \qquad \frac{dq}{dt} = \frac{p}{m}. \tag{1.19}$$

Let us take two functions $A(q, p)$ and $B(q, p)$ defined on the phase space of a Hamiltonian H. Then the **Poisson bracket** $[A, B]$ is defined by [6]

$$[A, B] = \sum_k \left(\frac{\partial A}{\partial q_k} \frac{\partial B}{\partial p_k} - \frac{\partial A}{\partial p_k} \frac{\partial B}{\partial q_k} \right). \tag{1.20}$$

Exercise 1.1. Show that the Poisson bracket is a **Lie bracket**, namely it satisfies

$$[A, c_1 B_1 + c_2 B_2] = c_1[A, B_1] + c_2[A, B_2] \qquad \text{linearity} \tag{1.21a}$$

$$[A, B] = -[B, A] \qquad \text{skew-symmetry} \tag{1.21b}$$

$$[[A, B], C] + [[C, A], B] + [[B, C], A] = 0 \qquad \textbf{Jacobi identity.} \tag{1.21c}$$

The fundamental Poisson brackets are

$$[p_i, p_j] = [q_i, q_j] = 0 \qquad [q_i, p_j] = \delta_{ij}. \tag{1.22}$$

It is important to notice that the time development of a physical quantity $A(q, p)$ is expressed in terms of the Poisson bracket as

$$\begin{aligned}
\frac{dA}{dt} &= \sum_k \left(\frac{dA}{dq_k} \frac{dq_k}{dt} + \frac{dA}{dp_k} \frac{dp_k}{dt} \right) \\
&= \sum_k \left(\frac{dA}{dq_k} \frac{\partial H}{\partial p_k} - \frac{dA}{dp_k} \frac{\partial H}{\partial q_k} \right) \\
&= [A, H].
\end{aligned} \tag{1.23}$$

If it happens that $[A, H] = 0$, the quantity A is conserved, namely $dA/dt = 0$. The Hamilton equations of motion themselves are written as

$$\frac{dp_k}{dt} = [p_k, H] \qquad \frac{dq_k}{dt} = [q_k, H]. \tag{1.24}$$

Theorem 1.1. (**Noether's theorem**) Let $H(q_k, p_k)$ be a Hamiltonian which is invariant under an infinitesimal coordinate transformation $q_k \rightarrow q_k' = q_k + \varepsilon f_k(q)$. Then

$$Q = \sum_k p_k f_k(q) \tag{1.25}$$

is conserved.

Proof. One has $H(q_k, p_k) = H(q_k', p_k')$ by definition. It follows from $q_k' = q_k + \varepsilon f_k(q)$ that the Jacobian associated with the coordinate change is

$$\Lambda_{ij} = \frac{\partial q_i'}{\partial q_j} \simeq \delta_{ij} + \varepsilon \frac{\partial f_i(q)}{\partial q_j}$$

[6] When the commutation relation $[A, B]$ of operators is introduced later, the Poisson bracket will be denoted as $[A, B]_{PB}$ to avoid confusion.

up to $\mathcal{O}(\varepsilon)$. The momentum transforms under this coordinate change as

$$p_i \rightarrow \sum_j p_j \Lambda_{ji}^{-1} \simeq p_i - \varepsilon \sum_j p_j \frac{\partial f_j}{\partial q_i}.$$

Then, it follows that

$$
\begin{aligned}
0 &= H(q_k', p_k') - H(q_k, p_k) \\
&= \frac{\partial H}{\partial q_k} \varepsilon f(q) - \frac{\partial H}{\partial p_j} \varepsilon p_i \frac{\partial f_i}{\partial q_j} \\
&= \varepsilon \left[\frac{\partial H}{\partial q_k} f_k(q) - \frac{\partial H}{\partial p_j} p_i \frac{\partial f_i}{\partial q_j} \right] \\
&= \varepsilon [H, Q] = \varepsilon \frac{dQ}{dt},
\end{aligned}
$$

which shows that Q is conserved.   □

This theorem shows that to find a conserved quantity is equivalent to finding a transformation which leaves the Hamiltonian invariant.

A conserved quantity Q is the 'generator' of the transformation under discussion. In fact,

$$[q_i, Q] = \sum_k \left[\frac{\partial q_i}{\partial q_k} \frac{\partial Q}{\partial p_k} - \frac{\partial q_i}{\partial p_k} \frac{\partial Q}{\partial q_k} \right] = \sum_k \delta_{ik} f_k(q) = f_i(q)$$

which shows that $\delta q_i = \varepsilon f_i(q) = \varepsilon [q_i, Q]$.

A few examples are in order. Let $H = p^2/2m$ be the Hamiltonian of a free particle. Since H does not depend on q, it is invariant under $q \mapsto q + \varepsilon \cdot 1$, $p \mapsto p$. Therefore, $Q = p \cdot 1 = p$ is conserved. The conserved quantity Q is identified with the linear momentum.

Example 1.4. Let us consider a paticle m moving in a two-dimensional plane with the axial symmetric potential $V(r)$. The Lagrangian is

$$L(r, \theta) = \tfrac{1}{2} m(\dot{r}^2 + r^2 \dot{\phi}^2) - V(r).$$

The canonical conjugate momenta are:

$$p_r = m\dot{r} \qquad p_\theta = mr^2 \dot{\theta}.$$

The Hamiltonian is

$$H = p_r \dot{r} + p_\theta \dot{\theta} - L = \frac{p_r^2}{2m} + \frac{p_\theta^2}{2mr^2} + V(r).$$

This Hamiltonian is clearly independent of θ and, hence, invariant under the transformation

$$\theta \mapsto \theta + \varepsilon \cdot 1, \qquad p_\theta \mapsto p_\theta.$$

The corresponding conserved quantity is

$$Q = p_\theta \cdot 1 = mr^2 \dot{\theta}$$

that is the angular momentum.

1.2 Canonical quantization

It was known by the end of the 19th century that classical physics, namely Newtonian mechanics and classical electromagnetism, contains serious inconsistencies. Later at the beginning of the 20th century, these were resolved by the discoveries of special and general relativities and quantum mechanics. So far, there is no single experiment which contradicts quantum theory. It is surprising, however, that there is no *proof* for quantum theory. What one can say is that quantum theory is not in contradiction to Nature. Accordingly, we do not prove quantum mechanics here but will be satisfied with outlining some 'rules' on which quantum theory is based.

1.2.1 Hilbert space, bras and kets

Let us consider a complex Hilbert space[7]

$$\mathcal{H} = \{|\phi\rangle, |\psi\rangle, \ldots\}. \tag{1.26}$$

An element of $\mathcal{H}$ is called a **ket** or a **ket vector**.

A linear function $\alpha : \mathcal{H} \to \mathbb{C}$ is defined by

$$\alpha(c_1|\psi_1\rangle + c_2|\psi_2\rangle) = c_1\alpha(|\psi_1\rangle) + c_2\alpha(|\psi_2\rangle) \qquad \forall c_i \in \mathbb{C}, |\psi_i\rangle \in \mathcal{H}.$$

We employ a special notation introduced by Dirac and write the linear function as $\langle\alpha|$ and the action as $\langle\alpha|\psi\rangle \in \mathbb{C}$. The set of linear functions is itself a vector space called the **dual vector space** of $\mathcal{H}$, denoted $\mathcal{H}^*$. An element of $\mathcal{H}$ is called a **bra** or a **bra vector**.

Let $\{|e_1\rangle, |e_2\rangle, \ldots\}$ be a basis of $\mathcal{H}$.[8] Any vector $|\psi\rangle \in \mathcal{H}$ is then expanded as $|\psi\rangle = \sum_k \psi_k |e_k\rangle$, where $\psi_k \in \mathbb{C}$ is called the kth component of $|\psi\rangle$. Now let us introduce a basis $\{\langle\varepsilon_1|, \langle\varepsilon_2|, \ldots\}$ in $\mathcal{H}^*$. We require that this basis be a **dual basis** of $\{|e_k\rangle\}$, that is

$$\langle\varepsilon_i|e_j\rangle = \delta_{ij}. \tag{1.27}$$

---

[7] In quantum mechanics, a Hilbert space often means the space of square integrable functions $L^2(M)$ on a space (manifold) M. In the following, however, we need to deal with such functions as $\delta(x)$ and e^{ikx} with infinite norm. An extended Hilbert space which contains such functions is called the rigged Hilbert space. The treatment of Hilbert spaces here is not mathematically rigorous but it will not cause any inconvenience.

[8] We assume $\mathcal{H}$ is separable and there are, at most, a countably infinite number of vectors in the basis. Note that we cannot impose an orthonormal condition since we have not defined the norm of a vector.

Then an arbitrary linear function $\langle\alpha|$ is expanded as $\langle\alpha| = \sum_k \alpha_k \langle\varepsilon_k|$, where $\alpha_k \in \mathbb{C}$ is the kth component of $\langle\alpha|$. The action of $\langle\alpha| \in \mathcal{H}^*$ on $|\psi\rangle \in \mathcal{H}$ is now expressed in terms of their components as

$$\langle\alpha|\psi\rangle = \sum_{ij} \alpha_i \psi_j \langle\varepsilon_i|e_j\rangle = \sum_{ij} \alpha_i \psi_j \delta_{ij} = \sum_i \alpha_i \psi_i. \tag{1.28}$$

One may consider $|\psi\rangle$ as a column vector and $\langle\alpha|$ as a row vector so that $\langle\alpha|\psi\rangle$ is regarded as just a matrix multiplication of a row vector and a column vector, yielding a scalar.

It is possible to introduce a one-to-one correspondence between elements in $\mathcal{H}$ and $\mathcal{H}^*$. Let us fix a basis $\{|e_k\rangle\}$ of $\mathcal{H}$ and $\{\langle\varepsilon_k|\}$ of $\mathcal{H}^*$. Then corresponding to $|\psi\rangle = \sum_k \psi_k |e_k\rangle$, there exists an element $\langle\psi| = \sum_k \psi_k^* \langle\varepsilon_k| \in \mathcal{H}^*$. The reason for the complex conjugation of ψ_k becomes clear shortly. Then it is possible to introduce an **inner product** between two elements of $\mathcal{H}$. Let $|\phi\rangle, |\psi\rangle \in \mathcal{H}$. Their inner product is defined by

$$(|\phi\rangle, |\psi\rangle) \equiv \langle\phi|\psi\rangle = \sum_k \phi_k^* \psi_k. \tag{1.29}$$

We customarily use the same letter to denote corresponding bras and kets. The **norm** of a vector $|\psi\rangle$ is naturally defined by the inner product. Let $\||\psi\rangle\| = \sqrt{\langle\psi|\psi\rangle}$. It is easy to show that this definition satisfies all the axioms of the norm. Note that the norm is real and non-negative thanks to the complex conjugation in the components of the bra vector.

By using the inner product between two ket vectors, it becomes possible to construct an orthonormal basis $\{|e_k\rangle\}$ such that $(|e_i\rangle, |e_j\rangle) = \langle e_i|e_j\rangle = \delta_{ij}$. Suppose $|\psi\rangle = \sum_k \psi_k|e_k\rangle$. By multiplying $\langle e_k|$ from the left, one obtains $\langle e_k|\psi\rangle = \psi_k$. Then $|\psi\rangle$ is expressed as $|\psi\rangle = \sum_k \langle e_k|\psi\rangle|e_k\rangle = \sum_k |e_k\rangle\langle e_k|\psi\rangle$. Since this is true for any $|\psi\rangle$, we have obtained the **completeness relation**

$$\sum_k |e_k\rangle\langle e_k| = I, \tag{1.30}$$

I being the identity operator in $\mathcal{H}$ (the unit matrix when $\mathcal{H}$ is finite dimensional).

1.2.2 Axioms of canonical quantization

Given an isolated classical dynamical system such as a harmonic oscillator, we can construct a corresponding quantum system following a set of axioms.

A1. There exists a Hilbert space $\mathcal{H}$ for a quantum system and the state of the system is required to be described by a vector $|\psi\rangle \in \mathcal{H}$. In this sense, $|\psi\rangle$ is also called the **state** or a **state vector**. Moreover, two states $|\psi\rangle$ and $c|\psi\rangle$ ($c \in \mathbb{C}, c \neq 0$) describe the same state. The state can also be described as a **ray representation** of $\mathcal{H}$.

A2. A physical quantity A in classical mechanics is replaced by a Hermitian operator $\hat{A}$ acting on $\mathcal{H}$.[9] The operator $\hat{A}$ is often called an **observable**. The result obtained when A is measured is one of the eigenvalues of $\hat{A}$. (The Hermiticity of $\hat{A}$ has been assumed to guarantee real eigenvalues.)

A3. The Poisson bracket in classical mechanics is replaced by the **commutator**

$$[\hat{A}, \hat{B}] \equiv \hat{A}\hat{B} - \hat{B}\hat{A} \tag{1.31}$$

multiplied by $-i/\hbar$. The unit in which $\hbar = 1$ will be employed hereafter unless otherwise stated explicitly. The fundamental commutation relations are (cf (1.22))

$$[\hat{q}_i, \hat{q}_j] = [\hat{p}_i, \hat{p}_j] = 0 \qquad [\hat{q}_i, \hat{p}_j] = i\delta_{ij}. \tag{1.32}$$

Under this replacement, Hamilton's equations of motion become

$$\frac{d\hat{q}_i}{dt} = \frac{1}{i}[\hat{q}_i, H] \qquad \frac{d\hat{p}_i}{dt} = \frac{1}{i}[\hat{p}_i, H]. \tag{1.33}$$

When a classical quantity A is independent of t explicitly, A satisifies the same equation as Hamilton's equation. By analogy, for $\hat{A}$ which does not depend on t explicitly, one has **Heisenberg's equation of motion**:

$$\frac{d\hat{A}}{dt} = \frac{1}{i}[\hat{A}, \hat{H}]. \tag{1.34}$$

A4. Let $|\psi\rangle \in \mathcal{H}$ be an arbitrary state. Suppose one prepares many systems, each of which is in this state. Then, observation of A in these systems at time t yields random results in general. Then the expectation value of the results is given by

$$\langle A \rangle_t = \frac{\langle \psi | \hat{A}(t) | \psi \rangle}{\langle \psi | \psi \rangle}. \tag{1.35}$$

A5. For any physical state $|\psi\rangle \in \mathcal{H}$, there exists an operator for which $|\psi\rangle$ is one of the eigenstates.[10]

These five axioms are adopted as the rules of the game. A few comments are in order. Let us examine axiom A4 more carefully. Let us assume that $|\psi\rangle$ is normalized as $\||\psi\rangle\|^2 = \langle \psi | \psi \rangle = 1$ for simplicity. Suppose $\hat{A}(t)$ has the set of discrete eigenvalues $\{a_n\}$ with the corresponding normalized eigenvectors $\{|n\rangle\}$:[11]

$$\hat{A}(t)|n\rangle = a_n|n\rangle \qquad \langle n|n\rangle = 1.$$

[9] An operator on $\mathcal{H}$ is denoted by $\hat{\ }$. This symbol will be dropped later unless this may cause confusion.

[10] This axiom is often ignored in the literature. The *raison d'etre* of this axiom will be clarified later.

[11] Since $\hat{A}(t)$ is Hermitian, it is always possible to choose $\{|n\rangle\}$ to be orthonormal.

Then the expectation value of $\hat{A}(t)$ with respect to an arbitrary state

$$|\psi\rangle = \sum_n \psi_n |n\rangle \qquad \psi_n = \langle n|\psi\rangle$$

is

$$\langle\psi|\hat{A}(t)|\psi\rangle = \sum_{m,n} \psi_m^* \psi_n \langle m|\hat{A}(t)|n\rangle = \sum_n a_n |\psi_n|^2.$$

From the fact that the result of the measurement of A in state $|n\rangle$ is always a_n, it follows that the probability of the outcome of the measurement being a_n, that is the probability of $|\psi\rangle$ being in $|n\rangle$, is

$$|\psi_n|^2 = |\langle n|\psi\rangle|^2.$$

The number $\langle n|\psi\rangle$ represents the 'weight' of the state $|n\rangle$ in the state $|\psi\rangle$ and is called the **probability amplitude**.

If $\hat{A}$ has a continuous spectrum a, the state $|\psi\rangle$ is expanded as

$$|\psi\rangle = \int da\, \psi(a)|a\rangle.$$

The completeness relation now takes the form

$$\int da\, |a\rangle\langle a| = I. \tag{1.36}$$

Then, from the identity $\int da'\, |a'\rangle\langle a'|a\rangle = |a\rangle$, one must have the normalization

$$\langle a'|a\rangle = \delta(a' - a), \tag{1.37}$$

where $\delta(a)$ is the **Dirac δ-function**. The expansion coefficient $\psi(a)$ is obtained from this normalization condition as $\psi(a) = \langle a|\psi\rangle$. If $|\psi\rangle$ is normalized as $\langle\psi|\psi\rangle = 1$, one should have

$$1 = \int da\, da'\, \psi^*(a)\psi(a')\langle a|a'\rangle = \int da\, |\psi(a)|^2.$$

It also follows from the relation

$$\langle\psi|\hat{A}|\psi\rangle = \int a|\psi(a)|^2\, da$$

that the probability with which the measured value of A is found in the interval $[a, a + da]$ is $|\psi(a)|^2\, da$. Therefore, the probability density is given by

$$\rho(a) = |\langle a|\psi\rangle|^2. \tag{1.38}$$

Finally let us clarify why axiom A5 is required. Suppose that the system is in the state $|\psi\rangle$ and assume that the probability of the state to be in $|\phi\rangle$ simultaneously is $|\langle\psi|\phi\rangle|^2$. This has already been mentioned, when $|\psi\rangle$ is an eigenstate of some observable. Axiom A5 asserts that this is true for an arbitrary state $|\psi\rangle$.

1.2.3 Heisenberg equation, Heisenberg picture and Schrödinger picture

The formal solution to the Heisenberg equation of motion

$$\frac{d\hat{A}}{dt} = \frac{1}{i}[\hat{A}, \hat{H}]$$

is easily obtained as

$$\hat{A}(t) = e^{i\hat{H}t}\hat{A}(0)e^{-i\hat{H}t}. \tag{1.39}$$

Therefore, the operators $\hat{A}(t)$ and $\hat{A}(0)$ are related by the unitary operator

$$\hat{U}(t) = e^{-i\hat{H}t} \tag{1.40}$$

and, hence, are unitary equivalent. This formalism, in which operators depend on t, while states do not, is called the **Heisenberg picture**.

It is possible to introduce another picture which is equivalent to the Heisenberg picture. Let us write down the expectation value of $\hat{A}$ with respect to the state $|\psi\rangle$ as

$$\langle \hat{A}(t) \rangle = \langle \psi | e^{i\hat{H}t}\hat{A}(0)e^{-i\hat{H}t} | \psi \rangle$$
$$= ((\langle \psi | e^{i\hat{H}t})\hat{A}(0)(e^{-i\hat{H}t}|\psi\rangle)).$$

If we write $|\psi(t)\rangle \equiv e^{-i\hat{H}t}|\psi\rangle$, we find that the expectation value at t is also expressed as

$$\langle \hat{A}(t) \rangle = \langle \psi(t) | \hat{A}(0) | \psi(t) \rangle. \tag{1.41}$$

Thus, states depend on t while operators do not in this formalism. This formalism is called the **Schrödinger picture**.

Our next task is to find the equation of motion for $|\psi(t)\rangle$. To avoid confusion, quantities associated with the Schrödinger picture (the Heisenberg picture) are denoted with the subscript S (H), respectively. Thus, $|\psi(t)\rangle_S = e^{-i\hat{H}t}|\psi\rangle_H$ and $\hat{A}_S = \hat{A}_H(0)$. By differentiating $|\psi(t)\rangle_S$ with respect to t, one finds the **Schrödinger equation**:

$$i\frac{d}{dt}|\psi(t)\rangle_S = \hat{H}|\psi(t)\rangle_S. \tag{1.42}$$

Note that the Hamiltonian $\hat{H}$ is the same for both the Schrödinger picture and the Heisenberg picture. We will drop the subscripts S and H whenever this does not cause confusion.

1.2.4 Wavefunction

Let us consider a particle moving on the real line $\mathbb{R}$ and let $\hat{x}$ be the position operator with the eigenvalue y and the corresponding eigenvector $|y\rangle$; $\hat{x}|y\rangle = y|y\rangle$. The eigenvectors are normalized as $\langle x|y\rangle = \delta(x - y)$.

Similarly, let q be the eigenvalue of $\hat{p}$ with the eigenvector $|q\rangle$; $\hat{p}|q\rangle = q|q\rangle$ such that $\langle p|q\rangle = \delta(p - q)$.

Let $|\psi\rangle \in \mathcal{H}$ be a state. The inner product

$$\psi(x) \equiv \langle x|\psi\rangle \tag{1.43}$$

is the component of $|\psi\rangle$ in the basis $|x\rangle$,

$$|\psi\rangle = \int |x\rangle\langle x|\, dx\, |\psi\rangle = \int \psi(x)|x\rangle\, dx.$$

The coefficient $\psi(x) \in \mathbb{C}$ is called the **wavefunction**. According to the earlier axioms of quantum mechanics outlined, it is the probability amplitude of finding the particle at x in the state $|\psi\rangle$, namely $|\psi(x)|^2\, dx$ is the probability of finding the particle in the interval $[x, x + dx]$. Then it is natural to impose the normalization condition

$$\int dx\, |\psi(x)|^2 = \langle \psi|\psi\rangle = 1 \tag{1.44}$$

since the probability of finding the particle anywhere on the real line is always unity.

Similarly, $\psi(p) = \langle p|\psi\rangle$ is the probability amplitude of finding the particle in the state with the momentum p and the probability of finding the momentum of the particle in the interval $[p, p + dp]$ is $|\psi(p)|^2\, dp$.

The inner product of two states in terms of the wavefunctions is

$$\langle \psi|\phi\rangle = \int dx\, \langle \psi|x\rangle\langle x|\phi\rangle = \int dx\, \psi^*(x)\phi(x), \tag{1.45a}$$

$$= \int dp\, \langle \psi|p\rangle\langle p|\phi\rangle = \int dp\, \psi^*(p)\phi(p). \tag{1.45b}$$

An abstract ket vector is now expressed in terms of a more concrete wavefunction $\psi(x)$ or $\psi(p)$. What about the operators? Now we write down the operators in the basis $|x\rangle$. From the defining equation $\hat{x}|x\rangle = x|x\rangle$, one obtains $\langle x|\hat{x} = \langle x|x$, which yields after multiplication by $|\psi\rangle$ from the right,

$$\langle x|\hat{x}|\psi\rangle = x\langle x|\psi\rangle = x\psi(x). \tag{1.46}$$

This is often written as $(\hat{x}\psi)(x) = x\psi(x)$.

What about the momentum operator $\hat{p}$? Let us consider the unitary operator

$$\hat{U}(a) = e^{-ia\hat{p}}.$$

Lemma 1.1. The operator $\hat{U}(a)$ defined as before satisfies

$$\hat{U}(a)|x\rangle = |x + a\rangle. \tag{1.47}$$

Proof. It follows from $[\hat{x}, \hat{p}] = i$ that $[\hat{x}, \hat{p}^n] = in\hat{p}^{n-1}$ for $n = 1, 2, \ldots$. Accordingly, we have

$$[\hat{x}, \hat{U}(a)] = \left[\hat{x}, \sum_n \frac{(-ia)^n}{n!} \hat{p}^n\right] = a\hat{U}(a)$$

which can also be written as

$$\hat{x}\hat{U}(a)|x\rangle = \hat{U}(a)(\hat{x} + a)|x\rangle = (x + a)\hat{U}(a)|x\rangle.$$

This shows that $\hat{U}(a)|x\rangle \propto |x + a\rangle$. Since $\hat{U}(a)$ is unitary, it preseves the norm of a vector. Thus, $\hat{U}(a)|x\rangle = |x + a\rangle$. $\qquad\qquad\square$

Let us take an infinitesimal number ε. Then

$$\hat{U}(\varepsilon)|x\rangle = |x + \varepsilon\rangle \simeq (1 - i\varepsilon\hat{p})|x\rangle.$$

It follows from this that

$$\hat{p}|x\rangle = \frac{|x + \varepsilon\rangle - |x\rangle}{-i\varepsilon} \xrightarrow{\varepsilon \to 0} i\frac{d}{dx}|x\rangle \qquad (1.48)$$

and its dual

$$\langle x|\hat{p} = \frac{\langle x + \varepsilon| - \langle x|}{i\varepsilon} \xrightarrow{\varepsilon \to 0} -i\frac{d}{dx}\langle x|. \qquad (1.49)$$

Therefore, for any state $|\psi\rangle$, one obtains

$$\langle x|\hat{p}|\psi\rangle = -i\frac{d}{dx}\langle x|\psi\rangle = -i\frac{d}{dx}\psi(x). \qquad (1.50)$$

This is also written as $(\hat{p}\psi)(x) = -i\, d\psi(x)/dx$.

Similarly, if one uses a basis $|p\rangle$, one will have the momentum representation of the operators as

$$\hat{x}|p\rangle = -i\frac{d}{dp}|p\rangle \qquad (1.51)$$

$$\hat{p}|p\rangle = p|p\rangle \qquad (1.52)$$

$$\langle p|\hat{x}|\psi\rangle = i\frac{d}{dp}\psi(p) \qquad (1.53)$$

$$\langle p|\hat{p}|\psi\rangle = p\psi(p). \qquad (1.54)$$

Exercise 1.2. Prove (1.51)–(1.54).

Proposition 1.1.

$$\langle x|p\rangle = \frac{1}{\sqrt{2\pi}}e^{ipx} \qquad (1.55)$$

$$\langle p|x\rangle = \frac{1}{\sqrt{2\pi}}e^{-ipx} \qquad (1.56)$$

Proof. Take $|\psi\rangle = |p\rangle$ in the relation

$$(\hat{p}\psi)(x) = \langle x|\hat{p}|\psi\rangle = -i\frac{d}{dx}\psi(x)$$

to find

$$p\langle x|p\rangle = \langle x|\hat{p}|p\rangle = -i\frac{d}{dx}\langle x|p\rangle.$$

The solution is easily found to be

$$\langle x|p\rangle = Ce^{ipx}.$$

The normalization condition requires that

$$\delta(x - y) = \langle x|y\rangle = \langle x| \int |p\rangle\langle p| \, dp \, |y\rangle$$

$$= C^2 \int dp \, e^{ip(x-y)}$$

$$= C^2 2\pi \delta(x - y),$$

where C has been taken to be real. This shows that $C = 1/\sqrt{2\pi}$. The proof of (1.56) is left as an exercise. $\qquad\square$

Thus, $\psi(x)$ and $\psi(p)$ are related as

$$\psi(p) = \langle p|\psi\rangle = \int dx \, \langle p|x\rangle\langle x|\psi\rangle = \int \frac{dx}{\sqrt{2\pi}} e^{-ipx}\psi(x) \qquad (1.57)$$

which is nothing other than the Fourier transform of $\psi(x)$.

Let us next derive the Schrödinger equation which $\psi(x)$ satisfies. By applying $\langle x|$ on (1.42) from the left, we obtain

$$\langle x|i\frac{d}{dt}|\psi(t)\rangle = \langle x|\hat{H}|\psi(t)\rangle$$

where the subscript S has been dropped. For a Hamiltonian of the type $\hat{H} = \hat{p}^2/2m + V(\hat{x})$, we obtain the **time-dependent Schrödinger equation**:

$$i\frac{d}{dt}\psi(x, t) = \left\langle x \left| \frac{\hat{p}^2}{2m} + V(\hat{x}) \right| \psi(t) \right\rangle$$

$$= -\frac{1}{2m}\frac{d^2}{dx^2}\psi(x, t) + V(x)\psi(x, t), \qquad (1.58)$$

where $\psi(x, t) \equiv \langle x|\psi(t)\rangle$.

Suppose a solution of this equation is written in the form $\psi(x, t) = T(t)\phi(x)$. By substituting this into (1.58) and dividing the result by $\psi(x, t)$, we obtain

$$\frac{iT'(t)}{T(t)} = \frac{-\phi''(x)/2m + V(x)\phi(x)}{\phi(x)}$$

where the prime denotes the derivative with respect to a relevant variable. Since the LHS is a function of t only while the right-hand side (RHS) of x only, they must be a constant, which we label E. Accordingly, there are two equations, which should be solved simultaneously,

$$iT'(t) = ET(t) \tag{1.59}$$

$$-\frac{1}{2m}\frac{d^2}{dx^2}\phi(x) + V(x)\phi(x) = E\phi(x). \tag{1.60}$$

The first equation is easily solved to yield

$$T(t) = \exp(-iEt) \tag{1.61}$$

while the second one is the eigenvalue problem of the Hamiltonian operator and called the **time-independent Schrödinger equation**, the **stationary state Schrödinger equation** or, simply, the **Schrödinger equation**. For three-dimensional space, it is written as

$$-\frac{1}{2m}\nabla^2\phi(\boldsymbol{x}) + V(\boldsymbol{x})\phi(\boldsymbol{x}) = E\phi(\boldsymbol{x}). \tag{1.62}$$

1.2.5 Harmonic oscillator

It is instructive to stop here for the moment and work out some non-trivial example. We take a one-dimensional harmonic oscillator as an example since it is not trivial, it is still solvable exactly and it is very important in the folllowing applications.

The Hamiltonian operator is

$$\hat{H} = \frac{\hat{p}^2}{2m} + \frac{1}{2}m\omega^2\hat{x}^2 \qquad [\hat{x}, \hat{p}] = i. \tag{1.63}$$

The (time-independent) Schrödinger equation is

$$-\frac{1}{2m}\frac{d^2}{dx^2}\psi(x) + \frac{1}{2}m\omega^2 x^2\psi(x) = E\psi(x). \tag{1.64}$$

By rescaling the variables as $\xi = \sqrt{m\omega}x$, $\varepsilon = E/\hbar\omega$, one arrives at

$$\psi'' + (\varepsilon - \xi^2)\psi = 0. \tag{1.65}$$

The normalizable solution of this ordinary differential equation (ODE) exists only when $\varepsilon = \varepsilon_n \equiv (n + \frac{1}{2})$ $(n = 0, 1, 2, \ldots)$ namely

$$E = E_n \equiv (n + \tfrac{1}{2})\omega \qquad (n = 0, 1, 2, \ldots) \tag{1.66}$$

and the normalized solution is written in terms of the Hermite polynomial

$$H_n(\xi) = (-1)^n e^{\xi^2/2}\frac{d^n e^{-\xi^2/2}}{d\xi^n} \tag{1.67}$$

as

$$\psi(\xi) = \sqrt{\frac{m\omega}{2^n n! \sqrt{\pi}}} H_n(\xi) e^{-\xi^2/2}. \tag{1.68}$$

This eigenvalue problem can also be analysed by an algebraic method. Define the **annihilation operator** $\hat{a}$ and the **creation operator** $\hat{a}^\dagger$ by

$$\hat{a} = \sqrt{\frac{m\omega}{2}} \hat{x} + i\sqrt{\frac{1}{2m\omega}} \hat{p} \tag{1.69}$$

$$\hat{a}^\dagger = \sqrt{\frac{m\omega}{2}} \hat{x} - i\sqrt{\frac{1}{2m\omega}} \hat{p}. \tag{1.70}$$

The number operator $\hat{N}$ is defined by

$$\hat{N} = \hat{a}^\dagger \hat{a}. \tag{1.71}$$

Exercise 1.3. Show that

$$[\hat{a}, \hat{a}] = [\hat{a}^\dagger, \hat{a}^\dagger] = 0 \qquad [\hat{a}, \hat{a}^\dagger] = 1 \tag{1.72}$$

and

$$[\hat{N}, \hat{a}] = -\hat{a} \qquad [\hat{N}, \hat{a}^\dagger] = \hat{a}^\dagger. \tag{1.73}$$

Show also that

$$\hat{H} = (\hat{N} + \tfrac{1}{2})\omega. \tag{1.74}$$

Let $|n\rangle$ be a normalized eigenvector of $\hat{N}$,

$$\hat{N}|n\rangle = n|n\rangle.$$

Then it follows from the commutation relations proved in exercise 1.3 that

$$\hat{N}(\hat{a}|n\rangle) = (\hat{a}\hat{N} - \hat{a})|n\rangle = (n-1)(\hat{a}|n\rangle)$$
$$\hat{N}(\hat{a}^\dagger|n\rangle) = (\hat{a}^\dagger\hat{N} + \hat{a}^\dagger)|n\rangle = (n+1)(\hat{a}^\dagger|n\rangle).$$

Therefore, $\hat{a}$ decreases the eigenvalue by one while $\hat{a}^\dagger$ increases it by one, hence the name *annihilation* and *creation*. Note that the eigenvalue $n \geq 0$ since

$$n = \langle n|\hat{N}|n\rangle = ((\langle n|\hat{a}^\dagger)(\hat{a}|n\rangle)) = \|\hat{a}|n\rangle\|^2 \geq 0.$$

The equality holds if and only if $\hat{a}|n\rangle = 0$. Take a fixed $n_0 > 0$ and apply $\hat{a}$ many times on $|n_0\rangle$. Eventually the eigenvalue of $\hat{a}^k|n_0\rangle$ will be negative for some integer $k > n_0$, which is a contradiction. This can be avoided only when n_0 is a non-negative integer. Thus, there exists a state $|0\rangle$ which satisfies $\hat{a}|0\rangle = 0$. The state $|0\rangle$ is called the **ground state**. Since $\hat{N}|0\rangle = \hat{a}^\dagger\hat{a}|0\rangle = 0$, this state is the eigenvector of $\hat{N}$ with the eigenvalue 0. The wavefunction $\psi_0(x) \equiv \langle x|0\rangle$ is obtained by solving the *first-order* ODE

$$\langle x|\hat{a}|0\rangle = \sqrt{\frac{1}{2m\omega}} \left(\frac{d}{dx}\psi_0(x) + m\omega x \psi_0(x) \right) = 0. \qquad (1.75)$$

The solution is easily found to be

$$\psi_0(x) = C \exp(-m\omega x^2/2) \qquad (1.76)$$

where C is the normalization constant given in (1.68). An arbitrary vector $|n\rangle$ is obtained from $|0\rangle$ by a repeated application of $\hat{a}^\dagger$.

Exercise 1.4. Show that

$$|n\rangle = \frac{1}{\sqrt{n!}}(\hat{a}^\dagger)^n|0\rangle \qquad (1.77)$$

satisfies $\hat{N}|n\rangle = n|n\rangle$ and is normalized.

Thus, the spectrum of $\hat{N}$ turns out to be Spec $\hat{N} = \{0, 1, 2, \ldots\}$ and hence the spectrum of the Hamiltonian is

$$\text{Spec } \hat{H} = \{\tfrac{1}{2}, \tfrac{3}{2}, \tfrac{5}{2}, \ldots\}. \qquad (1.78)$$

1.3   Path integral quantization of a Bose particle

The canonical quantization of a classical system has been discussed in the previous section. There the main role was played by the Hamiltonian and the Lagrangian did not show up at all. In the present section, it will be shown that there exists a quantization process, called the path integral quantization, based heavily on the Lagrangian.

1.3.1   Path integral quantization

We start our analysis with one-dimensional systems. Let $\hat{x}(t)$ be the position operator in the Heisenberg picture. Suppose the particle is found at x_i at time t_i (>0). Then the probability amplitude of finding this particle at x_f at later time t_f ($>t_i$) is

$$\langle x_f, t_f|x_i, t_i\rangle \qquad (1.79)$$

where the vectors are defined in the Heisenberg picture, [12]

$$\hat{x}(t_i)|x_i, t_i\rangle = x_i|x_i, t_i\rangle \qquad (1.80)$$

$$\hat{x}(t_f)|x_f, t_f\rangle = x_f|x_f, t_f\rangle. \qquad (1.81)$$

---

[12] We have dropped S and H again to simplify the notation. Note that $|x_i, t_i\rangle$ is an instantaneous eigenvector and hence parametrized by the time t_i when the position is measured. This should not be confused with the dynamical time dependence of a wavefunction in the Schrödinger picture.

The probability amplitude (1.79) is also called the **transition amplitude**.

Let us rewrite the probability amplitude in terms of the Schrödinger picture. Let $\hat{x} = \hat{x}(0)$ be the position operator with the eigenvector

$$\hat{x}|x\rangle = x|x\rangle. \tag{1.82}$$

Since $\hat{x}$ has no time dependence, its eigenvector should be also time independent. If

$$\hat{x}(t_i) = e^{i\hat{H}t_i}\hat{x}e^{-i\hat{H}t_i} \tag{1.83}$$

is substituted into (1.80), we obtain

$$e^{i\hat{H}t_i}\hat{x}e^{-i\hat{H}t_i}|x_i, t_i\rangle = x_i|x_i, t_i\rangle.$$

By multiplying $e^{-i\hat{H}t_i}$ from the left, we find

$$\hat{x}[e^{-i\hat{H}t_i}|x_i, t_i\rangle] = x_i[e^{-i\hat{H}t_i}|x_i, t_i\rangle].$$

This shows that the two eigenvectors are related as

$$|x_i, t_i\rangle = e^{i\hat{H}t_i}|x_i\rangle. \tag{1.84}$$

Similarly, we have

$$|x_f, t_f\rangle = e^{i\hat{H}t_f}|x_f\rangle, \tag{1.85}$$

from which we obtain

$$\langle x_f, t_f| = \langle x_f|e^{-i\hat{H}t_f}. \tag{1.86}$$

From these results, we express the probability amplitude in the Schrödinger picture as

$$\langle x_f, t_f|x_i, t_i\rangle = \langle x_f|e^{-i\hat{H}(t_f - t_i)}|x_i\rangle. \tag{1.87}$$

In general, the function

$$h(x, y; \beta) \equiv \langle x|e^{-\hat{H}\beta}|y\rangle \tag{1.88}$$

is called the **heat kernel** of $\hat{H}$. This nomenclature originates from the similarity between the Schrödinger equation and the heat equation. The amplitude (1.87) is the heat kernel of $\hat{H}$ with imaginary β:

$$\langle x_f, t_f|x_i, t_i\rangle = h(x_f, x_i; i(t_f - t_i)). \tag{1.89}$$

Now the amplitude (1.87) is expressed in the path integral formalism. To this end, we consider the case in which $t_f - t_i = \varepsilon$ is an infinitesimal positive number. Let us put $x_i = x$ and $x_f = y$ to simplify the notation and suppose the Hamiltonian is of the form

$$\hat{H} = \frac{\hat{p}^2}{2m} + V(\hat{x}). \tag{1.90}$$

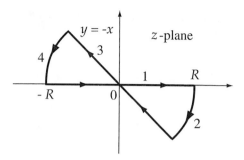

Figure 1.1. The integration contour.

We first prove the following lemma.

Lemma 1.2. Let a be a positive constant. Then

$$\int_{-\infty}^{\infty} e^{-iap^2} \, dp = \sqrt{\frac{\pi}{ia}}. \tag{1.91}$$

Proof. The integral is different from an ordinary Gaussian integral in that the coefficient of p^2 is a pure imaginary number. First replace p by $z = x + iy$. The integrand $\exp(-iaz^2)$ is analytic in the whole z-plane. Now change the integration contour from the real axis to the one shown in figure 1.1. Along path 1, we have $dz = dx$ and hence this path gives the same contribution as the original integration (1.91). The contribution from paths 2 and 4 vanishes as $R \to \infty$. Noting that the variable along path 3 is $z = (1 - i)x$, we evaluate the contribution from this path as

$$(1 - i) \int_{\infty}^{-\infty} e^{-2ax^2} \, dx = -e^{-i\pi/4} \sqrt{\frac{\pi}{a}}.$$

The summation of all the contribution must vanish due to Cauchy's theorem and, hence,

$$\int_{-\infty}^{\infty} dp \, e^{-iap^2} = e^{-i\pi/4} \sqrt{\frac{\pi}{a}} = \sqrt{\frac{\pi}{ia}}. \qquad \square$$

Now this lemma is employed to obtain the heat kernel for an infinitesimal time interval.

Proposition 1.2. Let $\hat{H}$ be a Hamiltonian of the form (1.90) and ε be an infinitesimal positive number. Then for any $x, y \in \mathbb{R}$, we find that

$$\langle x|e^{-i\hat{H}\varepsilon}|y\rangle = \frac{1}{\sqrt{2\pi i\varepsilon}} \exp\left[i\varepsilon \left\{ \frac{m}{2} \left(\frac{(x-y)^2}{\varepsilon} \right)^2 \right. \right.$$
$$\left. \left. - V\left(\frac{x+y}{2} \right) \right\} + \mathcal{O}(\varepsilon^2) + \mathcal{O}(\varepsilon(x-y)^2) \right]. \tag{1.92}$$

Proof. The completeness relation for the momentum eigenvectors is inserted into the LHS of (1.92) to yield

$$\langle x|e^{-i\hat{H}\varepsilon}|y\rangle = \int dk \langle x|e^{-i\varepsilon\hat{H}}|k\rangle\langle k|y\rangle$$

$$= \int \frac{dk}{2\pi} e^{-iky} e^{-i\varepsilon\hat{H}_x} e^{ikx}$$

where

$$\hat{H}_x = -\frac{1}{2m}\frac{d^2}{dx^2} + V(x).$$

Now we find from the commutation relation of $\partial_x \equiv d/dx$ and e^{ikx} that

$$\partial_x e^{ikx} = ike^{ikx} + e^{ikx}\partial_x = e^{ikx}(ik + \partial_x).$$

Repeated application of this commutation relation yields

$$\partial_x^n e^{ikx} = e^{ikx}(ik + \partial_x)^n \qquad (n = 0, 1, 2, \ldots)$$

from which we obtain

$$e^{-i\varepsilon[-\partial_x^2/2m+V(x)]}e^{ikx} = e^{ikx}e^{-i\varepsilon[-(ik+\partial_x)^2/2m+V(x)]}.$$

Therefore,

$$\langle x|e^{-i\hat{H}\varepsilon}|y\rangle = \int \frac{dk}{2\pi} e^{ik(x-y)} e^{-i\varepsilon[-(ik+\partial_x)^2/2m+V(x)]}$$

$$= \int \frac{dk}{2\pi} e^{-i[\varepsilon k^2/2m - k(x-y)]} e^{-i\varepsilon[-ik\partial_x/m - \partial_x^2/2m+V(x)]} \cdot 1$$

where the '1' at the end of the last line is written explicitly to remind us of the fact $\partial_x 1 = 0$. If we further put $p = \sqrt{\varepsilon/2m}\,k$ and expands the last exponential function in the last line, we obtain

$$\langle x|e^{-i\varepsilon\hat{H}}|y\rangle = \sqrt{\frac{2m}{\varepsilon}} e^{im(x-y)^2/2\varepsilon} \int \frac{dp}{2\pi} e^{-i[p+\sqrt{m/2\varepsilon}(x-y)]^2}$$

$$\times \sum_{n=0}^{\infty} \frac{(-i\varepsilon)^n}{n!}\left[i\sqrt{\frac{2}{\varepsilon m}}p\partial_x - \frac{\partial_x^2}{2m} + V(x)\right]^n \cdot 1.$$

If we put $q = p + \sqrt{m/2\varepsilon}(x - y)$ and use lemma 1.2, we obtain:

$$
\begin{aligned}
\langle x|e^{-i\varepsilon\hat{H}}|y\rangle &= \sqrt{\frac{2m}{\varepsilon}} e^{im(x-y)^2/2\varepsilon} \int \frac{dq}{2\pi} e^{-iq^2} \\
&\quad \times \left[1 + (-i\varepsilon)V(x) + \frac{(-\varepsilon^2)}{2}\frac{(-i)}{\varepsilon}(x-y)\partial_x V(x) \right. \\
&\qquad \left. + \mathcal{O}(\varepsilon^2) + \mathcal{O}(\varepsilon|x-y|^2) \right] \\
&= \sqrt{\frac{m}{2\pi i\varepsilon}} e^{i\varepsilon(m/2)[(x-y)/\varepsilon]^2} \\
&\quad \times \exp\left[-i\varepsilon V\left(\frac{x+y}{2}\right) + \mathcal{O}(\varepsilon^2) + \mathcal{O}(\varepsilon|x-y|^2) \right].
\end{aligned}
$$

Thus, the proposition has been proved.   □

Note that the average value $(x+y)/2$ appeared as the variable of V in (1.92). This prescription is often called the **Weyl ordering**.

It is found from (1.92) that the integrand oscillates very rapidly for $|x - y| > \sqrt{\varepsilon}$ and it can be regarded as zero in the sense of distribution (the Riemann–Lebesgue theorem). Therefore, as $x - y < \varepsilon$, the exponent of (1.92) approaches the action for an infinitesimal time interval $[0, \varepsilon]$,

$$
\Delta S = \int_0^\varepsilon dt \left[\frac{m}{2}v^2 - V(x) \right] \simeq \left[\frac{m}{2}v^2 - V(x) \right]\varepsilon \tag{1.93}
$$

where $v = (x - y)/\varepsilon$ is the average velocity and x is the average position.

Equation (1.92) also satisfies the boundary condition for $\varepsilon \to 0$,

$$
\langle x|e^{-i\hat{H}\varepsilon}|y\rangle \xrightarrow{\varepsilon \to 0} \langle x|y\rangle = \delta(x - y). \tag{1.94}
$$

This can be shown by noting that

$$
\int_{-\infty}^{\infty} dx \sqrt{\frac{m}{2\pi i\varepsilon}} e^{im(x-y)^2/2\varepsilon} = 1.
$$

The transition amplitude (1.79) for a finite time interval is obtained by infinitely repeating the transition amplitude for an infinitesimal time interval one after another. Let us first divide the interval $t_f - t_i$ into n equal intervals,

$$
\varepsilon = \frac{t_f - t_i}{n}.
$$

Put $t_0 = t_i$ and $t_k = t_0 + \varepsilon k$ ($0 \leq k \leq n$). Clearly $t_n = t_f$. Insert the completeness relation

$$
1 = \int dx_k |x_k, t_k\rangle\langle x_k, t_k| \qquad (1 \leq k \leq n - 1)
$$

for each instant of time t_k into (1.79) to yield

$$\langle x_f, t_f | x_i, t_i \rangle = \langle x_f, t_f | \int dx_{n-1} | x_{n-1}, t_{n-1} \rangle \langle x_{n-1}, t_{n-1} |$$

$$\times \int dx_{n-2} | x_{n-2}, t_{n-2} \rangle \dots \int dx_1 | x_1, t_1 \rangle \langle x_1, t_1 | x_0, t_0 \rangle.$$

Let us consider here the limit $\varepsilon \to 0$, namely $n \to \infty$. Proposition 1.2 states that for an infinitesimal ε, we have

$$\langle x_k, t_k | x_{k-1}, t_{k-1} \rangle \simeq \sqrt{\frac{m}{2\pi i \varepsilon}} e^{i \Delta S_k}$$

where

$$\Delta S_k = \varepsilon \left[\frac{m}{2} \left(\frac{x_k - x_{k-1}}{\varepsilon} \right)^2 - V \left(\frac{x_{k-1} + x_k}{2} \right) \right].$$

Therefore, we find

$$\langle x_f, t_f | x_i, t_i \rangle = \lim_{n \to \infty} \left(\frac{m}{2\pi i \varepsilon} \right)^{n/2} \int \prod_{j=1}^{n-1} dx_j \exp \left(i \sum_{k=1}^{n} \Delta S_k \right). \tag{1.95}$$

If $n - 1$ points $x_1, x_2, \dots, x_{n-1}$ are fixed, we obtain a piecewise linear path from x_0 to x_n via these points. Then we define $S(\{x_k\}) = \sum_k \Delta S_k$, which in the limit $n \to \infty$ can be written as

$$S(\{x_k\}) \overset{n \to \infty}{\longrightarrow} S[x(t)] = \int_{t_i}^{t_f} dt \left[\frac{m}{2} v^2 - V(x) \right]. \tag{1.96}$$

Note, however, that the $S[x(t)]$ defined here is formal; the variables x_k and x_{k-1} need not be close to each other and hence $v = (x_k - x_{k-1})/\varepsilon$ may diverge. This transition amplitude is written *symbolically* as

$$\langle x_f, t_f | x_i, t_i \rangle = \int \mathcal{D}x \, \exp \left[i \int_{t_i}^{t_f} dt \left(\frac{m}{2} v^2 - V(x) \right) \right]$$

$$= \int \mathcal{D}x \, \exp \left[i \int_{t_i}^{t_f} dt \, L(x, \dot{x}) \right] \tag{1.97}$$

which is called the **path integral** representation of the transition amplitude. It should be stressed again that the 'v' is not well defined and that this expression is just a symbolic representation of the limit (1.95).

The integration measure is understood as

$$\int \mathcal{D}x = \text{summation over all paths } x(t) \text{ with } x(t_i) = x_i, x(t_f) = x_f \tag{1.98}$$

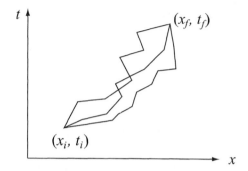

Figure 1.2. All the paths with fixed endpoints are considered in the path integral. The integrand $\exp[iS(\{x_k\})]$ is integrated over these paths.

see figure 1.2. Although $\mathcal{D}x$ or $S(\{x_k\})$ is ill defined in the limit $n \to \infty$, the amplitude $\langle x_f, t_f | x_i, t_i \rangle$ constructed from $\mathcal{D}x$ and $S(\{x_k\})$ together is well defined and hence meaningful. This point is clarified in the following example.

Example 1.5. Let us work out the transition amplitude of a free particle moving on the real axis with the Lagrangian

$$L = \tfrac{1}{2}m\dot{x}^2. \tag{1.99}$$

The canonical conjugate momentum is $p = \partial L/\partial \dot{x} = m\dot{x}$ and the Hamiltonian is

$$H = p\dot{x} - L = \frac{p^2}{2m}. \tag{1.100}$$

The transition amplitude is calculated within the canonical quantum theory as

$$\langle x_f, t_f | x_i, t_i \rangle = \langle x_f | e^{-i\hat{H}T} | x_i \rangle = \int dp \langle x_f | e^{-i\hat{H}T} | p \rangle \langle p | x_i \rangle$$

$$= \int \frac{dp}{2\pi} e^{ip(x_f - x_i)} e^{-iT(p^2/2m)}$$

$$= \sqrt{\frac{m}{2\pi iT}} \exp\left(\frac{im(x_f - x_i)^2}{2T}\right) \tag{1.101}$$

where $T = t_f - t_i$.

This result is obtained using the path integral formalism next. The amplitude is expressed as

$$\langle x_f, t_f | x_i, t_i \rangle = \lim_{n \to \infty} \left(\frac{m}{2\pi i\varepsilon}\right)^{n/2} \int dx_1 \ldots dx_{n-1}$$

$$\exp\left[i\varepsilon \sum_{k=1}^{n} \frac{m}{2} \left(\frac{x_k - x_{k-1}}{\varepsilon}\right)^2\right] \tag{1.102}$$

where $\varepsilon = T/n$. After scaling the coordinates as

$$y_k = \left(\frac{m}{2\varepsilon}\right)^{1/2} x_k$$

the amplitude becomes

$$\langle x_f, t_f | x_i, t_i \rangle = \lim_{n \to \infty} \left(\frac{m}{2\pi i \varepsilon}\right)^{n/2} \left(\frac{2\varepsilon}{m}\right)^{(n-1)/2}$$
$$\int dy_1 \ldots dy_{n-1} \exp\left[i \sum_{k=1}^{n} (y_k - y_{k-1})^2\right]. \quad (1.103)$$

It can be shown by induction (exercise) that

$$\int dy_1 \ldots dy_{n-1} \exp\left[i \sum_{k=1}^{n} (y_k - y_{k-1})^2\right] = \left[\frac{(i\pi)^{(n-1)}}{n}\right]^{1/2} e^{i(y_n - y_0)^2/n}.$$

Taking the limit $n \to \infty$, we finally obtain

$$\langle x_f, t_f | x_i, t_i \rangle = \lim_{n \to \infty} \left(\frac{m}{2\pi i \varepsilon}\right)^{n/2} \left(\frac{2\pi i \varepsilon}{m}\right)^{(n-1)/2} \frac{1}{\sqrt{n}} e^{im(x_f - x_i)^2/(2n\varepsilon)}$$
$$= \sqrt{\frac{m}{2\pi i T}} \exp\left[\frac{im(x_f - x_i)^2}{2T}\right]. \quad (1.104)$$

It should be noted here that the exponent is the classical action. In fact, if we note that the average velocity is $v = (x_f - x_i)/(t_f - t_i)$, the classical action is found to be

$$S_{cl} = \int_{t_i}^{t_f} dt \frac{1}{2} mv^2 = \frac{m(x_f - x_i)^2}{2(t_f - t_i)}.$$

It happens in many exactly solvable systems that the transition amplitude takes the form

$$\langle x_f, t_f | x_i, t_i \rangle = A e^{iS_{cl}}, \quad (1.105)$$

where all the effects of quantum fluctuation are taken into account in the prefactor A.

1.3.2 Imaginary time and partition function

Suppose the spectrum of a Hamiltonian $\hat{H}$ is bounded from below. Then it is always possible, by adding a positive constant to the Hamiltonian, to make $\hat{H}$ positive definite;

$$\text{Spec } \hat{H} = \{0 < E_0 \le E_1 \le E_2 \le \cdots\}. \quad (1.106)$$

It has been assumed for simplicity that the ground state is not degenerate. The spectral decomposition of $e^{-i\hat{H}t}$ given by

$$e^{-i\hat{H}t} = \sum_n e^{-iE_n t} |n\rangle\langle n| \tag{1.107}$$

is analytic in the lower half-plane of t, where $\hat{H}|n\rangle = E_n|n\rangle$. Introduce the **Wick rotation** by the replacement

$$t = -i\tau \qquad (\tau \in \mathbb{R}_+) \tag{1.108}$$

where $\mathbb{R}_+$ is the set of positive real numbers. The variable τ is regarded as imaginary time, which is also known as the Euclidean time since the world distance changes from $t^2 - x^2$ to $-(\tau^2 + x^2)$. Physical quantities change under this change of variable as

$$\dot{x} = \frac{dx}{dt} = i\frac{dx}{d\tau}$$
$$e^{-i\hat{H}t} = e^{-\hat{H}\tau}$$
$$i\int_{t_i}^{t_f} dt \left[\frac{1}{2}m\dot{x}^2 - V(x) \right] = i(-i)\int_{\tau_i}^{\tau_f} d\tau \left[-\frac{1}{2}m\left(\frac{dx}{d\tau}\right)^2 - V(x) \right]$$
$$= -\int_{\tau_i}^{\tau_f} d\tau \left[\frac{1}{2}m\left(\frac{dx}{d\tau}\right)^2 + V(x) \right].$$

Accordingly, the path integral is expressed in terms of the new variable as

$$\langle x_f, \tau_f | x_i, \tau_i \rangle = \langle x_f | e^{-\hat{H}(\tau_f - \tau_i)} | x_i \rangle$$
$$= \int \bar{D}x\, e^{-\int_{\tau_i}^{\tau_f} d\tau \left[\frac{1}{2}m\left(\frac{dx}{d\tau}\right)^2 + V(x) \right]}, \tag{1.109}$$

where $\bar{D}$ is the integration measure in the imaginary time τ.

For a given Hamiltonian $\hat{H}$, the **partition function** is defined as

$$Z(\beta) = \text{Tr}\, e^{-\beta\hat{H}} \qquad (\beta > 0), \tag{1.110}$$

where the trace is over the Hilbert space associated with $\hat{H}$.

Let us take the eigenstates $\{|E_n\rangle\}$ of $\hat{H}$ as the basis vectors of the Hilbert space;

$$\hat{H}|E_n\rangle = E_n|E_n\rangle, \qquad \langle E_m|E_n\rangle = \delta_{mn}.$$

Then the partition function is expressed as

$$Z(\beta) = \sum_n \langle E_n | e^{-\beta\hat{H}} | E_n \rangle = \sum_n \langle E_n | e^{-\beta E_n} | E_n \rangle$$
$$= \sum_n e^{-\beta E_n}. \tag{1.111}$$

The partition function is also expressed in terms of the eigenvector $|x\rangle$ of $\hat{x}$. Namely

$$Z(\beta) = \int dx \langle x | e^{-\beta \hat{H}} | x \rangle. \qquad (1.112)$$

If β is identified with the Euclidean time by putting $\beta = iT$, we find that

$$\langle x_f | e^{-i\hat{H}T} | x_i \rangle = \langle x_f | e^{-\beta \hat{H}} | x_i \rangle,$$

from which we obtain the path integral expression of the partition function

$$Z(\beta) = \int dy \int_{x(0)=x(\beta)=y} \bar{\mathcal{D}}x \, \exp\left\{ -\int_0^\beta d\tau \left(\frac{1}{2}m\dot{x}^2 + V(x) \right) \right\}$$

$$= \int_{\text{periodic}} \bar{\mathcal{D}}x \, \exp\left\{ -\int_0^\beta d\tau \left(\frac{1}{2}m\dot{x}^2 + V(x) \right) \right\}, \qquad (1.113)$$

where the integral in the last line is over all paths periodic in $[0, \beta]$.

1.3.3   Time-ordered product and generating functional

Define the **T-product** of Heisenberg operators $A(t)$ and $B(t)$ by

$$T[A(t_1)B(t_2)] = A(t_1)B(t_2)\theta(t_1 - t_2) + B(t_2)A(t_1)\theta(t_2 - t_1) \qquad (1.114)$$

$\theta(t)$ being the Heaviside function.[13] Generalization to the case with more than three operators should be trivial; operators in the bracket are rearranged so that the time parameters decrease from the left to the right. The T-product of n operators is expanded into $n!$ terms, each of which is proportional to the product of $n - 1$ Heaviside functions. An important quantity in quantum mechanics is the matrix element of the T-product,

$$\langle x_f, t_f | T[\hat{x}(t_1)\hat{x}(t_f) \cdots \hat{x}(t_n)] | x_i, t_i \rangle, \qquad (t_i < t_1, t_2, \ldots, t_n < t_f). \quad (1.115)$$

Suppose $t_i < t_1 \leq t_2 \leq \cdots \leq t_n < t_f$ in equation (1.115). By inserting the completeness relation

$$1 = \int_{-\infty}^{\infty} dx_k |x_k, t_k\rangle\langle x_k, t_k| \qquad (k = 1, 2, \ldots, n)$$

into equation (1.115), we obtain

$$\langle x_f, t_f | \hat{x}(t_n) \cdots \hat{x}(t_1) | x_i, t_i \rangle$$

$$= \langle x_f, t_f | \hat{x}(t_n) \int dx_n |x_n, t_n\rangle\langle x_n, t_n| \cdots \hat{x}(t_1) \int dx_1 |x_1, t_1\rangle\langle x_1, t_1 | x_i, t_i \rangle$$

$$= \int dx_1 \ldots dx_n \, x_1 \ldots x_n \langle x_f, t_f | x_n, t_n \rangle \cdots \langle x_1, t_1 | x_i, t_i \rangle \qquad (1.116)$$

---

[13] The Heaviside function is defined by

$$\theta(x) = \begin{cases} 0 & x < 0 \\ 1 & x \geq 0. \end{cases}$$

where use has been made of the eigenvalue equation $\hat{x}(t_k)|x_k, t_k\rangle = x_k|x_k, t_k\rangle$. If $\langle x_k, t_k|x_{k-1}, t_{k-1}\rangle$ in the last line is expressed in terms of a path integral, we find

$$\langle x_f, t_f|\hat{x}(t_n)\ldots\hat{x}(t_1)|x_i, t_i\rangle = \int \mathcal{D}x \, x(t_1)\ldots x(t_n)e^{iS}. \tag{1.117}$$

It is crucial to note that $\hat{x}(t_k)$ in the LHS is a Heisenberg operator, while $x(t_k) \, (=x_k)$ in the RHS is the real value of a classical path $x(t)$ at time t_k. Accordingly, the RHS remains true for any ordering of the time parameters in the LHS as long as the Heisenberg operators are arranged in a way defined by the T-product. Thus, the path integral expression automatically takes the T-product ordering into account to yield

$$\langle x_f, t_f|T[\hat{x}(t_n)\ldots\hat{x}(t_1)]|x_i, t_i\rangle = \int \mathcal{D}x \, x(t_1)\ldots x(t_n)e^{iS}. \tag{1.118}$$

The reader is encouraged to verify this result explicitly for $n = 2$.

It turns out to be convenient to define the **generating functional** $Z[J]$ to obtain the matrix elements of the T-products efficiently. We couple an **external field** $J(t)$ (also called the **source**) with the coordinate $x(t)$ as $x(t)J(t)$ in the Lagrangian, where $J(t)$ is defined on the interval $[t_i, t_f]$. Define the action with the source as

$$S[x(t), J(t)] = \int_{t_i}^{t_f} dt \, [\tfrac{1}{2}m\dot{x}^2 - V(x) + xJ]. \tag{1.119}$$

The transition amplitude in the presence of $J(t)$ is then given by

$$\langle x_f, t_f|x_i, t_i\rangle_J = \int \mathcal{D}x \exp\left[i\int_{t_i}^{t_f} dt \, (\tfrac{1}{2}m\dot{x}^2 - V(x) + xJ)\right]. \tag{1.120}$$

The functional derivative of this equation with respect to $J(t)$ $(t_i < t < t_f)$ yields

$$\frac{\delta}{\delta J(t)}\langle x_f, t_f|x_i, t_i\rangle_J = \int \mathcal{D}x \, ix(t) \exp\left[i\int_{t_i}^{t_f} dt \, (\tfrac{1}{2}m\dot{x}^2 - V(x) + xJ)\right]. \tag{1.121}$$

Higher functional derivatives are easy to obtain; the factor $ix(t_k)$ appears in the integrand of the path integral each time $\delta/\delta J(t)$ acts on $\langle x_f, t_f|x_i, t_i\rangle_J$. This is nothing but the matrix element of the T-product of the Heisenberg operator $\hat{x}(t)$ in the presence of the source $J(t)$. Accordingly, if we put $J(t) = 0$ in the end of the calculation, we obtain

$$\langle x_f, t_f|T[x(t_n)\ldots x(t_1)]|x_i, t_i\rangle$$
$$= (-i)^n \frac{\delta^n}{\delta J(t_1)\ldots\delta J(t_n)} \int \mathcal{D}x \, e^{iS[x(t), J(t)]}\bigg|_{J=0}. \tag{1.122}$$

It often happens in physical applications that the transition probability amplitude between general states, in particular the ground states, is required rather than those between coordinate eigenstates. Suppose the system under consideration is in the ground state $|0\rangle$ at t_i and calculate the probability amplitude with which the system is also in the ground state at later time t_f. Suppose $J(t)$ is non-vanishing only on an interval $[a, b] \subset [t_i, t_f]$. (The reason for this assumption will become clear later.) The transition amplitude in the presence of $J(t)$ may be obtained from the Hamiltonian $H^J = H - x(t)J(t)$ and the unitary operator $U^J(t_f, t_i)$ of the Hamiltonian. The transition probability amplitude between the coordinate eigenstates is

$$\langle x_f, t_f | x_i, t_i \rangle_J = \langle x_f | U^J(t_f, t_i) | x_i \rangle$$
$$= \langle x_f | e^{-iH(t_f - b)} U^J(b, a) e^{-iH(a - t_i)} | x_i \rangle, \qquad (1.123)$$

where use has been made of the fact $H^J = H$ outside the interval $[a, b]$. By inserting the completeness relations of the energy eigenvectors $\sum_n |n\rangle\langle n| = 1$ into this equation, we obtain

$$\langle x_f, t_f | x_i, t_i \rangle_J = \sum_{m,n} \langle x_f | e^{-iH(t_f - b)} | m \rangle \langle m | U^J(b, a) | n \rangle \langle n | e^{-iH(a - t_i)} | x_i \rangle$$
$$= \sum_{m,n} e^{-iE_m(t_f - b)} e^{-iE_n(a - t_i)} \langle x_f | m \rangle \langle n | x_i \rangle \langle m | U^J(b, a) | n \rangle.$$

$$(1.124)$$

Now let us Wick rotate the time variable $t \to -i\tau$ under which the exponential function changes as $e^{-iEt} \to e^{-E\tau}$. Then the limit $\tau_f \to \infty, \tau_i \to -\infty$ picks up only the ground states $m = n = 0$. Alternatively, we may introduce a small imaginary term $-i\varepsilon x^2$ in the Hamiltonian so that the eigenvalue has a small negative imaginary part. Then only the ground state survives in the summations over m and n under $\tau_f \to \infty, \tau_i \to -\infty$.

After all we have proved that

$$\lim_{\substack{t_f \to \infty \\ t_i \to -\infty}} \langle x_f, t_f | x_i, t_i \rangle_J = \langle x_f | 0 \rangle \langle 0 | x_i \rangle Z[J] \qquad (1.125)$$

where we have defined the **generating functional**

$$Z[J] = \langle 0 | U^J(b, a) | 0 \rangle = \lim_{\substack{t_f \to \infty \\ t_i \to -\infty}} \langle 0 | U^J(t_f, t_i) | 0 \rangle. \qquad (1.126)$$

The generating functional may be also expressed as

$$Z[J] = \lim_{\substack{t_f \to \infty \\ t_i \to -\infty}} \frac{\langle x_f, t_f | x_i, t_i \rangle_J}{\langle x_f | 0 \rangle \langle 0 | x_i \rangle}. \qquad (1.127)$$

Note that the denominator is just a constant independent of $Z[J]$. Now we have found the path integral representation for $Z[J]$,

$$Z[J] = \mathcal{N} \int \mathcal{D}x \, e^{iS[x,J]} \qquad (1.128)$$

where the path integral is over paths with arbitrarily fixed x_i and x_f. The normalization constant $\mathcal{N}$ is chosen so that $Z[0] = 1$, namely

$$\mathcal{N}^{-1} = \int \mathcal{D}x \, e^{iS[x,0]}.$$

It is readily shown that $Z[J]$ generates the matrix elements of the T-product between the ground states:

$$\langle 0|T\,[x(t_1)\cdots x(t_n)]\,|0\rangle = (-i)^n \frac{\delta^n}{\delta J(t_1)\cdots\delta J(t_n)} Z[J]\bigg|_{J=0}. \qquad (1.129)$$

1.4 Harmonic oscillator

We work out the path integral quantization of a harmonic oscillator, which is an example of systems for which the path integral may be evaluated exactly. We also introduce the zeta function regularization, which is a useful tool in many areas of theoretical physics.

1.4.1 Transition amplitude

The Lagrangian of a one-dimensional harmonic oscillator is

$$L = \tfrac{1}{2}m\dot{x}^2 - \tfrac{1}{2}m\omega^2 x^2. \qquad (1.130)$$

The transition amplitude is given by

$$\langle x_f, t_f | x_i, t_i \rangle = \int \mathcal{D}x \, e^{iS[x(t)]}, \qquad (1.131)$$

where $S[x(t)] = \int_{t_i}^{t_f} L \, dt$ is the action.

Let us expand $S[x]$ around its extremum $x_c(t)$ satisfying

$$\frac{\delta S[x]}{\delta x}\bigg|_{x=x_c(t)} = 0. \qquad (1.132)$$

Clearly $x_c(t)$ is the classical path connecting (x_i, t_i) and (x_f, t_f) and satifies the Euler–Lagrange equation

$$\ddot{x}_c + \omega^2 x_c = 0. \qquad (1.133)$$

The solution of equation (1.133) satifying $x_c(t_i) = x_i$ and $x_c(t_f) = x_f$ is easily obtained as

$$x_c(t) = \frac{1}{\sin \omega T}[x_f \sin \omega(t - t_i) + x_i \sin \omega(t_f - t)] \qquad (1.134)$$

where $T = t_f - t_i$. Substituting this solution into the action, we obtain (exercise)

$$\begin{aligned} S_c &\equiv S[x_c] \\ &= \frac{m\omega}{2\sin \omega T}[(x_f^2 + x_i^2)\cos \omega T - 2x_f x_i]. \end{aligned} \qquad (1.135)$$

Now the expansion of $S[x]$ around $x = x_c$ takes the form

$$S[x_c + y] = S[x_c] + \frac{1}{2!} \int dt_1 \, dt_2 \, y(t_1)y(t_2) \frac{\delta^2 S[x]}{\delta x(t_1)\delta x(t_2)}\bigg|_{x=x_c} \quad (1.136)$$

where $y(t)$ satisfies the boundary condition $y(t_i) = y(t_f) = 0$. Note that (1) the first-order term vanishes since $\delta S[x]/\delta x = 0$ at $x = x_c$ and (2) terms of order three and higher do not exist since the action is second order in x. Therefore, this expansion is *exact* and this problem is exactly solvable as we see later.

By noting that

$$\frac{\delta}{\delta x(t_1)} \int_{t_i}^{t_f} dt \left[\frac{1}{2}m\dot{x}(t)^2 - \frac{1}{2}m\omega^2 x(t)^2 \right] = -m\frac{d^2}{dt_1^2}x(t_1) - m\omega^2 x(t_1)$$

$$= -m\left(\frac{d^2}{dt_1^2} + \omega^2 \right) x(t_1)$$

and that

$$\frac{\delta x(t_1)}{\delta x(t_2)} = \delta(t_1 - t_2)$$

we obtain the second-order functional derivative

$$\frac{\delta^2 S[x]}{\delta x(t_1)\delta x(t_2)} = -m\left(\frac{d^2}{dt_1^2} + \omega^2 \right) \delta(t_1 - t_2). \quad (1.137)$$

Substituting this into equation (1.136) we find that

$$S[x_c + y] = S[x_c] - \frac{m}{2!} \int dt_1 \, dt_2 \, y(t_1)y(t_2) \left(\frac{d^2}{dt_1^2} + \omega^2 \right) \delta(t_1 - t_2)$$

$$= S[x_c] + \frac{m}{2} \int dt \, (\dot{y}^2 - \omega^2 y^2), \quad (1.138)$$

where the boundary condition $y(t_i) = y(t_f) = 0$ has been taken into account.

Since $\mathcal{D}x$ is translationally invariant,[14] we may replace $\mathcal{D}x$ by $\mathcal{D}y$ to obtain

$$\langle x_f, t_f | x_i, t_i \rangle = e^{iS[x_c]} \int_{y(t_i)=y(t_f)=0} \mathcal{D}y \, e^{i\frac{m}{2}\int_{t_i}^{t_f} dt\,(\dot{y}^2 - \omega^2 y^2)}. \quad (1.139)$$

Let us evaluate the fluctuation part

$$I_f = \int_{y(0)=y(T)=0} \mathcal{D}y \, e^{i\frac{m}{2}\int_0^T dt\,(\dot{y}^2 - \omega^2 y^2)} \quad (1.140)$$

[14] Integrating over all possible paths $x(t)$ with $x(t_i) = x_i$ and $x(t_f) = x_f$ is equivalent to integrating over all possible paths $y(t)$ with $y(t_i) = y(t_f) = 0$, where $x(t) = x_c(t) + y(t)$.

where we have shifted the t variable so that t_i now becomes $t = 0$. We expand $y(t)$ as

$$y(t) = \sum_{n \in \mathbb{N}} a_n \sin \frac{n\pi t}{T} \qquad (1.141)$$

in conformity with the boundary condition. Substitution of this expansion into the integral in the exponent yields

$$\int_0^T dt \, (\dot{y}^2 - \omega^2 y^2) = \frac{T}{2} \sum_{n \in \mathbb{N}} a_n^2 \left[\left(\frac{n\pi}{T} \right)^2 - \omega^2 \right].$$

The Fourier transform from $y(t)$ to $\{a_n\}$ may be regarded as a change of variables in the integration. For this transformation to be well defined, the number of variables must be the same. Suppose the number of the time slice is $N + 1$, including $t = 0$ and $t = T$, for which there are $N - 1$ independent y_k. Correspondingly, we must put $a_n = 0$ for $n > N - 1$. The Jacobian associated with this change of variables is

$$J_N = \det \frac{\partial y_k}{\partial a_n} = \det \left[\sin \left(\frac{n\pi t_k}{T} \right) \right] \qquad (1.142)$$

where t_k is the kth time step when $[0, T]$ is divided into N infinitesimal steps.

This Jacobian can be evaluated most easily for a free particle. Since the transformation $\{y_k\} \to \{a_n\}$ is independent of the potential, the Jacobian should be identical for both cases. The probability amplitude for a free particle has been obtained in (1.104) leading to

$$\langle x_f, T | x_i, 0 \rangle = \left(\frac{1}{2\pi i T} \right)^{1/2} \exp \left[i \frac{m}{2T} (x_f - x_i)^2 \right] = \left(\frac{1}{2\pi i T} \right)^{1/2} e^{iS[x_c]}. \qquad (1.143)$$

This is written in terms of a path integral as

$$e^{iS[x_c]} \int_{y(0)=y(T)=0} \mathcal{D}y \, e^{i\frac{m}{2} \int_0^T dt \, \dot{y}^2}. \qquad (1.144)$$

By comparing these two expressions and noting that

$$\frac{m}{2} \int_0^T dt \, \dot{y}^2 \to m \sum_{n=1}^N \frac{a_n^2 n^2 \pi^2}{4T}$$

we arrive at the equality

$$\left(\frac{1}{2\pi i T} \right)^{1/2} = \int_{y(0)=y(T)=0} \mathcal{D}y \, e^{i\frac{m}{2} \int_0^T dt \, \dot{y}^2}$$

$$= \lim_{N \to \infty} J_N \left(\frac{1}{2\pi i \varepsilon} \right)^{1/2} \int da_1 \ldots da_{N-1} \exp \left(im \sum_{n=1}^{N-1} \frac{a_n^2 \pi^2 n^2}{4T} \right).$$

By carrying out the Gaussian integrals, it is found that

$$\left(\frac{1}{2\pi iT}\right)^{1/2} = \lim_{N\to\infty} J_N \left(\frac{1}{2\pi i\varepsilon}\right)^{N/2} \prod_{n=1}^{N-1} \frac{1}{n} \left(\frac{4\pi iT}{\pi^2}\right)^{1/2}$$

$$= \lim_{N\to\infty} J_N \left(\frac{1}{2\pi i\varepsilon}\right)^{N/2} \frac{1}{(N-1)!} \left(\frac{4\pi iT}{\pi^2}\right)^{(N-1)/2}$$

from which we finally obtain, for *finite* N, that

$$J_N = N^{-N/2} 2^{-(N-1)/2} \pi^{N-1} (N-1)!. \tag{1.145}$$

The Jacobian J_N clearly diverges as $N \to \infty$. This does not matter at all, however, since we are not interested in J_N on its own but a combination with other (divergent) factors.

The transition amplitude of a harmonic oscillator is now given by

$$\langle x_f, T | x_i, 0 \rangle = \lim_{N\to\infty} J_N \left(\frac{1}{2\pi i\varepsilon}\right)^{N/2} e^{iS[x_c]}$$

$$\times \int da_1 \ldots da_{N-1} \exp\left[i\frac{mT}{4} \sum_{n=1}^{N-1} a_n^2 \left\{\left(\frac{n\pi}{T}\right)^2 - \omega^2\right\}\right]. \tag{1.146}$$

The integrals over a_n are simple Gaussian integrals and easily carried out to yield

$$\int da_n \exp\left[\frac{imT}{4} a_n^2 \left\{\left(\frac{n\pi}{T}\right)^2 - \omega^2\right\}\right] = \left(\frac{4iT}{\pi n^2}\right)^{1/2} \left[1 - \left(\frac{\omega T}{n\pi}\right)^2\right]^{-1/2}.$$

By substituting this result into equation (1.146), we obtain

$$\langle x_f, t_f | x_i, t_i \rangle = \lim_{N\to\infty} J_N \left(\frac{N}{2\pi iT}\right)^{N/2} e^{iS[x_c]}$$

$$\times \prod_{k=1}^{N-1} \left[\frac{1}{k} \left(\frac{4iT}{\pi}\right)^{1/2}\right] \prod_{n=1}^{N-1} \left[1 - \left(\frac{\omega T}{n\pi}\right)^2\right]^{-1/2}$$

$$= \left(\frac{1}{2\pi iT}\right)^{1/2} e^{iS[x_c]} \prod_{n=1}^{N-1} \left[1 - \left(\frac{\omega T}{n\pi}\right)^2\right]^{-1/2}. \tag{1.147}$$

The infinite product over n is well known and reduces to

$$\lim_{N\to\infty} \prod_{n=1}^{N} \left[1 - \left(\frac{\omega T}{n\pi}\right)^2\right] = \frac{\sin \omega T}{\omega T} \tag{1.148}$$

Note that the divergence of J_N cancelled with the divergence of the other terms to yield a finite value. Finally we have shown that

$$
\begin{aligned}
\langle x_f, t_f | x_i, t_i \rangle &= \left(\frac{\omega}{2\pi i \sin \omega T} \right)^{1/2} e^{iS[x_c]} \\
&= \left(\frac{\omega}{2\pi i \sin \omega T} \right)^{1/2} \exp \left[\frac{i\omega}{2 \sin \omega T} \{ (x_f^2 + x_i^2) \cos \omega T - 2x_i x_f \} \right].
\end{aligned}
\tag{1.149}
$$

1.4.2 Partition function

The partition function of a harmonic oscillator is easily obtained from the eigenvalue $E_n = (n + 1/2)\omega$,

$$
\mathrm{Tr}\, e^{-\beta \hat{H}} = \sum_{n=0}^{\infty} e^{-\beta(n+1/2)\omega} = \frac{1}{2 \sinh(\beta\omega/2)}.
\tag{1.150}
$$

The inverse temperature β can be regarded as the imaginary time by putting $iT = \beta$. Then the partition function may be evaluated from the path integral point of view.

Method 1: The trace may be taken over $\{|x\rangle\}$ to yield

$$
\begin{aligned}
Z(\beta) &= \int \mathrm{d}x \, \langle x | e^{-\beta \hat{H}} | x \rangle \\
&= \left(\frac{\omega}{2\pi i(-i \sinh \beta\omega)} \right)^{1/2} \\
&\quad \times \int \mathrm{d}x \, \exp i \left[\frac{\omega}{-2i \sinh \beta\omega} (2x^2 \cosh \beta\omega - 2x^2) \right] \\
&= \left(\frac{\omega}{2\pi \sinh \beta\omega} \right)^{1/2} \left[\frac{\pi}{\omega \tanh(\beta\omega/2)} \right]^{1/2} \\
&= \frac{1}{2 \sinh(\beta\omega/2)}
\end{aligned}
\tag{1.151}
$$

where use has been made of equation (1.149).

The following exercise serves as a preliminary to Method 2.

Exercise 1.5. (1) Let A be a symmetric positive-definite $n \times n$ matrix. Show that

$$
\int \mathrm{d}x_1 \ldots \mathrm{d}x_n \, \exp \left(-\sum_{i,j} x_i A_{ij} x_j \right) = \pi^{n/2} (\det A)^{-1/2} = \pi^{n/2} \prod_i \lambda_i^{-1/2}
\tag{1.152}
$$

where λ_i is the eigenvalue of A.

(2) Let A be a positive-definite $n \times n$ Hermite matrix. Show that

$$\int dz_1 \, d\bar{z}_1 \dots dz_n \, d\bar{z}_n \, \exp\left(-\sum_{i,j} \bar{z}_i A_{ij} z_j \right) = \pi^n (\det A)^{-1} = \pi^n \prod_i \lambda_i^{-1}.$$

(1.153)

Method 2: We next obtain the partition function by evaluating the path integral over the fluctuations with the help of the functional determinant and the ζ-function regularization. We introduce the imaginary time $\tau = it$ and rewrite the path integral as

$$\int_{y(0)=y(T)=0} \mathcal{D}y \, \exp\left[\frac{i}{2} \int dt \, y\left(-\frac{d^2}{dt^2} - \omega^2 \right) y \right]$$

$$\rightarrow \int_{y(0)=y(\beta)=0} \bar{\mathcal{D}}y \, \exp\left[-\frac{1}{2} \int d\tau \, y\left(-\frac{d^2}{d\tau^2} + \omega^2 \right) y \right],$$

where we noted the boundary condition $y(0) = y(\beta) = 0$. Here the bar on $\mathcal{D}$ implies the path integration measure with imaginary time.

Let A be an $n \times n$ Hermitian matrix with positive-definite eigenvalues λ_k ($1 \leq k \leq n$). Then for real variables x_k, we obtain from exercise 1.5 that

$$\prod_{k=1}^{n} \left(\int_{-\infty}^{\infty} dx_k \right) e^{-\frac{1}{2} \sum_{p,q} x_p A_{pq} x_q} = \prod_{k=1}^{n} \frac{1}{\sqrt{\lambda_k}} = \frac{1}{\sqrt{\det A}}$$

where we neglected numerical factors. This is a generalization of the well-known Gaussian integral

$$\int_{-\infty}^{\infty} dx \, e^{-\frac{1}{2}\lambda x^2} = \sqrt{\frac{2\pi}{\lambda}}$$

for $\lambda > 0$. We define the determinant of an operator $\mathcal{O}$ by the (properly regularized) infinite product of its eigenvalues λ_k as $\text{Det } \mathcal{O} = \prod_k \lambda_k$.[15] Then the previous path integral is written as

$$\int_{y(0)=y(\beta)=0} \bar{\mathcal{D}}y \, \exp\left[-\frac{1}{2} \int d\tau \, y\left(-\frac{d^2}{d\tau^2} + \omega^2 \right) y \right] = \frac{1}{\sqrt{\text{Det}_D(-d^2/d\tau^2 + \omega^2)}},$$

(1.154)

where the subscript 'D' implies that the eigenvalues are evaluated with the Dirichlet boundary condition $y(0) = y(\beta) = 0$.

The general solution $y(\tau)$ satisfying the boundary condition is written as

$$y(\tau) = \frac{1}{\sqrt{\beta}} \sum_{n \in \mathbb{N}} y_n \sin \frac{n\pi\tau}{\beta}.$$

(1.155)

---

[15] We will use 'det' for the determinant of a finite dimensional matrix while 'Det' for the (formal) determinant of an operator throughout this book. Similarly, the trace of a finite-dimensional matrix is denoted 'tr' while that of an operator is denoted 'Tr'.

Note that $y_n \in \mathbb{R}$ since $y(\tau)$ is a real function. Since the eigenvalue of the eigenfunction $\sin(n\pi\tau/\beta)$ is $\lambda_n = (n\pi/\beta)^2 + \omega^2$, the functional determinant is formally written as

$$\mathrm{Det}_D\left(-\frac{d^2}{d\tau^2} + \omega^2\right) = \prod_{n=1}^{\infty} \lambda_n = \prod_{n=1}^{\infty}\left[\left(\frac{n\pi}{\beta}\right)^2 + \omega^2\right]$$

$$= \prod_{n=1}^{\infty}\left(\frac{n\pi}{\beta}\right)^2 \prod_{p=1}^{\infty}\left[1 + \left(\frac{\beta\omega}{p\pi}\right)^2\right]. \qquad (1.156)$$

The first infinite product in the last line is written as

$$\mathrm{Det}_D\left(-\frac{d^2}{d\tau^2}\right).$$

We will evaluate this infinite product through the ζ-function regularization. Let $\mathcal{O}$ be an operator with positive-definite eigenvalues λ_n. Then we have *formally*

$$\log \mathrm{Det}\,\mathcal{O} = \mathrm{Tr}\log\mathcal{O} = \sum_n \log\lambda_n. \qquad (1.157)$$

Now we define the **spectral ζ-function** as

$$\zeta_{\mathcal{O}}(s) \equiv \sum_n \frac{1}{\lambda_n^s}. \qquad (1.158)$$

The RHS converges for sufficiently large Re s and $\zeta_{\mathcal{O}}(s)$ is analytic with respect to s in this region. Moreover, it can be analytically continued to the whole s-plane except at a possible finite number of points. By noting that

$$\left.\frac{d\zeta_{\mathcal{O}}(s)}{ds}\right|_{s=0} = -\sum_n \log\lambda_n$$

we arrive at the expression

$$\mathrm{Det}\,\mathcal{O} = \exp\left[-\left.\frac{d\zeta_{\mathcal{O}}(s)}{ds}\right|_{s=0}\right]. \qquad (1.159)$$

We replace $\mathcal{O}$ by $-d^2/d\tau^2$ in the case at hand to find

$$\zeta_{-d^2/d\tau^2}(s) = \sum_{n\geq 1}\left(\frac{n\pi}{\beta}\right)^{-2s} = \left(\frac{\beta}{\pi}\right)^{2s}\zeta(2s) \qquad (1.160)$$

where $\zeta(2s)$ is the celebrated **Riemann ζ-function**. It is analytic over the whole s-plane except at the simple pole at $s = 1$. From the well-known values

$$\zeta(0) = -\tfrac{1}{2} \qquad \zeta'(0) = -\tfrac{1}{2}\log(2\pi) \qquad (1.161)$$

we obtain

$$\zeta'_{-d^2/d\tau^2}(0) = 2\log\left(\frac{\beta}{\pi}\right)\zeta(0) + 2\zeta'(0) = -\log(2\beta).$$

We have finally shown that

$$\mathrm{Det_D}\left(-\frac{d^2}{d\tau^2}\right) = e^{\log(2\beta)} = 2\beta \tag{1.162}$$

and that

$$\mathrm{Det_D}\left(-\frac{d^2}{d\tau^2} + \omega^2\right) = 2\beta \prod_{p=1}^{\infty}\left[1 + \left(\frac{\beta\omega}{p\pi}\right)^2\right]. \tag{1.163}$$

The infinite product in this equation is well known but let us pretend that we are ignorant about this product.

The partition function is now expressed as

$$\mathrm{Tr}\,e^{-\beta H} = \left[2\beta \prod_{p=1}^{\infty}\left\{1 + \left(\frac{\beta\pi}{p\pi}\right)^2\right\}\right]^{-1/2}\left[\frac{\pi}{\omega\tanh(\beta\omega/2)}\right]^{1/2}. \tag{1.164}$$

By comparing this with the result (1.151), we have *proved* the formula

$$\prod_{n=1}^{\infty}\left[1 + \left(\frac{\beta\omega}{n\pi}\right)^2\right] = \frac{\pi}{\beta\omega}\sinh(\beta\omega)$$

namely

$$\prod_{n=1}^{\infty}\left(1 + \frac{x^2}{n^2}\right) = \frac{\sinh(\pi x)}{\pi x}. \tag{1.165}$$

What about the infinite product expansion of the cosh function? This is given by using the path integral with respect to the fermion, which we will work out in the next section.

1.5   Path integral quantization of a Fermi particle

The particles observed in Nature are not necessarily Bose particles whose position and momentum operators obey the commutation relation $[p, x] = -i$. There are particles called fermions whose operators satisfy anti-commutation relations. A classical description of a fermion requires anti-commuting numbers called the **Grassmann numbers**.

1.5.1 Fermionic harmonic oscillator

The bosonic harmonic oscillator in the previous section is described by the Hamiltonian[16]

$$H = \tfrac{1}{2}(a^\dagger a + a a^\dagger)$$

where a and $a^\dagger$ satisfy the commutation relations

$$[a, a^\dagger] = 1 \qquad [a, a] = [a^\dagger, a^\dagger] = 0.$$

The Hamiltonian has eigenvalues $(n + 1/2)\omega$ $(n \in \mathbb{N})$ with the eigenvector $|n\rangle$:

$$H|n\rangle = (n + \tfrac{1}{2})\omega|n\rangle.$$

Now suppose there is a Hamiltonian

$$H = \tfrac{1}{2}(c^\dagger c - c c^\dagger)\omega. \tag{1.166}$$

This is called the **fermionic harmonic oscillator**, which may be regarded as a Fourier component of the Dirac Hamiltonian, which describes relativistic fermions. If the operators c and $c^\dagger$ should satisfy the same commutation relations as those satisfied by bosons, the Hamiltonian would be a constant $H = -\omega/2$. Suppose, in contrast, they satisfy the *anti*-commutation relations

$$\{c, c^\dagger\} \equiv c c^\dagger + c^\dagger c = 1 \qquad \{c, c\} = \{c^\dagger, c^\dagger\} = 0. \tag{1.167}$$

The Hamiltonian takes the form

$$H = \tfrac{1}{2}[c^\dagger c - (1 - c c^\dagger)]\omega = (N - \tfrac{1}{2})\omega \tag{1.168}$$

where $N = c^\dagger c$. It is easy to see that the eigenvalue of N must be either 0 or 1. In fact, N satisfies $N^2 = c^\dagger c c^\dagger c = N$, namely $N(N - 1) = 0$. This is nothing other than the Pauli principle.

Let us study the Hilbert space of the Hamiltonian H. Let $|n\rangle$ be an eigenvector of H with the eigenvalue n, where $n = 0, 1$ as shown earlier. It is easy to verify the following relations;

$$H|0\rangle = -\frac{\omega}{2}|0\rangle \qquad H|1\rangle = \frac{\omega}{2}|1\rangle$$

$$c^\dagger|0\rangle = |1\rangle \qquad c|0\rangle = 0 \qquad c^\dagger|1\rangle = 0 \qquad c|1\rangle = |0\rangle.$$

It is convenient to introduce the component expressions

$$|0\rangle = \begin{pmatrix} 0 \\ 1 \end{pmatrix} \qquad |1\rangle = \begin{pmatrix} 1 \\ 0 \end{pmatrix}.$$

---

[16] We will drop ˆ on operators from now on unless this may cause confusion.

Exercise 1.6. Suppose the basis vectors have this form. Show that the operators have the following matrix representations

$$c = \begin{pmatrix} 0 & 0 \\ 1 & 0 \end{pmatrix}, \qquad c^\dagger = \begin{pmatrix} 0 & 1 \\ 0 & 0 \end{pmatrix},$$

$$N = \begin{pmatrix} 1 & 0 \\ 0 & 0 \end{pmatrix}, \qquad H = \frac{\omega}{2} \begin{pmatrix} 1 & 0 \\ 0 & -1 \end{pmatrix}.$$

The commutation relation $[x, p] = i$ for a boson has been replaced by $[x, p] = 0$ in the path integral formalism of a boson. For a fermion, the anti-commutation relation $\{c, c^\dagger\} = 1$ should be replaced by $\{\theta, \theta^*\} = 0$, where θ and θ^* are anti-commuting classical numbers called Grassmann numbers.

1.5.2   Calculus of Grassmann numbers

To distinguish anti-commuting Grassmann numbers from commuting real and complex numbers, the latter will be called the 'c-number', where c stands for commuting. Let n generators $\{\theta_1, \ldots, \theta_n\}$ satisfy the anti-commutation relations

$$\{\theta_i, \theta_j\} = 0 \qquad \forall i, j. \tag{1.169}$$

Then the set of the linear combinations of $\{\theta_i\}$ with the c-number coefficients is called the **Grassmann number** and the algebra generated by $\{\theta_i\}$ is called the **Grassmann algebra**, denoted by Λ^n. An arbitrary element f of Λ^n is expanded as

$$f(\theta) = f_0 + \sum_{i=1}^{n} f_i \theta_i + \sum_{i<j} f_{ij} \theta_i \theta_j + \cdots$$

$$= \sum_{0 \le k \le n} \frac{1}{k!} \sum_{\{i\}} f_{i_1, \ldots, i_k} \theta_{i_1} \ldots \theta_{i_k}, \tag{1.170}$$

where $f_0, f_i, f_{ij}, \ldots$ and $f_{i_1, \ldots, i_k}$ are c-numbers that are anti-symmetric under the exchange of any two indices. The element f is also written as

$$f(\theta) = \sum_{k_i = 0,1} \tilde{f}_{k_1, \ldots, k_n} \theta_1^{k_1} \ldots \theta_n^{k_n}. \tag{1.171}$$

Take $n = 2$ for example. Then

$$f(\theta) = f_0 + f_1 \theta_1 + f_2 \theta_2 + f_{12} \theta_1 \theta_2$$
$$= \tilde{f}_{00} + \tilde{f}_{10} \theta_1 + \tilde{f}_{01} \theta_2 + \tilde{f}_{11} \theta_1 \theta_2.$$

The subset of λ^n which is generated by monomials of even (resp. odd) power in θ_k is denoted by Λ_+^n (Λ_-^n):

$$\Lambda^n = \Lambda_+^n \oplus \Lambda_-^n. \tag{1.172}$$

The separation of Λ^n into these two subspaces is called $\mathbb{Z}_2$-**grading**. We call an element of Λ^n_+ (Λ^n_-) G-even (G-odd). Note that $\dim \lambda^n = 2^n$ while $\dim \Lambda^n_+ = \dim \Lambda^n_- = 2^{(n-1)}$.

The generator θ_k does not have a magnitude and hence the set of Grassmann numbers is not an ordered set. Zero is the only number that is a c-number as well as a Grassmann number simultaneously. A Grassmann number commutes with a c-number. It should be clear that the generators satisfy the following relations:

$$\theta_k^2 = 0$$
$$\theta_{k_1} \theta_{k_2} \dots \theta_{k_n} = \varepsilon_{k_1 k_2 \dots k_n} \theta_1 \theta_2 \dots \theta_n \qquad (1.173)$$
$$\theta_{k_1} \theta_{k_2} \dots \theta_{k_m} = 0 \qquad (m > n),$$

where

$$\varepsilon_{k_1 \dots k_n} = \begin{cases} +1 & \text{if } \{k_1 \dots k_n\} \text{ is an even permutation of } \{1 \dots n\} \\ -1 & \text{if } \{k_1 \dots k_n\} \text{ is an odd permutation of } \{1 \dots n\} \\ 0 & \text{otherwise.} \end{cases}$$

A function of Grassmann numbers is defined as a Taylor expansion of the function. When $n = 1$, for example, we have

$$e^\theta = 1 + \theta$$

since higher-order terms in θ vanish identically.

1.5.3 Differentiation

It is assumed that the differential operator acts on a function from the left:

$$\frac{\partial \theta_j}{\partial \theta_i} = \frac{\partial}{\partial \theta_i} \theta_j = \delta_{ij}. \qquad (1.174)$$

It is also assumed that the differential operator anti-commutes with θ_k. The Leibnitz rule then takes the form

$$\frac{\partial}{\partial \theta_i} (\theta_j \theta_k) = \frac{\partial \theta_j}{\partial \theta_i} \theta_k - \theta_j \frac{\partial \theta_k}{\partial \theta_i} = \delta_{ij} \theta_k - \delta_{ik} \theta_j. \qquad (1.175)$$

Exercise 1.7. Show that

$$\frac{\partial}{\partial \theta_i} \frac{\partial}{\partial \theta_j} + \frac{\partial}{\partial \theta_j} \frac{\partial}{\partial \theta_i} = 0. \qquad (1.176)$$

It is easily shown from this exercise that the differential operator is nilpotent

$$\frac{\partial^2}{\partial \theta_i^2} = 0. \qquad (1.177)$$

Exercise 1.8. Show that

$$\frac{\partial}{\partial \theta_i} \theta_j + \theta_j \frac{\partial}{\partial \theta_i} = \delta_{ij}. \qquad (1.178)$$

1.5.4 Integration

Supprisingly enough, integration with respect to a Grassmann variable is equivalent to differentiation. Let D denote differentiation with respect to a Grassmann variable and let I denote integration, where integration is understood as a definite integral. Suppose they satisfy the relations

(1) $ID = 0,$
(2) $DI = 0,$
(3) $D(A) = 0 \Rightarrow I(BA) = I(B)A,$

where A and B are arbitrary functions of Grassmann variables. The first relation states that the integration of a derivative of any function yields the surface term and it is set to zero. The second relation states that a derivative of a definite integral vanishes. The third relation implies that A is a constant if $D(A) = 0$ and hence it can be taken out of the integral. These relations are satified if we take $I \propto D$. Here we adopt the normalization $I = D$ and put

$$\int d\theta \, f(\theta) = \frac{\partial f(\theta)}{\partial \theta}. \tag{1.179}$$

We find from the previous definition that

$$\int d\theta = \frac{\partial 1}{\partial \theta} = 0 \qquad \int d\theta \, \theta = \frac{\partial \theta}{\partial \theta} = 1.$$

If there are n generators $\{\theta_k\}$, equation (1.179) is generalized as

$$\int d\theta_1 \, d\theta_2 \ldots d\theta_n \, f(\theta_1, \theta_2, \ldots, \theta_n) = \frac{\partial}{\partial \theta_1} \frac{\partial}{\partial \theta_2} \cdots \frac{\partial}{\partial \theta_n} f(\theta_1, \theta_2, \ldots, \theta_n). \tag{1.180}$$

Note the order of $d\theta_k$ and $\partial/\partial \theta_k$.

The equivalence of differentiation and integration leads to an odd behaviour of integration under the change of integration variables. Let us consider the case $n = 1$ first. Under the change of variable $\theta' = a\theta$ ($a \in \mathbb{C}$), we obtain

$$\int d\theta \, f(\theta) = \frac{\partial f(\theta)}{\partial \theta} = \frac{\partial f(\theta'/a)}{\partial \theta'/a} = a \int d\theta' \, f(\theta'/a)$$

which leads to $d\theta' = (1/a)d\theta$. This is readily extended to the case of n variables. Let $\theta_i \rightarrow \theta'_i = a_{ij}\theta_j$. Then

$$
\begin{aligned}
\int d\theta_1 \ldots \theta_n f(\theta) &= \frac{\partial}{\partial \theta_1} \cdots \frac{\partial}{\partial \theta_n} f(\theta) \\
&= \sum_{k_i=1}^{n} \frac{\partial \theta'_{k_1}}{\partial \theta_1} \cdots \frac{\partial \theta'_{k_n}}{\partial \theta_n} \frac{\partial}{\partial \theta'_{k_1}} \cdots \frac{\partial}{\partial \theta'_{k_n}} f(a^{-1}\theta') \\
&= \sum_{k_i=1}^{n} \varepsilon_{k_1 \ldots k_n} a_{k_1 1} \ldots a_{k_n n} \frac{\partial}{\partial \theta'_{k_1}} \cdots \frac{\partial}{\partial \theta'_{k_n}} f(a^{-1}\theta') \\
&= \det a \int d\theta'_1 \ldots \theta'_n f(a^{-1}\theta').
\end{aligned}
$$

Accordingly, the integral measure transforms as

$$
d\theta_1 \, d\theta_2 \ldots \theta_n = \det a \, d\theta'_1 \, d\theta'_2 \ldots d\theta'_n. \tag{1.181}
$$

1.5.5 Delta-function

The δ-function of a Grassmann variable is introduced as

$$
\int d\theta \, \delta(\theta - \alpha) f(\theta) = f(\alpha) \tag{1.182}
$$

for a single variable. If we substitute the expansion $f(\theta) = a + b\theta$ into this definition, we obtain

$$
\int d\theta \, \delta(\theta - \alpha)(a + b\theta) = a + b\alpha
$$

from which we find that the δ-function is explicitly given by

$$
\delta(\theta - \alpha) = \theta - \alpha. \tag{1.183}
$$

Extension of this result to n variables is easily verified to be (note the order of variables)

$$
\delta^n(\theta - \alpha) = (\theta_n - \alpha_n) \ldots (\theta_2 - \alpha_2)(\theta_1 - \alpha_1). \tag{1.184}
$$

The integral form of the δ-function is obtained from

$$
\int d\xi \, e^{i\xi\theta} = \int d\xi \, (1 + i\xi\theta) = i\theta
$$

as

$$
\delta(\theta) = \theta = -i \int d\xi \, e^{i\xi\theta}. \tag{1.185}
$$

1.5.6 Gaussian integral

Let us consider the integral

$$I = \int d\theta_1^* \, d\theta_1 \ldots d\theta_n^* \, d\theta_n \, e^{-\sum_{ij} \theta_i^* M_{ij} \theta_j} \tag{1.186}$$

where $\{\theta_i\}$ and $\{\theta_i^*\}$ are two sets of independent Grassmann variables. The $n \times n$ c-number matrix M is taken to be anti-symmetric since θ_i and θ_i^* anti-commute. The integral is evaluated with the help of the change of variables $\theta_i' = \sum_j M_{ij} \theta_j$ as

$$\begin{aligned}
I &= \det M \int d\theta_1^* \, d\theta_1' \ldots d\theta_n^* \, d\theta_n' e^{-\sum_i \theta_i^* \theta_i'} \\
&= \det M \left[\int d\theta^* \, d\theta (1 + \theta'\theta^*) \right]^n \\
&= \det M. \tag{1.187}
\end{aligned}$$

We prove an interesting formula as an application of the Gaussian integral.

Proposition 1.3. Let a be an anti-symmetric matrix of order $2n$ and define the **Pfaffian** of a by

$$Pf(a) = \frac{1}{2^n n!} \sum_{\substack{\text{Permutations of} \\ \{i_1,\ldots,i_{2n}\}}} \text{sgn}(P) a_{i_1 i_2} \ldots a_{i_{2n-1} i_{2n}}. \tag{1.188}$$

Then

$$\det a = Pf(a)^2. \tag{1.189}$$

Proof. Observe that

$$I = \int d\theta_{2n} \ldots d\theta_1 \exp\left[\frac{1}{2} \sum_{ij} \theta_i a_{ij} \theta_j \right] = \frac{1}{2^n n!} \int d\theta_{2n} \ldots d\theta_1 \left(\sum_{ij} \theta_i a_{ij} \theta_j \right)^n$$

$$= Pf(a).$$

Note also that

$$I^2 = \int d\theta_{2n} \ldots d\theta_1 \, d\theta_{2n}' \ldots d\theta_1' \exp\left[\frac{1}{2} \sum_{ij} (\theta_i a_{ij} \theta_j + \theta_i' a_{ij} \theta_j') \right].$$

Under the change of variables

$$\eta_k = \frac{1}{\sqrt{2}} (\theta_k + \theta_k'), \qquad \eta_k^* = \frac{1}{\sqrt{2i}} (\theta_k - \theta_k'),$$

we obtain the Jacobian $= (-1)^n$ and

$$\theta_i \theta_j + \theta'_i \theta'_j = \eta_i \eta^*_j - \eta^*_j \eta_i$$

$$d\eta_{2n} \ldots d\eta_i \, d\eta^*_{2n} \ldots d\eta^*_1 = (-1)^{n^2} d\eta_1 \, d\eta^*_1 \ldots d\eta_{2n} \, d\eta^*_{2n},$$

from which we verify that

$$\mathrm{Pf}(a)^2 = \int d\eta_1 \, d\eta^*_1 \ldots d\eta_{2n} \, d\eta^*_{2n} \exp\left[\sum_{ij} \eta^*_i a_{ij} \eta_j\right] = \det a. \qquad \square$$

Exercise 1.9. (1) Let M be a skewsymmetric matrix and K_i be Grassmann numbers. Show that

$$\int d\theta_1 \ldots d\theta_n \, e^{-\frac{1}{2}{}^t\theta \cdot M \cdot \theta + {}^t K \cdot \theta} = 2^{n/2}\sqrt{\det M} \, e^{-{}^t K \cdot M^{-1} \cdot K/4}. \qquad (1.190)$$

(2) Let M be a skew-Hermitian matrix and K_i and K^*_i be Grassmann numbers. Show that

$$\int d\theta^*_1 \, d\theta_1 \ldots d\theta^*_n \, d\theta_n \, e^{-\theta^\dagger \cdot M \cdot \theta + K^\dagger \cdot \theta + \theta^\dagger \cdot K} = \det M \, e^{K^\dagger \cdot M^{-1} \cdot K}. \qquad (1.191)$$

1.5.7 Functional derivative

The functional derivative with respect to a Grassmann variable can be defined similarly to that for a commuting variable. Let $\psi(t)$ be a Grassmann variable depending on a c-number parameter t and $F[\psi(t)]$ be a functional of ψ. Then we define

$$\frac{\delta F[\psi(t)]}{\delta \psi(s)} = \frac{1}{\varepsilon}\{F[\psi(t) + \varepsilon\delta(t-s)] - F[\psi(t)]\}, \qquad (1.192)$$

where ε is a Grassmann parameter. The Taylor expansion of $F[\psi(t) - \varepsilon\delta(t-s)]$ with respect to ε is linear in ε since $\varepsilon^2 = 0$. Accordingly, the limit $\varepsilon \to 0$ is not necessary. A word of caution: division by a Grassmann number is not well defined in general. Here, however, the numerator is proportional to ε and division by ε simply means picking up the coefficient of ε in the numerator.

1.5.8 Complex conjugation

Let $\{\theta_i\}$ and $\{\theta^*_i\}$ be two sets of the generators of Grassmann numbers. Define the complex conjugation of θ_i by $(\theta_i)^* = \theta^*_i$ and $(\theta^*_i)^* = \theta_i$. We define

$$(\theta_i \theta_j)^* = \theta^*_j \theta^*_i. \qquad (1.193)$$

Otherwise, the *real* c-number $\theta_i \theta^*_i$ does not satisfy the reality condition $(\theta_i \theta^*_i)^* = \theta_i \theta^*_i$.

1.5.9 Coherent states and completeness relation

The fermion annihilation and creation operators c and $c^\dagger$ satisfy the anti-commutation relations $\{c, c\} = \{c^\dagger, c^\dagger\} = 0$ and $\{c, c^\dagger\} = 1$ and the number operator $N = c^\dagger c$ has the eigenvectors $|0\rangle$ and $|1\rangle$. Let us consider the Hilbert space spanned by these vectors

$$\mathcal{H} = \mathrm{Span}\{|0\rangle, |1\rangle\}.$$

An arbitrary vector $|f\rangle$ in $\mathcal{H}$ may be written in the form

$$|f\rangle = |0\rangle f_0 + |1\rangle f_1,$$

where $f_0, f_1 \in \mathbb{C}$.

Now we consider the states

$$|\theta\rangle = |0\rangle + |1\rangle\theta \tag{1.194}$$
$$\langle\theta| = \langle 0| + \theta^*\langle 1| \tag{1.195}$$

where θ and θ^* are Grassmann numbers. These states are called the **coherent states** and are eigenstates of c and $c^\dagger$ respectively,

$$c|\theta\rangle = |0\rangle\theta = |\theta\rangle\theta, \qquad \langle\theta|c^\dagger = \theta^*\langle 0| = \theta^*\langle\theta|.$$

Exercise 1.10. Verify the following identities;

$$\langle\theta'|\theta\rangle = 1 + \theta'^*\theta = e^{\theta'^*\theta},$$
$$\langle\theta|f\rangle = f_0 + \theta^* f_1,$$
$$\langle\theta|c^\dagger|f\rangle = \langle\theta|1\rangle f_0 = \theta^* f_0 = \theta^*\langle\theta|f\rangle,$$
$$\langle\theta|c|f\rangle = \langle\theta|0\rangle f_1 = \frac{\partial}{\partial\theta^*}\langle\theta|f\rangle.$$

Let
$$h(c, c^\dagger) = h_{00} + h_{10}c^\dagger + h_{01}c + h_{11}c^\dagger c \qquad h_{ij} \in \mathbb{C}$$

be an arbitrary function of c and $c^\dagger$. Then the matrix elements of h are

$$\langle 0|h|0\rangle = h_{00} \qquad \langle 0|h|1\rangle = h_{01} \qquad \langle 1|h|0\rangle = h_{10} \qquad \langle 1|h|1\rangle = h_{00} + h_{11}.$$

It is easily found from these matrix elements that

$$\langle\theta|h|\theta'\rangle = (h_{00} + \theta^* h_{10} + h_{01}\theta' + \theta^*\theta' h_{11})e^{\theta^*\theta'}. \tag{1.196}$$

Lemma 1.3. Let $|\theta\rangle$ and $\langle\theta|$ be defined as before. Then the completeness relation takes the form

$$\int d\theta^* \, d\theta \, |\theta\rangle\langle\theta|e^{-\theta^*\theta} = I. \tag{1.197}$$

Proof. Straightforward calculation yields

$$\int d\theta^* \, d\theta \, |\theta\rangle\langle\theta| e^{-\theta^*\theta}$$

$$= \int d\theta^* \, d\theta \, (|0\rangle + |1\rangle\theta)(\langle 0| + \theta^*\langle 1|)(1 - \theta^*\theta)$$

$$= \int d\theta^* \, d\theta \, \left(|0\rangle\langle 0| + |1\rangle\theta\langle 0| + |0\rangle\theta^*\langle 1| + |1\rangle\theta\theta^*\langle 1|\right)(1 - \theta^*\theta)$$

$$= |0\rangle\langle 0| + |1\rangle\langle 1| = I. \qquad \square$$

1.5.10   Partition function of a fermionic oscillator

We obtain here the partition fuction of a fermionic harmonic oscillator as an application of the path integral formalism of fermions. The Hamiltonian is $H = (c^\dagger c - 1/2)\omega$, which has eigenvalues $\pm\omega/2$. The partition function is then

$$Z(\beta) = \text{Tr} \, e^{-\beta H} = \sum_{n=0}^{1} \langle n|e^{-\beta H}|n\rangle = e^{\beta\omega/2} + e^{-\beta\omega/2} = 2\cosh(\beta\omega/2).$$

(1.198)

Now we evaluate $Z(\beta)$ in two different ways using a path integral. We start our exposition with the following lemma.

Lemma 1.4. Let H be the Hamiltonian of a fermionic harmonic oscillator. Then the partition function is written as

$$\text{Tr} \, e^{-\beta H} = \int d\theta^* \, d\theta \, \langle -\theta|e^{-\beta H}|\theta\rangle e^{-\theta^*\theta}. \qquad (1.199)$$

Proof. Let us insert the completeness relation (1.197) into the definition of a partition function to obtain

$$Z(\beta) = \sum_{n=0,1} \langle n|e^{-\beta H}|n\rangle$$

$$= \sum_{n} \int d\theta^* \, d\theta e^{-\theta^*\theta} \langle n|\theta\rangle\langle\theta|e^{-\beta H}|n\rangle$$

$$= \sum_{n} \int d\theta^* \, d\theta \, (1 - \theta^*\theta)(\langle n|0\rangle + \langle n|1\rangle\theta)(\langle 0|e^{-\beta H}|n\rangle + \theta^*\langle 1|e^{-\beta H}|n\rangle)$$

$$= \sum_{n} \int d\theta^* \, d\theta (1 - \theta^*\theta)[\langle 0|e^{-\beta H}|n\rangle\langle n|0\rangle$$

$$- \theta^*\theta\langle 1|e^{-\beta H}|n\rangle\langle n|1\rangle + \theta\langle 0|e^{-\beta H}|n\rangle\langle n|1\rangle + \theta^*\langle 1|e^{-\beta H}|n\rangle\langle n|0\rangle].$$

The last term of the last line does not contribute to the integral and hence we may change θ^* to $-\theta^*$. Then

$$
\begin{aligned}
Z(\beta) &= \sum_n \int d\theta^* \, d\theta (1 - \theta^*\theta) [\langle 0|e^{-\beta H}|n\rangle \langle n|0\rangle \\
&\quad - \theta^*\theta \langle 1|e^{-\beta H}|n\rangle \langle n|1\rangle + \theta \langle 0|e^{-\beta H}|n\rangle \langle n|1\rangle - \theta^* \langle 1|e^{-\beta H}|n\rangle \langle n|0\rangle] \\
&= \int d\theta^* \, d\theta \, e^{-\theta^*\theta} \langle -\theta|e^{-\beta H}|\theta\rangle. \qquad \square
\end{aligned}
$$

Accordingly, the coordinate in the trace is over *anti-periodic* orbits. The Grassmann variable is θ at $\tau = 0$ while $-\theta$ at $\tau = \beta$ and we have to impose an anti-periodic boundary condition over $[0, \beta]$ in the trace.

Use the expression

$$
e^{-\beta H} = \lim_{N\to\infty} (1 - \beta H/N)^N
$$

and insert the completeness relation at each time step to find

$$
\begin{aligned}
Z(\beta) &= \lim_{N\to\infty} \int d\theta^* \, d\theta \, e^{-\theta^*\theta} \langle -\theta|(1 - \beta H/N)^N|\theta\rangle \\
&= \lim_{N\to\infty} \int d\theta^* \, d\theta \prod_{k=1}^{N-1} \int d\theta_k^* \, d\theta_k \, e^{-\sum_{n=1}^{N-1} \theta_n^*\theta_n} \\
&\quad \times \langle -\theta|(1 - \varepsilon H)|\theta_{N-1}\rangle \langle \theta_{N-1}|\dots|\theta_1\rangle \langle \theta_1|(1 - \varepsilon H)|\theta\rangle \\
&= \lim_{N\to\infty} \int \prod_{k=1}^{N} d\theta_k^* \, d\theta_k \, e^{-\sum_{n=1}^{N} \theta_n^*\theta_n} \\
&\quad \times \langle \theta_N|(1 - \varepsilon H)|\theta_{N-1}\rangle \langle \theta_{N-1}|\dots|\theta_1\rangle \langle \theta_1|(1 - \varepsilon H)| - \theta_N\rangle
\end{aligned}
$$

where we have put $\varepsilon = \beta/N$ and $\theta = -\theta_N = \theta_0$, $\theta^* = -\theta_N^* = \theta_0^*$.

Each matrix element is evaluated as

$$
\begin{aligned}
\langle \theta_k|(1 - \varepsilon H)|\theta_{k-1}\rangle &= \langle \theta_k|\theta_{k-1}\rangle \left[1 - \varepsilon \frac{\langle \theta_k|H|\theta_{k-1}\rangle}{\langle \theta_k|\theta_{k-1}\rangle}\right] \\
&\simeq \langle \theta_k|\theta_{k-1}\rangle e^{-\varepsilon \langle \theta_k|H|\theta_{k-1}\rangle / \langle \theta_k|\theta_{k-1}\rangle} \\
&= e^{\theta_k^*\theta_{k-1}} e^{-\varepsilon\omega(\theta_k^*\theta_{k-1} - 1/2)} \\
&= e^{\varepsilon\omega/2} e^{(1-\varepsilon\omega)\theta_k^*\theta_{k-1}}.
\end{aligned}
$$

The partition function is now expressed in terms of the path integral as

$$Z(\beta) = \lim_{N\to\infty} e^{\beta\omega/2} \prod_{k=1}^{N} \int d\theta_k^* d\theta_k e^{-\sum_{n=1}^{N} \theta_n^* \theta_n} e^{(1-\varepsilon\omega) \sum_{n=1}^{N} \theta_n^* \theta_{n-1}}$$

$$= e^{\beta\omega/2} \lim_{N\to\infty} \prod_{k=1}^{N} \int d\theta_k^* d\theta_k e^{-\sum_{n=1}^{N} [\theta_n^*(\theta_n - \theta_{n-1}) + \varepsilon\omega\theta_n^* \theta_{n-1}]}$$

$$= e^{\beta\omega/2} \lim_{N\to\infty} \prod_{k=1}^{N} \int d\theta_k^* d\theta_k e^{-\theta^\dagger \cdot B \cdot \theta}, \tag{1.200}$$

where

$$\theta = \begin{pmatrix} \theta_1 \\ \theta_2 \\ \vdots \\ \theta_N \end{pmatrix} \qquad \theta^\dagger = \begin{pmatrix} \theta_1^*, \theta_2^*, \ldots, \theta_N^* \end{pmatrix}$$

$$B_N = \begin{pmatrix} 1 & 0 & \ldots & 0 & -y \\ y & 1 & 0 & \ldots 0 \\ 0 & y & 1 & \ldots 0 \\ \vdots & & \ddots & & \vdots \\ 0 & 0 & \ldots & y & 1 \end{pmatrix}$$

with $y = -1 + \varepsilon\omega$ in the last line. We finally find from the definition of the Gaussian integral of Grassmann numbers that

$$Z(\beta) = e^{\beta\omega/2} \lim_{N\to\infty} \det B_N = e^{\beta\omega/2} \lim_{N\to\infty} [1 + (1 - \beta\omega/N)^N]$$

$$= e^{\beta\omega/2}(1 + e^{-\beta\omega}) = 2\cosh \tfrac{1}{2}\beta\omega. \tag{1.201}$$

This should be compared with the partition function (1.151) of the bosonic harmonic oscillator.

This partition function is also obtained by making use of the ζ-function regularization. It follows from the second line of equation (1.200) that

$$Z(\beta) = e^{\beta\omega/2} \lim_{N\to\infty} \prod_{k=1}^{N} \int d\theta_k^* d\theta_k e^{-\sum_n [(1-\varepsilon\omega)\theta_n^*(\theta_n - \theta_{n-1})/\varepsilon + \omega\theta_n^* \theta_n]}$$

$$= e^{\beta\omega/2} \int \mathcal{D}\theta^* \mathcal{D}\theta \exp\left[-\int_0^\beta d\tau\, \theta^* \left((1 - \varepsilon\omega)\frac{d}{d\tau} + \omega \right)\theta \right]$$

$$= e^{\beta\omega/2} \mathrm{Det}_{\mathrm{APBC}} \left((1 - \varepsilon\omega)\frac{d}{d\tau} + \omega \right).$$

Here the subscript APBC implies that the eigenvalue should be evaluated for the solutions that satisfy the anti-periodic boundary condition $\theta(\beta) = -\theta(0)$. It might seem odd that the differential operator contains ε. We find later that this gives a finite contribution to the infinite product of eigenvalues. Let us expand the orbit $\theta(\tau)$ in the Fourier modes. The eigenmodes and the corresponding eigenvalues are

$$\exp\left(\frac{\pi i(2n+1)\tau}{\beta}\right), \qquad (1-\varepsilon\omega)\frac{\pi i(2n+1)}{\beta}+\omega,$$

where $n = 0, \pm1, \pm2, \ldots$. It should be noted that the coherent states are overcomplete and that the actual number of degrees of freedom is N, which is related to ε as $\varepsilon = \beta/N$. Then we have to truncate the product at $-N/4 \le k \le N/4$ since one complex variable has two real degrees of freedom. Accordingly, the partition function takes the form

$$Z(\beta) = e^{\beta\omega/2} \lim_{N\to\infty} \prod_{k=-N/4}^{N/4} \left[i(1-\varepsilon\omega)\frac{\pi(2n-1)}{\beta}+\omega\right]$$

$$= e^{\beta\omega/2}e^{-\beta\omega/2} \prod_{k=1}^{\infty}\left[\left(\frac{2\pi(n-1/2)}{\beta}\right)^2+\omega^2\right]$$

$$= \prod_{k=1}^{\infty}\left[\frac{\pi(2k-1)}{\beta}\right]^2 \prod_{n=1}^{\infty}\left[1+\left(\frac{\beta\omega}{\pi(2n-1)}\right)^2\right].$$

The first infinite product, which we call P, is divergent and requires regularization. Note, first, that

$$\log P = \sum_{k=1}^{\infty} 2\log\left[\frac{2\pi(k-1/2)}{\beta}\right].$$

Define the corresponding ζ-function by

$$\tilde{\zeta}(s) = \sum_{k=1}^{\infty}\left[\frac{2\pi(k-1/2)}{\beta}\right]^{-s} = \left(\frac{\beta}{2\pi}\right)^s \zeta(s,1/2)$$

with which we obtain $P = e^{-2\tilde{\zeta}'(0)}$. Here

$$\zeta(s,a) = \sum_{k=0}^{\infty}\frac{1}{(k+a)^s} \qquad (0 < a < 1) \qquad\qquad (1.202)$$

is the **generalized ζ-function** (the **Hurwitz ζ-function**). The derivative of $\tilde{\zeta}(s)$ at $s=0$ yields

$$\tilde{\zeta}'(0) = \log\left(\frac{\beta}{2\pi}\right)\zeta(0,1/2)+\zeta'(0,1/2) = -\frac{1}{2}\log 2,$$

where use has been made of the values [17]

$$\zeta(0, 1/2) = 0 \qquad \zeta'(0, 1/2) = -\tfrac{1}{2} \log 2.$$

Finally we obtain

$$P = e^{-2\tilde{\zeta}'(0)} = e^{\log 2} = 2. \tag{1.203}$$

Note that P is independent of β after regularization.

Putting them all together, we arrive at the partition function

$$Z(\beta) = 2 \prod_{n=1}^{\infty} \left[1 + \left(\frac{\beta\omega}{\pi(2n-1)} \right)^2 \right]. \tag{1.204}$$

By making use of the well-known formula

$$\cosh \frac{x}{2} = \prod_{n=1}^{\infty} \left[1 + \frac{x^2}{\pi^2(2n-1)^2} \right] \tag{1.205}$$

we obtain

$$Z(\beta) = 2 \cosh \frac{\beta\omega}{2}. \tag{1.206}$$

Suppose, alternatively, we are ignorant about the formula (1.205). Then, by equating equation (1.201) with equation (1.204), we have *proved* the formula (1.205) with the help of path integrals. This is a typical application of physics to mathematics: evaluate some physical quantity by two different methods and equate the results. Then we often obtain a non-trivial relation which is mathematically useful.

1.6   Quantization of a scalar field

1.6.1   Free scalar field

The analysis made in the previous sections may be easily generalized to a case with many degrees of freedom. We are interested, in particular, in a system with infinitely many degrees of freedom; the **quantum field theory** (QFT). Let us start our exposition with the simplest case, that is, the scalar field theory. Let $\phi(x)$ be a real scalar field at the spacetime coordinates $x = (\boldsymbol{x}, x^0)$ where $\boldsymbol{x}$ is the space coordinate while x^0 is the time coordinate. The action depends on ϕ and its derivatives $\partial_\mu \phi(x) = \partial\phi(x)/\partial x^\mu$:

$$S = \int dx \, \mathcal{L}(\phi, \partial_\mu \phi). \tag{1.207}$$

---

[17] The first formula follows from the relation $\zeta(s, 1/2) = (2^s - 1)\zeta(s)$, which is derived from the identity $\zeta(s, 1/2) + \zeta(s) = 2^s \sum_{n=1}^{\infty}[1/(2n-1)^s + 1/(2n)^s] = 2^s \zeta(s)$. The second formula is obtained by differentiating $\zeta(s, 1/2) = (2^s - 1)\zeta(s)$ with respect to s and using the formula $\zeta(0) = -1/2$.

Here $\mathcal{L}$ is the Lagrangian density. The Euler–Lagrange equation now takes the form

$$\frac{\partial}{\partial x^\mu}\left(\frac{\partial \mathcal{L}}{\partial(\partial_\mu \phi)}\right) - \frac{\partial \mathcal{L}}{\partial \phi} = 0. \tag{1.208}$$

The Lagrangian density of a free scalar field is

$$\mathcal{L}_0(\phi, \partial_\mu \phi) = -\tfrac{1}{2}(\partial_\mu \phi \partial^\mu \phi + m^2 \phi^2). \tag{1.209}$$

The Euler–Lagrange equation derived from this Lagrangian density is the Klein–Gordon equation

$$(\Box - m^2)\phi = 0, \tag{1.210}$$

where $\Box = \partial^\mu \partial_\mu = -\partial_0^2 + \nabla^2$.

The vacuum-to-vacuum amplitude in the presence of a source J has the path integral representation $\langle 0, \infty | 0, -\infty \rangle^J \propto Z_0[J]$, where

$$Z_0[J] = \int \mathcal{D}\phi \exp\left[i \int dx \left(\mathcal{L}_0 + J\phi + \frac{i}{2}\varepsilon \phi^2\right)\right] \tag{1.211}$$

where the $i\varepsilon$ term has been added to regularize the path integral.[18] Integration by parts yields

$$Z_0[J] = \int \mathcal{D}\phi \exp\left[i \int dx \left(\tfrac{1}{2}\{\phi(\Box - m^2)\phi + i\varepsilon \phi^2\} + J\phi\right)\right]. \tag{1.212}$$

Let ϕ_c be the classical solution to the Klein–Gordon equation in the presence of the source,

$$(\Box - m^2 + i\varepsilon)\phi_c = -J. \tag{1.213}$$

The solution is easily found to be

$$\phi_c(x) = -\int dy\, \Delta(x - y) J(y) \tag{1.214}$$

where $\Delta(x - y)$ is the Feynman propagator

$$\Delta(x - y) = \frac{-1}{(2\pi)^d} \int d^d k \frac{e^{ik(x-y)}}{k^2 + m^2 - i\varepsilon}. \tag{1.215}$$

Here d denotes the spacetime dimension. Note that $\Delta(x - y)$ satisfies

$$(\Box - m^2 + i\varepsilon)\Delta(x - y) = \delta^d(x - y).$$

It is easy to show that (exercise) the functional $Z_0[J]$ is now written as

$$Z_0[J] = Z_0[0] \exp\left[-\frac{i}{2}\int dx\, dy\, J(x)\Delta(x - y)J(y)\right]. \tag{1.216}$$

[18] Alternatively, we can introduce the imaginary time $\tau = ix^0$ to Wick rotate the time axis.

It is instructive to note that the propagator is conversely obtained by the functional derivative of $Z_0[J]$,

$$\Delta(x - y) = \frac{i}{Z_0[0]} \frac{\delta^2 Z_0[J]}{\delta J(x)\delta J(y)}\bigg|_{J=0}. \tag{1.217}$$

The amplitude $Z_0[0]$ is the vacuum-to-vacuum amplitude in the absence of the source and may be evaluated as follows. Let us introduce the imaginary time $x^4 = \tau = ix^0$. Then, we obtain

$$Z_0[0] = \int \bar{\mathcal{D}}\phi \exp\left[\frac{1}{2}\int dx\, \phi(\bar{\Box} - m^2)\phi\right]$$
$$= [\text{Det}(\bar{\Box} - m^2)]^{-1/2}, \tag{1.218}$$

where $\bar{\Box} = \partial_\tau^2 + \nabla^2$ and the deteminant is understood in the sense of section 1.4, namely it is the product of eigenvalues with a relevant boundary condition.

A free complex scalar field theory has a Lagrangian density

$$\mathcal{L}_0 = -\partial_\mu\phi^*\partial^\mu\phi - m^2|\phi|^2 + J\phi^* + J^*\phi \tag{1.219}$$

where the source terms have been included. The generating functional is now given by

$$Z_0[J, J^*] = \int \mathcal{D}\phi\mathcal{D}\phi^* \exp\left[i\int dx\,(\mathcal{L}_0 - i\varepsilon|\phi|^2)\right]$$
$$= \int \mathcal{D}\phi\mathcal{D}\phi^* \exp\left[i\int dx\,\{\phi^*(\Box - m^2 + i\varepsilon)\phi + J^*\phi + J\phi^*\}\right]. \tag{1.220}$$

The propagator is now given by

$$\Delta(x - y) = \frac{i}{Z_0[0, 0]} \frac{\delta^2 Z_0[J, J^*]}{\delta J^*(x)\delta J(y)}\bigg|_{J=J^*=0}. \tag{1.221}$$

By substituting the Klein–Gordon equations

$$(\Box - m^2)\phi_c = -J \qquad (\Box - m^2)\phi_c^* = -J^* \tag{1.222}$$

we separate the generating functional as

$$Z_0[J, J^*] = Z_0[0, 0]\exp\left[-i\int dx\, dy\, J^*(x)\Delta(x - y)J(y)\right] \tag{1.223}$$

where

$$Z_0[0, 0] = \int \mathcal{D}\phi\,\mathcal{D}\phi^* \exp\left[-i\int dx\phi^*(\Box - m^2 - i\varepsilon)\phi\right]$$
$$= [\text{Det}(\bar{\Box} - m^2)]^{-1}. \tag{1.224}$$

Wick rotation has been made to occur at the last line.

1.6.2   Interacting scalar field

It is possible to add interaction terms to the free field Lagrangian (1.209),

$$\mathcal{L}(\phi, \partial_\mu\phi) = \mathcal{L}_0(\phi, \partial_\mu\phi) - V(\phi). \tag{1.225}$$

The possible form of $V(\phi)$ is restricted by the symmetry and renormalizability of the theory. A typical form of V is a polynomial

$$V(\phi) = \frac{g}{n!}\phi^n \qquad (n \geq 3, n \in \mathbb{N})$$

where the constant $g \in \mathbb{R}$ controls the strength of the interaction. The generating functional is defined similarly to the free theory as

$$Z[J] = \int \mathcal{D}\phi \exp\left[i\int dx\, \{\tfrac{1}{2}\phi(\square - m^2)\phi - V(\phi) + J\phi\}\right]. \tag{1.226}$$

The presence of $V(\phi)$ makes things slightly more complicated. It can be handled at least perturbatively as

$$
\begin{aligned}
Z[J] &= \int \mathcal{D}\phi \exp\left[-i\int dx\, V(\phi)\right]\exp\left[i\int dx\, \{\mathcal{L}_0 + J\phi\}\right]\\
&= \exp\left[-i\int dx\, V\left(\frac{1}{i}\frac{\delta}{\delta J(x)}\right)\right]\int \mathcal{D}\phi \exp\left[i\int dx\, \{\mathcal{L}_0 + J\phi\}\right]\\
&= \exp\left[-i\int dx\, V\left(\frac{1}{i}\frac{\delta}{\delta J(x)}\right)\right]Z_0[J]\\
&= \sum_{k=0}^{\infty}\int dx_1 \ldots \int dx_k \frac{(-i)^k}{k!}\\
&\quad \times V\left(\frac{1}{i}\frac{\delta}{\delta J(x_1)}\right)\ldots V\left(\frac{1}{i}\frac{\delta}{\delta J(x_k)}\right)Z_0[J].
\end{aligned}\tag{1.227}
$$

The generating functional $Z[J]$ generates the vacuum expectation value of the T-product of field operators, also known as the **Green function** $G_n(x_1, \ldots, x_n)$, as

$$
\begin{aligned}
G_n(x_1, \ldots, x_n) &\equiv \langle 0|T[\phi(x_1)\ldots\phi(x_n)]|0\rangle\\
&= \frac{(-i)^n \delta^n}{\delta J(x_1)\ldots\delta J(x_n)}Z[J]\bigg|_{J=0}.
\end{aligned}\tag{1.228}
$$

Since this is the nth functional derivative of $Z[J]$ around $J = 0$, we obtain the functional Taylor expansion of $Z[J]$ as

$$
\begin{aligned}
Z[J] &= \sum_{n=1}^{\infty}\frac{1}{n!}\left[\prod_{i=1}^{n}\int dx_i\, J(x_i)\right]\langle 0|T[\phi(x_1)\ldots\phi(x_n)]|0\rangle\\
&= \langle 0|Te^{\int dx\, J(x)\phi(x)}|0\rangle.
\end{aligned}\tag{1.229}
$$

The connected n-point functions are generated by $W[J]$ defined by

$$Z[J] = e^{-W[J]}. \tag{1.230}$$

The **effective action** $\Gamma[\phi_{cl}]$ is defined by the Legendre transformation

$$\Gamma[\phi_{cl}] \equiv W[J] - \int d\tau \, d\mathbf{x} \, J\phi_{cl} \tag{1.231}$$

where

$$\phi_{cl} \equiv \langle \phi \rangle^J = \frac{\delta W[J]}{\delta J}. \tag{1.232}$$

The functional $\Gamma[\phi_{cl}]$ generates **one-particle irreducible** diagrams.

1.7 Quantization of a Dirac field

The Lagrangian of the free **Dirac field** ψ is

$$\mathcal{L}_0 = \bar{\psi}(i\slashed{\partial} - m)\psi, \tag{1.233}$$

where $\slashed{\partial} = \gamma^\mu \partial_\mu$. In general $\slashed{A} \equiv \gamma^\mu A_\mu$. Variation with respect to $\bar{\psi}$ yields the **Dirac equation**

$$(i\slashed{\partial} - m)\psi = 0. \tag{1.234}$$

The Dirac field, in canonical quantization, satisifes the anti-commutation relation

$$\{\bar{\psi}(x^0, \mathbf{x}), \psi(x^0, \mathbf{y})\} = \delta(\mathbf{x} - \mathbf{y}). \tag{1.235}$$

Accordingly, it is expressed as a Grassmann number function in path integrals. The generating functional is

$$Z_0[\bar{\eta}, \eta] = \int \mathcal{D}\bar{\psi}\mathcal{D}\psi \exp\left[i \int d\mathbf{x} \left(\bar{\psi}(i\slashed{\partial} - m)\psi + \bar{\psi}\eta + \bar{\eta}\psi\right)\right] \tag{1.236}$$

where η, $\bar{\eta}$ are Grassmannian sources.

The propagator is given by the functional derivative with respect to the sources,

$$\begin{aligned} S(x - y) &= -\frac{\delta^2 Z_0[\bar{\eta}, \eta]}{\delta\bar{\eta}(x)\delta\eta(y)} \\ &= \frac{1}{(2\pi)^d} \int d^d k \frac{e^{ikx}}{\slashed{k} - m - i\varepsilon} = (i\slashed{\partial} + m + i\varepsilon)\Delta(x - y) \end{aligned} \tag{1.237}$$

where $\Delta(x - y)$ is the scalar field propagator.

By making use of the Dirac equations

$$(i\slashed{\partial} - m)\psi = -\eta \qquad \bar{\psi}(i\overleftarrow{\slashed{\partial}} + m) = \bar{\eta} \tag{1.238}$$

the generating functional is cast into the form

$$Z_0[\bar{\eta}, \eta] = Z_0[0, 0] \exp\left[-i \int dx\, dy\, \bar{\eta}(x) S(x - y) \eta(y)\right]. \qquad (1.239)$$

After Wick rotation $\tau = ix^0$, the normalization factor is obtained as

$$Z_0[0, 0] = \text{Det}(i\slashed{\partial} - m) = \prod_i \lambda_i \qquad (1.240)$$

where λ_i is the ith eigenvalue of the Dirac operator $i\slashed{\partial} - m$.

1.8 Gauge theories

At present, physically sensible theories of fundamental interactions are based on gauge theories. The gauge principle—*physics should not depend on how we describe it*—is in harmony with the principle of general relativity. Here we give a brief summary of classical aspects of gauge theories. For further references, the reader should consult those books listed at the beginning of this chapter.

1.8.1 Abelian gauge theories

The reader should be familiar with Maxwell's equations:

$$\text{div } \boldsymbol{B} = 0 \qquad (1.241a)$$

$$\frac{\partial \boldsymbol{B}}{\partial t} + \text{curl } \boldsymbol{E} = 0 \qquad (1.241b)$$

$$\text{div } \boldsymbol{E} = \rho \qquad (1.241c)$$

$$\frac{\partial \boldsymbol{E}}{\partial t} - \text{curl } \boldsymbol{E} = -\boldsymbol{j}. \qquad (1.241d)$$

The magnetic field $\boldsymbol{B}$ and the electric field $\boldsymbol{E}$ are expressed in terms of the vector potential $A_\mu = (\phi, \boldsymbol{A})$ as

$$\boldsymbol{B} = \text{curl } \boldsymbol{A} \qquad \boldsymbol{E} = \frac{\partial \boldsymbol{A}}{\partial t} - \text{grad } \phi. \qquad (1.242)$$

Maxwell's equations are invariant under the **gauge transformation**

$$A_\mu \to A_\mu + \partial_\mu \chi \qquad (1.243)$$

where χ is a scalar function. This invariance is manifest if we define the **electromagnetic field tensor** $F_{\mu\nu}$ by

$$F_{\mu\nu} \equiv \partial_\mu A_\nu - \partial_\nu A_\mu = \begin{pmatrix} 0 & -E_x & -E_y & -E_z \\ E_x & 0 & B_z & -B_y \\ E_y & -B_z & 0 & B_z \\ E_z & B_y & -B_x & 0 \end{pmatrix}. \qquad (1.244)$$

From the construction, F is invariant under (1.243). The Lagrangian of the electromagnetic fields is given by

$$\mathcal{L}_{\text{EM}} = -\tfrac{1}{4} F_{\mu\nu} F^{\mu\nu} + A_\mu j^\mu \tag{1.245}$$

where $j^\mu = (\rho, \boldsymbol{j})$.

Exercise 1.11. Show that (1.241a) and (1.241b) are written as

$$\partial_\xi F_{\mu\nu} + \partial_\mu F_{\nu\xi} + \partial_\nu F_{\xi\mu} = 0 \tag{1.246a}$$

while (1.241c) and (1.241d) are

$$\partial_\nu F^{\mu\nu} = j^\mu \tag{1.246b}$$

where the raising and lowering of spacetime indices are carried out with the Minkowski metric $\eta = \text{diag}(-1, 1, 1, 1)$. Verify that (1.246b) is the Euler–Lagrange equation derived from (1.245).

Let ψ be a Dirac field with electric charge e. The free Dirac Lagrangian

$$\mathcal{L}_0 = \bar{\psi}(\mathrm{i}\gamma^\mu \partial_\mu + m)\psi \tag{1.247}$$

is clearly invariant under the *global* gauge transformation

$$\psi \to \mathrm{e}^{-\mathrm{i}e\alpha}\psi \qquad \bar{\psi} \to \bar{\psi}\mathrm{e}^{\mathrm{i}e\alpha} \tag{1.248}$$

where $\alpha \in \mathbb{R}$ is a constant. We elevate this symmetry to invariance under the *local* gauge transformation,

$$\psi \to \mathrm{e}^{-\mathrm{i}e\alpha(x)}\psi \qquad \bar{\psi} \to \bar{\psi}\mathrm{e}^{\mathrm{i}e\alpha(x)}. \tag{1.249}$$

The Lagrangian transforms under (1.249) as

$$\bar{\psi}(\mathrm{i}\gamma^\mu \partial_\mu + m)\psi \to \bar{\psi}(\mathrm{i}\gamma^\mu \partial_\mu + e\gamma^\mu \partial_\mu\alpha + m)\psi. \tag{1.250}$$

Since the extra term $e\partial_\mu\alpha$ looks like a gauge transformation of the vector potential, we couple the gauge field A_μ with ψ so that the Lagrangian has a local gauge symmetry. We find that

$$\mathcal{L} = \bar{\psi}[\mathrm{i}\gamma^\mu(\partial_\mu - \mathrm{i}eA_\mu) + m]\psi \tag{1.251}$$

is invariant under the combined gauge transformation,

$$\psi \to \psi' = \mathrm{e}^{-\mathrm{i}e\alpha(x)}\psi \qquad \bar{\psi} \to \bar{\psi}' = \bar{\psi}\mathrm{e}^{\mathrm{i}e\alpha(x)}$$
$$A_\mu \to A'_\mu = A_\mu - \partial_\mu\alpha(x). \tag{1.252}$$

Let us introduce the **covariant derivatives**,

$$\nabla_\mu \equiv \partial_\mu - ieA_\mu \qquad \nabla'_\mu \equiv \partial_\mu - ieA'_\mu. \qquad (1.253)$$

The reader should verify that $\nabla_\mu \psi$ transforms in a nice way,

$$\nabla'_\mu \psi' = e^{-ie\alpha(x)} \nabla_\mu \psi. \qquad (1.254)$$

The total quantum electrodynamic (QED) Lagrangian is

$$\mathcal{L}_{QED} = -\tfrac{1}{4}F^{\mu\nu}F_{\mu\nu} + \bar\psi(i\gamma^\mu\nabla_\mu + m)\psi. \qquad (1.255)$$

Exercise 1.12. Let $\phi = (\phi_1 + i\phi_2)/\sqrt{2}$ be a complex scalar field with electric charge e. Show that the Lagrangian

$$\mathcal{L} = \eta^{\mu\nu}(\nabla_\mu\phi)^\dagger(\nabla_\nu\phi) + m^2\phi^\dagger\phi \qquad (1.256)$$

is invariant under the gauge transformation

$$\phi \to e^{-ie\alpha(x)}\phi \qquad \phi^\dagger \to \phi^\dagger e^{ie\alpha(x)} \qquad A_\mu \to A_\mu - \partial_\mu\alpha(x). \qquad (1.257)$$

1.8.2 Non-Abelian gauge theories

The gauge transformation just described is a member of a U(1) group, that is a complex number of modulus 1, which happens to be an Abelian group. A few decades ago, Yang and Mills (1954) introduced non-Abelian gauge transformations. At that time, non-Abelian gauge theories were studied from curiosity. Nowadays, they play a central role in elementary particle physics.

Let G be a compact semi-simple Lie group such as SO(N) or SU(N). The anti-Hermitian generators $\{T_\alpha\}$ satisfy the commutation relations

$$[T_\alpha, T_\beta] = f_{\alpha\beta}{}^\gamma T_\gamma \qquad (1.258)$$

where the numbers $f_{\alpha\beta}{}^\gamma$ are called the **structure constants** of G. An element U of G near the unit element can be expressed as

$$U = \exp(-\theta^\alpha T_\alpha). \qquad (1.259)$$

We suppose a Dirac field ψ transforms under $U \in G$ as

$$\psi \to U\psi \qquad \bar\psi \to \bar\psi U^\dagger. \qquad (1.260)$$

[*Remark*: Strictly speaking, we have to specify the representation of G to which ψ belongs. If readers feel uneasy about (1.260), they may consider ψ is in the fundamental representation, for example.]

Consider the Lagrangian

$$\mathcal{L} = \bar\psi[i\gamma^\mu(\partial_\mu + g\mathcal{A}_\mu) + m]\psi \qquad (1.261)$$

where the **Yang–Mills gauge field** $\mathcal{A}_\mu$ takes its values in the Lie algebra of G, that is, $\mathcal{A}_\mu$ can be expanded in terms of T_α as $\mathcal{A}_\mu = A_\mu{}^\alpha T_\alpha$. (Script fields are anti-Hermitian.) The constant g is the coupling constant which controls the strength of the coupling between the Dirac field and the gauge field. It is easily verified that $\mathcal{L}$ is invariant under

$$\psi \to \psi' = U\psi \qquad \bar{\psi} \to \bar{\psi}' = \bar{\psi}U^\dagger$$
$$\mathcal{A}_\mu \to \mathcal{A}'_\mu = U\mathcal{A}_\mu U^\dagger + g^{-1}U\partial_\mu U^\dagger. \tag{1.262}$$

The covariant derivative is defined by $\nabla_\mu = \partial_\mu + g\mathcal{A}_\mu$ as before. The covariant derivative $\nabla_\mu \psi$ transforms covariantly under the gauge transformation

$$\nabla'_\mu \psi' = U\nabla_\mu \psi. \tag{1.263}$$

The **Yang–Mills field tensor** is

$$\mathcal{F}_{\mu\nu} \equiv \partial_\mu \mathcal{A}_\nu - \partial_\nu \mathcal{A}_\mu + g[\mathcal{A}_\mu, \mathcal{A}_\nu]. \tag{1.264}$$

The component $F_{\mu\nu}{}^\alpha$ is

$$F_{\mu\nu}{}^\alpha = \partial_\mu A_\nu{}^\alpha - \partial_\nu A_\mu{}^\alpha + g f_{\beta\gamma}{}^\alpha A_\mu{}^\beta A_\nu{}^\gamma. \tag{1.265}$$

If we define the **dual field tensor** $*\mathcal{F}_{\mu\nu} \equiv \frac{1}{2}\varepsilon_{\mu\nu\kappa\lambda}\mathcal{F}^{\kappa\lambda}$, it satisfies the **Bianchi identity**,

$$\mathcal{D}_\mu * \mathcal{F}^{\mu\nu} \equiv \partial_\mu * \mathcal{F}^{\mu\nu} + g[\mathcal{A}_\mu, *\mathcal{F}^{\mu\nu}] = 0. \tag{1.266}$$

Exercise 1.13. Show that $\mathcal{F}_{\mu\nu}$ transforms under (1.262) as

$$\mathcal{F}_{\mu\nu} \to U\mathcal{F}_{\mu\nu}U^\dagger. \tag{1.267}$$

From this exercise, we find a gauge-invariant action

$$\mathcal{L}_{\text{YM}} = -\frac{1}{2}\operatorname{tr}(\mathcal{F}^{\mu\nu}\mathcal{F}_{\mu\nu}) \tag{1.268a}$$

where the trace is over the group matrix. The component form is

$$\mathcal{L}_{\text{YM}} = -\frac{1}{2}F^{\mu\nu\alpha}F_{\mu\nu}{}^\beta \operatorname{tr}(T_\alpha T_\beta) = \frac{1}{4}F^{\mu\nu\alpha}F_{\mu\nu\alpha} \tag{1.268b}$$

where we have normalized $\{T_\alpha\}$ so that $\operatorname{tr}(T_\alpha T_\beta) = -\frac{1}{2}\delta_{\alpha\beta}$. The field equation derived from (1.268) is

$$\mathcal{D}_\mu \mathcal{F}_{\mu\nu} = \partial_\mu \mathcal{F}_{\mu\nu} + g[\mathcal{A}_\mu, \mathcal{F}_{\mu\nu}] = 0. \tag{1.269}$$

1.8.3 Higgs fields

If the gauge symmetry is manifest in our world, there would be many observable massless vector fields. The absence of such fields, except for the electromagnetic field, forces us to break the gauge symmetry. The theory is left renormalizable if the symmetry is broken spontaneously.

Let us consider a U(1) gauge field coupled to a complex scalar field ϕ, whose Lagrangian is given by

$$\mathcal{L} = -\tfrac{1}{4}F^{\mu\nu}F_{\mu\nu} + (\nabla_\mu\phi)^\dagger(\nabla_\mu\phi) - \lambda(\phi^\dagger\phi - v^2)^2. \tag{1.270}$$

The potential $V(\phi) = \lambda(\phi^\dagger\phi - v^2)^2$ has minima $V = 0$ at $|\phi| = v$. The Lagrangian (1.270) is invariant under the local gauge transformation

$$A_\mu \to A_\mu - \partial_\mu\alpha \qquad \phi \to e^{-i e\alpha}\phi \qquad \phi^\dagger \to e^{i e\alpha}\phi^\dagger. \tag{1.271}$$

This symmetry is spontaneously broken due to the **vacuum expectation value** (VEV) $\langle\phi\rangle$ of the **Higgs field** ϕ. We expand ϕ as

$$\phi = \frac{1}{\sqrt{2}}[v + \rho(x)]e^{i\alpha(x)/v} \sim \frac{1}{\sqrt{2}}[v + \rho(x) + i\alpha(x)]$$

assuming $v \neq 0$. If $v \neq 0$, we may take the **unitary gauge** in which the phase of ϕ is 'gauged away' so that ϕ has only the real part,

$$\phi(x) = \frac{1}{\sqrt{2}}(v + \rho(x)). \tag{1.272}$$

If we substitute (1.272) into (1.270) and expand in ρ, we have

$$\mathcal{L} = -\tfrac{1}{4}F_{\mu\nu}F^{\mu\nu} + \tfrac{1}{2}\partial_\mu\rho\partial^\mu\rho + \tfrac{1}{2}e^2 A_\mu A^\mu(v^2 + 2v\rho + \rho^2)$$
$$\qquad - \tfrac{1}{4}\lambda(4v^2\rho^2 + 4v\rho^3 + \rho^4). \tag{1.273}$$

The equations of motion for A_μ and ρ derived from the free parts are

$$\partial^\nu F_{\nu\mu} + 2e^2 v^2 A_\mu = 0 \qquad \partial_\mu\partial^\mu\rho + 2\lambda v^2\rho = 0. \tag{1.274}$$

From the first equation, we find A_μ must satisfy the Lorentz condition $\partial_\mu A^\mu = 0$. The apparent degrees of freedom of (1.270) are 2(photon) + 2(complex scalar) = 4. If VEV $\neq 0$, we have 3(massive vector) + 1(real scalar) = 4. The field A_0 has a mass term with the wrong sign and so cannot be a physical degree of freedom. The creation of massive fields out of a gauge field is called the **Higgs mechanism.**.

1.9 Magnetic monopoles

Maxwell's equations unify electricity and magnetism. In the history of physics they should be recognized as the first attempt to unify forces in Nature. In spite of their great success, Dirac (1931) noticed that there existed an asymmetry in Maxwell's equations: the equation div $\boldsymbol{B} = 0$ denies the existence of magnetic charges. He introduced the magnetic monopole, a point magnetic charge, to make the theory symmetric.

1.9.1 Dirac monopole

Consider a monopole of strength g sitting at $r = 0$,

$$\text{div } \boldsymbol{B} = 4\pi g \delta^3(\boldsymbol{r}). \tag{1.275}$$

It follows from $\Delta(1/r) = -4\pi \delta^3(\boldsymbol{r})$ and $\nabla(1/r) = -\boldsymbol{r}/r^3$ that the solution of this equation is

$$\boldsymbol{B} = g\boldsymbol{r}/r^3. \tag{1.276}$$

The magnetic flux Φ is obtained by integrating $\boldsymbol{B}$ over a sphere S of radius R so that

$$\Phi = \oint_S \boldsymbol{B} \cdot d\boldsymbol{S} = 4\pi g. \tag{1.277}$$

What about the vector potential which gives the monopole field (1.276)? If we define the vector potential $\boldsymbol{A}^{\mathrm{N}}$ by

$$A^{\mathrm{N}}{}_x = \frac{-gy}{r(r+z)} \qquad A^{\mathrm{N}}{}_y = \frac{gx}{r(r+z)} \qquad A^{\mathrm{N}}{}_z = 0 \tag{1.278a}$$

we easily verify that

$$\text{curl } \boldsymbol{A}^{\mathrm{N}} = g\boldsymbol{r}/r^3 + 4\pi g \delta(x)\delta(y)\theta(-z). \tag{1.279}$$

We have curl $\boldsymbol{A}^{\mathrm{N}} = \boldsymbol{B}$ except along the negative z-axis ($\theta = \pi$). The singularity along the z-axis is called the **Dirac string** and reflects the poor choice of the coordinate system. If, instead, we define another vector potential

$$A^{\mathrm{S}}{}_x = \frac{gy}{r(r-z)} \qquad A^{\mathrm{S}}{}_y = \frac{-gx}{r(r-z)} \qquad A^{\mathrm{S}}{}_z = 0 \tag{1.278b}$$

we have curl $\boldsymbol{A}^{\mathrm{S}} = \boldsymbol{B}$ except along the positive z-axis ($\theta = 0$) this time. The existence of a singularity is a natural consequence of (1.277). If there were a vector $\boldsymbol{A}$ such that $\boldsymbol{B} = \text{curl } \boldsymbol{A}$ with no singularity, we would have, from Gauss' law,

$$\Phi = \oint_S \boldsymbol{B} \cdot d\boldsymbol{S} = \oint_S \text{curl } \boldsymbol{A} \cdot d\boldsymbol{S} = \int_V \text{div}(\text{curl } \boldsymbol{A}) \, dV = 0$$

where V is the volume inside the surface S. This problem is avoided only when we abandon the use of a single vector potential.

Exercise 1.14. Let us introduce the polar coordinates (r, θ, ϕ). Show that the vector potentials $\boldsymbol{A}^{\mathrm{N}}$ and $\boldsymbol{A}^{\mathrm{S}}$ are expressed as

$$\boldsymbol{A}^{\mathrm{N}}(\boldsymbol{r}) = \frac{g(1 - \cos\theta)}{r\sin\theta} \hat{\boldsymbol{e}}_\phi \tag{1.280a}$$

$$\boldsymbol{A}^{\mathrm{S}}(\boldsymbol{r}) = -\frac{g(1 + \cos\theta)}{r\sin\theta} \hat{\boldsymbol{e}}_\phi \tag{1.280b}$$

where $\hat{\boldsymbol{e}}_\phi = -\sin\phi \, \hat{\boldsymbol{e}}_x + \cos\phi \, \hat{\boldsymbol{e}}_y$.

1.9.2 The Wu–Yang monopole

Wu and Yang (1975) noticed that the geometrical and topological structures behind the Dirac monopole are best described by fibre bundles. In chapters 9 and 10, we give an account of the Dirac monopole in terms of fibre bundles and their connections. Here we outline the idea of Wu and Yang without introducing the fibre bundle. Wu and Yang noted that we may employ more than one vector potential to describe a monopole. For example, we may avoid singularities if we adopt A^N in the northern hemisphere and A^S in the southern hemisphere of the sphere S surrounding the monopole. These vector potentials yield the magnetic field $B = gr/r^3$, which is non-singular everywhere on the sphere. On the equator of the sphere, which is the boundary between the northern and southern hemispheres, A^N and A^S are related by the gauge transformation, $A^N - A^S = \text{grad } \Lambda$. To compute this quantity Λ, we employ the result of exercise 1.14,

$$A^N - A^S = \frac{2g}{r \sin \theta} \hat{e}_\phi = \text{grad}(2g\phi) \tag{1.281}$$

where use has been made of the expression

$$\text{grad } f = \frac{\partial f}{\partial r} \hat{e}_r + \frac{1}{r} \frac{\partial f}{\partial \theta} \hat{e}_\theta + \frac{1}{r \sin \theta} \frac{\partial f}{\partial \phi} \hat{e}_\phi.$$

Accordingly, the gauge transformation function connecting A^N and A^S is

$$\Lambda = 2g\phi. \tag{1.282}$$

Note that Λ is ill defined at $\theta = 0$ and $\theta = \pi$. Since we perform the gauge transformation only at $\theta = \pi/2$, these singularities do not show up in our analysis. The total flux is

$$\Phi = \oint_S \text{curl } A \cdot dS = \int_{U_N} \text{curl } A^N \cdot dS + \int_{U_S} \text{curl } A^S \cdot dS \tag{1.283}$$

where U_N and U_S stand for the northern and southern hemispheres respectively. Stokes' theorem yields

$$\Phi = \oint_{\text{equator}} A^N \cdot ds - \oint_{\text{equator}} A^S \cdot ds = \oint_{\text{equator}} (A^N - A^S) \cdot ds$$

$$= \oint_{\text{equator}} \text{grad}(2g\phi) \cdot ds = 4g\pi \tag{1.284}$$

in agreement with (1.277).

1.9.3 Charge quantization

Consider a point particle with electric charge e and mass m moving in the field of a magnetic monopole of charge g. If the monopole is heavy enough, the

assume $a^i \boldsymbol{g}_i + b^i \boldsymbol{h}_i = \mathbf{0}$. Then $\mathbf{0} = f(\mathbf{0}) = f(a^i \boldsymbol{g}_i + b^i \boldsymbol{h}_i) = b^i f(\boldsymbol{h}_i) = b^i \boldsymbol{h}'_i$, which implies that $b^i = 0$. Then it follows from $a^i \boldsymbol{g}_i = \mathbf{0}$ that $a^i = 0$, and the set $\{\boldsymbol{g}_1, \ldots, \boldsymbol{g}_r, \boldsymbol{h}_1, \ldots, \boldsymbol{h}_s\}$ is linearly independent in V. Finally we find $\dim V = r + s = \dim(\ker f) + \dim(\mathrm{im}\, f)$. □

[*Remark*: The vector space spanned by $\{\boldsymbol{h}_1, \ldots, \boldsymbol{h}_s\}$ is called the **orthogonal complement** of $\ker f$ and is denoted by $(\ker f)^{\perp}$.]

Exercise 2.8. (1) Let $f : V \to W$ be a linear map. Show that both $\ker f$ and $\mathrm{im}\, f$ are vector spaces.

(2) Show that a linear map $f : V \to V$ is an isomorphism if and only if $\ker f = \{\mathbf{0}\}$.

2.2.3 Dual vector space

The dual vector space has already been introduced in section 1.2 in the context of quantum mechanics. The exposition here is more mathematical and complements the materials presented there.

Let $f : V \to K$ be a linear function on a vector space $V(n, K)$ over a field K. Let $\{e_i\}$ be a basis and take an arbitrary vector $\boldsymbol{v} = v^1 e_1 + \cdots + v^n e_n$. From the linearity of f, we have $f(\boldsymbol{v}) = v^1 f(e_1) + \cdots + v^n f(e_n)$. Thus, if we know $f(e_i)$ for all i, we know the result of the operation of f on any vector. It is remarkable that the set of linear functions is made into a vector space, namely a linear combination of two linear functions is also a linear function.

$$(a_1 f_1 + a_2 f_2)(\boldsymbol{v}) = a_1 f_1(\boldsymbol{v}) + a_2 f_2(\boldsymbol{v}) \tag{2.9}$$

This linear space is called the **dual vector space** to $V(n, K)$ and is denoted by $V^*(n, K)$ or simply by V^*. If $\dim V$ is finite, $\dim V^*$ is equal to $\dim V$. Let us introduce a basis $\{e^{*i}\}$ of V^*. Since e^{*i} is a linear function it is completely specified by giving $e^{*i}(e_j)$ for all j. Let us choose the **dual basis**,

$$e^{*i}(e_j) = \delta^i_j. \tag{2.10}$$

Any linear function f, called a **dual vector** in this context, is expanded in terms of $\{e^{*i}\}$,

$$f = f_i e^{*i}. \tag{2.11}$$

The action of f on $\boldsymbol{v}$ is interpreted as an **inner product** between a column vector and a row vector,

$$f(\boldsymbol{v}) = f_i e^{*i}(v^j e_j) = f_i v^j e^{*i}(e_j) = f_i v^i. \tag{2.12}$$

We sometimes use the notation $\langle \, , \, \rangle : V^* \times V \to K$ to denote the inner product.

Let V and W be vector spaces with a linear map $f : V \to W$ and let $g : W \to K$ be a linear function on W ($g \in W^*$). It is easy to see that the

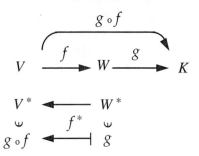

Figure 2.9. The pullback of a function g is a function $f^*(g) = g \circ f$.

composite map $g \circ f$ is a linear function on V. Thus, f and g give rise to an element $h \in V^*$ defined by

$$h(v) \equiv g(f(v)) \qquad v \in V. \tag{2.13}$$

Given $g \in W^*$, a map $f : V \to W$ has induced a map $h \in V^*$. Accordingly, we have an induced map $f^* : W^* \to V^*$ defined by $f^* : g \mapsto h = f^*(g)$, see figure 2.9. The map h is called the **pullback** of g by f^*.

Since $\dim V^* = \dim V$, there exists an isomorphism between V and V^*. However, this isomorphism is not canonical; we have to specify an inner product in V to define an isomorphism between V and V^* and *vice versa*, see the next section. The equivalence of a vector space and its dual vector space will appear recurrently in due course.

Exercise 2.9. Suppose $\{f_j\}$ is another basis of V and $\{f^{*i}\}$ the dual basis. In terms of the old basis, f_i is written as $f_i = A_i{}^j e_j$ where $A \in GL(n, K)$. Show that the dual bases are related by $e^{*i} = f^{*j} A_j{}^i$.

2.2.4 Inner product and adjoint

Let $V = V(m, K)$ be a vector space with a basis $\{e_i\}$ and let g be a vector space isomorphism $g : V \to V^*$, where g is an arbitrary element of $GL(m, K)$. The component representation of g is

$$g : v^j \to g_{ij} v^j. \tag{2.14}$$

Once this isomorphism is given, we may define the **inner product** of two vectors $v_1, v_2 \in V$ by

$$g(v_1, v_2) \equiv \langle g v_1, v_2 \rangle. \tag{2.15}$$

Let us assume that the field K is a real number $\mathbb{R}$. for definiteness. Then equation (2.15) has a component expression,

$$g(v_1, v_2) = v_1{}^i g_{ji} v_2{}^j. \tag{2.16}$$

We require that the matrix (g_{ij}) be positive definite so that the inner product $g(v, v)$ has the meaning of the squared norm of v. We also require that the metric be symmetric: $g_{ij} = g_{ji}$ so that $g(v_1, v_2) = g(v_2, v_1)$.

Next, let $W = W(n, \mathbb{R})$ be a vector space with a basis $\{f_\alpha\}$ and a vector space isomorphism $G : W \to W^*$. Given a map $f : V \to W$, we may define the **adjoint** of f, denoted by $\tilde{f}$, by

$$G(w, fv) = g(v, \tilde{f}w) \qquad (2.17)$$

where $v \in V$ and $w \in W$. It is easy to see that $(\tilde{\tilde{f}}) = f$. The component expression of equation (2.17) is

$$w^\alpha G_{\alpha\beta} f^\beta{}_i v^i = v^i g_{ij} \tilde{f}^j{}_\alpha w^\alpha \qquad (2.18)$$

where $f^\beta{}_i$ and $\tilde{f}^j{}_\alpha$ are the matrix representations of f and $\tilde{f}$ respectively. If $g_{ij} = \delta_{ij}$ and $G_{\alpha\beta} = \delta_{\alpha\beta}$, the adjoint $\tilde{f}$ reduces to the transpose f^t of the matrix f.

Let us show that $\dim \operatorname{im} f = \dim \operatorname{im} \tilde{f}$. Since (2.18) holds for any $v \in V$ and $w \in W$, we have $G_{\alpha\beta} f^\beta{}_i = g_{ij} \tilde{f}^j{}_\alpha$, that is

$$\tilde{f} = g^{-1} f^t G^t. \qquad (2.19)$$

Making use of the result of the following exercise, we obtain $\operatorname{rank} f = \operatorname{rank} \tilde{f}$, where the rank of a map is defined by that of the corresponding matrix (note that $g \in GL(m, \mathbb{R})$ and $G \in GL(n, \mathbb{R})$). It is obvious that $\dim \operatorname{im} f$ is the rank of a matrix representing the map f and we conclude $\dim \operatorname{im} f = \dim \operatorname{im} \tilde{f}$.

Exercise 2.10. Let $V = V(m, \mathbb{R})$ and $W = W(n, \mathbb{R})$ and let f be a matrix corresponding to a linear map from V to W. Verify that $\operatorname{rank} f = \operatorname{rank} f^t = \operatorname{rank}(M f^t N)$, where $M \in GL(m, \mathbb{R})$ and $N \in GL(n, \mathbb{R})$.

Exercise 2.11. Let V be a vector space over $\mathbb{C}$. The inner product of two vectors v_1 and v_2 is defined by

$$g(v_1, v_2) = \bar{v}_1{}^i g_{ij} v_2{}^j \qquad (2.20)$$

where $\bar{}$ denotes the complex conjugate. From the positivity and symmetry of the inner product, $g(v_1, v_2) = \overline{g(v_2, v_1)}$, the vector space isomorphism $g : V \to V^*$ is required to be a positive-definite Hermitian matrix. Let $f : V \to W$ be a (complex) linear map and $G : W \to W^*$ be a vector space isomorphism. The adjoint of f is defined by $g(v, \tilde{f}w) = \overline{G(w, fv)}$. Repeat the analysis to show that

(a) $\tilde{f} = g^{-1} f^\dagger G^\dagger$, where $\dagger$ denotes the Hermitian conjugate, and
(b) $\dim \operatorname{im} f = \dim \operatorname{im} \tilde{f}$.

Theorem 2.2. (**Toy index theorem**) Let V and W be finite-dimensional vector spaces over a field K and let $f : V \to W$ be a linear map. Then

$$\dim \ker f - \dim \ker \tilde{f} = \dim V - \dim W. \qquad (2.21)$$

Proof. Theorem 2.1 tells us that

$$\dim V = \dim \ker f + \dim \operatorname{im} f$$

and, if applied to $\tilde{f} : W \to V$,

$$\dim W = \dim \ker \tilde{f} + \dim \operatorname{im} \tilde{f}.$$

We saw earlier that $\dim \operatorname{im} f = \dim \operatorname{im} \tilde{f}$, from which we obtain

$$\dim V - \dim \ker f = \dim W - \dim \ker \tilde{f}. \qquad \qquad \square$$

Note that in (2.21), each term on the LHS depends on the details of the map f. The RHS states, however, that the *difference* in the two terms is independent of f! This may be regarded as a finite-dimensional analogue of the index theorems, see chapter 12.

2.2.5 Tensors

A dual vector is a linear object that maps a vector to a scalar. This may be generalized to multilinear objects called **tensors**, which map several vectors and dual vectors to a scalar. A tensor T of type (p, q) is a multilinear map that maps p dual vectors and q vectors to $\mathbb{R}$,

$$T : \overset{p}{\bigotimes} V^* \overset{q}{\bigotimes} V \to \mathbb{R}. \tag{2.22}$$

For example, a tensor of type $(0, 1)$ maps a vector to a real number and is identified with a dual vector. Similarly, a tensor of type $(1, 0)$ is a vector. If ω maps a dual vector and two vectors to a scalar, $\omega : V^* \times V \times V \to \mathbb{R}$, ω is of type $(1, 2)$.

The set of all tensors of type (p, q) is called the **tensor space** of type (p, q) and denoted by $\mathcal{T}_q^p$. The **tensor product** $\tau = \mu \otimes \nu \in \mathcal{T}_q^p \otimes \mathcal{T}_{q'}^{p'}$ is an element of $\mathcal{T}_{q+q'}^{p+p'}$ defined by

$$\tau(\omega_1, \ldots, \omega_p, \xi_1, \ldots, \xi_{p'}; u_1, \ldots, u_q, v_1, \ldots, v_{q'})$$
$$= \mu(\omega_1, \ldots, \omega_p; u_1, \ldots, u_q)\nu(\xi_1, \ldots, \xi_{p'}; v_1, \ldots, v_{q'}). \tag{2.23}$$

Another operation in a tensor space is the **contraction**, which is a map from a tensor space of type (p, q) to type $(p - 1, q - 1)$ defined by

$$\tau(\ldots, e^{*i}, \ldots; \ldots, e_i, \ldots) \tag{2.24}$$

where $\{e_i\}$ and $\{e^{*i}\}$ are the dual bases.

Exercise 2.12. Let V and W be vector spaces and let $f : V \to W$ be a linear map. Show that f is a tensor of type $(1, 1)$.

2.3 Topological spaces

The most general structure with which we work is a topological space. Physicists often tend to think that all the spaces they deal with are equipped with metrics. However, this is not always the case. In fact, metric spaces form a subset of manifolds and manifolds form a subset of topological spaces.

2.3.1 Definitions

Definition 2.3. Let X be any set and $\mathcal{T} = \{U_i | i \in I\}$ denote a certain collection of subsets of X. The pair $(X, \mathcal{T})$ is a **topological space** if $\mathcal{T}$ satisfies the following requirements.

(i) $\emptyset, X \in \mathcal{T}$.
(ii) If J is any (maybe infinite) subcollection of I, the family $\{U_j | j \in J\}$ satisfies $\cup_{j \in J} U_j \in \mathcal{T}$.
(iii) If K is any *finite* subcollection of I, the family $\{U_k | k \in K\}$ satisfies $\cap_{k \in K} U_k \in \mathcal{T}$.

X alone is sometimes called a topological space. The U_i are called the **open sets** and $\mathcal{T}$ is said to give a **topology** to X.

Example 2.7. (a) If X is a set and $\mathcal{T}$ is the collection of *all* the subsets of X, then (i)–(iii) are automatically satisfied. This topology is called the **discrete topology**.
 (b) Let X be a set and $\mathcal{T} = \{\emptyset, X\}$. Clearly $\mathcal{T}$ satisfies (i)–(iii). This topology is called the **trivial topology**. In general the discrete topology is too stringent while the trivial topology is too trivial to give any interesting structures on X.
 (c) Let X be the real line $\mathbb{R}$. All open intervals (a, b) and their unions define a topology called the **usual topology**; a and b may be $-\infty$ and ∞ respectively. Similarly, the usual topology in $\mathbb{R}^n$ can be defined. [Take a product $(a_1, b_1) \times \cdots \times (a_n, b_n)$ and their unions....]

Exercise 2.13. In definition 2.3, axioms (ii) and (iii) look somewhat unbalanced. Show that, if we allow infinite intersection in (iii), the usual topology in $\mathbb{R}$ reduces to the discrete topology (and is thus not very interesting).

A **metric** $d : X \times X \to \mathbb{R}$ is a function that satisfies the conditions:

(i) $d(x, y) = d(y, x)$
(ii) $d(x, y) \geq 0$ where the equality holds if and only if $x = y$
(iii) $d(x, y) + d(y, z) \geq d(x, z)$

for any $x, y, z \in X$. If X is endowed with a metric d, X is made into a topological space whose open sets are given by 'open discs',

$$U_\varepsilon(X) = \{y \in X | d(x, y) < \varepsilon\} \tag{2.25}$$

and all their possible unions. The topology $\mathcal{T}$ thus defined is called the **metric topology** determined by d. The topological space $(X, \mathcal{T})$ is called a **metric space**. [*Exercise*: Verify that a metric space $(X, \mathcal{T})$ is indeed a topological space.]

Let $(X, \mathcal{T})$ be a topological space and A be any subset of X. Then $\mathcal{T} = \{U_i\}$ induces the **relative topology** in A by $\mathcal{T}' = \{U_i \cap A | U_i \in \mathcal{T}\}$.

Example 2.8. Let $X = \mathbb{R}^{n+1}$ and take the n-sphere S^n,

$$(x^0)^2 + (x^1)^2 + \cdots + (x^n)^2 = 1. \tag{2.26}$$

A topology in S^n may be given by the relative topology induced by the usual topology on $\mathbb{R}^{n+1}$.

2.3.2 Continuous maps

Definition 2.4. Let X and Y be topological spaces. A map $f : X \to Y$ is **continuous** if the *inverse* image of an open set in Y is an open set in X.

This definition is in agreement with our intuitive notion of continuity. For instance, let $f : \mathbb{R} \to \mathbb{R}$ be defined by

$$f(x) = \begin{cases} -x + 1 & x \le 0 \\ -x + \frac{1}{2} & x > 0. \end{cases} \tag{2.27}$$

We take the usual topology in $\mathbb{R}$, hence any open interval (a, b) is an open set. In the usual calculus, f is said to have a discontinuity at $x = 0$. For an open set $(3/2, 2) \subset Y$, we find $f^{-1}((3/2, 2)) = (-1, -1/2)$ which is an open set in X. If we take an open set $(1 - 1/4, 1 + 1/4) \subset Y$, however, we find $f^{-1}((1 - 1/4, 1 + 1/4)) = (-1/4, 0]$ which is not an open set in the usual topology.

Exercise 2.14. By taking a continuous function $f : \mathbb{R} \to \mathbb{R}$, $f(x) = x^2$ as an example, show that the reverse definition, '*a map f is continuous if it maps an open set in X to an open set in Y*', does not work. [*Hint*: Find where $(-\varepsilon, +\varepsilon)$ is mapped to under f.]

2.3.3 Neighbourhoods and Hausdorff spaces

Definition 2.5. Suppose $\mathcal{T}$ gives a topology to X. N is a **neighbourhood** of a point $x \in X$ if N is a subset of X and N contains some (at least one) open set U_i to which x belongs. (The subset N need not be an open set. If N happens to be an open set in $\mathcal{T}$, it is called an **open neighbourhood**.)

Example 2.9. Take $X = \mathbb{R}$ with the usual topology. The interval $[-1, 1]$ is a neighbourhood of an arbitrary point $x \in (-1, 1)$.

Definition 2.6. A topological space $(X, \mathcal{T})$ is a **Hausdorff space** if, for an arbitrary pair of distinct points $x, x' \in X$, there always exist neighbourhoods U_x of x and $U_{x'}$ of x' such that $U_x \cap U_{x'} = \emptyset$.

Exercise 2.15. Let $X = \{\text{John, Paul, Ringo, George}\}$ and $U_0 = \emptyset, U_1 = \{\text{John}\}, U_2 = \{\text{John, Paul}\}, U_3 = \{\text{John, Paul, Ringo, George}\}$. Show that $\mathcal{T} = \{U_0, U_1, U_2, U_3\}$ gives a topology to X. Show also that $(X, \mathcal{T})$ is not a Hausdorff space.

Unlike this exercise, most spaces that appear in physics satisfy the Hausdorff property. In the rest of the present book we always assume this is the case.

Exercise 2.16. Show that $\mathbb{R}$ with the usual topology is a Hausdorff space. Show also that any metric space is a Hausdorff space.

2.3.4 Closed set

Let $(X, \mathcal{T})$ be a topological space. A subset A of X is **closed** if its complement in X is an open set, that is $X - A \in \mathcal{T}$. According to the definition, X and $\emptyset$ are both open *and* closed. Consider a set A (either open or closed). The **closure** of A is the smallest closed set that contains A and is denoted by $\bar{A}$. The **interior** of A is the largest open subset of A and is denoted by A°. The **boundary** $b(A)$ of A is the complement of A° in A; $b(A) = A - A^\circ$. An open set is always disjoint from its boundary while a closed set always contains its boundary.

Example 2.10. Take $X = \mathbb{R}$ with the usual topology and take a pair of open intervals $(-\infty, a)$ and (b, ∞) where $a < b$. Since $(-\infty, a) \cup (b, \infty)$ is open under the usual topology, the complement $[a, b]$ is closed. Any closed interval is a closed set under the usual topology. Let $A = (a, b)$, then $\bar{A} = [a, b]$. The boundary $b(A)$ consists of two points $\{a, b\}$. The sets $(a, b), [a, b], (a, b],$ and $[a, b)$ all have the same boundary, closure and interior. In $\mathbb{R}^n$, the product $[a_1, b_1] \times \cdots \times [a_n, b_n]$ is a closed set under the usual topology.

Exercise 2.17. Whether a set $A \subset X$ is open or closed depends on X. Let us take an interval $I = (0, 1)$ in the x-axis. Show that I is open in the x-axis $\mathbb{R}$ while it is neither closed nor open in the xy-plane $\mathbb{R}^2$.

2.3.5 Compactness

Let $(X, \mathcal{T})$ be a topological space. A family $\{A_i\}$ of subsets of X is called a **covering** of X, if

$$\bigcup_{i \in I} A_i = X.$$

If all the A_i happen to be the open sets of the topology $\mathcal{T}$, the covering is called an **open covering**.

Definition 2.7. Consider a set X and all possible coverings of X. The set X is **compact** if, for every open covering $\{U_i | i \in I\}$, there exists a *finite* subset J of I such that $\{U_j | j \in J\}$ is also a covering of X.

In general, if a set is compact in $\mathbb{R}^n$, it must be bounded. What else is needed? We state the result without the proof.

Theorem 2.3. Let X be a subset of $\mathbb{R}^n$. X is compact if and only if it is *closed* and *bounded*.

Example 2.11. (a) A point is compact.

(b) Take an open interval (a, b) in $\mathbb{R}$ and choose an open covering $U_n = (a, b - 1/n), n \in \mathbb{N}$. Evidently

$$\bigcup_{n \in \mathbb{Z}} U_n = (a, b).$$

However, no finite subfamily of $\{U_n\}$ covers (a, b). Thus, an open interval (a, b) is non-compact in conformity with theorem 2.3.

(c) S^n in example 2.8 with the relative topology is compact, since it is closed and bounded in $\mathbb{R}^{n+1}$.

The reader might not appreciate the significance of compactness from the definition and the few examples given here. It should be noted, however, that some mathematical analyses as well as physics become rather simple on a compact space. For example, let us consider a system of electrons in a solid. If the solid is non-compact with infinite volume, we have to deal with quantum statistical mechanics in an infinite volume. It is known that this is mathematically quite complicated and requires knowledge of the advanced theory of Hilbert spaces. What we usually do is to confine the system in a finite volume V surrounded by hard walls so that the electron wavefunction vanishes at the walls, or to impose periodic boundary conditions on the walls, which amounts to putting the system in a torus, see example 2.5(b). In any case, the system is now put in a compact space. Then we may construct the Fock space whose excitations are labelled by discrete indices. Another significance of compactness in physics will be found when we study extended objects such as instantons and Belavin–Polyakov monopoles, see section 4.8. In field theories, we usually assume that the field approaches some asymptotic form corresponding to the vacuum (or one of the vacua) at spatial infinities. Similarly, a class of order parameter distributions in which the spatial infinities have a common order parameter is an interesting class to study from various points of view as we shall see later. Since all points at infinity are mapped to a point, we have effectively compactified the non-compact space $\mathbb{R}^n$ to a compact space $S^n = \mathbb{R}^n \cup \{\infty\}$. This procedure is called the **one-point compactification**.

2.3.6   Connectedness

Definition 2.8. (a) A topological space X is **connected** if it cannot be written as $X = X_1 \cup X_2$, where X_1 and X_2 are both open and $X_1 \cap X_2 = \emptyset$. Otherwise X is called **disconnected**.

(b) A topological space X is called **arcwise connected** if, for any points $x, y \in X$, there exists a continuous map $f : [0, 1] \to X$ such that $f(0) = x$ and $f(1) = y$. With a few pathological exceptions, arcwise connectedness is practically equivalent to connectedness.

(c) A **loop** in a topological space X is a continuous map $f : [0, 1] \to X$ such that $f(0) = f(1)$. If any loop in X can be continuously shrunk to a point, X is called **simply connected**.

Example 2.12. (a) The real line $\mathbb{R}$ is arcwise connected while $\mathbb{R} - \{0\}$ is not. $\mathbb{R}^n$ ($n \geq 2$) is arcwise connected and so is $\mathbb{R}^n - \{0\}$.

(b) S^n is arcwise connected. The circle S^1 is not simply connected. If $n \geq 2$, S^n is simply connected. The n-dimensional torus

$$T^n = \underbrace{S^1 \times S^1 \times \cdots \times S^1}_{n} \qquad (n \geq 2)$$

is arcwise connected but not simply connected.

(c) $\mathbb{R}^2 - \mathbb{R}$ is not arcwise connected. $\mathbb{R}^2 - \{0\}$ is arcwise connected but not simply connected. $\mathbb{R}^3 - \{0\}$ is arcwise connected and simply connected.

2.4   Homeomorphisms and topological invariants

2.4.1   Homeomorphisms

As we mentioned at the beginning of this chapter, the main purpose of topology is to classify spaces. Suppose we have several figures and ask ourselves which are equal and which are different. Since we have not defined what is meant by *equal* or *different*, we may say 'they are all different from each other' or 'they are all the same figures'. Some of the definitions of equivalence are too stringent and some are too loose to produce any sensible classification of the figures or spaces. For example, in elementary geometry, the equivalence of figures is given by congruence, which turns out to be too stringent for our purpose. In topology, we define two figures to be equivalent if it is possible to deform one figure into the other by *continuous deformation*. Namely we introduce the equivalence relation under which geometrical objects are classified according to whether it is possible to deform one object into the other by continuous deformation. To be more mathematical, we need to introduce the following notion of homeomorphism.

Definition 2.9. Let X_1 and X_2 be topological spaces. A map $f : X_1 \to X_2$ is a **homeomorphism** if it is continuous and has an inverse $f^{-1} : X_2 \to X_1$ which is

(a) (b)

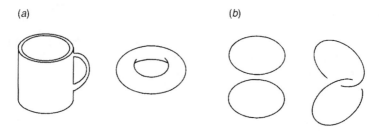

Figure 2.10. (*a*) A coffee cup is homeomorphic to a doughnut. (*b*) The linked rings are homeomorphic to the separated rings.

also continuous. If there exists a homeomorphism between X_1 and X_2, X_1 is said to be **homeomorphic** to X_2 and *vice versa*.

In other words, X_1 is homeomorphic to X_2 if there exist maps $f : X_1 \to X_2$ and $g : X_2 \to X_1$ such that $f \circ g = \mathrm{id}_{X_2}$, and $g \circ f = \mathrm{id}_{X_1}$. It is easy to show that a homeomorphism is an equivalence relation. Reflectivity follows from the choice $f = \mathrm{id}_X$, while symmetry follows since if $f : X_1 \to X_2$ is a homeomorphism so is $f^{-1} : X_2 \to X_1$ by definition. Transitivity follows since, if $f : X_1 \to X_2$ and $g : X_2 \to X_3$ are homeomorphisms so is $g \circ f : X_1 \to X_3$. Now we divide all topological spaces into equivalence classes according to whether it is possible to deform one space into the other by a homeomorphism. Intuitively speaking, we suppose the topological spaces are made out of ideal rubber which we can deform at our will. Two topological spaces are homeomorphic to each other if we can deform one into the other *continuously*, that is, without tearing them apart or pasting.

Figure 2.10 shows some examples of homeomorphisms. It seems impossible to deform the left figure in figure 2.10(*b*) into the right one by continuous deformation. However, this is an artefact of the embedding of these objects in $\mathbb{R}^3$. In fact, they are continuously deformable in $\mathbb{R}^4$, see problem 2.3. To distinguish one from the other, we have to embed them in S^3, say, and compare the complements of these objects in S^3. This approach is, however, out of the scope of the present book and we will content ourselves with homeomorphisms.

2.4.2 Topological invariants

Now our main question is: '*How can we characterize the equivalence classes of homeomorphism?*' In fact, we do not know the complete answer to this question yet. Instead, we have a rather modest statement, that is, if two spaces have different '**topological invariants**', they are not homeomorphic to each other. Here topological invariants are those quantities which are conserved under homeomorphisms. A topological invariant may be a number such as the number of connected components of the space, an algebraic structure such as a group or a ring which is constructed out of the space, or something like connectedness, compactness or the Hausdorff property. (Although it seems to be intuitively clear that these are topological invariants, we have to prove that they indeed are. We omit the proofs. An interested reader may consult any text book on topology.) If we knew the complete set of topological invariants we could specify the equivalence class by giving these invariants. However, so far we know a partial set of topological invariants, which means that even if all the known topological invariants of two topological spaces coincide, they may not be homeomorphic to each other. Instead, what we can say at most is: *if two topological spaces have different topological invariants they cannot be homeomorphic to each other.*

Example 2.13. (a) A closed line $[-1, 1]$ is not homeomorphic to an open line $(-1, 1)$, since $[-1, 1]$ is compact while $(-1, 1)$ is not.

(b) A circle S^1 is not homeomorphic to $\mathbb{R}$, since S^1 is compact in $\mathbb{R}^2$ while $\mathbb{R}$ is not.

(c) A parabola $(y = x^2)$ is not homeomorphic to a hyperbola $(x^2 - y^2 = 1)$ although they are both non-compact. A parabola is (arcwise) connected while a hyperbola is not.

(d) A circle S^1 is not homeomorphic to an interval $[-1, 1]$, although they are both compact and (arcwise) connected. $[-1, 1]$ is simply connected while S^1 is not. Alternatively $S^1 - \{p\}$, p being any point in S^1 is connected while $[-1, 1] - \{0\}$ is not, which is more evidence against their equivalence.

(e) Surprisingly, an interval without the endpoints is homeomorphic to a line $\mathbb{R}$. To see this, let us take $X = (-\pi/2, \pi/2)$ and $Y = \mathbb{R}$ and let $f : X \to Y$ be $f(x) = \tan x$. Since $\tan x$ is one to one on X and has an inverse, $\tan^{-1} x$, which is one to one on $\mathbb{R}$, this is indeed a homeomorphism. Thus, *boundedness* is not a topological invariant.

(f) An open disc $D^2 = \{(x, y) \in \mathbb{R}^2 | x^2 + y^2 < 1\}$ is homeomorphic to $\mathbb{R}^2$. A homeomorphism $f : D^2 \to \mathbb{R}^2$ may be

$$f(x, y) = \left(\frac{x}{\sqrt{1 - x^2 - y^2}}, \frac{y}{\sqrt{1 - x^2 - y^2}} \right) \tag{2.28}$$

while the inverse $f^{-1} : \mathbb{R}^2 \to D^2$ is

$$f^{-1}(x, y) = \left(\frac{x}{\sqrt{1 + x^2 + y^2}}, \frac{y}{\sqrt{1 + x^2 + y^2}} \right). \tag{2.29}$$

The reader should verify that $f \circ f^{-1} = \mathrm{id}_{\mathbb{R}^2}$, and $f^{-1} \circ f = \mathrm{id}_{D^2}$. As we saw in example 2.5(e), a closed disc whose boundary S^1 corresponds to a point is homeomorphic to S^2. If we take this point away, we have an open disc. The present analysis shows that this open disc is homeomorphic to $\mathbb{R}^2$. By reversing the order of arguments, we find that if we add a point (infinity) to $\mathbb{R}^2$, we obtain a compact space S^2. This procedure is the one-point compactification $S^2 = \mathbb{R}^2 \cup \{\infty\}$ introduced in the previous section. We similarly have $S^n = \mathbb{R}^n \cup \{\infty\}$.

(g) A circle $S^1 = \{(x, y) \in \mathbb{R}^2 \,|\, x^2 + y^2 = 1\}$ is homeomorphic to a square $I^2 = \{(x, y) \in \mathbb{R}^2 \,|\, (|x| = 1, |y| \le 1), (|x| \le 1, |y| = 1)\}$. A homeomorphism $f : I^2 \to S^1$ may be given by

$$f(x, y) = \left(\frac{x}{r}, \frac{y}{r}\right) \qquad r = \sqrt{x^2 + y^2}. \tag{2.30}$$

Since r cannot vanish, (2.27) is invertible.

Exercise 2.18. Find a homeomorphism between a circle $S^1 = \{(x, y) \in \mathbb{R}^2 \,|\, x^2 + y^2 = 1\}$ and an ellipse $E = \{(x, y) \in \mathbb{R}^2 \,|\, (x/a)^2 + (y/b)^2 = 1\}$.

2.4.3 Homotopy type

An equivalence class which is somewhat coarser than homeomorphism but which is still quite useful is 'of the **same homotopy type**'. We relax the conditions in definition 2.9 so that the continuous functions f or g need not have inverses. For example, take $X = (0, 1)$ and $Y = \{0\}$ and let $f : X \to Y$, $f(x) = 0$ and $g : Y \to X$, $g(0) = \frac{1}{2}$. Then $f \circ g = \mathrm{id}_Y$, while $g \circ f \ne \mathrm{id}_X$. This shows that an open interval $(0, 1)$ is of the same homotopy type as a point $\{0\}$, although it is not homeomorphic to $\{0\}$. We have more on this topic in section 4.2.

Example 2.14. (a) S^1 is of the same homotopy type as a cylinder, since a cylinder is a direct product $S^1 \times \mathbb{R}$ and we can shrink $\mathbb{R}$ to a point at each point of S^1. By the same reason, the Möbius strip is of the same homotopy type as S^1.

(b) A disc $D^2 = \{(x, y) \in \mathbb{R}^2 \,|\, x^2 + y^2 < 1\}$ is of the same homotopy type as a point. $D^2 - \{(0, 0)\}$ is of the same homotopy type as S^1. Similarly, $\mathbb{R}^2 - \{\mathbf{0}\}$ is of the same homotopy type as S^1 and $\mathbb{R}^3 - \{\mathbf{0}\}$ as S^2.

2.4.4 Euler characteristic: an example

The Euler characteristic is one of the most useful topological invariants. Moreover, we find the prototype of the algebraic approach to topology in it. To avoid unnecessary complication, we restrict ourselves to points, lines and surfaces in $\mathbb{R}^3$. A **polyhedron** is a geometrical object surrounded by faces. The boundary of two faces is an edge and two edges meet at a vertex. We extend the definition of a polyhedron a bit to include polygons and the boundaries of polygons, lines or points. We call the faces, edges and vertices of a polyhedron **simplexes**. Note that the boundary of two simplexes is either empty or another simplex. (For example, the boundary of two faces is an edge.) Formal definitions of a simplex and a polyhedron in a general number of dimensions will be given in chapter 3. We are now ready to define the Euler characteristic of a figure in $\mathbb{R}^3$.

Definition 2.10. Let X be a subset of $\mathbb{R}^3$, which is homeomorphic to a polyhedron K. Then the **Euler characteristic** $\chi(X)$ of X is defined by

$$\chi(X) = (\text{number of verticies in } K) - (\text{number of edges in } K)$$
$$+ (\text{number of faces in } K). \tag{2.31}$$

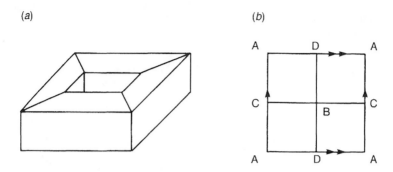

Figure 2.11. Example of a polyhedron which is homeomorphic to a torus.

The reader might wonder if $\chi(X)$ depends on the polyhedron K or not. The following theorem due to Poincaré and Alexander guarantees that it is, in fact, independent of the polyhedron K.

Theorem 2.4. (**Poincaré–Alexander**) The Euler characteristic $\chi(X)$ is independent of the polyhedron K as long as K is homeomorphic to X.

Examples are in order. The Euler characteristic of a point is $\chi(\cdot) = 1$ by definition. The Euler characteristic of a line is $\chi(\text{———}) = 2 - 1 = 1$, since a line has two vertices and an edge. For a triangular disc, we find $\chi(\text{triangle}) = 3 - 3 + 1 = 1$. An example which is a bit non-trivial is the Euler characteristic of S^1. The simplest polyhedron which is homeomorphic to S^1 is made of three edges of a triangle. Then $\chi(S^1) = 3 - 3 = 0$. Similarly, the sphere S^2 is homeomorphic to the surface of a tetrahedron, hence $\chi(S^2) = 4 - 6 + 4 = 2$. It is easily seen that S^2 is also homeomorphic to the surface of a cube. Using a cube to calculate the Euler characteristic of S^2, we have $\chi(S^2) = 8 - 12 + 6 = 2$, in accord with theorem 2.4. Historically this is the conclusion of **Euler's theorem**: if K is any polyhedron homeomorphic to S^2, with v vertices, e edges and f two-dimensional faces, then $v - e + f = 2$.

Example 2.15. Let us calculate the Euler characteristic of the torus T^2. Figure 2.11(a) is an example of a polyhedron which is homeomorphic to T^2. From this polyhedron, we find $\chi(T^2) = 16 - 32 + 16 = 0$. As we saw in example 2.5(b), T^2 is equivalent to a rectangle whose edges are identified; see figure 2.4. Taking care of this identification, we find an example of a polyhedron made of rectangular faces as in figure 2.11(b), from which we also have $\chi(T^2) = 0$. This approach is quite useful when the figure cannot be realized (embedded) in $\mathbb{R}^3$. For example, the Klein bottle (figure 2.5(a)) cannot be realized in $\mathbb{R}^3$ without intersecting itself. From the rectangle of figure 2.5(a), we find $\chi(\text{Klein bottle}) = 0$. Similarly, we have $\chi(\text{projective plane}) = 1$.

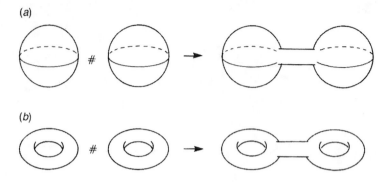

(a)

(b)

Figure 2.12. The connected sum. (a) $S^2 \sharp S^2 = S2$, (b) $T^2 \sharp T^2 = \Sigma_2$.

Exercise 2.19. (a) Show that χ(Möbius strip) $= 0$.

(b) Show that $\chi(\Sigma_2) = -2$, where Σ_2 is the torus with two handles (see example 2.5). The reader may either construct a polyhedron homeomorphic to Σ_2 or make use of the octagon in figure 2.6(a). We show later that $\chi(\Sigma_g) = 2 - 2g$, where Σ_g is the torus with g handles.

The **connected sum** $X \sharp Y$ of two surfaces X and Y is a surface obtained by removing a small disc from each of X and Y and connecting the resulting holes with a cylinder; see figure 2.12. Let X be an arbitrary surface. Then it is easy to see that

$$S^2 \sharp X = X \tag{2.32}$$

since S^2 and the cylinder may be deformed so that they fill in the hole on X; see figure 2.12(a). If we take a connected sum of two tori we get (figure 2.12(b))

$$T^2 \sharp T^2 = \Sigma_2. \tag{2.33}$$

Similarly, Σ_g may be given by the connected sum of g tori,

$$\underbrace{T^2 \sharp T^2 \sharp \cdots \sharp T^2}_{g \text{ factors}} = \Sigma_g. \tag{2.34}$$

The connected sum may be used as a trick to calculate an Euler characteristic of a complicated surface from those of known surfaces. Let us prove the following theorem.

Theorem 2.5. Let X and Y be two surfaces. Then the Euler characteristic of the connected sum $X \sharp Y$ is given by

$$\chi(X \sharp Y) = \chi(X) + \chi(Y) - 2.$$

Proof. Take polyhedra K_X and K_Y homeomorphic to X and Y, respectively. We assume, without loss of generality, that each of K_Y and K_Y has a triangle in it. Remove the triangles from them and connect the resulting holes with a trigonal cylinder. Then the number of vertices does not change while the number of edges increases by three. Since we have removed two faces and added three faces, the number of faces increases by $-2 + 3 = 1$. Thus, the change of the Euler characteristic is $0 - 3 + 1 = -2$. □

From the previous theorem and the equality $\chi(T^2) = 0$, we obtain $\chi(\Sigma_2) = 0 + 0 - 2 = -2$ and $\chi(\Sigma_g) = g \times 0 - 2(g - 1) = 2 - 2g$, cf exercise 2.19(b).

The significance of the Euler characteristic is that it is a topological invariant, which is calculated relatively easily. We accept, without proof, the following theorem.

Theorem 2.6. Let X and Y be two figures in $\mathbb{R}^3$. If X is homeomorphic to Y, then $\chi(X) = \chi(Y)$. In other words, if $\chi(X) \neq \chi(Y)$, X cannot be homeomorphic to Y.

Example 2.16. (a) S^1 is not homeomorphic to S^2, since $\chi(S^1) = 0$ while $\chi(S^2) = 2$.

(b) Two figures, which are not homeomorphic to each other, may have the same Euler characteristic. A point ($\cdot$) is not homeomorphic to a line (———) but $\chi(\cdot) = \chi(———) = 1$. This is a general consequence of the following fact: *if a figure X is of the same homotopy type as a figure Y, then $\chi(X) = \chi(Y)$.*

The reader might have noticed that the Euler characteristic is different from other topological invariants such as compactness or connectedness in character. Compactness and connectedness are geometrical properties of a figure or a space while the Euler characteristic is an *integer* $\chi(X) \in \mathbb{Z}$. Note that $\mathbb{Z}$ is an algebraic object rather than a geometrical one. Since the work of Euler, many mathematicians have worked out the relation between geometry and algebra and elaborated this idea, in the last century, to establish combinatorial topology and algebraic topology. We may compute the Euler characteristic of a smooth surface by the celebrated Gauss–Bonnet theorem, which relates the integral of the Gauss curvature of the surface with the Euler characteristic calculated from the corresponding polyhedron. We will give the generalized form of the Gauss–Bonnet theorem in chapter 12.

Problems

2.1 Show that the $4g$-gon in figure 2.13(a), with the boundary identified, represents the torus with genus g of figure 2.13(b). The reader may use equation (2.34).

2.2 Let $X = \{1, 1/2, \ldots, 1/n, \ldots\}$ be a subset of $\mathbb{R}$. Show that X is not closed in $\mathbb{R}$. Show that $Y = \{1, 1/2, \ldots, 1/n, \ldots, 0\}$ is closed in $\mathbb{R}$, hence compact.

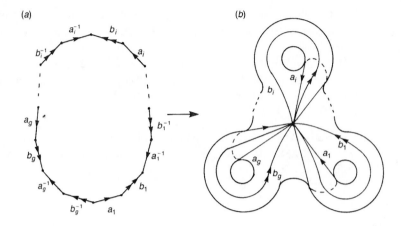

Figure 2.13. The polygon (a) whose edges are identified is the torus Σ_g with genus g.

2.3 Show that two figures in figure 2.109(b) are homeomorphic to each other. Find how to unlink the right figure in $\mathbb{R}^4$.

2.4 Show that there are only five regular polyhedra: a tetrahedron, a hexahedron, an octahedron, a dodecahedron and an icosahedron. [*Hint*: Use Euler's theorem.]

3

HOMOLOGY GROUPS

Among the topological invariants the Euler characteristic is a quantity readily computable by the 'polyhedronization' of space. The homology groups are *refinements*, so to speak, of the Euler characteristic. Moreover, we can easily read off the Euler characteristic from the homology groups. Let us look at figure 3.1. In figure 3.1(a), the interior is included but not in figure 3.1(b). How do we characterize this difference? An obvious observation is that the three edges of figure 3.1(a) form a boundary of the interior while the edges of figure 3.1(b) do not (the interior is *not* a part of figure 3.1(b)). Clearly the edges in both cases form a closed path (loop), having no boundary. In other words, the existence of a loop that is not a boundary of some area implies the existence of a hole within the loop. This is our guiding principle in classifying spaces here: *find a region without boundaries, which is not itself a boundary of some region*. This principle is mathematically elaborated into the theory of homology groups.

Our exposition follows Armstrong (1983), Croom (1978) and Nash and Sen (1983). An introduction to group theory is found in Fraleigh (1976).

3.1 Abelian groups

The mathematical structures underlying homology groups are *finitely generated Abelian groups*. Throughout this chapter, the group operation is denoted by $+$ since all the groups considered here are Abelian (commutative). The unit element is denoted by 0.

3.1.1 Elementary group theory

Let G_1 and G_2 be Abelian groups. A map $f : G_1 \to G_2$ is said to be a **homomorphism** if

$$f(x + y) = f(x) + f(y) \tag{3.1}$$

for any $x, y \in G_1$. If f is also a *bijection*, f is called an **isomorphism**. If there exists an isomorphism $f : G_1 \to G_2$, G_1 is said to be **isomorphic** to G_2, denoted by $G_1 \cong G_2$. For example, a map $f : \mathbb{Z} \to \mathbb{Z}_2 = \{0, 1\}$ defined by

$$f(2n) = 0 \qquad f(2n + 1) = 1$$

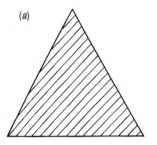 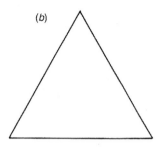

Figure 3.1. (*a*) is a solid triangle while (*b*) is the edges of a triangle without an interior.

is a homomorphism. Indeed

$$f(2m + 2n) = f(2(m + n)) = 0 = 0 + 0 = f(2m) + f(2n)$$
$$f(2m + 1 + 2n + 1) = f(2(m + n + 1)) = 0 = 1 + 1$$
$$= f(2m + 1) + f(2n + 1)$$
$$f(2m + 1 + 2n) = f(2(m + n) + 1) = 1 = 1 + 0$$
$$= f(2m + 1) + f(2n).$$

A subset $H \subset G$ is a subgroup if it is a group with respect to the group operation of G. For example,

$$k\mathbb{Z} \equiv \{kn | n \in \mathbb{Z}\} \qquad k \in \mathbb{N}$$

is a subgroup of $\mathbb{Z}$, while $\mathbb{Z}_2 = \{0, 1\}$ is not.

Let H be a subgroup of G. We say $x, y \in G$ are equivalent if

$$x - y \in H \tag{3.2}$$

and write $x \sim y$. Clearly $\sim$ is an equivalence relation. The equivalence class to which x belongs is denoted by $[x]$. Let G/H be the quotient space. The group operation $+$ in G naturally induces the group operation $+$ in G/H by

$$[x] + [y] = [x + y]. \tag{3.3}$$

Note that $+$ on the LHS is an operation in G/H while $+$ on the RHS is that in G. The operation in G/H should be independent of the choice of representatives. In fact, if $[x'] = [x]$, $[y'] = [y]$, then $x - x' = h$, $y - y' = g$ for some $h, g \in H$ and we find that

$$x' + y' = x + y - (h + g) \in [x + y]$$

Furthermore, G/H becomes a group with this operation, since H is always a normal subgroup of G; see example 2.6. The unit element of G/H is $[0] = [h]$,

$h \in H$. If $H = G$, $0 - x \in G$ for any $x \in G$ and G/G has just one element [0]. If $H = \{0\}$, G/H is G itself since $x - y = 0$ if and only if $x = y$.

Example 3.1. Let us work out the quotient group $\mathbb{Z}/2\mathbb{Z}$. For even numbers we have $2n - 2m = 2(n - m) \in 2\mathbb{Z}$ and $[2m] = [2n]$. For odd numbers $(2n+1) - (2m+1) = 2(n-m) \in 2\mathbb{Z}$ and $[2m+1] = [2n+1]$. Even numbers and odd numbers never belong to the same equivalence class since $2n - (2m+1) \notin 2\mathbb{Z}$. Thus, it follows that

$$\mathbb{Z}/2\mathbb{Z} = \{[0], [1]\}. \tag{3.4}$$

If we define an isomorphism $\varphi : \mathbb{Z}/2\mathbb{Z} \to \mathbb{Z}_2$ by $\varphi([0]) = 0$ and $\varphi([1]) = 1$, we find $\mathbb{Z}/2\mathbb{Z} \cong \mathbb{Z}_2$. For general $k \in \mathbb{N}$, we have

$$\mathbb{Z}/k\mathbb{Z} \cong \mathbb{Z}_k. \tag{3.5}$$

Lemma 3.1. Let $f : G_1 \to G_2$ be a homomorphism. Then
(a) ker $f = \{x | x \in G_1, f(x) = 0\}$ is a subgroup of G_1,
(b) im $f = \{x | x \in f(G_1) \subset G_2\}$ is a subgroup of G_2.

Proof. (a) Let $x, y \in \ker f$. Then $x + y \in \ker f$ since $f(x + y) = f(x) + f(y) = 0 + 0 = 0$. Note that $0 \in \ker f$ for $f(0) = f(0) + f(0)$. We also have $-x \in \ker f$ since $f(0) = f(x - x) = f(x) + f(-x) = 0$.
(b) Let $y_1 = f(x_1)$, $y_2 = f(x_2) \in \text{im } f$ where $x_1, x_2 \in G_1$. Since f is a homomorphism we have $y_1 + y_2 = f(x_1) + f(x_2) = f(x_1 + x_2) \in \text{im } f$. Clearly $0 \in \text{im } f$ since $f(0) = 0$. If $y = f(x)$, $-y \in \text{im } f$ since $0 = f(x - x) = f(x) + f(-x)$ implies $f(-x) = -y$. $\qquad\square$

Theorem 3.1. (**Fundamental theorem of homomorphism**) Let $f : G_1 \to G_2$ be a homomorphism. Then

$$G_1/\ker f \cong \text{im } f. \tag{3.6}$$

Proof. Both sides are groups according to lemma 3.1. Define a map $\varphi : G_1/\ker f \to \text{im } f$ by $\varphi([x]) = f(x)$. This map is well defined since for $x' \in [x]$, there exists $h \in \ker f$ such that $x' = x + h$ and $f(x') = f(x + h) = f(x) + f(h) = f(x)$. Now we show that φ is an isomorphism. First, φ is a homomorphism,

$$\varphi([x] + [y]) = \varphi([x + y]) = f(x + y)$$
$$= f(x) + f(y) = \varphi([x]) + \varphi([y]).$$

Second, φ is one to one: if $\varphi([x]) = \varphi([y])$, then $f(x) = f(y)$ or $f(x) - f(y) = f(x - y) = 0$. This shows that $x - y \in \ker f$ and $[x] = [y]$. Finally, φ is onto: if $y \in \text{im } f$, there exists $x \in G_1$ such that $f(x) = y = \varphi([x])$. $\qquad\square$

Example 3.2. Let $f : \mathbb{Z} \to \mathbb{Z}_2$ be defined by $f(2n) = 0$ and $f(2n+1) = 1$. Then ker $f = 2\mathbb{Z}$ and im $f = \mathbb{Z}_2$ are groups. Theorem 3.1 states that $\mathbb{Z}/2\mathbb{Z} \cong \mathbb{Z}_2$, in agreement with example 3.1.

3.1.2   Finitely generated Abelian groups and free Abelian groups

Let x be an element of a group G. For $n \in \mathbb{Z}, nx$ denotes

$$\underbrace{x + \cdots + x}_{n} \qquad (\text{if } n > 0)$$

and

$$\underbrace{(-x) + \cdots + (-x)}_{|n|} \qquad (\text{if } n < 0).$$

If $n = 0$, we put $0x = 0$. Take r elements $x_1, \ldots, x_r$ of G. The elements of G of the form

$$n_1 x_1 + \cdots + n_r x_r \qquad (n_i \in \mathbb{Z}, 1 \leq i \leq r) \tag{3.7}$$

form a subgroup of G, which we denote H. H is called a subgroup of G **generated** by the **generators** $x_1, \ldots, x_r$. If G itself is generated by finite elements $x_1, \ldots, x_r$, G is said to be **finitely generated**. If $n_1 x_1 + \cdots + n_r x_r = 0$ is satisfied only when $n_1 = \cdots = n_r = 0$, $x_1, \ldots, x_r$ are said to be **linearly independent**.

Definition 3.1. If G is finitely generated by r *linearly independent* elements, G is called a **free Abelian group** of **rank** r.

Example 3.3. $\mathbb{Z}$ is a free Abelian group of rank 1 finitely generated by 1 (or -1). Let $\mathbb{Z} \oplus \mathbb{Z}$ be the set of pairs $\{(i, j)|i, j \in \mathbb{Z}\}$. It is a free Abelian group of rank 2 finitely generated by generators $(1, 0)$ and $(0, 1)$. More generally

$$\underbrace{\mathbb{Z} \oplus \mathbb{Z} \oplus \cdots \oplus \mathbb{Z}}_{r}$$

is a free Abelian group of rank r. The group $\mathbb{Z}_2 = \{0, 1\}$ is finitely generated by 1 but is *not* free since 1 is not linearly independent (note $1 + 1 = 0$).

3.1.3   Cyclic groups

If G is generated by one element x, $G = \{0, \pm x, \pm 2x, \ldots\}$, G is called a **cyclic group**. If $nx \neq 0$ for any $n \in \mathbb{Z} - \{0\}$, it is an **infinite** cyclic group while if $nx = 0$ for some $n \in \mathbb{Z} - \{0\}$, a **finite cyclic group**. Let G be a cyclic group generated by x and let $f : \mathbb{Z} \to G$ be a homomorphism defined by $f(n) = nx$. f maps $\mathbb{Z}$ onto G but not necessarily one to one. From theorem 3.1, we have $G = \text{im} f \cong \mathbb{Z}/\ker f$. Let N be the smallest positive integer such that $Nx = 0$. Clearly

$$\ker f = \{0, \pm N, \pm 2N, \ldots\} = N\mathbb{Z} \tag{3.8}$$

and we have

$$G \cong \mathbb{Z}/N\mathbb{Z} \cong \mathbb{Z}_N. \tag{3.9}$$

If G is an infinite cyclic group, then ker $f = \{0\}$ and $G \cong \mathbb{Z}$. Any infinite cyclic group is isomorphic to $\mathbb{Z}$ while a finite cyclic group is isomorphic to some $\mathbb{Z}_N$.

We will need the following lemma and theorem in due course. We first state the lemma without proof.

Lemma 3.2. Let G be a free Abelian group of rank r and let H ($\neq \emptyset$) be a subgroup of G. We may always choose p generators $x_1, \ldots, x_p$, out of r generators of G so that $k_1 x_1, \ldots, k_p x_p$ generate H. Thus, $H \cong k_1 \mathbb{Z} \oplus \ldots \oplus k_p \mathbb{Z}$ and H is of rank p.

Theorem 3.2. (**Fundamental theorem of finitely generated Abelian groups**) Let G be a finitely generated Abelian group (not necessarily free) with m generators. Then G is isomorphic to the direct sum of cyclic groups,

$$G \cong \underbrace{\mathbb{Z} \oplus \cdots \oplus \mathbb{Z}}_{r} \oplus \mathbb{Z}_{k_1} \oplus \cdots \oplus \mathbb{Z}_{k_p} \tag{3.10}$$

where $m = r + p$. The number r is called the **rank** of G.

Proof. Let G be generated by m elements $x_1, \ldots, x_m$ and let

$$f : \underbrace{\mathbb{Z} \oplus \cdots \oplus \mathbb{Z}}_{m} \to G$$

be a surjective homomorphism,

$$f(n_1, \ldots, n_m) = n_1 x_1 + \cdots + n_m x_m.$$

Theorem 3.1 states that

$$\underbrace{\mathbb{Z} \oplus \cdots \oplus \mathbb{Z}}_{m} / \ker f \cong G.$$

Since ker f is a subgroup of

$$\underbrace{\mathbb{Z} \oplus \cdots \oplus \mathbb{Z}}_{m}$$

lemma 3.2 claims that if we choose the generators properly, we have

$$\ker f \cong k_1 \mathbb{Z} \oplus \cdots \oplus k_p \mathbb{Z}.$$

We finally obtain

$$G \cong \underbrace{\mathbb{Z} \oplus \cdots \oplus \mathbb{Z}}_{m} / \ker f \cong \underbrace{\mathbb{Z} \oplus \cdots \oplus \mathbb{Z}}_{m} / (k_1 \mathbb{Z} \oplus \cdots \oplus k_p \mathbb{Z})$$

$$\cong \underbrace{\mathbb{Z} \oplus \cdots \oplus \mathbb{Z}}_{m-p} \oplus \mathbb{Z}_{k_1} \oplus \cdots \oplus \mathbb{Z}_{k_p}. \qquad \square$$

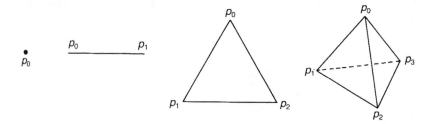

Figure 3.2. 0-, 1-, 2- and 3-simplexes.

3.2 Simplexes and simplicial complexes

Let us recall how the Euler characteristic of a surface is calculated. We first construct a polyhedron homeomorphic to the given surface, then count the numbers of vertices, edges and faces. The Euler characteristic of the polyhedron, and hence of the surface, is then given by equation (2.31). We abstract this procedure so that we may represent each part of a figure by some *standard* object. We take triangles and their analogues in other dimensions, called simplexes, as the standard objects. By this standardization, it becomes possible to assign to each figure Abelian group structures.

3.2.1 Simplexes

Simplexes are building blocks of a polyhedron. A 0-simplex $\langle p_0 \rangle$ is a point, or a vertex, and a 1-simplex $\langle p_0 p_1 \rangle$ is a line, or an edge. A 2-simplex $\langle p_0 p_1 p_2 \rangle$ is defined to be a triangle with its interior included and a 3-simplex $\langle p_0 p_1 p_2 p_3 \rangle$ is a solid tetrahedron (figure 3.2). It is common to denote a 0-simplex without the bracket; $\langle p_0 \rangle$ may be also written as p_0. It is easy to continue this construction to any r-simplex $\langle p_0 p_1 \ldots p_r \rangle$. Note that for an r-simplex to represent an r-dimensional object, the vertices p_i must be *geometrically independent*, that is, no $(r - 1)$-dimensional hyperplane contains all the $r + 1$ points. Let $p_0, \ldots, p_r$ be points geometrically independent in $\mathbb{R}^m$ where $m \geq r$. The r-simplex $\sigma_r = \langle p_0, \ldots, p_r \rangle$ is expressed as

$$\sigma^r = \left\{ x \in \mathbb{R}^m \,\middle|\, x = \sum_{i=0}^{r} c_i p_i, c_i \geq 0, \sum_{i=0}^{r} c_i = 1 \right\}. \qquad (3.11)$$

$(c_0, \ldots, c_r)$ is called the **barycentric coordinate** of x. Since σ_r is a bounded and closed subset of $\mathbb{R}^m$, it is compact.

Let q be an integer such that $0 \leq q \leq r$. If we choose $q + 1$ points $p_{i_0}, \ldots, p_{i_q}$ out of $p_0, \ldots, p_r$, these $q + 1$ points define a q-simplex $\sigma_q = \langle p_{i_0}, \ldots, p_{i_q} \rangle$, which is called a **$q$-face** of σ_r. We write $\sigma_q \leq \sigma_r$ if σ_q is a face of

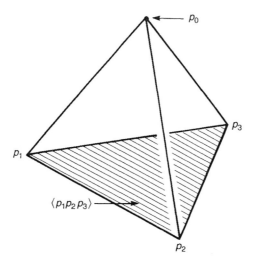

Figure 3.3. A 0-face p_0 and a 2-face $\langle p_1 p_2 p_3 \rangle$ of a 3-simplex $\langle p_0 p_1 p_2 p_3 \rangle$.

σ_r. If $\sigma_q \neq \sigma_r$, we say σ_q is a **proper face** of σ_r, denoted as $\sigma_q < \sigma_r$. Figure 3.3 shows a 0-face p_0 and a 2-face $\langle p_1 p_2 p_3 \rangle$ of a 3-simplex $\langle p_0 p_1 p_2 p_3 \rangle$. There are one 3-face, four 2-faces, six 1-faces and four 0-faces. The reader should verify that the number of q-faces in an r-simplex is $\begin{pmatrix} r+1 \\ q+1 \end{pmatrix}$. A 0-simplex is defined to have no proper faces.

3.2.2 Simplicial complexes and polyhedra

Let K be a set of finite number of simplexes in $\mathbb{R}^m$. If these simplexes are *nicely* fitted together, K is called a **simplicial complex**. By 'nicely' we mean:

(i)   an arbitrary face of a simplex of K belongs to K, that is, if $\sigma \in K$ and $\sigma' \leq \sigma$ then $\sigma' \in K$; and

(ii) if σ and σ' are two simplexes of K, the intersection $\sigma \cap \sigma'$ is either empty or a common face of σ and σ', that is, if $\sigma, \sigma' \in K$ then either $\sigma \cap \sigma' = \emptyset$ or $\sigma \cap \sigma' \leq \sigma$ and $\sigma \cap \sigma' \leq \sigma'$.

For example, figure 3.4(*a*) is a simplicial complex but figure 3.4(*b*) is not. The dimension of a simplicial complex K is defined to be the largest dimension of simplexes in K.

Example 3.4. Let σ_r be an r-simplex and $K = \{\sigma' | \sigma' \leq \sigma_r\}$ be the set of faces of σ_r. K is an r-dimensional simplicial complex. For example, take

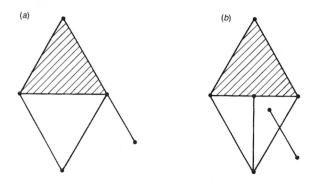

Figure 3.4. (*a*) is a simplicial complex but (*b*) is not.

$\sigma_3 = \langle p_0 p_1 p_2 p_3 \rangle$ (figure 3.3). Then

$$K = \{p_0,\, p_1,\, p_2,\, p_3,\, \langle p_0 p_1 \rangle,\, \langle p_0 p_2 \rangle,\, \langle p_0 p_3 \rangle,$$
$$\langle p_1 p_2 \rangle,\, \langle p_1 p_3 \rangle,\, \langle p_2 p_3 \rangle,\, \langle p_0 p_1 p_2 \rangle,\, \langle p_0 p_1 p_3 \rangle,$$
$$\langle p_0 p_2 p_3 \rangle,\, \langle p_1 p_2 p_3 \rangle,\, \langle p_0 p_1 p_2 p_3 \rangle\}. \tag{3.12}$$

A simplicial complex K is a *set* whose elements are simplexes. If each simplex is regarded as a subset of $\mathbb{R}^m$ ($m \geq \dim K$), the union of all the simplexes becomes a subset of $\mathbb{R}^m$. This subset is called the **polyhedron** $|K|$ of a simplicial complex K. The dimension of $|K|$ as a subset of $\mathbb{R}^m$ is the same as that of K; $\dim |K| = \dim K$.

Let X be a topological space. If there exists a simplicial complex K and a homeomorphism $f : |K| \to X$, X is said to be **triangulable** and the pair (K, f) is called a **triangulation** of X. Given a topological space X, its triangulation is far from unique. We will be concerned with triangulable spaces only.

Example 3.5. Figure 3.5(*a*) is a triangulation of a cylinder $S^1 \times [0, 1]$. The reader might think that somewhat simpler choices exist, figure 3.5(*b*), for example. This is, however, not a triangulation since, for $\sigma_2 = \langle p_0 p_1 p_2 \rangle$ and $\sigma_2' = \langle p_2 p_3 p_0 \rangle$, we find $\sigma_2 \cap \sigma_2' = \langle p_0 \rangle \cup \langle p_2 \rangle$, which is neither empty nor a simplex.

3.3 Homology groups of simplicial complexes

3.3.1 Oriented simplexes

We may assign *orientations* to an r-simplex for $r \geq 1$. Instead of $\langle \ldots \rangle$ for an unoriented simplex, we will use $(\ldots)$ to denote an oriented simplex. The symbol σ_r is used to denote both types of simplex. An oriented 1-simplex $\sigma_1 = (p_0 p_1)$ is a directed line segment traversed in the direction $p_0 \to p_1$ (figure 3.6(*a*)). Now

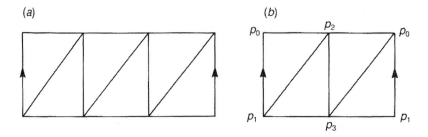

Figure 3.5. (*a*) is a triangulation of a cylinder while (*b*) is not.

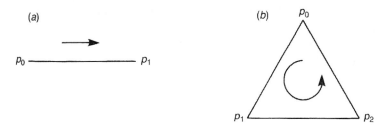

Figure 3.6. An oriented 1-simplex (*a*) and an oriented 2-simplex (*b*).

$(p_0 p_1)$ should be distinguished from $(p_1 p_0)$. We require that

$$(p_0 p_1) = -(p_1 p_0). \tag{3.13}$$

Here '$-$' in front of $(p_1 p_0)$ should be understood in the sense of a finitely generated Abelian group. In fact, $(p_1 p_0)$ is regarded as the *inverse* of $(p_0 p_1)$. Going from p_0 to p_1 followed by going from p_1 to p_0 means going nowhere, $(p_0 p_1) + (p_1 p_0) = 0$, hence $-(p_1 p_0) = (p_0 p_1)$.

Similarly, an oriented 2-simplex $\sigma_2 = (p_0 p_1 p_2)$ is a triangular region $p_0 p_1 p_2$ with a prescribed orientation along the edges (figure 3.6(*b*)). Observe that the orientation given by $p_0 p_1 p_2$ is the same as that given by $p_2 p_0 p_1$ or $p_1 p_2 p_0$ but opposite to $p_0 p_2 p_1$, $p_2 p_1 p_0$ or $p_1 p_0 p_2$. We require that

$$(p_0 p_1 p_2) = (p_2 p_0 p_1) = (p_1 p_2 p_0)$$
$$= -(p_0 p_2 p_1) = -(p_2 p_1 p_0) = -(p_1 p_0 p_2).$$

Let P be a permutation of $0, 1, 2$

$$P = \begin{pmatrix} 0 & 1 & 2 \\ i & j & k \end{pmatrix}.$$

These relations are summarized as

$$(p_i p_j p_k) = \mathrm{sgn}(P)(p_0 p_1 p_2)$$

where $\text{sgn}(P) = +1 \, (-1)$ if P is an even (odd) permutation.

An oriented 3-simplex $\sigma_3 = (p_0 p_1 p_2 p_3)$ is an ordered sequence of four vertices of a tetrahedron. Let

$$P = \begin{pmatrix} 0 & 1 & 2 & 3 \\ i & j & k & l \end{pmatrix}$$

be a permutation. We define

$$(p_i p_j p_k p_l) = \text{sgn}(P)(p_0 p_1 p_2 p_3).$$

It is now easy to construct an oriented r-simplex for any $r \geq 1$. The formal definition goes as follows. Take $r + 1$ geometrically independent points $p_0, p_1, \ldots, p_r$ in $\mathbb{R}^m$. Let $\{p_{i_0}, p_{i_1}, \ldots, p_{i_r}\}$ be a sequence of points obtained by a permutation of the points $p_0, \ldots, p_r$. We define $\{p_0, \ldots, p_r\}$ and $\{p_{i_0}, \ldots, p_{i_r}\}$ to be equivalent if

$$P = \begin{pmatrix} 0 & 1 & \ldots & r \\ i_0 & i_1 & \ldots & i_r \end{pmatrix}$$

is an even permutation. Clearly this is an equivalence relation, the equivalence class of which is called an **oriented r-simplex**. There are two equivalence classes, one consists of even permutations of $p_0, \ldots, p_r$, the other of odd permutations. The equivalence class (oriented r-simplex) which contains $\{p_0, \ldots, p_r\}$ is denoted by $\sigma_r = (p_0 p_1 \ldots p_r)$, while the other is denoted by $-\sigma_r = -(p_0 p_1 \ldots p_r)$. In other words,

$$(p_{i_0} p_{i_1} \ldots p_{i_r}) = \text{sgn}(P)(p_0 p_1 \ldots p_r). \tag{3.14}$$

For $r = 0$, we formally define an oriented 0-simplex to be just a point $\sigma_0 = p_0$.

3.3.2 Chain group, cycle group and boundary group

Let $K = \{\sigma_\alpha\}$ be an n-dimensional simplicial complex. We regard the simplexes σ_α in K as oriented simplexes and denote them by the same symbols σ_α as remarked before.

Definition 3.2. The **r-chain group** $C_r(K)$ of a simplicial complex K is a free Abelian group generated by the oriented r-simplexes of K. If $r > \dim K$, $C_r(K)$ is defined to be 0. An element of $C_r(K)$ is called an **r-chain**.

Let there be I_r r-simplexes in K. We denote each of them by $\sigma_{r,i}$ $(1 \leq i \leq I_r)$. Then $c \in C_r(K)$ is expressed as

$$c = \sum_{i=1}^{I_r} c_i \sigma_{r,i} \qquad c_i \in \mathbb{Z}. \tag{3.15}$$

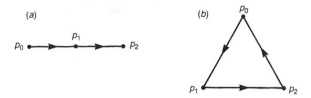

Figure 3.7. (a) An oriented 1-simplex with a fictitious boundary p_1. (b) A simplicial complex without a boundary.

The integers c_i are called the coefficients of c. The group structure is given as follows. The addition of two r-chains, $c = \sum_i c_i \sigma_{r,i}$ and $c' = \sum_i c'_i \sigma_{r,i}$, is

$$c + c' = \sum_i (c_i + c'_i) \sigma_{r,i}. \tag{3.16}$$

The unit element is $0 = \sum_i 0 \cdot \sigma_{r,i}$, while the inverse element of c is $-c = \sum_i (-c_i) \sigma_{r,i}$. [*Remark*: An oppositely oriented r-simplex $-\sigma_r$ is identified with $(-1)\sigma_r \in C_r(K)$.] Thus, $C_r(K)$ is a free Abelian group of rank I_r,

$$C_r(K) \cong \underbrace{\mathbb{Z} \oplus \mathbb{Z} \oplus \cdots \oplus \mathbb{Z}}_{I_r}. \tag{3.17}$$

Before we define the cycle group and the boundary group, we need to introduce the boundary operator. Let us denote the boundary of an r-simplex σ_r by $\partial_r \sigma_r$. ∂_r should be understood as an *operator* acting on σ_r to produce its boundary. This point of view will be elaborated later. Let us look at the boundaries of lower-dimensional simplexes. Since a 0-simplex has no boundary, we define

$$\partial_0 p_0 = 0. \tag{3.18}$$

For a 1-simplex $(p_0 p_1)$, we define

$$\partial_1 (p_0 p_1) = p_1 - p_0. \tag{3.19}$$

The reader might wonder about the appearance of a minus sign in front of p_0. This is again related to the orientation. The following examples will clarify this point. In figure 3.7(a), an oriented 1-simplex $(p_0 p_2)$ is divided into two, $(p_0 p_1)$ and $(p_1 p_2)$. We agree that the boundary of $(p_0 p_2)$ is $\{p_0\} \cup \{p_2\}$ and so should be that of $(p_0 p_1) + (p_1 p_2)$. If $\partial_1 (p_0 p_2)$ were defined to be $p_0 + p_2$, we would have $\partial_1 (p_0 p_1) + \partial_1 (p_1 p_2) = p_0 + p_1 + p_1 + p_2$. This is not desirable since p_1 is a *fictitious* boundary. If, instead, we take $\partial_1 (p_0 p_2) = p_2 - p_0$, we will have $\partial_1 (p_0 p_1) + \partial_1 (p_1 p_2) = p_1 - p_0 + p_2 - p_1 = p_2 - p_0$ as expected. The next example is the triangle of figure 3.7(b). It is the sum of three oriented 1-simplexes,

$(p_0 p_1) + (p_1 p_2) + (p_2 p_0)$. We agree that it has no boundary. If we insisted on the rule $\partial_1(p_0 p_1) = p_0 + p_1$, we would have

$$\partial_1(p_0 p_1) + \partial_1(p_1 p_2) + \partial_1(p_2 p_0) = p_0 + p_1 + p_1 + p_2 + p_2 + p_0$$

which contradicts our intuition. If, on the other hand, we take $\partial_1(p_0 p_1) = p_1 - p_0$, we have

$$\partial_1(p_0 p_1) + \partial_1(p_1 p_2) + \partial_1(p_2 p_0) = p_1 - p_0 + p_2 - p_1 + p_0 - p_2 = 0$$

as expected. Hence, we put a plus sign if the first vertex is omitted and a minus sign if the second is omitted. We employ this fact to define the boundary of a general r-simplex.

Let $\sigma_r(p_0 \dots p_r)$ $(r > 0)$ be an oriented r-simplex. The **boundary** $\partial_r \sigma_r$ of σ_r is an $(r-1)$-chain defined by

$$\partial_r \sigma_r \equiv \sum_{i=0}^{r} (-1)^i (p_0 p_1 \dots \hat{p}_i \dots p_r) \tag{3.20}$$

where the point p_i under $\hat{}$ is omitted. For example,

$$\partial_2(p_0 p_1 p_2) = (p_1 p_2) - (p_0 p_2) + (p_0 p_1)$$
$$\partial_3(p_0 p_1 p_2 p_3) = (p_1 p_2 p_3) - (p_0 p_2 p_3) + (p_0 p_1 p_3) - (p_0 p_1 p_2).$$

We formally define $\partial_0 \sigma_0 = 0$ for $r = 0$.

The operator ∂_r acts linearly on an element $c = \sum_i c_i \sigma_{r,i}$ of $C_r(K)$,

$$\partial_r c = \sum_i c_i \partial_r \sigma_{r,i}. \tag{3.21}$$

The RHS of (3.21) is an element of $C_{r-1}(K)$. Accordingly, ∂_r defines a map

$$\partial_r : C_r(K) \to C_{r-1}(K). \tag{3.22}$$

∂_r is called the **boundary operator**. It is easy to see that the boundary operator is a homomorphism.

Let K be an n-dimensional simplicial complex. There exists a sequence of free Abelian groups and homomorphisms,

$$0 \xrightarrow{i} C_n(K) \xrightarrow{\partial_n} C_{n-1}(K) \xrightarrow{\partial_{n-1}} \cdots \xrightarrow{\partial_2} C_1(K) \xrightarrow{\partial_1} C_0(K) \xrightarrow{\partial_0} 0 \tag{3.23}$$

where $i : 0 \hookrightarrow C_n(K)$ is an inclusion map (0 is regarded as the unit element of $C_n(K)$). This sequence is called the **chain complex** associated with K and is denoted by $C(K)$. It is interesting to study the *image* and *kernel* of the homomorphisms ∂_r.

Definition 3.3. If $c \in C_r(K)$ satisfies

$$\partial_r c = 0 \tag{3.24}$$

c is called an **r-cycle**. The set of r-cycles $Z_r(K)$ is a subgroup of $C_r(K)$ and is called the **r-cycle group**. Note that $Z_r(K) = \ker \partial_r$. [*Remark:* If $r = 0$, $\partial_0 c$ vanishes identically and $Z_0(K) = C_0(K)$, see (3.23).]

Definition 3.4. Let K be an n-dimensional simplicial complex and let $c \in C_r(K)$. If there exists an element $d \in C_{r+1}(K)$ such that

$$c = \partial_{r+1} d \tag{3.25}$$

then c is called an **r-boundary**. The set of r-boundaries $B_r(K)$ is a subgroup of $C_r(K)$ and is called the **r-boundary group**. Note that $B_r(K) = \operatorname{im} \partial_{r+1}$. [*Remark:* $B_n(K)$ is defined to be 0.]

From lemma 3.1, it follows that $Z_r(K)$ and $B_r(K)$ are subgroups of $C_r(K)$. We now prove an important relation between $Z_r(K)$ and $B_r(K)$, which is crucial in the definition of homology groups.

Lemma 3.3. The composite map $\partial_r \circ \partial_{r+1} : C_{r+1}(K) \to C_{r-1}(K)$ is a zero map; that is, $\partial_r(\partial_{r+1} c) = 0$ for any $c \in C_{r+1}(K)$.

Proof. Since ∂_r is a linear operator on $C_r(K)$, it is sufficient to prove the identity $\partial_r \circ \partial_{r+1} = 0$ for the generators of $C_{r+1}(K)$. If $r = 0$, $\partial_0 \circ \partial_1 = 0$ since ∂_0 is a zero operator. Let us assume $r > 0$. Take $\sigma = (p_0 \dots p_r p_{r+1}) \in C_{r+1}(K)$. We find

$$\partial_r(\partial_{r+1} \sigma) = \partial_r \sum_{i=0}^{r+1} (-1)^i (p_0 \dots \hat{p}_i \dots p_{r+1})$$

$$= \sum_{i=0}^{r+1} (-1)^i \partial_r (p_0 \dots \hat{p}_i \dots p_{r+1})$$

$$= \sum_{i=0}^{r+1} (-1)^i \left(\sum_{j=0}^{i-1} (-1)^j (p_0 \dots \hat{p}_j \dots \hat{p}_i \dots p_{r+1}) \right.$$

$$\left. + \sum_{j=i+1}^{r+1} (-1)^{j-1} (p_0 \dots \hat{p}_i \dots \hat{p}_j \dots p_{r+1}) \right)$$

$$= \sum_{j<i} (-1)^{i+j} (p_0 \dots \hat{p}_j \dots \hat{p}_i \dots p_{r+1})$$

$$- \sum_{j>i} (-1)^{i+j} (p_0 \dots \hat{p}_i \dots \hat{p}_j \dots p_{r+1}) = 0 \tag{3.26}$$

which proves the lemma.   □

Theorem 3.3. Let $Z_r(K)$ and $B_r(K)$ be the r-cycle group and the r-boundary group of $C_r(K)$, then

$$B_r(K) \subset Z_r(K) \qquad (\subset C_r(K)). \qquad (3.27)$$

Proof. This is obvious from lemma 3.3. Any element c of $B_r(K)$ is written as $c = \partial_{r+1} d$ for some $d \in C_{r+1}(K)$. Then we find $\partial_r c = \partial_r(\partial_{r+1} d) = 0$, that is, $c \in Z_r(K)$. This implies $Z_r(K) \supset B_r(K)$.   □

What are the geometrical pictures of r-cycles and r-boundaries? With our definitions, ∂_r picks up the boundary of an r-chain. If c is an r-cycle, $\partial_r c = 0$ tells us that c has no boundary. If $c = \partial_{r+1} d$ is an r-boundary, c is the boundary of d whose dimension is higher than c by one. Our intuition tells us that a boundary has no boundary, hence $Z_r(K) \supset B_r(K)$. Those elements of $Z_r(K)$ that are *not* boundaries play the central role in this chapter.

3.3.3 Homology groups

So far we have defined three groups $C_r(K)$, $Z_r(K)$ and $B_r(K)$ associated with a simplicial complex K. How are they related to topological properties of K or to the topological space whose triangulation is K? Is it possible for $C_r(K)$ to express any property which is conserved under homeomorphism? We all know that the edges of a triangle and those of a square are homeomorphic to each other. What about their chain groups? For example, the 1-chain group associated with a triangle is

$$C_1(K_1) = \{i(p_0 p_1) + j(p_1 p_2) + k(p_2 p_0) | i, j, k \in \mathbb{Z}\}$$
$$\cong \mathbb{Z} \oplus \mathbb{Z} \oplus \mathbb{Z}$$

while that associated with a square is

$$C_1(K_2) \cong \mathbb{Z} \oplus \mathbb{Z} \oplus \mathbb{Z} \oplus \mathbb{Z}.$$

Clearly $C_1(K_1)$ is not isomorphic to $C_1(K_2)$, hence $C_r(K)$ cannot be a candidate of a topological invariant. The same is true for $Z_r(K)$ and $B_r(K)$. It turns out that the homology groups defined in the following provide the desired topological invariants.

Definition 3.5. Let K be an n-dimensional simplicial complex. The **rth homology group** $H_r(K), 0 \le r \le n$, associated with K is defined by

$$H_r(K) \equiv Z_r(K)/B_r(K). \qquad (3.28)$$

[*Remarks*: If necessary, we define $H_r(K) = 0$ for $r > n$ or $r < 0$. If we want to stress that the group structure is defined with integer coefficients, we write $H_r(K; \mathbb{Z})$. We may also define the homology groups with $\mathbb{R}$-coefficients, $H_r(K; \mathbb{R})$ or those with $\mathbb{Z}_2$-coefficients, $H_r(K; \mathbb{Z}_2)$.]

Since $B_r(K)$ is a subgroup of $Z_r(K)$, $H_r(K)$ is well defined. The group $H_r(K)$ is the set of equivalence classes of r-cycles,

$$H_r(K) \equiv \{[z]|z \in Z_r(K)\} \tag{3.29}$$

where each equivalence class $[z]$ is called a **homology class**. Two r-cycles z and z' are in the same equivalence class if and only if $z - z' \in B_r(K)$, in which case z is said to be **homologous** to z' and denoted by $z \sim z'$ or $[z] = [z']$. Geometrically $z - z'$ is a boundary of some space. By definition, any boundary $b \in B_r(K)$ is homologous to 0 since $b - 0 \in B_r(K)$. We accept the following theorem without proof.

Theorem 3.4. Homology groups are topological invariants.   Let X be homeomorphic to Y and let (K, f) and (L, g) be triangulations of X and Y respectively. Then we have

$$H_r(K) \cong H_r(L) \qquad r = 0, 1, 2, \ldots. \tag{3.30}$$

In particular, if (K, f) and (L, g) are two triangulations of X, then

$$H_r(K) \cong H_r(L) \qquad r = 0, 1, 2, \ldots. \tag{3.31}$$

Accordingly, it makes sense to talk of homology groups of a topological space X which is not necessarily a polyhedron but which is triangulable. For an arbitrary triangulation (K, f), $H_r(X)$ is defined to be

$$H_r(X) \equiv H_r(K) \qquad r = 0, 1, 2, \ldots. \tag{3.32}$$

Theorem 3.4 tells us that this is independent of the choice of the triangulation (K, f).

Example 3.6. Let $K = \{p_0\}$. The 0-chain is $C_0(K) = \{ip_0|i \in \mathbb{Z}\} \cong \mathbb{Z}$. Clearly $Z_0(K) = C_0(K)$ and $B_0(K) = \{0\}$ ($\partial_0 p_0 = 0$ and p_0 cannot be a boundary of anything). Thus

$$H_0(K) \equiv Z_0(K)/B_0(K) = C_0(K) \cong \mathbb{Z}. \tag{3.33}$$

Exercise 3.1. Let $K = \{p_0, p_1\}$ be a simplicial complex consisting of two 0-simplexes. Show that

$$H_r(K) = \begin{cases} \mathbb{Z} \oplus \mathbb{Z} & (r = 0) \\ \{0\} & (r \neq 0). \end{cases} \tag{3.34}$$

Example 3.7. Let $K = \{p_0, p_1, (p_0 p_1)\}$. We have

$$C_0(K) = \{i p_0 + j p_1 | i, j \in \mathbb{Z}\}$$
$$C_1(K) = \{k(p_0 p_1) | k \in \mathbb{Z}\}.$$

Since $(p_0 p_1)$ is not a boundary of any simplex in K, $B_1(K) = \{0\}$ and

$$H_1(K) = Z_1(K)/B_1(K) = Z_1(K).$$

If $z = m(p_0 p_1) \in Z_1(K)$, it satisfies

$$\partial_1 z = m \partial_1 (p_0 p_1) = m\{p_1 - p_0\} = m p_1 - m p_0 = 0.$$

Thus, m has to vanish and $Z_1(K) = 0$, hence

$$H_1(K) = 0. \tag{3.35}$$

As for $H_0(K)$, we have $Z_0(K) = C_0(K) = \{i p_0 + j p_1\}$ and

$$B_0(K) = \text{im } \partial_1 = \{\partial_1 i(p_0 p_1) | i \in \mathbb{Z}\} = \{i(p_0 - p_1) | i \in \mathbb{Z}\}.$$

Define a surjective (onto) homomorphism $f : Z_0(K) \to \mathbb{Z}$ by

$$f(i p_0 + j p_1) = i + j.$$

Then we find

$$\ker f = f^{-1}(0) = B_0(K).$$

Theorem 3.1 states that $Z_0(K)/\ker f \cong \text{im } f = \mathbb{Z}$, or

$$H_0(K) = Z_0(K)/B_0(K) \cong \mathbb{Z}. \tag{3.36}$$

Example 3.8. Let $K = \{p_0, p_1, p_2, (p_0 p_1), (p_1 p_2), (p_2 p_0)\}$, see figure 3.7(*b*). This is a triangulation of S^1. Since there are no 2-simplexes in K, we have $B_1(K) = 0$ and $H_1(K) = Z_1(K)/B_1(K) = Z_1(K)$. Let $z = i(p_0 p_1) + j(p_1 p_2) + k(p_2 p_0) \in Z_1(K)$ where $i, j, k \in \mathbb{Z}$. We require that

$$\partial_1 z = i(p_1 - p_0) + j(p_2 - p_1) + k(p_0 - p_2)$$
$$= (k - i) p_0 + (i - j) p_1 + (j - k) p_2 = 0.$$

This is satisfied only when $i = j = k$. Thus, we find that

$$Z_1(K) = \{i\{(p_0 p_1) + (p_1 p_2) + (p_2 p_0)\} | i \in \mathbb{Z}\}.$$

This shows that $Z_1(K)$ is isomorphic to $\mathbb{Z}$ and

$$H_1(K) = Z_1(K) \cong \mathbb{Z}. \tag{3.37}$$

Let us compute $H_0(K)$. We have $Z_0(K) = C_0(K)$ and

$$B_0(K) = \{\partial_1[l(p_0p_1) + m(p_1p_2) + n(p_2p_0)]|l, m, n \in \mathbb{Z}\}$$
$$= \{(n - l)p_0 + (l - m)p_1 + (m - n)p_2 \mid l, m, n \in \mathbb{Z}\}.$$

Define a surjective homomorphism $f : Z_0(K) \to \mathbb{Z}$ by

$$f(ip_0 + jp_1 + kp_2) = i + j + k.$$

We verify that

$$\ker f = f^{-1}(0) = B_0(K).$$

From theorem 3.1 we find $Z_0(K)/\ker f \cong \operatorname{im} f = \mathbb{Z}$, or

$$H_0(K) = Z_0(K)/B_0(K) \cong \mathbb{Z}. \tag{3.38}$$

K is a triangulation of a circle S^1, and (3.37) and (3.38) are the homology groups of S^1.

Exercise 3.2. Let $K = \{p_0, p_1, p_2, p_3, (p_0p_1), (p_1p_2), (p_2p_3), (p_3p_0)\}$ be a simplicial complex whose polyhedron is a square. Verify that the homology groups are the same as those of example 3.8 above.

Example 3.9. Let $K = \{p_0, p_1, p_2, (p_0p_1), (p_1p_2), (p_2p_0), (p_0p_1p_2)\}$; see figure 3.6(b). Since the structure of 0-simplexes and 1-simplexes is the same as that of example 3.8, we have

$$H_0(K) \cong \mathbb{Z}. \tag{3.39}$$

Let us compute $H_1(K) = Z_1(K)/B_1(K)$. From the previous example, we have

$$Z_1(K) = \{i\{(p_0p_1) + (p_1p_2) + (p_2p_0)\}|i \in \mathbb{Z}\}.$$

Let $c = m(p_0p_1p_2) \in C_2(K)$. If $b = \partial_2 c \in B_1(K)$, we have

$$b = m\{(p_1p_2) - (p_0p_2) + (p_0p_1)\}$$
$$= m\{(p_0p_1) + (p_1p_2) + (p_2p_0)\} \qquad m \in \mathbb{Z}.$$

This shows that $Z_1(K) \cong B_1(K)$, hence

$$H_1(K) = Z_1(K)/B_1(K) \cong \{0\}. \tag{3.40}$$

Since there are no 3-simplexes in K, we have $B_2(K) = \{0\}$. Then $H_2(K) = Z_2(K)/B_2(K) = Z_2(K)$. Let $z = m(p_0p_1p_2) \in Z_2(K)$. Since $\partial_2 z = m\{(p_1p_2) - (p_0p_2) + (p_0p_1)\} = 0$, m must vanish. Hence, $Z_1(K) = \{0\}$ and we have

$$H_2(K) \cong \{0\}. \tag{3.41}$$

Exercise 3.3. Let

$$K = \{p_0, p_1, p_2, p_3, (p_0p_1), (p_0p_2), (p_0p_3), (p_1p_2), (p_1p_3), (p_2p_3),$$
$$(p_0p_1p_2), (p_0p_1p_3), (p_0p_2p_3), (p_1p_2p_3)\}$$

be a simplicial complex whose polyhedron is the surface of a tetrahedron. Verify that

$$H_0(K) \cong \mathbb{Z} \qquad H_1(K) \cong \{0\} \qquad H_2(K) \cong \mathbb{Z}. \qquad (3.42)$$

K is a triangulation of the sphere S^2 and (3.42) gives the homology groups of S^2.

3.3.4 Computation of $H_0(K)$

Examples 3.6–3.9 and exercises 3.2, 3.3 share the same zeroth homology group, $H_0(K) \cong \mathbb{Z}$. What is common to these simplicial complexes? We have the following answer.

Theorem 3.5. Let K be a *connected* simplicial complex. Then

$$H_0(K) \cong \mathbb{Z}. \qquad (3.43)$$

Proof. Since K is connected, for any pair of 0-simplexes p_i and p_j, there exists a sequence of 1-simplexes $(p_i p_k), (p_k p_l), \ldots, (p_m p_j)$ such that $\partial_1((p_i p_k) + (p_k p_l) + \cdots + (p_m p_j)) = p_j - p_i$. Then it follows that p_i is homologous to p_j, namely $[p_i] = [p_j]$. Thus, any 0-simplex in K is homologous to p_1 say. Suppose

$$z = \sum_{i=1}^{I_0} n_i p_i \in Z_0(K)$$

where I_0 is the number of 0-simplexes in K. Then the homology class $[z]$ is generated by a single point,

$$[z] = \left[\sum_i n_i p_i\right] = \sum_i n_i[p_i] = \sum_i n_i[p_1].$$

It is clear that $[z] = 0$, namely $z \in B_0(K)$, if $\sum n_i = 0$.

Let $\sigma_j = (p_{j,1} p_{j,2})$ $(1 \le j \le I_1)$ be 1-simplexes in K, I_1 being the number of 1-simplexes in K, then

$$B_0(K) = \mathrm{im}\,\partial_1$$
$$= \{\partial_1(n_1\sigma_1 + \cdots + n_{I_1}\sigma_{I_1})|n_1, \ldots, n_{I_1} \in \mathbb{Z}\}$$
$$= \{n_1(p_{1,2} - p_{1,1}) + \cdots + n_{I_1}(p_{I_1,2} - p_{I_1,1})|n_1, \ldots, n_{I_1} \in \mathbb{Z}\}.$$

Note that n_j $(1 \le j \le I_1)$ always appears as a pair $+n_j$ and $-n_j$ in an element of $B_0(K)$. Thus, if

$$z = \sum_j n_j p_j \in B_0(K) \qquad \text{then} \sum_j n_j = 0.$$

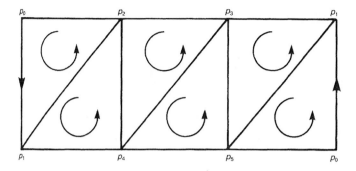

Figure 3.8. A triangulation of the Möbius strip.

Now we have proved for a connected complex K that $z = \sum n_i p_i \in B_0(K)$ if and only if $\sum n_i = 0$.

Define a surjective homomorphism $f : Z_0(K) \to \mathbb{Z}$ by

$$f(n_1 p_1 + \cdots + n_{I_0} p_{I_0}) = \sum_{i=1}^{I_0} n_i.$$

We then have $\ker f = f^{-1}(0) = B_0(K)$. It follows from theorem 3.1 that $H_0(K) = Z_0(K)/B_0(K) = Z_0(K)/\ker f \cong \mathrm{im}\, f = \mathbb{Z}$.   □

3.3.5 More homology computations

Example 3.10. This and the next example deal with homology groups of non-orientable spaces. Figure 3.8 is a triangulation of the Möbius strip. Clearly $B_2(K) = 0$. Let us take a cycle $z \in Z_2(K)$,

$$z = i(p_0 p_1 p_2) + j(p_2 p_1 p_4) + k(p_2 p_4 p_3)$$
$$+ l(p_3 p_4 p_5) + m(p_3 p_5 p_1) + n(p_1 p_5 p_0).$$

z satisfies

$$\begin{aligned}
\partial_2 z = \ & i\{(p_1 p_2) - (p_0 p_2) + (p_0 p_1)\} \\
& + j\{(p_1 p_4) - (p_2 p_4) + (p_2 p_1)\} \\
& + k\{(p_4 p_3) - (p_2 p_3) + (p_2 p_4)\} \\
& + l\{(p_4 p_5) - (p_3 p_5) + (p_3 p_4)\} \\
& + m\{(p_5 p_1) - (p_3 p_1) + (p_3 p_5)\} \\
& + n\{(p_5 p_0) - (p_1 p_0) + (p_1 p_5)\} = 0.
\end{aligned}$$

Since each of $(p_0 p_2)$, $(p_1 p_4)$, $(p_2 p_3)$, $(p_4 p_5)$, $(p_3 p_1)$ and $(p_5 p_0)$ appears once and only once in $\partial_2 z$, all the coefficients must vanish, $i = j = k = l = m = n =$

0. Thus, $Z_2(K) = \{0\}$ and

$$H_2(K) = Z_2(K)/B_2(K) \cong \{0\}. \tag{3.44}$$

To find $H_1(K)$, we use our intuition rather than doing tedious computations. Let us find the loops which make complete circuits. One such loop is

$$z = (p_0 p_1) + (p_1 p_4) + (p_4 p_5) + (p_5 p_0).$$

Then all the other complete circuits are homologous to multiples of z. For example, let us take

$$z' = (p_1 p_2) + (p_2 p_3) + (p_3 p_5) + (p_5 p_1).$$

We find that $z \sim z'$ since

$$z - z' = \partial_2 \{(p_2 p_1 p_4) + (p_2 p_4 p_3) + (p_3 p_4 p_5) + (p_1 p_5 p_0)\}.$$

If, however, we take

$$z'' = (p_1 p_4) + (p_4 p_5) + (p_5 p_0) + (p_0 p_2) + (p_2 p_3) + (p_3 p_1)$$

we find that $z'' \sim 2z$ since

$$\begin{aligned}
2z - z'' &= 2(p_0 p_1) + (p_1 p_4) + (p_4 p_5) + (p_5 p_0) - (p_0 p_2) \\
&\quad - (p_2 p_3) - (p_3 p_1) \\
&= \partial_2 \{(p_0 p_1 p_2) + (p_1 p_4 p_2) + (p_2 p_4 p_3) + (p_3 p_4 p_5) \\
&\quad + (p_3 p_5 p_1) + (p_0 p_1 p_5)\}.
\end{aligned}$$

We easily verify that all the closed circuits are homologous to nz, $n \in \mathbb{Z}$. $H_1(K)$ is generated by just one element $[z]$,

$$H_1(K) = \{i[z] | i \in \mathbb{Z}\} \cong \mathbb{Z}. \tag{3.45}$$

Since K is connected, it follows from theorem 3.5 that $H_0(K) = \{i[p_a] | i \in \mathbb{Z}\} \cong \mathbb{Z}$, p_a being any 0-simplex of K.

Example 3.11. The projective plane $\mathbb{R}P^2$ has been defined in example 2.5(c) as the sphere S^2 whose antipodal points are identified. As a coset space, we may take the hemisphere (or the disc D^2) whose opposite points on the boundary S^1 are identified, see figure 2.5(b). Figure 3.9 is a triangulation of the projective plane. Clearly $B_2(K) = \{0\}$. Take a cycle $z \in Z_2(K)$,

$$\begin{aligned}
z &= m_1(p_0 p_1 p_2) + m_2(p_0 p_4 p_1) + m_3(p_0 p_5 p_4) \\
&\quad + m_4(p_0 p_3 p_5) + m_5(p_0 p_2 p_3) + m_6(p_2 p_4 p_3) \\
&\quad + m_7(p_2 p_5 p_4) + m_8(p_2 p_1 p_5) + m_9(p_1 p_3 p_5) + m_{10}(p_1 p_4 p_3).
\end{aligned}$$

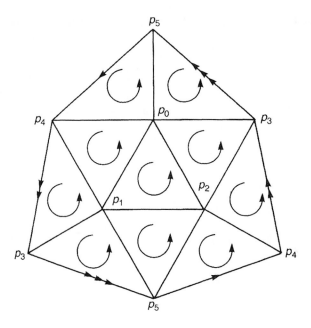

Figure 3.9. A triangulation of the projective plane.

The boundary of z is

$$
\begin{aligned}
\partial_2 z = {} & m_1\{(p_1 p_2) - (p_0 p_2) + (p_0 p_1)\} \\
& + m_2\{(p_4 p_1) - (p_0 p_1) + (p_0 p_4)\} \\
& + m_3\{(p_5 p_4) - (p_0 p_4) + (p_0 p_5)\} \\
& + m_4\{(p_3 p_5) - (p_0 p_5) + (p_0 p_3)\} \\
& + m_5\{(p_2 p_3) - (p_0 p_3) + (p_0 p_2)\} \\
& + m_6\{(p_4 p_3) - (p_2 p_3) + (p_2 p_4)\} \\
& + m_7\{(p_5 p_4) - (p_2 p_4) + (p_2 p_5)\} \\
& + m_8\{(p_1 p_5) - (p_2 p_5) + (p_2 p_1)\} \\
& + m_9\{(p_3 p_5) - (p_1 p_5) + (p_1 p_3)\} \\
& + m_{10}\{(p_4 p_3) - (p_1 p_3) + (p_1 p_4)\} = 0.
\end{aligned}
$$

Let us look at the coefficient of each 1-simplex. For example, we have $(m_1 - m_2)(p_0 p_1)$, hence $m_1 - m_2 = 0$. Similarly,

$$
\begin{aligned}
& -m_1 + m_5 = 0, \, m_4 - m_5 = 0, \, m_2 - m_3 = 0, \, m_1 - m_8 = 0, \\
& m_9 - m_{10} = 0, \, -m_2 + m_{10} = 0, \, m_5 - m_6 = 0, \, m_6 - m_7 = 0, \\
& m_6 + m_{10} = 0.
\end{aligned}
$$

These ten conditions are satisfied if and only if $m_i = 0$, $1 \leq i \leq 10$. This means that the cycle group $Z_2(K)$ is trivial and we have

$$H_2(K) = Z_2(K)/B_2(K) \cong \{0\}. \tag{3.46}$$

Before we calculate $H_1(K)$, we examine $H_2(K)$ from a slightly different viewpoint. Let us add all the 2-simplexes in K with the same coefficient,

$$z \equiv \sum_{i=1}^{10} m\sigma_{2,i} \qquad m \in \mathbb{Z}.$$

Observe that each 1-simplex of K is a common face of exactly two 2-simplexes. As a consequence, the boundary of z is

$$\partial_2 z = 2m(p_3 p_5) + 2m(p_5 p_4) + 2m(p_4 p_3). \tag{3.47}$$

Thus, if $z \in Z_2(K)$, m must vanish and we find $Z_2(K) = \{0\}$ as before. This observation remarkably simplifies the computation of $H_1(K)$. Note that any 1-cycle is homologous to a multiple of

$$z = (p_3 p_5) + (p_5 p_4) + (p_4 p_3)$$

cf example 3.10. Furthermore, equation (3.47) shows that an even multiple of z is a boundary of a 2-chain. Thus, z is a cycle and $z + z$ is homologous to 0. Hence, we find that

$$H_1(K) = \{[z]|[z] + [z] \sim [0]\} \cong \mathbb{Z}_2. \tag{3.48}$$

This example shows that a homology group is not necessarily free Abelian but may have the full structure of a finitely generated Abelian group. Since K is connected, we have $H_0(K) \cong \mathbb{Z}$.

It is interesting to compare example 3.11 with the following examples. In these examples, we shall use the intuition developed in this section on boundaries and cycles to obtain results rather than giving straightforward but tedious computations.

Example 3.12. Let us consider the torus T^2. A formal derivation of the homology groups of T^2 is left as an exercise to the reader: see Fraleigh (1976), for example. This is an appropriate place to recall the intuitive meaning of the homology groups. The rth homology group is generated by those boundaryless r-chains that are not, by themselves, boundaries of some $(r + 1)$-chains. For example, the surface of the torus has no boundary but it is not a boundary of some 3-chain. Thus, $H_2(T^2)$ is freely generated by one generator, the surface itself, $H_2(T^2) \cong \mathbb{Z}$. Let us look at $H_1(T^2)$ next. Clearly the loops a and b in figure 3.10 have no boundaries but are not boundaries of some 2-chains. Take another loop a'. a' is homologous to a since $a' - a$ bounds the shaded area of figure 3.10.

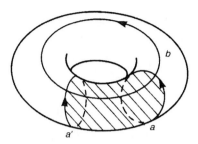

Figure 3.10. a' is homologous to a but b is not. a and b generate $H_1(T^2)$.

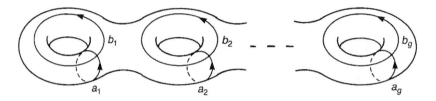

Figure 3.11. a_i and b_i $(1 \le i \le g)$ generate $H_1(\Sigma_g)$.

Hence, $H_1(T^2)$ is freely generated by a and b and $H_1(T^2) \cong \mathbb{Z} \oplus \mathbb{Z}$. Since T^2 is connected, we have $H_0(T^2) \cong \mathbb{Z}$.

Now it is easy to extend our analysis to the torus Σ_g of genus g. Since Σ_g has no boundary and there are no 3-simplexes, the surface Σ_g itself freely generates $H_2(T^2) \cong \mathbb{Z}$. The first homology group $H_1(\Sigma_g)$ is generated by those loops which are not boundaries of some area. Figure 3.11 shows the standard choice for the generators. We find

$$H_1(\Sigma_g) = \{i_1[a_1] + j_1[b_1] + \cdots + i_g[a_g] + j_g[b_g]\}$$
$$\cong \underbrace{\mathbb{Z} \oplus \mathbb{Z} \oplus \cdots \oplus \mathbb{Z}}_{2g}. \tag{3.49}$$

Since Σ_g is connected, $H_0(\Sigma_g) \cong \mathbb{Z}$. Observe that $a_i(b_i)$ is homologous to the edge $a_i(b_i)$ of figure 2.12. The $2g$ curves $\{a_i, b_i\}$ are called the **canonical system of curves** on Σ_g.

Example 3.13. Figure 3.12 is a triangulation of the Klein bottle. Computations of the homology groups are much the same as those of the projective plane. Since $B_2(K) = 0$, we have $H_2(K) = Z_2(K)$. Let $z \in Z_2(K)$. If z is a combination of all the 2-simplexes of K with the same coefficient, $z = \sum m\sigma_{2,i}$, the inner 1-simplexes cancel out to leave only the outer 1-simplexes

$$\partial_2 z = -2ma$$

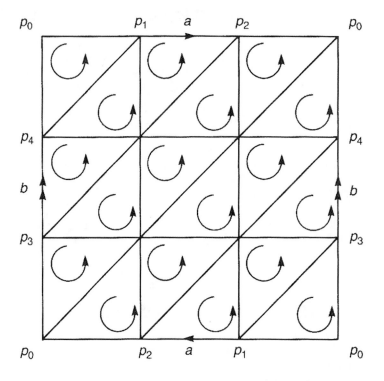

Figure 3.12. A triangulation of the Klein bottle.

where $a = (p_0 p_1) + (p_1 p_2) + (p_2 p_0)$. For $\partial_2 z$ to be 0, the integer m must vanish and we have

$$H_2(K) = Z_2(K) \cong \{0\}. \tag{3.50}$$

To compute $H_1(K)$ we first note, from our experience with the torus, that every 1-cycle is homologous to $ia + jb$ for some $i, j \in \mathbb{Z}$. For a 2-chain to have a boundary consisting of a and b only, all the 2-simplexes in K must be added with the same coefficient. As a result, for such a 2-chain $z = \sum m\sigma_{2,i}$, we have $\partial z = 2ma$. This shows that $2ma \sim 0$. Thus, $H_1(K)$ is generated by two cycles a and b such that $a + a = 0$, namely

$$H_1(K) = \{i[a] + j[b] | i, j \in \mathbb{Z}\} \cong \mathbb{Z}_2 \oplus \mathbb{Z}. \tag{3.51}$$

We obtain $H_0(K) \cong \mathbb{Z}$ since K is connected.

3.4 General properties of homology groups

3.4.1 Connectedness and homology groups

Let $K = \{p_0\}$ and $L = \{p_0, p_1\}$. From example 3.6 and exercise 3.1, we have $H_0(K) = \mathbb{Z}$ and $H_0(L) = \mathbb{Z} \oplus \mathbb{Z}$. More generally, we have the following theorem.

Theorem 3.6. Let K be a disjoint union of N connected components, $K = K_1 \cup K_2 \cup \cdots \cup K_N$ where $K_i \cap K_j = \emptyset$. Then

$$H_r(K) = H_r(K_1) \oplus H_r(K_2) \oplus \cdots \oplus H_r(K_N). \tag{3.52}$$

Proof. We first note that an r-chain group is consistently separated into a direct sum of N r-chain subgroups. Let

$$C_r(K) = \left\{ \sum_{i=1}^{I_r} c_i \sigma_{r,i} \,\middle|\, c_i \in \mathbb{Z} \right\}$$

where I_r is the number of linearly independent r-simplexes in K. It is always possible to rearrange σ_i so that those r-simplexes in K_1 come first, those in K_2 next and so on. Then $C_r(K)$ is separated into a direct sum of subgroups,

$$C_r(K) = C_r(K_1) \oplus C_r(K_2) \oplus \cdots \oplus C_r(K_N).$$

This separation is also carried out for $Z_r(K)$ and $B_r(K)$ as

$$Z_r(K) = Z_r(K_1) \oplus Z_r(K_2) \oplus \cdots \oplus Z_r(K_N)$$
$$B_r(K) = B_r(K_1) \oplus B_r(K_2) \oplus \cdots \oplus B_r(K_N).$$

We now define the homology groups of each component K_i by

$$H_r(K_i) = Z_r(K_i)/B_r(K_i).$$

This is well defined since $Z_r(K_i) \supset B_r(K_i)$. Finally, we have

$$\begin{aligned} H_r(K) &= Z_r(K)/B_r(K) \\ &= Z_r(K_1) \oplus \cdots \oplus Z_r(K_N)/B_r(K_1) \oplus \cdots \oplus B_r(K_N) \\ &= \{Z_r(K_1)/B_r(K_1)\} \oplus \cdots \oplus \{Z_r(K_N)/B_r(K_N)\} \\ &= H_r(K_1) \oplus \cdots \oplus H_r(K_N). \end{aligned} \qquad \square$$

Corollary 3.1. (a) Let K be a disjoint union of N connected components, $K_1, \ldots, K_N$. Then it follows that

$$H_0(K) \cong \underbrace{\mathbb{Z} \oplus \cdots \oplus \mathbb{Z}}_{N \text{ factors}}. \tag{3.53}$$

(b) If $H_0(K) \cong \mathbb{Z}$, K is connected. [Together with theorem 3.5 we conclude that $H_0(K) \cong \mathbb{Z}$ if and only if K is connected.]

3.4.2 Structure of homology groups

$Z_r(K)$ and $B_r(K)$ are free Abelian groups since they are subgroups of a free Abelian group $C_r(K)$. It does not mean that $H_r(K) = Z_r(K)/B_r(K)$ is also free Abelian. In fact, according to theorem 3.2, the most general form of $H_r(K)$ is

$$H_r(K) \cong \underbrace{\mathbb{Z} \oplus \cdots \oplus \mathbb{Z}}_{f} \oplus \mathbb{Z}_{k_1} \oplus \cdots \oplus \mathbb{Z}_{k_p}. \tag{3.54}$$

It is clear from our experience that the number of generators of $H_r(K)$ counts the number of $(r + 1)$-dimensional holes in $|K|$. The first f factors form a free Abelian group of rank f and the next p factors are called the **torsion subgroup** of $H_r(K)$. For example, the projective plane has $H_1(K) \cong \mathbb{Z}_2$ and the Klein bottle has $H_1(K) \cong \mathbb{Z} \oplus \mathbb{Z}_2$. In a sense, the torsion subgroup detects the 'twisting' in the polyhedron $|K|$. We now clarify why the homology groups with $\mathbb{Z}$-coefficients are preferable to those with $\mathbb{Z}_2$- or $\mathbb{R}$-coefficients. Since $\mathbb{Z}_2$ has no non-trivial subgroups, the torsion subgroup can never be recognized. Similarly, if $\mathbb{R}$-coefficients are employed, we cannot see the torsion subgroup either, since $\mathbb{R}/m\mathbb{R} \cong \{0\}$ for any $m \in \mathbb{Z} - \{0\}$. [For any $a, b \in \mathbb{R}$, there exists a number $c \in \mathbb{R}$ such that $a - b = mc$.] If $H_r(K; \mathbb{Z})$ is given by (3.54), $H_r(K; \mathbb{R})$ is

$$H_r(K; \mathbb{R}) \cong \underbrace{\mathbb{R} \oplus \mathbb{R} \oplus \cdots \oplus \mathbb{R}}_{f}. \tag{3.55}$$

3.4.3 Betti numbers and the Euler–Poincaré theorem

Definition 3.6. Let K be a simplicial complex. The rth **Betti number** $b_r(K)$ is defined by

$$b_r(K) \equiv \dim H_r(K; \mathbb{R}). \tag{3.56}$$

In other words, $b_r(K)$ is the rank of the free Abelian part of $H_r(K; \mathbb{Z})$.

For example, the Betti numbers of the torus T^2 are (see example 3.12)

$$b_0(K) = 1, \qquad b_1(K) = 2, \qquad b_2(K) = 1$$

and those of the sphere S^2 are (exercise 3.3)

$$b_0(K) = 1, \qquad b_1(K) = 0, \qquad b_2(K) = 1.$$

The following theorem relates the Euler characteristic to the Betti numbers.

Theorem 3.7. (**The Euler–Poincaré theorem**) Let K be an n-dimensional simplicial complex and let I_r be the number of r-simplexes in K. Then

$$\chi(K) \equiv \sum_{r=0}^{n} (-1)^r I_r = \sum_{r=0}^{n} (-1)^r b_r(K). \tag{3.57}$$

[*Remark:* The first equality *defines* the Euler characteristic of a general polyhedron $|K|$. Note that this is the generalization of the Euler characteristic defined for surfaces in section 2.4.]

Proof. Consider the boundary homomorphism,

$$\partial_r : C_r(K; \mathbb{R}) \to C_{r-1}(K; \mathbb{R})$$

where $C_{-1}(K; \mathbb{R})$ is defined to be $\{0\}$. Since both $C_{r-1}(K; \mathbb{R})$ and $C_r(K; \mathbb{R})$ are vector spaces, theorem 2.1 can be applied to yield

$$
\begin{aligned}
I_r &= \dim C_r(K; \mathbb{R}) = \dim(\ker \partial_r) + \dim(\operatorname{im} \partial_r) \\
&= \dim Z_r(K; \mathbb{R}) + \dim B_{r-1}(K; \mathbb{R})
\end{aligned}
$$

where $B_{-1}(K)$ is defined to be trivial. We also have

$$
\begin{aligned}
b_r(K) &= \dim H_r(K; \mathbb{R}) = \dim(Z_r(K; \mathbb{R})/B_r(K; \mathbb{R})) \\
&= \dim Z_r(K; \mathbb{R}) - \dim B_r(K; \mathbb{R}).
\end{aligned}
$$

From these relations, we obtain

$$
\begin{aligned}
\chi(K) &= \sum_{r=0}^{n} (-1)^r I_r = \sum_{r=0}^{n} (-1)^r (\dim Z_r(K; \mathbb{R}) + \dim B_{r-1}(K; \mathbb{R})) \\
&= \sum_{r=0}^{n} \{(-1)^r \dim Z_r(K; \mathbb{R}) - (-1)^r \dim B_r(K; \mathbb{R})\} \\
&= \sum_{r=0}^{n} (-1)^r b_r(K). \qquad \square
\end{aligned}
$$

Since the Betti numbers are topological invariants, $\chi(K)$ is also conserved under a homeomorphism. In particular, if $f : |K| \to X$ and $g : |K'| \to X$ are two triangulations of X, we have $\chi(K) = \chi(K')$. Thus, it makes sense to define the Euler characteristic of X by $\chi(K)$ for any triangulation (K, f) of X.

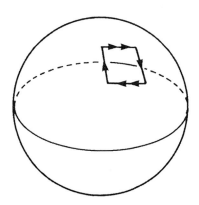

Figure 3.13. A hole in S^2, whose edges are identified as shown. We may consider S^2 with q such holes.

Problems

3.1 The most general orientable two-dimensional surface is a 2-sphere with h handles and q holes. Compute the homology groups and the Euler characteristic of this surface.

3.2 Consider a sphere with a hole and identify the edges of the hole as shown in figure 3.13. The surface we obtained was simply the projective plane $\mathbb{R}P^2$. More generally, consider a sphere with q such 'crosscaps' and compute the homology groups and the Euler characteristic of this surface.

4

HOMOTOPY GROUPS

The idea of homology groups in the previous chapter was to assign a group structure to cycles that are not boundaries. In homotopy groups, however, we are interested in continuous deformation of maps one to another. Let X and Y be topological spaces and let $\mathcal{F}$ be the set of continuous maps, from X to Y. We introduce an equivalence relation, called 'homotopic to', in $\mathcal{F}$ by which two maps $f, g \in \mathcal{F}$ are identified if the image $f(X)$ is continuously deformed to $g(X)$ in Y. We choose X to be some *standard* topological spaces whose structures are well known. For example, we may take the n-sphere S^n as the standard space and study all the maps from S^n to Y to see how these maps are classified according to homotopic equivalence. This is the basic idea of homotopy groups.

We will restrict ourselves to an elementary study of homotopy groups, which is sufficient for the later discussion. Nash and Sen (1983) and Croom (1978) complement this chapter.

4.1 Fundamental groups

4.1.1 Basic ideas

Let us look at figure 4.1. One disc has a hole in it, the other does not. What characterizes the difference between these two discs? We note that any loop in figure 4.1(*b*) can be continuously shrunk to a point. In contrast, the loop α in figure 4.1(*a*) cannot be shrunk to a point due to the existence of a hole in it. Some loops in figure 4.1(*a*) may be shrunk to a point while others cannot. We say a loop α is homotopic to β if α can be obtained from β by a *continuous* deformation. For example, any loop in Y is homotopic to a point. It turns out that 'homotopic to' is an equivalence relation, the equivalence class of which is called the homotopy class. In figure 4.1, there is only one homotopy class associated with Y. In X, each homotopy class is characterized by $n \in \mathbb{Z}$, n being the number of times the loop encircles the hole; $n < 0$ if it winds clockwise, $n > 0$ if counterclockwise, $n = 0$ if the loop does not wind round the hole. Moreover, $\mathbb{Z}$ is an additive group and the group operation (addition) has a geometrical meaning; $n + m$ corresponds to going round the hole first n times and then m times. The set of homotopy classes is endowed with a group structure called the fundamental group.

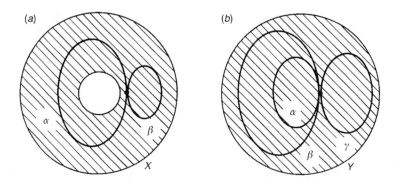

Figure 4.1. A disc with a hole (*a*) and without a hole (*b*). The hole in (*a*) prevents the loop α from shrinking to a point.

4.1.2 Paths and loops

Definition 4.1. Let X be a topological space and let $I = [0, 1]$. A continuous map $\alpha : I \to X$ is called a **path** with an initial point x_0 and an end point x_1 if $\alpha(0) = x_0$ and $\alpha(1) = x_1$. If $\alpha(0) = \alpha(1) = x_0$, the path is called a **loop** with **base point** x_0 (or a loop at x_0).

For $x \in X$, a **constant path** $c_x : I \to X$ is defined by $c_x(s) = x$, $s \in I$. A constant path is also a constant loop since $c_x(0) = c_x(1) = x$. The set of paths or loops in a topological space X may be endowed with an algebraic structure as follows.

Definition 4.2. Let $\alpha, \beta : I \to X$ be paths such that $\alpha(1) = \beta(0)$. The product of α and β, denoted by $\alpha * \beta$, is a path in X defined by

$$
\alpha * \beta(s) = \begin{cases} \alpha(2s) & 0 \le s \le \frac{1}{2} \\ \beta(2s - 1) & \frac{1}{2} \le s \le 1 \end{cases} \tag{4.1}
$$

see figure 4.2. Since $\alpha(1) = \beta(0)$, $\alpha * \beta$ is a continuous map from I to X. [Geometrically, $\alpha * \beta$ corresponds to traversing the image $\alpha(I)$, in the first half, then followed by $\beta(I)$ in the remaining half. Note that the velocity is doubled.]

Definition 4.3. Let $\alpha : I \to X$ be a path from x_0 to x_1. The inverse path α^{-1} of α is defined by

$$
\alpha^{-1}(s) \equiv \alpha(1 - s) \qquad s \in I. \tag{4.2}
$$

[The inverse path α^{-1} corresponds to traversing the image of α in the opposite direction from x_1 to x_0.]

Since a loop is a special path for which the initial point and end point agree, the product of loops and the inverse of a loop are defined in exactly the same way.

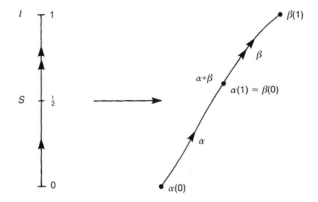

Figure 4.2. The product $\alpha * \beta$ of paths α and β with a common end point.

It seems that a constant map c_x is the unit element. However, it is not: $\alpha * \alpha^{-1}$ is not equal to c_x! We need a concept of homotopy to define a group operation in the space of loops.

4.1.3 Homotopy

The algebraic structure of loops introduced earlier is not so useful as it is. For example, the constant path is not exactly the unit element. We want to classify the paths and loops according to a neat equivalence relation so that the equivalence classes admit a group structure. It turns out that if we identify paths or loops that can be deformed continuously one into another, the equivalence classes form a group. Since we are primarily interested in loops, most definitions and theorems are given for loops. However, it should be kept in mind that many statements are also applied to paths with proper modifications.

Definition 4.4. Let $\alpha, \beta : I \to X$ be loops at x_0. They are said to be **homotopic**, written as $\alpha \sim \beta$, if there exists a continuous map $F : I \times I \to X$ such that

$$F(s, 0) = \alpha(s), \qquad F(s, 1) = \beta(s) \qquad \forall s \in I$$
$$F(0, t) = F(1, t) = x_0 \qquad \forall t \in I. \tag{4.3}$$

The connecting map F is called a **homotopy** between α and β.

It is helpful to represent a homotopy as figure 4.3(a). The vertical edges of the square $I \times I$ are mapped to x_0. The lower edge is $\alpha(s)$ while the upper edge is $\beta(s)$. In the space X, the image is continuously deformed as in figure 4.3(b).

Proposition 4.1. The relation $\alpha \sim \beta$ is an equivalence relation.

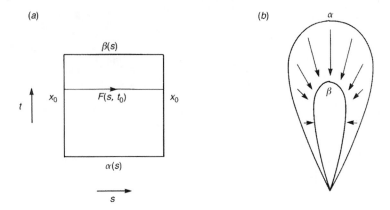

Figure 4.3. (*a*) The square represents a homotopy F interpolating the loops α and β. (*b*) The image of α is continuously deformed to the image of β in real space X.

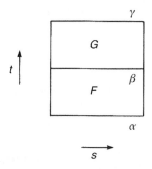

Figure 4.4. A homotopy H between α and γ via β.

Proof. Reflectivity: $\alpha \sim \alpha$. The homotopy may be given by $F(s, t) = \alpha(s)$ for any $t \in I$.

Symmetry: Let $\alpha \sim \beta$ with the homotopy $F(s, t)$ such that $F(s, 0) = \alpha(s)$, $F(s, 1) = \beta(s)$. Then $\beta \sim \alpha$, where the homotopy is given by $F(s, 1 - t)$.

Transitivity: Let $\alpha \sim \beta$ and $\beta \sim \gamma$. Then $\alpha \sim \gamma$. If $F(s, t)$ is a homotopy between α and β and $G(s, t)$ is a homotopy between β and γ, a homotopy between α and γ may be (figure 4.4)

$$H(s, t) = \begin{cases} F(s, 2t) & 0 \le t \le \frac{1}{2} \\ G(s, 2t - 1) & \frac{1}{2} \le t \le 1. \end{cases} \qquad \square$$

4.1.4 Fundamental groups

The equivalence class of loops is denoted by $[\alpha]$ and is called the **homotopy class** of α. The product between loops naturally defines the product in the set of homotopy classes of loops.

Definition 4.5. Let X be a topological space. The set of homotopy classes of loops at $x_0 \in X$ is denoted by $\pi_1(X, x_0)$ and is called the **fundamental group** (or the **first homotopy group**) of X at x_0. The product of homotopy classes $[\alpha]$ and $[\beta]$ is defined by

$$[\alpha] * [\beta] = [\alpha * \beta]. \tag{4.4}$$

Lemma 4.1. The product of homotopy classes is independent of the representative, that is, if $\alpha \sim \alpha'$ and $\beta \sim \beta'$, then $\alpha * \beta \sim \alpha' * \beta'$.

Proof. Let $F(s, t)$ be a homotopy between α and α' and $G(s, t)$ be a homotopy between β and β'. Then

$$H(s, t) = \begin{cases} F(2s, t) & 0 \le s \le \frac{1}{2} \\ G(2s - 1, t) & \frac{1}{2} \le s \le 1 \end{cases}$$

is a homotopy between $\alpha * \beta$ and $\alpha' * \beta'$, hence $\alpha * \beta \sim \alpha' * \beta'$ and $[\alpha] * [\beta]$ is well defined. $\qquad\square$

Theorem 4.1. The fundamental group is a group. Namely, if $\alpha, \beta, \ldots$ are loops at $x \in X$, the following group properties are satisfied:

(1) $([\alpha] * [\beta]) * [\gamma] = [\alpha] * ([\beta] * [\gamma])$
(2) $[\alpha] * [c_x] = [\alpha]$ and $[c_x] * [\alpha] = [\alpha]$ (unit element)
(3) $[\alpha] * [\alpha^{-1}] = [c_x]$, hence $[\alpha]^{-1} = [\alpha^{-1}]$ (inverse).

Proof. (1) Let $F(s, t)$ be a homotopy between $(\alpha * \beta) * \gamma$ and $\alpha * (\beta * \gamma)$. It may be given by (figure 4.5(a))

$$F(s, t) = \begin{cases} \alpha\left(\dfrac{4s}{1+t}\right) & 0 \le s \le \dfrac{1+t}{4} \\[2mm] \beta(4s - t - 1) & \dfrac{1+t}{4} \le s \le \dfrac{2+t}{4} \\[2mm] \gamma\left(\dfrac{4s - t - 2}{2 - t}\right) & \dfrac{2+t}{4} \le s \le 1. \end{cases}$$

Thus, we may simply write $[\alpha * \beta * \gamma]$ to denote $[(\alpha * \beta) * \gamma]$ or $[\alpha * (\beta * \gamma)]$.

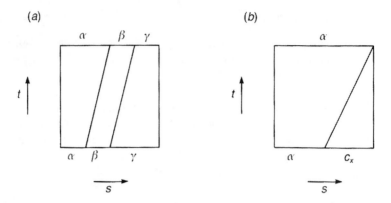

Figure 4.5. (*a*) A homotopy between $(\alpha * \beta) * \gamma$ and $\alpha * (\beta * \gamma)$. (*b*) A homotopy between $\alpha * c_x$ and α.

(2) Define a homotopy $F(s, t)$ by (figure 4.5(*b*))

$$
F(s, t) = \begin{cases} \alpha\left(\dfrac{2s}{1+t}\right) & 0 \le s \le \dfrac{t+1}{2} \\[2mm] x & \dfrac{t+1}{2} \le s \le 1. \end{cases}
$$

Clearly this is a homotopy between $\alpha * c_x$ and α. Similarly, a homotopy between $c_x * \alpha$ and α is given by

$$
F(s, t) = \begin{cases} x & 0 \le s \le \dfrac{1-t}{2} \\[2mm] \alpha\left(\dfrac{2s-1+t}{1+t}\right) & \dfrac{1-t}{2} \le s \le 1. \end{cases}
$$

This shows that $[\alpha] * [c_x] = [\alpha] = [c_x] * [\alpha]$.

(3) Define a map $F : I \times I \to X$ by

$$
F(s, t) = \begin{cases} \alpha(2s(1-t)) & 0 \le s \le \frac{1}{2} \\ \alpha(2(1-s)(1-t)) & \frac{1}{2} \le s \le 1. \end{cases}
$$

Clearly $F(s, 0) = \alpha * \alpha^{-1}$ and $F(s, 1) = c_x$, hence

$$
[\alpha * \alpha^{-1}] = [\alpha] * [\alpha^{-1}] = [c_x].
$$

This shows that $[\alpha^{-1}] = [\alpha]^{-1}$.   □

In summary, $\pi_1(X, x)$ is a group whose unit element is the homotopy class of the constant loop c_x. The product $[\alpha] * [\beta]$ is well defined and satisfies the

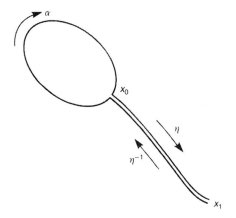

Figure 4.6. From a loop α at x_0, a loop $\eta^{-1} * \alpha * \eta$ at x_1 is constructed.

group axioms. The inverse of $[\alpha]$ is $[\alpha]^{-1} = [\alpha^{-1}]$. In the next section we study the general properties of fundamental groups, which simplify the actual computations.

4.2 General properties of fundamental groups

4.2.1 Arcwise connectedness and fundamental groups

In section 2.3 we defined a topological space X to be arcwise connected if, for any $x_0, x_1 \in X$, there exists a path α such that $\alpha(0) = x_0$ and $\alpha(1) = x_1$.

Theorem 4.2. Let X be an arcwise connected topological space and let $x_0, x_1 \in X$. Then $\pi_1(X, x_0)$ is isomorphic to $\pi_1(X, x_1)$.

Proof. Let $\eta : I \to X$ be a path such that $\eta(0) = x_0$ and $\eta(1) = x_1$. If α is a loop at x_0, then $\eta^{-1} * \alpha * \eta$ is a loop at x_1 (figure 4.6). Given an element $[\alpha] \in \pi_1(X, x_0)$, this correspondence induces a unique element $[\alpha'] = [\eta^{-1} * \alpha * \eta] \in \pi_1(X, x_1)$. We denote this map by $P_\eta : \pi_1(X, x_0) \to \pi_1(X, x_1)$ so that $[\alpha'] = P_\eta([\alpha])$.

We show that P_η is an isomorphism. First, P_η is a *homomorphism*, since for $[\alpha], [\beta] \in \pi_1(X, x_0)$, we have

$$P_\eta([\alpha] * [\beta]) = [\eta^{-1}] * [\alpha] * [\beta] * [\eta]$$
$$= [\eta^{-1}] * [\alpha] * [\eta] * [\eta^{-1}] * [\beta] * [\eta]$$
$$= P_\eta([\alpha]) * P_\eta([\beta]).$$

To show that P_η is *bijective*, we introduce the inverse of P_η. Define a map $P_\eta^{-1} : \pi_1(X, x_1) \to \pi_1(X, x_0)$ whose action on $[\alpha']$ is $P_\eta^{-1}([\alpha']) = [\eta * \alpha * \eta^{-1}]$.

Clearly P^{-1} is the inverse of P_η since

$$P_\eta^{-1} \circ P_\eta([\alpha]) = P_\eta^{-1}([\eta^{-1} * \alpha * \eta]) = [\eta * \eta^{-1} * \alpha * \eta * \eta^{-1}] = [\alpha].$$

Thus, $P_\eta^{-1} \circ P_\eta = \mathrm{id}_{\pi_1(X, x_0)}$. From the symmetry, we have $P_\eta \circ P_\eta^{-1} = \mathrm{id}_{\pi_1(X, x_1)}$. We find from exercise 2.3 that P_η is one to one and onto. □

Accordingly, if X is arcwise connected, we do not need to specify the base point since $\pi_1(X, x_0) \cong \pi_1(X, x_1)$ for any $x_0, x_1 \in X$, and we may simply write $\pi_1(X)$.

Exercise 4.1. (1) Let η and ζ be paths from x_0 to x_1, such that $\eta \sim \zeta$. Show that $P_\eta = P_\zeta$.
  (2) Let η and ζ be paths such that $\eta(1) = \zeta(0)$. Show that $P_{\eta * \zeta} = P_\zeta \circ P_\eta$.

4.2.2 Homotopic invariance of fundamental groups

The homotopic equivalence of paths and loops is easily generalized to arbitrary maps. Let $f, g : X \to Y$ be continuous maps. If there exists a continuous map $F : X \times I \to Y$ such that $F(x, 0) = f(x)$ and $F(x, 1) = g(x)$, f is said to be **homotopic** to g, denoted by $f \sim g$. The map F is called a **homotopy** between f and g.

Definition 4.6. Let X and Y be topological spaces. X and Y are of the same **homotopy type**, written as $X \simeq Y$, if there exist continuous maps $f : X \to Y$ and $g : Y \to X$ such that $f \circ g \sim \mathrm{id}_Y$ and $g \circ f \sim \mathrm{id}_X$. The map f is called the **homotopy equivalence** and g, its **homotopy inverse**. [*Remark*: If X is homeomorphic to Y, X and Y are of the same homotopy type but the converse is not necessarily true. For example, a point $\{p\}$ and the real line $\mathbb{R}$ are of the same homotopy type but $\{p\}$ is not homeomorphic to $\mathbb{R}$.]

Proposition 4.2. 'Of the same homotopy type' is an equivalence relation in the set of topological spaces.

Proof. Reflectivity: $X \simeq X$ where id_X is a homotopy equivalence. *Symmetry*: Let $X \simeq Y$ with the homotopy equivalence $f : X \to Y$. Then $Y \simeq X$, the homotopy equivalence being the homotopy inverse of f. *Transitivity*: Let $X \simeq Y$ and $Y \simeq Z$. Suppose $f : X \to Y$, $g : Y \to Z$ are homotopy equivalences and $f' : Y \to X$, $g' : Z \to Y$, their homotopy inverses. Then

$$(g \circ f)(f' \circ g') = g(f \circ f')g' \sim g \circ \mathrm{id}_Y \circ g' = g \circ g' \sim \mathrm{id}_Z$$
$$(f' \circ g')(g \circ f) = f'(g' \circ g)f \sim f' \circ \mathrm{id}_Y \circ f = f' \circ f \sim \mathrm{id}_X$$

from which it follows $X \simeq Z$. □

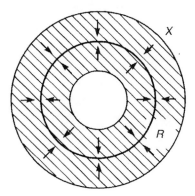

Figure 4.7. The circle R is a retract of the annulus X. The arrows depict the action of the retraction.

One of the most remarkable properties of the fundamental groups is that two topological spaces of the same homotopy type have the same fundamental group.

Theorem 4.3. Let X and Y be topological spaces of the same homotopy type. If $f : X \to Y$ is a homotopy equivalence, $\pi_1(X, x_0)$ is isomorphic to $\pi_1(Y, f(x_0))$.

The following corollary follows directly from theorem 4.3.

Corollary 4.1. A fundamental group is invariant under homeomorphisms, and hence is a topological invariant.

In this sense, we must admit that fundamental groups classify topological spaces in a less strict manner than homeomorphisms. What we claim at most is that if topological spaces X and Y have different fundamental groups, X cannot be homeomorphic to Y. Note, however, that the homotopy groups including the fundamental groups have many applications to physics as we shall see in due course. We should stress that the main usage of the homotopy groups in physics is not to classify spaces but to classify maps or field configurations.

It is rather difficult to appreciate what is meant by 'of the same homotopy type' for an arbitrary pair of X and Y. In practice, however, it often happens that Y is a subspace of X. We then claim that $X \simeq Y$ if Y is obtained by a continuous deformation of X.

Definition 4.7. Let R ($\neq \emptyset$) be a subspace of X. If there exists a continuous map $f : X \to R$ such that $f|_R = \mathrm{id}_R$, R is called a **retract** of X and f a **retraction**.

Note that the whole of X is mapped onto R keeping points in R fixed. Figure 4.7 is an example of a retract and retraction.

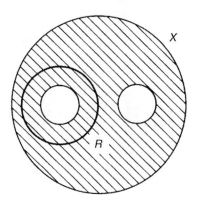

Figure 4.8. The circle R is not a deformation retract of X.

Definition 4.8. Let R be a subspace of X. If there exists a continuous map $H : X \times I \to X$ such that

$$H(x, 0) = x \qquad H(x, 1) \in R \qquad \text{for any } x \in X \qquad (4.5)$$
$$H(x, t) = x \qquad \text{for any } x \in R \text{ and any } t \in I. \qquad (4.6)$$

The space R is said to be a **deformation retract** of X. Note that H is a homotopy between id_X and a retraction $f : X \to R$, which leaves all the points in R fixed during deformation.

A retract is not necessarily a deformation retract. In figure 4.8, the circle R is a retract of X but not a deformation retract, since the hole in X is an obstruction to continuous deformation of id_X to the retraction. Since X and R are of the same homotopy type, we have

$$\pi_1(X, a) \cong \pi_1(R, a) \qquad a \in R. \qquad (4.7)$$

Example 4.1. Let X be the unit circle and Y be the annulus,

$$X = \{e^{i\theta} | 0 \le \theta < 2\pi\} \qquad (4.8)$$
$$Y = \{re^{i\theta} | 0 \le \theta < 2\pi, \tfrac{1}{2} \le r \le \tfrac{2}{3}\} \qquad (4.9)$$

see figure 4.7. Define $f : X \hookrightarrow Y$ by $f(e^{i\theta}) = e^{i\theta}$ and $g : Y \to X$ by $g(re^{i\theta}) = e^{i\theta}$. Then $f \circ g : re^{i\theta} \mapsto e^{i\theta}$ and $g \circ f : e^{i\theta} \mapsto e^{i\theta}$. Observe that $f \circ g \sim \mathrm{id}_Y$ and $g \circ f = \mathrm{id}_X$. There exists a homotopy

$$H(re^{i\theta}, t) = \{1 + (r - 1)(1 - t)\}e^{i\theta}$$

which interpolates between id_X and $f \circ g$, keeping the points on X fixed. Hence, X is a deformation retract of Y. As for the fundamental groups we have $\pi_1(X, a) \cong \pi_1(Y, a)$ where $a \in X$.

Definition 4.9. If a point $a \in X$ is a deformation retract of X, X is said to be **contractible**.

Let $c_a : X \to \{a\}$ be a constant map. If X is contractible, there exists a homotopy $H : X \times I \to X$ such that $H(x, 0) = c_a(x) = a$ and $H(x, 1) = \mathrm{id}_X(x) = x$ for any $x \in X$ and, moreover, $H(a, t) = a$ for any $t \in I$. The homotopy H is called the **contraction**.

Example 4.2. $X = \mathbb{R}^n$ is contractible to the origin 0. In fact, if we define $H : \mathbb{R}^n \times I \to \mathbb{R}$ by $H(x, t) = tx$, we have (i) $H(x, 0) = 0$ and $H(x, 1) = x$ for any $x \in X$ and (ii) $H(0, 1) = 0$ for any $t \in I$. Now it is clear that any convex subset of $\mathbb{R}^n$ is contractible.

Exercise 4.2. Let $D^2 = \{(x, y) \in \mathbb{R}^2 | x^2 + y^2 \leq 1\}$. Show that the unit circle S^1 is a deformation retract of $D^2 - \{0\}$. Show also that the unit sphere S^n is a deformation retract of $D^{n+1} - \{0\}$, where $D^{n+1} = \{x \in \mathbb{R}^{n+1} | |x| \leq 1\}$.

Theorem 4.4. The fundamental group of a contractible space X is trivial, $\pi_1(X, x_0) \cong \{e\}$. In particular, the fundamental group of $\mathbb{R}^n$ is trivial, $\pi_1(\mathbb{R}^n, x_0) \cong \{e\}$.

Proof. A contractible space has the same fundamental group as a point $\{p\}$ and a point has a trivial fundamental group. $\square$

If an arcwise connected space X has a trivial fundamental group, X is said to be **simply connected**, see section 2.3.

4.3 Examples of fundamental groups

There does not exist a routine procedure to compute the fundamental groups, in general. However, in certain cases, they are obtained by relatively simple considerations. Here we look at the fundamental groups of the circle S^1 and related spaces.

Let us express S^1 as $\{z \in \mathbb{C} | |z| = 1\}$. Define a map $p : \mathbb{R} \to S^1$ by $p : x \mapsto \exp(\mathrm{i}x)$. Under p, the point $0 \in \mathbb{R}$ is mapped to $1 \in S^1$, which is taken to be the base point. We imagine that $\mathbb{R}$ wraps around S^1 under p, see figure 4.9. If $x, y \in \mathbb{R}$ satisfies $x - y = 2\pi m (m \in \mathbb{Z})$, they are mapped to the same point in S^1. Then we write $x \sim y$. This is an equivalence relation and the equivalence class $[x] = \{y | x - y = 2\pi m$ for some $m \in \mathbb{Z}\}$ is identified with a point $\exp(\mathrm{i}x) \in S^1$. It then follows that $S^1 \cong \mathbb{R}/2\pi\mathbb{Z}$. Let $\tilde{f} : \mathbb{R} \to \mathbb{R}$ be a continuous map such that $\tilde{f}(0) = 0$ and $\tilde{f}(x + 2\pi) \sim \tilde{f}(x)$. It is obvious that $\tilde{f}(x + 2\pi) = \tilde{f}(x) + 2n\pi$ for any $x \in \mathbb{R}$, where n is a fixed integer. If $x \sim y$ $(x - y = 2\pi m)$, we have

$$\tilde{f}(x) - \tilde{f}(y) = \tilde{f}(y + 2\pi m) - \tilde{f}(y)$$
$$= \tilde{f}(y) + 2\pi mn - \tilde{f}(y) = 2\pi mn$$

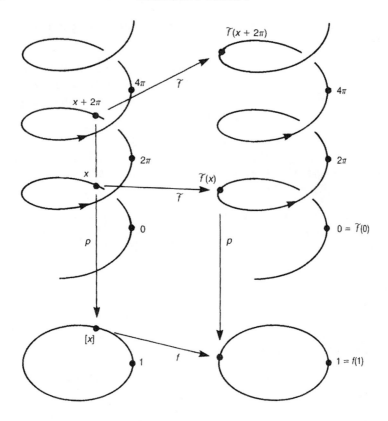

Figure 4.9. The map $p : \mathbb{R} \to S^1$ defined by $x \mapsto \exp(\mathrm{i}x)$ projects $x + 2m\pi$ to the same point on S^1, while $\tilde{f} : \mathbb{R} \to \mathbb{R}$, such that $\tilde{f}(0) = 0$ and $\tilde{f}(x + 2\pi) = \tilde{f}(x) + 2n\pi$ for fixed n, defines a map $f : S^1 \to S^1$. The integer n specifies the homotopy class to which f belongs.

hence $\tilde{f}(x) \sim \tilde{f}(y)$. Accordingly, $\tilde{f} : \mathbb{R} \to \mathbb{R}$ uniquely defines a continuous map $f : \mathbb{R}/2\pi\mathbb{Z} \to \mathbb{R}/2\pi\mathbb{Z}$ by $f([x]) = p \circ \tilde{f}(x)$, see figure 4.9. Note that f keeps the base point $1 \in S^1$ fixed. Conversely, given a map $f : S^1 \to S^1$, which leaves $1 \in S^1$ fixed, we may define a map $\tilde{f} : \mathbb{R} \to \mathbb{R}$ such that $\tilde{f}(0) = 0$ and $\tilde{f}(x + 2\pi) = \tilde{f}(x) + 2\pi n$.

In summary, there is a one-to-one correspondence between the set of maps from S^1 to S^1 with $f(1) = 1$ and the set of maps from $\mathbb{R}$ to $\mathbb{R}$ such that $\tilde{f}(0) = 0$ and $\tilde{f}(x + 2\pi) = \tilde{f}(x) + 2\pi n$. The integer n is called the **degree** of f and is denoted by $\deg(f)$. While x encircles S^1 once, $f(x)$ encircles S^1 n times.

Lemma 4.2. (1) Let $f, g : S^1 \to S^1$ such that $f(1) = g(1) = 1$. Then $\deg(f) = \deg(g)$ if and only if f is homotopic to g.

(2) For any $n \in \mathbb{Z}$, there exists a map $f : S^1 \to S^1$ such that $\deg(f) = n$.

Proof. (1) Let $\deg(f) = \deg(g)$ and $\tilde{f}, \tilde{g} : \mathbb{R} \to \mathbb{R}$ be the corresponding maps. Then $\tilde{F}(x, t) \equiv t\tilde{f}(x) + (1 - t)\tilde{g}(x)$ is a homotopy between $\tilde{f}(x)$ and $\tilde{g}(x)$. It is easy to verify that $F \equiv p \circ \tilde{F}$ is a homotopy between f and g. Conversely, if $f \sim g : S^1 \to S^1$, there exists a homotopy $F : S^1 \times I \to S^1$ such that $F(1, t) = 1$ for any $t \in I$. The corresponding homotopy $\tilde{F} : \mathbb{R} \times I \to \mathbb{R}$ between $\tilde{f}$ and $\tilde{g}$ satisfies $\tilde{F}(x + 2\pi, t) = \tilde{F}(x, t) + 2n\pi$ for some $n \in \mathbb{Z}$. Thus, $\deg(f) = \deg(g)$.

(2) $\tilde{f} : x \mapsto nx$ induces a map $f : S^1 \to S^1$ with $\deg(f) = n$. $\qquad \square$

Lemma 4.2 tells us that by assigning an integer $\deg(f)$ to a map $f : S^1 \to S^1$ such that $f(1) = 1$, there is a bijection between $\pi_1(S^1, 1)$ and $\mathbb{Z}$. Moreover, this is an isomorphism. In fact, for $f, g : S^1 \to S^1$, $f * g$, defined as a product of loops, satisfies $\deg(f * g) = \deg(f) + \deg(g)$. [Let $\tilde{f}(x + 2\pi) = \tilde{f}(x) + 2\pi n$ and $\tilde{g}(x + 2\pi) = \tilde{g}(x) + 2\pi m$. Then $f * g(x + 2\pi) = f * g(x) + 2\pi(m + n)$. Note that $*$ is not a composite of maps but a product of paths.] We have finally proved the following theorem.

Theorem 4.5. The fundamental group of S^1 is isomorphic to $\mathbb{Z}$,

$$\pi_1(S^1) \cong \mathbb{Z}. \tag{4.10}$$

[Since S^1 is arcwise connected, we may drop the base point.]

Although the proof of the theorem is not too obvious, the statement itself is easily understood even by children. Suppose we encircle a cylinder with an elastic band. If it encircles the cylinder n times, the configuration cannot be continuously deformed into that with m ($\neq n$) encirclements. If an elastic band encircles a cylinder first n times and then m times, it encircles the cylinder $n + m$ times in total.

4.3.1 Fundamental group of torus

Theorem 4.6. Let X and Y be arcwise connected topological spaces. Then $\pi_1(X \times Y, (x_0, y_0))$ is isomorphic to $\pi_1(X, x_0) \oplus \pi_1(Y, y_0)$.

Proof. Define projections $p_1 : X \times Y \to X$ and $p_2 : X \times Y \to Y$. If α is a loop in $X \times Y$ at (x_0, y_0), $\alpha_1 \equiv p_1(\alpha)$ is a loop in X at x_0, and $\alpha_2 \equiv p_2(\alpha)$ is a loop in Y at y_0. Conversely, any pair of loops α_1 of X at x_0 and α_2 of Y at y_0 determines a unique loop $\alpha = (\alpha_1, \alpha_2)$ of $X \times Y$ at (x_0, y_0). Define a homomorphism $\varphi : \pi_1(X \times Y, (x_0, y_0)) \to \pi_1(X, x_0) \oplus \pi_1(Y, y_0)$ by

$$\varphi([\alpha]) = ([\alpha_1], [\alpha_2]).$$

By construction φ has an inverse, hence it is the required isomorphism and $\pi_1(X \times Y, (x_0, y_0)) \cong \pi_1(X, x_0) \oplus \pi_1(Y, y_0)$. $\qquad \square$

Example 4.3. (1) Let $T^2 = S^1 \times S^1$ be a torus. Then

$$\pi_1(T^2) \cong \pi_1(S^1) \oplus \pi_1(S^1) \cong \mathbb{Z} \oplus \mathbb{Z}. \qquad (4.11)$$

Similarly, for the n-dimensional torus

$$T^n = \underbrace{S^1 \times S^1 \times \cdots \times S^1}_{n}$$

we have

$$\pi_1(T^n) \cong \underbrace{\mathbb{Z} \oplus \mathbb{Z} \oplus \cdots \oplus \mathbb{Z}}_{n}. \qquad (4.12)$$

(2) Let $X = S^1 \times \mathbb{R}$ be a cylinder. Since $\pi_1(\mathbb{R}) \cong \{e\}$, we have

$$\pi_1(X) \cong \mathbb{Z} \oplus \{e\} \cong \mathbb{Z}. \qquad (4.13)$$

4.4 Fundamental groups of polyhedra

The computation of fundamental groups in the previous section was, in a sense, *ad hoc* and we certainly need a more systematic way of computing the fundamental groups. Fortunately if a space X is triangulable, we can compute the fundamental group of the polyhedron K, and hence that of X by a routine procedure. Let us start with some aspects of group theories.

4.4.1 Free groups and relations

The free groups that we define here are not necessarily Abelian and we employ multiplicative notation for the group operation. A subset $X = \{x_j\}$ of a group G is called a **free set of generators** of G if any element $g \in G - \{e\}$ is *uniquely* written as

$$g = x_1^{i_1} x_2^{i_2} \cdots x_n^{i_n} \qquad (4.14)$$

where n is finite and $i_k \in \mathbb{Z}$. We assume no adjacent x_j are equal; $x_j \neq x_{j+1}$. If $i_j = 1$, x_j^1 is simply written as x_j. If $i_j = 0$, the term x_j^0 should be dropped from g. For example, $g = a^3 b^{-2} c b^3$ is acceptable but $h = a^3 a^{-2} c b^0$ is not. If each element is to be written uniquely, h must be reduced to $h = ac$. If G has a free set of generators, it is called a **free group**.

Conversely, given a set X, we can construct a free group G whose free set of generators is X. Let us call each element of X a **letter**. The product

$$w = x_1^{i_1} x_2^{i_2} \cdots x_n^{i_n} \qquad (4.15)$$

is called a **word**, where $x_j \in X$ and $i_j \in \mathbb{Z}$. If $i_j \neq 0$ and $x_j \neq x_{j+1}$ the word is called a **reduced word**. It is always possible to reduce a word by finite steps. For example,

$$a^{-2} b^{-3} b^3 a^4 b^3 c^{-2} c^4 = a^{-2} b^0 a^4 b^3 c^2 = a^2 b^3 c^2.$$

A word with no letters is called an **empty word** and denoted by 1. For example, it is obtained by reducing $w = a^0$.

A product of words is defined by simply juxtaposing two words. Note that a juxtaposition of reduced words is not necessarily reduced but it is always possible to reduce it. For example, if $v = a^2 c^{-3} b^2$ and $w = b^{-2} c^2 b^3$, the product vw is reduced as

$$vw = a^2 c^{-3} b^2 b^{-2} c^2 b^3 = a^2 c^{-3} c^2 b^3 = a^2 c^{-1} b^3.$$

Thus, the set of all reduced words form a well-defined free group called the free group generated by X, denoted by $F[X]$. The multiplication is the juxtaposition of two words followed by reduction, the unit element is the empty word and the inverse of

$$w = x_1^{i_1} x_2^{i_2} \cdots x_n^{i_n}$$

is

$$w^{-1} = x_n^{-i_n} \cdots x_2^{-i_2} x_1^{-i_1}.$$

Exercise 4.3. Let $X = \{a\}$. Show that the free group generated by X is isomorphic to $\mathbb{Z}$.

In general, an arbitrary group G is specified by the generators and certain constraints that these must satisfy. If $\{x_k\}$ is the set of generators, the constraints are most commonly written as

$$r = x_{k_1}^{i_1} x_{k_2}^{i_2} \cdots x_{k_n}^{i_n} = 1 \tag{4.16}$$

and are called **relations**. For example, the cyclic group of order n generated by x (in multiplicative notation) satisfies a relation $x^n = 1$.

More formally, let G be a group which is generated by $X = \{x_k\}$. Any element $g \in G$ is written as $g = x_1^{i_1} x_2^{i_2} \cdots x_n^{i_n}$, where we do not require that the expression be unique (G is not necessarily free). For example, we have $x^i = x^{n+1}$ in $\mathbb{Z}$. Let $F[X]$ be the free group generated by X. Then there is a natural homomorphism φ from $F[X]$ onto G defined by

$$x_1^{i_1} x_2^{i_2} \cdots x_n^{i_n} \xrightarrow{\varphi} x_1^{i_1} x_2^{i_2} \cdots x_n^{i_n} \in G. \tag{4.17}$$

Note that this is not an isomorphism since the LHS is not unique. φ is onto since X generates both $F[X]$ and G. Although $F[X]$ is not isomorphic to G, $F[X]/\ker \varphi$ is (see theorem 3.1),

$$F[X]/\ker \varphi \cong G. \tag{4.18}$$

In this sense, the set of generators X and $\ker \varphi$ completely determine the group G. [$\ker \varphi$ is a normal subgroup. Lemma 3.1 claims that $\ker \varphi$ is a subgroup of $F[X]$. Let $r \in \ker \varphi$, that is, $r \in F[X]$ and $\varphi(r) = 1$. For any element $x \in F[X]$, we have $\varphi(x^{-1} r x) = \varphi(x^{-1}) \varphi(r) \varphi(x) = \varphi(x)^{-1} \varphi(r) \varphi(x) = 1$, hence $x^{-1} r x \in \ker \varphi$.]

In this way, a group G generated by X is specified by the relations. The juxtaposition of generators and relations

$$(x_1, \ldots, x_p; r_1, \ldots, r_q) \qquad (4.19)$$

is called the **presentation** of G. For example, $\mathbb{Z}_n = (x; x^n)$ and $\mathbb{Z} = (x; \emptyset)$.

Example 4.4. Let $\mathbb{Z} \oplus \mathbb{Z} = \{x^n y^m | n, m \in \mathbb{Z}\}$ be a free Abelian group generated by $X = \{x, y\}$. Then we have $xy = yx$. Since $xyx^{-1}y^{-1} = 1$, we have a relation $r = xyx^{-1}y^{-1}$. The presentation of $\mathbb{Z} \oplus \mathbb{Z}$ is $(x, y : xyx^{-1}y^{-1})$.

4.4.2 Calculating fundamental groups of polyhedra

We shall be sketchy here to avoid getting into the technical details. We shall follow Armstrong (1983); the interested reader should consult this book or any textbook on algebraic topology. As noted in the previous chapter, a polyhedron $|K|$ is a nice approximation of a given topological space X within a homeomorphism. Since fundamental groups are topological invariants, we have $\pi_1(X) = \pi_1(|K|)$. We assume X is an arcwise connected space and drop the base point. Accordingly, if we have a systematic way of computing $\pi_1(|K|)$, we can also find $\pi_1(X)$.

We first define the edge group of a simplicial complex, which corresponds to the fundamental group of a topological space, then introduce a convenient way of computing it. Let $f : |K| \to X$ be a triangulation of a topological space X. If we note that an element of the fundamental group of X can be represented by loops in X, we expect that similar loops must exist in $|K|$ as well. Since any loop in $|K|$ is made up of 1-simplexes, we look at the set of all 1-simplexes in $|K|$, which can be endowed with a group structure called the edge group of K.

An **edge path** in a simplicial complex K is a sequence $v_0 v_1 \ldots v_k$ of vertices of $|K|$, in which the consecutive pair $v_i v_{i+1}$ is a 0- or 1-simplex of $|K|$. [For technical reasons, we allow the possibility $v_i = v_{i+1}$, in which case the relevant simplex is a 0-simplex $v_i = v_{i+1}$.] If $v_0 = v_k\ (=v)$, the edge path is called an **edge loop** at v. We classify these loops into equivalence classes according to some equivalence relation. We define two edge loops α and β to be equivalent if one is obtained from the other by repeating the following operations a finite number of times.

(1) If the vertices u, v and w span a 2-simplex in K, the edge path uvw may be replaced by uw and *vice versa*; see figure 4.10(a).

(2) As a special case, if $u = w$ in (1), the edge path uvw corresponds to traversing along uv first then reversing backwards from v to $w = u$. This edge path uvu may be replaced by a 0-simplex u and *vice versa*, see figure 4.10(b).

Let us denote the equivalence class of edge loops at v, to which $vv_1 \ldots v_{k-1}v$ belongs, by $\{vv_1 \ldots v_{k-1}v\}$. The set of equivalence classes forms a group under the product operation defined by

$$\{vu_1 \ldots u_{k-1}v\} * \{vv_1 \ldots v_{i-1}v\} = \{vu_1 \ldots u_{k-1}vv_1 \ldots v_{i-1}v\}. \qquad (4.20)$$

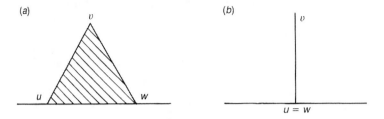

Figure 4.10. Possible deformations of the edge loops. In (a), uvw is replaced by uw. In (b), uvu is replaced by u.

The unit element is an equivalence class $\{v\}$ while the inverse of $\{vv_1 \ldots v_{k-1}v\}$ is $\{vv_{k-1} \ldots v_1 v\}$. This group is called the **edge group** of K at v and denoted by $E(K; v)$.

Theorem 4.7. $E(K; v)$ is isomorphic to $\pi_1(|K|; v)$.

The proof is found in Armstrong (1983), for example. This isomorphism $\varphi : E(K; v) \to \pi_1(|K|; v)$ is given by identifying an edge loop in K with a loop in $|K|$. To find $E(K; v)$, we need to read off the generators and relations. Let L be a simplicial subcomplex of K, such that

(a) L contains *all the vertices* (0-simplexes) of K;
(b) the polyhedron $|L|$ is *arcwise connected* and *simply connected*.

Given an arcwise-connected simplicial complex K, there always exists a subcomplex L that satisfies these conditions. A one-dimensional simplicial complex that is arcwise connected and simply connected is called a **tree**. A tree T_M is called the **maximal tree** of K if it is not a proper subset of other trees.

Lemma 4.3. A maximal tree T_M contains all the vertices of K and hence satisfies conditions (a) and (b) above.

Proof. Suppose T_M does not contain some vertex w. Since K is arcwise connected, there is a 1-simplex vw in K such that $v \in T_M$ and $w \notin T_M$. $T_M \cup \{vw\} \cup \{w\}$ is a one-dimensional subcomplex of K which is arcwise connected, simply connected and contains T_M, which contradicts the assumption. □

Suppose we have somehow obtained the subcomplex L. Since $|L|$ is simply connected, the edge loops in $|L|$ do not contribute to $E(K; v)$. Thus, we can effectively ignore the simplexes in L in our calculations. Let $v_0 (=v), v_1, \ldots, v_n$ be the vertices of K. Assign an 'object' g_{ij} for each ordered pair of vertices v_i, v_j if $\langle v_i v_j \rangle$ is a 1-simplex of K. Let $G(K; L)$ be a group that is generated by all g_{ij}. What about the relations? We have the following.

(1) Since we ignore those simplexes in L, we assign $g_{ij} = 1$ if $\langle v_i v_j \rangle \in L$.
(2) If $\langle v_i v_j v_k \rangle$ is a 2-simplex of K, there are no non-trivial loops around $v_i v_j v_k$ and we have the relation $g_{ij} g_{jk} g_{ki} = 1$.

The generators $\{g_{ij}\}$ and the set of relations completely determine the group $G(K; L)$.

Theorem 4.8. $G(K; L)$ is isomorphic to $E(K; v) \simeq \pi_1(|K|; v)$.

In fact, we can be more efficient than is apparent. For example, g_{ii} should be set equal to 1 since g_{ii} corresponds to the vertex v_i which is an element of L. Moreover, from $g_{ij} g_{ji} = g_{ii} = 1$, we have $g_{ij} = g_{ji}^{-1}$. Therefore, we only need to introduce those generators g_{ij} for each pair of vertices v_i, v_j such that $\langle v_i v_j \rangle \in K - L$ and $i < j$. Since there are no generators g_{ij} such that $\langle v_i v_j \rangle \in L$, we can ignore the first type of relation. If $\langle v_i v_j v_k \rangle$ is a 2-simplex of $K - L$ such that $i < j < k$, the corresponding relation is *uniquely* given by $g_{ij} g_{jk} = g_{ik}$ since we are only concerned with simplexes $\langle v_i v_j \rangle$ such that $i < j$.

To summarize, the rules of the game are as follows.

(1) First, find a triangulation $f : |K| \to X$.
(2) Find the subcomplex L that is arcwise connected, simply connected and contains all the vertices of K.
(3) Assign a generator g_{ij} to each 1-simplex $\langle v_i v_j \rangle$ of $K - L$, for which $i < j$.
(4) Impose a relation $g_{ij} g_{jk} = g_{ik}$ if there is a 2-simplex $\langle v_i v_j v_k \rangle$ such that $i < j < k$. If two of the vertices v_i, v_j and v_k form a 1-simplex of L, the corresponding generator should be set equal to 1.
(5) Now $\pi_1(X)$ is isomorphic to $G(K; L)$ which is a group generated by $\{g_{ij}\}$ with the relations obtained in (4).

Let us work out several examples.

Example 4.5. From our construction, it should be clear that $E(K; v)$ and $G(K; L)$ involve only the 0-, 1- and 2-simplexes of K. Accordingly, if $K^{(2)}$ denotes a **2-skeleton** of K, which is defined to be the set of all 0-, 1- and 2-simplexes in K, we should have

$$\pi_1(|K|) \cong \pi_1(|K^{(2)}|). \tag{4.21}$$

This is quite useful in actual computations. For example, a 3-simplex and its boundary have the same 2-skeleton. A 3-simplex is a polyhedron $|K|$ of the solid ball D^3, while its boundary $|L|$ is a polyhedron of the sphere S^2. Since D^3 is contractible, $\pi_1(|K|) \cong \{e\}$. From (4.21) we find $\pi_1(S^2) \cong \pi_1(|K|) \cong \{e\}$. In general, for $n \geq 2$, the $(n + 1)$-simplex σ_{n+1} and the boundary of σ_{n+1} have the same 2-skeleton. If we note that σ_{n+1} is contractible and the boundary of σ_{n+1} is a polyhedron of S^n, we find the formula

$$\pi_1(S^n) \cong \{e\} \qquad n \geq 2. \tag{4.22}$$

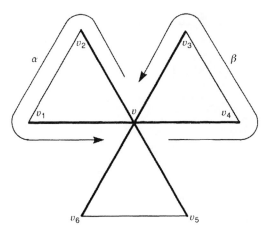

Figure 4.11. A triangulation of a 3-bouquet. The bold lines denote the maximal tree L.

Example 4.6. Let $K \equiv \{v_1, v_2, v_3, \langle v_1 v_2 \rangle, \langle v_1 v_3 \rangle, \langle v_2 v_3 \rangle\}$ be a simplicial complex of a circle S^1. We take v_1 as the base point. A maximal tree may be $L = \{v_1, v_2, v_3, \langle v_1 v_2 \rangle, \langle v_1 v_3 \rangle\}$. There is only one generator g_{23}. Since there are no 2-simplexes in K, the relation is empty. Hence,

$$\pi_1(S^1) \cong G(K; L) = (g_{23}; \emptyset) \cong \mathbb{Z} \tag{4.23}$$

in agreement with theorem 4.5.

Example 4.7. An **n-bouquet** is defined by the one-point union of n circles. For example, figure 4.11 is a triangulation of a 3-bouquet. Take the common point v as the base point. The bold lines in figure 4.11 form a maximal tree L. The generators of $G(K; L)$ are g_{12}, g_{34} and g_{56}. There are no relations and we find

$$\pi_1(\text{3-bouquet}) = G(K; L) = (x, y, z; \emptyset). \tag{4.24}$$

Note that this is a free group but not free *Abelian*. The non-commutativity can be shown as follows. Consider loops α and β at v encircling different holes. Obviously the product $\alpha * \beta * \alpha^{-1}$ cannot be continuously deformed into β, hence $[\alpha] * [\beta] * [\alpha]^{-1} \neq [\beta]$, or

$$[\alpha] * [\beta] \neq [\beta] * [\alpha]. \tag{4.25}$$

In general, an n-bouquet has n generators $g_{12}, \ldots, g_{2n-1\ 2n}$ and the fundamental group is isomorphic to the free group with n generators with no relations.

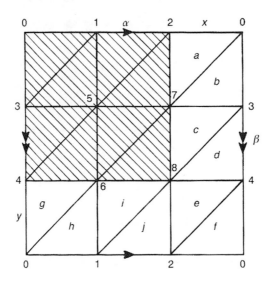

Figure 4.12. A triangulation of the torus.

Example 4.8. Let D^2 be a two-dimensional disc. A triangulation K of D^2 is given by a triangle with its interior included. Clearly K itself may be L and $K - L$ is empty. Thus, we find $\pi_1(K) \cong \{e\}$.

Example 4.9. Figure 4.12 is a triangulation of the torus T^2. The shaded area is chosen to be the subcomplex L. [Verify that it contains all the vertices and is both arcwise and simply connected.] There are 11 generators with ten relations. Let us take $x = g_{02}$ and $y = g_{04}$ and write down the relations

$$
\begin{array}{lllll}
\text{(a)} & \dfrac{g_{02}}{x} \; \dfrac{g_{27}}{1} & = & g_{07} & \rightarrow & g_{07} = x \\[2ex]
\text{(b)} & \dfrac{g_{03}}{1} \; \dfrac{g_{37}}{} & = & \dfrac{g_{07}}{x} & \rightarrow & g_{37} = x \\[2ex]
\text{(c)} & \dfrac{g_{37}}{x} \; \dfrac{g_{78}}{1} & = & g_{38} & \rightarrow & g_{38} = x \\[2ex]
\text{(d)} & \dfrac{g_{34}}{1} \; \dfrac{g_{48}}{} & = & \dfrac{g_{38}}{x} & \rightarrow & g_{48} = x \\[2ex]
\text{(e)} & g_{24} \; \dfrac{g_{48}}{x} & = & g_{28} & \rightarrow & g_{24}x = g_{28} \\[2ex]
\text{(f)} & \dfrac{g_{02}}{x} \; g_{24} & = & \dfrac{g_{04}}{y} & \rightarrow & xg_{24} = y \\
\end{array}
$$

(g) $\quad \begin{matrix} g_{04} & g_{46} \\ y & 1 \end{matrix} \ = \ g_{06} \quad \rightarrow \quad g_{06} = y$

(h) $\quad \begin{matrix} g_{01} & g_{16} \\ 1 & \end{matrix} \ = \ \begin{matrix} g_{06} \\ y \end{matrix} \quad \rightarrow \quad g_{16} = y$

(i) $\quad \begin{matrix} g_{16} & g_{68} \\ y & 1 \end{matrix} \ = \ g_{18} \quad \rightarrow \quad g_{18} = y$

(j) $\quad \begin{matrix} g_{12} & g_{28} \\ 1 & \end{matrix} \ = \ \begin{matrix} g_{18} \\ y \end{matrix} \quad \rightarrow \quad g_{28} = y \quad .$

It follows from (e) and (f) that $x^{-1}yx = g_{28}$. We finally have

$$g_{02} = g_{07} = g_{37} = g_{38} = g_{48} = x$$
$$g_{04} = g_{06} = g_{16} = g_{18} = g_{28} = y$$
$$g_{24} = x^{-1}y$$

with a relation $x^{-1}yx = y$ or

$$xyx^{-1}y^{-1} = 1. \tag{4.26}$$

This shows that $G(K; L)$ is generated by two commutative generators (note $xy = yx$), hence (cf example 4.4)

$$G(K; L) = (x, y; xyx^{-1}y^{-1}) \cong \mathbb{Z} \oplus \mathbb{Z} \tag{4.27}$$

in agreement with (4.11).

We have the following intuitive picture. Consider loops $\alpha = 0 \to 1 \to 2 \to 0$ and $\beta = 0 \to 3 \to 4 \to 0$. The loop α is identified with $x = g_{02}$ since $g_{12} = g_{01} = 1$ and β with $y = g_{04}$. They generate $\pi_1(T^2)$ since α and β are independent non-trivial loops. In terms of these, the relation is written as

$$\alpha * \beta * \alpha^{-1} * \beta^{-1} \sim c_v \tag{4.28}$$

where c_v is a constant loop at v, see figure 4.13.

More generally, let Σ_g be the torus with genus g. As we have shown in problem 2.1, Σ_g is expressed as a subset of $\mathbb{R}^2$ with proper identifications at the boundary. The fundamental group of Σ_g is generated by $2g$ loops α_i, β_i ($1 \le i \le g$). Similarly, to (4.28), we verify that

$$\prod_{i=1}^{g} (\alpha_i * \beta_i * \alpha_i^{-1} * \beta_i^{-1}) \sim c_v \tag{4.29}$$

If we denote the generators corresponding to α_i by x_i and β_i by y_i, there is only one relation among them,

$$\prod_{i=1}^{g} (x_i y_i x_i^{-1} y_i^{-1}) = 1. \tag{4.30}$$

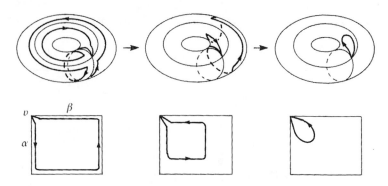

Figure 4.13. The loops α and β satisfy the relation $\alpha * \beta * \alpha^{-1} * \beta^{-1} \sim c_v$.

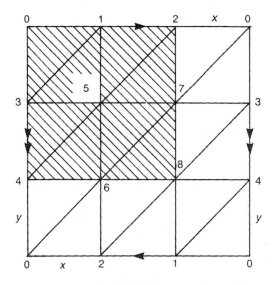

Figure 4.14. A triangulation of the Klein bottle.

Exercise 4.4. Figure 4.14 is a triangulation of the Klein bottle. The shaded area is the subcomplex L. There are 11 generators and ten relations. Take $x = g_{02}$ and $y = g_{04}$ and write down the relations for 2-simplexes to show that

$$\pi_1(\text{Klein bottle}) \cong (x, y; xyxy^{-1}). \tag{4.31}$$

Example 4.10. Figure 4.15 is a triangulation of the projective plane $\mathbb{R}P^2$. The shaded area is the subcomplex L. There are seven generators and six relations.

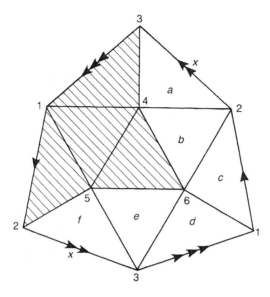

Figure 4.15. A triangulation of the projective plane.

Let us take $x = g_{23}$ and write down the relations

$$
\begin{array}{llllll}
\text{(a)} & \underset{x}{g_{23}} & \underset{1}{g_{34}} & = & g_{24} & \rightarrow & g_{24} = x \\[2mm]
\text{(b)} & \underset{x}{g_{24}} & \underset{1}{g_{46}} & = & g_{26} & \rightarrow & g_{26} = x \\[2mm]
\text{(c)} & \underset{1}{g_{12}} & \underset{x}{g_{26}} & = & g_{16} & \rightarrow & g_{16} = x \\[2mm]
\text{(d)} & \underset{1}{g_{13}} & g_{36} & = & \underset{x}{g_{16}} & \rightarrow & g_{36} = x \\[2mm]
\text{(e)} & g_{35} & \underset{1}{g_{56}} & = & \underset{x}{g_{36}} & \rightarrow & g_{35} = x \\[2mm]
\text{(f)} & \underset{x}{g_{23}} & \underset{x}{g_{35}} & = & \underset{1}{g_{25}} & \rightarrow & x^2 = 1.
\end{array}
$$

Hence, we find that

$$\pi_1(\mathbb{R}P^2) \cong (x; x^2) \cong \mathbb{Z}_2. \tag{4.32}$$

Intuitively, the appearance of a cyclic group is understood as follows. Figure 4.16(a) is a schematic picture of $\mathbb{R}P^2$. Take loops α and β. It is easy to see that α is continuously deformed to a point, and hence is a trivial element of $\pi_1(\mathbb{R}P^2)$. Since diametrically opposite points are identified in $\mathbb{R}P^2$, β is actually

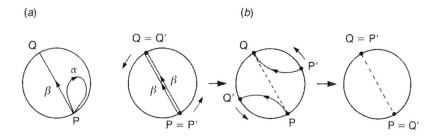

Figure 4.16. (*a*) α is a trivial loop while the loop β cannot be shrunk to a point. (*b*) $\beta * \beta$ is continuously shrunk to a point.

a closed loop. Since it cannot be shrunk to a point, it is a non-trivial element of $\pi_1(\mathbb{R}P^2)$. What about the product? $\beta * \beta$ is a loop which traverses from P to Q $\sim$ P twice. It can be read off from figure 4.16(*b*) that $\beta * \beta$ is continuously shrunk to a point, and thus belongs to the trivial class. This shows that the generator x, corresponding to the homotopy class of the loop β, satisfies the relation $x^2 = 1$, which verifies our result.

The same pictures can be used to show that

$$\pi_1(\mathbb{R}P^3) \cong \mathbb{Z}_2 \qquad (4.33)$$

where $\mathbb{R}P^3$ is identified as S^3 with diametrically opposite points identified, $\mathbb{R}P^3 = S^3/(x \sim -x)$. If we take the hemisphere of S^3 as the representative, $\mathbb{R}P^3$ can be expressed as a solid ball D^3 with diametrically opposite points on the surface identified. If the discs D^2 in figure 4.16 are interpreted as solid balls D^3, the same pictures verify (4.33).

Exercise 4.5. A triangulation of the Möbius strip is given by figure 3.8. Find the maximal tree and show that

$$\pi_1(\text{Möbius strip}) \cong \mathbb{Z}. \qquad (4.34)$$

[*Note*: Of course the Möbius strip is of the same homotopy type as S^1, hence (4.34) is trivial. The reader is asked to obtain this result through routine procedures.]

4.4.3 Relations between $H_1(K)$ and $\pi_1(|K|)$

The reader might have noticed that there is a certain similarity between the first homology group $H_1(K)$ and the fundamental group $\pi_1(|K|)$. For example, the fundamental groups of many spaces (circle, disc, n-spheres, torus and many more) are identical to the corresponding first homology group. In some cases, however, they are different: $H_1(2\text{-bouquet}) \cong \mathbb{Z} \oplus \mathbb{Z}$ and $\pi_1(2\text{-bouquet}) = (x, y : \emptyset)$, for example. Note that H_1(2-bouquet) is a free *Abelian* group while π_1(2-bouquet) is a free group. The following theorem relates $\pi_1(|K|)$ to $H_1(K)$.

Theorem 4.9. Let K be a connected simplicial complex. Then $H_1(K)$ is isomorphic to $\pi_1(|K|)/F$, where F is the commutator subgroup (see later) of $\pi_1(|K|)$.

Let G be a group whose presentation is $(x_i; r_m)$. The **commutator subgroup** F of G is a group generated by the elements of the form $x_i x_j x_i^{-1} x_j^{-1}$. Thus, G/F is a group generated by $\{x_i\}$ with the set of relations $\{r_m\}$ and $\{x_i x_j x_i^{-1} x_j^{-1}\}$. The theorem states that if $\pi_1(|K|) = (x_i : r_m)$, then $H_1(K) \cong (x_i : r_m, x_i x_j x_i^{-1} x_j^{-1})$. For example, from π_1(2-bouquet) $= (x, y : \emptyset)$, we find

$$\pi_1(\text{2-bouquet})/F \cong (x, y; xyx^{-1}y^{-1}) \cong \mathbb{Z} \oplus \mathbb{Z}$$

which is isomorphic to H_1(2-bouquet).

The proof of theorem 4.9 is found in Greenberg and Harper (1981) and also outlined in Croom (1978).

Example 4.11. From π_1(Klein bottle) $\cong (x, y; xyxy^{-1})$, we have

$$\pi_1(\text{Klein bottle})/F \cong (x, y; xyxy^{-1}, xyx^{-1}y^{-1}).$$

Two relations are replaced by $x^2 = 1$ and $xyx^{-1}y^{-1} = 1$ to yield

$$\pi_1(\text{Klein bottle})/F \cong (x, y; xyx^{-1}y^{-1}, x^2) \cong \mathbb{Z} \oplus \mathbb{Z}_2$$
$$\cong H_1(\text{Klein bottle})$$

where the factor $\mathbb{Z}$ is generated by y and $\mathbb{Z}_2$ by x.

Corollary 4.2. Let X be a connected topological space. Then $\pi_1(X)$ is isomorphic to $H_1(X)$ if and only if $\pi_1(X)$ is commutative. In particular, if $\pi_1(X)$ is generated by one generator, $\pi_1(X)$ is always isomorphic to $H_1(X)$. [Use theorem 4.9.]

Corollary 4.3. If X and Y are of the same homotopy type, their first homology groups are identical: $H_1(X) = H_1(Y)$. [Use theorems 4.9 and 4.3.]

4.5 Higher homotopy groups

The fundamental group classifies the homotopy classes of loops in a topological space X. There are many ways to assign other groups to X. For example, we may classify homotopy classes of the spheres in X or homotopy classes of the tori in X. It turns out that the homotopy classes of the sphere S^n ($n \geq 2$) form a group similar to the fundamental group.

4.5.1 Definitions

Let I^n ($n \geq 1$) denote the unit n-cube $I \times \cdots \times I$,

$$I^n = \{(s_1, \ldots, s_n) | 0 \leq s_i \leq 1 \, (1 \leq i \leq n)\}. \tag{4.35}$$

The boundary ∂I^n is the geometrical boundary of I^n,

$$\partial I^n = \{(s_1, \ldots, s_n) \in I^n | \text{ some } s_i = 0 \text{ or } 1\}. \tag{4.36}$$

We recall that in the fundamental group, the boundary ∂I of $I = [0, 1]$ is mapped to the base point x_0. Similarly, we assume here that we shall be concerned with continuous maps $\alpha : I^n \to X$, which map the boundary ∂I^n to a point $x_0 \in X$. Since the boundary is mapped to a single point x_0, we have effectively obtained S^n from I^n; cf figure 2.8. If $I^n/\partial I^n$ denotes the cube I^n whose boundary ∂I^n is shrunk to a point, we have $I^n/\partial I^n \cong S^n$. The map α is called an **n-loop** at x_0. A straightforward generalization of definition 4.4 is as follows.

Definition 4.10. Let X be a topological space and $\alpha, \beta : I^n \to X$ be n-loops at $x_0 \in X$. The map α is **homotopic** to β, denoted by $\alpha \sim \beta$, if there exists a continuous map $F : I^n \times I \to X$ such that

$$F(s_1, \ldots, s_n, 0) = \alpha(s_1, \ldots, s_n) \tag{4.37a}$$
$$F(s_1, \ldots, s_n, 1) = \beta(s_1, \ldots, s_n) \tag{4.37b}$$
$$F(s_1, \ldots, s_n, t) = x_0 \qquad \text{for } (s_1, \ldots, s_n) \in \partial I^n, t \in I. \tag{4.37c}$$

F is called a **homotopy** between α and β.

Exercise 4.6. Show that $\alpha \sim \beta$ is an equivalence relation. The equivalence class to which α belongs is called the **homotopy class** of α and is denoted by $[\alpha]$.

Let us define the group operations. The product $\alpha * \beta$ of n-loops α and β is defined by

$$\alpha * \beta(s_1, \ldots, s_n) = \begin{cases} \alpha(2s_1, \ldots, s_n) & 0 \leq s_1 \leq \frac{1}{2} \\ \beta(2s_1 - 1, \ldots, s_n) & \frac{1}{2} \leq s_1 \leq 1. \end{cases} \tag{4.38}$$

The product $\alpha * \beta$ looks like figure 4.17(a) in X. It is helpful to express it as figure 4.17(b). If we define α^{-1} by

$$\alpha^{-1}(s_1, \ldots, s_n) \equiv \alpha(1 - s_1, \ldots, s_n) \tag{4.39}$$

it satisfies

$$\alpha^{-1} * \alpha(s_1, \ldots, s_n) \sim \alpha * \alpha^{-1}(s_1, \ldots, s_n) \sim c_{x_0}(s_1, \ldots, s_n) \tag{4.40}$$

where c_{x_0} is a constant n-loop at $x_0 \in X$, $c_{x_0} : (s_1, \ldots, s_n) \mapsto x_0$. Verify that both $\alpha * \beta$ and α^{-1} are n-loops at x_0.

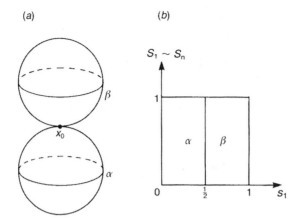

Figure 4.17. A product $\alpha * \beta$ of n-loops α and β.

Definition 4.11. Let X be a topological space. The set of homotopy classes of n-loops $(n \geq 1)$ at $x_0 \in X$ is denoted by $\pi_n(X, x_0)$ and called the **nth homotopy group** at x_0. $\pi_n(x, x_0)$ is called the *higher* homotopy group if $n \geq 2$.

The product $\alpha * \beta$ just defined naturally induces a product of homotopy classes defined by

$$[\alpha] * [\beta] \equiv [\alpha * \beta] \tag{4.41}$$

where α and β are n-loops at x_0. The following exercises verify that this product is well defined and satisfies the group axioms.

Exercise 4.7. Show that the product of n-loops defined by (4.41) is independent of the representatives: cf lemma 4.1.

Exercise 4.8. Show that the nth homotopy group is a group. To prove this, the following facts may be verified; cf theorem 4.1.

(1) $([\alpha] * [\beta]) * [\gamma] = [\alpha] * ([\beta] * [\gamma])$.
(2) $[\alpha] * [c_x] = [c_x] * [\alpha] = [\alpha]$.
(3) $[\alpha] * [\alpha^{-1}] = [c_x]$, which defines the inverse $[\alpha]^{-1} = [\alpha^{-1}]$.

We have excluded $\pi_0(X, x_0)$ so far. Let us classify maps from I^0 to X. We note $I^0 = \{0\}$ and $\partial I^0 = \emptyset$. Let $\alpha, \beta : \{0\} \to X$ be such that $\alpha(0) = x$ and $\beta(0) = y$. We define $\alpha \sim \beta$ if there exists a continuous map $F : \{0\} \times I \to X$ such that $F(0, 0) = x$ and $F(0, 1) = y$. This shows that $\alpha \sim \beta$ if and only if x and y are connected by a curve in X, namely they are in the same (arcwise) connected component. Clearly this equivalence relation is independent of x_0 and we simply denote the zeroth homology group by $\pi_0(X)$. Note, however, that $\pi_0(X)$ is not a group and denotes the number of (arcwise) connected components of X.

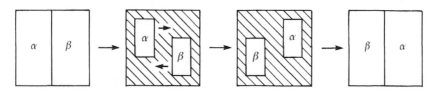

Figure 4.18. Higher homotopy groups are always commutative, $\alpha * \beta \sim \beta * \alpha$.

4.6 General properties of higher homotopy groups

4.6.1 Abelian nature of higher homotopy groups

Higher homotopy groups are always Abelian; for any n-loops α and β at $x_0 \in X$, $[\alpha]$ and $[\beta]$ satisfy

$$[\alpha] * [\beta] = [\beta] * [\alpha]. \tag{4.42}$$

To verify this assertion let us observe figure 4.18. Clearly the deformation is homotopic at each step of the sequence. This shows that $\alpha * \beta \sim \beta * \alpha$, namely $[\alpha] * [\beta] = [\beta] * [\alpha]$.

4.6.2 Arcwise connectedness and higher homotopy groups

If a topological space X is arcwise connected, $\pi_n(X, x_0)$ is isomorphic to $\pi_n(X, x_1)$ for any pair $x_0, x_1 \in X$. The proof is parallel to that of theorem 4.2. Accordingly, if X is arcwise connected, the base point need not be specified.

4.6.3 Homotopy invariance of higher homotopy groups

Let X and Y be topological spaces of the same homotopy type; see definition 4.6. If $f : X \to Y$ is a homotopy equivalence, the homotopy group $\pi_n(X, x_0)$ is isomorphic to $\pi_n(Y, f(x_0))$; cf theorem 4.3. Topological invariance of higher homotopy groups is the direct consequence of this fact. In particular, if X is contractible, the homotopy groups are trivial: $\pi_n(X, x_0) = \{e\}, n > 1$.

4.6.4 Higher homotopy groups of a product space

Let X and Y be arcwise connected topological spaces. Then

$$\pi_n(X \times Y) \cong \pi_n(X) \oplus \pi_n(Y) \tag{4.43}$$

cf theorem 4.6.

4.6.5 Universal covering spaces and higher homotopy groups

There are several cases in which the homotopy groups of one space are given by the known homotopy groups of the other space. There is a remarkable property between the higher homotopy groups of a topological space and its *universal covering space*.

Definition 4.12. Let X and $\widetilde{X}$ be connected topological spaces. The pair $(\widetilde{X}, p)$, or simply $\widetilde{X}$, is called the **covering space** of X if there exists a continuous map $p : \widetilde{X} \to X$ such that

(1) p is surjective (onto)
(2) for each $x \in X$, there exists a connected open set $U \subset X$ containing x, such that $p^{-1}(U)$ is a disjoint union of open sets in $\widetilde{X}$, each of which is mapped homeomorphically onto U by p.

In particular, if $\widetilde{X}$ is *simply* connected, $(\widetilde{X}, p)$ is called the **universal covering space** of X. [*Remarks*: Certain groups are known to be topological spaces. They are called topological groups. For example $SO(n)$ and $SU(n)$ are topological groups. If X and $\widetilde{X}$ in definition 4.12 happen to be topological groups and $p : \widetilde{X} \to X$ to be a group homomorphism, the (universal) covering space is called the (**universal**) **covering group**.]

For example, $\mathbb{R}$ is the universal covering space of S^1, see section 4.3. Since S^1 is identified with $U(1)$, $\mathbb{R}$ is a universal covering group of $U(1)$ if $\mathbb{R}$ is regarded as an additive group. The map $p : \mathbb{R} \to U(1)$ may be $p : x \to e^{i2\pi x}$. Clearly p is surjective and if $U = \{e^{i2\pi x} | x \in (x_0 - 0.1, x_0 + 0.1)\}$, then

$$p^{-1}(U) = \bigcup_{n \in \mathbb{Z}} (x_0 - 0.1 + n, x_0 + 0.1 + n)$$

which is a disjoint union of open sets of $\mathbb{R}$. It is easy to show that p is also a homomorphism with respect to addition in $\mathbb{R}$ and multiplication in $U(1)$. Hence, $(\mathbb{R}, p)$ is the universal covering group of $U(1) = S^1$.

Theorem 4.10. Let $(\widetilde{X}, p)$ be the universal covering space of a connected topological space X. If $x_0 \in X$ and $\tilde{x}_0 \in \widetilde{X}$ are base points such that $p(\tilde{x}_0) = x_0$, the induced homomorphism

$$p_* : \pi_n(\widetilde{X}, \tilde{x}_0) \to \pi_n(X, x_0) \tag{4.44}$$

is an isomorphism for $n \geq 2$. [*Warning*: This theorem cannot be applied if $n = 1$; $\pi_1(\mathbb{R}) = \{e\}$ while $\pi_1(S^1) = \mathbb{Z}$.]

The proof is given in Croom (1978). For example, we have $\pi_n(\mathbb{R}) = \{e\}$ since $\mathbb{R}$ is contractible. Then we find

$$\pi_n(S^1) \cong \pi_n(U(1)) = \{e\} \qquad n \geq 2. \tag{4.45}$$

Example 4.12. Let $S^n = \{x \in \mathbb{R}^{n+1} | |x|^2 = 1\}$. The real projective space $\mathbb{R}P^n$ is obtained from S^n by identifying the pair of antipodal points $(x, -x)$. It is easy to see that S^n is a covering space of $\mathbb{R}P^n$ for $n \geq 2$. Since $\pi_1(S^n) = \{e\}$ for $n \geq 2$, S^n is the universal covering space of $\mathbb{R}P^n$ and we have

$$\pi_n(\mathbb{R}P^m) \cong \pi_n(S^m). \tag{4.46}$$

It is interesting to note that $\mathbb{R}P^3$ is identified with SO(3). To see this let us specify an element of SO(3) by a rotation about an axis $\boldsymbol{n}$ by an angle θ $(0 < \theta < \pi)$ and assign a 'vector' $\boldsymbol{\Omega} \equiv \theta\boldsymbol{n}$ to this element. Ω takes its value in the disc D^3 of radius π. Moreover, $\pi\boldsymbol{n}$ and $-\pi\boldsymbol{n}$ represent the same rotation and should be identified. Thus, the space to which Ω belongs is a disc D^3 whose anti-podal points on the surface S^2 are identified. Note also that we may express $\mathbb{R}P^3$ as the northern hemisphere D^3 of S^3, whose anti-podal points on the boundary S^2 are identified. This shows that $\mathbb{R}P^3$ is identified with SO(3).

It is also interesting to see that S^3 is identified with SU(2). First note that any element $g \in$ SU(2) is written as

$$g = \begin{pmatrix} a & -\bar{b} \\ b & \bar{a} \end{pmatrix} \qquad |a|^2 + |b|^2 = 1. \tag{4.47}$$

If we write $a = u + iv$ and $b = x + iy$, this becomes S^3,

$$u^2 + v^2 + x^2 + y^2 = 1.$$

Collecting these results, we find

$$\pi_n(\mathrm{SO}(3)) = \pi_n(\mathbb{R}P^3) = \pi_n(S^3) = \pi_n(\mathrm{SU}(2)) \qquad n \geq 2. \tag{4.48}$$

More generally, the universal covering group Spin(n) of SO(n) is called the **spin group**. For small n, they are

$$\mathrm{Spin}(3) = \mathrm{SU}(2) \tag{4.49}$$
$$\mathrm{Spin}(4) = \mathrm{SU}(2) \times \mathrm{SU}(2) \tag{4.50}$$
$$\mathrm{Spin}(5) = \mathrm{USp}(4) \tag{4.51}$$
$$\mathrm{Spin}(6) = \mathrm{SU}(4). \tag{4.52}$$

Here USp($2N$) stands for the compact group of $2N \times 2N$ matrices A satisfying $A^t J A = J$, where

$$J = \begin{pmatrix} 0 & I_N \\ -I_N & 0 \end{pmatrix}.$$

4.7   Examples of higher homotopy groups

In general, there are no algorithms to compute higher homotopy groups $\pi_n(X)$. An *ad hoc* method is required for each topological space for $n \geq 2$. Here, we study several examples in which higher homotopy groups may be obtained by intuitive arguments. We also collect useful results in table 4.1.

Table 4.1. Useful homotopy groups.

		π_1	π_2	π_3	π_4	π_5	π_6
$SO(3)$		$\mathbb{Z}_2$	0	$\mathbb{Z}$	$\mathbb{Z}_2$	$\mathbb{Z}_2$	$\mathbb{Z}_{12}$
$SO(4)$		$\mathbb{Z}_2$	0	$\mathbb{Z}+\mathbb{Z}$	$\mathbb{Z}_2+\mathbb{Z}_2$	$\mathbb{Z}_2+\mathbb{Z}_2$	$\mathbb{Z}_{12}+\mathbb{Z}_{12}$
$SO(5)$		$\mathbb{Z}_2$	0	$\mathbb{Z}$	$\mathbb{Z}_2$	$\mathbb{Z}_2$	0
$SO(6)$		$\mathbb{Z}_2$	0	$\mathbb{Z}$	0	$\mathbb{Z}$	0
$SO(n)$	$n>6$	$\mathbb{Z}_2$	0	$\mathbb{Z}$	0	0	0
$U(1)$		$\mathbb{Z}$	0	0	0	0	0
$SU(2)$		0	0	$\mathbb{Z}$	$\mathbb{Z}_2$	$\mathbb{Z}_2$	$\mathbb{Z}_{12}$
$SU(3)$		0	0	$\mathbb{Z}$	0	$\mathbb{Z}$	$\mathbb{Z}_6$
$SU(n)$	$n>3$	0	0	$\mathbb{Z}$	0	$\mathbb{Z}$	0
S^2		0	$\mathbb{Z}$	$\mathbb{Z}$	$\mathbb{Z}_2$	$\mathbb{Z}_2$	$\mathbb{Z}_{12}$
S^3		0	0	$\mathbb{Z}$	$\mathbb{Z}_2$	$\mathbb{Z}_2$	$\mathbb{Z}_{12}$
S^4		0	0	0	$\mathbb{Z}$	$\mathbb{Z}_2$	$\mathbb{Z}_2$
G_2		0	0	$\mathbb{Z}$	0	0	$\mathbb{Z}_3$
F_4		0	0	$\mathbb{Z}$	0	0	0
E_6		0	0	$\mathbb{Z}$	0	0	0
E_7		0	0	$\mathbb{Z}$	0	0	0
E_8		0	0	$\mathbb{Z}$	0	0	0

Example 4.13. If we note that $\pi_n(X, x_0)$ is the set of the homotopy classes of n-loops S^n in X, we immediately find that

$$\pi_n(S^n, x_0) \cong \mathbb{Z} \qquad n \geq 1. \tag{4.53}$$

If α maps S^n onto a point $x_0 \in S^n$, $[\alpha]$ is the unit element $0 \in \mathbb{Z}$. Since both $I^n/\partial I^n$ and S^n are orientable, we may assign orientations to them. If α maps $I^n/\partial I^n$ homeomorphically to S^n in the same sense of orientation, then $[\alpha]$ is assigned an element $1 \in \mathbb{Z}$. If a homeomorphism α maps $I^n/\partial I^n$ onto S^n in an orientation of opposite sense, $[\alpha]$ corresponds to an element -1. For example, let $n = 2$. Since $I^2/\partial I^2 \cong S^2$, the point in I^2 can be expressed by the polar coordinate (θ, ϕ), see figure 4.19. Similarly, $X = S^2$ can be expressed by the polar coordinate (θ', ϕ'). Let $\alpha : (\theta, \phi) \to (\theta', \phi')$ be a 2-loop in X. If $\theta' = \theta$ and $\phi' = \phi$, the point (θ', ϕ') sweeps S^2 once while the point (θ, ϕ) scans I^2 once in the same orientation. This 2-loop belongs to the class $+1 \in \pi_2(S^2, x_0)$. If $\alpha : (\theta, \phi) \to (\theta', \phi')$ is given by $\theta' = \theta$ and $\phi' = 2\phi$, the point (θ', ϕ') sweeps S^2 twice while (θ, ϕ) scans I^2 once. This 2-loop belongs to the class $2 \in \pi_2(S^2, x_0)$. In general, the map $(\theta, \phi) \mapsto (\theta, k\phi), k \in \mathbb{Z}$, corresponds to the class k of $\pi_2(S^2, x_0)$. A similar argument verifies (4.53) for general $n > 2$.

Example 4.14. Noting that S^n is a universal covering space of $\mathbb{R}P^n$ for $n > 2$, we find

$$\pi_n(\mathbb{R}P^n) \cong \pi_n(S^n) \cong \mathbb{Z} \qquad n \geq 2. \tag{4.54}$$

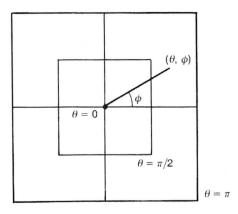

Figure 4.19. A point in I^2 may be expressed by polar coordinates (θ, ϕ).

[Of course this happens to be true for $n = 1$, since $\mathbb{R}P^1 = S^1$.] For example, we have $\pi_2(\mathbb{R}P^2) \cong \pi_2(S^2) \cong \mathbb{Z}$. Since $SU(2) = S^3$ is the universal covering group of $SO(3) = \mathbb{R}P^3$, it follows from theorem 4.10 that (see also (4.48))

$$\pi_3(SO(3)) \cong \pi_3(SU(2)) \cong \pi_3(S^3) \cong \mathbb{Z}. \qquad (4.55)$$

Shankar's monopoles in superfluid ^{3}He-A correspond to non-trivial elements of these homotopy classes, see section 4.10. $\pi_3(SU(2))$ is also employed in the classification of instantons in example 9.8.

In summary, we have table 4.1. In this table, other useful homotopy groups are also listed. We comment on several interesting facts.

(a) Since $Spin(4) = SU(2) \times SU(2)$ is the universal covering group of $SO(4)$, we have $\pi_n(SO(4)) = \pi_n(SU(2)) \oplus \pi_n(SU(2))$ for $n > 2$.
(b) There exists a map J called the **J-homomorphism** $J : \pi_k(SO(n)) \to \pi_{k+n}(S^n)$, see Whitehead (1978). In particular, if $k = 1$, the homomorphism is known to be an isomorphism and we have $\pi_1(SO(n)) = \pi_{n+1}(S^n)$. For example, we find

$$\pi_1(SO(2)) \cong \pi_3(S^2) \cong \mathbb{Z}$$
$$\pi_1(SO(3)) \cong \pi_4(S^3) \cong \pi_4(SU(2)) \cong \pi_4(SO(3)) \cong \mathbb{Z}_2.$$

(c) The **Bott periodicity theorem** states that

$$\pi_k(U(n)) \cong \pi_k(SU(n)) \cong \begin{cases} \{e\} & \text{if } k \text{ is even} \\ \mathbb{Z} & \text{if } k \text{ is odd} \end{cases} \qquad (4.56)$$

for $n \geq (k+1)/2$. Similarly,

$$\pi_k(O(n)) \cong \pi_k(SO(n)) \cong \begin{cases} \{e\} & \text{if } k \equiv 2, 4, 5, 6 \pmod{8} \\ \mathbb{Z}_2 & \text{if } k \equiv 0, 1 \pmod{8} \\ \mathbb{Z} & \text{if } k \equiv 3, 7 \pmod{8} \end{cases} \qquad (4.50)$$

for $n \geq k + 2$. Similar periodicity holds for symplectic groups which we shall not give here.

Many more will be found in appendix A, table 6 of Ito (1987).

4.8 Orders in condensed matter systems

Recently topological methods have played increasingly important roles in condensed matter physics. For example, homotopy theory has been employed to classify possible forms of extended objects, such as solitons, vortices, monopoles and so on, in condensed systems. These classifications will be studied in sections 4.8–4.10. Here, we briefly look at the order parameters of condensed systems that undergo phase transitions.

4.8.1 Order parameter

Let H be a Hamiltonian describing a condensed matter system. We assume H is invariant under a certain symmetry operation. The ground state of the system need not preserve the symmetry of H. If this is the case, we say the system undergoes **spontaneous symmetry breakdown**.

To illustrate this phenomenon, we consider the **Heisenberg Hamiltonian**

$$H = -J \sum_{(i,j)} S_i \cdot S_j + h \cdot \sum_i S_i \qquad (4.57)$$

which describes N ferromagnetic Heisenberg spins $\{S_i\}$. The parameter J is a positive constant, the summation is over the pair of the nearest-neighbour sites (i, j) and h is the uniform external magnetic field. The partition function is $Z = \mathrm{tr}\, e^{-\beta H}$, where $\beta = 1/T$ is the inverse temperature. The free energy F is defined by $\exp(-\beta F) = Z$. The average magnetization per spin is

$$m \equiv \frac{1}{N} \sum_i \langle S_i \rangle = \frac{1}{N\beta} \frac{\partial F}{\partial h} \qquad (4.58)$$

where $\langle \ldots \rangle \equiv \mathrm{tr}(\ldots e^{-\beta H})/Z$. Let us consider the limit $h \to 0$. Although H is invariant under the $SO(3)$ rotations of all S_i in this limit, it is well known that m does not vanish for large enough β and the system does not observe the $SO(3)$ symmetry. It is said that the system exhibits **spontaneous magnetization** and the maximum temperature, such that $m \neq 0$ is called the **critical temperature**.

The vector m is the **order parameter** describing the phase transition between the ordered state ($m \neq 0$) and the disordered state ($m = 0$). The system is still symmetric under SO(2) rotations around the magnetization axis m.

What is the mechanism underlying the phase transition? The free energy is $F = \langle H \rangle - TS$, S being the entropy. At low temperature, the term TS in F may be negligible and the minimum of F is attained by minimizing $\langle H \rangle$, which is realized if all S_i align in the same direction. At high temperature, however, the entropy term dominates F and the minimum of F is attained by maximizing S, which is realized if the directions of S_i are totally random.

If the system is at a uniform temperature, the magnitude $|m|$ is independent of the position and m is specified by its direction only. In the ground state, m itself is expected to be independent of position. It is convenient to introduce the polar coordinate (θ, ϕ) to specify the direction of m. There is a one-to-one correspondence between m and a point on the sphere S^2. Suppose m varies as a function of position: $m = m(x)$. At each point x of the space, a point (θ, ϕ) of S^2 is assigned and we have a map $(\theta(x), \phi(x))$ from the space to S^2. Besides the ground state (and excited states that are described by small oscillations (spin waves) around the ground state) the system may carry various excited states that cannot be obtained from the ground state by small perturbations. What kinds of excitation are possible depends on the dimension of the space and the order parameter. For example, if the space is two dimensional, the Heisenberg ferromagnet may admit an excitation called the **Belavin–Polyakov monopole** shown in figure 4.20 (Belavin and Polyakov 1975). Observe that m approaches a constant vector ($\hat{z}$ in this case) so the energy does not diverge. This condition guarantees the stability of this excitation; it is impossible to deform this configuration into the uniform one with m far from the origin kept fixed. These kinds of excitation whose stability depends on topological arguments are called **topological excitations**. Note that the field $m(x)$ defines a map $m : S^2 \to S^2$ and, hence, are classified by the homotopy group $\pi_2(S^2) = \mathbb{Z}$.

4.8.2 Superfluid ^{4}He and superconductors

In Bogoliubov's theory, the order parameter of superfluid ^{4}He is the expectation value

$$\langle \phi(x) \rangle = \Psi(r) = \Delta_0(x) e^{i\alpha(x)} \qquad (4.59)$$

where $\phi(x)$ is the field operator. In the operator formalism,

$$\phi(x) \sim \text{(creation operator)} + \text{(annihilation operator)}$$

from which we find the number of particles is not conserved if $\Psi(x) \neq 0$. This is related to the spontaneous breakdown of the global gauge symmetry. The

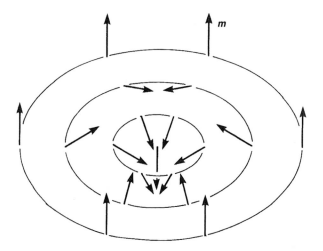

Figure 4.20. A sketch of the Belavin–Polyakov monopole. The vector m approaches $\hat{z}$ as $|x| \to \infty$.

Hamiltonian of ^{4}He is

$$H = \int dx\, \phi^\dagger(x) \left(-\frac{\nabla^2}{2m} - \mu \right) \phi(x)$$

$$+ \frac{1}{2} \int dx\, dy\, \phi^\dagger(y)\phi(y) V(|x-y|)\phi^\dagger(x)\phi(x). \qquad (4.60)$$

Clearly H is invariant under the global gauge transformation

$$\phi(x) \to e^{i\chi}\phi(x). \qquad (4.61)$$

The order parameter, however, transforms as

$$\Psi(x) \to e^{i\chi}\Phi(x) \qquad (4.62)$$

and hence does not observe the symmetry of the Hamiltonian. The phenomenological free energy describing ^{4}He is made up of two contributions. The main contribution is the **condensation energy**

$$\mathcal{F}_0 \equiv \frac{\alpha}{2!}|\Psi(x)|^2 + \frac{\beta}{4!}|\Psi(x)|^4 \qquad (4.63a)$$

where $\alpha \sim \alpha_0(T - T_c)$ changes sign at the critical temperature $T \sim 4$ K. Figure 4.21 sketches $\mathcal{F}_0$ for $T > T_c$ and $T < T_c$. If $T > T_c$, the minimum of $\mathcal{F}_0$ is attained at $\Psi(x) = 0$ while if $T < T_c$ at $|\Psi| = \Delta_0 \equiv [-(6\alpha/\beta)]^{1/2}$. If $\Psi(x)$ depends on x, we have an additional contribution called the **gradient energy**

$$\mathcal{F}_{grad} \equiv \frac{1}{2}K\overline{\nabla\Psi(x)} \cdot \nabla\Psi(x) \qquad (4.63b)$$

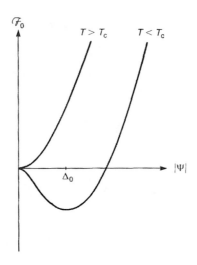

Figure 4.21. The free energy has a minimum at $|\Psi| = 0$ for $T > T_c$ and at $|\Psi| = \Delta_0$ for $T < T_c$.

K being a positive constant. If the spatial variation of $\Psi(x)$ is mild enough, we may assume Δ_0 is constant (the London limit).

In the BCS theory of superconductors, the order parameter is given by (Tsuneto 1982)

$$\Psi_{\alpha\beta} \equiv \langle \psi_\alpha(x)\psi_\beta(x) \rangle \qquad (4.64)$$

$\psi_\alpha(x)$ being the (non-relativistic) electron field operator of spin $\alpha = (\uparrow, \downarrow)$. It should be noted, however, that (4.64) is not an irreducible representation of the spin algebra. To see this, we examine the behaviour of $\Psi_{\alpha\beta}$ under a spin rotation. Consider an infinitesimal spin rotation around an axis n by an angle θ, whose matrix representation is

$$R = I_2 + \mathrm{i}\frac{\theta}{2}n^\mu \sigma_\mu,$$

σ_μ being the Pauli matrices. Since ψ_α transforms as $\psi_\alpha \to R_\alpha{}^\beta \psi_\beta$ we have

$$\Psi_{\alpha\beta} \to R_\alpha{}^{\alpha'} \Psi_{\alpha'\beta'} R_\beta{}^{\beta'} = (R \cdot \Psi \cdot R^t)_{\alpha\beta}$$

$$= \left[\Psi + \mathrm{i}\frac{\delta}{2}n(\sigma\Psi\sigma_2 - \Psi\sigma_2\sigma) \right]_{\alpha\beta}$$

where we note that $\sigma_\mu^t = -\sigma_2\sigma_\mu\sigma_2$. Suppose $\Psi_{\alpha\beta} \propto \mathrm{i}(\sigma_2)_{\alpha\beta}$. Then Ψ does not change under this rotation, hence it represents the spin-singlet pairing. We write

$$\Psi_{\alpha\beta}(x) = \Delta(x)(\mathrm{i}\sigma_2)_{\alpha\beta} = \Delta_0(x)\mathrm{e}^{\mathrm{i}\varphi(x)}(\mathrm{i}\sigma_2)_{\alpha\beta}. \qquad (4.65a)$$

If, however, we take

$$\Psi_{\alpha\beta}(x) = \Delta^\mu(x)\mathrm{i}(\sigma_\mu \cdot \sigma_2)_{\alpha\beta} \qquad (4.65b)$$

we have

$$\Psi_{\alpha\beta} \rightarrow [\Delta^{\mu} + \delta\varepsilon^{\mu\nu\lambda}n_{\nu}\Delta_{\lambda}](\mathrm{i}\sigma_{\mu} \cdot \sigma_{2})_{\alpha\beta}.$$

This shows that Δ^{μ} is a vector in spin space, hence (4.65b) represents the spin-triplet pairing.

The order parameter of a conventional superconductor is of the form (4.65a) and we restrict the analysis to this case for the moment. In (4.65a), $\Delta(x)$ assumes the same form as $\Psi(x)$ of superfluid ^{4}He and the free energy is again given by (4.63). This similarity is attributed to the Cooper pair. In the superfluid state, a macroscopic number of ^{4}He atoms occupy the ground state (Bose–Einstein condensation) which then behaves like a huge molecule due to the quantum coherence. In this state creating elementary excitations requires a finite amount of energy and the flow cannot decay unless this critical energy is supplied. Since an electron is a fermion there is, at first sight, no Bose–Einstein condensation. The key observation is the Cooper pair. By the exchange of phonons, a pair of electrons feels an attractive force that barely overcomes the Coulomb repulsion. This tiny attractive force makes it possible for electrons to form a pair (in momentum space) that obeys Bose statistics. The pairs then condense to form the superfluid state of the Cooper pairs of electric charge $2e$.

An electromagnetic field couples to the system through the minimal coupling

$$\mathcal{F}_{\mathrm{grad}} = \tfrac{1}{2}K \left| (\partial_{\mu} - \mathrm{i}2eA_{\mu})\Delta(x) \right|^2. \tag{4.66}$$

(The term $2e$ is used since the Cooper pair carries charge $2e$.) Superconductors are roughly divided into two types according to their behaviour in applied magnetic fields. The type-I superconductor forms an intermediate state in which normal and superconducting regions coexist in strong magnetic fields. The type-II superconductor forms a vortex lattice (**Abrikosov lattice**) to confine the magnetic fields within the cores of the vortices with other regions remaining in the superconducting state. A similar vortex lattice has been observed in rotating superfluid ^{4}He in a cylinder.

4.8.3 General consideration

In the next two sections, we study applications of homotopy groups to the classification of defects in ordered media. The analysis of this section is based on Toulouse and Kléman (1976), Mermin (1979) and Mineev (1980).

As we saw in the previous subsections, when a condensed matter system undergoes a phase transition, the symmetry of the system is reduced and this reduction is described by the order parameter. For definiteness, let us consider the three-dimensional medium of a superconductor. The order parameter takes the form $\psi(x) = \Delta_0(x)e^{i\varphi(x)}$. Let us consider a homogeneous system under uniform external conditions (temperature, pressure etc). The amplitude Δ_0 is uniquely fixed by minimizing the condensation free energy. Note that there are still a large number of degrees of freedom left. ψ may take any value in the circle $S^1 \cong U(1)$

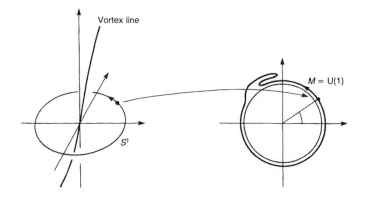

Figure 4.22. A circle S^1 surrounding a line defect (vortex) is mapped to $U(1) = S^1$. This map is classified by the fundamental group $\pi_1(U(1))$.

determined by the phase $e^{i\varphi}$. In this way, a uniform system takes its value in a certain region M called the **order parameter space**. For a superconductor, $M = U(1)$. For the Heisenberg spin system, $M = S^2$. The nematic liquid crystal has $M = \mathbb{R}P^2$ while $M = S^2 \times SO(3)$ for the superfluid ^{3}He-A, see sections 4.9–4.10.

If the system is in an inhomogeneous state, the gradient free energy cannot be negligible and ψ may not be in M. If the characteristic size of the variation of the order parameter is much larger than the coherence length, however, we may still assume that the order parameter takes its value in M, where the value is a function of position this time. If this is the case, there may be points, lines or surfaces in the medium on which the order parameter is not uniquely defined. They are called the **defects**. We have **point defects (monopoles)**, **line defects (vortices)** and **surface defects (domain walls)** according to their dimensionalities. These defects are classified by the homotopy groups.

To be more mathematical, let X be a space which is filled with the medium under consideration. The order parameter is a classical field $\psi(x)$, which is also regarded as a *map* $\psi : X \to M$. Suppose there is a defect in the medium. For concreteness, we consider a line defect in the three-dimensional medium of a superconductor. Imagine a circle S^1 which encircles the line defect. If each part of S^1 is far from the line defect, much further than the coherence length ξ, we may assume the order parameter along S^1 takes its value in the order parameter space $M = U(1)$, see figure 4.22. This is how the fundamental group comes into the problem; we talk of loops in a topological space $U(1)$. The map $S^1 \to U(1)$ is classified by the homotopy classes. Take a point $r_0 \in S^1$ and require that r_0 be mapped to $x_0 \in M$. By noting that $\pi_1(U(1), x_0) = \mathbb{Z}$, we may assign an integer to the line defect. This integer is called the **winding number** since it counts how many times the image of S^1 winds the space $U(1)$. If two line defects have the same winding number, one can be continuously deformed to the other. If two line defects A and B merge together, the new line defect belongs to the homotopy class of the product of the homotopy classes to which A and B belonged before coalescence. Since the group operation in $\mathbb{Z}$ is an addition, the new winding number is a sum of the old winding numbers. A uniform distribution of the order parameter corresponds to the constant map $\psi(x) = x_0 \in M$, which belongs to the unit element $0 \in \mathbb{Z}$. If two line defects of opposite winding numbers merge together, the new line defect can be continuously deformed into the defect-free configuration.

What about the other homotopy groups? We first consider the dimensionality of the defect and the sphere S^n which *surrounds* it. For example, consider a point defect in a three-dimensional medium. It can be surrounded by S^2 and the defect is classified by $\pi_2(M, x_0)$. If M has many components, $\pi_0(M)$ is non-trivial. Let us consider a three-dimensional Ising model for which $M = \{\downarrow\} \cup \{\uparrow\}$. Then there is a domain wall on which the order parameter is not defined. For example, if $S = \uparrow$ for $x < 0$ and $S = \downarrow$ for $x > 0$, there is a domain wall in the yz-plane at $x = 0$. In general, an m-dimensional defect in a d-dimensional medium is classified by the homotopy group $\pi_n(M, x_0)$ where

$$n = d - m - 1. \tag{4.67}$$

In the case of the Ising model, $d = 3, m = 2$; hence $n = 0$.

4.9 Defects in nematic liquid crystals

4.9.1 Order parameter of nematic liquid crystals

Certain organic crystals exhibit quite interesting optical properties when they are in their fluid phases. They are called liquid crystals and they are characterized by their optical anisotropy. Here we are interested in so-called nematic liquid crystals. An example of this is *octyloxy-cyanobiphenyl* whose molecular structure is

$$N \equiv C - \langle\!\!\bigcirc\!\!\rangle - \langle\!\!\bigcirc\!\!\rangle - O - C_8H_{17}.$$

The molecule of a nematic liquid crystal is very much like a rod and the order parameter, called the **director**, is given by the average direction of the rod. Even though the molecule itself has a head and a tail, the director has an inversion symmetry; it does not make sense to distinguish the directors $n = \rightarrow$ and $-n = \leftarrow$. We are tempted to assign a point on S^2 to specify the director. This works except for one point. Two antipodal points $n = (\theta, \phi)$ and $-n = (\pi - \theta, \pi + \phi)$ represent the same state; see figure 4.23. Accordingly, the order parameter of the nematic liquid crystal is the **projective plane** $\mathbb{R}P^2$. The director field in general

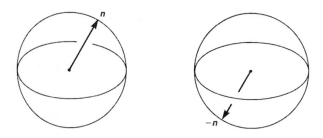

Figure 4.23. Since the director n has no head or tail, one cannot distinguish n from $-n$. Therefore, these two pictures correspond to the same order-parameter configuration.

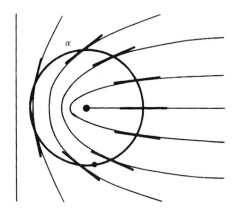

Figure 4.24. A vortex in a nematic liquid crystal, which corresponds to the non-trivial element of $\pi_1(\mathbb{R}P^2) = \mathbb{Z}_2$.

depends on the position r. Then we may define a map $f : \mathbb{R}^3 \to \mathbb{R}P^2$. This map is called the **texture**. The actual order-parameter configuration in $\mathbb{R}^3$ is also called the texture.

4.9.2 Line defects in nematic liquid crystals

From example 4.10 we have $\pi_1(\mathbb{R}P^2) \cong \mathbb{Z}_2 = \{0, 1\}$. There exist two kinds of line defect in nematic liquid crystals; one can be continuously deformed into a uniform configuration while the other cannot. The latter represents a stable vortex, whose texture is sketched in figure 4.24. The reader should observe how the loop α is mapped to $\mathbb{R}P^2$ by this texture.

Exercise 4.9. Show that the line 'defect' in figure 4.25 is fictitious, namely the singularity at the centre may be eliminated by a continuous deformation of directors with directors at the boundary fixed. This corresponds to the operation $1 + 1 = 0$.

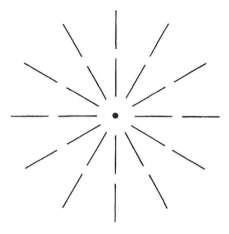

Figure 4.25. A line defect which may be continuously deformed into a uniform configuration.

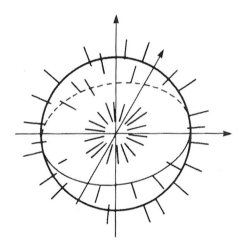

Figure 4.26. The texture of a point defect in a nematic liquid crystal.

4.9.3 Point defects in nematic liquid crystals

From example 4.14, we have $\pi_2(\mathbb{R}P^2) = \mathbb{Z}$. Accordingly, there are stable point defects in the nematic liquid crystal. Figure 4.26 shows the texture of the point defects that belong to the class $1 \in \mathbb{Z}$.

It is interesting to point out that a line defect and a point defect may be combined into a **ring defect**, which is specified by both $\pi_1(\mathbb{R}P^2)$ and $\pi_2(\mathbb{R}P^2)$, see Mineev (1980). If the ring defect is observed from far away, it looks like

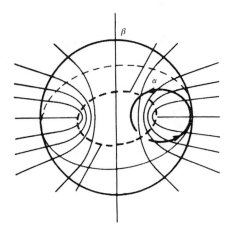

Figure 4.27. The texture of a ring defect in a nematic liquid crystal. The loop α classifies $\pi_1(\mathbb{R}P^2)$ while the sphere (2-loop) β classifies $\pi_2(\mathbb{R}P^2)$.

a point defect, while its local structure along the ring is specified by $\pi_1(\mathbb{R}P^2)$. Figure 4.27 is an example of such a ring defect. The loop α classifies $\pi_1(\mathbb{R}P^2) \cong \mathbb{Z}_2$ while the sphere (2-loop) β classifies $\pi_2(\mathbb{R}P^2) = \mathbb{Z}$.

4.9.4 Higher dimensional texture

The third homotopy group $\pi(\mathbb{R}P^2) \cong \mathbb{Z}$ leads to an interesting singularity-free texture in a three-dimensional medium of nematic liquid crystal. Suppose the director field approaches an asymptotic configuration, say $n = (1, 0, 0)^t$, as $|r| \rightarrow \infty$. Then the medium is effectively compactified into the three-dimensional sphere S^3 and the topological structure of the texture is classified by $\pi_3(\mathbb{R}P^2) \cong \mathbb{Z}$. What is the texture corresponding to a non-trivial element of the homotopy group?

An arbitrary rotation in $\mathbb{R}^3$ is specified by a unit vector e, around which the rotation is carried out, and the rotation angle α. It is possible to assign a 'vector' $\Omega = \alpha e$ to this rotation. It is not exactly a vector since $\Omega = \pi e$ and $-\Omega = -\pi e$ are the same rotation and hence should be identified. Therefore, Ω belongs to the real projective space $\mathbb{R}P^3$. Suppose we take $n_0 = (1, 0, 0)^t$ as a standard director. Then an arbitrary director configuration is specified by rotating n_0 around some axis e by an angle α: $n = R(e, \alpha)n_0$, where $R(e, \alpha)$ is the corresponding rotation matrix in SO(3). Suppose a texture field is given by applying the rotation

$$\alpha e(r) = f(r)\hat{r} \tag{4.68}$$

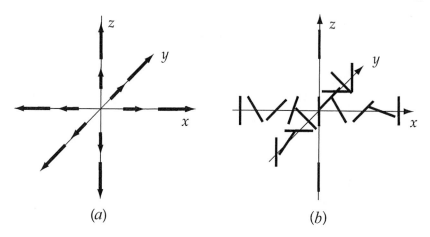

Figure 4.28. The texture of the non-trivial element of $\pi_3(\mathbb{R}P^2) \cong \mathbb{Z}$. (*a*) shows the rotation 'vector' αe. The length α approaches π as $|r| \to \infty$. (*b*) shows the corresponding director field.

to $\boldsymbol{n}_0$, where $\hat{\boldsymbol{r}}$ is the unit vector in the direction of the position vector $\boldsymbol{r}$ and

$$f(r) = \begin{cases} 0 & r = 0 \\ \pi & r \to \infty. \end{cases}$$

Figure 4.28 shows the director field of this texture. Note that although there is no singularity in the texture, it is impossible to 'wind off' this to a uniform configuration.

4.10 Textures in superfluid ^{3}He-A

4.10.1 Superfluid ^{3}He-A

Here comes the last and most interesting example. Before 1972 the only example of the BCS superfluid was the conventional superconductor (apart from indirect observations of superfluid neutrons in neutron stars). Figure 4.29 is the phase diagram of superfluid ^{3}He without an external magnetic field. From NMR and other observations, it turns out that the superfluid is in the spin-triplet p-wave state. Instead of the field operators (see (4.65b)), we define the order parameter in terms of the creation and annihilation operators. The most general form of the triplet superfluid order parameter is

$$\langle c_{\alpha,\boldsymbol{k}} c_{\beta,-\boldsymbol{k}} \rangle \propto \sum_{\mu=1}^{3} (i\sigma_2 \sigma_\mu)_{\alpha\beta} d_\mu(\boldsymbol{k}) \tag{4.69a}$$

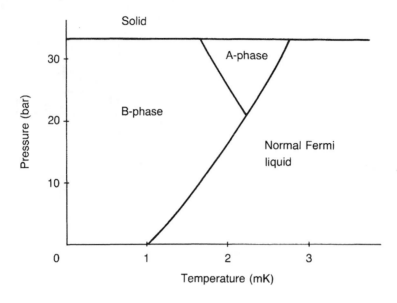

Figure 4.29. The phase diagram of superfluid ^{3}He.

where α and β are spin indices. The Cooper pair forms in the p-wave state hence $d_\mu(\mathbf{k})$ is proportional to $Y_{1m} \sim k_i$,

$$d_\mu(\mathbf{k}) = \sum_{i=1}^{3} \Delta_0 A_{\mu i} k_i. \qquad (4.69\text{b})$$

The bulk energy has several minima. The absolute minimum depends on the pressure and the temperature. We are particularly interested in the A phase in figure 4.29.

The A-phase order parameter takes the form

$$A_{\mu i} = d_\mu(\mathbf{\Delta}_1 + \mathrm{i}\mathbf{\Delta}_2)_i \qquad (4.70)$$

where $\mathbf{d}$ is a unit vector along which the spin projection of the Cooper pair vanishes and $(\mathbf{\Delta}_1, \mathbf{\Delta}_2)$ is a pair of orthonormal unit vectors. The vector $\mathbf{d}$ takes its value in S^2. If we define $\mathbf{l} \equiv \mathbf{\Delta}_1 \times \mathbf{\Delta}_2$, the triad $(\mathbf{\Delta}_1, \mathbf{\Delta}_2, \mathbf{l})$ forms an orthonormal frame at each point of the medium. Since any orthonormal frame can be obtained from a standard orthonormal frame $(\mathbf{e}_1, \mathbf{e}_2, \mathbf{e}_3)$ by an application of a three-dimensional rotation matrix, we conclude that the order parameter of ^{3}He-A is $S^2 \times SO(3)$. The vector $\mathbf{l}$ introduced here is the axis of the angular momentum of the Cooper pair.

For simplicity, we neglect the variation of the $\hat{\mathbf{d}}$-vector. [In fact, $\hat{\mathbf{d}}$ is locked along $\hat{l}$ due to the dipole force.] The order parameter assumes the form

$$A_i = \Delta_0(\hat{\mathbf{\Delta}}_1 + \hat{\mathbf{\Delta}}_2)_i \tag{4.71}$$

where $\hat{\mathbf{\Delta}}_1$, $\hat{\mathbf{\Delta}}_2$ and $\hat{l} \equiv \hat{\mathbf{\Delta}}_1 \times \hat{\mathbf{\Delta}}_2$ form an orthonormal frame at each point of the medium. Let us take a standard orthonormal frame (e_1, e_2, e_2). The frame $(\hat{\mathbf{\Delta}}_1, \hat{\mathbf{\Delta}}_2, \hat{l})$ is obtained by applying an element $g \in SO(3)$ to the standard frame,

$$g : (e_1, e_2, e_2) \rightarrow (\hat{\mathbf{\Delta}}_1, \hat{\mathbf{\Delta}}_2, \hat{l}). \tag{4.72}$$

Since g depends on the coordinate x, the configuration $(\hat{\mathbf{\Delta}}_1(x), \hat{\mathbf{\Delta}}_2(x), \hat{l}(x))$ defines a map $\psi : X \rightarrow SO(3)$ as $x \mapsto g(x)$. The map ψ is called the **texture** of a superfluid ^{3}He.[1] The relevant homotopy groups for classifying defects in superfluid ^{3}He-A are $\pi_n(SO(3))$.

If a container is filled with ^{3}He-A, the boundary poses certain conditions on the texture. The vector $\hat{l}$ is understood as the direction of the angular momentum of the Cooper pair. The pair should rotate in the plane parallel to the boundary wall, thus $\hat{l}$ should be perpendicular to the wall. [*Remark:* If the wall is *diffuse*, the orbital motion of Cooper pairs is disturbed and there is a depression in the amplitude of the order parameter in the vicinity of the wall. We assume, for simplicity, that the wall is *specularly smooth* so that Cooper pairs may execute orbital motion with no disturbance.] There are several kinds of free energy and the texture is determined by solving the Euler–Lagrange equation derived from the total free energy under given boundary conditions.

Reviews on superfluid ^{3}He are found in Anderson and Brinkman (1975), Leggett (1975) and Mermin (1978).

4.10.2   Line defects and non-singular vortices in ^{3}He-A

The fundamental group of $SO(3) \cong \mathbb{R}P^3$ is $\pi_1(\mathbb{R}P^3) \cong \mathbb{Z}_2 \cong \{0, 1\}$. Textures which belong to class 0 can be continuously deformed into the uniform configuration. Configurations in class 1 are called **disgyrations** and have been analysed by Maki and Tsuneto (1977) and Buchholtz and Fetter (1977). Figure 4.30 describes these disgyrations in their lowest free energy configurations.

A remarkable property of $\mathbb{Z}_2$ is the addition $1 + 1 = 0$; the coalescence of two disgyrations produces a trivial texture. By merging two disgyrations, we may construct a texture that looks like a vortex of double vorticity (homotopy class '2') without a singular core; see figure 4.31(*a*). It is easy to verify that the image of the loop α traverses $\mathbb{R}P^3$ twice while that of the smaller loop β may be shrunk to a point. This texture is called the **Anderson–Toulouse vortex** (Anderson and Toulouse 1977). Mermin and Ho (1976) pointed out that if the medium is in a cylinder, the boundary imposes the condition $\hat{l} \perp$ (boundary) and the vortex is cut at the surface, see figure 4.31(*b*) (the **Mermin–Ho vortex**).

---

[1] The name 'texture' is, in fact, borrowed from the order-parameter configuration in liquid crystals, see section 4.9.

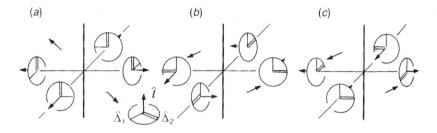

Figure 4.30. Disgyrations in ^{3}He-A.

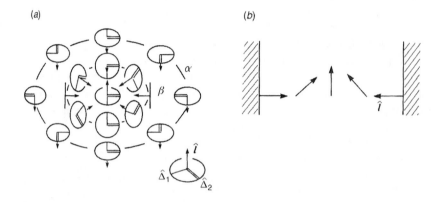

Figure 4.31. The Anderson–Toulouse vortex (a) and the Mermin–Ho vortex (b). In (b) the boundary forces $\hat{l}$ to be perpendicular to the wall.

Since $\pi_2(\mathbb{R}P^3) \cong \{e\}$, there are no point defects in ^{3}He-A. However, $\pi_3(\mathbb{R}P^3) \cong \mathbb{Z}$ introduces a new type of pointlike structure called the Shankar monopole, which we will study next.

4.10.3 Shankar monopole in ^{3}He-A

Shankar (1977) pointed out that there exists a pointlike singularity-free object in ^{3}He-A. Consider an infinite medium of ^{3}He-A. We assume the medium is asymptotically uniform, that is, $(\hat{\mathbf{\Delta}}_1, \hat{\mathbf{\Delta}}_2, \hat{l})$ approaches a standard orthonormal frame (e_1, e_2, e_3) as $|x| \to \infty$. Since all the points far from the origin are mapped to a single point, we have compactified $\mathbb{R}^3$ to S^3. Then the texture is classified according to $\pi_3(\mathbb{R}P^3) = \mathbb{Z}$. Let us specify an element of SO(3) by a 'vector' $\mathbf{\Omega} = \theta n$ in $\mathbb{R}P^3$ as before (example 4.12). Shankar (1977) proposed a texture,

$$\mathbf{\Omega}(r) = \frac{r}{r} \cdot f(r) \qquad (4.73)$$

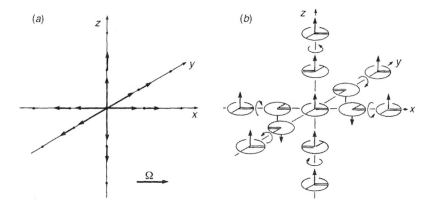

Figure 4.32. The Shankar monopole: (*a*) shows the 'vectors' $\Omega(r)$ and (*b*) shows the triad $(\hat{\Delta}_1, \hat{\Delta}_2, \hat{l})$. Note that as $|r| \to \infty$ the triad approaches the same configuration.

where $f(r)$ is a monotonically decreasing function such that

$$f(r) = \begin{cases} 2\pi & r = 0 \\ 0 & r = \infty. \end{cases} \tag{4.74}$$

We formally extend the radius of $\mathbb{R}P^3$ to 2π and define the rotation angle modulo 2π. This texture is called the **Shankar monopole**, see figure 4.32(*a*). At first sight it appears that there is a singularity at the origin. Note, however, that the length of Ω is 2π there and it is equivalent to the unit element of SO(3). Figure 4.32(*b*) describes the triad field. Since $\Omega(r) = 0$ as $r \to \infty$, irrespective of the direction, the space $\mathbb{R}^3$ is compactified to S^3. As we scan the whole space, $\Omega(r)$ sweeps SO(3) twice and this texture corresponds to class 1 of $\pi_3(\text{SO}(3)) \cong \mathbb{Z}$.

Exercise 4.10. Sketch the Shankar monopole which belongs to the class -1 of $\pi_3(\mathbb{R}P^3)$. [You cannot simply reverse the arrows in figure 4.32.]

Exercise 4.11. Consider classical Heisenberg spins defined in $\mathbb{R}^2$, see section 4.8. Suppose spins take the asymptotic value

$$n(x) \to e_z \qquad |x| \geq L \tag{4.75}$$

for the total energy to be finite, see figure 4.20. Show that the extended objects in this system are classified by $\pi_2(S^2)$. Sketch examples of spin configurations for the classes -1 and $+2$.

Problems

4.1 Show that the n-sphere S^n is a deformation retract of punctured Euclidean space $R^{n+1} - \{0\}$. Find a retraction.

4.2 Let D^2 be the two-dimensional closed disc and $S^1 = \partial D^2$ be its boundary. Let $f : D^2 \to D^2$ be a smooth map. Suppose f has no fixed points, namely $f(p) \neq p$ for any $p \in D^2$. Consider a semi-line starting at p through $f(p)$ (this semi-line is always well defined if $p \neq f(p)$). The line crosses the boundary at some point $q \in S^1$. Then define $\tilde{f} : D^2 \to S^1$ by $\tilde{f}(p) = q$. Use $\pi_1(S^1) = \mathbb{Z}$ and $\pi_1(D^2) = \{0\}$ to show that such an $\tilde{f}$ does not exist and hence, that f must have fixed points. [*Hint*: Show that if such an $\tilde{f}$ existed, D^2 and S^1 would be of the same homotopy type.] This is the two-dimensional version of the **Brouwer fixed-point theorem**.

4.3 Construct a map $f : S^3 \to S^2$ which belongs to the elements 0 and 1 of $\pi_3(S^2) \cong \mathbb{Z}$. See also example 9.9.

5

MANIFOLDS

Manifolds are generalizations of our familiar ideas about curves and surfaces to arbitrary dimensional objects. A curve in three-dimensional Euclidean space is parametrized locally by a single number t as $(x(t), y(t), z(t))$, while two numbers u and v parametrize a surface as $(x(u, v), y(u, v), z(u, v))$. A curve and a surface are considered locally homeomorphic to $\mathbb{R}$ and $\mathbb{R}^2$, respectively. A manifold, in general, is a topological space which is homeomorphic to $\mathbb{R}^m$ *locally*; it may be different from $\mathbb{R}^m$ *globally*. The local homeomorphism enables us to give each point in a manifold a set of m numbers called the (local) coordinate. If a manifold is not homeomorphic to $\mathbb{R}^m$ globally, we have to introduce several local coordinates. Then it is possible that a single point has two or more coordinates. We require that the transition from one coordinate to the other be *smooth*. As we will see later, this enables us to develop the usual calculus on a manifold. Just as topology is based on continuity, so the theory of manifolds is based on *smoothness*.

Useful references on this subject are Crampin and Pirani (1986), Matsushima (1972), Schutz (1980) and Warner (1983). Chapter 2 and appendices B and C of Wald (1984) are also recommended. Flanders (1963) is a beautiful introduction to differential forms. Sattinger and Weaver (1986) deals with Lie groups and Lie algebras and contains many applications to problems in physics.

5.1 Manifolds

5.1.1 Heuristic introduction

To clarify these points, consider the usual sphere of unit radius in $\mathbb{R}^3$. We parametrize the surface of S^2, among other possibilities, by two coordinate systems—polar coordinates and stereographic coordinates. Polar coordinates θ and ϕ are usually defined by (figure 5.1)

$$x = \sin\theta\cos\phi \qquad y = \sin\theta\sin\phi \qquad z = \cos\theta, \qquad (5.1)$$

where ϕ runs from 0 to 2π and θ from 0 to π. They may be inverted on the sphere to yield

$$\theta = \tan^{-1}\frac{\sqrt{x^2 + y^2}}{z} \qquad \phi = \tan^{-1}\frac{y}{x}. \qquad (5.2)$$

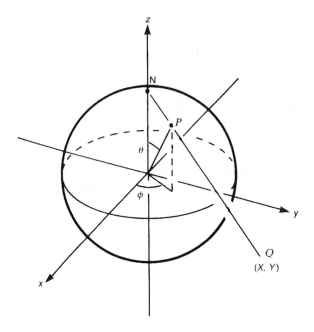

Figure 5.1. Polar coordinates (θ, ϕ) and stereographic coordinates (X, Y) of a point P on the sphere S^2.

Stereographic coordinates, however, are defined by the projection from the North Pole onto the equatorial plane as in figure 5.1. First, join the North Pole $(0, 0, 1)$ to the point $P(x, y, z)$ on the sphere and then continue in a straight line to the equatorial plane $z = 0$ to intersect at $Q(X, Y, 0)$. Then X and Y are the stereographic coordinates of P. We find

$$X = \frac{x}{1-z} \qquad Y = \frac{y}{1-z}. \qquad (5.3)$$

The two coordinate systems are related as

$$X = \cot \tfrac{1}{2}\theta \cos \phi \qquad Y = \cot \tfrac{1}{2}\theta \sin \phi. \qquad (5.4)$$

Of course, other systems, polar coordinates with different polar axes or projections from different points on S^2, could be used. The coordinates on the sphere may be kept arbitrary until some specific calculation is to be carried out. [The longitude is historically measured from Greenwich. However, there is no reason why it cannot be measured from New York or Kyoto.] This arbitrariness of the coordinate choice underlies the theory of manifolds: *all coordinate systems are equally good*. It is also in harmony with the basic principle of physics: *a physical system behaves in the same way whatever coordinates we use to describe it*.

Another point which can be seen from this example is that *no coordinate system may be usable everywhere at once*. Let us look at the polar coordinates on S^2. Take the equator ($\theta = \frac{1}{2}\pi$) for definiteness. If we let ϕ range from 0 to 2π, then it changes continuously as we go round the equator until we get all the way to $\phi = 2\pi$. There the ϕ-coordinate has a discontinuity from 2π to 0 and nearby points have quite different ϕ-values. Alternatively we could continue ϕ through 2π. Then we will encounter another difficulty: at each point we must have infinitely many ϕ-values, differing from one another by an integral multiple of 2π. A further difficulty arises at the poles, where ϕ is not determined at all. [An explorer on the Pole is in a state of timelessness since time is defined by the longitude.] Stereographic coordinates also have difficulties at the North Pole or at any projection point that is not projected to a point on the equatorial plane; and nearby points close to the Pole have widely different stereographic coordinates.

Thus, we cannot label the points on the sphere with a single coordinate system so that both of the following conditions are satisfied.

(i)   Nearby points always have nearby coordinates.
(ii) Every point has unique coordinates.

Note, however, that there are infinitely many ways to introduce coordinates that satisfy these requirements on a *part* of S^2. We may take advantage of this fact to define coordinates on S^2: introduce two or more overlapping coordinate systems, each covering a part of the sphere whose points are to be labelled so that the following conditions hold.

(i′) Nearby points have nearby coordinates in at least one coordinate system.
(ii′) Every point has unique coordinates in each system that contains it.

For example, we may introduce two stereographic coordinates on S^2, one a projection from the North Pole, the other from the South Pole. Are these conditions (i′) and (ii′) enough to develop sensible theories of the manifold? In fact, we need an extra condition on the coordinate systems.

(iii) If two coordinate systems overlap, they are related to each other in a sufficiently smooth way.

Without this condition, a differentiable function in one coordinate system may not be differentiable in the other system.

5.1.2 Definitions

Definition 5.1. M is an m-dimensional differentiable manifold if

(i) M is a topological space;
(ii) M is provided with a family of pairs $\{(U_i, \varphi_i)\}$;
(iii) $\{U_i\}$ is a family of open sets which covers M, that is, $\cup_i U_i = M$. φ_i is a homeomorphism from U_i onto an open subset U_i' of $\mathbb{R}^m$ (figure 5.2); and

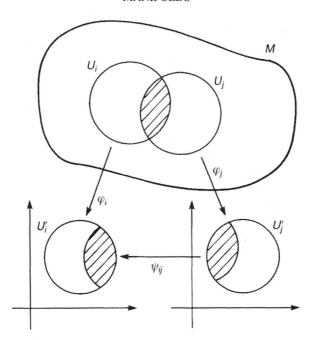

Figure 5.2. A homeomorphism φ_i maps U_i onto an open subset $U_i' \subset \mathbb{R}^m$, providing coordinates to a point $p \in U_i$. If $U_i \cap U_j \neq \emptyset$, the transition from one coordinate system to another is smooth.

(iv) given U_i and U_j such that $U_i \cap U_j \neq \emptyset$, the map $\psi_{ij} = \varphi_i \circ \varphi_j^{-1}$ from $\varphi_j(U_i \cap U_j)$ to $\varphi_i(U_i \cap U_j)$ is infinitely differentiable.

The pair (U_i, φ_i) is called a **chart** while the whole family $\{(U_i, \varphi_i)\}$ is called, for obvious reasons, an **atlas**. The subset U_i is called the **coordinate neighbourhood** while φ_i is the **coordinate function** or, simply, the **coordinate**. The homeomorphism φ_i is represented by m functions $\{x^1(p), \ldots, x^m(p)\}$. The set $\{x^\mu(p)\}$ is also called the **coordinate**. A point $p \in M$ exists independently of its coordinates; it is up to us how we assign coordinates to a point. We sometimes employ the rather sloppy notation x to denote a point whose coordinates are $\{x^1, \ldots, x^m\}$, unless several coordinate systems are in use. From (ii) and (iii), M is locally Euclidean. In each coordinate neighbourhood U_i, M looks like an open subset of $\mathbb{R}^m$ whose element is $\{x^1, \ldots, x^m\}$. Note that we do not require that M be $\mathbb{R}^m$ *globally*. We are living on the earth whose surface is S^2, which does not look like $\mathbb{R}^2$ globally. However, it looks like an open subset of $\mathbb{R}^2$ *locally*. Who can tell that we live on the sphere by just looking at a map of London, which, of course, looks like a part of $\mathbb{R}^2$?[1]

[1] Strictly speaking the distance between two longitudes in the northern part of the city is slightly

If U_i and U_j overlap, two coordinate systems are assigned to a point in $U_i \cap U_j$. Axiom (iv) asserts that the transition from one coordinate system to another be *smooth* (C^∞). The map φ_i assigns m coordinate values x^μ ($1 \le \mu \le m$) to a point $p \in U_i \cap U_j$, while φ_j assigns y^ν ($1 \le \nu \le m$) to the same point and the transition from y to x, $x^\mu = x^\mu(y)$, is given by m functions of m variables. The coordinate transformation functions $x^\mu = x^\mu(y)$ are the explicit form of the map $\psi_{ji} = \varphi_j \circ \varphi_i^{-1}$. Thus, the differentiability has been defined in the usual sense of calculus: the coordinate transformation is differentiable if each function $x^\mu(y)$ is differentiable with respect to each y^ν. We may restrict ourselves to the differentiability up to kth order (C^k). However, this does not bring about any interesting conclusions. We simply require, instead, that the coordinate transformations be infinitely differentiable, that is, of class C^∞. Now coordinates have been assigned to M in such a way that if we move over M in whatever fashion, the coordinates we use vary in a smooth manner.

If the union of two atlases $\{(U_i, \varphi_i)\}$ and $\{(V_j, \psi_j)\}$ is again an atlas, these two atlases are said to be **compatible**. The compatibility is an equivalence relation, the equivalence class of which is called the **differentiable structure**. It is also said that mutually compatible atlases define the same differentiable structure on M.

Before we give examples, we briefly comment on manifolds *with boundaries*. So far, we have assumed that the coordinate neighbourhood U_i is homeomorphic to an open set of $\mathbb{R}^m$. In some applications, however, this turns out to be too restrictive and we need to relax this condition. If a topological space M is covered by a family of open sets $\{U_i\}$ each of which is homeomorphic to an open set of $H^m \equiv \{(x^1, \ldots, x^m) \in \mathbb{R}^m | x^m \ge 0\}$, M is said to be a **manifold with a boundary**, see figure 5.3. The set of points which are mapped to points with $x^m = 0$ is called the **boundary** of M, denoted by ∂M. The coordinates of ∂M may be given by $m - 1$ numbers $(x^1, \ldots, x^{m-1}, 0)$. Now we have to be careful when we define the smoothness. The map $\psi_{ij} : \varphi_j(U_i \cap U_j) \to \varphi_i(U_i \cap U_j)$ is defined on an open set of H^m in general, and ψ_{ij} is said to be smooth if it is C^∞ in an open set of $\mathbb{R}^m$ which contains $\varphi_j(U_i \cap U_j)$. Readers are encouraged to use their imagination since our definition is in harmony with our intuitive notions about boundaries. For example, the boundary of the solid ball D^3 is the sphere S^2 and the boundary of the sphere is an empty set.

5.1.3 Examples

We now give several examples to develop our ideas about manifolds. They are also of great relevance to physics.

Example 5.1. The Euclidean space $\mathbb{R}^m$ is the most trivial example, where a single chart covers the whole space and φ may be the identity map.

---

shorter than that in the southern part and one may suspect that one lives on a curved surface. Of course, it is the other way around if one lives in a city in the southern hemisphere.

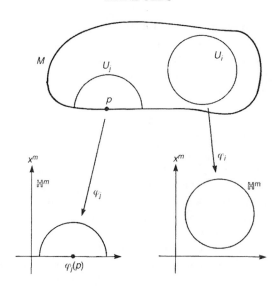

Figure 5.3. A manifold with a boundary. The point p is on the boundary.

Example 5.2. Let $m = 1$ and require that M be connected. There are only two manifolds possible: a real line $\mathbb{R}$ and the circle S^1. Let us work out an atlas of S^1. For concreteness take the circle $x^2 + y^1 = 1$ in the xy-plane. We need at least two charts. We may take them as in figure 5.4. Define $\varphi_1^{-1} : (0, 2\pi) \to S^1$ by

$$\varphi_1^{-1} : \theta \mapsto (\cos\theta, \sin\theta) \tag{5.5a}$$

whose image is $S^1 - \{(1, 0)\}$. Define also $\psi_2^{-1} : (-\pi, \pi) \to S^1$ by

$$\varphi_2^{-1} : \theta \mapsto (\cos\theta, \sin\theta) \tag{5.5b}$$

whose image is $S^1 - \{(-1, 0)\}$. Clearly φ_1^{-1} and φ_2^{-1} are invertible and all the maps $\varphi_1, \varphi_2, \varphi_1^{-1}$ and φ_2^{-1} are continuous. Thus, φ_1 and φ_2 are homeomorphisms. Verify that the maps $\psi_{12} = \varphi_1 \circ \varphi_2^{-1}$ and $\psi_{21} = \varphi_2 \circ \varphi_1^{-1}$ are smooth.

Example 5.3. The n-dimensional sphere S^n is a differentiable manifold. It is realized in $\mathbb{R}^{n+1}$ as

$$\sum_{i=0}^{n}(x^i)^2 = 1. \tag{5.6}$$

Let us introduce the coordinate neighbourhoods

$$U_{i+} \equiv \{(x^0, x^1, \ldots, x^n) \in S^n | x^i > 0\} \tag{5.7a}$$

$$U_{i-} \equiv \{(x^0, x^1, \ldots, x^n) \in S^n | x^i < 0\}. \tag{5.7b}$$

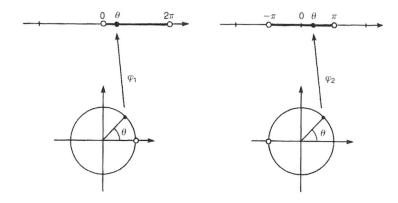

Figure 5.4. Two charts of a circle S^1.

Define the coordinate map $\varphi_{i+} : U_{i+} \to \mathbb{R}^n$ by

$$\varphi_{i+}(x^0, \ldots, x^n) = (x^0, \ldots, x^{i-1}, x^{i+1}, \ldots, x^n) \tag{5.8a}$$

and $\varphi_{i-} : U_{i-} \to \mathbb{R}^n$ by

$$\varphi_{i-}(x^0, \ldots, x^n) = (x^0, \ldots, x^{i-1}, x^{i+1}, \ldots, x^n). \tag{5.8b}$$

Note that the domains of φ_{i+} and φ_{i-} are different. $\varphi_{i\pm}$ are the projections of the hemispheres $U_{i\pm}$ to the plane $x^i = 0$. The transition functions are easily obtained from (5.8). Take S^2 as an example. The coordinate neighbourhoods are $U_{x\pm}$, $U_{y\pm}$ and $U_{z\pm}$. The transition function $\psi_{y-x+} \equiv \varphi_{y-} \circ \varphi_{x+}^{-1}$ is given by

$$\psi_{y-x+} : (y, z) \mapsto \left(\sqrt{1 - y^2 - z^2}, z \right) \tag{5.9}$$

which is infinitely differentiable on $U_{x+} \cap U_{y-}$.

Exercise 5.1. At the beginning of this chapter, we introduced the stereographic coordinates on S^2. We may equally define the stereographic coordinates projected from points other than the North Pole. For example, the stereographic coordinates (U, V) of a point in $S^2 - \{\text{South Pole}\}$ projected from the South Pole and (X, Y) for a point in $S^2 - \{\text{North Pole}\}$ projected from the North Pole are shown in figure 5.5. Show that the transition functions between (U, V) and (X, Y) are C^∞ and that they define a differentiable structure on M. See also example 8.1.

Example 5.4. The real projective space $\mathbb{R}P^n$ is the set of lines through the origin in $\mathbb{R}^{n+1}$. If $x = (x^0, \ldots, x^n) \neq 0$, x defines a line through the origin. Note that $y \in \mathbb{R}^{n+1}$ defines the same line as x if there exists a real number $a \neq 0$ such that $y = ax$. Introduce an equivalence relation $\sim$ by $x \sim y$ if there

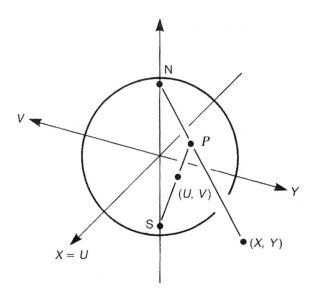

Figure 5.5. Two stereographic coordinate systems on S^2. The point P may be projected from the North Pole N giving (X, Y) or from the South Pole S giving (U, V).

exists $a \in \mathbb{R} - \{0\}$ such that $y = ax$. Then $\mathbb{R}P^n = (\mathbb{R}^{n+1} - \{0\})/ \sim$. The $n + 1$ numbers $x^0, x^1, \ldots, x^n$ are called the **homogeneous coordinates**. The homogeneous coordinates cannot be a good coordinate system, since $\mathbb{R}P^n$ is an n-dimensional manifold (an $(n + 1)$-dimensional space with a one-dimensional degree of freedom killed). The charts are defined as follows. First we take the coordinate neighbourhood U_i as the set of lines with $x^i \neq 0$, and then introduce the **inhomogeneous coordinates** on U_i by

$$\xi^j_{(i)} = x^j/x^i. \tag{5.10}$$

The inhomogeneous coordinates

$$\xi_{(i)} = (\xi^0_{(i)}, \xi^1_{(i)}, \ldots, \xi^{i-1}_{(i)}, \xi^{i+1}_{(i)}, \ldots, \xi^n_{(i)})$$

with $\xi^i_{(i)} = 1$ omitted, are well defined on U_i since $x^i \neq 0$, and furthermore they are independent of the choice of the representative of the equivalence class since $x^j/x^i = y^j/y^i$ if $y = ax$. The inhomogeneous coordinate $\xi_{(i)}$ gives the coordinate map $\varphi_i : U_i \to \mathbb{R}^n$, that is

$$\varphi_i : (x^0, \ldots, x^n) \mapsto (x^0/x^i, \ldots, x^{i-1}/x^i, x^{i+1}/x^i, \ldots, x^n/x^i)$$

where $x^i/x^i = 1$ is omitted. For $x = (x^0, x^1, \ldots, x^n) \in U_i \cap U_j$ we assign two inhomogeneous coordinates, $\xi^k_{(i)} = x^k/x^i$ and $\xi^k_{(j)} = x^k/x^j$. The coordinate transformation $\psi_{ij} = \varphi_i \circ \varphi_j^{-1}$ is

$$\psi_{ij} : \xi_{(j)}^k \mapsto \xi_{(i)}^k = (x^j/x^i)\xi_{(j)}^k. \tag{5.11}$$

This is a multiplication by x^j/x^i.

In example 4.12, we defined $\mathbb{R}P^n$ as the sphere S^n with antipodal points identified. This picture is in conformity with the definition here. As a representative of the equivalence class $[x]$, we may take points $|x| = 1$ on a line through the origin. These are points on the unit sphere. Since there are two points on the intersection of a line with S^n we have to take one of them consistently, that is nearby lines are represented by nearby points in S^n. This amounts to taking the hemisphere. Note, however, that the antipodal points on the boundary (the equator of S^n) are identified by definition, $(x^0, \ldots, x^n) \sim -(x^0, \ldots, x^n)$. This 'hemisphere' is homeomorphic to the ball D^n with antipodal points on the boundary S^{n-1} identified.

Example 5.5. A straightforward generalization of $\mathbb{R}P^n$ is the **Grassmann manifold**. An element of $\mathbb{R}P^n$ is a one-dimensional subspace in $\mathbb{R}^{n+1}$. The Grassmann manifold $G_{k,n}(\mathbb{R})$ is the set of k-dimensional planes in $\mathbb{R}^n$. Note that $G_{1,n+1}(\mathbb{R})$ is nothing but $\mathbb{R}P^n$. The manifold structure of $G_{k,n}(\mathbb{R})$ is defined in a manner similar to that of $\mathbb{R}P^n$.

Let $M_{k,n}(\mathbb{R})$ be the set of $k \times n$ matrices of rank k ($k \le n$). Take $A = (a_{ij}) \in M_{k,n}(\mathbb{R})$ and define k vectors $\boldsymbol{a}_i$ ($1 \le i \le k$) in $\mathbb{R}^n$ by $\boldsymbol{a}_i = (a_{ij})$. Since rank $A = k$, k vectors $\boldsymbol{a}_i$ are linearly independent and span a k-dimensional plane in $\mathbb{R}^n$. Note, however, that there are infinitely many matrices in $M_{k,n}(\mathbb{R})$ that yield the same k-plane. Take $g \in \mathrm{GL}(k, \mathbb{R})$ and consider a matrix $\bar{A} = gA \in M_{k,n}(\mathbb{R})$. $\bar{A}$ defines the same k-plane as A, since g simply rotates the basis within the k-plane. Introduce an equivalence relation $\sim$ by $\bar{A} \sim A$ if there exists $g \in \mathrm{GL}(k, \mathbb{R})$ such that $\bar{A} = gA$. We identify $G_{k,n}(\mathbb{R})$ with the coset space $M_{k,n}(\mathbb{R})/\mathrm{GL}(k, \mathbb{R})$.

Let us find the charts of $G_{k,n}(\mathbb{R})$. Take $A \in M_{k,n}(\mathbb{R})$ and let $\{A_1, \ldots, A_l\}$, $l = \binom{n}{k}$, be the collection of all $k \times k$ minors of A. Since rank $A = k$, there exists some A_α ($1 \le \alpha \le l$) such that $\det A \ne 0$. For example, let us assume the minor A_1 made of the first k columns has non-vanishing determinant,

$$A = (A_1, \widetilde{A_1}) \tag{5.12}$$

where $\widetilde{A_1}$ is a $k \times (n - k)$ matrix. Let us take the representative of the class to which A belongs to be

$$A_1^{-1} \cdot A = (I_k, A_1^{-1} \cdot \widetilde{A_1}) \tag{5.13}$$

where I_k is the $k \times k$ unit matrix. Note that A_1^{-1} always exists since $\det A_1 \ne 0$. Thus, the real degrees of freedom are given by the entries of the $k \times (n - k)$ matrix $A_1^{-1} \cdot \widetilde{A_1}$. We denote this subset of $G_{k,n}(\mathbb{R})$ by U_1. U_1 is a coordinate neighbourhood whose coordinates are given by $k(n - k)$ entries of $A_1^{-1} \cdot \widetilde{A_1}$. Since U_1 is homeomorphic to $\mathbb{R}^{k(n-k)}$ we find that

$$\dim G_{k,n}(\mathbb{R}) = k(n - k). \tag{5.14}$$

In the case where $\det A_\alpha \neq 0$, where A_α is composed of the columns $(i_1, i_2, \ldots, i_k)$, we multiply A_α^{-1} to obtain the representative

$$
A_\alpha^{-1} \cdot A =
\begin{matrix}
\text{column} \rightarrow & i_1 & i_2 & \cdots & i_k \\
\end{matrix}
\begin{pmatrix}
\cdots & 1 & \cdots & 0 & \cdots\cdots & 0 & \cdots \\
\cdots & 0 & \cdots & 1 & \cdots\cdots & 0 & \cdots \\
\cdots & . & \cdots & . & \cdots\cdots & . & \cdots \\
\cdots & 0 & \cdots & 0 & \cdots\cdots & 1 & \cdots
\end{pmatrix}
\tag{5.15}
$$

where the entries not written explicitly form a $k \times (n - k)$ matrix. We denote this subset of $M_{k,n}(\mathbb{R})$ with $\det A_\alpha \neq 0$ by U_α. The entries of the $k \times (n - k)$ matrix are the coordinates of U_α.

The relation between the projective space and the Grassmann manifold is evident. An element of $M_{1,n+1}(\mathbb{R})$ is a vector $A = (x^0, x^1, \ldots, x^n)$. Since the αth minor A_α of A is a number x^α, the condition $\det A_\alpha \neq 0$ becomes $x^\alpha \neq 0$. The representative (5.15) is just the inhomogeneous coordinate

$$
(x^\alpha)^{-1}(x^0, x^1, \ldots, x^\alpha, \ldots, x^n) \\
= (x^0/x^\alpha, x^1/x^\alpha, \ldots, x^\alpha/x^\alpha = 1, \ldots, x^n/x^\alpha).
$$

Let M be an m-dimensional manifold with an atlas $\{(U_i, \varphi_i)\}$ and N be an n-dimensional manifold with $\{(V_j, \psi_j)\}$. A **product manifold** $M \times N$ is an $(m+n)$-dimensional manifold whose atlas is $\{(U_i \times V_j), (\varphi_i, \psi_j)\}$. A point in $M \times N$ is written as (p, q), $p \in M$, $q \in N$, and the coordinate function (φ_i, ψ_j) acts on (p, q) to yield $(\varphi_i(p), \psi_j(p)) \in \mathbb{R}^{m+n}$. The reader should verify that a product manifold indeed satisfies the axioms of definition 5.1.

Example 5.6. The torus T^2 is a product manifold of two circles, $T^2 = S^1 \times S^1$. If we denote the polar angle of each circle as $\theta_i \bmod 2\pi$ $(i = 1, 2)$, the coordinates of T^2 are (θ_1, θ_2). Since each S^1 is embedded in $\mathbb{R}^2$, T^2 may be embedded in $\mathbb{R}^4$. We often imagine T^2 as the surface of a doughnut in $\mathbb{R}^3$, in which case, however, we inevitably have to introduce bending of the surface. This is an extrinsic feature brought about by the 'embedding'. When we say 'a torus is a flat manifold', we refer to the flat surface embedded in $\mathbb{R}^4$. See definition 5.3 for further details.

We may also consider a direct product of n circles,

$$
T^n = \underbrace{S^1 \times S^1 \times \cdots \times S^1}_{n}.
$$

Clearly T^n is an n-dimensional manifold with the coordinates $(\theta_1, \theta_2, \ldots, \theta_n)$ $\bmod 2\pi$. This may be regarded as an n-cube whose opposite faces are identified, see figure 2.4 for $n = 2$.

5.2 The calculus on manifolds

The significance of differentiable manifolds resides in the fact that we may use the usual calculus developed in $\mathbb{R}^n$. Smoothness of the coordinate transformations

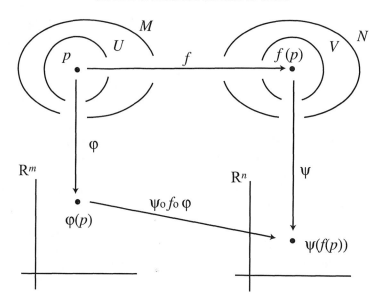

Figure 5.6. A map $f : M \to N$ has a coordinate presentation $\psi \circ f \circ \varphi^{-1} : \mathbb{R}^m \to \mathbb{R}^n$.

ensures that the calculus is independent of the coordinates chosen.

5.2.1 Differentiable maps

Let $f : M \to N$ be a map from an m-dimensional manifold M to an n-dimensional manifold N. A point $p \in M$ is mapped to a point $f(p) \in N$, namely $f : p \mapsto f(p)$, see figure 5.6. Take a chart (U, φ) on M and (V, ψ) on N, where $p \in U$ and $f(p) \in V$. Then f has the following coordinate presentation:

$$\psi \circ f \circ \varphi^{-1} : \mathbb{R}^m \to \mathbb{R}^n. \tag{5.16}$$

If we write $\varphi(p) = \{x^\mu\}$ and $\psi(f(p)) = \{y^\alpha\}$, $\psi \circ f \circ \varphi^{-1}$ is just the usual vector-valued function $y = \psi \circ f \circ \varphi^{-1}(x)$ of m variables. We sometimes use (in fact, abuse!) the notation $y = f(x)$ or $y^\alpha = f^\alpha(x^\mu)$, when we know which coordinate systems on M and N are in use. If $y = \psi \circ f \circ \varphi^{-1}(x)$, or simply $y^\alpha = f^\alpha(x^\mu)$, is C^∞ with respect to each x^μ, f is said to be **differentiable** at p or at $x = \varphi(p)$. Differentiable maps are also said to be **smooth**. Note that we require infinite (C^∞) differentiability, in harmony with the smoothness of the transition functions ψ_{ij}.

The differentiability of f is independent of the coordinate system. Consider two overlapping charts (U_1, φ_1) and (U_2, φ_2). Take a point $p \in U_1 \cap U_2$, whose coordinates by φ_1 are $\{x_1^\mu\}$, while those by φ_2 are $\{x_2^\nu\}$. When expressed in terms of $\{x_1^\mu\}$, f takes the form $\psi \circ f \circ \varphi_1^{-1}$, while in $\{x_2^\nu\}$, $\psi \circ f \circ \varphi_2^{-1} =$

$\psi \circ f \circ \varphi_1^{-1}(\varphi_1 \circ \varphi_2^{-1})$. By definition, $\psi_{12} = \varphi_1 \circ \varphi_2^{-1}$ is C^∞. In the simpler expressions, they correspond to $y = f(x_1)$ and $y = f(x_1(x_2))$. It is clear that if $f(x_1)$ is C^∞ with respect to x_1^μ and $x_1(x_2)$ is C^∞ with respect to x_2^ν, then $y = f(x_1(x_2))$ is also C^∞ with respect to x_2^ν.

Exercise 5.2. Show that the differentiability of f is also independent of the chart in N.

Definition 5.2. Let $f : M \to N$ be a homeomorphism and ψ and φ be coordinate functions as previously defined. If $\psi \circ f \circ \varphi^{-1}$ is invertible (that is, there exists a map $\varphi \circ f^{-1} \circ \psi^{-1}$) and both $y = \psi \circ f \circ \varphi^{-1}(x)$ and $x = \varphi \circ f^{-1} \circ \psi^{-1}(y)$ are C^∞, f is called a **diffeomorphism** and M is said to be **diffeomorphic** to N and *vice versa*, denoted by $M \equiv N$.

Clearly $\dim M = \dim N$ if $M \equiv N$. In chapter 2, we noted that homeomorphisms classify spaces according to whether it is possible to deform one space into another *continuously*. Diffeomorphisms classify spaces into equivalence classes according to whether it is possible to deform one space to another *smoothly*. Two diffeomorphic spaces are regarded as the same manifold. Clearly a diffeomorphism is a homeomorphism. What about the converse? Is a homeomorphism a diffeomorphism? In the previous section, we defined the differentiable structure as an equivalence class of atlases. Is it possible for a topological space to carry many differentiable structures? It is rather difficult to give examples of 'diffeomorphically inequivalent homeomorphisms' since it is known that this is possible only in higher-dimensional spaces ($\dim M \geq 4$). It was believed before 1956 that a topological space admits only one differentiable structure. However, Milnor (1956) pointed out that S^7 admits 28 differentiable structures. A recent striking discovery in mathematics is that $\mathbb{R}^4$ admits an infinite number of differentiable structures. Interested readers should consult Donaldson (1983) and Freed and Uhlenbeck (1984). Here we assume that a manifold admits a unique differentiable structure, for simplicity.

The set of diffeomorphisms $f : M \to M$ is a group denoted by $\mathrm{Diff}(M)$. Take a point p in a chart (U, φ) such that $\varphi(p) = x^\mu(p)$. Under $f \in \mathrm{Diff}(M)$, p is mapped to $f(p)$ whose coordinates are $\varphi(f(p)) = y^\mu(f(p))$ (we have assumed $f(p) \in U$). Clearly y is a differentiable function of x; this is an *active* point of view to the coordinate transformation. However, if (U, φ) and (V, ψ) are overlapping charts, we have two coordinate values $x^\mu = \varphi(p)$ and $y^\mu = \psi(p)$ for a point $p \in U \cap V$. The map $x \mapsto y$ is differentiable by the assumed smoothness of the manifold; this reparametrization is a *passive* point of view to the coordinate transformation. We also denote the group of reparametrizations by $\mathrm{Diff}(M)$.

Now we look at special classes of mappings, namely **curves** and **functions**. An open curve in an m-dimensional manifold M is a map $c : (a, b) \to M$ where (a, b) is an open interval such that $a < 0 < b$. We assume that the curve does not intersect with itself (figure 5.7). The number a (b) may be $-\infty$ ($+\infty$) and we have included 0 in the interval for later convenience. If a curve is closed, it is

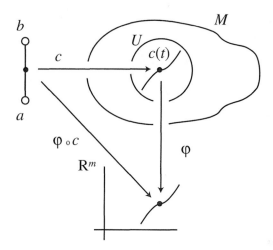

Figure 5.7. A curve c in M and its coordinate presentation $\varphi \circ c$.

regarded as a map $c : S^1 \to M$. In both cases, c is locally a map from an open interval to M. On a chart (U, φ), a curve $c(t)$ has the coordinate presentation $x = \varphi \circ c : \mathbb{R} \to \mathbb{R}^m$.

A function f on M is a smooth map from M to $\mathbb{R}$, see figure 5.8. On a chart (U, φ), the coordinate presentation of f is given by $f \circ \varphi^{-1} : \mathbb{R}^m \to \mathbb{R}$ which is a real-valued function of m variables. We denote the set of smooth functions on M by $\mathcal{F}(M)$.

5.2.2 Vectors

Now that we have defined maps on a manifold, we are ready to define other geometrical objects: vectors, dual vectors and tensors. In general, an elementary picture of a vector as an arrow connecting a point and the origin does not work in a manifold. [Where is the origin? What is a straight arrow? How do we define a straight arrow that connects London and Los Angeles on the *surface* of the Earth?] On a manifold, a vector is defined to be a **tangent vector** to a curve in M.

To begin with, let us look at a tangent line to a curve in the xy-plane. If the curve is differentiable, we may approximate the curve in the vicinity of x_0 by

$$y - y(x_0) = a(x - x_0) \tag{5.17}$$

where $a = \mathrm{d}y/\mathrm{d}x|_{x=x_0}$. The tangent vectors on a manifold M generalize this tangent line. To define a tangent vector we need a curve $c : (a, b) \to M$ and a function $f : M \to \mathbb{R}$, where (a, b) is an open interval containing $t = 0$, see figure 5.9. We define the tangent vector at $c(0)$ as a directional derivative of a function $f(c(t))$ along the curve $c(t)$ at $t = 0$. The rate of change of $f(c(t))$ at

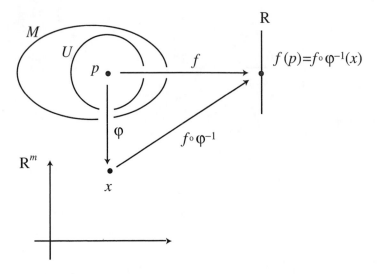

Figure 5.8. A function $f : M \to \mathbb{R}$ and its coordinate presentation $f \circ \varphi^{-1}$.

$t = 0$ along the curve is

$$\frac{\mathrm{d} f(c(t))}{\mathrm{d} t}\bigg|_{t=0}. \tag{5.18}$$

In terms of the local coordinate, this becomes

$$\frac{\partial f}{\partial x^\mu} \frac{\mathrm{d} x^\mu(c(t))}{\mathrm{d} t}\bigg|_{t=0}. \tag{5.19}$$

[Note the abuse of the notation! The derivative $\partial f/\partial x^\mu$ really means $\partial(f \circ \varphi^{-1}(x))/\partial x^\mu$.] In other words, $\mathrm{d} f(c(t))/\mathrm{d} t$ at $t = 0$ is obtained by applying the differential operator X to f, where

$$X = X^\mu \left(\frac{\partial}{\partial x^\mu} \right) \qquad \left(X^\mu = \frac{\mathrm{d} x^\mu(c(t))}{\mathrm{d} t}\bigg|_{t=0} \right) \tag{5.20}$$

that is,

$$\frac{\mathrm{d} f(c(t))}{\mathrm{d} t}\bigg|_{t=0} = X^\mu \left(\frac{\partial f}{\partial x^\mu} \right) \equiv X[f]. \tag{5.21}$$

Here the last equality defines $X[f]$. It is $X = X^\mu \partial/\partial x^\mu$ which we now define as the tangent vector to M at $p = c(0)$ along the direction given by the curve $c(t)$.

Example 5.7. If X is applied to the coordinate functions $\varphi(c(t)) = x^\mu(t)$, we have

$$X[x^\mu] = \left(\frac{\mathrm{d} x^\nu}{\mathrm{d} t} \right) \left(\frac{\partial x^\mu}{\partial x^\nu} \right) = \frac{\mathrm{d} x^\mu(t)}{\mathrm{d} t}\bigg|_{t=0}$$

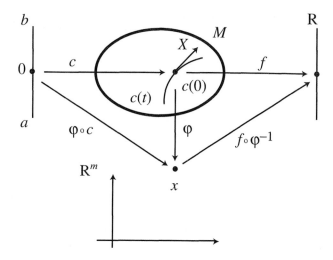

Figure 5.9. A curve c and a function f define a tangent vector along the curve in terms of the directional derivative.

which is the μth component of the velocity vector if t is understood as time.

To be more mathematical, we introduce an equivalence class of curves in M. If two curves $c_1(t)$ and $c_2(t)$ satisfy

(i) $c_1(0) = c_2(0) = p$

(ii) $\left. \dfrac{dx^\mu(c_1(t))}{dt} \right|_{t=0} = \left. \dfrac{dx^\mu(c_2(t))}{dt} \right|_{t=0}$

$c_1(t)$ and $c_2(t)$ yield the same differential operator X at p, in which case we define $c_1(t) \sim c_2(t)$. Clearly $\sim$ is an equivalence relation and defines the equivalence classes. We identify the *tangent vector* X with the *equivalence class of curves*

$$[c(t)] = \left\{ \tilde{c}(t) \,\middle|\, \tilde{c}(0) = c(0) \text{ and } \left. \frac{dx^\mu(\tilde{c}(t))}{dt} \right|_{t=0} = \left. \frac{dx^\mu(c(t))}{dt} \right|_{t=0} \right\} \tag{5.22}$$

rather than a curve itself.

All the equivalence classes of curves at $p \in M$, namely all the tangent vectors at p, form a vector space called the **tangent space** of M at p, denoted by $T_p M$. To analyse $T_p M$, we may use the theory of vector spaces developed in section 2.2. Evidently, $e_\mu = \partial/\partial x^\mu$ $(1 \le \mu \le m)$ are the basis vectors of $T_p M$, see (5.20), and $\dim T_p M = \dim M$. The basis $\{e_\mu\}$ is called the **coordinate basis**. If a vector $V \in T_p M$ is written as $V = V^\mu e_\mu$, the numbers V^μ are called the components of V with respect to e_μ. By construction, it is obvious that a vector X exists without specifying the coordinate, see (5.21). The assignment of the coordinate is simply for our convenience. This coordinate independence of a vector enables us to find the transformation property of the *components* of the vector. Let $p \in U_i \cap U_j$ and $x = \varphi_i(p)$, $y = \varphi_j(p)$. We have two expressions for $X \in T_p M$,

$$X = X^\mu \frac{\partial}{\partial x^\mu} = \widetilde{X}^\mu \frac{\partial}{\partial y^\mu}.$$

This shows that X^μ and $\widetilde{X}^\mu$ are related as

$$\widetilde{X}^\mu = X^\nu \frac{\partial y^\mu}{\partial x^\nu}. \tag{5.23}$$

Note again that the components of the vector transform in such a way that the vector itself is left invariant.

The basis of $T_p M$ need not be $\{e_\mu\}$, and we may think of the linear combinations $\hat{e}_i \equiv A_i{}^\mu e_\mu$, where $A = (A_i{}^\mu) \in \mathrm{GL}(m, \mathbb{R})$. The basis $\{\hat{e}_i\}$ is known as the **non-coordinate basis**.

5.2.3   One-forms

Since $T_p M$ is a vector space, there exists a dual vector space to $T_p M$, whose element is a linear function from $T_p M$ to $\mathbb{R}$, see section 2.2. The dual space is called the **cotangent space** at p, denoted by $T_p^* M$. An element $\omega : T_p M \to \mathbb{R}$ of $T_p^* M$ is called a **dual vector**, **cotangent vector** or, in the context of differential forms, a **one-form**. The simplest example of a one-form is the differential $\mathrm{d}f$ of a function $f \in \mathcal{F}(M)$. The action of a vector V on f is $V[f] = V^\mu \partial f / \partial x^\mu \in \mathbb{R}$. Then the action of $\mathrm{d}f \in T_p^* M$ on $V \in T_p M$ is defined by

$$\langle \mathrm{d}f, V \rangle \equiv V[f] = V^\mu \frac{\partial f}{\partial x^\mu} \in \mathbb{R}. \tag{5.24}$$

Clearly $\langle \mathrm{d}f, V \rangle$ is $\mathbb{R}$-linear in both V and f.

Noting that $\mathrm{d}f$ is expressed in terms of the coordinate $x = \varphi(p)$ as $\mathrm{d}f = (\partial f / \partial x^\mu)\mathrm{d}x^\mu$, it is natural to regard $\{\mathrm{d}x^\mu\}$ as a basis of $T_p^* M$. Moreover, this is a dual basis, since

$$\left\langle \mathrm{d}x^\mu, \frac{\partial}{\partial x^\mu} \right\rangle = \frac{\partial x^\nu}{\partial x^\mu} = \delta_\mu^\nu. \tag{5.25}$$

An arbitrary one-form ω is written as

$$\omega = \omega_\mu \, \mathrm{d}x^\mu \tag{5.26}$$

where the ω_μ are the components of ω. Take a vector $V = V^\mu \partial / \partial x^\mu$ and a one-form $\omega = \omega_\mu \mathrm{d}x^\mu$. The **inner product** $\langle \ , \ \rangle : T_p^* M \times T_p M \to \mathbb{R}$ is defined by

$$\langle \omega, V \rangle = \omega_\mu V^\nu \left\langle \mathrm{d}x^\mu, \frac{\partial}{\partial x^\nu} \right\rangle = \omega_\mu V^\nu \delta_\nu^\mu = \omega_\mu V^\mu. \tag{5.27}$$

Note that the inner product is defined between a vector and a dual vector and not between two vectors or two dual vectors.

Since ω is defined without reference to any coordinate system, for a point $p \in U_i \cap U_j$, we have

$$\omega = \omega_\mu dx^\mu = \widetilde{\omega}_\nu \, dy^\nu$$

where $x = \varphi_i(p)$ and $y = \varphi_j(p)$. From $dy^\nu = (\partial y^\nu / \partial x^\mu) dx^\mu$ we find that

$$\widetilde{\omega}_\nu = \omega_\mu \frac{\partial x^\mu}{\partial y^\nu}. \tag{5.28}$$

5.2.4 Tensors

A **tensor** of type (q, r) is a multilinear object which maps q elements of $T_p^* M$ and r elements of $T_p M$ to a real number. $\mathcal{T}^q_{r,p}(M)$ denotes the set of type (q, r) tensors at $p \in M$. An element of $\mathcal{T}^q_{r,p}(M)$ is written in terms of the bases described earlier as

$$T = T^{\mu_1 \dots \mu_q}{}_{\nu_1 \dots \nu_r} \frac{\partial}{\partial x^{\mu_1}} \cdots \frac{\partial}{\partial x^{\mu_q}} dx^{\nu_1} \dots dx^{\nu_r}. \tag{5.29}$$

Clearly this is a linear function from

$$\otimes^q T_p^* M \otimes^r T_p M$$

to $\mathbb{R}$. Let $V_i = V_i^\mu \partial / \partial x^\mu$ $(1 \leq i \leq r)$ and $\omega_i = \omega_{i\mu} dx^\mu$ $(1 \leq i \leq q)$. The action of T on them yields a number

$$T(\omega_1, \dots, \omega_q; V_1, \dots, V_r) = T^{\mu_1 \dots \mu_q}{}_{\nu_1 \dots \nu_r} \omega_{1\mu_1} \dots \omega_{q\mu_q} V_1^{\nu_1} \dots V_r^{\nu_r}.$$

In the present notation, the inner product is $\langle \omega, X \rangle = \omega(X)$.

5.2.5 Tensor fields

If a vector is assigned *smoothly* to each point of M, it is called a **vector field** over M. In other words, V is a vector field if $V[f] \in \mathcal{F}(M)$ for any $f \in \mathcal{F}(M)$. Clearly each component of a vector field is a smooth function from M to $\mathbb{R}$. The set of the vector fields on M is denoted as $\mathcal{X}(M)$. A vector field X at $p \in M$ is denoted by $X|_p$, which is an element of $T_p M$. Similarly, we define a **tensor field** of type (q, r) by a smooth assignment of an element of $\mathcal{T}^q_{r,p}(M)$ at each point $p \in M$. The set of the tensor fields of type (q, r) on M is denoted by $\mathcal{T}^q_r(M)$. For example, $\mathcal{T}^0_1(M)$ is the set of the dual vector fields, which is also denoted by $\Omega^1(M)$ in the context of differential forms, see section 5.4. Similarly, $\mathcal{T}^0_0(M) = \mathcal{F}(M)$ is denoted by $\Omega^0(M)$ in the same context.

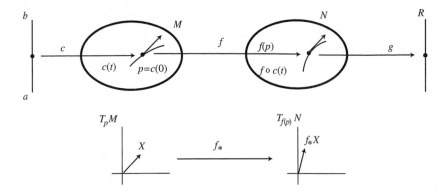

Figure 5.10. A map $f : M \to N$ induces the differential map $f_* : T_pM \to T_{f(p)}N$.

5.2.6 Induced maps

A smooth map $f : M \to N$ naturally induces a map f_* called the **differential map** (figure 5.10),

$$f_* : T_pM \to T_{f(p)}N. \tag{5.30}$$

The explicit form of f_* is obtained by the definition of a tangent vector as a directional derivative along a curve. If $g \in \mathcal{F}(N)$, then $g \circ f \in \mathcal{F}(M)$. A vector $V \in T_pM$ acts on $g \circ f$ to give a number $V[g \circ f]$. Now we define $f_*V \in T_{f(p)}N$ by

$$(f_*V)[g] \equiv V[g \circ f] \tag{5.31}$$

or, in terms of charts (U, φ) on M and $(V. \psi)$ on N,

$$(f_*V)[g \circ \psi^{-1}(y)] \equiv V[g \circ f \circ \varphi^{-1}(x)] \tag{5.32}$$

where $x = \varphi(p)$ and $y = \psi(f(p))$. Let $V = V^\mu \partial/\partial x^\mu$ and $f_*V = W^\alpha \partial/\partial y^\alpha$. Then (5.32) yields

$$W^\alpha \frac{\partial}{\partial y^\alpha}[g \circ \psi^{-1}(y)] = V^\mu \frac{\partial}{\partial x^\mu}[g \circ f \circ \varphi^{-1}(x)].$$

If we take $g = y^\alpha$, we obtain the relation between W^α and V^μ,

$$W^\alpha = V^\mu \frac{\partial}{\partial x^\mu} y^\alpha(x). \tag{5.33}$$

Note that the matrix $(\partial y^\alpha/\partial x^\mu)$ is nothing but the Jacobian of the map $f : M \to N$. The differential map f_* is naturally extended to tensors of type $(q, 0)$, $f_* : \mathcal{T}_{0,p}^q(M) \to \mathcal{T}_{0,f(p)}^q(N)$.

Example 5.8. Let (x^1, x^2) and (y^1, y^2, y^3) be the coordinates in M and N, respectively, and let $V = a\partial/\partial x^1 + b\partial/\partial x^2$ be a tangent vector at (x^1, x^2).

Let $f : M \to N$ be a map whose coordinate presentation is $y = (x^1, x^2, \sqrt{1 - (x^1)^2 - (x^2)^2})$. Then

$$f_* V = V^\mu \frac{\partial y^\alpha}{\partial x^\mu} \frac{\partial}{\partial y^\alpha} = a \frac{\partial}{\partial y^1} + b \frac{\partial}{\partial y^2} - \left(a \frac{y^1}{y^3} + b \frac{y^2}{y^3} \right) \frac{\partial}{\partial y^3}.$$

Exercise 5.3. Let $f : M \to N$ and $g : N \to P$. Show that the differential map of the composite map $g \circ f : M \to P$ is

$$(g \circ f)_* = g_* \circ f_*. \tag{5.34}$$

A map $f : M \to N$ also induces a map

$$f^* : T^*_{f(p)} N \to T^*_p M. \tag{5.35}$$

Note that f_* goes in the same direction as f, while f^* goes backward, hence the name **pullback**, see section 2.2. If we take $V \in T_p M$ and $\omega \in T^*_{f(p)} N$, the pullback of ω by f^* is defined by

$$\langle f^* \omega, V \rangle = \langle \omega, f_* V \rangle. \tag{5.36}$$

The pullback f^* naturally extends to tensors of type $(0, r)$, $f^* : T^0_{r, f(p)}(N) \to T^0_{r, p}(M)$. The component expression of f^* is given by the Jacobian matrix $(\partial y^\alpha / \partial x^\mu)$, see exercise 5.4.

Exercise 5.4. Let $f : M \to N$ be a smooth map. Show that for $\omega = \omega_\alpha \, \mathrm{d} y^\alpha \in T^*_{f(p)} N$, the induced one-form $f^* \omega = \xi_\mu \, \mathrm{d} x^\mu \in T^*_p M$ has components

$$\xi_\mu = \omega_\alpha \frac{\partial y^\alpha}{\partial x^\mu}. \tag{5.37}$$

Exercise 5.5. Let f and g be as in exercise 5.3. Show that the pullback of the composite map $g \circ f$ is

$$(g \circ f)^* = f^* \circ g^*. \tag{5.38}$$

There is no natural extension of the induced map for a tensor of mixed type. The extension is only possible if $f : M \to N$ is a diffeomorphism, where the Jacobian of f^{-1} is also defined.

Exercise 5.6. Let

$$T^\mu{}_\nu \frac{\partial}{\partial x^\mu} \otimes \mathrm{d} x^\nu$$

be a tensor field of type $(1, 1)$ on M and let $f : M \to N$ be a diffeomorphism. Show that the induced tensor on N is

$$f_* \left(T^\mu{}_\nu \frac{\partial}{\partial x^\mu} \otimes \mathrm{d} x^\nu \right) = T^\mu{}_\nu \left(\frac{\partial y^\alpha}{\partial x^\mu} \right) \left(\frac{\partial x^\nu}{\partial y^\beta} \right) \frac{\partial}{\partial y^\alpha} \otimes \mathrm{d} y^\beta$$

where x^μ and y^α are local coordinates in M and N, respectively.

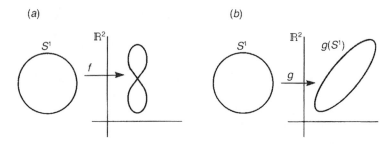

Figure 5.11. (*a*) An immersion f which is not an embedding. (*b*) An embedding g and the submanifold $g(S^1)$.

5.2.7 Submanifolds

Before we close this section, we define a submanifold of a manifold. The meaning of embedding is also clarified here.

Definition 5.3. (**Immersion, submanifold, embedding**) Let $f : M \to N$ be a smooth map and let dim $M \le$ dim N.

   (a) The map f is called an **immersion** of M into N if $f_* : T_p M \to T_{f(p)} N$ is an injection (one to one), that is rank $f_* = $ dim M.
   (b) The map f is called an **embedding** if f is an injection and an immersion. The image $f(M)$ is called a **submanifold** of N. [In practice, $f(M)$ thus defined is diffeomorphic to M.]

   If f is an immersion, f^* maps $T_p M$ isomorphically to an m-dimensional vector subspace of $T_{f(p)} N$ since rank $f_* = $ dim M. From theorem 2.1, we also find ker $f_* = \{0\}$. If f is an embedding, M is diffeomorphic to $f(M)$. Examples will clarify these rather technical points. Consider a map $f : S^1 \to \mathbb{R}^2$ in figure 5.11(*a*). It is an immersion since a one-dimensional tangent space of S^1 is mapped by f_* to a subspace of $T_{f(p)} \mathbb{R}^2$. The image $f(S^1)$ is not a submanifold of $\mathbb{R}^2$ since f is not an injection. The map $g : S^1 \to \mathbb{R}^2$ in figure 5.11(*b*) is an embedding and $g(S^1)$ is a submanifold of $\mathbb{R}^2$. Clearly, an embedding is an immersion although the converse is not necessarily true. In the previous section, we occasionally mentioned the embedding of S^n into $\mathbb{R}^{n+1}$. Now this meaning is clear; if S^n is embedded by $f : S^n \to \mathbb{R}^{n+1}$ then S^n is diffeomorphic to $f(S^n)$.

5.3 Flows and Lie derivatives

Let X be a vector field in M. An integral curve $x(t)$ of X is a curve in M, whose tangent vector at $x(t)$ is $X|_x$. Given a chart (U, φ), this means

$$\frac{dx^\mu}{dt} = X^\mu(x(t)) \tag{5.39}$$

where $x^\mu(t)$ is the μth component of $\varphi(x(t))$ and $X = X^\mu \partial/\partial x^\mu$. Note the abuse of the notation: x is used to denote a point in M as well as its coordinates. [For later convenience we assume the point $x(0)$ is included in U.] Put in another way, finding the integral curve of a vector field X is equivalent to solving the autonomous system of ordinary differential equations (ODEs) (5.39). The initial condition $x_0^\mu = x^\mu(0)$ corresponds to the coordinates of an integral curve at $t = 0$. The existence and uniqueness theorem of ODEs guarantees that there is a unique solution to (5.39), at least locally, with the initial data x_0^μ. It may happen that the integral curve is defined only on a subset of $\mathbb{R}$, in which case we have to pay attention so that the parameter t does not exceed the given interval. In the following we assume that t is maximally extended. It is known that if M is a compact manifold, the integral curve exists for all $t \in \mathbb{R}$.

Let $\sigma(t, x_0)$ be an integral curve of X which passes a point x_0 at $t = 0$ and denote the coordinate by $\sigma^\mu(t, x_0)$. Equation (5.39) then becomes

$$\frac{\mathrm{d}}{\mathrm{d}t} \sigma^\mu(t, x_0) = X^\mu(\sigma(t, x_0)) \tag{5.40a}$$

with the initial condition

$$\sigma^\mu(0, x_0) = x_0^\mu. \tag{5.40b}$$

The map $\sigma : \mathbb{R} \times M \to M$ is called a **flow** generated by $X \in \mathfrak{X}(M)$. A flow satisfies the rule

$$\sigma(t, \sigma^\mu(s, x_0)) = \sigma(t + s, x_0) \tag{5.41}$$

for any $s, t \in \mathbb{R}$ such that both sides of (5.41) make sense. This can be seen from the uniqueness of ODEs. In fact, we note that

$$\frac{\mathrm{d}}{\mathrm{d}t} \sigma^\mu(t, \sigma^\mu(s, x_0)) = X^\mu(\sigma(t, \sigma^\mu(s, x_0)))$$
$$\sigma(0, \sigma(s, x_0)) = \sigma(s, x_0)$$

and

$$\frac{\mathrm{d}}{\mathrm{d}t} \sigma^\mu(t + s, x_0) = \frac{\mathrm{d}}{\mathrm{d}(t + s)} \sigma^\mu(t + s, x_0) = X^\mu(\sigma(t + s, x_0))$$
$$\sigma(0 + s, x_0) = \sigma(s, x_0).$$

Thus, both sides of (5.41) satisfy the same ODE and the same initial condition. From the uniqueness of the solution, they should be the same. We have obtained the following theorem.

Theorem 5.1. For any point $x \in M$, there exists a differentiable map $\sigma : \mathbb{R} \times M \to M$ such that

(i) $\sigma(0, x) = x$;
(ii) $t \mapsto \sigma(t, x)$ is a solution of (5.40a) and (5.40b); and

(iii) $\sigma(t, \sigma^{\mu}(s, x)) = \sigma(t + s, x)$.

[*Note:* We denote the initial point by x instead of x_0 to emphasize that σ is a map $\mathbb{R} \times M \to M$.]

We may imagine a flow as a (steady) stream flow. If a particle is observed at a point x at $t = 0$, it will be found at $\sigma(t, x)$ at later time t.

Example 5.9. Let $M = \mathbb{R}^2$ and let $X((x, y)) = -y\partial/\partial x + x\partial/\partial y$ be a vector field in M. It is easy to verify that

$$\sigma(t, (x, y)) = (x \cos t - y \sin t, x \sin t + y \cos t)$$

is a flow generated by X. The flow through (x, y) is a circle whose centre is at the origin. Clearly, $\sigma(t, (x, y)) = (x, y)$ if $t = 2n\pi, n \in \mathbb{Z}$. If $(x, y) = (0, 0)$, the flow stays at $(0, 0)$.

Exercise 5.7. Let $M = \mathbb{R}^2$, and let $X = y\partial/\partial x + x\partial/\partial y$ be a vector field in M. Find the flow generated by X.

5.3.1   One-parameter group of transformations

For fixed $t \in \mathbb{R}$, a flow $\sigma(t, x)$ is a diffeomorphism from M to M, denoted by $\sigma_t : M \to M$. It is important to note that σ_t is made into a *commutative group* by the following rules.

(i) $\sigma_t(\sigma_s(x)) = \sigma_{t+s}(x)$, that is, $\sigma_t \circ \sigma_s = \sigma_{t+s}$;
(ii) $\sigma_0 = $ the identity map $(= $ unit element); and
(iii) $\sigma_{-t} = (\sigma_t)^{-1}$.

This group is called the **one-parameter group of transformations**. The group locally looks like the additive group $\mathbb{R}$, although it may not be isomorphic to $\mathbb{R}$ globally. In fact, in example 5.9, $\sigma_{2\pi n+t}$ was the same map as σ_t and we find that the one-parameter group is isomorphic to SO(2), the multiplicative group of 2×2 real matrices of the form

$$\begin{pmatrix} \cos \theta & -\sin \theta \\ \sin \theta & \cos \theta \end{pmatrix}$$

or U(1), the multiplicative group of complex numbers of unit modulus $e^{i\theta}$.

Under the action of σ_ε, with an infinitesimal ε, we find from (5.40a) and (5.40b) that a point x whose coordinate is x^{μ} is mapped to

$$\sigma_\varepsilon^{\mu}(x) = \sigma^{\mu}(\varepsilon, x) = x^{\mu} + \varepsilon X^{\mu}(x). \tag{5.42}$$

The vector field X is called, in this context, the **infinitesimal generator** of the transformation σ_t.

Given a vector field X, the corresponding flow σ is often referred to as the **exponentiation** of X and is denoted by

$$\sigma^{\mu}(t, x) = \exp(tX)x^{\mu}. \tag{5.43}$$

The name 'exponentiation' is justified as we shall see now. Let us take a parameter t and evaluate the coordinate of a point which is separated from the initial point $x = \sigma(0, x)$ by the parameter distance t along the flow σ. The coordinate corresponding to the point $\sigma(t, x)$ is

$$
\begin{aligned}
\sigma^{\mu}(t, x) &= x^{\mu} + t \frac{d}{ds} \sigma^{\mu}(s, x)\Big|_{s=0} + \frac{t^2}{2!} \left(\frac{d}{ds}\right)^2 \sigma^{\mu}(s, x)\Big|_{s=0} + \cdots \\
&= \left[1 + t\frac{d}{ds} + \frac{t^2}{2!}\left(\frac{d}{ds}\right)^2 + \cdots\right] \sigma^{\mu}(s, x)\Big|_{s=0} \\
&\equiv \exp\left(t\frac{d}{ds}\right)\sigma^{\mu}(s, x)\Big|_{s=0}.
\end{aligned}
\tag{5.44}
$$

The last expression can also be written as $\sigma^{\mu}(t, x) = \exp(tX)x^{\mu}$, as in (5.43). The flow σ satisfies the following exponential properties.

(i) $\quad \sigma(0, x) = x = \exp(0X)x$ \hfill (5.45a)

(ii) $\quad \dfrac{d\sigma(t, x)}{dt} = X\exp(tX)x = \dfrac{d}{dt}[\exp(tX)x]$ \hfill (5.45b)

(iii) $\quad \sigma(t, \sigma(s, x)) = \sigma(t, \exp(sX)x) = \exp(tX)\exp(sX)x$

$$= \exp\{(t+s)X\}x = \sigma(t+s, x). \tag{5.45c}$$

5.3.2 Lie derivatives

Let $\sigma(t, x)$ and $\tau(t, x)$ be two flows generated by the vector fields X and Y,

$$\frac{d\sigma^{\mu}(s, x)}{ds} = X^{\mu}(\sigma(s, x)) \tag{5.46a}$$

$$\frac{d\tau^{\mu}(t, x)}{dt} = Y^{\mu}(\tau(t, x)). \tag{5.46b}$$

Let us evaluate the change of the vector field Y along $\sigma(s, x)$. To do this, we have to compare the vector Y at a point x with that at a nearby point $x' = \sigma_{\varepsilon}(x)$, see figure 5.12. However, we cannot simply take the difference between the components of Y at two points since they belong to different tangent spaces T_pM and $T_{\sigma_{\varepsilon}(x)}M$; the naive difference between vectors at different points is ill defined. To define a sensible derivative, we first map $Y|_{\sigma_{\varepsilon}(x)}$ to T_xM by $(\sigma_{-\varepsilon})_* : T_{\sigma_{\varepsilon}(x)}M \to T_xM$, after which we take a difference between two vectors $(\sigma_{-\varepsilon})_* Y|_{\sigma_{\varepsilon}(x)}$ and $Y|_x$, both of which are vectors in T_xM. The **Lie derivative** of a vector field Y along the flow σ of X is defined by

$$\mathcal{L}_X Y = \lim_{\varepsilon \to 0} \frac{1}{\varepsilon}[(\sigma_{-\varepsilon})_* Y|_{\sigma_{\varepsilon}(x)} - Y|_x]. \tag{5.47}$$

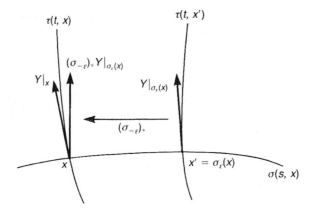

Figure 5.12. To compare a vector $Y|_x$ with $Y|_{\sigma_\varepsilon(x)}$, the latter must be transported back to x by the differential map $(\sigma_{-\varepsilon})_*$.

Exercise 5.8. Show that $\mathcal{L}_X Y$ is also written as

$$\mathcal{L}_X Y = \lim_{\varepsilon \to 0} \frac{1}{\varepsilon}[Y|_x - (\sigma_\varepsilon)_* Y|_{\sigma_{-\varepsilon}(x)}]$$

$$= \lim_{\varepsilon \to 0} \frac{1}{\varepsilon}[Y|_{\sigma_\varepsilon(x)} - (\sigma_\varepsilon)_* Y|_x].$$

Let (U, φ) be a chart with the coordinates x and let $X = X^\mu \partial/\partial x^\mu$ and $Y = Y^\mu \partial/\partial x^\mu$ be vector fields defined on U. Then $\sigma_\varepsilon(x)$ has the coordinates $x^\mu + \varepsilon X^\mu(x)$ and

$$Y|_{\sigma_\varepsilon(x)} = Y^\mu(x^\nu + \varepsilon X^\nu(x)) e_\mu|_{x+\varepsilon X}$$
$$\simeq [Y^\mu(x) + \varepsilon X^\nu(x)\partial_\nu Y^\mu(x)] e_\mu|_{x+\varepsilon X}$$

where $\{e_\mu\} = \{\partial/\partial x^\mu\}$ is the coordinate basis and $\partial_\nu \equiv \partial/\partial x^\nu$. If we map this vector defined at $\sigma_\varepsilon(x)$ to x by $(\sigma_{-\varepsilon})_*$, we obtain

$$[Y^\mu(x) + \varepsilon X^\lambda(x)\partial_\lambda Y^\mu(x)]\partial_\mu[x^\nu - \varepsilon X^\nu(x)] e_\nu|_x$$
$$= [Y^\mu(x) + \varepsilon X^\lambda(x)\partial_\lambda Y^\mu(x)][\delta_\mu^\nu - \varepsilon \partial_\mu X^\nu(x)] e_\nu|_x$$
$$= Y^\mu(x) e_\mu|_x + \varepsilon[X^\mu(x)\partial_\mu Y^\nu(x) - Y^\mu(x)\partial_\mu X^\nu(x)] e_\nu|_x + O(\varepsilon^2).$$

$$(5.48)$$

From (5.47) and (5.48), we find that

$$\mathcal{L}_X Y = (X^\mu \partial_\mu Y^\nu - Y^\mu \partial_\mu X^\nu) e_\nu. \qquad (5.49a)$$

Exercise 5.9. Let $X = X^\mu \partial/\partial x^\mu$ and $Y = Y^\mu \partial/\partial x^\mu$ be vector fields in M. Define the **Lie bracket** $[X, Y]$ by

$$[X, Y]f = X[Y[f]] - Y[X[f]] \tag{5.50}$$

where $f \in \mathcal{F}(M)$. Show that $[X, Y]$ is a vector field given by

$$(X^\mu \partial_\mu Y^\nu - Y^\mu \partial_\mu X^\nu)e_\nu.$$

This exercise shows that the Lie derivative of Y along X is

$$\mathcal{L}_X Y = [X, Y]. \tag{5.49b}$$

[*Remarks:* Note that neither XY nor YX is a vector field since they are second-order derivatives. The combination $[X, Y]$ is, however, a first-order derivative and indeed a vector field.]

Exercise 5.10. Show that the Lie bracket satisfies

(a) bilinearity

$$[X, c_1 Y_1 + c_2 Y_2] = c_1[X, Y_1] + c_2[X, Y_2]$$
$$[c_1 X_1 + c_2 X_2, Y] = c_1[X_1, Y] + c_2[X_2, Y]$$

    for any constants c_1 and c_2,

(b) skew-symmetry

$$[X, Y] = -[YX]$$

(c) the Jacobi identity

$$[[X, Y], Z] + [[Z, X], Y] + [[Y, Z], X] = 0.$$

Exercise 5.11. (a) Let $X, Y \in \mathfrak{X}(M)$ and $f \in \mathcal{F}(M)$. Show that

$$\mathcal{L}_{fX} Y = f[X, Y] - Y[f]X \tag{5.51a}$$
$$\mathcal{L}_X(fY) = f[X, Y] + X[f]Y. \tag{5.51b}$$

(b) Let $X, Y \in \mathfrak{X}(M)$ and $f : M \to N$. Show that

$$f_*[X, Y] = [f_* X, f_* Y]. \tag{5.52}$$

Geometrically, the Lie bracket shows the non-commutativity of two flows. This is easily observed from the following consideration. Let $\sigma(s, x)$ and $\tau(t, x)$ be two flows generated by vector fields X and Y, as before, see figure 5.13. If we move by a small parameter distance ε along the flow σ first, then by δ along τ, we shall be at the point whose coordinates are

$$\tau^\mu(\delta, \sigma(\varepsilon, x)) \simeq \tau^\mu(\delta, x^\nu + \varepsilon X^\nu(x))$$
$$\simeq x^\mu + \varepsilon X^\mu(x) + \delta Y^\mu(x^\nu + \varepsilon X^\nu(x))$$
$$\simeq x^\mu + \varepsilon X^\mu(x) + \delta Y^\mu(x) + \varepsilon\delta X^\nu(x)\partial_\nu Y^\mu(x).$$

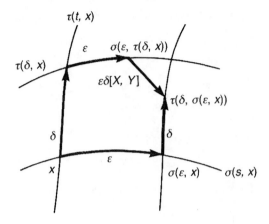

Figure 5.13. A Lie bracket $[X, Y]$ measures the failure of the closure of the parallelogram.

If, however, we move by δ along τ first, then by ε along σ, we will be at the point

$$
\begin{aligned}
\sigma^\mu(\varepsilon, \tau(\delta, x)) &\simeq \sigma^\mu(\varepsilon, x^\nu + \delta Y^\nu(x)) \\
&\simeq x^\mu + \delta Y^\mu(x) + \varepsilon X^\mu(x^\nu + \delta Y^\nu(x)) \\
&\simeq x^\mu + \delta Y^\mu(x) + \varepsilon X^\mu(x) + \varepsilon \delta Y^\nu(x)\partial_\nu X^\mu(x).
\end{aligned}
$$

The difference between the coordinates of these two points is proportional to the Lie bracket,

$$
\tau^\mu(\delta, \sigma(\varepsilon, x)) - \sigma^\mu(\varepsilon, \tau(\delta, x)) = \varepsilon\delta[X, Y]^\mu.
$$

The Lie bracket of X and Y measures the failure of the closure of the parallelogram in figure 5.13. It is easy to see $\mathcal{L}_X Y = [X, Y] = 0$ if and only if

$$
\sigma(s, \tau(t, x)) = \tau(t, \sigma(s, x)). \tag{5.53}
$$

We may also define the Lie derivative of a one-form $\omega \in \Omega^1(M)$ along $X \in \mathfrak{X}(M)$ by

$$
\mathcal{L}_X\omega \equiv \lim_{\varepsilon \to 0} \frac{1}{\varepsilon}[(\sigma_\varepsilon)^*\omega|_{\sigma_\varepsilon(x)} - \omega|_x] \tag{5.54}
$$

where $\omega|_x \in T_x^*M$ is ω at x. Put $\omega = \omega_\mu dx^\mu$. Repeating a similar analysis as before, we obtain

$$
(\sigma_\varepsilon)^*\omega|_{\sigma_\varepsilon(x)} = \omega_\mu(x)\,dx^\mu + \varepsilon[X^\nu(x)\partial_\nu\omega_\mu(x) + \partial_\mu X^\nu(x)\omega_\nu(x)]\,dx^\mu
$$

which leads to

$$
\mathcal{L}_X\omega = (X^\nu\partial_\nu\omega_\mu + \partial_\mu X^\nu\omega_\nu)\,dx^\mu. \tag{5.55}
$$

Clearly $\mathcal{L}_X\omega \in T_x^*(M)$, since it is a difference of two one-forms at the same point x.

The Lie derivative of $f \in \mathcal{F}(M)$ along a flow σ_s generated by a vector field X is

$$\mathcal{L}_X f \equiv \lim_{\varepsilon \to 0} \frac{1}{\varepsilon}[f(\sigma_\varepsilon(x)) - f(x)]$$

$$= \lim_{\varepsilon \to 0} \frac{1}{\varepsilon}[f(x^\mu + \varepsilon X^\mu(x)) - f(x^\mu)]$$

$$= X^\mu(x)\frac{\partial f}{\partial x^\mu} = X[f] \tag{5.56}$$

which is the usual directional derivative of f along X.

The Lie derivative of a general tensor is obtained from the following proposition.

Proposition 5.1. The Lie derivative satisfies

$$\mathcal{L}_X(t_1 + t_2) = \mathcal{L}_X t_1 + \mathcal{L}_X t_2 \tag{5.57a}$$

where t_1 and t_2 are tensor fields of the same type and

$$\mathcal{L}_X(t_1 \otimes t_2) = (\mathcal{L}_X t_1) \otimes t_2 + t_1 \otimes (\mathcal{L}_X t_2) \tag{5.57b}$$

where t_1 and t_2 are tensor fields of arbitrary types.

Proof. (a) is obvious. Rather than giving the general proof of (b), which is full of indices, we give an example whose extension to more general cases is trivial. Take $Y \in \mathfrak{X}(M)$ and $\omega \in \Omega^1(M)$ and construct the tensor product $Y \otimes \omega$. Then $(Y \otimes \omega)|_{\sigma_\varepsilon(x)}$ is mapped onto a tensor at x by the action of $(\sigma_{-\varepsilon})_* \otimes (\sigma_\varepsilon)^*$:

$$[(\sigma_{-\varepsilon})_* \otimes (\sigma_\varepsilon)^*](Y \otimes \omega)|_{\sigma_\varepsilon(x)} = [(\sigma_{-\varepsilon})_* Y \otimes (\sigma_\varepsilon)^* \omega]|_x.$$

Then there follows (the Leibniz rule):

$$\mathcal{L}_X(Y \otimes \omega) = \lim_{\varepsilon \to 0} \frac{1}{\varepsilon}[\{(\sigma_{-\varepsilon})_* Y \otimes (\sigma_\varepsilon)^* \omega\}|_x - (Y \otimes \omega)|_x]$$

$$= \lim_{\varepsilon \to 0} \frac{1}{\varepsilon}[(\sigma_{-\varepsilon})_* Y \otimes \{(\sigma_\varepsilon)^* \omega - \omega\} + \{(\sigma_{-\varepsilon})_* Y - Y\} \otimes \omega]$$

$$= Y \otimes (\mathcal{L}_X \omega) + (\mathcal{L}_X Y) \otimes \omega.$$

Extensions to more general cases are obvious.   □

This proposition enables us to calculate the Lie derivative of a general tensor field. For example, let $t = t_\mu{}^\nu \, dx^\mu \otimes e_\nu \in T^1_1(M)$. Proposition 5.1 gives

$$\mathcal{L}_X t = X[t_\mu{}^\nu] \, dx^\mu \otimes e_\nu + t_\mu{}^\nu (\mathcal{L}_X dx^\mu) \otimes e_\nu + t_\mu{}^\nu \, dx^\mu \otimes (\mathcal{L}_X e_\nu).$$

Exercise 5.12. Let t be a tensor field. Show that

$$\mathcal{L}_{[X,Y]} t = \mathcal{L}_X \mathcal{L}_Y t - \mathcal{L}_Y \mathcal{L}_X t. \tag{5.58}$$

5.4 Differential forms

Before we define differential forms, we examine the symmetry property of tensors. The symmetry operation on a tensor $\omega \in \mathcal{T}^0_{r,p}(M)$ is defined by

$$P\omega(V_1, \ldots, V_r) \equiv \omega(V_{P(1)}, \ldots, V_{P(r)}) \qquad (5.59)$$

where $V_i \in T_pM$ and P is an element of S_r, the **symmetric group** of order r. Take the coordinate basis $\{e_\mu\} = \{\partial/\partial x^\mu\}$. The component of ω in this basis is

$$\omega(e_{\mu_1}, e_{\mu_2}, \ldots, e_{\mu_r}) = \omega_{\mu_1 \mu_2 \ldots \mu_r}.$$

The component of $P\omega$ is obtained from (5.59) as

$$P\omega(e_{\mu_1}, e_{\mu_2}, \ldots, e_{\mu_r}) = \omega_{\mu_{P(1)} \mu_{P(2)} \ldots \mu_{P(r)}}.$$

For a general tensor of type (q, r), the symmetry operations are defined for q indices and r indices separately.

For $\omega \in \mathcal{T}^0_{r,p}(M)$, the **symmetrizer** $\mathcal{S}$ is defined by

$$\mathcal{S}\omega = \frac{1}{r!} \sum_{P \in S_r} P\omega \qquad (5.60)$$

while the **anti-symmetrizer** $\mathcal{A}$ is

$$\mathcal{A}\omega = \frac{1}{r!} \sum_{P \in S_r} \text{sgn}(P) P\omega \qquad (5.61)$$

where $\text{sgn}(P) = +1$ for even permutations and -1 for odd permutations. $\mathcal{S}\omega$ is *totally symmetric* (that is, $P\mathcal{S}\omega = \mathcal{S}\omega$ for any $P \in S_r$) and $\mathcal{A}\omega$ is *totally antisymmetric* ($P\mathcal{A}\omega = \text{sgn}(P)\mathcal{A}\omega$).

5.4.1 Definitions

Definition 5.4. A **differential form** of order r or an **r-form** is a totally antisymmetric tensor of type $(0, r)$.

Let us define the **wedge product** $\wedge$ of r one-forms by the totally antisymmetric tensor product

$$dx^{\mu_1} \wedge dx^{\mu_2} \wedge \ldots \wedge dx^{\mu_r} = \sum_{P \in S_r} \text{sgn}(P)\, dx^{\mu_{P(1)}} \wedge dx^{\mu_{P(2)}} \wedge \ldots \wedge dx^{\mu_{P(r)}}. \quad (5.62)$$

For example,

$$dx^\mu \wedge dx^\nu = dx^\mu \otimes dx^\nu - dx^\nu \otimes dx^\mu$$
$$dx^\lambda \wedge dx^\mu \wedge dx^\nu = dx^\lambda \otimes dx^\mu \otimes dx^\nu + dx^\nu \otimes dx^\lambda \otimes dx^\mu$$
$$+ dx^\mu \otimes dx^\nu \otimes dx^\lambda - dx^\lambda \otimes dx^\nu \otimes dx^\mu$$
$$- dx^\nu \otimes dx^\mu \otimes dx^\lambda - dx^\mu \otimes dx^\lambda \otimes dx^\nu.$$

It is readily verified that the wedge product satisfies the following.

(i) $dx^{\mu_1} \wedge \ldots \wedge dx^{\mu_r} = 0$ if some index μ appears at least twice.
(ii) $dx^{\mu_1} \wedge \ldots \wedge dx^{\mu_r} = \text{sgn}(P) dx^{\mu_{P(1)}} \wedge \ldots \wedge dx^{\mu_{P(r)}}$.
(iii) $dx^{\mu_1} \wedge \ldots \wedge dx^{\mu_r}$ is linear in each dx^{μ}.

If we denote the vector space of r-forms at $p \in M$ by $\Omega_p^r(M)$, the set of r-forms (5.62) forms a basis of $\Omega_p^r(M)$ and an element $\omega \in \Omega_p^r(M)$ is expanded as

$$\omega = \frac{1}{r!} \omega_{\mu_1\mu_2\ldots\mu_r} dx^{\mu_1} \wedge dx^{\mu_2} \wedge \ldots \wedge dx^{\mu_r} \tag{5.63}$$

where $\omega_{\mu_1\mu_2\ldots\mu_r}$ are taken *totally anti-symmetric*, reflecting the anti-symmetry of the basis. For example, the components of any second-rank tensor $\omega_{\mu\nu}$ are decomposed into the symmetric part $\sigma_{\mu\nu}$ and the anti-symmetric part $\alpha_{\mu\nu}$:

$$\sigma_{\mu\nu} = \omega_{(\mu\nu)} \equiv \tfrac{1}{2}(\omega_{\mu\nu} + \omega_{\nu\mu}) \tag{5.64a}$$

$$\alpha_{\mu\nu} = \omega_{[\mu\nu]} \equiv \tfrac{1}{2}(\omega_{\mu\nu} - \omega_{\nu\mu}). \tag{5.64b}$$

Observe that $\sigma_{\mu\nu} dx^{\mu} \wedge dx^{\nu} = 0$, while $\alpha_{\mu\nu} dx^{\mu} \wedge dx^{\nu} = \omega_{\mu\nu} dx^{\mu} \wedge dx^{\nu}$.

Since there are $\binom{m}{r}$ choices of the set $(\mu_1, \mu_2, \ldots, \mu_r)$ out of $(1, 2, \ldots, m)$ in (5.62), the dimension of the vector space $\Omega_p^r(M)$ is

$$\binom{m}{r} = \frac{m!}{(m-r)!r!}.$$

For later convenience we define $\Omega_p^0(M) = \mathbb{R}$. Clearly $\Omega_p^1(M) = T_p^*M$. If r in (5.62) exceeds m, it vanishes identically since some index appears at least twice in the anti-symmetrized summation. The equality $\binom{m}{r} = \binom{m}{m-r}$ implies $\dim \Omega_p^r(M) = \dim \Omega_p^{m-r}(M)$. Since $\Omega_p^r(M)$ is a vector space, $\Omega_p^r(M)$ is isomorphic to $\Omega_p^{m-r}(M)$ (see section 2.2).

Define the **exterior product** of a q-form and an r-form $\wedge : \Omega_p^q(M) \times \Omega_p^r(M) \to \Omega_p^{q+r}(M)$ by a trivial extension. Let $\omega \in \Omega_p^q(M)$ and $\xi \in \Omega_p^r(M)$, for example. The action of the $(q+r)$-form $\omega \wedge \xi$ on $q+r$ vectors is defined by

$$(\omega \wedge \xi)(V_1, \ldots, V_{q+r})$$

$$= \frac{1}{q!r!} \sum_{P \in S_{q+r}} \text{sgn}(P)\omega(V_{P(1)}, \ldots, V_{P(q)})\xi(V_{P(q+1)}, \ldots, V_{P(q+r)}) \tag{5.65}$$

where $V_i \in T_pM$. If $q + r > m$, $\omega \wedge \xi$ vanishes identically. With this product, we define an algebra

$$\Omega_p^*(M) \equiv \Omega_p^0(M) \oplus \Omega_p^1(M) \oplus \ldots \oplus \Omega_p^m(M). \tag{5.66}$$

Table 5.1.

r-forms	Basis	Dimension
$\Omega^0(M) = \mathcal{F}(M)$	$\{1\}$	1
$\Omega^1(M) = T^*M$	$\{dx^\mu\}$	m
$\Omega^2(M)$	$\{dx^{\mu_1} \wedge dx^{\mu_2}\}$	$m(m-1)/2$
$\Omega^3(M)$	$\{dx^{\mu_1} \wedge dx^{\mu_2} \wedge dx^{\mu_3}\}$	$m(m-1)(m-2)/6$
$\vdots$	$\vdots$	$\vdots$
$\Omega^m(M)$	$\{dx^1 \wedge dx^2 \wedge \ldots dx^m\}$	1

$\Omega_p^*(M)$ is the space of all differential forms at p and is closed under the exterior product.

Exercise 5.13. Take the Cartesian coordinates (x, y) in $\mathbb{R}^2$. The two-form $dx \wedge dy$ is the oriented area element (the vector product in elementary vector algebra). Show that, in polar coordinates, this becomes $r dr \wedge d\theta$.

Exercise 5.14. Let $\xi \in \Omega_p^q(M)$, $\eta \in \Omega_p^r(M)$ and $\omega \in \Omega_p^s(M)$. Show that

$$\xi \wedge \xi = 0 \qquad \text{if } q \text{ is odd} \tag{5.67a}$$

$$\xi \wedge \eta = (-1)^{qr} \eta \wedge \xi \tag{5.67b}$$

$$(\xi \wedge \eta) \wedge \omega = \xi \wedge (\eta \wedge \omega). \tag{5.67c}$$

We may assign an r-form smoothly at each point on a manifold M. We denote the space of smooth r-forms on M by $\Omega^r(M)$. We also define $\Omega^0(M)$ to be the algebra of smooth functions, $\mathcal{F}(M)$. In summary we have table 5.1.

5.4.2 Exterior derivatives

Definition 5.5. The exterior derivative d_r is a map $\Omega^r(M) \to \Omega^{r+1}(M)$ whose action on an r-form

$$\omega = \frac{1}{r!} \omega_{\mu_1 \ldots \mu_r} \, dx^{\mu_1} \wedge \ldots \wedge dx^{\mu_r}$$

is defined by

$$d_r \omega = \frac{1}{r!} \left(\frac{\partial}{\partial x^\nu} \omega_{\mu_1 \ldots \mu_r} \right) dx^\nu \wedge dx^{\mu_1} \wedge \ldots \wedge dx^{\mu_r}. \tag{5.68}$$

It is common to drop the subscript r and write simply d. The wedge product automatically anti-symmetrizes the coefficient.

Example 5.10. The *r*-forms in three-dimensional space are:

(i) $\omega_0 = f(x, y, z)$,

(ii) $\omega_1 = \omega_x(x, y, z)\,dx + \omega_y(x, y, z)\,dy + \omega_z(x, y, z)\,dz$,

(iii) $\omega_2 = \omega_{xy}(x, y, z)\,dx \wedge dy + \omega_{yz}(x, y, z)\,dy \wedge dz + \omega_{zx}(x, y, z)\,dz \wedge dx$

and

(iv) $\omega_3 = \omega_{xyz}(x, y, z)\,dx \wedge dy \wedge dz$.

If we define an *axial vector* α^μ by $\varepsilon^{\mu\nu\lambda}\omega_{\nu\lambda}$, a two-form may be regarded as a 'vector'. The **Levi-Civita symbol** $\varepsilon^{\mu\nu\lambda}$ is defined by $\varepsilon^{P(1)P(2)P(3)} = \text{sgn}(P)$ and provides the isomorphism between $\mathfrak{X}(M)$ and $\Omega^2(M)$. [Note that both of these are of dimension three.]

The action of d is

(i) $d\omega_0 = \dfrac{\partial f}{\partial x}\,dx + \dfrac{\partial f}{\partial y}\,dy + \dfrac{\partial f}{\partial z}\,dz$,

(ii) $d\omega_1 = \left(\dfrac{\partial \omega_y}{\partial x} - \dfrac{\partial \omega_x}{\partial y}\right) dx \wedge dy + \left(\dfrac{\partial \omega_z}{\partial y} - \dfrac{\partial \omega_y}{\partial z}\right) dy \wedge dz$
$+ \left(\dfrac{\partial \omega_x}{\partial z} - \dfrac{\partial \omega_z}{\partial x}\right) dz \wedge dx$,

(iii) $d\omega_2 = \left(\dfrac{\partial \omega_{yz}}{\partial x} + \dfrac{\partial \omega_{zx}}{\partial y} + \dfrac{\partial \omega_{xy}}{\partial z}\right) dx \wedge dy \wedge dz$ and

(iv) $d\omega_3 = 0$.

Hence, the action of d on ω_0 is identified with 'grad', on ω_1 with 'rot' and on ω_2 with 'div' in the usual vector calculus.

Exercise 5.15. Let $\xi \in \Omega^q(M)$ and $\omega \in \Omega^r(M)$. Show that

$$d(\xi \wedge \omega) = d\xi \wedge \omega + (-1)^q \xi \wedge d\omega. \tag{5.69}$$

A useful expression for the exterior derivative is obtained as follows. Let us take $X = X^\mu \partial/\partial x^\mu$, $Y = Y^\nu \partial/\partial x^\nu \in \mathfrak{X}(M)$ and $\omega = \omega_\mu\,dx^\mu \in \Omega^1(M)$. It is easy to see that the combination

$$X[\omega(Y)] - Y[\omega(X)] - \omega([X, Y]) = \frac{\partial \omega_\mu}{\partial x^\nu}(X^\nu Y^\mu - X^\mu Y^\nu)$$

is equal to $d\omega(X, Y)$, and we have the coordinate-free expression

$$d\omega(X, Y) = X[\omega(Y)] - Y[\omega(X)] - \omega([X, Y]). \tag{5.70}$$

For an *r*-form $\omega \in \Omega^r(M)$, this becomes

$$d\omega(X_1, \ldots, X_{r+1})$$
$$= \sum_{i=1}^{r} (-1)^{i+1} X_i \omega(X_1, \ldots, \hat{X}_i, \ldots, X_{r+1})$$
$$+ \sum_{i<j} (-1)^{i+j} \omega([X_i, X_j], X_1, \ldots, \hat{X}_i, \ldots, \hat{X}_j, \ldots, X_{r+1}) \tag{5.71}$$

where the entry below $\hat{\ }$ has been omitted. As an exercise, the reader should verify (5.71) explicitly for $r = 2$.

We now prove an important formula:

$$d^2 = 0 \qquad (\text{or } d_{r+1} d_r = 0). \tag{5.72}$$

Take

$$\omega = \frac{1}{r!}\omega_{\mu_1\ldots\mu_r} \, dx^{\mu_1} \wedge \ldots \wedge dx^{\mu_r} \in \Omega^r(M).$$

The action of d^2 on ω is

$$d^2\omega = \frac{1}{r!} \frac{\partial^2 \omega_{\mu_1\ldots\mu_r}}{\partial x^\lambda \partial x^\nu} \, dx^\lambda \wedge dx^\nu \wedge dx^{\mu_1} \wedge \ldots \wedge dx^{\mu_r}.$$

This vanishes identically since $\partial^2 \omega_{\mu_1\ldots\mu_r}/\partial x^\lambda \partial x^\nu$ is symmetric with respect to λ and ν while $dx^\lambda \wedge dx^\nu$ is anti-symmetric.

Example 5.11. It is known that the electromagnetic potential $A = (\phi, A)$ is a one-form, $A = A_\mu dx^\mu$ (see chapter 10). The electromagnetic tensor is defined by $F = dA$ and has the components

$$\begin{pmatrix} 0 & -E_x & -E_y & -E_x \\ E_x & 0 & B_z & -B_y \\ E_y & -B_z & 0 & B_x \\ E_z & B_y & -B_x & 0 \end{pmatrix} \tag{5.73}$$

where

$$\mathbf{E} = -\nabla\phi - \frac{\partial}{\partial x^0}\mathbf{A} \qquad \text{and} \qquad \mathbf{B} = \nabla \times \mathbf{A}$$

as usual. Two Maxwell equations, $\nabla \cdot \mathbf{B} = 0$ and $\partial\mathbf{B}/\partial t = -\nabla \times \mathbf{E}$ follow from the identity $dF = d(dA) = 0$, which is known as the **Bianchi identity**, while the other set is the equation of motion derived from the Lagrangian (1.245).

A map $f : M \to N$ induces the pullback $f^* : T^*_{f(p)}N \to T^*_p M$ and f^* is naturally extended to tensors of type $(0, r)$; see section 5.2. Since an r-form is a tensor of type $(0, r)$, this applies as well. Let $\omega \in \Omega^r(N)$ and let f be a map $M \to N$. At each point $f(p) \in N$, f induces the pullback $f^* : \Omega^r_{f(p)}N \to \Omega^r_p M$ by

$$(f^*\omega)(X_1, \ldots, X_r) \equiv \omega(f_*X_1, \ldots, f_*X_r) \tag{5.74}$$

where $X_i \in T_p M$ and f_* is the differential map $T_p M \to T_{f(p)}N$.

Exercise 5.16. Let $\xi, \omega \in \Omega^r(N)$ and let $f : M \to N$. Show that

$$d(f^*\omega) = f^*(d\omega) \tag{5.75}$$

$$f^*(\xi \wedge \omega) = (f^*\xi) \wedge (f^*\omega). \tag{5.76}$$

The exterior derivative d_r induces the sequence

$$0 \xrightarrow{i} \Omega^0(M) \xrightarrow{d_0} \Omega^1(M) \xrightarrow{d_1} \cdots \xrightarrow{d_{m-2}} \Omega^{m-1}(M) \xrightarrow{d_{m-1}} \Omega^m(M) \xrightarrow{d_m} 0 \quad (5.77)$$

where i is the inclusion map $0 \hookrightarrow \Omega^0(M)$. This sequence is called the **de Rham complex**. Since $d^2 = 0$, we have im $d_r \subset \ker d_{r+1}$. [Take $\omega \in \Omega^r(M)$. Then $d_r \omega \in$ im d_r and $d_{r+1}(d_r \omega) = 0$ imply $d_r \omega \in \ker d_{r+1}$.] An element of $\ker d_r$ is called a **closed r-form**, while an element of im d_{r-1} is called an **exact r-form**. Namely, $\omega \in \Omega^r(M)$ is closed if $d\omega = 0$ and exact if there exists an $(r-1)$-form ψ such that $\omega = d\psi$. The quotient space $\ker d_r / \text{im } d_{r-1}$ is called the rth **de Rham cohomology group** which is made into the dual space of the homology group; see chapter 6.

5.4.3 Interior product and Lie derivative of forms

Another important operation is the **interior product** $i_X : \Omega^r(M) \to \Omega^{r-1}(M)$, where $X \in \mathfrak{X}(M)$. For $\omega \in \Omega^r(M)$, we define

$$i_X \omega(X_1, \ldots, X_{r-1}) \equiv \omega(X, X_1, \ldots, X_{r-1}). \quad (5.78)$$

For $X = X^\mu \partial/\partial x^\mu$ and $\omega = (1/r!)\omega_{\mu_1 \ldots \mu_r} dx^{\mu_1} \wedge \ldots \wedge dx^{\mu_r}$ we have

$$i_X \omega = \frac{1}{(r-1)!} X^\nu \omega_{\nu \mu_2 \ldots \mu_r} dx^{\mu_2} \wedge \ldots \wedge dx^{\mu_r}$$

$$= \frac{1}{r!} \sum_{s=1}^r X^{\mu_s} \omega_{\mu_1 \ldots \mu_s \ldots \mu_r} (-1)^{s-1} dx^{\mu_1} \wedge \ldots \wedge \widehat{dx^{\mu_s}} \wedge \ldots \wedge dx^{\mu_r}$$

$$(5.79)$$

where the entry below $\hat{}$ has been omitted. For example, let (x, y, z) be the coordinates of $\mathbb{R}^3$. Then

$$i_{e_x}(dx \wedge dy) = dy, \qquad i_{e_x}(dy \wedge dz) = 0, \qquad i_{e_x}(dz \wedge dx) = -dz.$$

The Lie derivative of a form is most neatly written with the interior product. Let $\omega = \omega_\mu dx^\mu$ be a one-form. Consider the combination

$$\begin{aligned}
(di_X + i_X d)\omega &= d(X^\mu \omega_\mu) + i_X[\tfrac{1}{2}(\partial_\mu \omega_\nu - \partial_\nu \omega_\mu) dx^\mu \wedge dx^\nu] \\
&= (\omega_\mu \partial_\nu X^\mu + X^\mu \partial_\nu \omega_\mu) dx^\nu + X^\mu (\partial_\mu \omega_\nu - \partial_\nu \omega_\mu) dx^\nu \\
&= (\omega_\mu \partial_\nu X^\mu + X^\mu \partial_\mu \omega_\nu) dx^\nu.
\end{aligned}$$

Comparing this with (5.55), we find that

$$\mathcal{L}_X \omega = (di_X + i_X d)\omega. \quad (5.80)$$

For a general r-form $\omega = (1/r!)\omega_{\mu_1...\mu_r}\,dx^{\mu_1} \wedge ... \wedge dx^{\mu_r}$, we have

$$\mathcal{L}_X\omega = \lim_{\varepsilon \to 0} \frac{1}{\varepsilon}((\sigma_\varepsilon)^*\omega|_{\sigma_\varepsilon(x)} - \omega|_x)$$

$$= X^\nu \frac{1}{r!}\partial_\nu\omega_{\mu_1...\mu_r}\,dx^{\mu_1} \wedge ... \wedge dx^{\mu_r}$$

$$+ \sum_{s=1}^{r} \partial_{\mu_s} X^\nu \frac{1}{r!}\omega_{\mu_1...\overset{s}{\underset{\downarrow}{\nu}}...\mu_r}\,dx^{\mu_1} \wedge ... \wedge dx^{\mu_r}. \qquad (5.81)$$

We also have

$(d i_X + i_X d)\omega$

$$= \frac{1}{r!}\sum_{s=1}^{r}[\partial_\nu X^{\mu_s}\omega_{\mu_1...\mu_s...\mu_r} + X^{\mu_s}\partial_\nu\omega_{\mu_1...\mu_s...\mu_r}]$$

$$\times (-1)^{s-1}\,dx^\nu \wedge dx^{\mu_1} \wedge ... \wedge \widehat{dx^{\mu_s}} \wedge dx^{\mu_r}$$

$$+ \frac{1}{r!}[X^\nu\partial_\nu\omega_{\mu_1...\mu_r}\,dx^{\mu_1} \wedge ... \wedge dx^{\mu_r}$$

$$+ \sum_{s=1}^{r} X^{\mu_s}\omega_{\mu_1...\mu_s...\mu_r}(-1)^s\,dx^\nu \wedge dx^{\mu_1} \wedge ... \wedge \widehat{dx^{\mu_s}} \wedge ... \wedge dx^{\mu_r}]$$

$$= \frac{1}{r!}\sum_{s=1}^{r}[\partial_\nu X^{\mu_s}\omega_{\mu_1...\mu_s...\mu_r}(-1)^{s-1}\,dx^\nu \wedge dx^{\mu_1} \wedge ... \wedge \widehat{dx^{\mu_s}} \wedge ... \wedge dx^{\mu_r}$$

$$+ \frac{1}{r!}X^\nu\partial_\nu\omega_{\mu_1...\mu_r}\,dx^{\mu_1} \wedge ... \wedge dx^{\mu_r}.$$

If we interchange the roles of μ_s and ν in the first term of the last expression and compare it with (5.81), we verify that

$$(d i_X + i_X d)\omega = \mathcal{L}_X\omega \qquad (5.82)$$

for any r-form ω.

Exercise 5.17. Let $X, Y \in \mathfrak{X}(M)$ and $\omega \in \Omega^r(M)$. Show that

$$i_{[X,Y]}\omega = X(i_Y\omega) - Y(i_X\omega). \qquad (5.83)$$

Show also that i_X is an anti-derivation,

$$i_X(\omega \wedge \eta) = i_X\omega \wedge \eta + (-1)^r\omega \wedge i_X\eta \qquad (5.84)$$

and nilpotent,

$$i_X^2 = 0. \qquad (5.85)$$

Use the nilpotency to prove

$$\mathcal{L}_X i_X\omega = i_X\mathcal{L}_X\omega. \qquad (5.86)$$

Exercise 5.18. Let $t \in \mathcal{T}^n_m(M)$. Show that

$$(\mathcal{L}_X t)^{\mu_1 \ldots \mu_n}_{\nu_1 \ldots \nu_m} = X^\lambda \partial_\lambda t^{\mu_1 \ldots \mu_n}_{\nu_1 \ldots \nu_m} + \sum_{s=1}^{n} \partial_{\nu_s} X^\lambda t^{\mu_1 \ldots \mu_n}_{\nu_1 \ldots \lambda \ldots \nu_m} - \sum_{s=1}^{n} \partial_\lambda X^{\mu_s} t^{\mu_1 \ldots \lambda \ldots \mu_n}_{\nu_1 \ldots \nu_m}. \quad (5.87)$$

Example 5.12. Let us reformulate Hamiltonian mechanics (section 1.1) in terms of differential forms. Let H be a Hamiltonian and (q^μ, p_μ) be its phase space. Define a two-form

$$\omega = \mathrm{d}p_\mu \wedge \mathrm{d}q^\mu \qquad (5.88)$$

called the **symplectic two-form**. If we introduce a one-form

$$\theta = q^\mu \, \mathrm{d}p_\mu, \qquad (5.89)$$

the symplectic two-form is expressed as

$$\omega = \mathrm{d}\theta. \qquad (5.90)$$

Given a function $f(q, p)$ in the phase space, one can define the **Hamiltonian vector field**

$$X_f = \frac{\partial f}{\partial p_\mu} \frac{\partial}{\partial q^\mu} - \frac{\partial f}{\partial q^\mu} \frac{\partial}{\partial p_\mu}. \qquad (5.91)$$

Then it is easy to verify that

$$i_{X_f} \omega = -\frac{\partial f}{\partial p_\mu} \, \mathrm{d}p^\mu - \frac{\partial f}{\partial q^\mu} \, \mathrm{d}q^\mu = -\mathrm{d}f.$$

Consider a vector field generated by the Hamiltonian

$$X_H = \frac{\partial H}{\partial p_\mu} \frac{\partial}{\partial q^\mu} - \frac{\partial H}{\partial q^\mu} \frac{\partial}{\partial p_\mu}. \qquad (5.92)$$

For the solution (q^μ, p_μ) to Hamilton's equation of motion

$$\frac{\mathrm{d}q^\mu}{\mathrm{d}t} = \frac{\partial H}{\partial p_\mu} \qquad \frac{\mathrm{d}p_\mu}{\mathrm{d}t} = -\frac{\partial H}{\partial q^\mu}, \qquad (5.93)$$

we also obtain

$$X_H = \frac{\mathrm{d}p_\mu}{\mathrm{d}t} \frac{\partial}{\partial p_\mu} \frac{\mathrm{d}q^\mu}{\mathrm{d}t} \frac{\partial}{\partial q^\mu} = \frac{\mathrm{d}}{\mathrm{d}t}. \qquad (5.94)$$

The symplectic two-form ω is left invariant along the flow generated by X_H,

$$\mathcal{L}_{X_H} \omega = \mathrm{d}(i_{X_H} \omega) + i_{X_H}(\mathrm{d}\omega)$$
$$= \mathrm{d}(i_{X_H} \omega) = -\mathrm{d}^2 H = 0 \qquad (5.95)$$

where use has been made of (5.82). Conversely, if X satisifes $\mathcal{L}_X \omega = 0$, there exists a Hamiltonian H such that Hamilton's equation of motion is satisfied along the flow generated by X. This follows from the previous observation that $\mathcal{L}_X \omega = \mathrm{d}(\mathrm{i}_X \omega) = 0$ and hence by Poincaré's lemma, there exists a function $H(q, p)$ such that

$$\mathrm{i}_X \omega = -\mathrm{d}H.$$

The Poisson bracket is cast into a form independent of the special coordinates chosen with the help of the Hamiltonian vector fields. In fact,

$$\mathrm{i}_{X_f}(\mathrm{i}_{X_g} \omega) = -\mathrm{i}_{X_f}(\mathrm{d}g) = \frac{\partial f}{\partial q^\mu} \frac{\partial g}{\partial p_\mu} - \frac{\partial f}{\partial q^\mu} \frac{\partial g}{\partial p_\mu} = [f, g]_{\mathrm{PB}}. \qquad (5.96)$$

5.5 Integration of differential forms

5.5.1 Orientation

An integration of a differential form over a manifold M is defined only when M is 'orientable'. So we first define an **orientation** of a manifold. Let M be a connected m-dimensional differentiable manifold. At a point $p \in M$, the tangent space $T_p M$ is spanned by the basis $\{e_\mu\} = \{\partial/\partial x^\mu\}$, where x^μ is the local coordinate on the chart U_i to which p belongs. Let U_j be another chart such that $U_i \cap U_j \neq \emptyset$ with the local coordinates y^α. If $p \in U_i \cap U_j$, $T_p M$ is spanned by either $\{e_\mu\}$ or $\{\tilde{e}_\alpha\} = \{\partial/\partial y^\alpha\}$. The basis changes as

$$\tilde{e}_\alpha = \left(\frac{\partial x^\mu}{\partial y^\alpha} \right) e_\mu. \qquad (5.97)$$

If $J = \det(\partial x^\mu / \partial y^\alpha) > 0$ on $U_i \cap U_j$, $\{e_\mu\}$ and $\{\tilde{e}_\alpha\}$ are said to define the *same orientation* on $U_i \cap U_j$ and if $J < 0$, they define the *opposite orientation*.

Definition 5.6. Let M be a connected manifold covered by $\{U_i\}$. The manifold M is **orientable** if, for any overlapping charts U_i and U_j, there exist local coordinates $\{x^\mu\}$ for U_i and $\{y^\alpha\}$ for U_j such that $J = \det(\partial x^\mu / \partial y^\alpha) > 0$.

If M is non-orientable, J cannot be positive in all intersections of charts. For example, the Möbius strip in figure 5.14(a) is non-orientable since we have to choose J to be negative in the intersection B.

If an m-dimensional manifold M is orientable, there exists an m-form ω which vanishes nowhere. This m-form ω is called a **volume element**, which plays the role of a measure when we integrate a function $f \in \mathcal{F}(M)$ over M. Two volume elements ω and ω' are said to be *equivalent* if there exists a strictly positive function $h \in \mathcal{F}(M)$ such that $\omega = h\omega'$. A negative-definite function $h' \in \mathcal{F}(M)$ gives an inequivalent orientation to M. Thus, any orientable manifold admits *two* inequivalent orientations, one of which is called **right handed**, the other **left handed**. Take an m-form

$$\omega = h(p) \, \mathrm{d}x^1 \wedge \ldots \wedge \mathrm{d}x^m \qquad (5.98)$$

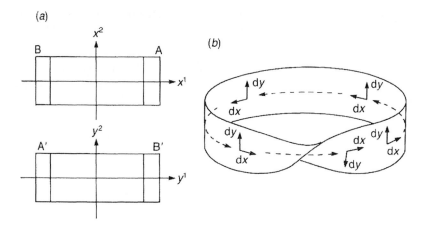

Figure 5.14. (*a*) The Möbius strip is obtained by twisting the part B′ of the second strip by π before pasting A with A′ and B with B′. The coordinate change on B is $y^1 = x^1, y^2 = -x^2$ and the Jacobian is -1. (*b*) Basis frames on the Möbius strip.

with a positive-definite $h(p)$ on a chart (U, φ) whose coordinate is $x = \varphi(p)$. If M is orientable, we may extend ω throughout M such that the component h is positive definite on any chart U_i. If M is orientable, this ω is a volume element. Note that this positivity of h is independent of the choice of coordinates. In fact, let $p \in U_i \cap U_j \neq \emptyset$ and let x^μ and y^α be the coordinates of U_i and U_j, respectively. Then (5.98) becomes

$$\omega = h(p)\frac{\partial x^1}{\partial y^{\mu_1}}\, dy^{\mu_1} \wedge \ldots \wedge \frac{\partial x^m}{\partial y^{\mu_m}}\, dy^{\mu_m} = h(p) \det\left(\frac{\partial x^\mu}{\partial y^\nu}\right) dy^1 \wedge \ldots \wedge dy^m.$$

$$(5.99)$$

The determinant in (5.99) is the Jacobian of the coordinate transformation and must be positive by assumed orientability. If M is non-orientable, ω with a positive-definite component cannot be defined on M. Let us look at figure 5.14 again. If we circumnavigate the strip along the direction shown in the figure, $\omega = dx \wedge dy$ changes the signature $dx \wedge dy \to -dx \wedge dy$ when we come back to the starting point. Hence, ω cannot be defined uniquely on M.

5.5.2 Integration of forms

Now we are ready to define an integration of a function $f : M \to \mathbb{R}$ over an orientable manifold M. Take a volume element ω. In a coordinate neighbourhood U_i with the coordinate x, we define the integration of an m-form $f\omega$ by

$$\int_{U_i} f\omega \equiv \int_{\varphi(U_i)} f(\varphi_i^{-1}(x))h(\varphi_i^{-1}(x))\, dx^1 \ldots dx^m. \qquad (5.100)$$

The RHS is an ordinary multiple integration of a function of m variables. Once the integral of f over U_i is defined, the integral of f over the whole of M is given with the help of the 'partition of unity' defined now.

Definition 5.7. Take an open covering $\{U_i\}$ of M such that each point of M is covered with a finite number of U_i. [If this is always possible, M is called **paracompact**, which we assume to be the case.] If a family of differentiable functions $\varepsilon_i(p)$ satisfies

(i) $0 \le \varepsilon_i(p) \le 1$
(ii) $\varepsilon_i(p) = 0$ if $p \notin U_i$ and
(iii) $\varepsilon_1(p) + \varepsilon_2(p) + \ldots = 1$ for any point $p \in M$

the family $\{\varepsilon(p)\}$ is called a **partition of unity** subordinate to the covering $\{U_i\}$.

From condition (iii), it follows that

$$f(p) = \sum_i f(p)\varepsilon_i(p) = \sum_i f_i(p) \qquad (5.101)$$

where $f_i(p) \equiv f(p)\varepsilon_i(p)$ vanishes outside U_i by (ii). Hence, given a point $p \in M$, assumed paracompactness ensures that there are only finite terms in the summation over i in (5.101). For each $f_i(p)$, we may define the integral over U_i according to (5.100). Finally the integral of f on M is given by

$$\int_M f\omega \equiv \sum_i \int_{U_i} f_i\omega. \qquad (5.102)$$

Although a different atlas $\{(V_i, \psi_i)\}$ gives different coordinates and a different partition of unity, the integral defined by (5.102) remains the same.

Example 5.13. Let us take the atlas of S^1 defined in example 5.2. Let $U_1 = S^1 - \{(1, 0)\}$, $U_2 = S^1 - \{(-1, 0)\}$, $\varepsilon_1(\theta) = \sin^2(\theta/2)$ and $\varepsilon_2(\theta) = \cos^2(\theta/2)$. The reader should verify that $\{\varepsilon_i(\theta)\}$ is a partition of unity subordinate to $\{U_i\}$. Let us integrate a function $f = \cos^2\theta$, for example. [Of course we know

$$\int_0^{2\pi} d\theta \, \cos^2\theta = \pi$$

but let us use the partition of unity.] We have

$$\int_{S^1} d\theta \, \cos^2\theta = \int_0^{2\pi} d\theta \, \sin^2\frac{\theta}{2}\cos^2\theta + \int_{-\pi}^{\pi} d\theta \, \cos^2\frac{\theta}{2}\cos^2\theta$$

$$= \tfrac{1}{2}\pi + \tfrac{1}{2}\pi = \pi.$$

So far, we have left h arbitrary provided it is strictly positive. The reader might be tempted to choose h to he unity. However, as we found in (5.99), h is multiplied by the Jacobian under the change of coordinates and there is no canonical way to single out the component h; unity in one coordinate might not be unity in the other. The situation changes if the manifold is endowed with a metric, as we will see in chapter 7.

5.6 Lie groups and Lie algebras

A Lie group is a manifold on which the group manipulations, *product* and *inverse*, are defined. Lie groups play an extremely important role in the theory of fibre bundles and also find vast applications in physics. Here we will work out the geometrical aspects of Lie groups and Lie algebras.

5.6.1 Lie groups

Definition 5.8. A Lie group G is a differentiable manifold which is endowed with a group structure such that the group operations

(i) $\cdot : G \times G \to G$, $(g_1, g_2) \mapsto g_1 \cdot g_2$
(ii) $^{-1} : G \to G$, $g \mapsto g^{-1}$

are differentiable. [*Remark:* It can be shown that G has a unique analytic structure with which the product and the inverse operations are written as convergent power series.]

The unit element of a Lie group is written as e. The dimension of a Lie group G is defined to be the dimension of G as a manifold. The product symbol may be omitted and $g_1 \cdot g_2$ is usually written as $g_1 g_2$. For example, let $\mathbb{R}^* \equiv \mathbb{R} - \{0\}$. Take three elements $x, y, z \in \mathbb{R}^*$ such that $xy = z$. Obviously if we multiply a number close to x by a number close to y, we have a number close to z. Similarly, an inverse of a number close to x is close to $1/x$. In fact, we can differentiate these maps with respect to the relevant arguments and $\mathbb{R}^*$ is made into a Lie group with these group operations. If the product is commutative, namely $g_1 g_2 = g_2 g_1$, we often use the additive symbol $+$ instead of the product symbol.

Exercise 5.19.

(a) Show that $\mathbb{R}^+ = \{x \in \mathbb{R} | x > 0\}$ is a Lie group with respect to multiplication.
(b) Show that $\mathbb{R}$ is a Lie group with respect to addition.
(c) Show that $\mathbb{R}^2$ is a Lie group with respect to addition defined by $(x_1, y_1) + (x_2, y_2) = (x_1 + x_2, y_1 + y_2)$.

Example 5.14. Let S^1 be the unit circle on the complex plane,

$$S^1 = \{e^{i\theta} | \theta \in \mathbb{R} \quad (\text{mod } 2\pi)\}.$$

The group operations defined by $e^{i\theta} e^{i\varphi} = e^{i(\theta+\varphi)}$ and $(e^{i\theta})^{-1} = e^{-i\theta}$ are differentiable and S^1 is made into a Lie group, which we call U(1). It is easy to see that the group operations are the same as those in exercise 5.19(*b*) modulo 2π.

Of particular interest in physical applications are the matrix groups which are subgroups of general linear groups GL($n, \mathbb{R}$) or GL($n, \mathbb{C}$). The product of elements is simply the matrix multiplication and the inverse is given by the matrix inverse. The coordinates of $GL(n, \mathbb{R})$ are given by n^2 entries of $M = \{x_{ij}\}$. $GL(n, \mathbb{R})$ is a non-compact manifold of real dimension n^2.

Interesting subgroups of $GL(n, \mathbb{R})$ are the **orthogonal group** $O(n)$, the **special linear group** $SL(n, \mathbb{R})$ and the **special orthogonal group** $SO(n)$:

$$O(n) = \{M \in GL(n, \mathbb{R}) | MM^t = M^tM = I_n\} \tag{5.103}$$

$$SL(n, \mathbb{R}) = \{M \in GL(n, \mathbb{R}) | \det M = 1\} \tag{5.104}$$

$$SO(n) = O(n) \cap SL(n, \mathbb{R}) \tag{5.105}$$

where ^{t} denotes the transpose of a matrix. In special relativity, we are familiar with the **Lorentz group**

$$O(1, 3) = \{M \in GL(4, \mathbb{R}) | M\eta M^t = \eta\}$$

where η is the Minkowski metric, $\eta = \mathrm{diag}(-1, 1, 1, 1)$. Extension to higher-dimensional spacetime is trivial.

Exercise 5.20. Show that the group $O(1, 3)$ is non-compact and has four connected components according to the sign of the determinant and the sign of the $(0, 0)$ entry. The component that contains the unit matrix is denoted by $O^\uparrow_+(1, 3)$.

The group $GL(n, \mathbb{C})$ is the set of non-singular linear transformations in $\mathbb{C}^n$, which are represented by $n \times n$ non-singular matrices with complex entries. The **unitary group** $U(n)$, the **special linear group** $SL(n, \mathbb{C})$ and the **special unitary group** $SU(n)$ are defined by

$$U(n) = \{M \in GL(n, \mathbb{C}) | MM^\dagger = M^\dagger M = \mathbf{1}\} \tag{5.106}$$

$$SL(n, \mathbb{C}) = \{M \in GL(n, \mathbb{C}) | \det M = 1\} \tag{5.107}$$

$$SU(n) = U(n) \cap SL(n, \mathbb{C}) \tag{5.108}$$

where † is the Hermitian conjugate.

So far we have just mentioned that the matrix groups are subgroups of a Lie group $GL(n, \mathbb{R})$ (or $GL(n, \mathbb{C})$). The following theorem guarantees that they are Lie subgroups, that is, these subgroups are Lie groups by themselves. We accept this important (and difficult to prove) theorem without proof.

Theorem 5.2. Every closed subgroup H of a Lie group G is a Lie subgroup.

For example, $O(n)$, $SL(n, \mathbb{R})$ and $SO(n)$ are Lie subgroups of $GL(n, \mathbb{R})$. To see why $SL(n, \mathbb{R})$ is a closed subgroup, consider a map $f : GL(n, \mathbb{R}) \to \mathbb{R}$ defined by $A \mapsto \det A$. Obviously f is a continuous map and $f^{-1}(1) = SL(n, \mathbb{R})$. A point $\{1\}$ is a closed subset of $\mathbb{R}$, hence $f^{-1}(1)$ is closed in $GL(n, \mathbb{R})$. Then theorem 5.2 states that $SL(n, \mathbb{R})$ is a Lie subgroup. The reader should verify that $O(n)$ and $SO(n)$ are also Lie subgroups of $GL(n, \mathbb{R})$.

Let G be a Lie group and H a Lie subgroup of G. Define an equivalence relation $\sim$ by $g \sim g'$ if there exists an element $h \in H$ such that $g' = gh$. An equivalence class $[g]$ is a set $\{gh|h \in H\}$. The coset space G/H is a manifold (not necessarily a Lie group) with $\dim G/H = \dim G - \dim H$. G/H is a Lie group if H is a normal subgroup of G, that is, if $ghg^{-1} \in H$ for any $g \in G$ and $h \in H$. In fact, take equivalence classes $[g]$, $[g'] \in G/H$ and construct the product $[g][g']$. If the group structure is well defined in G/H, the product must be independent of the choice of the representatives. Let gh and $g'h'$ be the representatives of $[g]$ and $[g']$ respectively. Then $ghg'h' = gg'h''h' \in [gg']$ where the equality follows since there exists $h'' \in H$ such that $hg' = g'h''$. It is left as an exercise to the reader to show that $[g]^{-1}$ is also a well defined operation and $[g]^{-1} = [g^{-1}]$.

5.6.2 Lie algebras

Definition 5.9. Let a and g be elements of a Lie group G. The **right-translation** $R_a : G \to G$ and the **left-translation** $L_a : G \to G$ of g by a are defined by

$$R_a g = ga \tag{5.109a}$$
$$L_a g = ag. \tag{5.109b}$$

By definition, R_a and L_a are diffeomorphisms from G to G. Hence, the maps $L_a : G \to G$ and $R_a : G \to G$ induce $L_{a*} : T_g G \to T_{ag}G$ and $R_{a*} : T_g G \to T_{ga}G$; see section 5.2. Since these translations give equivalent theories, we are concerned mainly with the left-translation in the following. The analysis based on the right-translation can be carried out in a similar manner.

Given a Lie group G, there exists a special class of vector fields characterized by an invariance under group action. [On the usual manifold there is no canonical way of discriminating some vector fields from the others.]

Definition 5.10. Let X be a vector field on a Lie group G. X is said to be a **left-invariant vector field** if $L_{a*}X|_g = X|_{ag}$.

Exercise 5.21. Verify that a left-invariant vector field X satisfies

$$L_{a*}X|_g = X^\mu(g)\frac{\partial x^\nu(ag)}{\partial x^\mu(g)}\frac{\partial}{\partial x^\nu}\bigg|_{ag} = X^\nu(ag)\frac{\partial}{\partial x^\nu}\bigg|_{ag} \tag{5.110}$$

where $x^\mu(g)$ and $x^\mu(ag)$ are coordinates of g and ag, respectively.

A vector $V \in T_e G$ defines a unique left-invariant vector field X_V throughout G by

$$X_V|_g = L_{g*}V \qquad g \in G. \tag{5.111}$$

In fact, we verify from (5.34) that $X_V|_{ag} = L_{ag*}V = (L_a L_g)_* V = L_{a*}L_{g*}V = L_{a*}X_V|_g$. Conversely, a left-invariant vector field X defines a unique vector $V = X|_e \in T_e G$. Let us denote the set of left-invariant vector fields on G by

g. The map $T_eG \to \mathfrak{g}$ defined by $V \mapsto X_V$ is an isomorphism and it follows that the set of left-invariant vector fields is a vector space isomorphic to T_eG. In particular, $\dim \mathfrak{g} = \dim G$.

Since $\mathfrak{g}$ is a set of vector fields, it is a subset of $\mathfrak{X}(G)$ and the Lie bracket defined in section 5.3 is also defined on $\mathfrak{g}$. We show that $\mathfrak{g}$ is closed under the Lie bracket. Take two points g and $ag = L_a g$ in G. If we apply L_{a*} to the Lie bracket $[X, Y]$ of $X, Y \in \mathfrak{g}$, we have

$$L_{a*}[X, Y]|_g = [L_{a*}X|_g, L_{a*}Y|_g] = [X, Y]|_{ag} \tag{5.112}$$

where the left-invariances of X and Y and (5.52) have been used. Thus, $[X, Y] \in \mathfrak{g}$, that is $\mathfrak{g}$ is closed under the Lie bracket.

It is instructive to work out the left-invariant vector field of $GL(n, \mathbb{R})$. The coordinates of $GL(n, \mathbb{R})$ are given by n^2 entries x^{ij} of the matrix. The unit element is $e = I_n = (\delta^{ij})$. Let $g = \{x^{ij}(g)\}$ and $a = \{x^{ij}(a)\}$ be elements of $GL(n, \mathbb{R})$. The left-translation is

$$L_a g = ag = \sum x^{ik}(a) x^{kj}(g).$$

Take a vector $V = \sum V^{ij} \partial/\partial x^{ij}|_e \in T_eG$ where the V^{ij} are the entries of V. The left-invariant vector field generated by V is

$$X_V|_g = L_{g*}V = \sum_{ijklm} V^{ij} \left.\frac{\partial}{\partial x^{ij}}\right|_e x^{kl}(g) x^{lm}(e) \left.\frac{\partial}{\partial x^{km}}\right|_g$$

$$= \sum V^{ij} x^{kl}(g) \delta_i^l \delta_j^m \left.\frac{\partial}{\partial x^{km}}\right|_g$$

$$= \sum x^{ki}(g) V^{ij} \left.\frac{\partial}{\partial x^{kj}}\right|_g = \sum (gV)^{kj} \left.\frac{\partial}{\partial x^{kj}}\right|_g \tag{5.113}$$

where gV is the usual matrix multiplication of g and V. The vector $X_V|_g$ is often abbreviated as gV since it gives the components of the vector.

The Lie bracket of X_V and X_W generated by $V = V^{ij} \partial/\partial x^{ij}|_e$ and $W = W^{ij} \partial/\partial x^{ij}|_e$ is

$$[X_V, X_W]|_g = \sum x^{ki}(g) V^{ij} \left.\frac{\partial}{\partial x^{kj}}\right|_g x^{ca}(g) W^{ab} \left.\frac{\partial}{\partial x^{cb}}\right|_g - (V \leftrightarrow W)$$

$$= \sum x^{ij}(g) [V^{jk}W^{kl} - W^{jk}V^{kl}] \left.\frac{\partial}{\partial x^{il}}\right|_g$$

$$= \sum (g[V, W])^{ij} \left.\frac{\partial}{\partial x^{ij}}\right|_g. \tag{5.114}$$

Clearly, (5.113) and (5.114) remain true for any matrix group and we establish that

$$L_{g*}V = gV \tag{5.115}$$

$$[X_V, X_W]|_g = L_{g*}[V, W] = g[V, W]. \tag{5.116}$$

Now a Lie algebra is defined as the set of left-invariant vector fields $\mathfrak{g}$ with the Lie bracket.

Definition 5.11. The set of left-invariant vector fields $\mathfrak{g}$ with the Lie bracket $[\quad , \quad] : \mathfrak{g} \times \mathfrak{g} \to \mathfrak{g}$ is called the **Lie algebra** of a Lie group G.

We denote the Lie algebra of a Lie group by the corresponding lower-case German gothic letter. For example $\mathfrak{so}(n)$ is the Lie algebra of $SO(n)$.

Example 5.15.

(a) Take $G = \mathbb{R}$ as in exercise 5.19(b). If we define the left translation L_a by $x \mapsto x + a$, the left-invariant vector field is given by $X = \partial/\partial x$. In fact,

$$L_{a*}X\Big|_x = \frac{\partial(a+x)}{\partial x}\frac{\partial}{\partial(a+x)} = \frac{\partial}{\partial(x+a)} = X\Big|_{x+a}.$$

Clearly this is the only left-invariant vector field on $\mathbb{R}$. We also find that $X = \partial/\partial\theta$ is the unique left-invariant vector field on $G = SO(2) = \{e^{i\theta}|0 \le \theta \le 2\pi\}$. Thus, the Lie groups $\mathbb{R}$ and $SO(2)$ share the common Lie algebra.
(b) Let $\mathfrak{gl}(n, \mathbb{R})$ be the Lie algebra of $GL(n, \mathbb{R})$ and $c : (-\varepsilon, \varepsilon) \to GL(n, \mathbb{R})$ be a curve with $c(0) = I_n$. The curve is approximated by $c(s) = I_n + sA + O(s^2)$ near $s = 0$, where A is an $n \times n$ matrix of real entries. Note that for small enough s, $\det c(s)$ cannot vanish and $c(s)$ is, indeed, in $GL(n, \mathbb{R})$. The tangent vector to $c(s)$ at I_n is $c'(s)\big|_{s=0} = A$. This shows that $\mathfrak{gl}(n, \mathbb{R})$ is the set of $n \times n$ matrices. Clearly $\dim \mathfrak{gl}(n, \mathbb{R}) = n^2 = \dim GL(n, \mathbb{R})$. Subgroups of $GL(n, \mathbb{R})$ are more interesting.
(c) Let us find the Lie algebra $\mathfrak{sl}(n, \mathbb{R})$ of $SL(n, \mathbb{R})$. Following this prescription, we approximate a curve through I_n by $c(s) = I_n + sA + O(s^2)$. The tangent vector to $c(s)$ at I_n is $c'(s)\big|_{s=0} = A$. Now, for the curve $c(s)$ to be in $SL(n, \mathbb{R})$, $c(s)$ has to satisfy $\det c(s) = 1 + s\mathrm{tr}A = 1$, namely $\mathrm{tr}\,A = 0$. Thus, $\mathfrak{sl}(n, \mathbb{R})$ is the set of $n \times n$ traceless matrices and $\dim \mathfrak{sl}(n, \mathbb{R}) = n^2 - 1$.
(d) Let $c(s) = I_n + sA + O(s^2)$ be a curve in $SO(n)$ through I_n. Since $c(s)$ is a curve in $SO(n)$, it satisfies $c(s)^{\mathrm{t}}c(s) = I_n$. Differentiating this identity, we obtain $c'(s)^{\mathrm{t}}c(s) + c(s)^{\mathrm{t}}c'(s) = 0$. At $s = 0$, this becomes $A^{\mathrm{t}} + A = 0$. Hence, $\mathfrak{so}(n)$ is the set of skew-symmetric matrices. Since we are interested only in the vicinity of the unit element, the Lie algebra of $O(n)$ is the same as that of $SO(n)$: $\mathfrak{o}(n) = \mathfrak{so}(n)$. It is easy to see that $\dim \mathfrak{o}(n) = \dim \mathfrak{so}(n) = n(n-1)/2$.
(e) A similar analysis can be carried out for matrix groups of $GL(n, \mathbb{C})$. $\mathfrak{gl}(n, \mathbb{C})$ is the set of $n \times n$ matrices with complex entries and $\dim \mathfrak{gl}(n, \mathbb{C}) = 2n^2$ (the dimension here is a real dimension). $\mathfrak{sl}(n, \mathbb{C})$ is the set of traceless matrices with real dimension $2(n^2 - 1)$. To find $\mathfrak{u}(n)$, we consider a curve $c(s) = I_n + sA + O(s^2)$ in $U(n)$. Since $c(s)^{\dagger}c(s) = I_n$, we have $c'(s)^{\dagger}c(s) + c(s)^{\dagger}c'(s) = 0$. At $s = 0$, we have $A^{\dagger} + A = 0$.

Hence, $\mathfrak{u}(n)$ is the set of skew-Hermitian matrices with $\dim \mathfrak{u}(n) = n^2$. $\mathfrak{su}(n) = \mathfrak{u}(n) \cap \mathfrak{sl}(n)$ is the set of traceless skew-Hermitian matrices with $\dim \mathfrak{su}(n) = n^2 - 1$.

Exercise 5.22. Let

$$c(s) = \begin{pmatrix} \cos s & -\sin s & 0 \\ \sin s & \cos s & 0 \\ 0 & 0 & 1 \end{pmatrix}$$

be a curve in SO(3). Find the tangent vector to this curve at I_3.

5.6.3   The one-parameter subgroup

A vector field $X \in \mathfrak{X}(M)$ generates a flow in M (section 5.3). Here we are interested in the flow generated by a left-invariant vector field.

Definition 5.12. A curve $\phi : \mathbb{R} \to G$ is called a **one-parameter subgroup** of G if it satisfies the condition

$$\phi(t)\phi(s) = \phi(t + s). \tag{5.117}$$

It is easy to see that $\phi(0) = e$ and $\phi^{-1}(t) = \phi(-t)$. Note that the curve ϕ thus defined is a homomorphism from $\mathbb{R}$ to G. Although G may be non-Abelian, a one-parameter subgroup is an Abelian subgroup: $\phi(t)\phi(s) = \phi(t + s) = \phi(s + t) = \phi(s)\phi(t)$.

Given a one-parameter subgroup $\phi : \mathbb{R} \to G$, there exists a vector field X, such that

$$\frac{\mathrm{d}\phi^{\mu}(t)}{\mathrm{d}t} = X^{\mu}(\phi(t)). \tag{5.118}$$

We now show that the vector field X is left-invariant. First note that the vector field $\mathrm{d}/\mathrm{d}t$ is left-invariant on $\mathbb{R}$, see example 5.15(a). Thus, we have

$$\left. (L_t)_* \frac{\mathrm{d}}{\mathrm{d}t} \right|_0 = \left. \frac{\mathrm{d}}{\mathrm{d}t} \right|_t. \tag{5.119}$$

Next, we apply the induced map $\phi_* : T_t\mathbb{R} \to T_{\phi(t)}G$ on the vectors $\mathrm{d}/\mathrm{d}t|_0$ and $\mathrm{d}/\mathrm{d}t|_t$,

$$\left. \phi_* \frac{\mathrm{d}}{\mathrm{d}t} \right|_0 = \left. \frac{\mathrm{d}\phi^{\mu}(t)}{\mathrm{d}t} \right|_0 \left. \frac{\partial}{\partial g^{\mu}} \right|_e = X|_e \tag{5.120a}$$

$$\left. \phi_* \frac{\mathrm{d}}{\mathrm{d}t} \right|_t = \left. \frac{\mathrm{d}\phi^{\mu}(t)}{\mathrm{d}t} \right|_t \left. \frac{\partial}{\partial g^{\mu}} \right|_g = X|_g \tag{5.120b}$$

where we put $\phi(t) = g$. From (5.119) and (5.120b), we have

$$\left. (\phi L_t)_* \frac{\mathrm{d}}{\mathrm{d}t} \right|_0 = \left. \phi_* L_{t*} \frac{\mathrm{d}}{\mathrm{d}t} \right|_0 = X|_g. \tag{5.121a}$$

It follows from the commutativity $\phi L_t = L_g \phi$ that $\phi_* L_{t*} = L_{g*} \phi_*$. Then (5.121a) becomes

$$\phi_* L_{t*} \left. \frac{d}{dt} \right|_0 = L_{g*} \phi_* \left. \frac{d}{dt} \right|_0 = L_{g*} X |_e. \tag{5.121b}$$

From (5.121), we conclude that

$$L_{g*} X |_e = X |_g. \tag{5.122}$$

Thus, given a flow $\phi(t)$, there exists an associated left-invariant vector field $X \in \mathfrak{g}$.

Conversely, a left-invariant vector field X defines a one-parameter group of transformations $\sigma(t, g)$ such that $d\sigma(t, g)/dt = X$ and $\sigma(0, g) = g$. If we define $\phi : \mathbb{R} \to G$ by $\phi(t) \equiv \sigma(t, e)$, the curve $\phi(t)$ becomes a one-parameter subgroup of G. To prove this, we have to show $\phi(s + t) = \phi(s)\phi(t)$. By definition, σ satisfies

$$\frac{d}{dt} \sigma(t, \sigma(s, e)) = X(\sigma(t, \sigma(s, e))). \tag{5.123}$$

[We have omitted the coordinate indices for notational simplicity. If readers feel uneasy, they may supplement the indices as in (5.118).] If the parameter s is fixed, $\bar{\sigma}(t, \phi(s)) \equiv \phi(s)\phi(t)$ is a curve $\mathbb{R} \to G$ at $\phi(s)\phi(0) = \phi(s)$. Clearly σ and $\bar{\sigma}$ satisfy the same initial condition,

$$\sigma(0, \sigma(s, e)) = \bar{\sigma}(0, \phi(s)) = \phi(s). \tag{5.124}$$

$\bar{\sigma}$ also satisfies the same differential equation as σ:

$$\begin{aligned} \frac{d}{dt} \bar{\sigma}(t, \phi(t)) &= \frac{d}{dt} \phi(s)\phi(t) = (L_{\phi(s)})_* \frac{d}{dt} \phi(t) \\ &= (L_{\phi(s)})_* X(\phi(t)) \\ &= X(\phi(s)\phi(t)) \qquad \text{(left-invariance)} \\ &= X(\bar{\sigma}(t, \phi(s))). \end{aligned} \tag{5.125}$$

From the uniqueness theorem of ODEs, we conclude that

$$\phi(s + t) = \phi(s)\phi(t). \tag{5.126}$$

We have found that there is a one-to-one correspondence between a one-parameter subgroup of G and a left-invariant vector field. This correspondence becomes manifest if we define the exponential map as follows.

Definition 5.13. Let G be a Lie group and $V \in T_e G$. The exponential map $\exp : T_e G \to G$ is defined by

$$\exp V \equiv \phi_V(1) \tag{5.127}$$

where ϕ_V is a one-parameter subgroup of G generated by the left-invariant vector field $X_V|_g = L_{g*}V$.

Proposition 5.2. Let $V \in T_eG$ and let $t \in \mathbb{R}$. Then

$$\exp(tV) = \phi_V(t) \tag{5.128}$$

where $\phi_V(t)$ is a one-parameter subgroup generated by $X_V|_g = L_{g*}V$.

Proof. Let $a \neq 0$ be a constant. Then $\phi_V(at)$ satisfies

$$\frac{d}{dt}\phi_V(at)\bigg|_{t=0} = a\frac{d}{dt}\phi_V(t)\bigg|_{t=0} = aV$$

which shows that $\phi_V(at)$ is a one-parameter subgroup generated by $L_{g*}aV$. The left-invariant vector field $L_{g*}aV$ also generates $\phi_{aV}(t)$ and, from the uniqueness of the solution, we find that $\phi_V(at) = \phi_{aV}(t)$. From definition 5.13, we have

$$\exp(aV) = \phi_{aV}(1) = \phi_V(a).$$

The proof is completed if a is replaced by t.   □

For a matrix group, the exponential map is given by the exponential of a matrix. Take $G = GL(n, \mathbb{R})$ and $A \in \mathfrak{gl}(n, \mathbb{R})$. Let us define a one-parameter subgroup $\phi_A : \mathbb{R} \to GL(n, \mathbb{R})$ by

$$\phi_A(t) = \exp(tA) = I_n + tA + \frac{t^2}{2!}A^2 + \cdots + \frac{t^n}{n!}A^n + \cdots. \tag{5.129}$$

In fact, $\phi_A(t) \in GL(n, \mathbb{R})$ since $[\phi_A(t)]^{-1} = \phi_A(-t)$ exists. It is also easy to see $\phi_A(t)\phi_A(s) = \phi(t+s)$. Now the exponential map is given by

$$\phi_A(1) = \exp(A) = I_n + A + \frac{1}{2!}A^2 + \cdots + \frac{1}{n!}A^n + \cdots. \tag{5.130}$$

The curve $g\exp(tA)$ is a flow through $g \in G$. We find that

$$\frac{d}{dt}g\exp(tA)\bigg|_{t=0} = L_{g*}A = X_A|_g$$

where X_A is a left-invariant vector field generated by A. From (5.115), we find, for a matrix group G, that

$$L_{g*}A = X_A|_g = gA. \tag{5.131}$$

The curve $g\exp(tA)$ defines a map $\sigma_t : G \to G$ by $\sigma_t(g) \equiv g\exp(tA)$ which is also expressed as a right-translation,

$$\sigma_t = R_{\exp(tA)}. \tag{5.132}$$

5.6.4 Frames and structure equation

Let the set of n vectors $\{V_1, V_2, \ldots, V_n\}$ be a basis of T_eG where $n = \dim G$. [We assume throughout this book that n is finite.] The basis defines the set of n linearly independent left-invariant vector fields $\{X_1, X_2, \ldots, X_n\}$ at each point g in G by $X_\mu|_g = L_{g*}V_\mu$. Note that the set $\{X_\mu\}$ is a frame of a basis defined throughout G. Since $[X_\mu, X_\nu]|_g$ is again an element of $\mathfrak{g}$ at g, it can be expanded in terms of $\{X_\mu\}$ as

$$[X_\mu, X_\nu] = c_{\mu\nu}{}^\lambda X_\lambda \tag{5.133}$$

where $c_{\mu\nu}{}^\lambda$ are called the **structure constants** of the Lie group G. If G is a matrix group, the LHS of (5.133) at $g = e$ is precisely the commutator of matrices V_μ and V_ν; see (5.116). We show that the $c_{\mu\nu}{}^\lambda$ are, indeed, constants independent of g. Let $c_{\mu\nu}{}^\lambda(e)$ be the structure constants at the unit element. If L_{g*} is applied to the Lie bracket, we have

$$[X_\mu, X_\nu]|_g = c_{\mu\nu}{}^\lambda(e)X_\lambda|_g$$

which shows the g-independence of the structure constants. In a sense, the structure constants determine a Lie group completely (Lie's theorem).

Exercise 5.23. Show that the structure constants satisfy

(a) *skew-symmetry*

$$c_{\mu\nu}{}^\lambda = -c_{\nu\mu}{}^\lambda \tag{5.134}$$

(b) *Jacobi identity*

$$c_{\mu\nu}{}^\tau c_{\tau\rho}{}^\lambda + c_{\rho\mu}{}^\tau c_{\tau\nu}{}^\lambda + c_{\nu\rho}{}^\tau c_{\tau\mu}{}^\lambda = 0. \tag{5.135}$$

Let us introduce a dual basis to $\{X_\mu\}$ and denote it by $\{\theta^\mu\}$; $\langle\theta^\mu, X_\nu\rangle = \delta_\nu^\mu$. $\{\theta^\mu\}$ is a basis for the left-invariant one-forms. We will show that the dual basis satisfies **Maurer–Cartan's structure equation**,

$$d\theta^\mu = -\tfrac{1}{2}c_{\nu\lambda}{}^\mu\theta^\nu \wedge \theta^\lambda. \tag{5.136}$$

This can be seen by making use of (5.70):

$$d\theta^\mu(X_\nu, X_\lambda) = X_\nu[\theta^\mu(X_\lambda)] - X_\lambda[\theta^\mu(X_\nu)] - \theta^\mu([X_\nu, X_\lambda])$$
$$= X_\nu[\delta_\lambda^\mu] - X_\lambda[\delta_\nu^\mu] - \theta^\mu(c_{\nu\lambda}{}^\kappa X_\kappa) = -c_{\nu\lambda}{}^\mu$$

which proves (5.136).

We define a Lie-algebra-valued one-form $\theta : T_gG \to T_eG$ by

$$\theta : X \mapsto (L_{g^{-1}})_* X = (L_g)_*^{-1} X \qquad X \in T_gG. \tag{5.137}$$

θ is called the **canonical one-form** or **Maurer–Cartan form** on G.

Theorem 5.3.   (a) The canonical one-form θ is expanded as

$$\theta = V_\mu \otimes \theta^\mu \tag{5.138}$$

where $\{V_\mu\}$ is the basis of $T_e G$ and $\{\theta^\mu\}$ the dual basis of $T_e^* G$.
(b) The canonical one-form θ satisfies

$$d\theta + \tfrac{1}{2}[\theta \wedge \theta] = 0 \tag{5.139}$$

where $d\theta \equiv V_\mu \otimes d\theta^\mu$ and

$$[\theta \wedge \theta] \equiv [V_\mu, V_\nu] \otimes \theta^\mu \wedge \theta^\nu. \tag{5.140}$$

Proof.

(a) Take any vector $Y = Y^\mu X_\mu \in T_g G$, where $\{X_\mu\}$ is the set of frame vectors generated by $\{V_\mu\}$; $X_\mu|_g = L_{g*} V_\mu$. From (5.137), we find

$$\theta(Y) = Y^\mu \theta(X_\mu) = Y^\mu (L_{g*})^{-1}[L_{g*} V_\mu] = Y^\mu V_\mu.$$

However,

$$(V_\mu \otimes \theta^\mu)(Y) = Y^\nu V_\mu \theta^\mu(X_\nu) = Y^\nu V_\mu \delta^\mu_\nu = Y^\mu V_\mu.$$

Since Y is arbitrary, we have $\theta = V_\mu \otimes \theta^\mu$.
(b) We use the Maurer–Cartan structure equation (5.136):

$$d\theta + \tfrac{1}{2}[\theta \wedge \theta] = -\tfrac{1}{2} V_\mu \otimes c_{\nu\lambda}{}^\mu \theta^\nu \wedge \theta^\lambda + \tfrac{1}{2} c_{\nu\lambda}{}^\mu V_\mu \otimes \theta^\nu \wedge \theta^\lambda = 0$$

where the $c_{\nu\lambda}{}^\mu$ are the structure constants of G.   □

5.7 The action of Lie groups on manifolds

In physics, a Lie group often appears as the set of transformations acting on a manifold. For example, SO(3) is the group of rotations in $\mathbb{R}^3$, while the Poincaré group is the set of transformations acting on the Minkowski spacetime. To study more general cases, we abstract the action of a Lie group G on a manifold M. We have already encountered this interaction between a group and geometry. In section 5.3 we defined a flow in a manifold M as a map $\sigma : \mathbb{R} \times M \to M$, in which $\mathbb{R}$ acts as an additive group. We abstract this idea as follows.

5.7.1 Definitions

Definition 5.14. Let G be a Lie group and M be a manifold. The **action** of G on M is a differentiable map $\sigma : G \times M \to M$ which satisfies the conditions

(i)	$\sigma(e, p) = p$	for any $p \in M$	(5.141a)
(ii)	$\sigma(g_1, \sigma(g_2, p)) = \sigma(g_1 g_2, p).$		(5.141b)

[*Remark:* We often use the notation gp instead of $\sigma(g, p)$. The second condition in this notation is $g_1(g_2 p) = (g_1 g_2)p$.]

Example 5.16. (a) A flow is an action of $\mathbb{R}$ on a manifold M. If a flow is periodic with a period T, it may be regarded as an action of U(1) or SO(2) on M. Given a periodic flow $\sigma(t, x)$ with period T, we construct a new action $\bar{\sigma}(\exp(2\pi i t/T), x) \equiv \sigma(t, x)$ whose group G is U(1).

(b) Let $M \in$ GL$(n, \mathbb{R})$ and let $x \in \mathbb{R}^n$. The action of GL$(n, \mathbb{R})$ on $\mathbb{R}^n$ is defined by the usual matrix action on a vector:

$$\sigma(M, x) = M \cdot x. \tag{5.142}$$

The action of the subgroups of GL$(n, \mathbb{R})$ is defined similarly. They may also act on a smaller space. For example, O(n) acts on $S^{n-1}(r)$, an $(n-1)$-sphere of radius r,

$$\sigma : \text{O}(n) \times S^{n-1}(r) \to S^{n-1}(r). \tag{5.143}$$

(c) It is known that SL$(2, \mathbb{C})$ acts on a four-dimensional Minkowski space M_4 in a special manner. For $x = (x^0, x^1, x^2, x^3) \in M_4$, define a Hermitian matrix,

$$X(x) \equiv x^\mu \sigma_\mu = \begin{pmatrix} x^0 + x^3 & x^1 - ix^2 \\ x^1 + ix^2 & x^0 - x^3 \end{pmatrix} \tag{5.144}$$

where $\sigma_\mu = (I_2, \sigma_1, \sigma_2, \sigma_3)$, σ_i $(i = 1, 2, 3)$ being the Pauli matrices. Conversely, given a Hermitian matrix X, a unique vector $(x^\mu) \in M_4$ is defined as

$$x^\mu = \tfrac{1}{2} \text{tr}(\sigma_\mu X) \tag{5.130}$$

where tr is over the 2×2 matrix indices. Thus, there is an isomorphism between M_4 and the set of 2×2 Hermitian matrices. It is interesting to note that $\det X(x) = (x^0)^2 - (x^1)^2 - (x^2)^2 - (x^3)^2 = -X^t \eta X = -(\text{Minkowski norm})^2$. Accordingly

$$\begin{aligned} \det X(x) &> 0 && \text{if } x \text{ is a timelike vector} \\ &= 0 && \text{if } x \text{ is on the light cone} \\ &< 0 && \text{if } x \text{ is a spacelike vector.} \end{aligned}$$

Take $A \in$ SL$(2, \mathbb{C})$ and define an action of SL$(2, \mathbb{C})$ on M_4 by

$$\sigma(A, x) \equiv AX(x)A^\dagger. \tag{5.145}$$

The reader should verify that this action, in fact, satisfies the axioms of definition 5.14. The action of SL$(2, \mathbb{C})$ on M_4 represents the Lorentz transformation O$(1, 3)$. First we note that the action preserves the Minkowski norm,

$$\det \sigma(A, x) = \det[AX(x)A^\dagger] = \det X(x)$$

since $\det A = \det A^\dagger = 1$. Moreover, there is a homomorphism $\varphi : \mathrm{SL}(2, \mathbb{C}) \rightarrow$ $\mathrm{O}(1, 3)$ since

$$A(BXB^\dagger)A^\dagger = (AB)X(AB)^\dagger.$$

However, this homomorphism cannot be one to one, since $A \in \mathrm{SL}(2, \mathbb{C})$ and $-A$ give the same element of $\mathrm{O}(1, 3)$; see (5.145). We verify (exercise 5.24) that the following matrix is an explicit form of a rotation about the unit vector $\hat{n}$ by an angle θ,

$$A = \exp\left[-\mathrm{i}\frac{\theta}{2}(\hat{n} \cdot \boldsymbol{\sigma})\right] = \cos\frac{\theta}{2}I_2 - \mathrm{i}(\hat{n} \cdot \boldsymbol{\sigma})\sin\frac{\theta}{2}. \qquad (5.146a)$$

The appearance of $\theta/2$ ensures that the homomorphism between $\mathrm{SL}(2, \mathbb{C})$ and the $\mathrm{O}(3)$ subgroup of $\mathrm{O}(1, 3)$ is indeed two to one. In fact, rotations about $\hat{n}$ by θ and by $2\pi + \theta$ should be the same $\mathrm{O}(3)$ rotation, but $A(2\pi + \theta) = -A(\theta)$ in $\mathrm{SL}(2, \mathbb{C})$. This leads to the existence of spinors. [See Misner *et al* (1973) and Wald (1984).] A boost along the direction $\hat{n}$ with the velocity $v = \tanh \alpha$ is given by

$$A = \exp\left[\frac{\alpha}{2}(\hat{n} \cdot \boldsymbol{\sigma})\right] = \cosh\frac{\alpha}{2}I_2 + (\hat{n} \cdot \boldsymbol{\sigma})\sinh\frac{\alpha}{2}. \qquad (5.146b)$$

We show that φ maps $\mathrm{SL}(2, \mathbb{C})$ onto the proper orthochronous Lorentz group $\mathrm{O}_+^\uparrow(1, 3) = \{\Lambda \in \mathrm{O}(1, 3) | \det \Lambda = +1, \Lambda_{00} > 0\}$. Take any

$$A = \begin{pmatrix} a & b \\ c & d \end{pmatrix} \in \mathrm{SL}(2, \mathbb{C})$$

and suppose $x^\mu = (1, 0, 0, 0)$ is mapped to x'^μ. If we write $\varphi(A) = \Lambda$, we have

$$x'^0 = \frac{1}{2}\mathrm{tr}(AXA^\dagger) = \frac{1}{2}\mathrm{tr}\left[\begin{pmatrix} a & b \\ c & d \end{pmatrix}\begin{pmatrix} \bar{a} & \bar{c} \\ \bar{b} & \bar{d} \end{pmatrix}\right]$$

$$= \frac{1}{2}(|a|^2 + |b|^2 + |c|^2 + |d|^2) > 0$$

hence $\Lambda_{00} > 0$. To show $\det A = +1$, we note that any element of $\mathrm{SL}(2, \mathbb{C})$ may be written as

$$A = \begin{pmatrix} e^{\mathrm{i}\alpha} & 0 \\ 0 & e^{-\mathrm{i}\alpha} \end{pmatrix}\begin{pmatrix} \cos\beta & \sin\beta\, e^{\mathrm{i}\gamma} \\ -\sin\beta\, e^{-\mathrm{i}\gamma} & \cos\beta \end{pmatrix}B$$

$$= \begin{pmatrix} e^{\mathrm{i}\alpha/2} & 0 \\ 0 & e^{\mathrm{i}\alpha/2} \end{pmatrix}^2\begin{pmatrix} \cos(\beta/2) & \sin(\beta/2)e^{\mathrm{i}\gamma} \\ -\sin(\beta/2)e^{-\mathrm{i}\gamma} & \cos(\beta/2) \end{pmatrix}^2 B$$

$$\equiv M^2 N^2 B_0^2$$

where $B \equiv B_0^2$ is a positive-definite matrix. This shows that $\varphi(A)$ is positive definite:

$$\det \varphi(A) = (\det \varphi(M))^2(\det \varphi(N))^2(\det \varphi(B_0))^2 > 0.$$

Now we have established that $\varphi(SL(2, \mathbb{C})) \subset O_+^\uparrow(1, 3)$. Equations (5.146a) and (5.146b) show that for any element of $O_+^\uparrow(1, 3)$, there is a corresponding matrix $A \in SL(2, \mathbb{C})$, hence φ is onto. Thus, we have established that

$$\varphi(SL(2, \mathbb{C})) = O_+^\uparrow(1, 3). \qquad (5.147)$$

It can be shown that $SL(2, \mathbb{C})$ is simply connected and is the universal covering group $\text{SPIN}(1, 3)$ of $O_+^\uparrow(1, 3)$, see section 4.6.

Exercise 5.24. Verify by explicit calculations that

(a)

$$A = \begin{pmatrix} e^{-i\theta/2} & 0 \\ 0 & e^{i\theta/2} \end{pmatrix}$$

represents a rotation about the z-axis by θ;
(b)

$$A = \begin{pmatrix} \cosh(\alpha/2) + \sinh(\alpha/2) & 0 \\ 0 & \cosh(\alpha/2) - \sinh(\alpha/2) \end{pmatrix}$$

represents a boost along the z-axis with the velocity $v = \tanh \alpha$.

Definition 5.15. Let G be a Lie group that acts on a manifold M by $\sigma : G \times M \to M$. The action σ is said to be

(a) **transitive** if, for any $p_1, p_2 \in M$, there exists an element $g \in G$ such that $\sigma(g, p_1) = p_2$;

(b) **free** if every non-trivial element $g \neq e$ of G has no fixed points in M, that is, if there exists an element $p \in M$ such that $\sigma(g, p) = p$, then g must be the unit element e; and

(c) **effective** if the unit element $e \in G$ is the unique element that defines the trivial action on M, i.e. if $\sigma(g, p) = p$ for all $p \in M$, then g must be the unit element e.

Exercise 5.25. Show that the right translation $R : (a, g) \mapsto R_a g$ and left translation $L : (a, g) \mapsto L_a g$ of a Lie group are free and transitive.

5.7.2 Orbits and isotropy groups

Given a point $p \in M$, the action of G on p takes p to various points in M. The **orbit** of p under the action σ is the subset of M defined by

$$Gp = \{\sigma(g, p) | g \in G\}. \qquad (5.148)$$

If the action of G on M is transitive, the orbit of any $p \in M$ is M itself. Clearly the action of G on any orbit Gp is transitive.

Example 5.17. (a) A flow σ generated by a vector field $X = -y\partial/\partial x + x\partial/\partial y$ is periodic with period 2π, see example 5.9. The action $\sigma : \mathbb{R} \times \mathbb{R}^2 \to \mathbb{R}^2$ defined by $(t, (x, y)) \to \sigma(t, (x, y))$ is not effective since $\sigma(2\pi n, (x, y)) = (x, y)$ for all $(x, y) \in \mathbb{R}^2$. For the same reason, this flow is not free either. The orbit through $(x, y) \neq (0, 0)$ is a circle S^1 centred at the origin.

(b) The action of $O(n)$ on $\mathbb{R}^n$ is not transitive since if $|x| \neq |x'|$, no element of $O(n)$ takes x to x'. However, the action of $O(n)$ on S^{n-1} is obviously transitive. The orbit through x is the sphere S^{n-1} of radius $|x|$. Accordingly, given an action $\sigma : O(n) \times \mathbb{R}^n \to \mathbb{R}^n$, the orbits divide $\mathbb{R}^n$ into mutually disjoint spheres of different radii. Introduce a relation by $x \sim y$ if $y = \sigma(g, x)$ for some $g \in G$. It is easily verified that $\sim$ is an equivalence relation. The equivalence class $[x]$ is an orbit through x. The coset space $\mathbb{R}^n/O(n)$ is $[0, \infty)$ since each equivalence class is parametrized by the radius.

Definition 5.16. Let G be a Lie group that acts on a manifold M. The **isotropy group** of $p \in M$ is a subgroup of G defined by

$$H(p) = \{g \in G | \sigma(g, p) = p\}. \tag{5.149}$$

$H(p)$ is also called the **little group** or **stabilizer** of p.

It is easy to see that $H(p)$ is indeed a subgroup. Let $g_1, g_2 \in H(p)$, then $g_1 g_2 \in H(p)$ since $\sigma(g_1 g_2, p) = \sigma(g_1, \sigma(g_2, p)) = \sigma(g_1, p) = p$. Clearly $e \in H(p)$ since $\sigma(e, p) = p$ by definition. If $g \in H(p)$, then $g^{-1} \in H(p)$ since $p = \sigma(e, p) = \sigma(g^{-1}g, p) = \sigma(g^{-1}, \sigma(g, p)) = \sigma(g^{-1}, p)$.

Exercise 5.26. Suppose a Lie group G acts on a manifold M freely. Show that $H(p) = \{e\}$ for any $p \in M$.

Theorem 5.4. Let G be a Lie group which acts on a manifold M. Then the isotropy group $H(p)$ for any $p \in M$ is a Lie subgroup.

Proof. For fixed $p \in M$, we define a map $\varphi_p : G \to M$ by $\varphi_p(g) \equiv gp$. Then $H(p)$ is the inverse image $\varphi_p^{-1}(p)$ of a *point* p, and hence a closed set. The group properties have been shown already. It follows from theorem 5.2 that $H(p)$ is a Lie subgroup. $\square$

For example, let $M = \mathbb{R}^3$ and $G = SO(3)$ and take a point $p = (0, 0, 1) \in \mathbb{R}^3$. The isotropy group $H(p)$ is the set of rotations about the z-axis, which is isomorphic to $SO(2)$.

Let G be a Lie group and H any subgroup of G. The coset space G/H admits a differentiable structure and G/H becomes a manifold, called a **homogeneous space**. Note that $\dim G/H = \dim G - \dim H$. Let G be a Lie group which acts on a manifold M transitively and let $H(p)$ be an isotropy group of $p \in M$. $H(p)$ is a Lie subgroup and the coset space $G/H(p)$ is a homogeneous space.

In fact, if G, $H(p)$ and M satisfy certain technical requirements (for example, $G/H(p)$ compact) is, it can be shown that $G/H(p)$ is homeomorphic to M, see example 5.18.

Example 5.18. (a) Let $G = \mathrm{SO}(3)$ be a group acting on $\mathbb{R}^3$ and $H = \mathrm{SO}(2)$ be the isotropy group of $x \in \mathbb{R}^3$. The group $\mathrm{SO}(3)$ acts on S^2 transitively and we have $\mathrm{SO}(3)/\mathrm{SO}(2) \cong S^2$. What is the geometrical picture of this? Let $g' = gh$ where $g, g' \in G$ and $h \in H$. Since H is the set of rotations in a plane, g and g' must be rotations about the common axis. Then the equivalence class $[g]$ is specified by the polar angles (θ, ϕ). Thus, we again find that $G/H = S^2$. Since $\mathrm{SO}(2)$ is not a normal subgroup of $\mathrm{SO}(3)$, S^2 does not admit a group structure.

It is easy to generalize this result to higher-dimensional rotation groups and we have the useful result

$$\mathrm{SO}(n+1)/\mathrm{SO}(n) = S^n. \tag{5.150}$$

$\mathrm{O}(n+1)$ also acts on S^n transitively and we have

$$\mathrm{O}(n+1)/\mathrm{O}(n) = S^n. \tag{5.151}$$

Similar relations hold for $\mathrm{U}(n)$ and $\mathrm{SU}(n)$:

$$\mathrm{U}(n+1)/\mathrm{U}(n) = \mathrm{SU}(n+1)/\mathrm{SU}(n) = S^{2n+1}. \tag{5.152}$$

(b) The group $\mathrm{O}(n+1)$ acts on $\mathbb{R}P^n$ transitively from the left. Note, first, that $\mathrm{O}(n+1)$ acts on $\mathbb{R}^{n+1}$ in the usual manner and preserves the equivalence relation employed to define $\mathbb{R}P^n$ (see example 5.12). In fact, take $x, x' \in \mathbb{R}^{n+1}$ and $g \in \mathrm{O}(n+1)$. If $x \sim x'$ (that is if $x' = ax$ for some $a \in \mathbb{R} - \{0\}$), then it follows that $gx \sim gx'$ ($gx' = agx$). Accordingly, this action of $\mathrm{O}(n+1)$ on $\mathbb{R}^{n+1}$ induces the natural action of $\mathrm{O}(n+1)$ on $\mathbb{R}P^n$. Clearly this action is transitive on $\mathbb{R}P^n$. (Look at two representatives with the same norm.) If we take a point p in $\mathbb{R}P^n$, which corresponds to a point $(1, 0, \dots, 0) \in \mathbb{R}^{n+1}$, the isotropy group $H(p)$ is

$$H(p) = \begin{pmatrix} \pm 1 & 0 & 0 & \cdots & 0 \\ 0 & & & & \\ 0 & & & & \\ \vdots & & \mathrm{O}(n) & & \\ 0 & & & & \end{pmatrix} = \mathrm{O}(1) \times \mathrm{O}(n) \tag{5.153}$$

where $\mathrm{O}(1)$ is the set $\{-1, +1\} = \mathbb{Z}_2$. Now we find that

$$\mathrm{O}(n+1)/[\mathrm{O}(1) \times \mathrm{O}(n)] \cong S^n/\mathbb{Z}_2 \cong \mathbb{R}P^n. \tag{5.154}$$

(c) This result is easily generalized to the Grassmann manifolds: $G_{k,n}(\mathbb{R}) = \mathrm{O}(n)/[\mathrm{O}(k) \times \mathrm{O}(n-k)]$. We first show that $\mathrm{O}(n)$ acts on $G_{k,n}(\mathbb{R})$ transitively.

Let A be an element of $G_{k,n}(\mathbb{R})$, then A is a k-dimensional plane in $\mathbb{R}^n$. Define an $n \times n$ matrix P_A which projects a vector $v \in \mathbb{R}^n$ to the plane A. Let us introduce an orthonormal basis $\{e_1, \ldots, e_n\}$ in $\mathbb{R}^n$ and another orthonormal basis $\{f_1, \ldots, f_k\}$ in the plane A, where the orthonormality is defined with respect to the Euclidean metric in $\mathbb{R}^n$. In terms of $\{e_i\}$, f_a is expanded as $f_a = \sum_i f_{ai} e_i$ and the projected vector is

$$
\begin{aligned}
P_A v &= (v f_1) f_1 + \cdots + (v f_k) f_k \\
&= \sum_{i,j} (v_i f_{1i} f_{1j} + \cdots + v_i f_{ki} f_{kj}) e_j = \sum_{i,a,j} v_i f_{ai} f_{aj} e_j.
\end{aligned}
$$

Thus, P_A is represented by a matrix

$$
(P_A)_{ij} = \sum f_{ai} f_{aj}. \tag{5.155}
$$

Note that $P_A^2 = P_A$, $P_A^t = P_A$ and $\operatorname{tr} P_A = k$. [The last relation holds since it is always possible to choose a coordinate system such that

$$
P_A = \operatorname{diag}(\underbrace{1, 1, \ldots, 1}_{k}, \underbrace{0, \ldots, 0}_{n-k}).
$$

This guarantees that A is, indeed, a k-dimensional plane.] Conversely any matrix P that satisfies these three conditions determines a unique k-dimensional plane in $\mathbb{R}^n$, that is a unique element of $G_{k,n}(\mathbb{R})$.

We now show that $O(n)$ acts on $G_{k,n}(\mathbb{R})$ transitively. Take $A \in G_{k,n}(\mathbb{R})$ and $g \in O(n)$ and construct $P_B \equiv g P_A g^{-1}$. The matrix P_B determines an element $B \in G_{k,n}(\mathbb{R})$ since $P_B^2 = P_B$, $P_B^t = P_B$ and $\operatorname{tr} P_B = k$. Let us denote this action by $B = \sigma(g, A)$. Clearly this action is transitive since given a standard k-dimensional basis of A, $\{f_1, \ldots, f_k\}$ for example, any k-dimensional basis $\{\widetilde{f}_1, \ldots, \widetilde{f}_k\}$ can be reached by an action of $O(n)$ on this basis.

Let us take a special plane C_0 which is spanned by the standard basis $\{f_1, \ldots, f_k\}$. Then an element of the isotropy group $H(C_0)$ is of the form

$$
M = \begin{pmatrix} \overset{k}{g_1} & \overset{n-k}{0} \\ 0 & g_2 \end{pmatrix} \begin{matrix} k \\ n-k \end{matrix} \tag{5.156}
$$

where $g_1 \in O(k)$. Since $M \in O(n)$, an $(n-k) \times (n-k)$ matrix g_2 must be an element of $O(n-k)$. Thus, the isotropy group is isomorphic to $O(k) \times O(n-k)$. Finally we verified that

$$
G_{k,n}(\mathbb{R}) \cong O(n)/[O(k) \times O(n-k)]. \tag{5.157}
$$

The dimension of $G_{k,n}(\mathbb{R})$ is obtained from the general formula as

$$
\begin{aligned}
\dim G_{k,n}(\mathbb{R}) &= \dim O(n) - \dim[O(k) \times O(n-k)] \\
&= \tfrac{1}{2} n(n-1) - [\tfrac{1}{2} k(k-1) + \tfrac{1}{2}(n-k)(n-k-1)] \\
&= k(n-k) \tag{5.158}
\end{aligned}
$$

in agreement with the result of example 5.5. Equation (5.157) also shows that the Grassmann manifold is compact.

5.7.3 Induced vector fields

Let G be a Lie group which acts on M as $(g, x) \mapsto gx$. A left-invariant vector field X_V generated by $V \in T_e G$ naturally induces a vector field in M. Define a flow in M by

$$\sigma(t, x) = \exp(tV)x, \tag{5.159}$$

$\sigma(t, x)$ is a one-parameter group of transformations, and define a vector field called the **induced vector field** denoted by $V^\sharp$,

$$V^\sharp|_x = \frac{d}{dt} \exp(tV)x \bigg|_{t=0}. \tag{5.160}$$

Thus, we have obtained a map $\sharp : T_e G \to \mathfrak{X}(M)$ defined by $V \mapsto V^\sharp$.

Exercise 5.27. The Lie group $SO(2)$ acts on $M = \mathbb{R}^2$ in the usual way. Let

$$V = \begin{pmatrix} 0 & -1 \\ 1 & 0 \end{pmatrix}$$

be an element of $\mathfrak{so}(2)$.
   (a) Show that

$$\exp(tV) = \begin{pmatrix} \cos t & -\sin t \\ \sin t & \cos t \end{pmatrix}$$

and find the induced flow through

$$x = \begin{pmatrix} x \\ y \end{pmatrix} \in \mathbb{R}^2.$$

   (b) Show that $V^\sharp|_x = -y\partial/\partial x + x\partial/\partial y$.

Example 5.19. Let us take $G = SO(3)$ and $M = \mathbb{R}^3$. The basis vectors of $T_e G$ are generated by rotations about the x, y and z axes. We denote them by X_x, X_y and X_z, respectively (see exercise 5.22),

$$X_x = \begin{pmatrix} 0 & 0 & 0 \\ 0 & 0 & -1 \\ 0 & 1 & 0 \end{pmatrix}, \quad X_y = \begin{pmatrix} 0 & 0 & 1 \\ 0 & 0 & 0 \\ -1 & 0 & 0 \end{pmatrix}, \quad X_z = \begin{pmatrix} 0 & -1 & 0 \\ 1 & 0 & 0 \\ 0 & 0 & 0 \end{pmatrix}.$$

Repeating a similar analysis to the previous one, we obtain the corresponding induced vectors,

$$X_x^\sharp = -z\frac{\partial}{\partial y} + y\frac{\partial}{\partial z}, \qquad X_y^\sharp = -x\frac{\partial}{\partial z} + z\frac{\partial}{\partial x}, \qquad X_z^\sharp = -y\frac{\partial}{\partial x} + x\frac{\partial}{\partial y}.$$

5.7.4 The adjoint representation

A Lie group G acts on G itself in a special way.

Definition 5.17. Take any $a \in G$ and define a homomorphism $\mathrm{ad}_a : G \to G$ by the conjugation,

$$\mathrm{ad}_a : g \mapsto aga^{-1}. \tag{5.161}$$

This homomorphism is called the **adjoint representation** of G.

Exercise 5.28. Show that ad_a is a homomorphism. Define a map $\sigma : G \times G \to G$ by $\sigma(a, g) \equiv \mathrm{ad}_a g$. Show that $\sigma(a, g)$ is an action of G on itself.

Noting that $\mathrm{ad}_a e = e$, we restrict the induced map $\mathrm{ad}_{a*} : T_g G \to T_{\mathrm{ad}_a g} G$ to $g = e$,

$$\mathrm{Ad}_a : T_e G \to T_e G \tag{5.162}$$

where $\mathrm{Ad}_a \equiv \mathrm{ad}_{a*}|_{T_e G}$. If we identify $T_e G$ with the Lie algebra $\mathfrak{g}$, we have obtained a map $\mathrm{Ad} : G \times \mathfrak{g} \to \mathfrak{g}$ called the **adjoint map** of G. Since $\mathrm{ad}_{a*}\mathrm{ad}_{b*} = \mathrm{ad}_{ab*}$, it follows that $\mathrm{Ad}_a \mathrm{Ad}_b = \mathrm{Ad}_{ab}$. Similarly, $\mathrm{Ad}_{a^{-1}} = \mathrm{Ad}_a^{-1}$ follows from $\mathrm{ad}_{a^{-1}*}\mathrm{ad}_{a*}|_{T_e G} = \mathrm{id}_{T_e G}$.

If G is a matrix group, the adjoint representation becomes a simple matrix operation. Let $g \in G$ and $X_V \in \mathfrak{g}$, and let $\sigma_V(t) = \exp(tV)$ be a one-parameter subgroup generated by $V \in T_e G$. Then ad_g acting on $\sigma_V(t)$ yields $g \exp(tV)g^{-1} = \exp(tgVg^{-1})$. As for Ad_g we have $\mathrm{Ad}_g : V \mapsto gVg^{-1}$ since

$$\mathrm{Ad}_g V = \frac{\mathrm{d}}{\mathrm{d}t}[\mathrm{ad}_g \exp(tV)]\Big|_{t=0}$$

$$= \frac{\mathrm{d}}{\mathrm{d}t}\exp(tgVg^{-1})\Big|_{t=0} = gVg^{-1}. \tag{5.163}$$

Problems

5.1 The Stiefel manifold $V(m, r)$ is the set of orthonormal vectors $\{e_i\}$ ($1 \le i \le r$) in $\mathbb{R}^m$ ($r \le m$). We may express an element A of $V(m, r)$ by an $m \times r$ matrix $(e_1, \ldots, e_r)$. Show that $SO(m)$ acts transitively on $V(m, r)$. Let

$$A_0 \equiv \begin{pmatrix} 1 & 0 & \cdots & 0 \\ 0 & 1 & \cdots & 0 \\ \cdots & \cdots & \cdots & \cdots \\ 0 & 0 & \cdots & 1 \\ 0 & 0 & \cdots & 0 \\ 0 & 0 & \cdots & 0 \end{pmatrix}$$

be an element of $V(m, r)$. Show that the isotropy group of A_0 is $SO(m-r)$. Verify that $V(m, r) = SO(m)/SO(m - r)$ and $\dim V(m, r) = [r(r - 1)]/2 + r(m - r)$. [*Remark*: The Stiefel manifold is, in a sense, a generalization of a sphere. Observe that $V(m, 1) = S^{m-1}$.]

5.2 Let M be the Minkowski four-spacetime. Define the action of a linear operator $* : \Omega^r(M) \to \Omega^{4-r}(M)$ by

$$
\begin{array}{lll}
r = 0: & *1 = -dx^0 \wedge dx^1 \wedge dx^2 \wedge dx^3; \\
r = 1: & *dx^i = -dx^j \wedge dx^k \wedge dx^0 & *dx^0 = -dx^1 \wedge dx^2 \wedge dx^3; \\
r = 2: & *dx^i \wedge dx^j = dx^k \wedge dx^0 & *dx^i \wedge dx^0 = -dx^j \wedge dx^k; \\
r = 3: & *dx^1 \wedge dx^2 \wedge dx^3 = -dx^0 & *dx^i \wedge dx^j \wedge dx^0 = -dx^k; \\
r = 4: & *dx^0 \wedge dx^1 \wedge dx^2 \wedge dx^3 = 1;
\end{array}
$$

where (i, j, k) is an even permutation of $(1, 2, 3)$. The vector potential A and the electromagnetic tensor F are defined as in example 5.11. $J = J_\mu dx^\mu = \rho dx^0 + j_k dx^k$ is the current one-form.

(a) Write down the equation $d * F = *J$ and verify that it reduces to two of the Maxwell equations $\nabla \cdot E = \rho$ and $\nabla \times B - \partial E / \partial t = j$.

(b) Show that the identity $0 = d(d * F) = d * J$ reduces to the charge conservation equation

$$
\partial_\mu J^\mu = \frac{\partial \rho}{\partial t} + \nabla \cdot \mathbf{j} = 0.
$$

(c) Show that the Lorentz condition $\partial_\mu A^\mu = 0$ is expressed as $d * A = 0$.

6

DE RHAM COHOMOLOGY GROUPS

The homology groups of topological spaces have been defined in chapter 3. If a topological space M is a manifold, we may define the *dual* of the homology groups out of differential forms defined on M. The dual groups are called the de Rham cohomology groups. Besides physicists' familiarity with differential forms, cohomology groups have several advantages over homology groups.

We follow closely Nash and Sen (1983) and Flanders (1963). Bott and Tu (1982) contains more advanced topics.

6.1 Stokes' theorem

One of the main tools in the study of de Rham cohomology groups is Stokes' theorem with which most physicists are familiar from electromagnetism. Gauss' theorem and Stokes' theorem are treated in a unified manner here.

6.1.1 Preliminary consideration

Let us define an integration of an r-form over an r-simplex in a Euclidean space. To do this, we need first to define the **standard n-simplex** $\bar{\sigma}_r = (p_0 p_1 \ldots p_r)$ in $\mathbb{R}^r$ where

$$p_0 = (0, 0, \ldots, 0)$$
$$p_1 = (1, 0, \ldots, 0)$$
$$\ldots$$
$$p_r = (0, 0, \ldots, 1)$$

see figure 6.1. If $\{x^\mu\}$ is a coordinate of $\mathbb{R}^r$, $\bar{\sigma}_r$ is given by

$$\bar{\sigma}_r = \left\{ (x^1, \ldots, x^r) \in \mathbb{R}^r \middle| x^\mu \geq 0, \sum_{\mu=1}^{r} x^\mu \leq 1 \right\}. \tag{6.1}$$

An r-form ω (the volume element) in $\mathbb{R}^r$ is written as

$$\omega = a(x) \, dx^1 \wedge dx^2 \wedge \ldots \wedge dx^r.$$

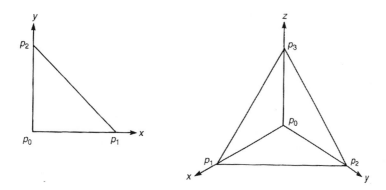

Figure 6.1. The standard 2-simplex $\bar{\sigma}_2 = (p_0 p_1 p_2)$ and the standard 3-simplex $\bar{\sigma}_3 = (p_0 p_1 p_2 p_3)$.

We define the integration of ω over $\bar{\sigma}_r$ by

$$\int_{\bar{\sigma}_r} \omega \equiv \int_{\bar{\sigma}_r} a(x)\,\mathrm{d}x^1\,\mathrm{d}x^2 \ldots \mathrm{d}x^r \tag{6.2}$$

where the RHS is the usual r-fold integration. For example, if $r = 2$ and $\omega = \mathrm{d}x \wedge \mathrm{d}y$, we have

$$\int_{\bar{\sigma}_2} \omega = \int_{\bar{\sigma}_2} \mathrm{d}x\,\mathrm{d}y = \int_0^1 \mathrm{d}x \int_0^{1-x} \mathrm{d}y = \tfrac{1}{2}.$$

Next we define an r-chain, an r-cycle and an r-boundary in an m-dimensional manifold M. Let σ_r be an r-simplex in $\mathbb{R}^r$ and let $f : \sigma_r \to M$ be a smooth map. [To avoid the subtlety associated with the differentiability of f at the boundary of σ_r, f may be defined over an open subset U of $\mathbb{R}^r$, which contains σ_r.] Here we assume f is not required to have an inverse. For example, im f may be a point in M. We denote the image of σ_r in M by s_r and call it a (**singular**) r-**simplex** in M. These simplexes are called singular since they do not provide a triangulation of M and, moreover, *geometrical independence* of points makes no sense in a manifold (see section 3.2). If $\{s_{r,i}\}$ is the set of r-simplexes in M, we define an r-**chain** in M by a formal sum of $\{s_{r,i}\}$ with $\mathbb{R}$-coefficients

$$c = \sum_i a_i s_{r,i} \qquad a_i \in \mathbb{R}. \tag{6.3}$$

In the following, we are concerned with $\mathbb{R}$-coefficients only and we omit the explicit quotation of $\mathbb{R}$. The r-chains in M form the **chain group** $C_r(M)$. Under $f : \sigma_r \to M$, the boundary $\partial \sigma_r$ is also mapped to a subset of M. Clearly, $\partial s_r \equiv f(\partial \sigma_r)$ is a set of $(r-1)$-simplexes in M and is called the **boundary** of

s_r. ∂s_r corresponds to the geometrical boundary of s_r with an induced orientation defined in section 3.3. We have a map

$$\partial : C_r(M) \to C_{r-1}(M). \tag{6.4}$$

The result of section 3.3 tells us that ∂ is nilpotent; $\partial^2 = 0$.

Cycles and boundaries are defined in exactly the same way as in section 3.3 (note, however, that $\mathbb{Z}$ is replaced by $\mathbb{R}$). If c_r is an **r-cycle**, $\partial c_r = 0$ while if c_r is an **r-boundary**, there exists an $(r + 1)$-chain c_{r+1} such that $c_r = \partial c_{r+1}$. The **boundary group** $B_r(M)$ is the set of r-boundaries and the **cycle group** $Z_r(M)$ is the set of r-cycles. There are infinitely many singular simplexes which make up $C_r(M)$, $B_r(M)$ and $Z_r(M)$. It follows from $\partial^2 = 0$ that $Z_r(M) \supset B_r(M)$; cf theorem 3.3. The singular homology group is defined by

$$H_r(M) \equiv Z_r(M)/B_r(M). \tag{6.5}$$

With mild topological assumptions, the singular homology group is isomorphic to the corresponding simplicial homology group with $\mathbb{R}$-coefficients and we employ the same symbol to denote both of them.

Now we are ready to define an integration of an r-form ω over an r-chain in M. We first define an integration of ω on an r-simplex s_r of M by

$$\int_{s_r} \omega = \int_{\bar{\sigma}_r} f^* \omega \tag{6.6}$$

where $f : \bar{\sigma}_r \to M$ is a smooth map such that $s_r = f(\bar{\sigma}_r)$. Since $f^* \omega$ is an r-form in $\mathbb{R}^r$, the RHS is the usual r-fold integral. For a general r-chain $c = \sum_i a_i s_{r,i} \in C_r(M)$, we define

$$\int_c \omega = \sum_i a_i \int_{s_{r,i}} \omega. \tag{6.7}$$

6.1.2   Stokes' theorem

Theorem 6.1. (**Stokes' theorem**) Let $\omega \in \Omega^{r-1}(M)$ and $c \in C_r(M)$. Then

$$\int_c d\omega = \int_{\partial c} \omega. \tag{6.8}$$

Proof. Since c is a linear combination of r-simplexes, it suffices to prove (6.8) for an r-simplex s_r in M. Let $f : \bar{\sigma}_r \to M$ be a map such that $f(\bar{\sigma}_r) = s_r$. Then

$$\int_{s_r} d\omega = \int_{\bar{\sigma}_r} f^*(d\omega) = \int_{\bar{\sigma}_r} d(f^*\omega)$$

where (5.75) has been used. We also have

$$\int_{\partial s_r} \omega = \int_{\partial \bar{\sigma}_r} f^* \omega.$$

Note that $f^*\omega$ is an $(r-1)$-form in $\mathbb{R}^r$. Thus, to prove Stokes' theorem

$$\int_{S_r} d\omega = \int_{\partial S_r} \omega \tag{6.9a}$$

it suffices to prove an alternative formula

$$\int_{\bar{\sigma}_r} d\psi = \int_{\partial \bar{\sigma}_r} \psi \tag{6.9b}$$

for an $(r-1)$-form ψ in $\mathbb{R}^r$. The most general form of ψ is

$$\psi = \sum a_\mu(x)\, dx^1 \wedge \ldots \wedge dx^{\mu-1} \wedge dx^{\mu+1} \wedge \ldots \wedge dx^r.$$

Since an integration is distributive, it suffices to prove (6.9b) for $\psi = a(x)dx^1 \wedge \ldots \wedge dx^{r-1}$. We note that

$$d\psi = \frac{\partial a}{\partial x^r}\, dx^r \wedge dx^1 \wedge \ldots \wedge dx^{r-1} = (-1)^{r-1}\frac{\partial a}{\partial x^r}\, dx^1 \wedge \ldots \wedge dx^{r-1} \wedge dx^r.$$

By direct computation, we find, from (6.2), that

$$\int_{\bar{\sigma}_r} d\psi = (-1)^{r-1}\int_{\bar{\sigma}_r}\frac{\partial a}{\partial x^r}\, dx^1 \ldots dx^{r-1}\, dx^r$$

$$= (-1)^{r-1}\int_{x^\mu \geq 0,\, \sum_{\mu=1}^{r-1} x^\mu \leq 1}\, dx^1 \ldots dx^{r-1}\int_0^{1-\sum_{\mu=1}^{r-1} x^\mu}\frac{\partial a}{\partial x^r}\, dx^r$$

$$= (-1)^{r-1}\int dx^1 \ldots dx^{r-1}$$

$$\times \left[a\left(x^1,\ldots,x^{r-1}, 1-\sum_{\mu=1}^{r-1}x^\mu\right) - a\left(x^1,\ldots,x^{r-1},0\right)\right].$$

For the boundary of $\bar{\sigma}_r$, we have

$$\partial\bar{\sigma}_r = (p_1, p_2, \ldots, p_r) - (p_0, p_2, \ldots, p_r)$$
$$+ \cdots + (-1)^r(p_0, p_1, \ldots, p_{r-1}).$$

Note that $\psi = a(x)dx^1 \wedge \ldots \wedge dx^{r-1}$ vanishes when one of $x^1,\ldots,x^{r-1}$ is constant. Then it follows that

$$\int_{(p_0,p_2,\ldots,p_r)} \psi = 0$$

since $x^1 \equiv 0$ on $(p_0, p_2, \ldots, p_r)$. In fact, most of the faces of $\partial\bar{\sigma}_r$ do not contribute to the RHS of (6.9b) and we are left with

$$\int_{\partial\bar{\sigma}_r} \psi = \int_{(p_1,p_2,\ldots,p_r)} \psi + (-1)^r\int_{(p_0,p_1,\ldots,p_{r-1})} \psi.$$

Since $(p_0, p_1, \ldots, p_{r-1})$ is the standard $(r-1)$-simplex $(x^\mu \geq 0, \sum_{\mu=1}^{r-1} x^\mu \leq 1)$, on which $x^r = 0$, the second term is

$$(-1)^r \int_{(p_0, p_1, \ldots, p_{r-1})} \psi = (-1)^r \int_{\bar{\sigma}_{r-1}} a(x^1, \ldots, x^{r-1}, 0) \, dx^1 \ldots dx^{r-1}.$$

The first term is

$$\int_{(p_1, p_2, \ldots, p_r)} \psi = \int_{(p_1, \ldots, p_{r-1}, p_0)} a\left(x^1, \ldots, x^{r-1}, 1 - \sum_{\mu=1}^{r-1} x^\mu\right) dx^1 \ldots dx^{r-1}$$

$$= (-1)^{r-1} \int_{\bar{\sigma}_{r-1}} a\left(x^1, \ldots, x^{r-1}, 1 - \sum_{\mu=1}^{r-1} x^\mu\right) dx^1 \ldots dx^{r-1}$$

where the integral domain $(p_1, \ldots, p_r)$ has been projected along x^r to the $(p_1, \ldots, p_{r-1}, p_0)$-plane, preserving the orientation. Collecting these results, we have proved (6.9b). [The reader is advised to verify this proof for $m = 3$ using figure 6.1.] $\square$

Exercise 6.1. Let $M = \mathbb{R}^3$ and $\omega = a \, dx + b \, dy + c \, dz$. Show that Stokes' theorem is written as

$$\int_S \text{curl} \, \omega \cdot d\mathbf{S} = \oint_C \omega \cdot d\mathbf{S} \qquad \text{(Stokes' theorem)} \qquad (6.10)$$

where $\omega = (a, b, c)$ and C is the boundary of a surface S. Similarly, for $\psi = \frac{1}{2}\psi_{\mu\nu} \, dx^\mu \wedge dx^\nu$, show that

$$\int_V \text{div} \, \psi \, dV = \oint_S \psi \cdot d\mathbf{S} \qquad \text{(Gauss' theorem)}$$

where $\psi^\lambda = \varepsilon^{\lambda\mu\nu}\psi_{\mu\nu}$ and S is the boundary of a volume V.

6.2 de Rham cohomology groups

6.2.1 Definitions

Definition 6.1. Let M be an m-dimensional differentiable manifold. The set of closed r-forms is called the rth **cocycle group**, denoted $Z^r(M)$. The set of exact r-forms is called the rth **coboundary group**, denoted $B^r(M)$. These are vector spaces with $\mathbb{R}$-coefficients. It follows from $d^2 = 0$ that $Z^r(M) \supset B^r(M)$.

Exercise 6.2. Show that

(a) if $\omega \in Z^r(M)$ and $\psi \in Z^s(M)$, then $\omega \wedge \psi \in Z^{r+s}(M)$;
(b) if $\omega \in Z^r(M)$ and $\psi \in B^s(M)$, then $\omega \wedge \psi \in B^{r+s}(M)$; and

(c) if $\omega \in B^r(M)$ and $\psi \in B^s(M)$, then $\omega \wedge \psi \in B^{r+s}(M)$.

Definition 6.2. The rth **de Rham cohomology group** is defined by

$$H^r(M; \mathbb{R}) \equiv Z^r(M)/B^r(M). \tag{6.11}$$

If $r \leq -1$ or $r \geq m+1$, $H^r(M; \mathbb{R})$ may be defined to be trivial. In the following, we omit the explicit quotation of $\mathbb{R}$-coefficients.

Let $\omega \in Z^r(M)$. Then $[\omega] \in H^r(M)$ is the equivalence class $\{\omega' \in Z^r(M)|\omega' = \omega + d\psi, \psi \in \Omega^{r-1}(M)\}$. Two forms which differ by an exact form are called **cohomologous**. We will see later that $H^r(M)$ is isomorphic to $H_r(M)$. The following examples will clarify the idea of de Rham cohomology groups.

Example 6.1. When $r = 0$, $B^0(M)$ has no meaning since there is no (-1)-form. We define $\Omega^{-1}(M)$ to be empty, hence $B^0(M) = 0$. Then $H^0(M) = Z^0(M) = \{f \in \Omega^0(M) = \mathcal{F}(M)|df = 0\}$. If M is connected, the condition $df = 0$ is satisfied if and only if f is constant over M. Hence, $H^0(M)$ is isomorphic to the vector space $\mathbb{R}$,

$$H^0(M) \cong \mathbb{R}. \tag{6.12}$$

If M has n connected components, $df = 0$ is satisfied if and only if f is constant on each connected component, hence it is specified by n real numbers,

$$H^0(M) \cong \underbrace{\mathbb{R} \oplus \mathbb{R} \oplus \cdots \oplus \mathbb{R}}_{n}. \tag{6.13}$$

Example 6.2. Let $M = \mathbb{R}$. From example 6.1, we have $H^0(\mathbb{R}) = \mathbb{R}$. Let us find $H^1(\mathbb{R})$ next. Let x be a coordinate of $\mathbb{R}$. Since $\dim \mathbb{R} = 1$, any one-form $\omega \in \Omega^1(\mathbb{R})$ is closed, $d\omega = 0$. Let $\omega = f \, dx$, where $f \in \mathcal{F}(\mathbb{R})$. Define a function $F(x)$ by

$$F(x) = \int_0^x f(s) \, ds \in \mathcal{F}(\mathbb{R}) = \Omega^0(\mathbb{R}).$$

Since $dF(x)/dx = f(x)$, ω is an exact form,

$$\omega = f \, dx = \frac{dF(x)}{dx} \, dx = dF.$$

Thus, any one-form is closed as well as exact. We have established

$$H^1(\mathbb{R}) = \{0\}. \tag{6.14}$$

Example 6.3. Let $S^1 = \{e^{i\theta}|0 \leq \theta < 2\pi\}$. Since S^1 is connected, we have $H^0(S^1) = \mathbb{R}$. We compute $H^1(S^1)$ next. Let $\omega = f(\theta) \, d\theta \in \Omega^1(S^1)$. Is it possible to write $\omega = dF$ for some $F \in \mathcal{F}(S^1)$? Let us repeat the analysis of the previous example. If $\omega = dF$, then $F \in \mathcal{F}(S^1)$ must be given by

$$F(\theta) = \int_0^\theta f(\theta') \, d\theta'.$$

For F to be defined uniquely on S^1, F must satisfy the periodicity $F(2\pi) = F(0)$ ($=0$). Namely F must satisfy

$$F(2\pi) = \int_0^{2\pi} f(\theta') \, d\theta' = 0.$$

If we define a map $\lambda : \Omega^1(S^1) \to \mathbb{R}$ by

$$\lambda : \omega = f \, d\theta \mapsto \int_0^{2\pi} f(\theta') \, d\theta' \qquad (6.15)$$

then $B^1(S^1)$ is identified with $\ker \lambda$. Now we have (theorem 3.1)

$$H^1(S^1) = \Omega^1(S^1)/\ker \lambda = \operatorname{im} \lambda = \mathbb{R}. \qquad (6.16)$$

This is also obtained from the following consideration. Let ω and ω' be closed forms that are not exact. Although $\omega - \omega'$ is not exact in general, we can show that there exists a number $a \in \mathbb{R}$ such that $\omega' - a\omega$ is exact. In fact, if we put

$$a = \int_0^{2\pi} \omega' \Big/ \int_0^{2\pi} \omega$$

we have

$$\int_0^{2\pi} (\omega' - a\omega) = 0.$$

This shows that, given a closed form ω which is not exact, any closed form ω' is cohomologous to $a\omega$ for some $a \in \mathbb{R}$. Thus, each cohomology class is specified by a real number a, hence $H^1(S^1) = \mathbb{R}$.

Exercise 6.3. Let $M = \mathbb{R}^2 - \{0\}$. Define a one-form ω by

$$\omega = \frac{-y}{x^2 + y^2} \, dx + \frac{x}{x^2 + y^2} \, dy. \qquad (6.17)$$

    (a) Show that ω is closed.
    (b) Define a 'function' $F(x, y) = \tan^{-1}(y/x)$. Show that $\omega = dF$. Is ω exact?

6.2.2 Duality of $H_r(M)$ and $H^r(M)$; de Rham's theorem

As the name itself suggests, the *co*homology group is a dual space of the homology group. The duality is provided by Stokes' theorem. We first define the inner product of an r-form and an r-chain in M. Let M be an m-dimensional manifold and let $C_r(M)$ be the chain group of M. Take $c \in C_r(M)$ and $\omega \in \Omega^r(M)$ where $1 \le r \le m$. Define an inner product $(\ ,\ ) : C_r(M) \times \Omega^r(M) \to \mathbb{R}$ by

$$c, \omega \mapsto (c, \omega) \equiv \int_c \omega. \tag{6.18}$$

Clearly, (c, ω) is linear in both c and ω and $(\ , \omega)$ may be regarded as a linear map acting on c and *vice versa*,

$$(c_1 + c_2, \omega) = \int_{c_1+c_2} \omega = \int_{c_1} \omega + \int_{c_2} \omega \tag{6.19a}$$

$$(c, \omega_1 + \omega_2) = \int_c (\omega_1 + \omega_2) = \int_c \omega_1 + \int_c \omega_2. \tag{6.19b}$$

Now Stokes' theorem takes a compact form:

$$(c, \mathrm{d}\omega) = (\partial c, \omega). \tag{6.20}$$

In this sense, the exterior derivative operator d is the adjoint of the boundary operator ∂ and *vice versa*.

Exercise 6.4. Let (i) $c \in B_r(M)$, $\omega \in Z^r(M)$ or (ii) $c \in Z_r(M)$, $\omega \in B^r(M)$. Show, in both cases, that $(c, \omega) = 0$.

The inner product $(\ ,\ )$ naturally induces an inner product λ between the elements of $H_r(M)$ and $H^r(M)$. We now show that $H_r(M)$ is the dual of $H^r(M)$. Let $[c] \in H_r(M)$ and $[\omega] \in H^r(M)$ and define an inner product $\Lambda : H_r(M) \times H^r(M) \to \mathbb{R}$ by

$$\Lambda([c], [\omega]) \equiv (c, \omega) = \int_c \omega. \tag{6.21}$$

This is well defined since (6.21) is independent of the choice of the representatives. In fact, if we take $c + \partial c'$, $c' \in C_{r+1}(M)$, we have, from Stokes' theorem,

$$(c + \partial c', \omega) = (c, \omega) + (c', \mathrm{d}\omega) = (c, \omega)$$

where $\mathrm{d}\omega = 0$ has been used. Similarly, for $\omega + \mathrm{d}\psi$, $\psi \in \Omega^{r-1}(M)$,

$$(c, \omega + \mathrm{d}\psi) = (c, \omega) + (\partial c, \psi) = (c, \omega)$$

since $\partial c = 0$. Note that $\Lambda(\ , [\omega])$ is a linear map $H_r(M) \to \mathbb{R}$, and $\Lambda([c], \ )$ is a linear map $H^r(M) \to \mathbb{R}$. To prove the duality of $H_r(M)$ and $H^r(M)$, we have to show that $\Lambda(\ , [\omega])$ has the maximal rank, that is, dim $H_r(M) = $ dim $H^r(M)$. We accept the following theorem due to de Rham without the proof which is highly non-trivial.

Theorem 6.2. (**de Rham's theorem**) If M is a compact manifold, $H_r(M)$ and $H^r(M)$ are finite dimensional. Moreover the map

$$\Lambda : H_r(M) \times H^r(M) \to \mathbb{R}$$

is bilinear and non-degenerate. Thus, $H^r(M)$ is the dual vector space of $H_r(M)$.

A **period** of a closed r-form ω over a cycle c is defined by $(c, \omega) = \int_c \omega$. Exercise 6.4 shows that the period vanishes if ω is exact or if c is a boundary. The following corollary is easily derived from de Rham's theorem.

Corollary 6.1. Let M be a compact manifold and let k be the rth Betti number (see section 3.4). Let $c_1, c_2, \ldots, c_k$ be properly chosen elements of $Z_r(M)$ such that $[c_i] \neq [c_j]$.

(a) A closed r-form ψ is exact if and only if

$$\int_{c_i} \psi = 0 \qquad (1 \le i \le k). \tag{6.22}$$

(b) For any set of real numbers $b_1, b_2, \ldots, b_k$ there exists a closed r-form ω such that

$$\int_{c_i} \omega = b_i \qquad (1 \le i \le k). \tag{6.23}$$

Proof. (a) de Rham's theorem states that the bilinear form $\Lambda([c], [\omega])$ is non-degenerate. Hence, if $\Lambda([c_i], \ )$ is regarded as a linear map acting on $H^r(M)$, the kernel consists of the trivial element, the cohomology class of exact forms. Accordingly, ψ is an exact form.

(b) de Rham's theorem ensures that corresponding to the homology basis $\{[c_i]\}$, we may choose the dual basis $\{[\omega_i]\}$ of $H^r(M)$ such that

$$\Lambda([c_i], [\omega_j]) = \int_{c_i} \omega_j = \delta_{ij}. \tag{6.24}$$

If we define $\omega \equiv \sum_{i=1}^{k} b_i \omega_i$, the closed r-form ω satisfies

$$\int_{c_i} \omega = b_i$$

as claimed.   □

For example. we observe the duality of the following groups.

(a) $H^0(M) \cong H_0(M) \cong \underbrace{\mathbb{R} \oplus \cdots \oplus \mathbb{R}}_{n}$ if M has n connected components.

(b) $H^1(S^1) \cong H_1(S^1) \cong \mathbb{R}$.

Since $H^r(M)$ is isomorphic to $H_r(M)$, we find that

$$b^r(M) \equiv \dim H^r(M) = \dim H_r(M) = b_r(M) \tag{6.25}$$

where $b_r(M)$ is the Betti number of M. The Euler characteristic is now written as

$$\chi(M) = \sum_{r=1}^{m} (-1)^r b^r(M). \tag{6.26}$$

This is quite an interesting formula; the LHS is purely *topological* while the RHS is given by an *analytic* condition (note that $d\omega = 0$ is a set of partial differential equations). We will frequently encounter this interplay between topology and analysis.

In summary, we have the chain complex $C(M)$ and the de Rham complex $\Omega^*(M)$,

$$\begin{aligned}
&\longleftarrow C_{r-1}(M) \xleftarrow{\partial_r} C_r(M) \xleftarrow{\partial_{r+1}} C_{r+1}(M) \longleftarrow \\
&\longrightarrow \Omega^{r-1}(M) \xrightarrow{d_r} \Omega^r(M) \xrightarrow{d_{r+1}} \Omega^{r+1}(M) \longleftarrow
\end{aligned} \tag{6.27}$$

for which the rth homology group is defined by

$$H_r(M) = Z_r(M)/B_r(M) = \ker \partial_r / \operatorname{im} \partial_{r+1}$$

and the rth de Rham cohomology group is defined by

$$H^r(M) = Z^r(M)/B^r(M) = \ker d_{r+1} / \operatorname{im} d_r.$$

6.3 Poincaré's lemma

An exact form is always closed but the converse is not necessarily true. However, the following theorem provides the situation in which the converse is also true.

Theorem 6.3. (**Poincaré's lemma**) If a coordinate neighbourhood U of a manifold M is contractible to a point $p_0 \in M$, any closed r-form on U is also exact.

Proof. We assume U is smoothly contractible to p_0, that is, there exists a smooth map $F : U \times I \to U$ such that

$$F(x, 0) = x, \qquad F(x, 1) = p_0 \qquad \text{for } x \in U.$$

Let us consider an r-form $\eta \in \Omega^r(U \times I)$,

$$
\begin{aligned}
\eta = a_{i_1 \dots i_r}(x, t)\, \mathrm{d}x^{i_1} \wedge \dots \wedge \mathrm{d}x^{i_r} \\
+ b_{j_1 \dots j_{r-1}}(x, t)\, \mathrm{d}t \wedge \mathrm{d}x^{j_1} \wedge \dots \wedge \mathrm{d}x^{j_{r-1}}
\end{aligned}
\tag{6.28}
$$

where x is the coordinate of U and t of I. Define a map $P : \Omega^r(U \times I) \to \Omega^{r-1}(U)$ by

$$
P\eta \equiv \left(\int_0^1 \mathrm{d}s\, b_{j_1 \dots j_{r-1}}(x, s) \right) \mathrm{d}x^{j_1} \wedge \dots \wedge \mathrm{d}x^{j_{r-1}}.
\tag{6.29}
$$

Next, define a map $f_t : U \to U \times I$ by $f_t(x) = (x, t)$. The pullback of the first term of (6.28) by f_t^* is an element of $\Omega^r(U)$,

$$
f_t^*\eta = a_{i_1 \dots i_r}(x, t)\, \mathrm{d}x^{i_1} \wedge \dots \wedge \mathrm{d}x^{i_r} \in \Omega^r(U).
\tag{6.30}
$$

We now prove the following identity,

$$
\mathrm{d}(P\eta) + P(\mathrm{d}\eta) = f_1^*\eta - f_0^*\eta.
\tag{6.31}
$$

Each term of the LHS is calculated to be

$$
\begin{aligned}
\mathrm{d}P\eta &= \mathrm{d}\left(\int_0^1 \mathrm{d}s\, b_{j_1 \dots j_{r-1}} \right) \mathrm{d}x^{j_1} \wedge \dots \wedge \mathrm{d}x^{j_{r-1}} \\
&= \int_0^1 \mathrm{d}s \left(\frac{\partial b_{j_1 \dots j_{r-1}}}{\partial x^{j_r}} \right) \mathrm{d}x^{j_r} \wedge \mathrm{d}x^{j_1} \wedge \dots \wedge \mathrm{d}x^{j_{r-1}} \\
P\,\mathrm{d}\eta &= P\left[\left(\frac{\partial a_{i_1 \dots i_r}}{\partial x^{i_{r+1}}} \right) \mathrm{d}x^{i_{r+1}} \wedge \mathrm{d}x^{i_1} \wedge \dots \wedge \mathrm{d}x^{i_r} \right. \\
&\quad + \left(\frac{\partial a_{i_1 \dots i_r}}{\partial t} \right) \mathrm{d}t \wedge \mathrm{d}x^{i_1} \wedge \dots \wedge \mathrm{d}x^{i_r} \\
&\quad \left. + \left(\frac{\partial b_{j_1 \dots j_{r-1}}}{\partial x^{j_r}} \right) \mathrm{d}x^{j_r} \wedge \mathrm{d}t \wedge \mathrm{d}x^{j_1} \wedge \dots \wedge \mathrm{d}x^{j_{r-1}} \right] \\
&= \left[\int_0^1 \mathrm{d}s \left(\frac{\partial a_{i_1 \dots i_r}}{\partial s} \right) \right] \mathrm{d}x^{i_1} \wedge \dots \wedge \mathrm{d}x^{i_r} \\
&\quad - \left[\int_0^1 \mathrm{d}s \left(\frac{\partial b_{j_1 \dots j_{r-1}}}{\partial x^{j_r}} \right) \right] \mathrm{d}x^{j_r} \wedge \mathrm{d}x^{j_1} \wedge \dots \wedge \mathrm{d}x^{j_{r-1}}.
\end{aligned}
$$

Collecting these results, we have

$$
\begin{aligned}
\mathrm{d}(P\eta) + P(\mathrm{d}\eta) &= \left[\int_0^1 \mathrm{d}s \left(\frac{\partial a_{i_1 \dots i_r}}{\partial s} \right) \right] \mathrm{d}x^{i_1} \wedge \dots \wedge \mathrm{d}x^{i_r} \\
&= [a_{i_1 \dots i_r}(x, 1) - a_{i_1 \dots i_r}(x, 0)]\, \mathrm{d}x^{i_1} \wedge \dots \wedge \mathrm{d}x^{i_r} \\
&= f_1^*\eta - f_0^*\eta.
\end{aligned}
$$

Poincaré's lemma readily follows from (6.31). Let ω be a closed r-form on a contractible chart U. We will show that ω is written as an exact form,

$$\omega = d(-PF^*\omega), \tag{6.32}$$

F being the smooth contraction map. In fact, if η in (6.31) is replaced by $F^*\omega \in \Omega^r(U \times I)$ we have

$$dPF^*\omega + P\,dF^*\omega = f_1{}^* \circ F^*\omega - f_0{}^* \circ F^*\omega$$
$$= (F \circ f_1)^*\omega - (F \circ f_0)^*\omega \tag{6.33}$$

where use has been made of the relation $(f \circ g)^* = g^* \circ f^*$. Clearly $F \circ f_1 : U \to U$ is a constant map $x \mapsto p_0$, hence $(F \circ f_1)^* = 0$. However, $F \circ f_0 = \mathrm{id}_U$, hence $(F \circ f_0)^* : \Omega^r(U) \to \Omega^r(U)$ is the identity map. Thus, the RHS of (6.33) is simply $-\omega$. The second term of the LHS vanishes since ω is closed; $dF^*\omega = F^*\,d\omega = 0$, where use has been made of (5.75). Finally, (6.33) becomes $\omega = -dP\,F^*\omega$, which proves the theorem. $\qquad\square$

Any closed form is exact at least locally. The de Rham cohomology group is regarded as an obstruction to the *global* exactness of closed forms.

Example 6.4. Since $\mathbb{R}^n$ is contractible, we have

$$H^r(\mathbb{R}^n) = 0 \qquad 1 \le r \le n. \tag{6.34}$$

Note, however, that $H^0(\mathbb{R}^n) = \mathbb{R}$.

6.4 Structure of de Rham cohomology groups

de Rham cohomology groups exhibit quite an interesting structure that is very difficult or even impossible to appreciate with homology groups.

6.4.1 Poincaré duality

Let M be a compact m-dimensional manifold and let $\omega \in H^r(M)$ and $\eta \in H^{m-r}(M)$. Noting that $\omega \wedge \eta$ is a volume element, we define an inner product $\langle\,,\,\rangle : H^r(M) \times H^{m-r}(M) \to \mathbb{R}$ by

$$\langle \omega, \eta \rangle \equiv \int_M \omega \wedge \eta. \tag{6.35}$$

The inner product is bilinear. Moreover, it is non-singular, that is, if $\omega \ne 0$ or $\eta \ne 0$, $\langle \omega, \eta \rangle$ cannot vanish identically. Thus, (6.35) defines the duality of $H^r(M)$ and $H^{m-r}(M)$,

$$H^r(M) \cong H^{m-r}(M) \tag{6.36}$$

called the **Poincaré duality**. Accordingly, the Betti numbers have a symmetry

$$b_r = b_{m-r}. \tag{6.37}$$

It follows from (6.37) that the Euler characteristic of an odd-dimensional space vanishes,

$$\chi(M) = \sum (-1)^r b_r = \tfrac{1}{2} \left\{ \sum (-1)^r b_r + \sum (-1)^{m-r} b_{m-r} \right\}$$

$$= \tfrac{1}{2} \left\{ \sum (-1)^r b_r - \sum (-1)^{-r} b_r \right\} = 0. \tag{6.38}$$

6.4.2 Cohomology rings

Let $[\omega] \in H^q(M)$ and $[\eta] \in H^r(M)$. Define a product of $[\omega]$ and $[\eta]$ by

$$[\omega] \wedge [\eta] \equiv [\omega \wedge \eta]. \tag{6.39}$$

It follows from exercise 6.2 that $\omega \wedge \eta$ is closed, hence $[\omega \wedge \eta]$ is an element of $H^{q+r}(M)$. Moreover, $[\omega \wedge \eta]$ is independent of the choice of the representatives of $[\omega]$ and $[\eta]$. For example, if we take $\omega' = \omega + d\psi$ instead of ω, we have

$$[\omega'] \wedge [\eta] \equiv [(\omega + d\psi) \wedge \eta] = [\omega \wedge \eta + d(\psi \wedge \eta)] = [\omega \wedge \eta].$$

Thus, the product $\wedge : H^q(M) \times H^r(M) \to H^{q+r}(M)$ is a well-defined map.
The **cohomology ring** $H^*(M)$ is defined by the direct sum,

$$H^*(M) \equiv \bigoplus_{r=1}^{m} H^r(M). \tag{6.40}$$

The product is provided by the exterior product defined earlier,

$$\wedge : H^*(M) \times H^*(M) \to H^*(M). \tag{6.41}$$

The addition is the formal sum of two elements of $H^*(M)$. One of the superiorities of cohomology groups over homology groups resides here. Products of chains are not well defined and homology groups cannot have a ring structure.

6.4.3 The Künneth formula

Let M be a product of two manifolds $M = M_1 \times M_2$. Let $\{\omega_i^p\}$ $(1 \le i \le b^p(M_1))$ be a basis of $H^p(M_1)$ and $\{\eta_i^p\}$ $(1 \le i \le b^p(M_2))$ be that of $H^p(M_2)$. Clearly $\omega_i^p \wedge \eta_j^{r-p}$ $(1 \le p \le r)$ is a closed r-form in M. We show that it is not exact. If it were exact, it would be written as

$$\omega_i^p \wedge \eta_j^{r-p} = d(\alpha^{p-1} \wedge \beta^{r-p} + \gamma^p \wedge \delta^{r-p-1}) \tag{6.42}$$

for some $\alpha^{p-1} \in \Omega^{p-1}(M_1)$, $\beta^{r-p} \in \Omega^{r-p}(M_2)$, $\gamma^p \in \Omega^p(M_1)$ and $\delta^{r-p-1} \in \Omega^{r-p-1}(M_2)$. [If $p = 0$, we put $\alpha^{p-1} = 0$.] By executing the exterior derivative in (6.42), we have

$$\omega_i^p \wedge \eta_j^{r-p} = d\alpha^{p-1} \wedge \beta^{r-p} + (-1)^{p-1}\alpha^{p-1} \wedge d\beta^{r-p}$$
$$+ d\gamma^p \wedge \delta^{r-p-1} + (-1)^p \gamma^p \wedge d\delta^{r-p-1}. \tag{6.43}$$

By comparing the LHS with the RHS, we find $\alpha^{p-1} = \delta^{r-p-1} = 0$, hence $\omega_i^p \wedge \eta_j^{r-p} = 0$ in contradiction to our assumption. Thus, $\omega_i^p \wedge \eta_j^{r-p}$ is a non-trivial element of $H^r(M)$. Conversely, any element of $H^r(M)$ can be decomposed into a sum of a product of the elements of $H^p(M_1)$ and $H^{r-p}(M_2)$ for $0 \le p \le r$. Now we have obtained the **Künneth formula**

$$H^r(M) = \bigoplus_{p+q=r} [H^p(M_1) \otimes H^q(M_2)]. \tag{6.44}$$

This is rewritten in terms of the Betti numbers as

$$b^r(M) = \sum_{p+q=r} b^p(M_1) b^q(M_2). \tag{6.45}$$

The Künneth formula also gives a relation between the cohomology rings of the respective manifolds,

$$H^*(M) = \sum_{r=1}^m H^r(M) = \sum_{r=1}^m \bigoplus_{p+q=r} H^p(M_1) \otimes H^q(M_2)$$
$$= \sum_p H^p(M_1) \otimes \sum_q H^q(M_2) = H^*(M_1) \otimes H^*(M_2). \tag{6.46}$$

Exercise 6.5. Let $M = M_1 \times M_2$. Show that

$$\chi(M) = \chi(M_1) \cdot \chi(M_2). \tag{6.47}$$

Example 6.5. Let $T^2 = S^1 \times S^1$ be the torus. Since $H^0(S^1) = \mathbb{R}$ and $H^1(S^1) = \mathbb{R}$, we have

$$H^0(T^2) = \mathbb{R} \otimes \mathbb{R} = \mathbb{R} \tag{6.48a}$$
$$H^1(T^2) = (\mathbb{R} \otimes \mathbb{R}) \oplus (\mathbb{R} \otimes \mathbb{R}) = \mathbb{R} \oplus \mathbb{R} \tag{6.48b}$$
$$H^2(T^2) = \mathbb{R} \otimes \mathbb{R} = \mathbb{R}. \tag{6.48c}$$

Observe the Poincaré duality $H^0(T^2) = H^2(T^2)$. [*Remark*: $\mathbb{R} \otimes \mathbb{R}$ is the tensor product and should not be confused with the direct product. Clearly the product of two real numbers is a real number.] Let us parametrize the coordinate of T^2

as (θ_1, θ_2) where θ_i is the coordinate of S^1. The groups $H^r(T^2)$ are generated by the following forms:

$$
\begin{aligned}
r = 0: & \quad \omega_0 = c_0 & c_0 \in \mathbb{R} \\
r = 1: & \quad \omega_1 = c_1 \, d\theta_1 + c_1' \, d\theta_2 & c_1, c_1' \in \mathbb{R} \\
r = 2: & \quad \omega_2 = c_2 \, d\theta_1 \wedge d\theta_2 & c_2 \in \mathbb{R}.
\end{aligned}
\tag{6.49a}
$$

Although the one-form $d\theta_i$ looks like an exact form, there is no *function* θ_i which is defined uniquely on S^1. Since $\chi(S^1) = 0$, we have $\chi(T^2) = 0$.

The de Rham cohomology groups of

$$
T^n = \underbrace{S^1 \times \cdots \times S^1}_{n}
$$

are obtained similarly. $H^r(T^n)$ is generated by r-forms of the form

$$
d\theta^{i_1} \wedge d\theta^{i_2} \wedge \ldots \wedge d\theta^{i_r}
\tag{6.50}
$$

where $i_1 < i_2 < \cdots < i_r$ are chosen from $1, \ldots, n$. Clearly

$$
b^r = \dim H^r(T^n) = \binom{n}{r}.
\tag{6.51}
$$

The Euler characteristic is directly obtained from (6.51) as

$$
\chi(T^n) = \sum (-1)^r \binom{n}{r} = (1 - 1)^n = 0.
\tag{6.52}
$$

6.4.4 Pullback of de Rham cohomology groups

Let $f : M \to N$ be a smooth map. Equation (5.75) shows that the pullback f^* maps closed forms to closed forms and exact forms to exact forms. Accordingly, we may define a pullback of the cohomology groups $f^* : H^r(N) \to H^r(M)$ by

$$
f^*[\omega] = [f^*\omega] \qquad [\omega] \in H^r(N).
\tag{6.53}
$$

The pullback f^* preserves the ring structure of $H^*(N)$. In fact, if $[\omega] \in H^p(N)$ and $[\eta] \in H^q(N)$, we find

$$
\begin{aligned}
f^*([\omega] \wedge [\eta]) = f^*[\omega \wedge \eta] &= [f^*(\omega \wedge \eta)] \\
&= [f^*\omega \wedge f^*\eta] = [f^*\omega] \wedge [f^*\eta].
\end{aligned}
\tag{6.54}
$$

6.4.5 Homotopy and $H^1(M)$

Let $f, g : M \to N$ be smooth maps. We assume f and g are homotopic to each other, that is, there exists a smooth map $F : M \times I \to N$ such that $F(p, 0) =$

$f(p)$ and $F(p, 1) = g(p)$. We now prove that $f^* : H^r(N) \to H^r(M)$ is equal to $g^* : H^r(N) \to H^r(M)$.

Lemma 6.1. Let f^* and g^* be defined as before. If $\omega \in \Omega^r(N)$ is a closed form, the difference of the pullback images is exact,

$$f^*\omega - g^*\omega = \mathrm{d}\psi \qquad \psi \in \Omega^{r-1}(M). \tag{6.55}$$

Proof. We first note that

$$f = F \circ f_0, \qquad g = F \circ f_1$$

where $f_t : M \to M \times I$ $(p \mapsto (p, t))$ has been defined in theorem 6.3. The LHS of (6.55) is

$$(F \circ f_0)^*\omega - (F \circ f_1)^*\omega = f_0^* \circ F^*\omega - f_1^* \circ F^*\omega$$
$$= -[\mathrm{d}P(F^*\omega) + P\,\mathrm{d}(F^*\omega)] = -\mathrm{d}P\,F^*\omega$$

where (6.33) has been used. This shows that $f^*\omega - g^*\omega = \mathrm{d}(-PF^*\omega)$.   □

Now it is easy to see that $f^* = g^*$ as the pullback maps $H^r(N) \to H^r(M)$. In fact, from the previous lemma,

$$[f^*\omega - g^*\omega] = [f^*\omega] - [g^*\omega] = [\mathrm{d}\psi] = 0.$$

We have established the following theorem.

Theorem 6.4. Let $f, g : M \to N$ be maps which are homotopic to each other. Then the pullback maps f^* and g^* of the de Rham cohomology groups $H^r(N) \to H^r(M)$ are identical.

Let M be a simply connected manifold, namely $\pi_1(M) \cong \{0\}$. Since $H_1(M) = \pi_1(M)$ modulo the commutator subgroup (theorem 4.9), it follows that $H_1(M)$ is also trivial. In terms of the de Rham cohomology group this can be expressed as follows.

Theorem 6.5. Let M be a simply connected manifold. Then its first de Rham cohomology group is trivial.

Proof. Let ω be a closed one-form on M. It is clear that if $\omega = \mathrm{d}f$, then a function f must be of the form

$$f(p) = \int_{p_0}^{p} \omega \tag{6.56}$$

$p_0 \in M$ being a fixed point.

We first prove that an integral of a closed form along a loop vanishes. Let $\alpha : I \to M$ be a loop at $p \in M$ and let $c_p : I \to M$ $(t \mapsto p)$ be a constant loop. Since M is simply connected, there exists a homotopy $F(s, t)$ such that $F(s, 0) = \alpha(s)$ and $F(s, 1) = c_p(s)$. We assume $F : I \times I \to M$ is smooth. Define the integral of a one-form ω over $\alpha(I)$ by

$$\int_{\alpha(I)} \omega = \int_{S^1} \alpha^* \omega \qquad (6.57)$$

where we have taken the integral domain in the RHS to be S^1 since $I = [0, 1]$ in the LHS is compactified to S^1. From lemma 6.1, we have, for a closed one-form ω,

$$\alpha^* \omega - c_p^* \omega = dg \qquad (6.58)$$

where $g = -PF^*\omega$. The pullback $c_p\omega$ vanishes since c_p is a constant map. Then (6.57) vanishes since ∂S^1 is empty,

$$\int_{S^1} \alpha^* \omega = \int_{S^1} dg = \int_{\partial S^1} g = 0. \qquad (6.59)$$

Let β and γ be two paths connecting p_0 and p. According to (6.59), integrals of ω along β and along γ are identical,

$$\int_{\beta(I)} \omega = \int_{\gamma(I)} \omega.$$

This shows that (6.56) is indeed well defined, hence ω is exact.   □

Example 6.6. The n-sphere S^n ($n \geq 2$) is simply connected, hence

$$H^1(S^n) = 0 \qquad n \geq 2. \qquad (6.60)$$

From the Poincaré duality, we find

$$H^0(S^n) \cong H^n(S^n) = \mathbb{R}. \qquad (6.61)$$

It can be shown that

$$H^r(S^n) = 0 \qquad 1 \leq r \leq n - 1. \qquad (6.62)$$

$H^n(S^n)$ is generated by the volume element Ω. Since there are no $(n + 1)$-forms on S^n, every n-form is closed. Ω cannot be exact since if $\Omega = d\psi$, we would have

$$\int_{S^n} \Omega = \int_{S^n} d\psi = \int_{\partial S^n} \psi = 0.$$

The Euler characteristic is

$$\chi(S^n) = 1 + (-1)^n = \begin{cases} 0 & n \text{ is odd,} \\ 2 & n \text{ is even.} \end{cases} \qquad (6.63)$$

Example 6.7. Take S^2 embedded in $\mathbb{R}^3$ and define

$$\Omega = \sin\theta \, d\theta \wedge d\phi \qquad (6.64)$$

where (θ, ϕ) is the usual polar coordinate. Verify that Ω is closed. We may *formally* write Ω as

$$\Omega = -d(\cos\theta) \wedge d\phi = -d(\cos\theta \, d\phi).$$

Note, however, that Ω is not exact.

7

RIEMANNIAN GEOMETRY

A manifold is a topological space which locally looks like $\mathbb{R}^n$. Calculus on a manifold is assured by the existence of smooth coordinate systems. A manifold may carry a further structure if it is endowed with a metric tensor, which is a natural generalization of the inner product between two vectors in $\mathbb{R}^n$ to an arbitrary manifold. With this new structure, we define an inner product between two vectors in a tangent space T_pM. We may also compare a vector at a point $p \in M$ with another vector at a different point $p' \in M$ with the help of the 'connection'.

There are many books about Riemannian geometry. Those which are accessible to physicists are Choquet-Bruhat *et al* (1982), Dodson and Poston (1977) and Hicks (1965). Lightman *et al* (1975) and chapter 3 of Wald (1984) are also recommended.

7.1 Riemannian manifolds and pseudo-Riemannian manifolds

7.1.1 Metric tensors

In elementary geometry, the inner product between two vectors U and V is defined by $U \cdot V = \sum_{i=1}^{m} U_i V_i$ where U_i and V_i are the components of the vectors in $\mathbb{R}^m$. On a manifold, an inner product is defined at each tangent space T_pM.

Definition 7.1. Let M be a differentiable manifold. A **Riemannian metric** g on M is a type $(0, 2)$ tensor field on M which satisfies the following axioms at each point $p \in M$:

(i) $g_p(U, V) = g_p(V, U)$,
(ii) $g_p(U, U) \geq 0$, where the equality holds only when $U = 0$.

Here $U, V \in T_pM$ and $g_p = g|_p$. In short, g_p is a symmetric positive-definite bilinear form.

A tensor field g of type $(0, 2)$ is a **pseudo-Riemannian metric** if it satisfies (i) and

(ii$'$) if $g_p(U, V) = 0$ for any $U \in T_pM$, then $V = 0$.

In chapter 5, we have defined the inner product between a vector $V \in T_M$ and a dual vector $\omega \in T_p^*M$ as a map $\langle \ , \ \rangle : T_p^*M \times T_pM \to \mathbb{R}$. If there exists a metric g, we define an inner product between two vectors $U, V \in T_pM$ by $g_p(U, V)$. Since g_p is a map $T_pM \otimes T_pM \to \mathbb{R}$ we may define a linear map $g_p(U, \ ) : T_pM \to \mathbb{R}$ by $V \mapsto g_p(U, V)$. Then $g_p(U, \ )$ is identified with a one-form $\omega_U \in T_p^*M$. Similarly, $\omega \in T_p^*M$ induces $V_\omega \in T_pM$ by $\langle \omega, U \rangle = g(V_\omega, U)$. Thus, the metric g_p gives rise to an isomorphism between T_pM and T_p^*M.

Let (U, φ) be a chart in M and $\{x^\mu\}$ the coordinates. Since $g \in \mathcal{T}_2^0(M)$, it is expanded in terms of $\mathrm{d}x^\mu \otimes \mathrm{d}x^\nu$ as

$$g_p = g_{\mu\nu}(p)\mathrm{d}x^\mu \otimes \mathrm{d}x^\nu. \tag{7.1a}$$

It is easily checked that

$$g_{\mu\nu}(p) = g_p\left(\frac{\partial}{\partial x^\mu}, \frac{\partial}{\partial x^\nu}\right) = g_{\nu\mu}(p) \qquad (p \in M). \tag{7.1b}$$

We usually omit p in $g_{\mu\nu}$ unless it may cause confusion. It is common to regard $(g_{\mu\nu})$ as a matrix whose (μ, ν)th entry is $g_{\mu\nu}$. Since $(g_{\mu\nu})$ has the maximal rank, it has an inverse denoted by $(g^{\mu\nu})$ according to the tradition: $g_{\mu\nu}g^{\nu\lambda} = g^{\lambda\nu}g_{\nu\mu} = \delta_\mu^\lambda$. The determinant $\det(g_{\mu\nu})$ is denoted by g. Clearly $\det(g^{\mu\nu}) = g^{-1}$. The isomorphism between T_pM and T_p^*M is now expressed as

$$\omega_\mu = g_{\mu\nu}U^\nu, \qquad U^\mu = g^{\mu\nu}\omega_\nu. \tag{7.2}$$

From (7.1a) and (7.1b) we recover the 'old-fashioned' definition of the metric as an infinitesimal distance squared. Take an infinitesimal displacement $\mathrm{d}x^\mu\partial/\partial x^\mu \in T_pM$ and plug it into g to find

$$\mathrm{d}s^2 = g\left(\mathrm{d}x^\mu\frac{\partial}{\partial x^\mu}, \mathrm{d}x^\nu\frac{\partial}{\partial x^\nu}\right) = \mathrm{d}x^\mu\,\mathrm{d}x^\nu g\left(\frac{\partial}{\partial x^\mu}, \frac{\partial}{\partial x^\nu}\right)$$
$$= g_{\mu\nu}\,\mathrm{d}x^\mu\,\mathrm{d}x^\nu. \tag{7.3}$$

We also call the quantity $\mathrm{d}s^2 = g_{\mu\nu}\,\mathrm{d}x^\mu\,\mathrm{d}x^\nu$ a metric, although in a strict sense the metric is a *tensor* $g = g_{\mu\nu}\,\mathrm{d}x^\mu \otimes \mathrm{d}x^\nu$.

Since $(g^{\mu\nu})$ is a symmetric matrix, the eigenvalues are real. If g is Riemannian, all the eigenvalues are strictly positive and if g is pseudo-Riemannian, some of them may be negative. If there are i positive and j negative eigenvalues, the pair (i, j) is called the **index** of the metric. If $j = 1$, the metric is called a **Lorentz metric**. Once a metric is diagonalized by an appropriate orthogonal matrix, it is easy to reduce all the diagonal elements to ± 1 by a suitable scaling of the basis vectors with positive numbers. If we start with a Riemannian metric we end up with the **Euclidean metric** $\delta = \mathrm{diag}(1, \ldots, 1)$ and if we start with a Lorentz metric, the **Minkowski metric** $\eta = \mathrm{diag}(-1, 1, \ldots, 1)$.

If (M, g) is Lorentzian, the elements of $T_p M$ are divided into three classes as follows,

(i) $g(U, U) > 0 \longrightarrow U$ is **spacelike**,

(ii) $g(U, U) = 0 \longrightarrow U$ is **lightlike** (or **null**),   (7.4)

(iii) $g(U, U) < 0 \longrightarrow U$ is **timelike**.

Exercise 7.1. Diagonalize the metric

$$(g_{\mu\nu}) = \begin{pmatrix} 0 & 1 & 0 & 0 \\ 1 & 0 & 0 & 0 \\ 0 & 0 & 1 & 0 \\ 0 & 0 & 0 & 1 \end{pmatrix}$$

to show that it reduces to the Minkowski metric. The frame on which the metric takes this form is known as the **light cone frame**. Let $\{e_0, e_1, e_2, e_3\}$ be the basis of the Minkowski frame in which the metric is $g_{\mu\nu} = \eta_{\mu\nu}$. Show that $\{e_+, e_-, e_2, e_3\}$ are the basis vectors in the light cone frame, where $e_\pm \equiv (e_1 \pm e_0)/\sqrt{2}$. Let $V = (V^+, V^-, V^2, V^3)$ be components of a vector V. Find the components of the corresponding one-form.

If a smooth manifold M admits a Riemannian metric g, the pair (M, g) is called a **Riemannian manifold**. If g is a pseudo-Riemannian metric, (M, g) is called a **pseudo-Riemannian manifold**. If g is Lorentzian, (M, g) is called a **Lorentz manifold**. Lorentz manifolds are of special interest in the theory of relativity. For example, an m-dimensional Euclidean space $(\mathbb{R}^m, \delta)$ is a Riemannian manifold and an m-dimensional Minkowski space $(\mathbb{R}^m, \eta)$ is a Lorentz manifold.

7.1.2   Induced metric

Let M be an m-dimensional submanifold of an n-dimensional Riemanian manifold N with the metric g_N. If $f : M \to N$ is the embedding which induces the submanifold structure of M (see section 5.2), the pullback map f^* induces the natural metric $g_M = f^* g_N$ on M. The components of g_M are given by

$$g_{M\mu\nu}(x) = g_{N\alpha\beta}(f(x)) \frac{\partial f^\alpha}{\partial x^\mu} \frac{\partial f^\beta}{\partial x^\nu} \qquad (7.5)$$

where f^α denote the coordinates of $f(x)$. For example, consider the metric of the unit sphere embedded in $(\mathbb{R}^3, \delta)$. Let (θ, ϕ) be the polar coordinates of S^2 and define f by the usual inclusion

$$f : (\theta, \phi) \mapsto (\sin\theta \cos\phi, \sin\theta \sin\phi, \cos\theta)$$

from which we obtain the **induced metric**

$$g_{\mu\nu}\, dx^{\mu} \otimes dx^{\nu} = \delta_{\alpha\beta} \frac{\partial f^{\alpha}}{\partial x^{\mu}} \frac{\partial f^{\beta}}{\partial x^{\nu}}\, dx^{\mu} \otimes dx^{\nu}$$

$$= d\theta \otimes d\theta + \sin^2\theta\, d\phi \otimes d\phi. \tag{7.6}$$

Exercise 7.2. Let $f : T^2 \to \mathbb{R}^3$ be an embedding of the torus into $(\mathbb{R}^3, \delta)$ defined by

$$f : (\theta, \phi) \mapsto ((R + r\cos\theta)\cos\phi, (R + r\cos\theta)\sin\phi, r\sin\theta)$$

where $R > r$. Show that the induced metric on T^2 is

$$g = r^2\, d\theta \otimes d\theta + (R + r\cos\theta)^2\, d\phi \otimes d\phi. \tag{7.7}$$

When a manifold N is pseudo-Riemannian, its submanifold $f : M \to N$ need not have a metric f^*g_N. The tensor f^*g_N is a metric only when it has a fixed index on M.

7.2 Parallel transport, connection and covariant derivative

A vector X is a directional derivative acting on $f \in \mathcal{F}(M)$ as $X : f \mapsto X[f]$. However, there is no directional derivative acting on a tensor field of type (p, q), which arises naturally from the differentiable structure of M. [Note that the Lie derivative $\mathcal{L}_V X = [V, X]$ is not a directional derivative since it depends on the *derivative* of V.] What we need is an extra structure called the **connection**, which specifies how tensors are transported along a curve.

7.2.1 Heuristic introduction

We first give a heuristic approach to parallel transport and covariant derivatives. As we have noted several times, two vectors defined at different points cannot be compared naively with each other. Let us see how the derivative of a vector field in a Euclidean space $\mathbb{R}^m$ is defined. The derivative of a vector field $V = V^{\mu}e_{\mu}$ with respect to x^{ν} has the μth component

$$\frac{\partial V^{\mu}}{\partial x^{\nu}} = \lim_{\Delta x \to 0} \frac{V^{\mu}(\ldots, x^{\nu} + \Delta x^{\nu}, \ldots) - V^{\mu}(\ldots, x^{\nu}, \ldots)}{\Delta x^{\nu}}.$$

The first term in the numerator of the LHS is defined at $x + \Delta x = (x^1, \ldots, x^{\nu} + \Delta x^{\nu}, \ldots, x^m)$, while the second term is defined at $x = (x^{\mu})$. To subtract $V^{\mu}(x)$ from $V^{\mu}(x + \Delta x)$, we have to transport $V^{\mu}(x)$ to $x + \Delta x$ *without change* and compute the difference. This transport of a vector is called a **parallel transport**. We have implicitly assumed that $V|_x$ parallel transported to $x + \Delta x$ has the same component $V^{\mu}(x)$. However, there is no natural way to parallel transport a vector in a manifold and we have to specify *how it is parallel transported* from one point to the other. Let $\widetilde{V}|_{x+\Delta x}$ denote a vector $V|_x$ parallel transported to $x + \Delta x$. We demand that the components satisfy

$$\widetilde{V}^\mu(x + \Delta x) - V^\mu(x) \propto \Delta x \qquad (7.8a)$$

$$\widetilde{(V^\mu + W^\mu)}(x + \Delta x) = \widetilde{V}^\mu(x + \Delta x) + \widetilde{W}^\mu(x + \Delta x). \qquad (7.8b)$$

These conditions are satisfied if we take

$$\widetilde{V}^\mu(x + \Delta x) = V^\mu(x) - V^\lambda(x)\Gamma^\mu{}_{\nu\lambda}(x)\Delta x^\nu. \qquad (7.9)$$

The covariant derivative of V with respect to x^ν is defined by

$$\lim_{\Delta x^\nu \to 0} \frac{V^\mu(x + \Delta x) - \widetilde{V}^\mu(x + \Delta x)}{\Delta x^\nu} \frac{\partial}{\partial x^\mu} = \left(\frac{\partial V^\mu}{\partial x^\nu} + V^\lambda \Gamma^\mu{}_{\nu\lambda} \right) \frac{\partial}{\partial x^\mu}. \qquad (7.10)$$

This quantity is a vector at $x + \Delta x$ since it is a difference of two vectors $V|_{x+\Delta x}$ and $\widetilde{V}|_{x+\Delta x}$ defined at the *same* point $x + \Delta x$. There are many distinct rules of parallel transport possible, one for each choice of Γ. If the manifold is endowed with a metric, there exists a preferred choice of Γ, called the Levi-Civita connection, see example 7.1 and section 7.4.

Example 7.1. Let us work out a simple example: two-dimensional Euclidean space $(\mathbb{R}^2, \delta)$. We define parallel transportation according to the usual sense in elementary geometry. In the Cartesian coordinate system (x, y), all the components of Γ vanish since $\widetilde{V}^\mu(x + \Delta x, y + \Delta y) = V^\mu(x, y)$ for any Δx and Δy. Next we take the polar coordinates (r, ϕ). If $(r, \phi) \mapsto (r\cos\phi, r\sin\phi)$ is regarded as an embedding, we find the induced metric,

$$g = dr \otimes dr + r^2 \, d\phi \otimes d\phi. \qquad (7.11)$$

Let $V = V^r \partial/\partial r + V^\phi \partial/\partial\phi$ be a vector defined at (r, ϕ). If we parallel transport this vector to $(r + \Delta r, \phi)$, we have a new vector $\widetilde{V} = \widetilde{V}^r \, \partial/\partial r|_{(r+\Delta r, \phi)} + \widetilde{V}^\phi \, \partial/\partial\phi|_{(r+\Delta r, \phi)}$ (figure 7.1(a)). Note that $V^r = V\cos\theta$ and $V^\phi = V(\sin\theta/r)$, where $V = \sqrt{g(V, V)}$ and θ is the angle between V and $\partial/\partial r$. Then we have $\widetilde{V}^r = V^r$ and

$$\widetilde{V}^\phi = \frac{r}{r + \Delta r} V^\phi \simeq V^\phi - \frac{\Delta r}{r} V^\phi.$$

By comparing these components with (7.9), we easily find that

$$\Gamma^r{}_{rr} = 0 \qquad \Gamma^r{}_{r\phi} = 0 \qquad \Gamma^\phi{}_{rr} = 0 \qquad \Gamma^\phi{}_{r\phi} = \frac{1}{r}. \qquad (7.12a)$$

Similarly, if V is parallel transported to $(r, \phi + \Delta\phi)$, it becomes

$$\widetilde{V} = \widetilde{V}^r \frac{\partial}{\partial r}\bigg|_{(r, \phi+\Delta\phi)} + \widetilde{V}^\phi \frac{\partial}{\partial\phi}\bigg|_{(r, \phi+\Delta\phi)}$$

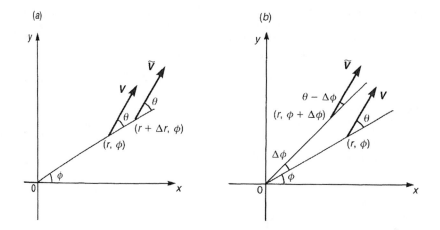

Figure 7.1. $\widetilde{V}$ is a vector V parallel transported to (a) $(r + \Delta r, \phi)$ and (b) $(r, \phi + \Delta\phi)$.

where

$$\widetilde{V}^r = V \cos(\theta - \Delta\phi) \simeq V \cos\theta + V \sin\theta \Delta\phi = V^r + V^\phi r \Delta\phi$$

and

$$\widetilde{V}^\phi = V \frac{\sin(\theta - \Delta\phi)}{r} \simeq V \frac{\sin\theta}{r} - V \cos\theta \frac{\Delta\phi}{r} = V^\phi - V^r \frac{\Delta\phi}{r}$$

(figure 7.1(b)). Then we find

$$\Gamma^r{}_{\phi r} = 0 \qquad \Gamma^r{}_{\phi\phi} = -r \qquad \Gamma^\phi{}_{\phi r} = \frac{1}{r} \qquad \Gamma^\phi{}_{\phi\phi} = 0. \qquad (7.12b)$$

Note that the Γ satisfy the symmetry $\Gamma^\lambda{}_{\mu\nu} = \Gamma^\lambda{}_{\nu\mu}$. It is also implicitly assumed that the norm of a vector is invariant under parallel transport. A rule of parallel transport which satisfies these two conditions is called a **Levi-Civita connection**, see section 7.4. Our intuitive approach leads us to the formal definition of the affine connection.

7.2.2 Affine connections

Definition 7.2. An affine connection ∇ is a map $\nabla : \mathfrak{X}(M) \times \mathfrak{X}(M) \to \mathfrak{X}(M)$, or $(X, Y) \mapsto \nabla_X Y$ which satisfies the following conditions:

$$\nabla_X(Y + Z) = \nabla_X Y + \nabla_X Z \qquad (7.13a)$$

$$\nabla_{(X+Y)}Z = \nabla_X Z + \nabla_Y Z \qquad (7.13b)$$

$$\nabla_{(fX)}Y = f\nabla_X Y \qquad (7.13c)$$

$$\nabla_X(fY) = X[f]Y + f\nabla_X Y \qquad (7.13d)$$

where $f \in \mathcal{F}(M)$ and $X, Y, Z \in \mathcal{X}(M)$.

Take a chart (U, φ) with the coordinate $x = \varphi(p)$ on M, and define m^3 functions $\Gamma^\lambda{}_{\nu\mu}$ called the **connection coefficients** by

$$\nabla_\nu e_\mu \equiv \nabla_{e_\nu} e_\mu = e_\lambda \Gamma^\lambda{}_{\nu\mu} \tag{7.14}$$

where $\{e_\mu\} = \{\partial/\partial x^\mu\}$ is the coordinate basis in $T_p M$. The connection coefficients specify how the basis vectors change from point to point. Once the action of ∇ on the basis vectors is defined, we can calculate the action of ∇ on any vectors. Let $V = V^\mu e_\mu$ and $W = W^\nu e_\nu$ be elements of $T_p(M)$. Then

$$\nabla_V W = V^\mu \nabla_{e_\mu}(W^\nu e_\nu) = V^\mu(e_\mu[W^\mu]e_\nu + W^\nu \nabla_{e_\mu} e_\nu)$$

$$= V^\mu \left(\frac{\partial W^\lambda}{\partial x^\mu} + W^\nu \Gamma^\lambda{}_{\mu\nu} \right) e_\lambda. \tag{7.15}$$

Note that this definition of the connection coefficient is in agreement with the previous heuristic result (7.10). By definition, ∇ maps two vectors V and W to a new vector given by the RHS of (7.15), whose λth component is $V^\mu \nabla_\mu W^\lambda$ where

$$\nabla_\mu W^\lambda \equiv \frac{\partial W^\lambda}{\partial x^\mu} + \Gamma^\lambda{}_{\mu\nu} W^\nu. \tag{7.16}$$

Note that $\nabla_\mu W^\lambda$ is the λth component of a vector $\nabla_\mu W = \nabla_\mu W^\lambda e_\lambda$ and should not be confused with the covariant derivative of a *component* W^λ. $\nabla_V W$ is independent of the derivative of V, unlike the Lie derivative $\mathcal{L}_V W = [V, W]$. In this sense, the covariant derivative is a proper generalization of the directional derivative of functions to tensors.

7.2.3 Parallel transport and geodesics

Given a curve in a manifold M, we may define the parallel transport of a vector along the curve. Let $c : (a, b) \to M$ be a curve in M. For simplicity, we assume the image is covered by a single chart (U, φ) whose coordinate is $x = \varphi(p)$. Let X be a vector field defined (at least) along $c(t)$,

$$X|_{c(t)} = X^\mu(c(t))e_\mu|_{c(t)} \tag{7.17}$$

where $e_\mu = \partial/\partial x^\mu$. If X satisfies the condition

$$\nabla_V X = 0 \qquad \text{for any } t \in (a, b) \tag{7.18a}$$

X is said to be **parallel transported** along $c(t)$ where $V = \mathrm{d}/\mathrm{d}t = (\mathrm{d}x^\mu(c(t))/\mathrm{d}t)e_\mu|_{c(t)}$ is the tangent vector to $c(t)$. The condition (7.18a) is written in terms of the components as

$$\frac{\mathrm{d}X^\mu}{\mathrm{d}t} + \Gamma^\mu{}_{\nu\lambda}\frac{\mathrm{d}x^\nu(c(t))}{\mathrm{d}t}X^\lambda = 0. \tag{7.18b}$$

If the tangent vector $V(t)$ itself is parallel transported along $c(t)$, namely if

$$\nabla_V V = 0 \tag{7.19a}$$

the curve $c(t)$ is called a **geodesic**. Geodesics are, in a sense, the *straightest possible curves* in a Riemannian manifold. In components, the geodesic equation (7.19a) becomes

$$\frac{d^2 x^\mu}{dt^2} + \Gamma^\mu{}_{\nu\lambda} \frac{dx^\nu}{dt} \frac{dx^\lambda}{dt} = 0 \tag{7.19b}$$

where $\{x^\mu\}$ are the coordinates of $c(t)$. We might say that (7.19a) is too strong to be the condition for the straightest possible curve, and instead require a weaker condition

$$\nabla_V V = f V \tag{7.20}$$

where $f \in \mathcal{F}(M)$. 'Change of V is parallel to V' is also a feature of a straight line. However, under the reparametrization $t \to t'$, the component of the tangent vector changes as

$$\frac{dx^\mu}{dt} \to \frac{dt}{dt'} \frac{dx^\mu}{dt}$$

and (7.20) reduces to (7.19a) if t' satisfies

$$\frac{d^2 t'}{dt^2} = f \frac{dt'}{dt}.$$

Thus, it is always possible to reparametrize the curve so that the geodesic equation takes the form (7.19a).

Exercise 7.3. Show that (7.19b) is left invariant under the affine reparametrization $t \to at + b$ $(a, b \in \mathbb{R})$.

7.2.4 The covariant derivative of tensor fields

Since ∇_X has the meaning of a derivative, it is natural to define the covariant derivative of $f \in \mathcal{F}(M)$ by the ordinary directional derivative:

$$\nabla_X f = X[f]. \tag{7.21}$$

Then (7.13d) looks exactly like the Leibnitz rule,

$$\nabla_X (fY) = (\nabla_X f)Y + f \nabla_X Y. \tag{7.13d'}$$

We require that this be true for any product of tensors,

$$\nabla_X (T_1 \otimes T_2) = (\nabla_X T_1) \otimes T_2 + T_1 \otimes (\nabla_X T_2) \tag{7.22}$$

where T_1 and T_2 are tensor fields of arbitrary types. Equation (7.22) is also true when some of the indices are contracted. With these requirements, we compute the covariant derivative of a one-form $\omega \in \Omega^1(M)$. Since $\langle \omega, Y \rangle \in \mathcal{F}(M)$ for $Y \in \mathfrak{X}(M)$, we should have

$$X[\langle \omega, Y \rangle] = \nabla_X [\langle \omega, Y \rangle] = \langle \nabla_X \omega, Y \rangle + \langle \omega, \nabla_X Y \rangle.$$

Writing down both sides in terms of the components we find

$$(\nabla_X \omega)_\nu = X^\mu \partial_\mu \omega_\nu - X^\mu \Gamma^\lambda_{\mu\nu} \omega_\lambda. \tag{7.23}$$

In particular, for $X = e_\mu$, we have

$$(\nabla_\mu \omega)_\nu = \partial_\mu \omega_\nu - \Gamma^\lambda_{\mu\nu} \omega_\lambda. \tag{7.24}$$

For $\omega = \mathrm{d}x^\nu$, we obtain (cf (7.14))

$$\nabla_\mu \mathrm{d}x^\nu = -\Gamma^\nu_{\mu\lambda} \mathrm{d}x^\lambda. \tag{7.25}$$

It is easy to generalize these results as

$$\nabla_\nu t^{\lambda_1 \dots \lambda_p}_{\mu_1 \dots \mu_q} = \partial_\nu t^{\lambda_1 \dots \lambda_p}_{\mu_1 \dots \mu_q} + \Gamma^{\lambda_1}_{\nu\kappa} t^{\kappa \lambda_2 \dots \lambda_p}_{\mu_1 \dots \mu_q} + \cdots$$
$$+ \Gamma^{\lambda_p}_{\nu\kappa} t^{\lambda_1 \dots \lambda_{p-1}\kappa}_{\mu_1 \dots \mu_q} - \Gamma^\kappa_{\nu\mu_1} t^{\lambda_1 \dots \lambda_p}_{\kappa\mu_2 \dots \mu_q} - \cdots$$
$$- \Gamma^\kappa_{\nu\mu_q} t^{\lambda_1 \dots \lambda_p}_{\mu_1 \dots \mu_{q-1}\kappa}. \tag{7.26}$$

Exercise 7.4. Let g be a metric tensor. Verify that

$$(\nabla_\nu g)_{\lambda\mu} = \partial_\nu g_{\lambda\mu} - \Gamma^\kappa_{\nu\lambda} g_{\kappa\mu} - \Gamma^\kappa_{\nu\mu} g_{\lambda\kappa}. \tag{7.27}$$

7.2.5 The transformation properties of connection coefficients

Introduce another chart (V, ψ) such that $U \cap V \neq \emptyset$, whose coordinates are $y = \psi(p)$. Let $\{e_\mu\} = \{\partial/\partial x^\mu\}$ and $\{f_\alpha\} = \{\partial/\partial y^\alpha\}$ be bases of the respective coordinates. Denote the connection coefficients with respect to the y-coordinates by $\tilde{\Gamma}^\alpha_{\beta\gamma}$. The basis vector f_α satisfies

$$\nabla_{f_\alpha} f_\beta = \tilde{\Gamma}^\gamma_{\alpha\beta} f_\gamma. \tag{7.28}$$

If we write $f_\alpha = (\partial x^\mu/\partial y^\alpha) e_\mu$, the LHS becomes

$$\nabla_{f_\alpha} f_\beta = \nabla_{f_\alpha} \left(\frac{\partial x^\mu}{\partial y^\beta} e_\mu \right) = \frac{\partial^2 x^\mu}{\partial y^\alpha \partial y^\beta} e_\mu + \frac{\partial x^\lambda}{\partial y^\alpha} \frac{\partial x^\mu}{\partial y^\beta} \nabla_{e_\lambda} e_\mu$$
$$= \left(\frac{\partial^2 x^\nu}{\partial y^\alpha \partial y^\beta} + \frac{\partial x^\lambda}{\partial y^\alpha} \frac{\partial x^\mu}{\partial y^\beta} \Gamma^\nu_{\lambda\mu} \right) e_\nu.$$

Since the RHS of (7.28) is equal to $\tilde{\Gamma}^{\gamma}{}_{\alpha\beta}(\partial x^{\nu}/\partial y^{\gamma})e_{\nu}$, the connection coefficients must transform as

$$\tilde{\Gamma}^{\gamma}{}_{\alpha\beta} = \frac{\partial x^{\lambda}}{\partial y^{\alpha}}\frac{\partial x^{\mu}}{\partial y^{\beta}}\frac{\partial y^{\gamma}}{\partial x^{\nu}}\Gamma^{\nu}{}_{\lambda\mu} + \frac{\partial^{2}x^{\nu}}{\partial y^{\alpha}\partial y^{\beta}}\frac{\partial y^{\gamma}}{\partial x^{\nu}}. \tag{7.29}$$

The reader should verify that this transformation rule indeed makes $\nabla_{X}Y$ a vector, namely

$$\tilde{X}^{\alpha}(\tilde{\partial}_{\alpha}\tilde{Y}^{\gamma} + \tilde{\Gamma}^{\gamma}{}_{\alpha\beta}\tilde{Y}^{\beta})f_{\gamma} = X^{\lambda}(\partial_{\lambda}Y^{\nu} + \Gamma^{\nu}{}_{\lambda\mu}Y^{\mu})e_{\nu}.$$

In the literature, connection coefficients are often defined as objects which transform as (7.29). From our viewpoint, however, they must transform according to (7.29) to make $\nabla_{X}Y$ independent of the coordinate chosen.

Exercise 7.5. Let Γ be an arbitrary connection coefficient. Show that $\Gamma^{\lambda}{}_{\mu\nu}+t^{\lambda}{}_{\mu\nu}$ is another connection coefficient provided that $t^{\lambda}{}_{\mu\nu}$ is a tensor field. Conversely, suppose $\Gamma^{\lambda}{}_{\mu\nu}$ and $\bar{\Gamma}^{\lambda}{}_{\mu\nu}$ are connection coefficients. Show that $\Gamma^{\lambda}{}_{\mu\nu} - \bar{\Gamma}^{\lambda}{}_{\mu\nu}$ is a component of a tensor of type $(1, 2)$.

7.2.6 The metric connection

So far we have left Γ arbitrary. Now that our manifold is endowed with a metric, we may put reasonable restrictions on the possible form of connections. We demand that the metric $g_{\mu\nu}$ be *covariantly constant*, that is, if two vectors X and Y are parallel transported along any curve, then the inner product between them remains constant under parallel transport. [In example 7.1, we have already assumed this reasonable condition.] Let V be a tangent vector to an arbitrary curve along which the vectors are parallel transported. Then we have

$$0 = \nabla_{V}[g(X, Y)] = V^{\kappa}[(\nabla_{\kappa}g)(X, Y) + g(\nabla_{\kappa}X, Y) + g(X, \nabla_{\kappa}Y)]$$
$$= V^{\kappa}X^{\mu}Y^{\nu}(\nabla_{\kappa}g)_{\mu\nu}$$

where we have noted that $\nabla_{\kappa}X = \nabla_{\kappa}Y = 0$. Since this is true for any curves and vectors, we must have

$$(\nabla_{\kappa}g)_{\mu\nu} = 0 \tag{7.30a}$$

or, from exercise 7.4,

$$\partial_{\lambda}g_{\mu\nu} - \Gamma^{\kappa}{}_{\lambda\mu}g_{\kappa\nu} - \Gamma^{\kappa}{}_{\lambda\nu}g_{\kappa\mu} = 0. \tag{7.30b}$$

If (7.30a) is satisfied, the affine connection ∇ is said to be **metric compatible** or simply a **metric connection**. We will deal with metric connections only. Cyclic permutations of (λ, μ, ν) yield

$$\partial_{\mu}g_{\nu\lambda} - \Gamma^{\kappa}{}_{\mu\nu}g_{\kappa\lambda} - \Gamma^{\kappa}{}_{\mu\lambda}g_{\kappa\nu} = 0 \tag{7.30c}$$

$$\partial_{\nu}g_{\lambda\mu} - \Gamma^{\kappa}{}_{\nu\lambda}g_{\kappa\mu} - \Gamma^{\kappa}{}_{\nu\mu}g_{\kappa\lambda} = 0. \tag{7.30d}$$

The combination $-(7.30b) + (7.30c) + (7.30d)$ yields

$$-\partial_\lambda g_{\mu\nu} + \partial_\mu g_{\nu\lambda} + \partial_\nu g_{\lambda\mu} + T^\kappa{}_{\lambda\mu} g_{\kappa\nu} + T^\kappa{}_{\lambda\nu} g_{\kappa\mu} - 2\Gamma^\kappa{}_{(\mu\nu)} g_{\kappa\lambda} = 0 \quad (7.31)$$

where $T^\kappa{}_{\lambda\mu} \equiv 2\Gamma^\kappa{}_{[\lambda\mu]} \equiv \Gamma^\kappa{}_{\lambda\mu} - \Gamma^\kappa{}_{\mu\lambda}$ and $\Gamma^\kappa{}_{(\mu\nu)} \equiv \frac{1}{2}(\Gamma^\kappa{}_{\nu\mu} + \Gamma^\kappa{}_{\mu\nu})$. The tensor $T^\kappa{}_{\lambda\mu}$ is anti-symmetric with respect to the lower indices $T^\kappa{}_{\lambda\mu} = -T^\kappa{}_{\mu\lambda}$ and called the **torsion tensor**, see exercise 7.6. The torsion tensor will be studied in detail in the next section. Equation (7.31) is solved for $\Gamma^\kappa{}_{(\mu\nu)}$ to yield

$$\Gamma^\kappa{}_{(\mu\nu)} = \left\{ \begin{matrix} \kappa \\ \mu\nu \end{matrix} \right\} + \frac{1}{2}\left(T_\nu{}^\kappa{}_\mu + T_\mu{}^\kappa{}_\nu\right) \quad (7.32)$$

where $\left\{ \begin{matrix} \kappa \\ \mu\nu \end{matrix} \right\}$ are the **Christoffel symbols** defined by

$$\left\{ \begin{matrix} \kappa \\ \mu\nu \end{matrix} \right\} = \frac{1}{2} g^{\kappa\lambda}\left(\partial_\mu g_{\nu\lambda} + \partial_\nu g_{\mu\lambda} - \partial_\lambda g_{\mu\nu}\right). \quad (7.33)$$

Finally, the connection coefficient Γ is given by

$$\begin{aligned} \Gamma^\kappa{}_{\mu\nu} &= \Gamma^\kappa{}_{(\mu\nu)} + \Gamma^\kappa{}_{[\mu\nu]} \\ &= \left\{ \begin{matrix} \kappa \\ \mu\nu \end{matrix} \right\} + \frac{1}{2}(T_\nu{}^\kappa{}_\mu + T_\mu{}^\kappa{}_\nu + T^\kappa{}_{\mu\nu}). \end{aligned} \quad (7.34)$$

The second term of the last expression of (7.34) is called the **contorsion**, denoted by $K^\kappa{}_{\mu\nu}$:

$$K^\kappa{}_{\mu\nu} \equiv \frac{1}{2}(T^\kappa{}_{\mu\nu} + T_\mu{}^\kappa{}_\nu + T_\nu{}^\kappa{}_\mu). \quad (7.35)$$

If the torsion tensor vanishes on a manifold M, the metric connection ∇ is called the **Levi-Civita connection**. Levi-Civita connections are natural generalizations of the connection defined in the classical geometry of surfaces, see section 7.4.

Exercise 7.6. Show that $T^\kappa{}_{\mu\nu}$ obeys the tensor transformation rule. [*Hint*: Use (7.29).] Show also that $K^\kappa{}_{[\mu\nu]} = \frac{1}{2}T^\kappa{}_{\mu\nu}$ and $K_{\kappa\mu\nu} = -K_{\nu\mu\kappa}$ where $K_{\kappa\mu\nu} = g_{\kappa\lambda}K^\lambda{}_{\mu\nu}$.

7.3 Curvature and torsion

7.3.1 Definitions

Since Γ is not a tensor, it cannot have an intrinsic geometrical meaning as a measure of how much a manifold is curved. For example, the connection coefficients in example 7.1 vanish if the Cartesian coordinate is employed while they do not in polar coordinates. As intrinsic objects, we define the **torsion tensor**

$T : \mathfrak{X}(M) \otimes \mathfrak{X}(M) \to \mathfrak{X}(M)$ and the **Riemann curvature tensor** (or **Riemann tensor**) $R : \mathfrak{X}(M) \otimes \mathfrak{X}(M) \otimes \mathfrak{X}(M) \to \mathfrak{X}(M)$ by

$$T(X, Y) \equiv \nabla_X Y - \nabla_Y X - [X, Y] \tag{7.36}$$

$$R(X, Y, Z) \equiv \nabla_X \nabla_Y Z - \nabla_Y \nabla_X Z - \nabla_{[X,Y]} Z. \tag{7.37}$$

It is common to write $R(X, Y)Z$ instead of $R(X, Y, Z)$, so that R looks like an operator acting on Z. Clearly, they satisfy

$$T(X, Y) = -T(Y, X), \qquad R(X, Y)Z = -R(Y, X)Z. \tag{7.38}$$

At first sight, T and R seem to be differential operators and it is not obvious that they are multilinear objects. We prove the tensorial property of R,

$$\begin{aligned}
R(fX, gY)hZ &= f\nabla_X\{g\nabla_Y(hZ)\} - g\nabla_Y\{f\nabla_X(hZ)\} - fX[g]\nabla_Y(hZ) \\
&\quad + gY[f]\nabla_X(hZ) - fg\nabla_{[X,Y]}(hZ) \\
&= fg\nabla_X\{Y[h]Z + h\nabla_Y Z\} - gf\nabla_Y\{X[h]Z + h\nabla_X Z\} \\
&\quad - fg[X, Y][h]Z - fgh\nabla_{[X,Y]}Z \\
&= fgh\{\nabla_X\nabla_Y Z - \nabla_Y\nabla_X Z - \nabla_{[X,Y]}Z\} \\
&= fgh\, R(X, Y)Z.
\end{aligned}$$

Now it is easy to see that R satisfies

$$R(X, Y)Z = X^\lambda Y^\mu Z^\nu R(e_\lambda, e_\mu)e_\nu \tag{7.39}$$

which verifies the tensorial property of R. Since R maps three vector fields to a vector field, it is a tensor field of type $(1, 3)$.

Exercise 7.7. Show that T defined by (7.36) is multilinear,

$$T(X, Y) = X^\mu Y^\nu T(e_\mu, e_\nu) \tag{7.40}$$

and hence a tensor field of type $(1, 2)$.

Since T and R are tensors, their operations on vectors are obtained once their actions on the basis vectors are known. With respect to the coordinate basis $\{e_\mu\}$ and the dual basis $\{dx^\mu\}$, the components of these tensors are given by

$$\begin{aligned}
T^\lambda{}_{\mu\nu} &= \langle dx^\lambda, T(e_\mu, e_\nu)\rangle = \langle dx^\lambda, \nabla_\mu e_\nu - \nabla_\nu e_\mu\rangle \\
&= \langle dx^\lambda, \Gamma^\eta{}_{\mu\nu}e_\eta - \Gamma^\eta{}_{\nu\mu}e_\eta\rangle = \Gamma^\lambda{}_{\mu\nu} - \Gamma^\lambda{}_{\nu\mu} \tag{7.41}
\end{aligned}$$

and

$$\begin{aligned}
R^\kappa{}_{\lambda\mu\nu} &= \langle dx^\kappa, R(e_\mu, e_\nu)e_\lambda\rangle = \langle dx^\kappa, \nabla_\mu \nabla_\nu e_\lambda - \nabla_\nu \nabla_\mu e_\lambda\rangle \\
&= \langle dx^\kappa, \nabla_\mu(\Gamma^\eta{}_{\nu\lambda}e_\eta) - \nabla_\nu(\Gamma^\eta{}_{\mu\nu}e_\eta)\rangle \\
&= \langle dx^\kappa, (\partial_\mu \Gamma^\eta{}_{\nu\lambda})e_\eta + \Gamma^\eta{}_{\nu\lambda}\Gamma^\xi{}_{\mu\eta}e_\xi - (\partial_\nu \Gamma^\eta{}_{\mu\lambda})e_\eta - \Gamma^\eta{}_{\mu\lambda}\Gamma^\xi{}_{\nu\eta}e_\xi\rangle \\
&= \partial_\mu \Gamma^\kappa{}_{\nu\lambda} - \partial_\nu \Gamma^\kappa{}_{\mu\lambda} + \Gamma^\eta{}_{\nu\lambda}\Gamma^\kappa{}_{\mu\eta} - \Gamma^\eta{}_{\mu\lambda}\Gamma^\kappa{}_{\nu\eta}. \tag{7.42}
\end{aligned}$$

We readily find (cf (7.38))

$$T^\lambda{}_{\mu\nu} = -T^\lambda{}_{\nu\mu} \qquad R^\kappa{}_{\lambda\mu\nu} = -R^\kappa{}_{\lambda\nu\mu}. \tag{7.43}$$

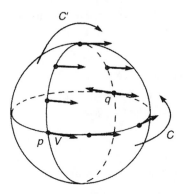

Figure 7.2. It is natural to define V parallel transported along a great circle if the angle V makes with the great circle is kept fixed. If V at p is parallel transported along great circles C and C', the resulting vectors at q point in opposite directions.

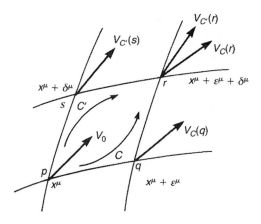

Figure 7.3. A vector V_0 at p is parallel transported along C and C' to yield $V_C(r)$ and $V_{C'}(r)$ at r. The curvature measures the difference between two vectors.

7.3.2 Geometrical meaning of the Riemann tensor and the torsion tensor

Before we proceed further, we examine the geometrical meaning of these tensors. We consider the Riemann tensor first. A crucial observation is that if we parallel transport a vector V at p to q along two different curves C and C', the resulting vectors at q are different in general (figure 7.2). If, however, we parallel transport a vector in a Euclidean space, where the parallel transport is defined in our usual sense, the resulting vector does not depend on the path along which it has been parallel transported. We expect that this non-integrability of parallel transport characterizes the intrinsic notion of curvature, which does not depend on the special coordinates chosen. Let us take an infinitesimal parallelogram $pqrs$ whose coordinates are $\{x^\mu\}$, $\{x^\mu + \varepsilon^\mu\}$, $\{x^\mu + \varepsilon^\mu + \delta^\mu\}$ and $\{x^\mu + \delta^\mu\}$ respectively, ε^μ and δ^μ being infinitesimal (figure 7.3). If we parallel transport a vector $V_0 \in T_p M$ along $C = pqr$, we will have a vector $V_C(r) \in T_r M$. The vector V_0 parallel transported to q along C is

$$V_C^\mu(q) = V_0^\mu - V_0^\kappa \Gamma^\mu{}_{\nu\kappa}(p)\varepsilon^\nu.$$

Then $V_C^\mu(r)$ is given by

$$\begin{aligned}
V_C^\mu(r) &= V_C^\mu(q) - V_C^\kappa(q)\Gamma^\mu{}_{\nu\kappa}(q)\delta^\nu \\
&= V_0^\mu - V_0^\kappa \Gamma^\mu{}_{\nu\kappa}\varepsilon^\nu - [V_0^\kappa - V_0^\rho \Gamma^\kappa{}_{\zeta\rho}(p)\varepsilon^\zeta] \\
&\quad \times [\Gamma^\mu{}_{\nu\kappa}(p) + \partial_\lambda \Gamma^\mu{}_{\nu\kappa}(p)\varepsilon^\lambda]\delta^\nu \\
&\simeq V_0^\mu - V_0^\kappa \Gamma^\mu{}_{\nu\kappa}(p)\varepsilon^\nu - V_0^\kappa \Gamma^\mu{}_{\nu\kappa}(p)\delta^\nu \\
&\quad - V_0^\kappa [\partial_\lambda \Gamma^\mu{}_{\nu\kappa}(p) - \Gamma^\rho{}_{\lambda\kappa}(p)\Gamma^\mu{}_{\nu\rho}(p)]\varepsilon^\lambda \delta^\nu
\end{aligned}$$

where we have kept terms of up to order two in ε and δ. Similarly, parallel transport of V_0 along $C' = psr$ yields another vector $V_{C'}(r) \in T_r M$, given by

$$\begin{aligned}
V_{C'}^\mu(r) &\simeq V_0^\mu - V_0^\kappa \Gamma^\mu{}_{\nu\kappa}(p)\delta^\nu - V_0^\kappa \Gamma^\mu{}_{\nu\kappa}(p)\varepsilon^\nu \\
&\quad - V_0^\kappa [\partial_\nu \Gamma^\mu{}_{\lambda\kappa}(p) - \Gamma^\rho{}_{\nu\kappa}(p)\Gamma^\mu{}_{\lambda\rho}(p)]\varepsilon^\lambda \delta^\nu.
\end{aligned}$$

The two vectors at r differ by

$$\begin{aligned}
V_{C'}(r) - V_C(r) &= V_0^\kappa [\partial_\lambda \Gamma^\mu{}_{\nu\kappa}(p) - \partial_\nu \Gamma^\mu{}_{\lambda\kappa}(p) \\
&\quad - \Gamma^\rho{}_{\lambda\kappa}(p)\Gamma^\mu{}_{\nu\rho}(p) + \Gamma^\rho{}_{\nu\kappa}(p)\Gamma^\mu{}_{\lambda\rho}(p)]\varepsilon^\lambda \delta^\nu \\
&= V_0^\kappa R^\mu{}_{\kappa\lambda\nu}\varepsilon^\lambda \delta^\nu.
\end{aligned} \tag{7.44}$$

We next look at the geometrical meaning of the torsion tensor. Let $p \in M$ be a point whose coordinates are $\{x^\mu\}$. Let $X = \varepsilon^\mu e_\mu$ and $Y = \delta^\mu e_\mu$ be infinitesimal vectors in $T_p M$. If these vectors are regarded as small displacements, they define two points q and s near p, whose coordinates are $\{x^\mu + \varepsilon^\mu\}$ and $\{x^\mu + \delta^\mu\}$ respectively (figure 7.4). If we parallel transport X along the line ps, we obtain a vector sr_1 whose component is $\varepsilon^\mu - \varepsilon^\lambda \Gamma^\mu{}_{\nu\lambda}\delta^\nu$. The displacement vector connecting p and r_1 is

$$pr_1 = ps + sr_1 = \delta^\mu + \varepsilon^\mu - \Gamma^\mu{}_{\nu\lambda}\varepsilon^\lambda \delta^\nu.$$

Similarly, the parallel transport of δ^μ along pq yields a vector

$$pr_2 = pq + qr_2 = \varepsilon^\mu + \delta^\mu - \Gamma^\mu{}_{\lambda\nu}\varepsilon^\lambda \delta^\nu.$$

In general, r_1 and r_2 do not agree and the difference is

$$r_2 r_1 = pr_2 - pr_1 = (\Gamma^\mu{}_{\nu\lambda} - \Gamma^\mu{}_{\lambda\nu})\varepsilon^\lambda \delta^\nu = T^\mu{}_{\nu\lambda}\varepsilon^\lambda \delta^\nu. \tag{7.45}$$

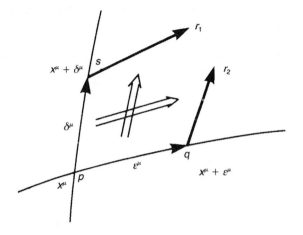

Figure 7.4. The vector qr_2 (sr_1) is the vector ps (pq) parallel transported to q (s). In general, $r_1 \neq r_2$ and the torsion measures the difference $r_2 r_1$.

Thus, the torsion tensor measures the failure of the closure of the parallelogram made up of the small displacement vectors and their parallel transports.

Example 7.2. Suppose we are navigating on the surface of the Earth. We define a vector to be parallel transported if the angle between the vector and the latitude is kept fixed during the navigation. [*Remarks*: This definition of parallel transport is not the usual one. For example, the geodesic is not a great circle but a straight line on Mercator's projection. See example 7.5.] Suppose we navigate along a small quadrilateral $pqrs$ made up of latitudes and longitudes (figure 7.5(a)). We parallel transport a vector at p along pqr and psr, separately. According to our definition of parallel transport, two vectors at r should agree, hence the curvature tensor vanishes. To find the torsion, we parametrize the points p, q, r and s as in figure 7.5(b). We find the torsion by evaluating the difference between pr_1 and pr_2 as in (7.45). If we parallel transport the vector pq along ps, we obtain a vector sr_1, whose length is $R \sin \theta d\phi$. However, a parallel transport of the vector ps along pq yields a vector $qr_2 = qr$. Since sr has a length $R \sin(\theta - d\theta) d\phi \simeq R \sin \theta d\phi - R \cos \theta d\theta d\phi$, we find that $r_1 r_2$ has a length $R \cos \theta d\theta d\phi$. Since $r_1 r_2$ is parallel to $-\partial/\partial\phi$, the connection has a torsion $T^{\phi}{}_{\theta\phi}$, see (7.45). From $g_{\phi\phi} = R^2 \sin^2 \theta$, we find that $r_1 r_2$ has components $(0, -\cot \theta d\theta d\phi)$. Since the ϕ-component of $r_1 r_2$ is equal to $T^{\phi}{}_{\theta\phi} d\theta d\phi$, we obtain $T^{\phi}{}_{\theta\phi} = -\cot \theta$.

Note that the basis $\{\partial/\partial\theta, \partial/\partial\phi\}$ is not well defined at the poles. It is known that the sphere S^2 does not admit two vector fields which are linearly independent everywhere on S^2. Any vector field on S^2 must vanish somewhere on S^2 and

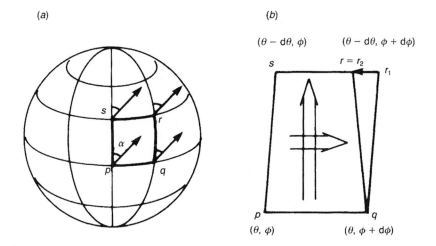

Figure 7.5. (*a*) If a vector makes an angle α with the longitude at p, this angle is kept fixed during parallel transport. (*b*) The vector sr_1 (qr_2) is the vector pq (ps) parallel transported to s (q). The torsion does not vanish.

hence cannot be linearly independent of the other vector field there. If an m-dimensional manifold M admits m vector fields which are linearly independent everywhere, M is said to be **parallelizable**. On a parallelizable manifold, we can use these m vector fields to define a tangent space at each point of M. A vector $V_p \in T_pM$ is defined to be parallel to $V_q \in T_qM$ if all the *components* of V_p at T_pM are equal to those of V_q at T_qM. Since the vector fields are defined throughout M, this parallelism should be independent of the path connecting p and q, hence the Riemann curvature tensor vanishes although the torsion tensor may not in general. For S^m, this is possible only when $m = 1, 3$ and 7, which is closely related to the existence of complex numbers, quaternions and octonions, respectively. For definiteness, let us consider

$$S^3 = \left\{ (x^1, x^2, x^3, x^4) \, \middle| \, \sum_{i=1}^{4} (x^i)^2 = 1 \right\}$$

embedded in $(\mathbb{R}^4, \delta)$. Three orthonormal vectors

$$
\begin{aligned}
e_1(x) &= (-x^2, x^1, -x^4, x^3) \\
e_2(x) &= (-x^3, x^4, x^1, -x^2) \\
e_3(x) &= (-x^4, -x^3, x^2, x^1)
\end{aligned}
\tag{7.46}
$$

are orthogonal to $x = (x^1, x^2, x^3, x^4)$ and linearly independent everywhere on S^3, hence define the tangent space T_xS^3. Two vectors $V_1(x)$ and $V_2(y)$

are parallel if $V_1(x) = \sum c^i e_i(x)$ and $V_2(y) = \sum c^i e_i(y)$. The connection coefficients are computed from (7.14). Let $\varepsilon e_1(x)$ be a small displacement under which $x = (x^1, x^2, x^3, x^4)$ changes to $x' = x + \varepsilon e_1(x) = \{x^1 - \varepsilon x^2, x^2 + \varepsilon x^1, x^3 - \varepsilon x^4, x^4 + \varepsilon x^3\}$. The difference between the basis vectors at x and x' is $e_2(x') - e_2(x) = (-x^3 - \varepsilon x^4, x^4 + \varepsilon x^3, x^1 - \varepsilon x^2, -x^2 - \varepsilon x^1) - (-x^3, x^4, x^1, -x^2) = -\varepsilon e_3(x) = \varepsilon \Gamma^\mu{}_{12} e_\mu(x)$, hence $\Gamma^3{}_{12} = -1, \Gamma^1{}_{12} = \Gamma^2{}_{12} = 0$. Similarly, $\Gamma^3{}_{21} = 1$ hence we find $T^3{}_{12} = -2$. The reader should complete the computation of the connection coefficients and verify that $T^\lambda{}_{\mu\nu} = -2$ $(+2)$ if $(\lambda\mu\nu)$ is an even (odd) permutation of (123) and vanishes otherwise.

Let us see how this parallelizability of S^3 is related to the existence of quaternions. The multiplication rule of quaternions is

$$(x^1, x^2, x^3, x^4) \cdot (y^1, y^2, y^3, y^4)$$
$$= (x^1 y^1 - x^2 y^2 - x^3 y^3 - x^4 y^4, x^1 y^2 + x^2 y^1 + x^3 y^4 - x^4 y^3,$$
$$x^1 y^3 - x^2 y^4 + x^3 y^1 + x^4 y^2, x^1 y^4 + x^2 y^3 - x^3 y^2 + x^4 y^1). \quad (7.47)$$

S^3 may be defined by the set of unit quaternions

$$S^3 = \{(x^1, x^2, x^3, x^4) | x \cdot \bar{x} = 1\}$$

where the conjugate of x is defined by $\bar{x} = (x^1, -x^2, -x^3, -x^4)$. According to (7.46), the tangent space at $x_0 = (1, 0, 0, 0)$ is spanned by

$$e_1 = (0, 1, 0, 0) \qquad e_2 = (0, 0, 1, 0) \qquad e_3 = (0, 0, 0, 1).$$

Then the basis vectors (7.46) of the tangent space at $x = (x^1, x^2, x^3, x^4)$ are expressed as the quaternion products

$$e_1(x) = e_1 \cdot x \qquad e_2(x) = e_2 \cdot x \qquad e_3(x) = e_3 \cdot x. \quad (7.48)$$

Because of this algebra, it is *always* possible to give a set of basis vectors at an arbitrary point of S^3 once it is given at some point, $x_0 = (1, 0, 0, 0)$, for example.

By the same token, a Lie group is parallelizable. If the set of basis vectors $\{V_1, \ldots, V_m\}$ at the unit element e of a Lie group G is given, we can always find a set of basis vectors of $T_g G$ by the left translation of $\{V_\mu\}$ (see section 5.6),

$$\{V_1, \ldots, V_n\} \xrightarrow{L_{g*}} \{X_1|_g, \ldots, X_n|_g\}. \quad (7.49)$$

7.3.3 The Ricci tensor and the scalar curvature

From the Riemann curvature tensor, we construct new tensors by contracting the indices. The **Ricci tensor** Ric is a type $(0, 2)$ tensor defined by

$$Ric(X, Y) \equiv \langle dx^\mu, R(e_\mu, Y)X \rangle \quad (7.50a)$$

whose component is

$$Ric_{\mu\nu} = Ric(e_\mu, e_\nu) = R^\lambda{}_{\mu\lambda\nu}. \tag{7.50b}$$

The **scalar curvature** $\mathcal{R}$ is obtained by further contracting indices,

$$\mathcal{R} \equiv g^{\mu\nu} Ric(e_\mu, e_\nu) = g^{\mu\nu} Ric_{\mu\nu}. \tag{7.51}$$

7.4 Levi-Civita connections

7.4.1 The fundamental theorem

Among affine connections, there is a special connection called the **Levi-Civita connection**, which is a natural generalization of the connection in the classical differential geometry of surfaces. A connection ∇ is called a **symmetric connection** if the torsion tensor vanishes. In the coordinate basis, connection coefficients of a symmetric connection satisfy

$$\Gamma^\lambda{}_{\mu\nu} = \Gamma^\lambda{}_{\nu\mu}. \tag{7.52}$$

Theorem 7.1. (**The fundamental theorem of (pseudo-)Riemannian geometry**) On a (pseudo-)Riemannian manifold (M, g), there exists a unique *symmetric* connection which is *compatible* with the metric g. This connection is called the **Levi-Civita connection**.

Proof. This follows directly from (7.34). Let ∇ be an arbitrary connection such that

$$\tilde{\Gamma}^\kappa{}_{\mu\nu} = \left\{ {\kappa \atop \mu\nu} \right\} + K^\kappa{}_{\mu\nu}$$

where $\left\{ {\kappa \atop \mu\nu} \right\}$ is the Christoffel symbol and K the contorsion tensor. It was shown in exercise 7.5 that $\Gamma^\kappa{}_{\mu\nu} \equiv \tilde{\Gamma}^\kappa{}_{\mu\nu} + t^\kappa{}_{\mu\nu}$ is another connection coefficient if t is a tensor field of type $(1, 2)$. Now we choose $t^\kappa{}_{\mu\nu} = -K^\kappa{}_{\mu\nu}$ so that

$$\Gamma^\kappa{}_{\mu\nu} = \left\{ {\kappa \atop \mu\nu} \right\} = \frac{1}{2} g^{\kappa\lambda}(\partial_\mu g_{\lambda\nu} + \partial_\nu g_{\lambda\mu} - \partial_\lambda g_{\mu\nu}). \tag{7.53}$$

By construction, this is symmetric and certainly unique given a metric.   □

Exercise 7.8. Let V be a Levi-Civita connection.

(a) Let $f \in \mathcal{F}(M)$. Show that

$$\nabla_\mu \nabla_\nu f = \nabla_\nu \nabla_\mu f. \tag{7.54}$$

(b) Let $\omega \in \Omega^1(M)$. Show that

$$d\omega = (\nabla_\mu \omega)_\nu \, dx^\mu \wedge dx^\nu. \tag{7.55}$$

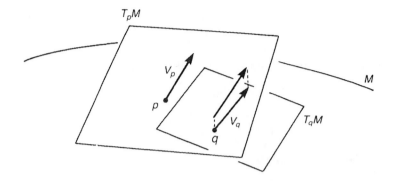

Figure 7.6. On a surface M, a vector $V_p \in T_pM$ is defined to be parallel to $V_q \in T_qM$ if the projection of V_q onto T_pM is parallel to V_p in our ordinary sense of parallelism in $\mathbb{R}^2$.

(c) Let $\omega \in \Omega^1(M)$ and let $U \in \mathfrak{X}(M)$ be the corresponding vector field: $U^\mu = g^{\mu\nu}\omega_\nu$. Show that, for any $V \in \mathfrak{X}(M)$,

$$g(\nabla_X U, V) = \langle \nabla_X \omega, V \rangle. \tag{7.56}$$

Example 7.3.

(a) The metric on $\mathbb{R}^2$ in polar coordinates is $g = \mathrm{d}r \otimes \mathrm{d}r + r^2\,\mathrm{d}\phi \otimes \mathrm{d}\phi$. The non-vanishing components of the Levi-Civita connection coefficients are $\Gamma^\phi{}_{r\phi} = \Gamma^\phi{}_{\phi r} = r^{-1}$ and $\Gamma^r{}_{\phi\phi} = -r$. This is in agreement with the result obtained in example 7.1.

(b) The induced metric on S^2 is $g = \mathrm{d}\theta \otimes \mathrm{d}\theta + \sin^2\theta\,\mathrm{d}\phi \otimes \mathrm{d}\phi$. The non-vanishing components of the Levi-Civita connection are

$$\Gamma^\theta{}_{\phi\phi} = -\cos\theta\sin\theta \qquad \Gamma^\phi{}_{\theta\phi} = \Gamma^\phi{}_{\phi\theta} = \cot\theta. \tag{7.57}$$

7.4.2 The Levi-Civita connection in the classical geometry of surfaces

In the classical differential geometry of surfaces embedded in $\mathbb{R}^3$, Levi-Civita defined the parallelism of vectors at the nearby points p and q in the following sense (figure 7.6). First, take the tangent plane at p and a vector V_p at p, which lies in the tangent plane. A vector V_q at q is defined to be parallel to V_p if the projection of V_q to the tangent plane at p is parallel to V_p in our usual sense. Now take two points q and s near p as in figure 7.7 and parallel transport the displacement vectors pq along ps and ps along pq. If the parallelism is defined in the sense of Levi-Civita, the displacement vectors projected to the tangent plane at p form a closed parallelogram, hence this parallelism has vanishing torsion. As has been proved in theorem 7.1, there exists a unique connection which has vanishing torsion, which generalizes the parallelism defined here to arbitrary manifolds.

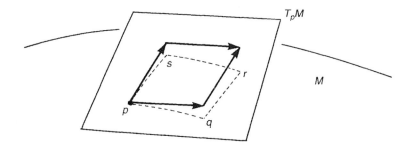

Figure 7.7. If the parallelism is defined in the sense of Levi-Civita, the torsion vanishes identically.

7.4.3 Geodesics

When the Levi-Civita connection is employed, we can compute the connection coefficients, Riemann tensors and many relations involving these by simple routines. Besides this simplicity, the Levi-Civita connection provides a geodesic (defined as the *straightest* possible curve) with another picture, namely the *shortest* possible curve connecting two given points. In Newtonian mechanics, the trajectory of a free particle is the straightest possible as well as the shortest possible curve, that is, a straight line. Einstein proposed that this property should be satisfied in general relativity as well; if gravity is understood as a part of the geometry of spacetime, a freely falling particle should follow the straightest as well as the shortest possible curve. [*Remark*: To be precise, the shortest possible curve is too strong a condition. As we see later, a geodesic defined with respect to the Levi-Civita connection gives the local extremum of the length of a curve connecting two points.]

Example 7.4. In a flat manifold $(\mathbb{R}^m, \delta)$ or $(\mathbb{R}^m, \eta)$, the Levi-Civita connection coefficients Γ vanish identically. Hence, the geodesic equation (7.19b) is easily solved to yield $x^\mu = A^\mu t + B^\mu$, where A^μ and B^μ are constants.

Exercise 7.9. A metric on a cylinder $S^1 \times \mathbb{R}$ is given by $g = \mathrm{d}\phi \otimes \mathrm{d}\phi + \mathrm{d}z \otimes \mathrm{d}z$, where ϕ is the polar angle of S^1 and z the coordinate of $\mathbb{R}$. Show that the geodesics given by the Levi-Civita connection are helices.

The equivalence of the straightest possible curve and the local extremum of the distance is proved as follows. First we parametrize the curve by the distance s along the curve, $x^\mu = x^\mu(s)$. The length of a path c connecting two points p and q is

$$I(c) = \int_c \mathrm{d}s = \int_c \sqrt{g_{\mu\nu} x'^\mu x'^\nu}\, \mathrm{d}s \tag{7.58}$$

where $x'^\mu = \mathrm{d}x^\mu/\mathrm{d}s$. Instead of deriving the Euler–Lagrange equation from (7.58), we will solve a slightly easier problem. Let $F \equiv \frac{1}{2} g_{\mu\nu} x'^\mu x'^\nu$ and write

(7.58) as $I(c) = \int_c L(F)ds$. The Euler–Lagrange equation for the original problem takes the form

$$\frac{d}{ds}\left(\frac{\partial L}{\partial x'^\lambda}\right) - \frac{\partial L}{\partial x^\lambda} = 0. \tag{7.59}$$

Then $F = L^2/2$ satisfies

$$\frac{d}{ds}\left(\frac{\partial F}{\partial x'^\lambda}\right) - \frac{\partial F}{\partial x^\lambda} = L\left[\frac{d}{ds}\left(\frac{\partial L}{\partial x'^\lambda}\right) - \frac{\partial L}{\partial x^\lambda}\right] + \frac{\partial L}{\partial x'^\lambda}\frac{dL}{ds} = \frac{\partial L}{\partial x'^\lambda}\frac{dL}{ds}. \tag{7.60}$$

The last expression vanishes since $L \equiv 1$ along the curve; $dL/ds = 0$. Now we have proved that F also satisfies the Euler–Lagrange equation provided that L does so. We then have

$$\frac{d}{ds}(g_{\lambda\mu}x'^\mu) - \frac{1}{2}\frac{\partial g_{\mu\nu}}{\partial x^\lambda}x'^\mu x'^\nu$$

$$= \frac{\partial g_{\lambda\mu}}{\partial x^\nu}x'^\mu x'^\nu + g_{\lambda\mu}\frac{d^2 x^\mu}{ds^2} - \frac{1}{2}\frac{\partial g_{\mu\nu}}{\partial x^\lambda}x'^\mu x'^\nu$$

$$= g_{\lambda\mu}\frac{d^2 x^\mu}{ds^2} + \frac{1}{2}\left(\frac{\partial g_{\lambda\mu}}{\partial x^\nu} + \frac{\partial g_{\lambda\nu}}{\partial x^\mu} - \frac{\partial g_{\mu\nu}}{\partial x^\lambda}\right)\frac{dx^\mu}{ds}\frac{dx^\nu}{ds} = 0. \tag{7.61}$$

If (7.61) is multiplied by $g^{\kappa\lambda}$, we reproduce the geodesic equation (7.19b).

Having proved that L and F satisfy the same variational problem, we take advantage of this to compute the Christoffel symbols. Take S^2, for example. F is given by $\frac{1}{2}(\theta'^2 + \sin^2\theta\phi'^2)$ and the Euler–Lagrange equations are

$$\frac{d^2\theta}{ds^2} - \sin\theta\cos\theta\left(\frac{d\phi}{ds}\right)^2 = 0 \tag{7.62a}$$

$$\frac{d^2\phi}{ds^2} + 2\cot\theta\frac{d\phi}{ds}\frac{d\theta}{ds} = 0. \tag{7.62b}$$

It is easy to read off the connection coefficients $\Gamma^\theta{}_{\phi\phi} = -\sin\theta\cos\theta$ and $\Gamma^\phi{}_{\phi\theta} = \Gamma^\phi{}_{\theta\phi} = \cot\theta$, see (7.57).

Example 7.5. Let us compute the geodesics of S^2. Rather than solving the geodesic equations (7.62) we find the geodesic by minimizing the length of a curve connecting two points on S^2. Without loss of generality, we may assign coordinates (θ_1, ϕ_0) and (θ_2, ϕ_0) to these points. Let $\phi = \phi(\theta)$ be a curve connecting these points. Then the length of the curve is

$$I(c) = \int_{\theta_1}^{\theta_2}\sqrt{1 + \sin^2\left(\frac{d\phi}{d\theta}\right)^2}\,d\theta \tag{7.63}$$

which is minimized when $d\phi/d\theta \equiv 0$, that is $\phi \equiv \phi_0$. Thus, the geodesic is a great circle (θ, ϕ_0), $\theta_1 \le \theta \le \theta_2$. [*Remark*: Solving (7.62) is not very difficult. Let $\theta = \theta(\phi)$ be the equation of the geodesic. Then

$$\frac{d\theta}{ds} = \frac{d\theta}{d\phi}\frac{d\phi}{ds} \qquad \frac{d^2\theta}{ds^2} = \frac{d^2\theta}{d\phi^2}\left(\frac{d\phi}{ds}\right)^2 + \frac{d\theta}{d\phi}\frac{d^2\phi}{ds^2}.$$

Substituting these into the first equation of (7.62), we obtain

$$\frac{d^2\theta}{d\phi^2}\left(\frac{d\phi}{ds}\right)^2 + \frac{d\theta}{d\phi}\frac{d^2\phi}{ds^2} - \sin\theta\cos\theta\left(\frac{d\phi}{ds}\right)^2 = 0. \qquad (7.64)$$

The second equation of (7.62) and (7.64) yields

$$\frac{d^2\theta}{d\phi^2} - 2\cot\theta\left(\frac{d\theta}{d\phi}\right)^2 - \sin\theta\cos\theta = 0. \qquad (7.65)$$

If we define $f(\theta) \equiv \cot\theta$, (7.65) becomes

$$\frac{d^2 f}{d\phi^2} + f = 0$$

whose general solution is $f(\theta) = \cot\theta = A\cos\phi + B\sin\phi$ or

$$A\sin\theta\cos\phi + B\sin\theta\sin\phi - \cos\theta = 0. \qquad (7.66)$$

Equation (7.66) is the equation of a great circle which lies in a plane whose normal vector is $(A, B, -1)$.]

Example 7.6. Let U be the upper half-plane $U \equiv \{(x, y)|y > 0\}$ and introduce the **Poincaré metric**

$$g = \frac{dx \otimes dx + dy \otimes dy}{y^2}. \qquad (7.67)$$

The geodesic equations are

$$x'' - \frac{2}{y}x'y' = 0 \qquad (7.68a)$$

$$y'' - \frac{1}{y}[x'^2 + 3y'^2] = 0 \qquad (7.68b)$$

where $x' \equiv dx/ds$ etc. The first equation of (7.68) is easily integrated, if divided by x', to yield

$$\frac{x'}{y^2} = \frac{1}{R} \qquad (7.69)$$

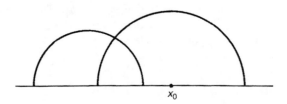

Figure 7.8. Geodesics defined by the Poincaré metric in the upper half-plane. The geodesic has an infinite length.

where R is a constant. Since the parameter s is taken so that the vector (x', y') has unit length, it satisfies $(x'^2 + y'^2)/y^2 = 1$. From (7.69), this becomes $y^2/R^2 + (y'/y)^2 = 1$ or

$$ds = \frac{dy}{y\sqrt{1 - y^2/R^2}} = \frac{dt}{\sin t}$$

where we put $y = R \sin t$. Equation (7.69) then becomes

$$x' = \frac{y^2}{R} = R \sin^2 t.$$

Now x is solved for t to yield

$$x = \int x' \, ds = \int \frac{dx}{ds} \frac{ds}{dt} \, dt$$

$$= \int R \sin t \, dt = -R \cos t + x_0.$$

Finally, we obtain the solution

$$x = -R \cos t + x_0 \qquad y = R \sin t \qquad (y > 0) \qquad (7.70)$$

which is a circle with radius R centred at $(x_0, 0)$. Maximally extended geodesics are given by $0 < t < \pi$ (figure 7.8) whose length is infinite,

$$I = \int ds = \int_{0+\varepsilon}^{\pi-\varepsilon} \frac{ds}{dt} \, dt = \int_{0+\varepsilon}^{\pi-\varepsilon} \frac{1}{\sin t} \, dt$$

$$= -\frac{1}{2} \log \frac{1 + \cos t}{1 - \cos t} \Big|_{0+\varepsilon}^{\pi-\varepsilon} \xrightarrow{\varepsilon \to 0} \infty.$$

7.4.4 The normal coordinate system

The subject here is not restricted to Levi-Civita connections but it does take an especially simple form when the Levi-Civita connection is employed. Let $c(t)$ be a geodesic in (M, g) defined with respect to a connection ∇, which satisfies

$$c(0) = p, \qquad \left.\frac{d}{dt}\right|_p = X = X^\mu e_\mu \in T_p M \tag{7.71}$$

where $\{e_\mu\}$ is the coordinate basis at p. Any geodesic emanating from p is specified by giving $X \in T_p M$. Take a point q near p. There are many geodesics which connect p and q. However, there exists a *unique* geodesic c_q such that $c_q(1) = q$. Let $X_q \in T_p M$ be the tangent vector of this geodesic at p. As long as q is not far from p, q uniquely specifies $X_q = X_q^\mu e_\mu \in T_p M$ and $\varphi : q \to X_q^\mu$ serves as a good coordinate system in the neighbourhood of p. This coordinate system is called the **normal coordinate system** based on p with basis $\{e_\mu\}$. Obviously $\varphi(p) = 0$. We define a map EXP : $T_p M \to M$ by EXP : $X_q \mapsto q$. By definition, we have

$$\varphi(\text{EXP } X_q^\mu e_\mu) = X_q^\mu. \tag{7.72}$$

With respect to this coordinate system, a geodesic $c(t)$ with $c(0) = p$ and $c(1) = q$ has the coordinate presentation

$$\varphi(c(t)) = X^\mu = X_q^\mu t \tag{7.73}$$

where X_q^μ are the normal coordinates of q.

We now show that Levi-Civita connection coefficients vanish in the normal coordinate system. We write down the geodesic equation in the normal coordinate system,

$$0 = \frac{d^2 X^\mu}{dt^2} + \Gamma^\mu{}_{\nu\lambda}(X_q^\kappa t)\frac{dX^\nu}{dt}\frac{dX^\lambda}{dt} = \Gamma^\mu{}_{\nu\lambda}(X_q^\kappa t)X_q^\nu X_q^\lambda. \tag{7.74}$$

Since $\Gamma^\mu{}_{\nu\lambda}(p)X_q^\nu X_q^\lambda = 0$ for *any* X_q^ν at p for which $t = 0$, we find $\Gamma^\mu{}_{\nu\lambda}(p) + \Gamma^\mu{}_{\lambda\nu}(p) = 0$. Since our connection is symmetric we must have

$$\Gamma^\mu{}_{\nu\lambda}(p) = 0. \tag{7.75}$$

As a consequence, the covariant derivative of any tensor t in this coordinate system takes the extremely simple form at p,

$$\nabla_X t^{\cdots}_{\cdots} = X[t^{\cdots}_{\cdots}]. \tag{7.76}$$

Equation (7.75) does not imply that $\Gamma^\mu{}_{\nu\lambda}$ vanishes at q ($\neq p$). In fact, we find from (7.42) that

$$R^\kappa{}_{\lambda\mu\nu}(p) = \partial_\mu \Gamma^\kappa{}_{\nu\lambda}(p) - \partial_\nu \Gamma^\kappa{}_{\mu\lambda}(p) \tag{7.77}$$

hence $\partial_\mu \Gamma^\kappa{}_{\nu\lambda}(p) \neq 0$ if $R^\kappa{}_{\lambda\mu\nu}(p) \neq 0$.

7.4.5 Riemann curvature tensor with Levi-Civita connection

Let ∇ be the Levi-Civita connection. The components of the Riemann curvature tensor are given by (7.42) with

$$\Gamma^\lambda{}_{\mu\nu} = \left\{ \begin{matrix} \kappa \\ \mu\nu \end{matrix} \right\}$$

while the torsion tensor vanishes by definition. Many formulae are simplified if the Levi-Civita connections are employed.

Exercise 7.10.

(a) Let $g = \mathrm{d}r \otimes \mathrm{d}r + r^2(\mathrm{d}\theta \otimes \mathrm{d}\theta + \sin^2\theta\, \mathrm{d}\phi \otimes \mathrm{d}\phi)$ be the metric of $(\mathbb{R}^3, \delta)$, where $0 \le \theta \le \pi$, $0 \le \phi < 2\pi$. Show, by direct calculation, that all the components of the Riemann curvature tensor with respect to the Levi-Civita connection vanish.

(b) The spatially homogeneous and isotropic universe is described by the **Robertson–Walker metric**,

$$g = -\mathrm{d}t \otimes \mathrm{d}t + a^2(t) \left(\frac{\mathrm{d}r \otimes \mathrm{d}r}{1 - kr^2} + r^2(\mathrm{d}\theta \otimes \mathrm{d}\theta + \sin^2\theta\, \mathrm{d}\phi \otimes \mathrm{d}\phi) \right) \quad (7.78)$$

where k is a constant, which may be chosen to be $-1, 0$ or $+1$ by a suitable rescaling of r and $0 \le \theta \le \pi$, $0 \le \phi < 2\pi$. If $k = +1$, r is restricted to $0 \le r < 1$. Compute the Riemann tensor, the Ricci tensor and the scalar curvature.

(c) The **Schwarzschild metric** takes the from

$$g = -\left(1 - \frac{2M}{r}\right) \mathrm{d}t \otimes \mathrm{d}t$$

$$+ \frac{1}{1 - \dfrac{2M}{r}}\, \mathrm{d}r \otimes \mathrm{d}r + r^2(\mathrm{d}\theta \otimes \mathrm{d}\theta + \sin^2\theta\, \mathrm{d}\phi \otimes \mathrm{d}\phi) \quad (7.79)$$

where $0 < 2M < r$, $0 \le \theta \le \pi$, $0 \le \phi < 2\pi$. Compute the Riemann tensor, the Ricci tensor and the scalar curvature. [*Remark*: The metric (7.79) describes a spacetime of a spherically symmetric object with mass M.]

Exercise 7.11. Let R be the Riemann tensor defined with respect to the Levi-Civita connection. Show that

$$R_{\kappa\lambda\mu\nu} = \frac{1}{2}\left(\frac{\partial^2 g_{\kappa\mu}}{\partial x^\lambda \partial x^\nu} - \frac{\partial^2 g_{\lambda\mu}}{\partial x^\kappa \partial x^\nu} - \frac{\partial^2 g_{\kappa\nu}}{\partial x^\lambda \partial x^\mu} + \frac{\partial^2 g_{\mu\nu}}{\partial x^\kappa \partial x^\mu} \right)$$

$$+ g_{\zeta\eta}(\Gamma^\zeta{}_{\kappa\mu}\Gamma^\eta{}_{\lambda\nu} - \Gamma^\zeta{}_{\kappa\nu}\Gamma^\eta{}_{\lambda\mu})$$

where $R_{\kappa\lambda\mu\nu} \equiv g_{\kappa\zeta} R^{\zeta}{}_{\lambda\mu\nu}$. Verify the following symmetries,

$$R_{\kappa\lambda\mu\nu} = -R_{\kappa\lambda\nu\mu} \qquad \text{(cf (7.43))} \tag{7.80a}$$

$$R_{\kappa\lambda\mu\nu} = -R_{\lambda\kappa\mu\nu} \tag{7.80b}$$

$$R_{\kappa\lambda\mu\nu} = R_{\mu\nu\kappa\lambda} \tag{7.80c}$$

$$Ric_{\mu\nu} = Ric_{\nu\mu}. \tag{7.80d}$$

Theorem 7.2. (**Bianchi identities**) Let R be the Riemann tensor defined with respect to the Levi-Civita connection. Then R satisfies the following identities:

$$R(X, Y)Z + R(Z, X)Y + R(Y, Z)X = 0$$

(the **first Bianchi identity**) $\tag{7.81a}$

$$(\nabla_X R)(Y, Z)V + (\nabla_Z R)(X, Y)V + (\nabla_Y R)(Z, X)V = 0$$

(the **second Bianchi identity**). $\tag{7.81b}$

Proof. Our proof follows Nomizu (1981). Define the symmetrizor $\mathfrak{S}$ by $\mathfrak{S}\{f(X, Y, Z)\} = f(X, Y, Z) + f(Z, X, Y) + f(Y, Z, X)$. Let us prove the first Bianchi identity $\mathfrak{S}\{R(X, Y)Z\} = 0$. Covariant differentiation of the identity $T(X, Y) = \nabla_X Y - \nabla_Y X - [X, Y] = 0$ with respect to Z yields

$$0 = \nabla_Z\{\nabla_X Y - \nabla_Y X - [X, Y]\}$$
$$= \nabla_Z \nabla_X Y - \nabla_Z \nabla_Y X - \{\nabla_{[X,Y]} Z + [Z, [X, Y]]\}$$

where the torsion-free condition has been used again to derive the second equality. Symmetrizing this, we have

$$0 = \mathfrak{S}\{\nabla_Z \nabla_X Y - \nabla_Z \nabla_Y X - \nabla_{[X,Y]}Z - [Z, [X, Y]]\}$$
$$= \mathfrak{S}\{\nabla_Z \nabla_X Y - \nabla_Z \nabla_Y X - \nabla_{[X,Y]}Z\} = \mathfrak{S}\{R(X, Y)Z\}$$

where the Jacobi identity $\mathfrak{S}\{[X, [Y, Z]]\} = 0$ has been used.

The second Bianchi identity becomes $\mathfrak{S}\{(\nabla_X R)(Y, Z)\}V = 0$ where $\mathfrak{S}$ symmetrizes (X, Y, Z) only. If the identity $R(T(X, Y), Z)V = R(\nabla_X Y - \nabla_Y X - [X, Y], Z)V = 0$ is symmetrized, we have

$$0 = \mathfrak{S}\{R(\nabla_X Y, Z) - R(\nabla_Y X, Z) - R([X, Y], Z)\}V$$
$$= \mathfrak{S}\{R(\nabla_Z X, Y) - R(X, \nabla_Z Y) - R([X, Y], Z)\}V. \tag{7.82}$$

If we note the Leibnitz rule,

$$\nabla_Z\{R(X, Y)V\} = (\nabla_Z R)(X, Y)V$$
$$+ R(X, Y)\nabla_Z V + R(\nabla_Z X, Y)V + R(X, \nabla_Z Y)V$$

(7.82) becomes

$$0 = \mathfrak{S}\{-(\nabla_Z R)(X, Y) + [\nabla_Z, R(X, Y)] - R([X, Y], Z)\}V.$$

The last two terms vanish if $R(X, Y)V = \{[\nabla_X, \nabla_Y] - \nabla_{[X,Y]}\}V$ is substituted into them,

$$\mathfrak{S}\{[\nabla_Z, R(X, Y)] - R([X, Y], Z)\}V$$
$$= \mathfrak{S}\{[\nabla_Z, [\nabla_X, \nabla_Y]] - [\nabla_Z, \nabla_{[X,Y]}] - [\nabla_{[X,Y]}, \nabla_Z] + \nabla_{[[X,Y],Z]}\}V$$
$$= 0$$

where the Jacobi identities $\mathfrak{S}\{[\nabla_Z, [\nabla_X, \nabla_Y]]\} = \mathfrak{S}\{[[X, Y], Z]\} = 0$ have been used. We finally obtain $\mathfrak{S}\{(\nabla_X R)(Y, Z)\}V = 0$. $\qquad\square$

In components, the Bianchi identities are

$$R^\kappa{}_{\lambda\mu\nu} + R^\kappa{}_{\mu\nu\lambda} + R^\kappa{}_{\nu\lambda\mu} = 0$$
(the first Bianchi identity) $\qquad\qquad$ (7.83a)

$$(\nabla_\kappa R)^\xi{}_{\lambda\mu\nu} + (\nabla_\mu R)^\xi{}_{\lambda\nu\kappa} + (\nabla_\nu R)^\xi{}_{\lambda\kappa\mu} = 0$$
(the second Bianchi identity). $\qquad\qquad$ (7.83b)

By contracting the indices ξ and μ of the second Bianchi identity, we obtain an important relation:

$$(\nabla_\kappa Ric)_{\lambda\nu} + (\nabla_\mu R)^\mu{}_{\lambda\nu\kappa} - (\nabla_\nu Ric)_{\lambda\kappa} = 0. \qquad (7.84)$$

If the indices λ and ν are further contracted, we have $\nabla_\mu(\mathcal{R}\delta - 2Ric)^\mu{}_\kappa = 0$ or

$$\nabla_\mu G^{\mu\nu} = 0 \qquad\qquad (7.85)$$

where $G^{\mu\nu}$ is the **Einstein tensor** defined by

$$G^{\mu\nu} = Ric^{\mu\nu} - \tfrac{1}{2}g^{\mu\nu}\mathcal{R}. \qquad\qquad (7.86)$$

Historically, when Einstein formulated general relativity, he first equated the Ricci tensor $Ric^{\mu\nu}$ to the energy–momentum tensor $T^{\mu\nu}$. Later he realized that $T^{\mu\nu}$ satisfies the covariant conservation equation $\nabla_\mu T^{\mu\nu} = 0$ while $Ric^{\mu\nu}$ does not. To avoid this difficulty, he proposed that $G^{\mu\nu}$ should be equated to $T^{\mu\nu}$. This new equation is natural in the sense that it can be derived from a scalar action by variation, see section 7.10.

Exercise 7.12. Let (M, g) be a two-dimensional manifold with $g = -dt \otimes dt + R^2(t)dx \otimes dx$, where $R(t)$ is an arbitrary function of t. Show that the Einstein tensor vanishes.

The symmetry properties (7.80a)–(7.80c) restrict the number of independent components of the Riemann tensor. Let m be the dimension of a manifold (M, g). The anti-symmetry $R_{\kappa\lambda\mu\nu} = -R_{\lambda\kappa\nu\mu}$ implies that there are $N \equiv \binom{m}{2}$ independent choices of the pair (μ, ν). Similarly, from $R_{\kappa\lambda\mu\nu} = -R_{\lambda\kappa\mu\nu}$, we find there are

N independent pairs of (κ, λ). Since $R_{\kappa\lambda\mu\nu}$ is symmetric with respect to the interchange of the pairs (κ, λ) and (μ, ν), the number of independent choices of the pairs reduces from N^2 to $\binom{N+1}{2} = \frac{1}{2}N(N+1)$. The first Bianchi identity

$$R_{\kappa\lambda\mu\nu} + R_{\kappa\mu\nu\lambda} + R_{\kappa\nu\lambda\mu} = 0 \qquad (7.87)$$

further reduces the number of independent components. The LHS of (7.87) is totally anti-symmetric with respect to the interchange of the indices (λ, μ, ν). Furthermore, the anti-symmetry (7.80b) ensures that it is totally anti-symmetric in all the indices. If $m < 4$, (7.87) is trivially satisfied and it imposes no additional restrictions. If $m \geq 4$, (7.87) yields non-trivial constraints only when all the indices are different. The number of constraints is equal to the number of possible ways of choosing four different indices out of m indices, namely $\binom{m}{4}$. Noting that $\binom{m}{4} = m(m-1)(m-2)(m-3)/4!$ vanishes for $m < 4$, the number of independent components of the Riemann tensor is given by

$$F(m) = \frac{1}{2}\binom{m}{2}\left[\binom{m}{2} + 1\right] - \binom{m}{4} = \frac{1}{12}m^2(m^2 - 1). \qquad (7.88)$$

$F(1) = 0$ implies that one-dimensional manifolds are flat. Since $F(2) = 1$, there is only one independent component R_{1212} on a two-dimensional manifold, other components being either 0 or $\pm R_{1212}$. $F(4) = 20$ is a well-known fact in general relativity.

Exercise 7.13. Let (M, g) be a two-dimensional manifold. Show that the Riemann tensor is written as

$$R_{\kappa\lambda\mu\nu} = K(g_{\kappa\mu}g_{\lambda\nu} - g_{\kappa\nu}g_{\lambda\mu}) \qquad (7.89)$$

where $K \in \mathcal{F}(M)$. Compute the Ricci tensor to show $Ric_{\mu\nu} \propto g_{\mu\nu}$. Compute the scalar curvature to show $K = \mathcal{R}/2$.

7.5 Holonomy

Let (M, g) be an m-dimensional Riemannian manifold with an affine connection ∇. The connection naturally defines a transformation group at each tangent space T_pM as follows.

Definition 7.3. Let p be a point in (M, g) and consider the set of closed loops at p, $\{c(t)|0 \leq t \leq 1, c(0) = c(1) = p\}$. Take a vector $X \in T_pM$ and parallel transport X along a curve $c(t)$. After a trip along $c(t)$, we end up with a new vector $X_c \in T_pM$. Thus, the loop $c(t)$ and the connection ∇ induce a linear transformation

$$P_c : T_pM \to T_pM. \qquad (7.90)$$

The set of these transformations is denoted by $H(p)$ and is called the **holonomy group** at p.

We assume that $H(p)$ acts on T_pM from the right, $P_cX = Xh$ $(h \in H(p))$. In components, this becomes $P_cX = X^\mu h_\mu{}^\nu e_\nu$, $\{e_\nu\}$ being the basis of T_pM. It is easy to see that $H(p)$ is a group. The product $P_{c'}P_c$ corresponds to parallel transport along c first and then c'. If we write $P_d = P_{c'}P_c$, the loop d is given by

$$d(t) = \begin{cases} c(2t) & 0 \le t \le \frac{1}{2} \\ c'(2t - 1) & \frac{1}{2} \le t \le 1. \end{cases} \tag{7.91}$$

The unit element corresponds to the constant map $c_p(t) = p$ $(0 \le t \le 1)$ and the inverse of P_c is given by $P_{c^{-1}}$, where $c^{-1}(t) = c(1 - t)$. Note that $H(p)$ is a subgroup of $GL(m, \mathbb{R})$, which is the maximal holonomy group possible. $H(p)$ is trivial if and only if the Riemann tensor vanishes. In particular, if (M, g) is parallelizable (see example 7.2), we can make $H(p)$ trivial.

If M is (arcwise-)connected, any two points $p, q \in M$ are connected by a curve a. The curve a defines a map $\tau_a : T_pM \to T_qM$ by parallel transporting a vector in T_pM to T_qM along a. Then the holonomy groups $H(p)$ and $H(q)$ are related by

$$H(q) = \tau_a^{-1}H(p)\tau_a \tag{7.92}$$

hence $H(q)$ is isomorphic to $H(p)$.

In general, the holonomy group is a subgroup of $GL(m, \mathbb{R})$. If ∇ is a metric connection, ∇ preserves the length of a vector, $g_p(P_c(X), P_c(X)) = g_p(X, X)$ for $X \in T_pM$. Then the holonomy group must be a subgroup of $SO(m)$ if (M, g) is orientable and Riemannian and $SO(m - 1, 1)$ if it is orientable and Lorentzian.

Example 7.7. We work out the holonomy group of the Levi-Civita connection on S^2 with the metric $g = d\theta \otimes d\theta + \sin^2 d\phi \otimes d\phi$. The non-vanishing connection coefficients are $\Gamma^\theta{}_{\phi\phi} = -\sin\theta\cos\theta$ and $\Gamma^\phi{}_{\phi\theta} = \Gamma^\phi{}_{\theta\phi} = \cot\theta$. For simplicity, we take a vector $e_\theta = \partial/\partial\theta$ at a point $(\theta_0, 0)$ and parallel transport it along a circle $\theta = \theta_0$, $0 \le \phi \le 2\pi$. Let X be the vector e_θ parallel transported along the circle. The vector $X = X^\theta e_\theta + X^\phi e_\phi$ satisfies

$$\partial_\phi X^\theta - \sin\theta_0\cos\theta_0 X^\phi = 0 \tag{7.93a}$$

$$\partial_\phi X^\phi + \cot\theta_0 X^\theta = 0. \tag{7.93b}$$

Equations (7.93a) and (7.93b) represent the harmonic oscillations. Indeed if we take a ϕ-derivative of (7.93a) and use (7.93b), we have

$$\frac{d^2X^\theta}{d\phi^2} - \sin\theta_0\cos\theta_0\frac{dX^\phi}{d\phi} = \frac{d^2X^\theta}{d\phi^2} - \cos^2\theta_0 X^\theta = 0. \tag{7.94}$$

The general solution is $X^\theta = A\cos(C_0\phi) + B\sin(C_0\phi)$, where $C_0 \equiv \cos\theta_0$. Since $X^\theta = 1$ at $\phi = 0$ we have

$$X^\theta = \cos(C_0\phi) \qquad X^\phi = -\frac{\sin(C_0\phi)}{\sin\theta_0}.$$

After parallel transport along the circle, we end up with

$$X(\phi = 2\pi) = \cos(2\pi C_0)e_\theta - \frac{\sin(2\pi C_0\phi)}{\sin\theta_0}e_\phi. \tag{7.95}$$

Now the vector is rotated by $\Theta = 2\pi\cos\theta_0$, with its magnitude kept fixed. If we take a point $p \in S^2$ and a circle in S^2 which passes through p, we can always find a coordinate system such that the circle is given by $\theta = \theta_0$ $(0 \le \theta < \pi)$ and we can apply our previous calculation. The rotation angle is $-2\pi \le \Theta < 2\pi$ and we find that the holonomy group at $p \in S^2$ is $SO(2)$.

In general, S^m $(m \ge 2)$ admits the holonomy group $SO(m)$. Product manifolds admit more restricted holonomy groups. The following example is taken from Horowitz (1986). Consider six-dimensional manifolds made of the spheres with standard metrics. Examples are S^6, $S^3 \times S^3$, $S^2 \times S^2 \times S^2$, $T^6 = S^1 \times \cdots \times S^1$. Their holonomy groups are:

(i) S^6: $H(p) = SO(6)$.
(ii) $S^3 \times S^3$: $H(p) = SO(3) \times SO(3)$.
(iii) $S^2 \times S^2 \times S^2$: $H(p) = SO(2) \times SO(2) \times SO(2)$.
(iv) T^6: $H(p)$ is trivial since the Riemann tensor vanishes.

Exercise 7.14. Show that the holonomy group of the Levi-Civita connection of the Poincaré metric given in example 7.6 is $SO(2)$.

7.6 Isometries and conformal transformations

7.6.1 Isometries

Definition 7.4. Let (M, g) be a (pseudo-)Riemannian manifold. A diffeomorphism $f : M \to M$ is an **isometry** if it preserves the metric

$$f^*g_{f(p)} = g_p \tag{7.96a}$$

that is, if $g_{f(p)}(f_*X, f_*Y) = g_p(X, Y)$ for $X, Y \in T_pM$.

In components, the condition (7.96a) becomes

$$\frac{\partial y^\alpha}{\partial x^\mu}\frac{\partial y^\beta}{\partial x^\nu}g_{\alpha\beta}(f(p)) = g_{\mu\nu}(p) \tag{7.96b}$$

where x and y are the coordinates of p and $f(p)$, respectively. The identity map, the composition of the isometries and the inverse of an isometry are isometries; all these isometries form a group. Since an isometry preserves the *length* of a vector, in particular that of an infinitesimal displacement vector, it may be regarded as a rigid motion. For example, in $\mathbb{R}^n$, the Euclidean group E^n, that is the set of maps $f : x \mapsto Ax + T$ $(A \in SO(n), T \in \mathbb{R}^n)$, is the isometry group.

7.6.2   Conformal transformations

Definition 7.5. Let (M, g) be a (pseudo-)Riemannian manifold. A diffeomorphism $f : M \to M$ is called a **conformal transformation** if it preserves the metric *up to a scale*,

$$f^* g_{f(p)} = e^{2\sigma} g_p \qquad \sigma \in \mathcal{F}(M) \tag{7.97a}$$

namely, $g_{f(p)}(f_* X, f_* Y) = e^{2\sigma} g_p(X, Y)$ for $X, Y \in T_p M$.

In components, the condition (7.97a) becomes

$$\frac{\partial y^\alpha}{\partial x^\mu} \frac{\partial y^\beta}{\partial x^\nu} g_{\alpha\beta}(f(p)) = e^{2\sigma(p)} g_{\mu\nu}(p). \tag{7.97b}$$

The set of conformal transformations on M is a group, the **conformal group** denoted by $\mathrm{Conf}(M)$. Let us define the angle θ between two vectors $X = X^\mu \partial_\mu$, $Y = Y^\mu \partial_\mu \in T_p M$ by

$$\cos\theta = \frac{g_p(X, Y)}{\sqrt{g_p(X, X) g_p(Y, Y)}} = \frac{g_{\mu\nu} X^\mu Y^\nu}{\sqrt{g_{\zeta\eta} X^\zeta X^\eta g_{\kappa\lambda} Y^\kappa Y^\lambda}}. \tag{7.98}$$

If f is a conformal transformation, the angle θ' between $f_* X$ and $f_* Y$ is given by

$$\cos\theta' = \frac{e^{2\sigma} g_{\mu\nu} X^\mu Y^\nu}{\sqrt{e^{2\sigma} g_{\zeta\eta} X^\zeta X^\eta \cdot e^{2\sigma} g_{\kappa\lambda} Y^\kappa Y^\lambda}} = \cos\theta$$

hence f preserves the angle. In other words, f changes the *scale* but not the *shape*.

A concept related to conformal transformations is Weyl rescaling. Let g and $\bar{g}$ be metrics on a manifold M. $\bar{g}$ is said to be **conformally related** to g if

$$\bar{g}_p = e^{2\sigma(p)} g_p. \tag{7.99}$$

Clearly this is an equivalence relation among the set of metrics on M. The equivalence class is called the **conformal structure**. The transformation $g \to e^{2\sigma} g$ is called a **Weyl rescaling**. The set of Weyl rescalings on M is a group denoted by $\mathrm{Weyl}(M)$.

Example 7.8. Let $w = f(z)$ be a holomorphic function defined on the complex plane $\mathbb{C}$. [A C^∞-function regarded as a function of $z = x + iy$ and $\bar{z} = x - iy$ is holomorphic if $\partial_{\bar{z}} f(z, \bar{z}) = 0$.] We write the real part and the imaginary part of the respective variables as $z = x + iy$ and $w = u + iv$. The map $f : (x, y) \mapsto (u, v)$ is conformal since

$$du^2 + dv^2 = \left(\frac{\partial u}{\partial x} dx + \frac{\partial u}{\partial y} dy\right)^2 + \left(\frac{\partial v}{\partial x} dx + \frac{\partial v}{\partial y} dy\right)^2$$

$$= \left[\left(\frac{\partial u}{\partial x}\right)^2 + \left(\frac{\partial u}{\partial y}\right)^2\right] (dx^2 + dy^2) \tag{7.100}$$

where use has been made of the Cauchy–Riemann relations

$$\frac{\partial u}{\partial x} = \frac{\partial v}{\partial y} \qquad \frac{\partial u}{\partial y} = -\frac{\partial v}{\partial x}.$$

Exercise 7.15. Let $f : M \to M$ be a conformal transformation on a Lorentz manifold (M, g). Show that $f_* : T_p M \to T_{f(p)} M$ preserves the local light cone structure, namely

$$f_* : \begin{cases} \text{timelike vector} & \mapsto & \text{timelike vector} \\ \text{null vector} & \mapsto & \text{null vector} \\ \text{spacelike vector} & \mapsto & \text{spacelike vector.} \end{cases} \qquad (7.101)$$

Let $\bar{g}$ be a metric on M, which is conformally related to g as $\bar{g} = e^{2\sigma(p)} g$. Let us compute the Riemann tensor of $\bar{g}$. We could simply substitute $\bar{g}$ into the defining equation (7.42). However, we follow the elegant coordinate-free derivation of Nomizu (1981). Let K be the difference of the covariant derivatives $\bar{\nabla}$ with respect to $\bar{g}$ and ∇ with respect to g,

$$K(X, Y) \equiv \bar{\nabla}_X Y - \nabla_X Y. \qquad (7.102)$$

Proposition 7.1. Let U be a vector field which corresponds to the one-form $d\sigma$: $Z[\sigma] = \langle d\sigma, Z \rangle = g(U, Z)$. Then

$$K(X, Y) = X[\sigma]Y + Y[\sigma]X - g(X, Y)U. \qquad (7.103)$$

Proof. It follows from the torsion-free condition that $K(X, Y) = K(Y, X)$. Since $\bar{\nabla}_X \bar{g} = \nabla_X g = 0$, we have

$$X[\bar{g}(Y, Z)] = \bar{\nabla}_X[\bar{g}(Y, Z)] = \bar{g}(\bar{\nabla}_X, Z) + \bar{g}(Y, \bar{\nabla}_X Z)$$

and also

$$\begin{aligned} X[\bar{g}(Y, Z)] &= \nabla_X[e^{2\sigma} g(Y, Z)] \\ &= 2X[\sigma]e^{2\sigma} g(Y, Z) + e^{2\sigma}[g(\nabla_X, Z) + g(Y, \nabla_X Z)]. \end{aligned}$$

Taking the difference between these two expressions, we have

$$g(K(X, Y), Z) + g(Y, K(X, Z)) = 2X[\sigma]g(Y, Z). \qquad (7.104a)$$

Permutations of (X, Y, Z) yield

$$g(K(Y, X), Z) + g(X, K(Y, Z)) = 2Y[\sigma]g(X, Z) \qquad (7.104b)$$
$$g(K(Z, X), Y) + g(X, K(Z, Y)) = 2Z[\sigma]g(X, Y). \qquad (7.104c)$$

The combination (7.104a) + (7.104b) − (7.104c) yields

$$g(K(X, Y), Z) = X[\sigma]g(Y, Z) + Y[\sigma]g(X, Z) - Z[\sigma]g(X, Y). \qquad (7.105)$$

The last term is modified as

$$Z[\sigma]g(X, Y) = g(U, Z)g(X, Y) = g(g(Y, X)U, Z).$$

Substituting this into (7.105), we find

$$g(K(X, Y) - X[\sigma]Y - Y[\sigma]X + g(X, Y)U, Z) = 0.$$

Since this is true for any Z, we have (7.103).   □

The component expression for K is

$$\begin{aligned}
K(e_\mu, e_\nu) &= \bar{\nabla}_\mu e_\nu - \nabla_\mu e_\nu = (\bar{\Gamma}^\lambda{}_{\mu\nu} - \Gamma^\lambda{}_{\mu\nu})e_\lambda \\
&= e_\mu[\sigma]e_\nu + e_\nu[\sigma]e_\mu - g(e_\mu, e_\nu)g^{\kappa\lambda}\partial_\kappa\sigma e_\lambda
\end{aligned}$$

from which it is readily seen that

$$\bar{\Gamma}^\lambda{}_{\mu\nu} = \Gamma^\lambda{}_{\mu\nu} + \delta^\lambda{}_\nu\partial_\mu\sigma + \delta^\lambda{}_\mu\partial_\nu\sigma - g_{\mu\nu}g^{\kappa\lambda}\partial_\kappa\sigma. \tag{7.106}$$

To find the Riemann curvature tensor, we start from the definition,

$$\begin{aligned}
\bar{R}(X, Y)Z &= \bar{\nabla}_X\bar{\nabla}_Y Z - \bar{\nabla}_Y\bar{\nabla}_X Z - \bar{\nabla}_{[X,Y]}Z \\
&= \bar{\nabla}_X[\nabla_Y Z + K(Y, Z)] - \bar{\nabla}_Y[\nabla_X Z + K(X, Z)] \\
&\quad - \{\nabla_{[X,Y]}Z + K([X, Y], Z)\} \\
&= \nabla_X\{\nabla_Y Z + K(Y, Z)\} + K(X, \nabla_Y Z + K(Y, Z)) \\
&\quad - \nabla_Y\{\nabla_X Z + K(X, Z)\} - K(Y, \nabla_X Z + K(X, Z)) \\
&\quad - \{\nabla_{[X,Y]}Z + K([X, Y], Z)\}. \tag{7.107}
\end{aligned}$$

After a straightforward but tedious calculation, we find that

$$\begin{aligned}
\bar{R}(X, Y)Z &= R(X, Y)Z + \langle\nabla_X \, d\sigma, Z\rangle Y - \langle\nabla_Y \, d\sigma, Z\rangle X \\
&\quad - g(Y, Z)\nabla_X U + Y[\sigma]Z[\sigma]X \\
&\quad - g(Y, Z)U[\sigma]X + X[\sigma]g(Y, Z)U \\
&\quad + g(X, Z)\nabla_Y U - X[\sigma]Z[\sigma]Y \\
&\quad + g(X, Z)U[\sigma]Y - Y[\sigma]g(X, Z)U. \tag{7.108}
\end{aligned}$$

Let us define a type $(1, 1)$ tensor field B by

$$BX \equiv -X[\sigma]U + \nabla_X U + \tfrac{1}{2}U[\sigma]X. \tag{7.109}$$

Since $g(\nabla_Y U, Z) = \langle\nabla_Y d\sigma, Z\rangle$ (exercise 7.8(c)), (7.108) becomes

$$\begin{aligned}
\bar{R}(X, Y)Z &= R(X, Y)Z - [g(Y, Z)BX - g(BX, Z)Y \\
&\quad + g(BY, Z)X - g(X, Z)BY]. \tag{7.110}
\end{aligned}$$

In components, this becomes

$$\bar{R}^\kappa{}_{\lambda\mu\nu} = R^\kappa{}_{\lambda\mu\nu} - g_{\nu\lambda}B_\mu{}^\kappa + g_{\xi\lambda}B_\mu{}^\xi\delta^\kappa{}_\nu - g_{\xi\lambda}B_\nu{}^\xi\delta^\kappa{}_\mu + g_{\mu\lambda}B_\nu{}^\kappa \tag{7.111}$$

where the components of the tensor B are

$$\begin{aligned}B_\mu{}^\kappa &= -\partial_\mu\sigma U^\kappa + (\nabla_\mu U)^\kappa + \tfrac{1}{2}U[\sigma]\delta_\mu{}^\kappa\\ &= -\partial_\mu\sigma g^{\kappa\lambda}\partial_\lambda\sigma + g^{\kappa\lambda}(\partial_\mu\partial_\lambda\sigma - \Gamma^\xi{}_{\mu\lambda}\partial_\xi\sigma) + \tfrac{1}{2}g^{\lambda\xi}\partial_\lambda\sigma\partial_\xi\sigma\delta_\mu{}^\kappa.\end{aligned} \tag{7.112}$$

Note that $B_{\mu\nu} \equiv g_{\nu\lambda}B_\mu{}^\lambda = B_{\nu\mu}$.

By contracting the indices in (7.111), we obtain

$$\overline{Ric}_{\mu\nu} = Ric_{\mu\nu} - g_{\mu\nu}B_\lambda{}^\lambda - (m-2)B_{\nu\mu} \tag{7.113}$$

$$e^{2\sigma}\bar{\mathcal{R}} = \mathcal{R} - 2(m-1)B_\lambda{}^\lambda \tag{7.114a}$$

where $m = \dim M$. Equation (7.114a) is also written as

$$\bar{g}_{\mu\nu}\bar{\mathcal{R}} = [\mathcal{R} - 2(m-1)B_\lambda{}^\lambda]g_{\mu\nu}. \tag{7.114b}$$

If we eliminate $g_{\mu\nu}B_\lambda{}^\lambda$ and $B_{\mu\nu}$ in $\bar{R}^\kappa{}_{\lambda\mu\nu}$ in favour of $\overline{Ric}$ and $\bar{\mathcal{R}}$ and separate barred and unbarred terms, we find a combination which is independent of σ,

$$\begin{aligned}C_{\kappa\lambda\mu\nu} &= R_{\kappa\lambda\mu\nu} + \frac{1}{m-2}(Ric_{\kappa\mu}g_{\lambda\nu} - Ric_{\lambda\mu}g_{\kappa\nu} + Ric_{\lambda\nu}g_{\kappa\mu} - Ric_{\kappa\nu}g_{\lambda\mu})\\ &\quad + \frac{\mathcal{R}}{(m-2)(m-1)}(g_{\kappa\mu}g_{\lambda\nu} - g_{\kappa\nu}g_{\lambda\mu})\end{aligned} \tag{7.115}$$

where $m \geq 4$ (see problem 7.2 for $m = 3$). The tensor C is called the **Weyl tensor**. The reader should verify that $C_{\kappa\lambda\mu\nu} = \bar{C}_{\kappa\lambda\mu\nu}$.

If every point p of a (pseudo-)Riemannian manifold (M, g) has a chart (U, φ) containing p such that $g_{\mu\nu} = e^{2\sigma}\delta_{\mu\nu}$, then (M, g) is said to be **conformally flat**. Since the Weyl tensor vanishes for a flat metric, it also vanishes for a conformally flat metric. If $\dim M \geq 4$, then $C = 0$ is the necessary and sufficient condition for conformal flatness (Weyl–Schouten). If $\dim M = 3$, the Weyl tensor vanishes identically; see problem 7.2. If $\dim M = 2$, M is always conformally flat; see the next example.

Example 7.9. Any two-dimensional Riemannian manifold (M, g) is conformally flat. Let (x, y) be the original local coordinates with which the metric takes the form

$$ds^2 = g_{xx}\,dx^2 + 2g_{xy}\,dx\,dy + g_{yy}\,dy^2. \tag{7.116}$$

Let $g \equiv g_{xx}g_{yy} - g_{xy}^2$ and write (7.116) as

$$ds^2 = \left(\sqrt{g_{xx}}\, dx + \frac{g_{xy} + i\sqrt{g}}{\sqrt{g_{xx}}}\, dy \right) \left(\sqrt{g_{yy}}\, dx + \frac{g_{xy} - i\sqrt{g}}{\sqrt{g_{xx}}}\, dy \right).$$

According to the theory of differential equations, there exists an integrating factor $\lambda(x, y) = \lambda_1(x, y) + i\lambda_2(x, y)$ such that

$$\lambda \left(\sqrt{g_{xx}}\, dx + \frac{g_{xy} + i\sqrt{g}}{\sqrt{g_{xx}}} dy \right) = du + i\, dv \qquad (7.117a)$$

$$\bar{\lambda} \left(\sqrt{g_{yy}}\, dx + \frac{g_{xy} - i\sqrt{g}}{\sqrt{g_{xx}}} dy \right) = du - i\, dv. \qquad (7.117b)$$

Then $ds^2 = (du^2 + dv^2)/|\lambda|^2$ and by setting $|\lambda|^{-2} = e^{2\sigma}$, we have the desired coordinate system. The coordinates (u, v) are called the **isothermal coordinates**. [*Remark*: If the curve $u = $ a constant is regarded as an isothermal curve, $v = $ a constant corresponds to the line of heat flow.]

For example, let $ds^2 = d\theta^2 + \sin^2\theta\, d\phi^2$ be the standard metric of S^2. Noting that

$$\frac{d}{d\theta} \log \left| \tan \frac{\theta}{2} \right| = \frac{1}{\sin\theta}$$

we find that $f : (\theta, \phi) \mapsto (u, v)$ defined by $u = \log|\tan\frac{1}{2}\theta|$ and $v = \phi$ yields a conformally flat metric. In fact,

$$ds^2 = \sin^2\theta \left(\frac{d\theta^2}{\sin^2\theta} + d\phi^2 \right) = \sin^2\theta (du^2 + dv^2).$$

If (M, g) is a Lorentz manifold, we have integrating factors $\lambda(x, y)$ and $\mu(x, y)$ such that

$$\lambda \left(\sqrt{g_{xx}}\, dx + \frac{g_{xy} + \sqrt{-g}}{\sqrt{g_{xx}}} dy \right) = du + dv \qquad (7.118a)$$

$$\mu \left(\sqrt{g_{xx}}\, dx + \frac{g_{xy} - \sqrt{-g}}{\sqrt{g_{xx}}} dy \right) = du - dv. \qquad (7.118b)$$

In terms of the coordinates (u, v) the metric takes the form $ds^2 = \lambda^{-1}\mu^{-1}(du^2 - dv^2)$. The product $\lambda\mu$ is either positive definite or negative definite and we may set $1/|\lambda\mu| = e^{2\sigma}$ to obtain the form

$$ds^2 = \pm e^{2\sigma}(du^2 - dv^2). \qquad (7.119)$$

Exercise 7.16. Let (M, g) be a two-dimensional Lorentz manifold with $g = -dt \otimes dt + t^2 dx \otimes dx$ (the **Milne universe**). Use the transformation $|t| \mapsto e^\eta$ to show that g is conformally flat. In fact, it is further simplified by $(\eta, x) \mapsto (u = e^\eta \sinh x, v = e^\eta \cosh x)$. What is the resulting metric?

7.7 Killing vector fields and conformal Killing vector fields

7.7.1 Killing vector fields

Let (M, g) be a Riemannian manifold and $X \in \mathfrak{X}(M)$. If a displacement εX, ε being infinitesimal, generates an isometry, the vector field X is called a **Killing vector field**. The coordinates x^μ of a point $p \in M$ change to $x^\mu + \varepsilon X^\mu(p)$ under this displacement, see (5.42). If $f : x^\mu \mapsto x^\mu + \varepsilon X^\mu$ is an isometry, it satisfies (7.96b),

$$\frac{\partial(x^\kappa + \varepsilon X^\kappa)}{\partial x^\mu} \frac{\partial(x^\lambda + \varepsilon X^\lambda)}{\partial x^\nu} g_{\kappa\lambda}(x + \varepsilon X) = g_{\mu\nu}(x).$$

After a simple calculation, we find that $g_{\mu\nu}$ and X^μ satisfy the **Killing equation**

$$X^\xi \partial_\xi g_{\mu\nu} + \partial_\mu X^\kappa g_{\kappa\nu} + \partial_\nu X^\lambda g_{\mu\lambda} = 0. \tag{7.120a}$$

From the definition of the Lie derivative, this is written in a compact form as

$$(\mathcal{L}_X g)_{\mu\nu} = 0. \tag{7.120b}$$

Let $\phi_t : M \to M$ be a one-parameter group of transformations which generates the Killing vector field X. Equation (7.120b) then shows that the local geometry does not change as we move along ϕ_t. In this sense, the Killing vector fields represent the direction of the symmetry of a manifold.

A set of Killing vector fields are defined to be dependent if one of them is expressed as a linear combination of others with *constant* coefficients. Thus, there may be more Killing vector fields than the dimension of the manifold. [The number of independent symmetries has no direct connection with $\dim M$. The *maximum* number, however, has; see example 7.10.]

Exercise 7.17. Let ∇ be the Levi-Civita connection. Show that the Killing equation is written as

$$(\nabla_\mu X)_\nu + (\nabla_\nu X)_\mu = \partial_\mu X_\nu + \partial_\nu X_\mu - 2\Gamma^\lambda_{\mu\nu} X_\lambda = 0. \tag{7.121}$$

Exercise 7.18. Find three Killing vector fields of $(\mathbb{R}^2, \delta)$. Show that two of them correspond to translations while the third corresponds to a rotation; cf next example.

Example 7.10. Let us work out the Killing vector fields of the Minkowski spacetime $(\mathbb{R}^4, \eta)$, for which all the Levi-Civita connection coefficients vanish. The Killing equation becomes

$$\partial_\mu X_\nu + \partial_\nu X_\mu = 0. \tag{7.122}$$

It is easy to see that X_μ is, at most, of the first order in x. The constant solutions

$$X^\mu_{(i)} = \delta^\mu_i \qquad (0 \le i \le 3) \tag{7.123a}$$

correspond to spacetime translations. Next, let $X_\mu = a_{\mu\nu} x^\mu$, $a_{\mu\nu}$ being constant. Equation (7.122) implies that $a_{\mu\nu}$ is anti-symmetric with respect to $\mu \leftrightarrow \nu$. Since $\binom{4}{2} = 6$, there are six independent solutions of this form, three of which

$$X_{(j)0} = 0 \qquad X_{(j)m} = \varepsilon_{jmn} x^n \qquad (1 \leq j, m, n \leq 3) \qquad (7.123b)$$

correspond to spatial rotations about the x^j-axis, while the others

$$X_{(k)0} = x^k \qquad X_{(k)m} = -\delta_{km} x^0 \qquad (1 \leq k, m \leq 3) \qquad (7.123c)$$

correspond to Lorentz boosts along the x^k-axis.

In m-dimensional Minkowski spacetime ($m \geq 2$), there are $m(m + 1)/2$ Killing vector fields, m of which generate translations, $(m - 1)$, boosts and $(m - 1)(m - 2)/2$, space rotations. Those spaces (or spacetimes) which admit $m(m + 1)/2$ Killing vector fields are called **maximally symmetric spaces**.

Let X and Y be two Killing vector fields. We easily verify that

(i)   a linear combination $aX + bY$ ($a, b \in \mathbb{R}$) is a Killing vector field; and
(ii) the Lie bracket $[X, Y]$ is a Killing vector field.

(i) is obvious from the linearity of the covariant derivative. To prove (ii), we use (5.58). We have $\mathcal{L}_{[X,Y]}g = \mathcal{L}_X \mathcal{L}_Y g - \mathcal{L}_Y \mathcal{L}_X g = 0$, since $\mathcal{L}_X g = \mathcal{L}_Y g = 0$. Thus, all the Killing vector fields form a Lie algebra of the symmetric operations on the manifold M; see the next example.

Example 7.11. Let $g = d\theta \otimes d\theta + \sin^2 \theta \, d\phi \otimes d\phi$ be the standard metric of S^2. The Killing equations (7.121) are:

$$\partial_\theta X_\theta + \partial_\theta X_\theta = 0 \qquad (7.124a)$$
$$\partial_\phi X_\phi + \partial_\phi X_\phi + 2 \sin \theta \cos \theta X_\theta = 0 \qquad (7.124b)$$
$$\partial_\theta X_\phi + \partial_\phi X_\theta - 2 \cot \theta X_\phi = 0. \qquad (7.124c)$$

It follows from (7.124a) that X_θ is independent of θ: $X_\theta(\theta, \phi) = f(\phi)$. Substituting this into (7.124b), we have

$$X_\phi = -F(\phi) \sin \theta \cos \theta + g(\theta) \qquad (7.125)$$

where $F(\phi) = \int^\phi f(\phi) \, d\phi$. Substitution of (7.125) into (7.124c) yields

$$-F(\phi)(\cos^2 \theta - \sin^2 \theta) + \frac{dg}{d\theta} + \frac{df}{d\phi} + 2 \cot \theta (F(\phi) \sin \theta \cos \theta - g(\theta)) = 0.$$

This equation may be separated into

$$\frac{dg}{d\theta} - 2 \cot \theta g(\theta) = -\frac{df}{d\phi} - F(\phi).$$

Since both sides must be separately constant ($\equiv C$), we have

$$\frac{dg}{d\theta} - 2\cot\theta g(\theta) = C \tag{7.126a}$$

$$\frac{df}{d\phi} + F(\phi) = -C. \tag{7.126b}$$

Equation (7.126a) is solved if we multiply both sides by $\exp(-\int d\theta\, 2\cot\theta) = \sin^{-2}\theta$ to make the LHS a total derivative,

$$\frac{d}{d\theta}\left(\frac{g(\theta)}{\sin^2\theta}\right) = \frac{C}{\sin^2\theta}.$$

The solution is easily found to be

$$g(\theta) = (C_1 - C\cot\theta)\sin^2\theta.$$

Differentiating (7.126b) again, we find that f is harmonic,

$$X_\theta(\phi) = f(\phi) = A\sin\phi + B\cos\phi$$
$$F(\phi) = -A\cos\phi + B\sin\phi - C.$$

Substituting these results into (7.125), we have

$$X_\phi(\theta, \phi) = -(-A\cos\phi + B\sin\phi - C)\sin\theta\cos\theta + (C_1 - C\cot\theta)\sin^2\theta$$
$$= (A\cos\phi - B\sin\phi)\sin\theta\cos\theta + C_1\sin^2\theta.$$

A general Killing vector is given by

$$X = X^\theta \frac{\partial}{\partial\theta} + X^\phi \frac{\partial}{\partial\phi}$$

$$= A\left(\sin\phi\frac{\partial}{\partial\theta} + \cos\phi\cot\theta\frac{\partial}{\partial\phi}\right)$$

$$+ B\left(\cos\phi\frac{\partial}{\partial\theta} - \sin\phi\cot\theta\frac{\partial}{\partial\phi}\right) + C_1\frac{\partial}{\partial\phi}. \tag{7.127}$$

The basis vectors

$$L_x = -\cos\phi\frac{\partial}{\partial\theta} + \cot\theta\sin\phi\frac{\partial}{\partial\phi} \tag{7.128a}$$

$$L_y = \sin\phi\frac{\partial}{\partial\theta} + \cot\theta\cos\phi\frac{\partial}{\partial\phi} \tag{7.128b}$$

$$L_z = \frac{\partial}{\partial\phi} \tag{7.128c}$$

generate rotations round the x, y and z axes respectively.

These vectors generate the Lie algebra $\mathfrak{so}(3)$. This reflects the fact that S^2 is the homogeneous space $SO(3)/SO(2)$ and the metric on S^2 retains this $SO(3)$ symmetry (see example 5.18(a)). In general $S^n = SO(n + 1)/SO(n)$ with the usual metric has $\dim SO(n + 1) = n(n + 1)/2$ Killing vectors and they form the Lie algebra $\mathfrak{so}(n + 1)$. The sphere S^n with the usual metric is a maximally symmetric space. We may *squash* S^n so that it has fewer symmetries. For example, if S^2 considered here is squashed along the z-axis it has a rotational symmetry around the z-axis only and there exists one Killing vector field $L_z = \partial/\partial\phi$.

7.7.2 Conformal Killing vector fields

Let (M, g) be a Riemannian manifold and let $X \in \mathfrak{X}(M)$. If an infinitesimal displacement given by εX generates a conformal transformation, the vector field X is called a **conformal Killing vector field** (CKV). Under the displacement $x^\mu \to x^\mu + \varepsilon X^\mu$, this condition is written as

$$\frac{\partial(x^\kappa + \varepsilon X^\kappa)}{\partial x^\mu} \frac{\partial(x^\lambda + \varepsilon X^\lambda)}{\partial x^\nu} g_{\kappa\lambda}(x + \varepsilon X) = e^{2\sigma} g_{\mu\nu}(x).$$

Noting that $\sigma \propto \varepsilon$, we set $\sigma = \varepsilon\psi/2$, where $\psi \in \mathcal{F}(M)$. Then we find that $g_{\mu\nu}$ and X^μ satisfy

$$\mathcal{L}_X g_{\mu\nu} = X^\xi \partial_\xi g_{\mu\nu} + \partial_\mu X^\kappa g_{\kappa\nu} + \partial_\nu X^\lambda g_{\mu\lambda} = \psi g_{\mu\nu}. \tag{7.129a}$$

Equation (7.129a) is easily solved for ψ to yield

$$\psi = \frac{X^\xi g^{\mu\nu} \partial_\xi g_{\mu\nu} + 2\partial_\mu X^\mu}{m} \tag{7.129b}$$

where $m = \dim M$. We verify that

(i) a linear combination of CKVs is a CKV: $(\mathcal{L}_{aX+bY} g)_{\mu\nu} = (a\varphi + b\psi) g_{\mu\nu}$ where $a, b \in \mathbb{R}$, $\mathcal{L}_X g_{\mu\nu} = \varphi g_{\mu\nu}$ and $\mathcal{L}_Y g_{\mu\nu} = \psi g_{\mu\nu}$;
(ii) the Lie bracket $[X, Y]$ of a CKV is again a CKV: $\mathcal{L}_{[X,Y]} g_{\mu\nu} = (X[\psi] - Y[\varphi]) g_{\mu\nu}$.

Example 7.12. Let x^μ be the coordinates of $(\mathbb{R}^m, \delta)$. The vector

$$D \equiv x^\mu \frac{\partial}{\partial x^\mu} \tag{7.130}$$

(dilatation vector) is a CKV. In fact,

$$\mathcal{L}_D \delta_{\mu\nu} = \partial_\mu x^\kappa \delta_{\kappa\nu} + \partial_\nu x^\lambda \delta_{\mu\lambda} = 2\delta_{\mu\nu}.$$

7.8 Non-coordinate bases

7.8.1 Definitions

In the coordinate basis, $T_p M$ is spanned by $\{e_\mu\} = \{\partial/\partial x^\mu\}$ and $T_p^* M$ by $\{dx^\mu\}$. If, moreover, M is endowed with a metric g, there may be an alternative choice. Let us consider the linear combination,

$$\hat{e}_\alpha = e_\alpha{}^\mu \frac{\partial}{\partial x^\mu} \qquad \{e_\alpha{}^\mu\} \in GL(m, \mathbb{R}) \tag{7.131}$$

where $\det e_\alpha{}^\mu > 0$. In other words, $\{\hat{e}_\alpha\}$ is the frame of basis vectors which is obtained by a $GL(m, \mathbb{R})$-rotation of the basis $\{e_\mu\}$ preserving the orientation. We require that $\{\hat{e}_\alpha\}$ be orthonormal with respect to g,

$$g(\hat{e}_\alpha, \hat{e}_\beta) = e_\alpha{}^\mu e_\beta{}^\nu g_{\mu\nu} = \delta_{\alpha\beta}. \tag{7.132a}$$

If the manifold is Lorentzian, $\delta_{\alpha\beta}$ should be replaced by $\eta_{\alpha\beta}$. We easily reverse (7.132a),

$$g_{\mu\nu} = e^\alpha{}_\mu e^\beta{}_\nu \delta_{\alpha\beta} \tag{7.132b}$$

where $e^\alpha{}_\mu$ is the inverse of $e_\alpha{}^\mu$; $e^\alpha{}_\mu e_\alpha{}^\nu = \delta_\mu{}^\nu$, $e^\alpha{}_\mu e_\beta{}^\mu = \delta^\alpha{}_\beta$. [We have used the same symbols for a matrix and its inverse. So long as the indices are written explicitly it does not cause confusion.] Since a vector V is independent of the basis chosen, we have $V = V^\mu e_\mu = V^\alpha \hat{e}_\alpha = V^\alpha e_\alpha{}^\mu e_\mu$. It follows that

$$V^\mu = V^\alpha e_\alpha{}^\mu \qquad V^\alpha = e^\alpha{}_\mu V^\mu. \tag{7.133}$$

Let us introduce the dual basis $\{\hat{\theta}^\alpha\}$ defined by $\langle \hat{\theta}^\alpha, \hat{e}_\beta \rangle = \delta^\alpha{}_\beta$. $\hat{\theta}^\alpha$ is given by

$$\hat{\theta}^\alpha = e^\alpha{}_\mu dx^\mu. \tag{7.134}$$

In terms of $\{\hat{\theta}^\alpha\}$, the metric is

$$g = g_{\mu\nu} dx^\mu \otimes dx^\nu = \delta_{\alpha\beta} \hat{\theta}^\alpha \otimes \hat{\theta}^\beta. \tag{7.135}$$

The bases $\{\hat{e}_\alpha\}$ and $\{\hat{\theta}^\alpha\}$ are called the **non-coordinate bases**. We use κ, λ, μ, ν, ... (α, β, γ, δ, ...) to denote the coordinate (non-coordinate) basis. The coefficients $e_\alpha{}^\mu$ are called the **vierbeins** if the space is four dimensional and **vielbeins** if it is *many* dimensional. The non-coordinate basis has a non-vanishing Lie bracket. If the $\{\hat{e}_\alpha\}$ are given by (7.131), they satisfy

$$[\hat{e}_\alpha, \hat{e}_\beta]|_p = c_{\alpha\beta}{}^\gamma(p) \hat{e}_\gamma|_p \tag{7.136a}$$

where

$$c_{\alpha\beta}{}^\gamma(p) = e^\gamma{}_\nu [e_\alpha{}^\mu \partial_\mu e_\beta{}^\nu - e_\beta{}^\mu \partial_\mu e_\alpha{}^\nu](p). \tag{7.136b}$$

Example 7.13. The standard metric on S^2 is

$$g = \mathrm{d}\theta \otimes \mathrm{d}\theta + \sin^2\theta\,\mathrm{d}\phi \otimes \mathrm{d}\phi = \hat{\theta}^1 \otimes \hat{\theta}^1 + \hat{\theta}^2 \otimes \hat{\theta}^2 \qquad (7.137)$$

where $\hat{\theta}^1 = \mathrm{d}\theta$ and $\hat{\theta}^2 = \sin\theta\,\mathrm{d}\phi$. The 'zweibeins' are

$$\begin{aligned} e^1{}_\theta &= 1 & e^1{}_\phi &= 0 \\ e^2{}_\theta &= 0 & e^2{}_\phi &= \sin\theta. \end{aligned} \qquad (7.138)$$

The non-vanishing components of $c_{\alpha\beta}{}^\gamma$ are $c_{12}{}^2 = -c_{21}{}^2 = -\cot\theta$.

Exercise 7.19. (a) Verify the identities,

$$\delta^{\alpha\beta} = g^{\mu\nu}e^\alpha{}_\mu e^\beta{}_\nu \qquad g^{\mu\nu} = \delta^{\alpha\beta}e_\alpha{}^\mu e_\beta{}^\nu. \qquad (7.139)$$

(b) Let γ^α be the Dirac matrices in Minkowski spacetime, which satisfy $\{\gamma^\alpha, \gamma^\beta\} = 2\eta^{\alpha\beta}$. Define the curved spacetime counterparts of the Dirac matrices by $\gamma^\mu \equiv e_\alpha{}^\mu \gamma^\alpha$. Show that

$$\{\gamma^\mu, \gamma^\nu\} = 2g^{\mu\nu}. \qquad (7.140)$$

7.8.2 Cartan's structure equations

In section 7.3 the curvature tensor R and the torsion tensor T have been defined by

$$\begin{aligned} R(X, Y)Z &= \nabla_X \nabla_Y Z - \nabla_Y \nabla_X Z - \nabla_{[X,Y]}Z \\ T(X, Y) &= \nabla_X Y - \nabla_Y X - [X, Y]. \end{aligned}$$

Let $\{\hat{e}_\alpha\}$ be the non-coordinate basis and $\{\hat{\theta}^\alpha\}$ the dual basis. The vector fields $\{\hat{e}_\alpha\}$ satisfy $[\hat{e}_\alpha, \hat{e}_\beta] = c_{\alpha\beta}{}^\gamma \hat{e}_\gamma$. Define the connection coefficients with respect to the basis $\{\hat{e}_\alpha\}$ by

$$\nabla_\alpha \hat{e}_\beta \equiv \nabla_{\hat{e}_\alpha} \hat{e}_\beta = \Gamma^\gamma{}_{\alpha\beta}\hat{e}_\gamma. \qquad (7.141)$$

Let $\hat{e}_\alpha = e_\alpha{}^\mu e_\mu$. Then (7.141) becomes $e_\alpha{}^\mu(\partial_\mu e_\beta{}^\nu + e_\beta{}^\lambda \Gamma^\nu{}_{\mu\lambda})e_\nu = \Gamma^\gamma{}_{\alpha\beta}e_\gamma{}^\nu e_\nu$, from which we find that

$$\Gamma^\gamma{}_{\alpha\beta} = e^\gamma{}_\nu e_\alpha{}^\mu(\partial_\mu e_\beta{}^\nu + e_\beta{}^\lambda \Gamma^\nu{}_{\mu\lambda}) = e^\gamma{}_\nu e_\alpha{}^\mu \nabla_\mu e_\beta{}^\nu. \qquad (7.142)$$

The components of T and R in this basis are given by

$$\begin{aligned} T^\alpha{}_{\beta\gamma} &= \langle \hat{\theta}^\alpha, T(\hat{e}_\beta, \hat{e}_\gamma)\rangle = \langle \hat{\theta}^\alpha, \nabla_\beta \hat{e}_\gamma - \nabla_\gamma \hat{e}_\beta - [\hat{e}_\beta, \hat{e}_\gamma]\rangle \\ &= \Gamma^\alpha{}_{\beta\gamma} - \Gamma^\alpha{}_{\gamma\beta} - c_{\beta\gamma}{}^\alpha. \end{aligned} \qquad (7.143)$$

$$\begin{aligned} R^\alpha{}_{\beta\gamma\delta} &= \langle \hat{\theta}^\alpha, \nabla_\gamma \nabla_\delta \hat{e}_\beta - \nabla_\delta \nabla_\gamma \hat{e}_\beta - \nabla_{[\hat{e}_\gamma, \hat{e}_\delta]}\hat{e}_\beta\rangle \\ &= \langle \hat{\theta}^\alpha, \nabla_\gamma(\Gamma^\varepsilon{}_{\delta\beta}\hat{e}_\varepsilon) - \nabla_\delta \Gamma^\varepsilon{}_{\gamma\beta}\hat{e}_\varepsilon) - c_{\gamma\delta}{}^\varepsilon \nabla_\varepsilon \hat{e}_\beta\rangle \\ &= \hat{e}_\gamma[\Gamma^\alpha{}_{\delta\beta}] - \hat{e}_\delta[\Gamma^\alpha{}_{\gamma\beta}] + \Gamma^\varepsilon{}_{\delta\beta}\Gamma^\alpha{}_{\gamma\varepsilon} - \Gamma^\varepsilon{}_{\gamma\beta}\Gamma^\alpha{}_{\delta\varepsilon} - c_{\gamma\delta}{}^\varepsilon \Gamma^\alpha{}_{\varepsilon\beta}. \end{aligned} \qquad (7.144)$$

We define a matrix-valued one-form $\{\omega^\alpha{}_\beta\}$ called the **connection one-form** by

$$\omega^\alpha{}_\beta \equiv \Gamma^\alpha{}_{\gamma\beta}\hat{\theta}^\gamma. \tag{7.145}$$

Theorem 7.3. The connection one-form $\omega^\alpha{}_\beta$ satisfies **Cartan's structure equations**,

$$\mathrm{d}\hat{\theta}^\alpha + \omega^\alpha{}_\beta \wedge \hat{\theta}^\beta = T^\alpha \tag{7.146a}$$

$$\mathrm{d}\omega^\alpha{}_\beta + \omega^\alpha{}_\gamma \wedge \omega^\gamma{}_\beta = R^\alpha{}_\beta \tag{7.146b}$$

where we have introduced the **torsion two-form** $T^\alpha \equiv \frac{1}{2}T^\alpha{}_{\beta\gamma}\hat{\theta}^\beta \wedge \hat{\theta}^\gamma$ and the **curvature two-form** $R^\alpha{}_\beta \equiv \frac{1}{2}R^\alpha{}_{\beta\gamma\delta}\hat{\theta}^\gamma \wedge \hat{\theta}^\delta$.

Proof. Let the LHS of (7.146a) act on the basis vectors $\hat{e}_\gamma$ and $\hat{e}_\delta$,

$$\mathrm{d}\hat{\theta}^\alpha(\hat{e}_\gamma, \hat{e}_\delta) + [\langle\omega^\alpha{}_\beta, \hat{e}_\gamma\rangle\langle\hat{\theta}^\beta, \hat{e}_\delta\rangle - \langle\hat{\theta}^\beta, \hat{e}_\gamma\rangle\langle\omega^\alpha{}_\beta, \hat{e}_\delta\rangle]$$

$$= \{\hat{e}_\gamma[\langle\hat{\theta}^\alpha, \hat{e}_\delta\rangle] - \hat{e}_\delta[\langle\hat{\theta}^\alpha, \hat{e}_\gamma\rangle] - \langle\hat{\theta}^\alpha, [\hat{e}_\gamma, \hat{e}_\delta]\rangle\} + \{\langle\omega^\alpha{}_\delta, \hat{e}_\gamma\rangle - \langle\omega^\alpha{}_\gamma, \hat{e}_\delta\rangle\}$$

$$= -c_{\gamma\delta}{}^\alpha + \Gamma^\alpha{}_{\gamma\delta} - \Gamma^\alpha{}_{\delta\gamma} = T^\alpha{}_{\gamma\delta}$$

where use has been made of (5.70). The RHS acting on $\hat{e}_\gamma$ and $\hat{e}_\delta$ yields

$$\frac{1}{2}T^\alpha{}_{\beta\varepsilon}[\langle\hat{\theta}^\beta, \hat{e}_\gamma\rangle\langle\hat{\theta}^\varepsilon, \hat{e}_\delta\rangle - \langle\hat{\theta}^\varepsilon, \hat{e}_\gamma\rangle\langle\hat{\theta}^\beta, \hat{e}_\delta\rangle] = T^\alpha{}_{\gamma\delta}$$

which verifies (7.146a).

Equation (7.146b) may be proved similarly (exercise). $\qquad\square$

Taking the exterior derivative of (7.146a) and (7.146b), we have the **Bianchi identities**

$$\mathrm{d}T^\alpha + \omega^\alpha{}_\beta \wedge T^\beta = R^\alpha{}_\beta \wedge \hat{\theta}^\beta \tag{7.147a}$$

$$\mathrm{d}R^\alpha{}_\beta + \omega^\alpha{}_\gamma \wedge R^\gamma{}_\beta - R^\alpha{}_\gamma \wedge \omega^\gamma{}_\beta = 0. \tag{7.147b}$$

These are the non-coordinate basis versions of (7.81a) and (7.81b).

7.8.3 The local frame

In an m-dimensional Riemannian manifold, the metric tensor $g_{\mu\nu}$ has $m(m+1)/2$ degrees of freedom while the vielbein $e_\alpha{}^\mu$ has m^2 degrees of freedom. There are many non-coordinate bases which yield the same metric, g, each of which is related to the other by the *local* orthogonal rotation,

$$\hat{\theta}^\alpha \longrightarrow \hat{\theta}'^\alpha(p) = \Lambda^\alpha{}_\beta(p)\hat{\theta}^\beta(p) \tag{7.148}$$

at each point p. The vielbein transforms as

$$e^\alpha{}_\mu(p) \longrightarrow e'^\alpha{}_\mu(p) = \Lambda^\alpha{}_\beta(p)e^\beta{}_\mu(p). \tag{7.149}$$

Unlike $\kappa, \lambda, \mu, \nu, \ldots$ which transform under coordinate changes, the indices $\alpha, \beta, \gamma, \ldots$ transform under the local orthogonal rotation and are inert under coordinate changes. Since the metric tensor is invariant under the rotation, $\Lambda^{\alpha}{}_{\beta}$ satisfies

$$\Lambda^{\alpha}{}_{\beta} \delta_{\alpha\delta} \Lambda^{\delta}{}_{\gamma} = \delta_{\beta\gamma} \qquad \text{if } M \text{ is Riemannian} \qquad (7.150a)$$

$$\Lambda^{\alpha}{}_{\beta} \eta_{\alpha\delta} \Lambda^{\delta}{}_{\gamma} = \eta_{\beta\gamma} \qquad \text{if } M \text{ is Lorentzian.} \qquad (7.150b)$$

This implies that $\{\Lambda^{\alpha}{}_{\beta}(p)\} \in SO(m)$ if M is Riemannian with dim $M = m$ and $\{\Lambda^{\alpha}{}_{\beta}(p)\} \in SO(m-1, 1)$ if M is Lorentzian. The dimension of these Lie groups is $m(m-1)/2 = m^2 - m(m+1)/2$, that is the difference between the degrees of freedom of $e_{\alpha}{}^{\mu}$ and $g_{\mu\nu}$. Under the local frame rotation $\Lambda^{\alpha}{}_{\beta}(p)$, the indices $\alpha, \beta, \gamma, \delta, \ldots$ are rotated while $\kappa, \lambda, \mu, \nu, \ldots$ (world indices) are not affected. Under the rotation (7.148), the basis vector transforms as

$$\hat{e}_{\alpha} \longrightarrow \hat{e}'_{\alpha} = \hat{e}_{\beta} (\Lambda^{-1})^{\beta}{}_{\alpha}. \qquad (7.151)$$

Let $t = t^{\mu}{}_{\nu} e_{\mu} \otimes dx^{\nu}$ be a tensor field of type $(1, 1)$. In the bases $\{\hat{e}_{\alpha}\}$ and $\{\hat{\theta}^{\alpha}\}$, we have $t = t^{\alpha}{}_{\beta} \hat{e}_{\alpha} \otimes \hat{\theta}^{\beta}$, where $t^{\alpha}{}_{\beta} = e^{\alpha}{}_{\mu} e_{\beta}{}^{\nu} t^{\mu}{}_{\nu}$. If the new frames $\{\hat{e}'_{\alpha}\} = \{\hat{e}_{\beta} (\Lambda^{-1})^{\beta}{}_{\alpha}\}$ and $\{\hat{\theta}'^{\alpha}\} = \{\Lambda^{\alpha}{}_{\beta} \hat{\theta}^{\beta}\}$ are employed, the tensor t is expressed as

$$t = t'^{\alpha}{}_{\beta} \hat{e}'_{\alpha} \otimes \hat{\theta}'^{\beta} = t'^{\alpha}{}_{\beta} \hat{e}_{\gamma} (\Lambda^{-1})^{\gamma}{}_{\alpha} \otimes \Lambda^{\beta}{}_{\delta} \hat{\theta}^{\delta}$$

from which we find the transformation rule,

$$t^{\alpha}{}_{\beta} \longrightarrow t'^{\alpha}{}_{\beta} = \Lambda^{\alpha}{}_{\gamma} t^{\gamma}{}_{\delta} (\Lambda^{-1})^{\delta}{}_{\beta}.$$

To summarize, the upper (lower) non-coordinate indices are rotated by Λ (Λ^{-1}). The change from the coordinate basis to the non-coordinate basis is carried out by multiplications of vielbeins.

From these facts we find the transformation rule of the connection one-form $\omega^{\alpha}{}_{\beta}$. The torsion two-form transforms as

$$T^{\alpha} \longrightarrow T'^{\alpha} = d\hat{\theta}'^{\alpha} + \omega'^{\alpha}{}_{\beta} \wedge \hat{\theta}'^{\beta} = \Lambda^{\alpha}{}_{\beta} [d\hat{\theta}^{\beta} + \omega^{\beta}{}_{\gamma} \wedge \hat{\theta}^{\gamma}].$$

Substituting $\hat{\theta}'^{\alpha} = \Lambda^{\alpha}{}_{\beta} \hat{\theta}^{\beta}$ into this equation, we find that

$$\omega'^{\alpha}{}_{\beta} \Lambda^{\beta}{}_{\gamma} = \Lambda^{\alpha}{}_{\delta} \omega^{\delta}{}_{\gamma} - d\Lambda^{\alpha}{}_{\gamma}.$$

Multiplying both sides by Λ^{-1} from the right, we have

$$\omega'^{\alpha}{}_{\beta} = \Lambda^{\alpha}{}_{\gamma} \omega^{\gamma}{}_{\delta} (\Lambda^{-1})^{\delta}{}_{\beta} + \Lambda^{\alpha}{}_{\gamma} (d\Lambda^{-1})^{\gamma}{}_{\beta} \qquad (7.152)$$

where use has been made of the identity $d\Lambda\,\Lambda^{-1} + \Lambda\,d\Lambda^{-1} = 0$, which is derived from $\Lambda\Lambda^{-1} = I_m$.

The curvature two-form transforms homogeneously as

$$R^{\alpha}{}_{\beta} \longrightarrow R'^{\alpha}{}_{\beta} = \Lambda^{\alpha}{}_{\gamma} R^{\gamma}{}_{\delta} (\Lambda^{-1})^{\delta}{}_{\beta} \qquad (7.153)$$

under a local frame rotation Λ.

7.8.4   The Levi-Civita connection in a non-coordinate basis

Let ∇ be a Levi-Civita connection on (M, g), which is characterized by the metric compatibility $\nabla_X g = 0$, and the vanishing torsion $\Gamma^\lambda{}_{\mu\nu} - \Gamma^\lambda{}_{\nu\mu} = 0$. It is interesting to see how these conditions are expressed in the present approach. The components $\Gamma^\lambda{}_{\mu\nu}$ and $\Gamma^\alpha{}_{\beta\gamma}$ are related to each other by (7.142). Let (M, g) be a Riemannian manifold (if (M, g) is Lorentzian, we simply replace $\delta_{\alpha\beta}$ all below by $\eta_{\alpha\beta}$). If we define the **Ricci rotation coefficient** $\Gamma_{\alpha\beta\gamma}$ by $\delta_{\alpha\delta}\Gamma^\delta{}_{\beta\gamma}$ the metric compatibility is expressed as

$$\Gamma_{\alpha\beta\gamma} = \delta_{\alpha\delta}e^\delta{}_\lambda e_\beta{}^\mu \nabla_\mu e_\gamma{}^\lambda = -\delta_{\alpha\delta}e_\gamma{}^\lambda e_\beta{}^\mu \nabla_\mu e^\delta{}_\lambda$$
$$= -\delta_{\gamma\delta}e^\delta{}_\lambda e_\beta{}^\mu \nabla_\mu e_\alpha{}^\lambda = -\Gamma_{\gamma\beta\alpha} \tag{7.154}$$

where $\nabla_\mu g = 0$ has been used. In terms of the connection one-form $\omega_{\alpha\beta} \equiv \delta_{\alpha\gamma}\omega^\gamma{}_\beta$, this becomes

$$\omega_{\alpha\beta} = -\omega_{\beta\alpha}. \tag{7.155}$$

The torsion-free condition is

$$\mathrm{d}\hat\theta^\alpha + \omega^\alpha{}_\beta \wedge \hat\theta^\beta = 0. \tag{7.156}$$

The reader should verify that (7.156) implies the symmetry of the connection coefficient $\Gamma^\lambda{}_{\mu\nu} = \Gamma^\lambda{}_{\nu\mu}$ in the coordinate basis. The condition (7.156) enables us to compute the $c_{\alpha\beta}{}^\gamma$ of the basis $\{\hat e_\alpha\}$. Let us look at the commutation relation

$$c_{\alpha\beta}{}^\gamma \hat e_\gamma = [\hat e_\alpha, \hat e_\beta] = \nabla_\alpha \hat e_\beta - \nabla_\beta \hat e_\alpha \tag{7.157}$$

where the final equality follows from the torsion-free condition. From (7.141), we find that

$$c_{\alpha\beta}{}^\gamma = \Gamma^\gamma{}_{\alpha\beta} - \Gamma^\gamma{}_{\beta\alpha}. \tag{7.158}$$

Substituting (7.158) into (7.144) we may express the Riemaun curvature tensor in terms of Γ only,

$$R^\alpha{}_{\beta\gamma\delta} = \hat e_\gamma[\Gamma^\alpha{}_{\delta\beta}] - \hat e_\delta[\Gamma^\alpha{}_{\gamma\beta}] + \Gamma^\varepsilon{}_{\delta\beta}\Gamma^\alpha{}_{\gamma\varepsilon} - \Gamma^\varepsilon{}_{\gamma\beta}\Gamma^\alpha{}_{\delta\varepsilon}$$
$$- (\Gamma^\varepsilon{}_{\gamma\delta} - \Gamma^\varepsilon{}_{\delta\gamma})\Gamma^\alpha{}_{\varepsilon\beta}. \tag{7.159}$$

Example 7.14. Let us take the sphere S^2 of example 7.13. The components of $e^\alpha{}_\mu$ are

$$e^1{}_\theta = 1 \qquad e^1{}_\phi = 0 \qquad e^2{}_\theta = 0 \qquad e^2{}_\phi = \sin\theta. \tag{7.160}$$

We first note that the metric condition implies $\omega_{11} = \omega_{22} = 0$, hence $\omega^1{}_1 = \omega^2{}_2 = 0$. Other connection one-forms are obtained from the torsion-free conditions,

$$\mathrm{d}(\mathrm{d}\theta) + \omega^1{}_2 \wedge (\sin\theta\,\mathrm{d}\phi) = 0 \tag{7.161a}$$
$$\mathrm{d}(\sin\theta\,\mathrm{d}\phi) + \omega^2{}_1 \wedge \mathrm{d}\theta = 0. \tag{7.161b}$$

From the second equation of (7.161), we easily see that $\omega^2{}_1 = \cos\theta\,d\phi$ and the metric condition $\omega_{12} = -\omega_{21}$ implies $\omega^1{}_2 = -\cos\theta\,d\phi$. The Riemann tensor is also found from Cartan's structure equation,

$$\omega^1{}_2 \wedge \omega^2{}_1 = \tfrac{1}{2} R^1{}_{1\alpha\beta} \hat\theta^\alpha \wedge \hat\theta^\beta \tag{7.162a}$$

$$d\omega^1{}_2 = \tfrac{1}{2} R^1{}_{2\alpha\beta} \hat\theta^\alpha \wedge \hat\theta^\beta \tag{7.162b}$$

$$d\omega^2{}_1 = \tfrac{1}{2} R^2{}_{1\alpha\beta} \hat\theta^\alpha \wedge \hat\theta^\beta \tag{7.162c}$$

$$\omega^2{}_1 \wedge \omega^1{}_2 = \tfrac{1}{2} R^2{}_{2\alpha\beta} \hat\theta^\alpha \wedge \hat\theta^\beta. \tag{7.162d}$$

The non-vanishing components are $R^1{}_{212} = -R^1{}_{221} = \sin\theta$, $R^2{}_{112} = -R^2{}_{121} = -\sin\theta$. The transition to the coordinate basis expression is carried out with the help of $e_\alpha{}^\mu$ and $e^\alpha{}_\mu$. For example,

$$R^\theta{}_{\phi\theta\phi} = e_\alpha{}^\theta e^\beta{}_\phi e^\gamma{}_\theta e^\delta{}_\phi R^\alpha{}_{\beta\gamma\delta} = \frac{1}{\sin^2\theta} R^1{}_{212} = \frac{1}{\sin\theta}.$$

Example 7.15. The Schwarzschild metric is given by

$$ds^2 = -\left(1 - \frac{2M}{r}\right) dt^2 + \frac{1}{1 - \dfrac{2M}{r}} dr^2 + r^2(d\theta^2 + \sin^2\theta\,d\phi^2)$$

$$= -\hat\theta^0 \otimes \hat\theta^0 + \hat\theta^1 \otimes \hat\theta^1 + \hat\theta^2 \otimes \hat\theta^2 + \hat\theta^3 \otimes \hat\theta^3 \tag{7.163}$$

where

$$\hat\theta^0 = \left(1 - \frac{2M}{r}\right)^{1/2} dt \qquad \hat\theta^1 = \left(1 - \frac{2M}{r}\right)^{-1/2} dr \tag{7.164}$$

$$\hat\theta^2 = r\,d\theta \qquad \hat\theta^3 = r\sin\theta\,d\phi.$$

The parameters run over the range $0 < 2M < r$, $0 \le \theta \le \pi$ and $0 \le \phi < 2\pi$. The metric condition yields $\omega^0{}_0 = \omega^1{}_1 = \omega^2{}_2 = \omega^3{}_3 = 0$ and the torsion-free conditions are:

$$d[(1 - 2M/r)^{1/2}dt] + \omega^0{}_\beta \wedge \hat\theta^\beta = 0 \tag{7.165a}$$

$$d[(1 - 2M/r)^{-1/2}dr] + \omega^1{}_\beta \wedge \hat\theta^\beta = 0 \tag{7.165b}$$

$$d(r\,d\theta) + \omega^2{}_\beta \wedge \hat\theta^\beta = 0 \tag{7.165c}$$

$$d(r\sin\theta\,d\phi) + \omega^3{}_\beta \wedge \hat\theta^\beta = 0. \tag{7.165d}$$

The non-vanishing components of the connection one-forms are

$$\omega^0{}_1 = \omega^1{}_0 = \frac{M}{r^2} dt \qquad \omega^2{}_1 = -\omega^1{}_2 = \left(1 - \frac{2M}{r}\right)^{1/2} d\theta$$

$$\omega^3{}_1 = -\omega^1{}_3 = \left(1 - \frac{2M}{r}\right)^{1/2} \sin\theta\,d\phi \qquad \omega^3{}_2 = -\omega^2{}_3 = \cos\theta\,d\phi.$$

$$\tag{7.166}$$

The curvature two-forms are found from the structure equations to be

$$R^0{}_1 = R^1{}_0 = \frac{2M}{r^3}\hat{\theta}^0 \wedge \hat{\theta}^1 \qquad R^0{}_2 = R^2{}_0 = -\frac{2M}{r^3}\hat{\theta}^0 \wedge \hat{\theta}^2$$

$$R^0{}_3 = R^3{}_0 = -\frac{M}{r^3}\hat{\theta}^0 \wedge \hat{\theta}^3 \qquad R^1{}_2 = -R^2{}_1 = -\frac{M}{r^3}\hat{\theta}^1 \wedge \hat{\theta}^2 \qquad (7.167)$$

$$R^1{}_3 = -R^3{}_1 = -\frac{M}{r^3}\hat{\theta}^1 \wedge \hat{\theta}^3 \qquad R^2{}_3 = -R^3{}_2 = \frac{2M}{r^3}\hat{\theta}^2 \wedge \hat{\theta}^3.$$

7.9 Differential forms and Hodge theory

7.9.1 Invariant volume elements

We have defined the volume element as a non-vanishing m-form on an m-dimensional orientable manifold M in section 5.5. If M is endowed with a metric g, there exists a natural volume element which is invariant under coordinate transformation. Let us define the **invariant volume element** by

$$\Omega_M \equiv \sqrt{|g|}\, dx^1 \wedge dx^2 \wedge \ldots \wedge dx^m \qquad (7.168)$$

where $g = \det g_{\mu\nu}$ and x^μ are the coordinates of the chart (U, φ). The m-form Ω_M is, indeed, invariant under a coordinate change. Let y^λ be the coordinates of another chart (V, ψ) with $U \cap V \neq \emptyset$. The invariant volume element is

$$\sqrt{\left|\det\left(\frac{\partial x^\mu}{\partial y^\kappa}\frac{\partial x^\nu}{\partial y^\lambda}g_{\mu\nu}\right)\right|}\, dy^1 \wedge \ldots \wedge dy^m$$

in terms of the y-coordinates. Noting that $dy^\lambda = (\partial y^\lambda/\partial x^\mu)\, dx^\mu$, this becomes

$$\left|\det\left(\frac{\partial x^\mu}{\partial y^\kappa}\right)\right|\sqrt{|g|}\det\left(\frac{\partial y^\lambda}{\partial x^\nu}\right) dx^1 \wedge dx^2 \wedge \ldots \wedge dx^m$$

$$= \pm\sqrt{|g|}dx^1 \wedge dx^2 \wedge \ldots \wedge dx^m.$$

If x^μ and y^κ define the same orientation, $\det(\partial x^\mu/\partial y^\kappa)$ is strictly positive on $U \cap V$ and Ω_M is invariant under the coordinate change.

Exercise 7.20. Let $\{\hat{\theta}^\alpha\} = \{e^\alpha{}_\mu dx^\mu\}$ be the non-coordinate basis. Show that the invariant volume element is written as

$$\Omega_M = |e|\, dx^1 \wedge dx^2 \wedge \ldots \wedge dx^m = \hat{\theta}^1 \wedge \hat{\theta}^2 \wedge \ldots \wedge \hat{\theta}^m \qquad (7.169)$$

where $e = \det e^\alpha{}_\mu$.

Now that we have defined the invariant volume element, it is natural to define an integration of $f \in \mathcal{F}(M)$ over M by

$$\int_M f\Omega_M \equiv \int_M f\sqrt{|g|}\, dx^1\, dx^2 \ldots dx^m. \qquad (7.170)$$

Obviously (7.170) is invariant under a change of coordinates. In physics, there are many objects which are expressed as volume integrals of this type, see section 7.10.

7.9.2   Duality transformations (Hodge star)

As noted in section 5.4, $\Omega^r(M)$ is isomorphic to $\Omega^{m-r}(M)$ on an m-dimensional manifold M. If M is endowed with a metric g, we can define a natural isomorphism between them called the **Hodge $*$ operation**. Define the totally anti-symmetric tensor ε by

$$
\varepsilon_{\mu_1\mu_2\ldots\mu_m} = \begin{cases} +1 & \text{if } (\mu_1\mu_2\ldots\mu_m) \text{ is an even permutation of } (12\ldots m) \\ -1 & \text{if } (\mu_1\mu_2\ldots\mu_m) \text{ is an odd permutation of } (12\ldots m) \\ 0 & \text{otherwise.} \end{cases}
$$

(7.171a)

Note that

$$
\varepsilon^{\mu_1\mu_2\ldots\mu_m} = g^{\mu_1\nu_1}g^{\mu_2\nu_2}\ldots g^{\mu_m\nu_m}\varepsilon_{\nu_1\nu_2\ldots\nu_m} = g^{-1}\varepsilon_{\mu_1\mu_2\ldots\mu_m}. \tag{7.171b}
$$

The Hodge $*$ is a linear map $* : \Omega^r(M) \to \Omega^{m-r}(M)$ whose action on a basis vector of $\Omega^r(M)$ is defined by

$$
\begin{aligned}
&* (dx^{\mu_1} \wedge dx^{\mu_2} \wedge \ldots \wedge dx^{\mu_r}) \\
&= \frac{\sqrt{|g|}}{(m-r)!}\varepsilon^{\mu_1\mu_2\ldots\mu_r}{}_{\nu_{r+1}\ldots\nu_m} dx^{\nu_{r+1}} \wedge \ldots \wedge dx^{\nu_m}.
\end{aligned} \tag{7.172}
$$

It should be noted that $*1$ is the invariant volume element:

$$
*1 = \frac{\sqrt{|g|}}{m!}\varepsilon_{\mu_1\mu_2\ldots\mu_m} dx^{\mu_1} \wedge \ldots \wedge dx^{\mu_m} = \sqrt{|g|}\, dx^1 \wedge \ldots \wedge dx^m.
$$

For

$$
\omega = \frac{1}{r!}\omega_{\mu_1\mu_2\ldots\mu_r} dx^{\mu_1} \wedge dx^{\mu_2} \wedge \ldots \wedge dx^{\mu_r} \in \Omega^r(M)
$$

we have

$$
*\omega = \frac{\sqrt{|g|}}{r!(m-r)!}\omega_{\mu_1\mu_2\ldots\mu_r}\varepsilon^{\mu_1\mu_2\ldots\mu_r}{}_{\nu_{r+1}\ldots\nu_m} dx^{\nu_{r+1}} \wedge \ldots \wedge dx^{\nu_m}. \tag{7.173}
$$

If we take the non-coordinate basis $\{\theta^\alpha\} = \{e^\alpha{}_\mu dx^\mu\}$, the $*$ operation becomes

$$
*(\hat{\theta}^{\alpha_1} \wedge \ldots \wedge \hat{\theta}^{\alpha_r}) = \frac{1}{(m-r)!}\varepsilon^{\alpha_1\ldots\alpha_r}{}_{\beta_{r+1}\ldots\beta_m}\hat{\theta}^{\beta_{r+1}} \wedge \ldots \wedge \hat{\theta}^{\beta_m} \tag{7.174}
$$

where

$$
\varepsilon_{\alpha_1\ldots\alpha_m} = \begin{cases} +1 & \text{if } (\alpha_1\ldots\alpha_m) \text{ is an even permutation of } (12\ldots m) \\ -1 & \text{if } (\alpha_1\ldots\alpha_m) \text{ is an odd permutation of } (12\ldots m) \\ 0 & \text{otherwise} \end{cases} \tag{7.175}
$$

and the indices are raised by $\delta^{\alpha\beta}$ or $\eta^{\alpha\beta}$.

Theorem 7.4.

$$* * \omega = (-1)^{r(m-r)}\omega. \tag{7.176a}$$

if (M, g) is Riemannian and

$$* * \omega = (-1)^{1+r(m-r)}\omega \tag{7.176b}$$

if Lorentzian.

Proof. It is simpler to prove (7.176a) with a non-coordinate basis. Let

$$\omega = \frac{1}{r!}\omega_{\alpha_1...\alpha_r}\hat{\theta}^{\alpha_1} \wedge \ldots \wedge \hat{\theta}^{\alpha_r}.$$

Repeated applications of $*$ on ω yield

$$* * \omega = \frac{1}{r!}\omega_{\alpha_1...\alpha_r}\frac{1}{(m-r)!}\varepsilon^{\alpha_1...\alpha_r}{}_{\beta_{r+1}...\beta_m}$$

$$\times \frac{1}{r!}\varepsilon^{\beta_{r+1}...\beta_m}{}_{\gamma_1...\gamma_r}\hat{\theta}^{\gamma_1} \wedge \ldots \wedge \hat{\theta}^{\gamma_r}$$

$$= \frac{(-1)^{r(m-r)}}{r!r!(m-r)!}\sum_{\alpha\beta\gamma}\omega_{\alpha_1...\alpha_r}\varepsilon_{\alpha_1...\alpha_r\beta_{r+1}...\beta_m}\varepsilon_{\gamma_1...\gamma_r\beta_{r+1}...\beta_m}$$

$$\times \hat{\theta}^{\gamma_1} \wedge \ldots \wedge \hat{\theta}^{\gamma_r}$$

$$= \frac{(-1)^{r(m-r)}}{r!}\omega_{\alpha_1...\alpha_r}\hat{\theta}^{\alpha_1} \wedge \ldots \wedge \hat{\theta}^{\alpha_r} = (-1)^{r(m-r)}\omega$$

where use has been made of the identity

$$\sum_{\beta\gamma}\varepsilon_{\alpha_1...\alpha_r\beta_{r+1}...\beta_m}\varepsilon_{\gamma_1...\gamma_r\beta_{r+1}...\beta_m}\hat{\theta}^{\gamma_1} \wedge \ldots \wedge \hat{\theta}^{\gamma_r} = r!(m-r)!\hat{\theta}^{\alpha_1} \wedge \ldots \wedge \hat{\theta}^{\alpha_r}.$$

The proof of (7.176b) is left as an exercise to the reader (use $\det \eta = -1$). $\square$

Thus, we find that $(-1)^{r(m-r)} * *$ (or $(-1)^{1+r(m-r)} * *$) is an identity map on $\Omega^r(M)$. We define the inverse of $*$ by

$$*^{-1} = (-1)^{r(m-r)} * \qquad (M, g) \text{ is Riemannian} \tag{7.177a}$$

$$*^{-1} = (-1)^{1+r(m-r)} * \qquad (M, g) \text{ is Lorentzian.} \tag{7.177b}$$

7.9.3 Inner products of r-forms

Take

$$\omega = \frac{1}{r!}\omega_{\mu_1...\mu_r} dx^{\mu_1} \wedge \ldots \wedge dx^{\mu_r}$$

$$\eta = \frac{1}{r!}\eta_{\mu_1...\mu_r} dx^{\mu_1} \wedge \ldots \wedge dx^{\mu_r}.$$

The exterior product $\omega \wedge *\eta$ is an m-form:

$$
\begin{aligned}
\omega \wedge *\eta &= \frac{1}{(r!)^2} \omega_{\mu_1 \ldots \mu_r} \eta_{\nu_1 \ldots \nu_r} \frac{\sqrt{|g|}}{(m-r)!} \varepsilon^{\nu_1 \ldots \nu_r}{}_{\mu_{r+1} \ldots \mu_m} \\
&\quad \times dx^{\mu_1} \wedge \ldots \wedge dx^{\mu_r} \wedge dx^{\mu_{r+1}} \wedge \ldots \wedge dx^{\mu_m} \\
&= \frac{1}{r!} \sum_{\mu\nu} \omega_{\mu_1 \ldots \mu_r} \eta^{\nu_1 \ldots \nu_r} \frac{1}{r!(m-r)!} \varepsilon_{\nu_1 \ldots \nu_r \mu_{r+1} \ldots \mu_m} \\
&\quad \times \varepsilon_{\mu_1 \ldots \mu_r \mu_{r+1} \ldots \mu_m} \sqrt{|g|} \, dx^1 \wedge \ldots \wedge dx^m \\
&= \frac{1}{r!} \omega_{\mu_1 \ldots \mu_r} \eta^{\mu_1 \ldots \mu_r} \sqrt{|g|} \, dx^1 \wedge \ldots \wedge dx^m.
\end{aligned}
\tag{7.178}
$$

This expression shows that the product is symmetric:

$$
\omega \wedge *\eta = \eta \wedge *\omega.
\tag{7.179}
$$

Let $\{\hat{\theta}^\alpha\}$ be the non-coordinate basis and

$$
\omega = \frac{1}{r!} \omega_{\alpha_1 \ldots \alpha_r} \hat{\theta}^{\mu_1} \wedge \ldots \wedge \hat{\theta}^{\alpha_r}
$$

$$
\eta = \frac{1}{r!} \eta_{\alpha_1 \ldots \alpha_r} \hat{\theta}^{\alpha_1} \wedge \ldots \wedge \hat{\theta}^{\alpha_r}.
$$

Equation (7.178) is rewritten as

$$
\omega \wedge *\eta = \frac{1}{r!} \omega_{\alpha_1 \ldots \alpha_r} \eta^{\alpha_1 \ldots \alpha_r} \hat{\theta}^1 \wedge \ldots \wedge \hat{\theta}^m.
\tag{7.180}
$$

Since $\alpha \wedge *\beta$ is an m-form, its integral over M is well defined. Define the inner product (ω, η) of two r-forms by

$$
\begin{aligned}
(\omega, \eta) &\equiv \int \omega \wedge *\eta \\
&= \frac{1}{r!} \int_M \omega_{\mu_1 \ldots \mu_r} \eta^{\mu_1 \ldots \mu_r} \sqrt{|g|} \, dx^1 \ldots dx^m.
\end{aligned}
\tag{7.181}
$$

Since $\omega \wedge *\eta = \eta \wedge *\omega$, the inner product is symmetric,

$$
(\omega, \eta) = (\eta, \omega).
\tag{7.182}
$$

If (M, g) is Riemannian, the inner product is positive definite,

$$
(\alpha, \alpha) \geq 0.
\tag{7.183}
$$

where the equality holds only when $\alpha = 0$. This is not true if (M, g) is Lorentzian.

7.9.4   Adjoints of exterior derivatives

Definition 7.6. Let d $: \Omega^{r-1}(M) \rightarrow \Omega^r(M)$ be the exterior derivative operator. The adjoint exterior derivative operator $d^\dagger : \Omega^r(M) \rightarrow \Omega^{r-1}(M)$ is defined by

$$d^\dagger = (-1)^{mr+m+1} * d* \qquad (7.184a)$$

if (M, g) is Riemannian and

$$d^\dagger = (-1)^{mr+m} * d* \qquad (7.184b)$$

if Lorentzian, where $m = \dim M$.

In summary, we have the following diagram (for a Riemannian manifold),

$$
\begin{array}{ccc}
\Omega^{m-r}(M) & \xrightarrow{\ (-1)^{mr+m+1}d\ } & \Omega^{m-r+1}(M) \\
{\scriptstyle *}\big\uparrow & & \big\downarrow{\scriptstyle *} \\
\Omega^r(M) & \xrightarrow{\quad d^\dagger \quad} & \Omega^{r-1}(M).
\end{array} \qquad (7.185)
$$

The operator $d^\dagger$ is nilpotent since d is: $d^{\dagger 2} = *d * *d* \propto *d^2* = 0$.

Theorem 7.5. Let (M, g) be a compact orientable manifold without a boundary and $\alpha \in \Omega^r(M)$, $\beta \in \Omega^{r-1}(M)$. Then

$$(d\beta, \alpha) = (\beta, d^\dagger \alpha). \qquad (7.186)$$

Proof. Since both $d\beta \wedge *\alpha$ and $\beta \wedge *d^\dagger \alpha$ are m-forms, their integrals over M are well defined. Let d act on $\beta \wedge *\alpha$,

$$d(\beta \wedge *\alpha) = d\beta \wedge *\alpha - (-1)^r \beta \wedge d * \alpha.$$

Suppose (M, g) is Riemannian. Noting that $d * \alpha$ is an $(m - r + 1)$-form and inserting the identity map $(-1)^{(m-r+1)[m-(m-r+1)]} * * = (-1)^{mr+m+r+1} * *$ in front of $d * \alpha$ in the second term, we have

$$d(\beta \wedge *\alpha) = d\beta \wedge *\alpha - (-1)^{mr+m+1} \beta \wedge *(*d * \alpha).$$

Integrating this equation over M, we have

$$\int_M d\beta \wedge *\alpha - \int_M \beta \wedge *[(-1)^{mr+m+1} * d * \alpha] = \int_M d(\beta \wedge *\alpha)$$

$$= \int_{\partial M} \beta \wedge *\alpha = 0$$

where the last equality follows by assumption. This shows that $(d\beta, \alpha) = (\beta, d^\dagger \alpha)$. The reader should check how the proof is modified when (M, g) is Lorentzian. $\qquad \square$

7.9.5 The Laplacian, harmonic forms and the Hodge decomposition theorem

Definition 7.7. The **Laplacian** $\Delta : \Omega^r(M) \to \Omega^r(M)$ is defined by

$$\Delta = (d + d^\dagger)^2 = dd^\dagger + d^\dagger d. \tag{7.187}$$

As an example, we obtain the explicit form of $\Delta : \Omega^0(M) \to \Omega^0(M)$. Let $f \in \mathcal{F}(M)$. Since $d^\dagger f = 0$, we have

$$\begin{aligned}
\Delta f = d^\dagger df &= - *d * (\partial_\mu f\, dx^\mu) \\
&= - *d \left(\frac{\sqrt{|g|}}{(m-1)!} \partial_\mu f g^{\mu\lambda} \varepsilon_{\lambda\nu_2\ldots\nu_m}\, dx^{\nu_2} \wedge \ldots \wedge dx^{\nu_m} \right) \\
&= - * \frac{1}{(m-1)!} \partial_\nu [\sqrt{|g|} g^{\lambda\mu} \partial_\mu f] \varepsilon_{\lambda\nu_2\ldots\nu_m}\, dx^\nu \wedge dx^{\nu_2} \wedge \ldots \wedge dx^{\nu_m} \\
&= - * \partial_\nu [\sqrt{|g|} g^{\nu\mu} \partial_\mu f] g^{-1}\, dx^1 \wedge \ldots \wedge dx^m \\
&= - \frac{1}{\sqrt{|g|}} \partial_\nu [\sqrt{|g|} g^{\nu\mu} \partial_\mu f].
\end{aligned} \tag{7.188}$$

Exercise 7.21. Take a one-form $\omega = \omega_\mu\, dx^\mu$ in the Euclidean space $(\mathbb{R}^m, \delta)$. Show that

$$\Delta\omega = - \sum_{\mu=1}^{m} \frac{\partial^2 \omega_\nu}{\partial x^\mu \partial x^\mu}\, dx^\nu.$$

Example 7.16. In example 5.11, it was shown that half of the Maxwell equations are reduced to the identity, $dF = d^2 A = 0$, where $A = A_\mu\, dx^\mu$ is the vector potential one-form and $F = dA$ is the electromagnetic two-form. Let ρ be the electric charge density and $\boldsymbol{j}$ the electric current density and form the current one-form $j = \eta_{\mu\nu} j^\nu\, dx^\mu = -\rho\, dt + \boldsymbol{j} \cdot d\boldsymbol{x}$. Then the remaining Maxwell equations become

$$d^\dagger F = d^\dagger dA = j. \tag{7.189a}$$

The component expression is

$$\nabla \cdot \boldsymbol{E} = \rho \qquad \nabla \times \boldsymbol{B} - \frac{\partial \boldsymbol{E}}{\partial t} = \boldsymbol{j}. \tag{7.189b}$$

The vector potential A has a large number of degrees of freedom and we can always choose an A which satisfies the **Lorentz condition** $d^\dagger A = 0$. Then (7.189a) becomes $(dd^\dagger + d^\dagger d)A = \Delta A = j$.

Let (M, g) be a compact Riemannian manifold. The Laplacian Δ is a positive operator on M in the sense that

$$(\omega, \Delta\omega) = (\omega, (d^\dagger d + dd^\dagger)\omega) = (d\omega, d\omega) + (d^\dagger\omega, d^\dagger\omega) \geq 0 \tag{7.190}$$

where (7.183) has been used. An r-form ω is called **harmonic** if $\Delta\omega = 0$ and **closed** (**coclosed**) if $d\omega = 0$ $(d^\dagger\omega = 0)$. The following theorem is a direct consequence of (7.190).

Theorem 7.6. An r-form ω is harmonic if and only if ω is closed and coclosed.

An r-form ω is called **coexact** if it is written *globally* as

$$\omega_r = d^\dagger\beta_{r+1} \tag{7.191}$$

where $\beta_{r+1} \in \Omega^{r+1}(M)$ [cf a form $\omega_r \in \Omega^r(M)$ is exact if $\omega_r = d\alpha_{r-1}$, $\alpha_{r-1} \in \Omega^{r-1}(M)$]. We denote the set of harmonic r-forms on M by $\mathrm{Harm}^r(M)$ and the set of exact r-forms (coexact r-forms) by $d\Omega^{r-1}(M)$ $(d^\dagger\Omega^{r+1}(M))$. [*Note*: The set of exact r-forms has been denoted by $B^r(M)$ so far.]

Theorem 7.7. (**Hodge decomposition theorem**) Let (M, g) be a compact orientable Riemannian manifold without a boundary. Then $\Omega^r(M)$ is uniquely decomposed as

$$\Omega^r(M) = d\Omega^{r-1}(M) \oplus d^\dagger\Omega^{r+1}(M) \oplus \mathrm{Harm}^r(M). \tag{7.192a}$$

[That is, any r-form ω_r is written globally as

$$\omega_r = d\alpha_{r-1} + d^\dagger\beta_{r+1} + \gamma_r \tag{7.192b}$$

where $\alpha_{r-1} \in \Omega^{r-1}(M)$, $\beta_{r+1} \in \Omega^{r+1}(M)$ and $\gamma_r \in \mathrm{Harm}^r(M)$.]

If $r = 0$, we define $\Omega^{-1}(M) = \{0\}$. The proof of this theorem requires the results of the following two easy exercises.

Exercise 7.22. Let (M, g) be as given in theorem 7.7. Show that

$$(d\alpha_{r-1}, d^\dagger\beta_{r+1}) = (d\alpha_{r-1}, \gamma_r) = (d^\dagger\beta_{r+1}, \gamma_r) = 0. \tag{7.193}$$

Show also that if $\omega_r \in \Omega^r(M)$ satisfies

$$(d\alpha_{r-1}, \omega_r) = (d^\dagger\beta_{r+1}, \omega_r) = (\gamma_r, \omega_r) = 0 \tag{7.194}$$

for any $d\alpha_{r-1} \in d\Omega^{r-1}(M)$, $d^\dagger\beta_{r+1} \in d^\dagger\Omega^{r+1}(M)$ and $\gamma_r \in \mathrm{Harm}^r(M)$, then $\omega_r = 0$.

Exercise 7.23. Suppose $\omega_r \in \Omega^r(M)$ is written as $\omega_r = \Delta\psi_r$ for some $\psi_r \in \Omega^r(M)$. Show that $(\omega_r, \gamma_r) = 0$ for any $\gamma_r \in \mathrm{Harm}^r(M)$. The proof of the converse 'if ω_r is orthogonal to any harmonic r-form, then ω_r is written as $\Delta\psi_r$ for some $\psi_r \in \Omega^r(M)$' is highly technical and we just state that the operator Δ^{-1} (the Green function) is well defined in the present problem and ψ_r is given by $\Delta^{-1}\omega_r$.

Let $P : \Omega^r(M) \rightarrow \mathrm{Harm}^r(M)$ be a projection operator to the space of harmonic r-forms. Take an element $\omega_r \in \Omega^r(M)$. Since $\omega_r - P\omega_r$ is orthogonal to $\mathrm{Harm}^r(M)$, it can be written as $\Delta\psi_r$ for some $\psi_r \in \Omega^r(M)$. Then we have

$$\omega_r = d(d^\dagger\psi_r) + d^\dagger(d\psi_r) + P\omega_r. \tag{7.195}$$

This realizes the decomposition of theorem 7.7.

7.9.6 Harmonic forms and de Rham cohomology groups

We show that any element of the de Rham cohomology group has a unique harmonic representative. Let $[\omega_r] \in H^r(M)$. We first show that $\omega_r \in$ Harm$^r(M) \oplus d\Omega^{r-1}(M)$. According to (7.192b), ω_r is decomposed as $\omega_r = \gamma_r + d\alpha_{r-1} + d^\dagger \beta_{r+1}$. Since $d\omega_r = 0$, we have

$$0 = (d\omega_r, \beta_{r+1}) = (dd^\dagger \beta_{r+1}, \beta_{r+1}) = (d^\dagger \beta_{r+1}, d^\dagger \beta_{r+1}).$$

This is satisfied if and only if $d^\dagger \beta_{r+1} = 0$. Hence, $\omega_r = \gamma_r + d\alpha_{r-1}$. From (7.195) we have

$$\omega_r = P\omega_r + d(d^\dagger \psi) = P\omega_r + dd^\dagger \Delta^{-1}\omega_r. \tag{7.196a}$$

$\gamma_r \equiv P\omega_r$ is the harmonic representative of $[\omega_r]$. Let $\widetilde{\omega}_r$ be another representative of $[\omega_r]$: $\widetilde{\omega}_r - \omega_r = d\eta_{r-1}, \eta_{r-1} \in \Omega^{r-1}(M)$. Corresponding to (7.196a), we have

$$\widetilde{\omega}_r = P\widetilde{\omega}_r + d(d^\dagger \Delta^{-1}\widetilde{\omega}_r) = P\omega_r + d(\ldots) \tag{7.196b}$$

where the last equality follows since $d\eta_{r-1}$ is orthogonal to Harm$^r(M)$ and hence its projection to Harm$^r(M)$ vanishes. (7.196a) and (7.196b) show that $[\omega_r]$ has a unique harmonic representative $P\omega_r$.

This proof shows that $H^r(M) \subset$ Harm$^r(M)$. Now we prove that $H^r(M) \supset$ Harm$^r(M)$. Since $d\gamma_r = 0$ for any $\gamma_r \in$ Harm$^r(M)$, we find that $Z^r(M) \supset$ Harm$^r(M)$. We also have $B^r(M) \cap$ Harm$^r(M) = \emptyset$ since $B^r(M) = d\Omega^{r-1}(M)$, see (7.192a). Thus, every element of Harm$^r(M)$ is a non-trivial member of $H^r(M)$ and we find that Harm$^r(M)$ is a vector subspace of $H^r(M)$ and hence Harm$^r(M) \subset H^r(M)$. We have proved:

Theorem 7.8. (**Hodge's theorem**) On a compact orientable Riemannian manifold (M, g), $H^r(M)$ is isomorphic to Harm$^r(M)$:

$$H^r(M) \cong \text{Harm}^r(M). \tag{7.197}$$

The isomorphism is provided by identifying $[\omega] \in H^r(M)$ with $P\omega \in$ Harm$^r(M)$.

In particular, we have

$$\dim \text{Harm}^r(M) = \dim H^r(M) = b^r \tag{7.198}$$

b^r being the Betti number. The Euler characteristic is given by

$$\chi(M) = \sum(-1)^r b^r = \sum(-1)^r \dim \text{Harm}^r(M) \tag{7.199}$$

see theorem 3.7. We note that the LHS is a topological quantity while the RHS is an analytical quantity given by the eigenvalue problem of the Laplacian Δ.

7.10 Aspects of general relativity

7.10.1 Introduction to general relativity

The general theory of relativity is one of the most beautiful and successful theories in classical physics. There is no disagreement between the theory and astrophysical and cosmological observations such as solar system tests, gravitational radiation from pulsars, gravitational red shifts, the recently discovered gravitational lens effect and so on. Readers not very familiar with general relativity may consult Berry (1989) or the *primer* by Price (1982).

Einstein proposed the following principles to construct the general theory of relativity

(I) **Principle of General Relativity**: All laws in physics take the same forms in any coordinate system.

(II) **Principle of Equivalence**: There exists a coordinate system in which the effect of a gravitational field vanishes locally. (An observer in a freely falling lift does not feel gravity until it crashes.)

Any theory of gravity must reduce to Newton's theory of gravity in the weak-field limit. In Newton's theory, the gravitational potential Φ satisfies the Poisson equation

$$\Delta\Phi = 4\pi G\rho \tag{7.200}$$

where ρ is the mass density. The Einstein equation generalizes this classical result so that the principle of general relativity is satisfied.

In general relativity, the gravitational potential is replaced by the components of the metric tensor. Then, instead of the LHS of (7.200), we have the **Einstein tensor** defined by

$$G_{\mu\nu} \equiv Ric_{\mu\nu} - \tfrac{1}{2}g_{\mu\nu}\mathcal{R}. \tag{7.201}$$

Similarly, the mass density is replaced by a more general object called the **energy–momentum tensor** $T_{\mu\nu}$. The **Einstein equation** takes a very similar form to (7.200):

$$G_{\mu\nu} = 8\pi G T_{\mu\nu}. \tag{7.202}$$

The constant $8\pi G$ is chosen so that (7.202) reproduces the Newtonian result in the weak-field limit. The tensor $T_{\mu\nu}$ is obtained from the matter action by the variational principle. From Noether's theorem, $T_{\mu\nu}$ must satisfy a conservation equation of the form $\nabla_\mu T^{\mu\nu} = 0$. A similar conservation law holds for $G_{\mu\nu}$ (but not for $Ric_{\mu\nu}$). We shall see in the next subsection that the LHS of (7.202) is also obtained from the variational principle.

Exercise 7.24. Consider a metric

$$g_{00} = -1 - \frac{2\Phi}{c^2} \qquad g_{0i} = 0 \qquad g_{ij} = \delta_{ij} \qquad 1 \le i, j \le 3$$

and $T_{\mu\nu}$ given by $T_{00} = \rho c^2$, $T_{0i} = T_{ij} = 0$ which corresponds to dust at rest. Show that (7.202) reduces to the Poisson equation in the weak-field limit ($\Phi/c^2 \ll 1$).

7.10.2 Einstein–Hilbert action

This and the next example are taken from Weinberg (1972). The general theory of relativity describes the dynamics of the geometry, that is, the dynamics of $g_{\mu\nu}$. What is the action principle for this theory? As usual, we require that the relevant action should be a scalar. Moreover, it should contain the derivatives of $g_{\mu\nu}$: $\int \sqrt{|g|}\, \mathrm{d}^m x$ cannot describe the dynamics of the metric. The simplest guess will be $S_{\mathrm{EH}} \propto \int \mathcal{R}\sqrt{|g|}\, \mathrm{d}^m x$. Since $\mathcal{R}$ is a scalar and $\sqrt{|g|}\, \mathrm{d}x^1\, \mathrm{d}x^2 \ldots \mathrm{d}x^m$ is the invariant volume element, S_{EH} is a scalar. In the following, we show that S_{EH} indeed yields the Einstein equation under the variation with respect to the metric. Our connection is restricted to the Levi-Civita connection. We first prove a technical proposition.

Proposition 7.2. Let (M, g) be a (pseudo-)Riemannian manifold. Under the variation $g_{\mu\nu} \to g_{\mu\nu} + \delta g_{\mu\nu}$, $g^{\mu\nu}$, g and $Ric_{\mu\nu}$ change as

(a) $\delta g^{\mu\nu} = -g^{\mu\kappa} g^{\lambda\nu} \delta g_{\kappa\lambda}$ (7.203)

(b) $\delta g = g g^{\mu\nu} \delta g_{\mu\nu}$, $\delta\sqrt{|g|} = \frac{1}{2}\sqrt{|g|}\, g^{\mu\nu} \delta g_{\mu\nu}$ (7.204)

(c) $\delta Ric_{\mu\nu} = \nabla_\kappa \delta\Gamma^\kappa{}_{\nu\mu} - \nabla_\nu \delta\Gamma^\kappa{}_{\kappa\mu}$ (**Palatini identity**).   (7.205)

Proof. (a) From $g_{\kappa\lambda} g^{\lambda\nu} = \delta_\kappa{}^\nu$, it follows that

$$0 = \delta(g_{\kappa\lambda} g^{\lambda\nu}) = \delta g_{\kappa\lambda} g^{\lambda\nu} + g_{\kappa\lambda} \delta g^{\lambda\nu}.$$

Multiplying by $g^{\mu\kappa}$ we find that $\delta g^{\mu\nu} = -g^{\mu\kappa} g^{\lambda\nu} \delta g_{\kappa\lambda}$.

(b) We first note the matrix identity $\ln(\det g_{\mu\nu}) = \mathrm{tr}(\ln g_{\mu\nu})$. This can be proved by diagonalizing $g_{\mu\nu}$. Under the variation $\delta g_{\mu\nu}$, the LHS becomes $\delta g \cdot g^{-1}$ while the RHS yields $g^{\mu\nu} \cdot \delta g_{\mu\nu}$, hence $\delta g = g g^{\mu\nu} \delta g_{\mu\nu}$. The rest of (7.204) is easily derived from this.

(c) Let Γ and $\tilde{\Gamma}$ be two connections. The difference $\delta\Gamma \equiv \tilde{\Gamma} - \Gamma$ is a tensor of type $(1, 2)$, see exercise 7.5. In the present case, we take $\tilde{\Gamma}$ to be a connection associated with $g + \delta g$ and Γ with g. We will work in the normal coordinate system in which $\Gamma \equiv 0$ (of course $\partial\Gamma \neq 0$ in general); see section 7.4. We find

$$\delta Ric_{\mu\nu} = \partial_\kappa \delta\Gamma^\kappa{}_{\nu\mu} - \partial_\nu \delta\Gamma^\kappa{}_{\kappa\mu} = \nabla_\kappa \delta\Gamma^\kappa{}_{\nu\mu} - \nabla_\nu \Gamma^\kappa{}_{\kappa\mu}.$$

[The reader should verify the second equality.] Since both sides are tensors, this is valid in any coordinate system.   □

We define the **Einstein–Hilbert action** by

$$S_{\mathrm{EH}} \equiv \frac{1}{16\pi G} \int \mathcal{R}\sqrt{-g}\, \mathrm{d}^4 x. \tag{7.206}$$

The constant factor $1/16\pi G$ is introduced to reproduce the Newtonian limit when matter is added; see (7.214). We prove that $\delta S_{\text{EH}} = 0$ leads to the vacuum Einstein equation. Under the variation $g \to g + \delta g$ such that $\delta g \to 0$ as $|x| \to 0$, the integrand changes as

$$
\begin{aligned}
\delta(\mathcal{R}\sqrt{-g}) &= \delta(g^{\mu\nu} Ric_{\mu\nu}\sqrt{-g}) \\
&= \delta g^{\mu\nu} Ric_{\mu\nu}\sqrt{-g} + g^{\mu\nu}\delta Ric_{\mu\nu}\sqrt{-g} + \mathcal{R}\delta(\sqrt{-g}) \\
&= -g^{\mu\kappa}g^{\lambda\nu}\delta g_{\kappa\lambda} Ric_{\mu\nu}\sqrt{-g} \\
&\quad + g^{\mu\nu}(\nabla_\kappa\delta\Gamma^\kappa{}_{\nu\mu} - \nabla_\nu\Gamma^\kappa{}_{\kappa\mu})\sqrt{-g} + \tfrac{1}{2}\mathcal{R}\sqrt{-g}g^{\mu\nu}\delta g_{\mu\nu}.
\end{aligned}
$$

We note that the second term is a total divergence,

$$
\begin{aligned}
&\nabla_\kappa(g^{\mu\nu}\delta\Gamma^\kappa{}_{\nu\mu}\sqrt{-g}) - \nabla_\nu(g^{\mu\nu}\delta\Gamma^\kappa{}_{\kappa\mu}\sqrt{-g}) \\
&= \partial_\kappa(g^{\mu\nu}\delta\Gamma^\kappa{}_{\nu\mu}\sqrt{-g}) - \partial_\nu(g^{\mu\nu}\delta\Gamma^\kappa{}_{\kappa\mu}\sqrt{-g})
\end{aligned}
$$

and hence does not contribute to the variation. From the remaining terms we have

$$
\delta S_{\text{EH}} = \frac{1}{16\pi G}\int\left(-Ric^{\mu\nu} + \frac{1}{2}\mathcal{R}g^{\mu\nu}\right)\delta g_{\mu\nu}\sqrt{-g}\,\mathrm{d}^4x. \tag{7.207}
$$

If we require that $\delta S_{\text{EH}} = 0$ under any variation δg, we obtain the vacuum Einstein equation,

$$
G_{\mu\nu} = Ric_{\mu\nu} - \tfrac{1}{2}g_{\mu\nu}\mathcal{R} = 0 \tag{7.208}
$$

where the symmetric tensor G is called the **Einstein tensor**.

So far we have considered the gravitational field only. Suppose there exists matter described by an action

$$
S_{\text{M}} \equiv \int \mathcal{L}(\phi)\sqrt{-g}\,\mathrm{d}^4x \tag{7.209}
$$

where $\mathcal{L}(\phi)$ is the Lagrangian density of the theory. Typical examples are the real scalar field and the Maxwell fields,

$$
S_{\text{S}} \equiv -\tfrac{1}{2}\int[g^{\mu\nu}\partial_\mu\phi\partial_\nu\phi + m^2\phi^2]\sqrt{-g}\,\mathrm{d}^4x \tag{7.210a}
$$

$$
S_{\text{ED}} \equiv -\tfrac{1}{4}\int F_{\mu\nu}F^{\mu\nu}\sqrt{-g}\,\mathrm{d}^4x \tag{7.210b}
$$

where $F_{\mu\nu} = \partial_\mu A_\nu - \partial_\nu A_\mu = \nabla_\mu A_\nu - \nabla_\nu A_\mu$. If the matter action changes by δS_{M} under δg, the **energy–momentum tensor** $T^{\mu\nu}$ is defined by

$$
\delta S_{\text{M}} = \tfrac{1}{2}\int T^{\mu\nu}\delta g_{\mu\nu}\sqrt{-g}\,\mathrm{d}^4x. \tag{7.211}
$$

Since $\delta g_{\mu\nu}$ is symmetric, $T^{\mu\nu}$ is also taken to be so. For example, $T_{\mu\nu}$ of a real scalar field is given by

$$
\begin{aligned}
T_{\mu\nu}(x) &= 2\frac{1}{\sqrt{-g}}\frac{\delta}{\delta g^{\mu\nu}(x)}S_S \\
&= \partial_\mu\phi\partial_\nu\phi - \tfrac{1}{2}g_{\mu\nu}(g^{\kappa\lambda}\partial_\kappa\phi\partial_\lambda\phi + m^2\phi^2).
\end{aligned}
\tag{7.212}
$$

Suppose we have a gravitational field coupled with a matter field whose action is S_M. Now our action principle is

$$
\delta(S_{EH} + S_M) = 0
\tag{7.213}
$$

under $g \to g + \delta g$. From (7.207) and (7.211), we obtain the **Einstein equation**

$$
G_{\mu\nu} = 8\pi G T_{\mu\nu}.
\tag{7.214}
$$

Exercise 7.25. We may add an extra scalar to the scalar curvature without spoiling the invariance of the action. For example, we can add a constant called the **cosmological constant** Λ,

$$
\widetilde{S}_{EH} = \frac{1}{16\pi G}\int_M (\mathcal{R} + \Lambda)\sqrt{-g}\,\mathrm{d}^4 x.
\tag{7.215}
$$

Write down the vacuum Einstein equation. Other possible scalars may be such terms as $\mathcal{R}^2$, $Ric^{\mu\nu}Ric_{\mu\nu}$ or $R_{\kappa\lambda\mu\nu}R^{\kappa\lambda\mu\nu}$.

7.10.3   Spinors in curved spacetime

For concreteness, we consider a Dirac spinor ψ in a four-dimensional Lorentz manifold M. The vierbein $e^\alpha{}_\mu$ defined by

$$
g_{\mu\nu} = e^\alpha{}_\mu e^\beta{}_\nu \eta_{\alpha\beta}
\tag{7.216}
$$

defines an orthonormal frame $\{\hat{\theta}^\alpha = e^\alpha{}_\mu \mathrm{d}x^\mu\}$ at each point $p \in M$. As noted before, $\alpha, \beta, \gamma, \dots$ are the local orthonormal indices while $\mu, \nu, \lambda, \dots$ are the coordinate indices. With respect to this frame, the Dirac matrices $\gamma^\alpha = e^\alpha{}_\mu\gamma^\mu$ satisfy $\{\gamma^\alpha, \gamma^\beta\} = 2\eta^{\alpha\beta}$. Under a local Lorentz transformation $\Lambda^\alpha{}_\beta(p)$, the Dirac spinor transforms as

$$
\psi(p) \to \rho(\Lambda)\psi(p) \qquad \bar{\psi}(p) \to \bar{\psi}(p)\rho(\Lambda)^{-1}
\tag{7.217}
$$

where $\bar{\psi} \equiv \psi^\dagger\gamma^0$ and $\rho(\Lambda)$ is the spinor representation of Λ. To construct an invariant action, we seek a covariant derivative $\nabla_\alpha\psi$ which is a local Lorentz vector and transforms as a spinor,

$$
\nabla_\alpha\psi \to \rho(\Lambda)\Lambda_\alpha{}^\beta\nabla_\beta\psi.
\tag{7.218}
$$

If we find such a $\nabla_\alpha \psi$, an invariant Lagrangian may be given by

$$\mathcal{L} = \bar{\psi} \left(i\gamma^\alpha \nabla_\alpha + m \right) \psi \tag{7.219}$$

m being the mass of ψ. We note that $e_\alpha{}^\mu \partial_\mu \psi$ transforms under $\Lambda(p)$ as

$$e_\alpha{}^\mu \partial_\mu \psi \to \Lambda_\alpha{}^\beta e_\beta{}^\mu \partial_\mu \rho(\Lambda)\psi = \Lambda_\alpha{}^\eta e_\beta{}^\mu [\rho(\Lambda)\partial_\mu \psi + \partial_\mu \rho(\Lambda)\psi]. \tag{7.220}$$

Suppose ∇_α is of the form

$$\nabla_\alpha \psi = e_\alpha{}^\mu [\partial_\mu + \Omega_\mu]\psi. \tag{7.221}$$

From (7.218) and (7.220), we find that Ω_μ satisfies

$$\Omega_\mu \to \rho(\Lambda)\Omega_\mu \rho(\Lambda)^{-1} - \partial_\mu \rho(\Lambda)\rho(\Lambda)^{-1}. \tag{7.222}$$

To find the explicit form of Ω_μ, we consider an infinitesimal local Lorentz transformation $\Lambda_\alpha{}^\beta(p) = \delta_\alpha{}^\beta + \varepsilon_\alpha{}^\beta(p)$. The Dirac spinor transforms as

$$\psi \to \exp[\tfrac{1}{2} i\varepsilon^{\alpha\beta} \Sigma_{\alpha\beta}]\psi \simeq [1 + \tfrac{1}{2} i\varepsilon^{\alpha\beta} \Sigma_{\alpha\beta}]\psi \tag{7.223}$$

where $\Sigma_{\alpha\beta} \equiv \tfrac{1}{4} i [\gamma_\alpha, \gamma_\beta]$ is the spinor representation of the generators of the Lorentz transformation. $\Sigma_{\alpha\beta}$ satisfies the $o(1, 3)$ Lie algebra

$$i[\Sigma_{\alpha\beta}, \Sigma_{\gamma\delta}] = \eta_{\gamma\beta} \Sigma_{\alpha\delta} - \eta_{\gamma\alpha} \Sigma_{\beta\delta} + \eta_{\delta\beta} \Sigma_{\gamma\alpha} - \eta_{\delta\alpha} \Sigma_{\gamma\beta}. \tag{7.224}$$

Under the same Lorentz transformation, Ω_μ transforms as

$$\Omega_\mu \to (1 + \tfrac{1}{2} i\varepsilon^{\alpha\beta} \Sigma_{\alpha\beta})\Omega_\mu(1 - \tfrac{1}{2} i\varepsilon^{\gamma\delta} \Sigma_{\gamma\delta}) - \tfrac{1}{2} i\partial_\mu \varepsilon^{\alpha\beta} \Sigma_{\alpha\beta}(1 - \tfrac{1}{2} i\varepsilon^{\gamma\delta} \Sigma_{\gamma\delta})$$
$$= \Omega_\mu + \tfrac{1}{2} i\varepsilon^{\alpha\beta} [\Sigma_{\alpha\beta}, \Omega_\mu] - \tfrac{1}{2} i\partial_\mu \varepsilon^{\alpha\beta} \Sigma_{\alpha\beta}. \tag{7.225}$$

We recall that the connection one-form $\omega^\alpha{}_\beta$ transforms under an infinitesimal Lorentz transformation as (see (7.152))

$$\omega^\alpha{}_\beta \to \omega^\alpha{}_\beta + \varepsilon^\alpha{}_\gamma \omega^\gamma{}_\beta - \omega^\alpha{}_\gamma \varepsilon^\gamma{}_\beta - d\varepsilon^\alpha{}_\beta \tag{7.226a}$$

or in components,

$$\Gamma^\alpha{}_{\mu\beta} \to \Gamma^\alpha{}_{\mu\beta} + \varepsilon^\alpha{}_\gamma \Gamma^\gamma{}_{\mu\beta} - \Gamma^\alpha{}_{\mu\gamma} \varepsilon^\gamma{}_\beta - \partial_\mu \varepsilon^\alpha{}_\beta. \tag{7.226b}$$

From (7.224), (7.225) and (7.226b), we find that the combination

$$\Omega_\mu \equiv \tfrac{1}{2} i\Gamma^\alpha{}_\mu{}^\beta \Sigma_{\alpha\beta} = \tfrac{1}{2} ie^\alpha{}_\nu \nabla_\mu e^{\beta\nu} \Sigma_{\alpha\beta} \tag{7.227}$$

satisfies the transformation property (7.222). In fact,

$$\tfrac{1}{2} i\Gamma^\alpha{}_\mu{}^\beta \Sigma_{\alpha\beta} \to \tfrac{1}{2} i(\Gamma^\alpha{}_\mu{}^\beta + \varepsilon^\alpha{}_\gamma \Gamma^\gamma{}_\mu{}^\beta - \Gamma^\alpha{}_{\mu\gamma} \varepsilon^{\gamma\beta} - \partial_\mu \varepsilon^{\alpha\beta})\Sigma_{\alpha\beta}$$
$$= \tfrac{1}{2} i\Gamma^\alpha{}_\mu{}^\beta \Sigma_{\alpha\beta} + \tfrac{1}{2} i(\varepsilon^\alpha{}_\gamma \Gamma^\gamma{}_\mu{}^\beta \Sigma_{\alpha\beta} - \Gamma^\alpha{}_{\mu\gamma} \varepsilon^{\gamma\beta} \Sigma_{\alpha\beta})$$
$$\quad - \tfrac{1}{2} i\partial_\mu \varepsilon^{\alpha\beta} \Sigma_{\alpha\beta}$$
$$= \tfrac{1}{2} i\Gamma^\alpha{}_\mu{}^\beta \Sigma_{\alpha\beta} + \tfrac{1}{2} ie^{\alpha\beta} [\Sigma_{\alpha\beta}, \tfrac{1}{2} i\Gamma^\gamma{}_\mu{}^\delta \Sigma_{\gamma\delta}] - \tfrac{1}{2} i\partial_\mu \varepsilon^{\alpha\beta} \Sigma_{\alpha\beta}.$$

We finally obtain the Lagrangian which is a scalar both under coordinate changes and local Lorentz rotations,

$$\mathcal{L} \equiv \bar{\psi}[i\gamma^\alpha e_\alpha{}^\mu (\partial_\mu + \tfrac{1}{2}i\Gamma^\beta{}_\mu{}^\gamma \Sigma_{\beta\gamma}) + m]\psi \tag{7.228}$$

and the scalar action

$$S_\psi \equiv \int_M d^4x \sqrt{-g}\,\bar{\psi}[i\gamma^\alpha e_\alpha{}^\mu (\partial_\mu + \tfrac{1}{2}i\Gamma^\beta{}_\mu{}^\gamma \Sigma_{\beta\gamma}) + m]\psi. \tag{7.229a}$$

If ψ is coupled to the gauge field $\mathcal{A}$, the action is given by

$$S_\psi = \int_M d^4x \sqrt{-g}\,\bar{\psi}[i\gamma^\alpha e_\alpha{}^\mu (\partial_\mu + \mathcal{A}_\mu + \tfrac{1}{2}i\Gamma^\beta{}_\mu{}^\gamma \Sigma_{\beta\gamma}) + m]\psi. \tag{7.229b}$$

It is interesting to note that the spin connection term vanishes if dim $M = 2$. To see this, we rewrite (7.229a) as

$$S_\psi = \tfrac{1}{2}\int_M d^2x \sqrt{-g}\,\bar{\psi}[i\gamma^\mu \overleftrightarrow{\partial_\mu} + \tfrac{1}{2}i\Gamma^\beta{}_\mu{}^\gamma \{i\gamma^\mu, \Sigma_{\beta\gamma}\} + m]\psi \tag{7.229a'}$$

where $\gamma^\mu = \gamma^\alpha e_\alpha{}^\mu$ and we have added total derivatives to the Lagrangian to make it Hermitian. The non-vanishing components of Σ are $\Sigma_{01} \propto [\gamma_0, \gamma_1] \propto \gamma_3$, where γ_3 is the two-dimensional analogue of γ_5. Since $\{\gamma^\mu, \gamma_3\} = 0$, the spin connection term drops out from S_ψ.

7.11 Bosonic string theory

Quantum field theory (QFT) is occasionally called particle physics since it deals with the dynamics of particles. As far as high-energy processes whose typical energy is much smaller than the Planck energy ($\sim 10^{19}$ GeV) are concerned there is no objection to this viewpoint. However, once we try to quantize gravity in this framework, there exists an impenetrable barrier. We do not know how to renormalize the ultraviolet divergences that are ubiquitous in the QFT of gravity. In the early 1980s, physicists tried to construct a consistent theory of gravity by introducing supersymmetry. In spite of a partial improvement, the resulting supergravity could not tame the ultraviolet behaviour completely.

In the late 1960s and early 1970s, the dual resonance model was extensively studied as a candidate for a model of hadrons. In this, particles are replaced by one-dimensional objects called **strings**. Unfortunately, it turned out that the theory contained tachyons (imaginary mass particles) and spin-2 particles and, moreover, it is consistent only in 26-dimensional spacetime! Due to these difficulties, the theory was abandoned and taken over by quantum chromodynamics (QCD). However, a small number of people noticed that the theory must contain the graviton and they thought it could be a candidate for the quantum theory of gravity.

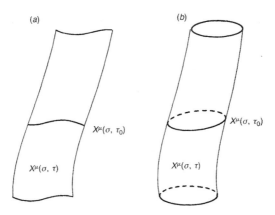

Figure 7.9. The trajectories of an open string (*a*) and a closed string (*b*). Slices of the trajectories at fixed parameter τ_0 are also shown.

Nowadays, supersymmetry has been built into string theory to form the **superstring theory**, which is free of tachyons and consistent in ten-dimensional spacetime. There are several candidates for consistent superstring theories. It is sometimes suggested that complete mathematical consistency will single out a unique *theory of everything* (TOE).

In this book, we study the elementary aspects of bosonic string theory in the final chapter. We also study some mathematical tools relevant for superstrings. The classical review is that of Scherk (1975). We give more references in chapter 14.

7.11.1 The string action

The trajectory of a particle in a D-dimensional Minkowski spacetime is given by the set of D functions $X^\mu(\tau)$, $1 \le \mu \le D$, where τ parametrizes the trajectory. A string is a one-dimensional object and its configuration is parametrized by two numbers (σ, τ), σ being spacelike and τ timelike. Its position in D-dimensional Minkowski spacetime is given by $X^\mu(\sigma, \tau)$, see figure 7.9. The parameter σ can be normalized as $\sigma \in [0, \pi]$. A string may be open or closed. We now seek an action that governs the dynamics of strings.

We first note that the action of a relativistic particle is the *length* of the *world line*,

$$S \equiv m \int_{s_i}^{s_f} ds = m \int_{\tau_i}^{\tau_f} d\tau \, (-\dot{X}^\mu \dot{X}_\mu)^{1/2} \tag{7.230}$$

where $\dot{X}^\mu \equiv dX^\mu/d\tau$. For some purposes, it is convenient to take another expression,

$$S = -\frac{1}{2} \int d\tau \, \sqrt{g}(g^{-1}\dot{X}^\mu \dot{X}_\mu - m^2) \tag{7.231}$$

where the auxiliary variable $g \equiv g_{\tau\tau}$ is regarded as a metric.

Exercise 7.26. Write down the Euler–Lagrange equations derived from (7.231). Eliminate g from (7.231) making use of the equation of motion to reproduce (7.230).

What is the advantage of (7.231) over (7.230)? We first note that (7.231) makes sense even when $m^2 = 0$, while (7.230) vanishes in this case. Second, (7.231) is quadratic in X while the X-dependence of (7.230) is rather complicated.

Nambu (1970) proposed an action describing the strings, which is proportional to the *area* of the **world sheet**, the surface spanned by the trajectory of a string. Clearly this is a generalization of the length of the world line of a particle. He proposed the **Nambu action**,

$$S = -\frac{1}{2\pi\alpha'} \int_0^\pi d\sigma \int_{\tau_i}^{\tau_f} d\tau \, [-\det(\partial_\alpha X^\mu \partial_\beta X_\mu)]^{1/2} \tag{7.232}$$

where $\xi^0 = \tau, \xi^1 = \sigma$ and $\partial_\alpha X^\mu \equiv \partial X^\mu / \partial \xi^\alpha$. The parameter τ_i (τ_f) is the initial (final) value of the parameter τ while α' is a parameter corresponding to the inverse string tension (the Regge slope).

Exercise 7.27. The action S is required to have no dimension. We take σ and τ to be dimensionless. Show that the dimension of α' is [length]^2.

Although the action provides a nice geometrical picture, it is not quadratic in X and it turned out that the quantization of the theory was rather difficult. Let us seek an equivalent action which is easier to quantize. We proceed analogously to the case of point particles. A quadratic action for strings is called the **Polyakov action** (Polyakov 1981) and is given by

$$S = -\frac{1}{4\pi\alpha'} \int_0^\pi d\sigma \int_{\tau_i}^{\tau_f} d\tau \, \sqrt{-g} g^{\alpha\beta} \partial_\alpha X^\mu \partial_\beta X_\mu \tag{7.233}$$

where $g = \det g_{\alpha\beta}$ and $g^{\alpha\beta} = (g^{-1})^{\alpha\beta}$. If the string is open, the trajectory is a sheet while if it is closed, it is a tube, see figure 7.9. It is shown here that the action (7.233) agrees with (7.232) upon eliminating g. It should be noted though that this is true only for the Lagrangian. There is no guarantee that this remains true at the quantum level. It has been shown that the quantum theory based on the respective Lagrangians agrees only for $D = 26$. The action (7.233) is invariant under

(i)   local reparametrization of the world sheet

$$\tau \to \tau'(\tau, \sigma) \qquad \sigma \to \sigma'(\tau, \sigma) \tag{7.234a}$$

(ii) Weyl rescaling

$$g_{\alpha\beta} \to g'_{\alpha\beta} \equiv e^{\phi(\sigma,\tau)} g_{\alpha\beta} \tag{7.234b}$$

(iii) global Poincaré invariance

$$X^\mu \to X^{\mu\prime} \equiv \Lambda^\mu{}_\nu X^\nu + a^\mu \qquad \Lambda \in SO(D-1, 1) \qquad a \in \mathbb{R}^D. \quad (7.234c)$$

These symmetries will be worked out later.

Exercise 7.28. Taking advantage of symmetries (i) and (iii), it is always possible to choose $g_{\alpha\beta}$ in the form $g_{\alpha\beta} = \eta_{\alpha\beta}$. Write down the equation of motion for X^μ to show that it obeys the equation

$$\eta^{\alpha\beta} \partial_\alpha \partial_\beta X^\mu = 0. \quad (7.235)$$

7.11.2 Symmetries of the Polyakov strings

The bosonic string theory is defined on a two-dimensional Lorentz manifold (M, g). The embedding $f : M \to \mathbb{R}^D$ is defined by $\xi^\alpha \mapsto X^\mu$ where $\{\xi^\alpha\} = (\tau, \sigma)$ are the local coordinates of M. We assume the physical spacetime is Minkowskian $(\mathbb{R}^D, \eta)$ for simplicity. The **Polyakov action**

$$S = -\tfrac{1}{2} \int d^2\xi \sqrt{-g} g^{\alpha\beta} \partial_\alpha X^\mu \partial_\beta X^\nu \eta_{\mu\nu} \quad (7.236)$$

is left invariant under the coordinate reparametrization Diff(M) since the volume element $\sqrt{-g} d^2\xi$ is invariant and $g^{\alpha\beta} \partial_\alpha X^\mu \partial_\beta X_\mu$ is a scalar.

Now we are ready to derive the equation of motion. Our variational parameters are the *embedding* X^μ and the geometry $g_{\alpha\beta}$. Under the variation δX^μ, we have the Euler–Lagrange equation

$$\partial_\alpha(\sqrt{-g} g^{\alpha\beta} \partial_\beta X_\mu) = 0. \quad (7.237a)$$

Under the variation $\delta g_{\alpha\beta}$, the integrand of S changes as

$$\begin{aligned}
\delta(\sqrt{-g} g^{\alpha\beta} \partial_\alpha X^\mu \partial_\beta X_\mu) &= \delta\sqrt{-g} g^{\alpha\beta} \partial_\alpha X^\mu \partial_\beta X_\mu + \sqrt{-g} \delta g^{\alpha\beta} \partial_\alpha X^\mu \partial_\beta X_\mu \\
&= -\tfrac{1}{2}\sqrt{-g} g_{\gamma\delta} \delta g^{\gamma\delta} g^{\alpha\beta} \partial_\alpha X^\mu \partial_\beta X_\mu \\
&\quad + \sqrt{-g} \delta g^{\alpha\beta} \partial_\alpha X^\mu \partial_\beta X_\mu
\end{aligned}$$

where proposition 7.2 has been used. Since this should vanish for any variation $\delta g_{\alpha\beta}$, we should have

$$T_{\alpha\beta} = \partial_\alpha X^\mu \partial_\beta X_\mu - \tfrac{1}{2} g_{\alpha\beta} (g^{\gamma\delta} \partial_\gamma X^\mu \partial_\delta X_\mu) = 0. \quad (7.237b)$$

This is solved for $g_{\alpha\beta}$ to yield

$$g_{\alpha\beta} = \partial_\alpha X^\mu \partial_\beta X^\nu \eta_{\mu\nu} \quad (7.238)$$

showing that the induced metric (the RHS) agrees with $g_{\alpha\beta}$. Substituting (7.238) into (7.236) to eliminate $g_{\alpha\beta}$, we recover the Nambu action,

$$S = -\tfrac{1}{2} \int d^2\xi \sqrt{-\det(\partial_\alpha X^\mu \partial_\beta X_\mu)}. \quad (7.239)$$

By construction, the action S is invariant under local reparametrization of M, $\{\xi^\alpha\} \rightarrow \{\xi'^\alpha(\xi)\}$. In addition to this, the action has extra invariances. Under the global **Poincaré transformation** in D-dimensional spacetime,

$$X^\mu \rightarrow X'^\mu \equiv \Lambda^\mu{}_\nu X^\nu + a^\mu \qquad (7.240)$$

the action S transforms as

$$S \rightarrow -\frac{1}{2} \int d^2\xi \sqrt{-g} g^{\alpha\beta} \partial_\alpha(\Lambda^\mu{}_\kappa X^\kappa + a^\mu) \partial_\beta(\Lambda^\nu{}_\lambda X^\lambda + a^\nu) \eta_{\mu\nu}$$

$$= -\frac{1}{2} \int d^2\xi \sqrt{-g} g^{\alpha\beta} \partial_\alpha X^\kappa \partial_\beta X^\lambda (\Lambda^\mu{}_\kappa \Lambda^\nu{}_\lambda \eta_{\mu\nu}).$$

From $\Lambda^\mu{}_\kappa \Lambda^\nu{}_\lambda \eta_{\mu\nu} = \eta_{\kappa\lambda}$, we find that S is invariant under global Poincaré transformations. The action S is also invariant under the **Weyl rescaling**, $g_{\alpha\beta}(\tau, \sigma) \rightarrow e^{2\sigma(\tau,\sigma)} g_{\alpha,\beta}(\tau, \sigma)$ keeping (τ, σ) fixed. In fact, S transforms as

$$S \rightarrow -\frac{1}{2} \int d^2\xi \sqrt{-e^{4\sigma} g} e^{-2\sigma} g^{\alpha\beta} \partial_\alpha X^\mu \partial_\beta X^\nu \eta_{\mu\nu}$$

and hence is left invariant. Note that the Weyl rescaling invariance exists only when M is two dimensional, making strings prominent among other extended objects such as membranes.

Since $\dim M = 2$, we can always parametrize the world sheet by the isothermal coordinate (example 7.9) so that

$$g_{\alpha\beta} = e^{2\sigma(\tau,\sigma)} \eta_{\alpha\beta}. \qquad (7.241)$$

Then the Weyl rescaling invariance allows us to choose the standard metric $\eta_{\alpha\beta}$ on the world sheet. The metric $g_{\alpha\beta}$ has three independent components while the reparametrization has two degrees of freedom and the Weyl scaling invariance has one. Thus, so long as we are dealing with strings, we can choose the standard metric $\eta_{\alpha\beta}$.

We end our analysis of Polyakov strings here. Polyakov strings will be quantized in the most elegant manner in chapter 14.

Exercise 7.29. Let (M, g) and (N, h) be Riemannian manifolds. Take a chart U of M in which the metric g takes the form

$$g = g_{\mu\nu}(x) \, dx^\mu \otimes dx^\nu.$$

Take a chart V of N on which h takes the form

$$h = G_{\alpha\beta}(\phi) d\phi^\alpha \otimes d\phi^\beta.$$

A map $\phi : M \rightarrow N$ defined by $x \mapsto \phi(x)$ is called a **harmonic map** if it satisfies

$$\frac{1}{\sqrt{g}} \partial_\mu[\sqrt{g} g^{\mu\nu} \partial_\nu \phi^\alpha] + \Gamma^\alpha{}_{\beta\gamma} \partial_\mu \phi^\alpha \partial_\nu \phi^\beta g^{\mu\nu} = 0. \qquad (7.242)$$

Show that this equation is obtained by the variation of the action

$$S \equiv \frac{1}{2} \int d^m x \sqrt{g} g^{\mu\nu} \partial_\mu \phi^\alpha \partial_\nu \phi^\beta h_{\alpha\beta}(\phi) \tag{7.243}$$

with respect to ϕ. Applications of harmonic maps to physics are found in Misner (1978) and Sánchez (1988). Mathematical aspects have been reviewed in Eells and Lemaire (1968).

Problems

7.1 Let ∇ be a general connection for which the torsion tensor does not vanish. Show that the first Bianchi identity becomes

$$\mathfrak{S}\{R(X, Y)Z\} = \mathfrak{S}\{T(X, [Y, Z])\} + \mathfrak{S}\{\nabla_X[T(Y, Z)]\}$$

where $\mathfrak{S}$ is the symmetrizer defined in theorem 7.2. Show also that the second Bianchi identity is given by

$$\mathfrak{S}\{(\nabla_X R)(Y, Z)\}V = \mathfrak{S}\{R(X, T(Y, Z))\}V$$

where $\mathfrak{S}$ symmetrizes X, Y and Z only.

7.2 Let (M, g) be a conformally flat three-dimensional manifold. Show that the **Weyl–Schouten tensor** defined by

$$C_{\lambda\mu\nu} \equiv \nabla_\nu Ric_{\lambda\mu} - \nabla_\mu Ric_{\lambda\nu} - \frac{1}{4}(g_{\lambda\mu}\partial_\nu \mathcal{R} - g_{\lambda\nu}\partial_\mu \mathcal{R})$$

vanishes. It is known that $C_{\lambda\mu\nu} = 0$ is the necessary and sufficient condition for conformal flatness if dim $M = 3$.

7.3 Consider a metric

$$g = -dt \otimes dt + dr \otimes dr + (1 - 4\mu^2)r^2 \, d\phi \otimes d\phi + dz \otimes dz$$

where $0 < \mu < 1/2$ and $\mu \neq 1/4$. Introduce a new variable

$$\tilde{\phi} \equiv (1 - 4\mu)\phi$$

and show that the metric g reduces to the Minkowski metric. Does this mean that g describes Minkowski spacetime? Compute the Riemann curvature tensor and show that there is a stringlike singularity at $r = 0$. This singularity is *conical* (the spacetime is flat except along the line). This metric models the spacetime of a cosmic string.

8

COMPLEX MANIFOLDS

A differentiable manifold is a topological space which admits differentiable structures. Here we introduce another structure which has relevance in physics. In elementary complex analysis, the partial derivatives are required to satisfy the Cauchy–Riemann relations. We talk not only of the differentiability but also of the analyticity of a function in this case. A complex manifold admits a complex structure in which each coordinate neighbourhood is homeomorphic to $\mathbb{C}^m$ and the transition from one coordinate system to the other is analytic.

The reader may consult Chern (1979), Goldberg (1962) or Greene (1987) for further details. Griffiths and Harris (1978), chapter 0 is a concise survey of the present topics. For applications to physics, see Horowitz (1986) and Candelas (1988).

8.1 Complex manifolds

To begin with, we define a holomorphic (or analytic) map on $\mathbb{C}^m$. A complex-valued function $f : \mathbb{C}^m \to \mathbb{C}$ is **holomorphic** if $f = f_1 + \mathrm{i}\, f_2$ satisfies the **Cauchy–Riemann relations** for each $z^\mu = x^\mu + \mathrm{i}\, y^\mu$,

$$\frac{\partial f_1}{\partial x^\mu} = \frac{\partial f_2}{\partial y^\mu} \qquad \frac{\partial f_2}{\partial x^\mu} = -\frac{\partial f_1}{\partial y^\mu}. \tag{8.1}$$

A map $(f^1, \ldots, f^n) : \mathbb{C}^m \to \mathbb{C}^n$ is called holomorphic if each function f^λ $(1 \le \lambda \le n)$ is holomorphic.

8.1.1 Definitions

Definition 8.1. M is a complex manifold if the following axioms hold,

(i) M is a topological space.

(ii) M is provided with a family of pairs $\{(U_i, \varphi_i)\}$.

(iii) $\{U_i\}$ is a family of open sets which covers M. The map φ_i is a homeomorphism from U_i to an open subset U of $\mathbb{C}^m$. [Hence, M is even dimensional.]

(iv) Given U_i and U_j such that $U_i \cap U_j \ne \emptyset$, the map $\psi_{ji} = \varphi_j \circ \varphi_i^{-1}$ from $\varphi_i(U_i \cap U_j)$ to $\varphi_j(U_i \cap U_j)$ is holomorphic.

The number m is called the complex dimension of M and is denoted as $\dim_{\mathbb{C}} M = m$. The real dimension $2m$ is denoted either by $\dim_{\mathbb{R}} M$ or simply by $\dim M$. Let $z^\mu = \varphi_i(p)$ and $w^\nu = \varphi_j(p)$ be the (complex) coordinates of a point $p \in U_i \cap U_j$ in the charts (U_i, φ_i) and (U_j, φ_j), respectively. Axiom (iv) asserts that the function $w^\nu = u^\nu + \mathrm{i}v^\nu$ ($1 \le \nu \le m$) is holomorphic in $z^\mu = x^\mu + \mathrm{i}y^\mu$, namely

$$\frac{\partial u^\nu}{\partial x^\nu} = \frac{\partial v^\nu}{\partial y^\nu} \qquad \frac{\partial u^\nu}{\partial y^\nu} = -\frac{\partial v^\nu}{\partial x^\nu} \qquad 1 \le \mu, \ \nu \le m.$$

These axioms ensure that calculus on complex manifolds can be carried out independently of the special coordinates chosen. For example, $\mathbb{C}^m$ is the simplest complex manifold. A single chart covers the whole space and φ is the identity map.

Let $\{(U_i, \varphi_i)\}$ and $\{(V_j, \psi_j)\}$ be atlases of M. If the union of two atlases is again an atlas which satisfies the axioms of definition 8.1, they are said to define the same complex structure. A complex manifold may carry a number of complex structures (see example 8.2).

8.1.2 Examples

Example 8.1. In exercise 5.1, it was shown that the stereographic coordinates of a point $P(x, y, z) \in S^2 - \{\text{North Pole}\}$ projected from the North Pole are

$$(X, Y) = \left(\frac{x}{1 - z}, \frac{y}{1 - z} \right)$$

while those of a point $P(x, y, z) \in S^2 - \{\text{South Pole}\}$ projected from the South Pole are

$$(U, V) = \left(\frac{x}{1 + z}, \frac{-y}{1 + z} \right).$$

[Note the orientation of (U, V) in figure 5.5.] Let us define complex coordinates

$$Z = X + \mathrm{i}Y, \qquad \overline{Z} = X - \mathrm{i}Y, \qquad W = U + \mathrm{i}V, \qquad \overline{W} = U - \mathrm{i}V.$$

W is a holomorphic function of Z,

$$W = \frac{x - \mathrm{i}y}{1 + z} = \frac{1 - z}{1 + z}(X - \mathrm{i}Y) = \frac{X - \mathrm{i}Y}{X^2 + Y^2} = \frac{1}{Z}.$$

Thus, S^2 is a complex manifold which is identified with the Riemann sphere $\mathbb{C} \cup \{\infty\}$.

Example 8.2. Take a complex plane $\mathbb{C}$ and define a lattice $L(\omega_1, \omega_2) \equiv \{\omega_1 m + \omega_2 n | m, n \in \mathbb{Z}\}$ where ω_1 and ω_2 are two non-vanishing complex numbers such

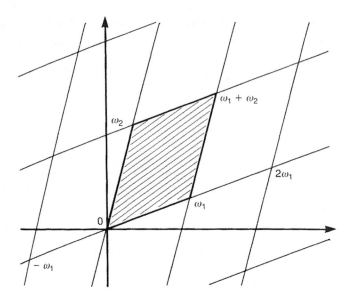

Figure 8.1. Two complex numbers ω_1 and ω_2 define a lattice $L(\omega_1, \omega_2)$ in the complex plane. $\mathbb{C}/L(\omega_1, \omega_2)$ is homeomorphic to the torus (the shaded area).

that $\omega_2/\omega_1 \notin \mathbb{R}$; see figure 8.1. Without loss of generality, we may take $\text{Im}(\omega_2/\omega_1) > 0$. The manifold $\mathbb{C}/L(\omega_1, \omega_2)$ is obtained by identifying the points $z_1, z_2 \in \mathbb{C}$ such that $z_1 - z_2 = \omega_1 m + \omega_2 n$ for some $m, n \in \mathbb{Z}$. Since the opposite sides of the shaded area of figure 8.1 are identified, $\mathbb{C}/L(\omega_1, \omega_2)$ is homeomorphic to the torus T^2. The complex structure of $\mathbb{C}$ naturally induces that of $\mathbb{C}/L(\omega_1, \omega_2)$. We say that the pair (ω_1, ω_2) defines a complex structure on T^2. There are many pairs (ω_1, ω_2) which give the same complex structure on T^2.

When do pairs (ω_1, ω_2) and (ω_1', ω_2') $(\text{Im}(\omega_2/\omega_1) > 0, \text{Im}(\omega_2'/\omega_1') > 0)$ define the same complex structure? We first note that two lattices $L(\omega_1, \omega_2)$ and $L(\omega_1', \omega_2')$ coincide if and only if there exists a matrix[1]

$$\begin{pmatrix} a & b \\ c & d \end{pmatrix} \in \text{PSL}(2, \mathbb{Z}) \equiv \text{SL}(2, \mathbb{Z})/\mathbb{Z}_2$$

such that

$$\begin{pmatrix} \omega_1' \\ \omega_2' \end{pmatrix} = \begin{pmatrix} a & b \\ c & d \end{pmatrix} \begin{pmatrix} \omega_1 \\ \omega_2 \end{pmatrix}. \tag{8.2}$$

This statement is proved as follows.
Suppose

$$\begin{pmatrix} \omega_1' \\ \omega_2' \end{pmatrix} = \begin{pmatrix} a & b \\ c & d \end{pmatrix} \begin{pmatrix} \omega_1 \\ \omega_2 \end{pmatrix} \qquad \text{where} \quad \begin{pmatrix} a & b \\ c & d \end{pmatrix} \in \text{SL}(2, \mathbb{Z}).$$

---

[1] The group $\text{SL}(2, \mathbb{Z})$ has been defined in (2.4). Two matrices A and $-A$ are identified in $\text{PSL}(2, \mathbb{Z})$.

Since $\omega_1', \omega_2' \in L(\omega_1, \omega_2)$, we find $L(\omega_1', \omega_2') \subset L(\omega_1, \omega_2)$. From

$$\begin{pmatrix} \omega_1 \\ \omega_2 \end{pmatrix} = \begin{pmatrix} d & -b \\ -c & a \end{pmatrix} \begin{pmatrix} \omega_1' \\ \omega_2' \end{pmatrix}$$

we also find $L(\omega_1, \omega_2) \subset L(\omega_1', \omega_2')$. Thus, $L(\omega_1, \omega_2) = L(\omega_1', \omega_2')$. Conversely, if $L(\omega_1, \omega_2) = L(\omega_1', \omega_2')$, ω_1' and ω_2' are lattice points of $L(\omega_1, \omega_2)$ and can be written as $\omega_1' = d\omega_1 + c\omega_2$ and $\omega_2' = b\omega_1 + a\omega_2$ where $a, b, c, d \in \mathbb{Z}$. Also ω_1 and ω_2 may be expressed as $\omega_1 = d'\omega_1' + c'\omega_2'$ and $\omega_2 = b'\omega_1' + a'\omega_2'$ where $a', b', c', d' \in \mathbb{Z}$. Then we have

$$\begin{pmatrix} \omega_1 \\ \omega_2 \end{pmatrix} = \begin{pmatrix} a' & b' \\ c' & d' \end{pmatrix} \begin{pmatrix} \omega_1' \\ \omega_2' \end{pmatrix} = \begin{pmatrix} a' & b' \\ c' & d' \end{pmatrix} \begin{pmatrix} a & b \\ c & d \end{pmatrix} \begin{pmatrix} \omega_1 \\ \omega_2 \end{pmatrix}$$

from which we find

$$\begin{pmatrix} a' & b' \\ c' & d' \end{pmatrix} \begin{pmatrix} a & b \\ c & d \end{pmatrix} = \begin{pmatrix} 1 & 0 \\ 0 & 1 \end{pmatrix}.$$

Equating the determinants of both sides, we have $(a'd' - b'c')(ad - bc) = 1$. All the entries being integers, this is possible only when $ad - bc = \pm 1$. Since

$$\operatorname{Im}\left(\frac{\omega_2'}{\omega_1'}\right) = \operatorname{Im}\left(\frac{b\omega_1 + a\omega_2}{d\omega_1 + c\omega_2}\right) = \frac{ad - bc}{|c(\omega_2/\omega_1) + d|^2} \operatorname{Im}\left(\frac{\omega_2'}{\omega_1'}\right) > 0$$

we must have $ad - bc > 0$, that is,

$$\begin{pmatrix} a & b \\ c & d \end{pmatrix} \in \operatorname{SL}(2, \mathbb{Z}).$$

In fact, it is clear that

$$\begin{pmatrix} a & b \\ c & d \end{pmatrix} \in \operatorname{SL}(2, \mathbb{Z})$$

defines the same lattice as

$$-\begin{pmatrix} a & b \\ c & d \end{pmatrix}$$

and we have to identify those matrices of $\operatorname{SL}(2, \mathbb{Z})$ which differ only by their overall signature. Thus, two lattices agree if they are related by $\operatorname{PSL}(2, \mathbb{Z}) \equiv \operatorname{SL}(2, \mathbb{Z})/\mathbb{Z}_2$.

Assume that there exists a one-to-one holomorphic map h of $\mathbb{C}/L(\omega_1, \omega_2)$ onto $\mathbb{C}/L(\tilde{\omega}_1, \tilde{\omega}_2)$ where $\operatorname{Im}(\omega_2/\omega_1) > 0$, $\operatorname{Im}(\tilde{\omega}_2/\tilde{\omega}_1) > 0$. Let $p : \mathbb{C} \to \mathbb{C}/L(\omega_1, \omega_2)$ and $\tilde{p} : \mathbb{C} \to \mathbb{C}/L(\tilde{\omega}_1, \tilde{\omega}_2)$ be the natural projections. For example, p maps a point in $\mathbb{C}$ to an equivalent point in $\mathbb{C}/L(\omega_1, \omega_2)$. Choose the origin 0 and define $h_*(0)$ to be a point such that $\tilde{p} \circ h_*(0) = h \circ p(0)$ (figure 8.2),

$$\begin{array}{ccc} \mathbb{C} & \xrightarrow{\quad h_* \quad} & \mathbb{C} \\ {\scriptstyle p}\downarrow & & \downarrow{\scriptstyle \tilde{p}} \\ \mathbb{C}/L(\omega_1, \omega_2) & \xrightarrow{\quad h \quad} & \mathbb{C}/L(\tilde{\omega}_1, \tilde{\omega}_2). \end{array} \qquad (8.3)$$

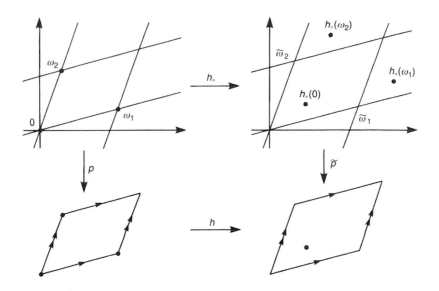

Figure 8.2. A holomorphic bijection $h : \mathbb{C}/L(\omega_1, \omega_2) \to \mathbb{C}/L(\tilde{\omega}_1, \tilde{\omega}_2)$ and the natural projections $p : \mathbb{C} \to \mathbb{C}/L(\omega_1, \omega_2)$, $\tilde{p} : \mathbb{C} \to \mathbb{C}/L(\tilde{\omega}_1, \tilde{\omega}_2)$ define a holomorphic bijection $h_* : \mathbb{C} \to \mathbb{C}$.

Then by analytic continuation from the origin, we obtain a one-to-one holomorphic map h_* of $\mathbb{C}$ onto itself satisfying

$$\tilde{p} \circ h_*(z) = h \circ p(z) \qquad \text{for all } z \in \mathbb{C} \tag{8.4}$$

so that the diagram (8.3) commutes. It is known that a one-to-one holomorphic map of $\mathbb{C}$ onto itself must be of the form $z \to h_*(z) = az + b$, where $a, b \in \mathbb{C}$ and $a \neq 0$. We then have $h_*(\omega_1) - h_*(0) = a\omega_1$ and $h_*(\omega_2) - h_*(0) = a\omega_2$. For h to be well defined as a map of $\mathbb{C}/L(\omega_1, \omega_2)$ onto $\mathbb{C}/L(\tilde{\omega}_1, \tilde{\omega}_2)$, we must have $a\omega_1, a\omega_2 \in L(\tilde{\omega}_1, \tilde{\omega}_2)$, see figure 8.2. By changing the roles of (ω_1, ω_2) and (ω_1', ω_2'), we have $\tilde{a}\tilde{\omega}_1, \tilde{a}\tilde{\omega}_1 \in L(\omega_1, \omega_2)$ where $\tilde{a} \neq 0$ is a complex number. Hence, we conclude that if $\mathbb{C}/L(\omega_1, \omega_2)$, $\mathbb{C}/L(\tilde{\omega}_1, \tilde{\omega}_2)$ have the same complex structure, there must be a matrix $M \in \mathrm{SL}(2, \mathbb{Z})$ and a complex number λ $(=\tilde{a}^{-1})$ such that

$$\begin{pmatrix} \tilde{\omega}_1 \\ \tilde{\omega}_2 \end{pmatrix} = \lambda M \begin{pmatrix} \omega_1 \\ \omega_2 \end{pmatrix}. \tag{8.5}$$

Conversely, we verify that (ω_1, ω_2) and (ω_1', ω_2') related by (8.5) define the same complex structure. In fact,

$$\begin{pmatrix} \omega_1 \\ \omega_2 \end{pmatrix} \qquad \text{and} \qquad M \begin{pmatrix} \omega_1 \\ \omega_2 \end{pmatrix}$$

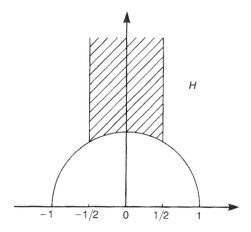

Figure 8.3. The quotient space $H/\mathrm{PSL}(2, \mathbb{Z})$.

define the same lattice (modulo translation) and we may take $h_* : \mathbb{C} \to \mathbb{C}$ to be $z \mapsto z + b$. $L(\omega_1, \omega_2)$ and $L(\lambda\omega_1, \lambda\omega_2)$ also define the same complex structure. We take, in this case, $h_* : z \mapsto \lambda z + b$.

We have shown that the complex structure on T^2 is defined by a pair of complex numbers (ω_1, ω_2) modulo a constant factor and $\mathrm{PSL}(2, \mathbb{Z})$. To get rid of the constant factor, we introduce the modular parameter $\tau \equiv \omega_2/\omega_1 \in H \equiv \{z \in \mathbb{C}| \,\mathrm{Im}\, z > 0\}$, to specify the complex structure of T^2. Without loss of generality, we take 1 and τ to be the generators of a lattice. Note, however, that not all of $\tau \in H$ are independent modular parameters. As was shown previously, τ and $\tau' = (a\tau + b)/(c\tau + d)$ define the same complex structure if

$$\begin{pmatrix} a & b \\ c & d \end{pmatrix} \in \mathrm{PSL}(2, \mathbb{Z}).$$

The quotient space $H/\mathrm{PSL}(2, \mathbb{Z})$ is shown in figure 8.3, the derivation of which can be found in Koblitz (1984) p 100, and Gunning (1962) p 4.

The change $\tau \to \tau'$ is called the **modular transformation** and is generated by $\tau \to \tau + 1$ and $\tau \to -1/\tau$. The transformation $\tau \to \tau + 1$ generates a **Dehn twist** along the meridian m as follows (figure 8.4(a)). (i) First, cut a torus along m. (ii) Then take one of the lips of the cut and rotate it by 2π with the other lip kept fixed. (iii) Then glue the lips together again. The other transformation $\tau \to -1/\tau$ corresponds to changing the roles of the longitude l and the meridian m (figure 8.4(b)).

Example 8.3. The **complex projective space** $\mathbb{C}P^n$ is defined similarly to $\mathbb{R}P^n$; see example 5.4. The *n*tuple $z = (z^0, \ldots, z^n) \in \mathbb{C}^{n+1}$ determines a complex line through the origin provided that $z \neq 0$. Define an equivalence relation

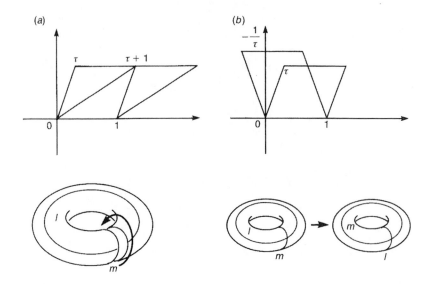

Figure 8.4. (*a*) Dehn twists generate modular transformations. (*b*) $\tau \to -1/\tau$ changes the roles of l and m.

by $z \sim w$ if there exists a complex number $a \neq 0$ such that $w = az$. Then $\mathbb{C}P^n \equiv (\mathbb{C}^{n+1} - \{0\})/ \sim$. The $(n + 1)$ numbers $z^0, z^1, \ldots, z^n$ are called the **homogeneous coordinates**, which is denoted by $[z^0, z^1, \ldots, z^n]$ where $(z^0, \ldots, z^n)$ is identified with $(\lambda z^0, \ldots, \lambda z^n)$ $(\lambda \neq 0)$. A chart U_μ is a subset of $\mathbb{C}^{n+1} - \{0\}$ such that $z^\mu \neq 0$. In a chart U_μ, the **inhomogeneous coordinates** are defined by $\xi^\nu_{(\mu)} = z^\nu/z^\mu$ $(\nu \neq \mu)$. In $U_\mu \cap U_\nu \neq \emptyset$, the coordinate transformation $\psi_{\mu\nu} : \mathbb{C}^n \to \mathbb{C}^n$ is

$$\xi^\lambda_{(\nu)} \mapsto \xi^\lambda_{(\mu)} = \frac{z^\nu}{z^\mu}\xi^\lambda_{(\nu)}. \tag{8.6}$$

Accordingly, $\psi_{\mu\nu}$ is a multiplication by z^ν/z^μ, which is, of course, holomorphic.

Example 8.4. The **complex Grassmann manifolds** $G_{k,n}(\mathbb{C})$ are defined similarly to the real Grassmann manifolds; see example 5.5. $G_{k,n}(\mathbb{C})$ is the set of complex k-dimensional subspaces of $\mathbb{C}^n$. Note that $\mathbb{C}P^n = G_{1,n+1}(\mathbb{C})$.

Let $M_{k,n}(\mathbb{C})$ be the set of $k \times n$ matrices of rank k $(k \leq n)$. Take $A, B \in M_{k,n}(\mathbb{C})$ and define an equivalence relation by $A \sim B$ if there exists $g \in \mathrm{GL}(k, \mathbb{C})$ such that $B = gA$. We identify $G_{k,n}(\mathbb{C})$ with $M_{k,n}(\mathbb{C})/\mathrm{GL}(k, \mathbb{C})$. Let $\{A_1, \ldots, A_l\}$ be the collection of all the $k \times k$ minors of $A \in M_{k,n}(\mathbb{C})$. We define the chart U_α to be a subset of $G_{k,n}(\mathbb{C})$ such that $\det A_\alpha \neq 0$. The $k(n - k)$ coordinates on U_α are given by the non-trivial entries of the matrix $A_\alpha^{-1}A$. See example 5.5 for details.

Example 8.5. The common zeros of a set of homogeneous polynomials are a compact submanifold of $\mathbb{C}P^n$ called an **algebraic variety**. For example, let $P(z^0, \ldots, z^n)$ be a homogeneous polynomial of degree d. If $a \neq 0$ is a complex number, P satisfies

$$P(az^0, \ldots, az^n) = a^d P(z^0, \ldots, z^n).$$

This shows that the zeros of P are defined on $\mathbb{C}P^n$; if $P(z^0, \ldots, z^n) = 0$ then $P([z^0, \ldots, z^n]) = 0$. For definiteness, consider

$$P(z^0, z^1, z^2) = (z^0)^2 + (z^1)^2 + (z^2)^2$$

and define N by

$$N = \{[z^0, z^1, z^2] \in \mathbb{C}P^2 \,|\, P(z^0, z^1, z^2) = 0\}. \tag{8.7}$$

We define U_μ as in example 8.3. In $N \cap U_0$, we have

$$[\xi^1_{(0)}]^2 + [\xi^2_{(0)}]^2 + 1 = 0$$

where $\xi^\mu_{(0)} = z^\mu/z^0$ (note that $z^0 \neq 0$). Consider a holomorphic change of coordinates $(\xi^1_{(0)}, \xi^2_{(0)}) \mapsto (\eta^1 = \xi^1_{(0)}, \eta^2 = [\xi^1_{(0)}]^2 + [\xi^2_{(0)}]^2 + 1)$. Note that $\partial(\eta^1, \eta^2)/\partial(\xi^1_{(0)}, \xi^2_{(0)}) \neq 0$ unless $\xi^2_{(0)} = z^2 = 0$. Then $N \cap U_0 \cap U_2 = \{(\eta^1, \eta^2) \in \mathbb{C}^2 \,|\, \eta^2 = 0\}$ is clearly a one-dimensional submanifold of $\mathbb{C}^2$. If $\xi^2_{(0)} = z^2 = 0$, we have $(\xi^1_{(0)}, \xi^2_{(0)}) \mapsto (\zeta^1 = [\xi^1_{(0)}]^2 + [\xi^2_{(0)}]^2 + 1, \zeta^2 = \xi^2_{(0)})$ for which the Jacobian does not vanish unless $\xi^1_{(0)} = z^1 = 0$. Then $N \cap U_0 \cap U_1 = \{(\zeta^1, \zeta^2) \in \mathbb{C}^2 \,|\, \zeta^1 = 0\}$ is a one-dimensional submanifold of $\mathbb{C}^2$. On $N \cap U_0 \cap U_1 \cap U_2$, the coordinate change $\eta^1 \mapsto \zeta^2$ is a multiplication by z^2/z^1 and is, hence, holomorphic. In this way, we may define a one-dimensional compact submanifold N of $\mathbb{C}P^2$.

A complex manifold is a differentiable manifold. For example, $\mathbb{C}^m$ is regarded as $\mathbb{R}^{2m}$ by the identification $z^\mu = x^\mu + iy^\mu$, $x^\mu, y^\mu \in \mathbb{R}$. Similarly, any chart U of a complex manifold has coordinates $(z^1, \ldots, z^m)$ which may be understood as real coordinates $(x^1, y^1, \ldots, x^m, y^m)$. The analytic property of the coordinate transformation functions ensures that they are differentiable when the manifold is regarded as a $2m$-dimensional differentiable manifold.

8.2 Calculus on complex manifolds

8.2.1 Holomorphic maps

Let $f : M \to N$, M and N being complex manifolds with $\dim_{\mathbb{C}} M = m$ and $\dim_{\mathbb{C}} N = n$. Take a point p in a chart (U, φ) of M. Let (V, ψ) be a chart of N such that $f(p) \in V$. If we write $\{z^\mu\} = \varphi(p)$ and $\{w^\nu\} = \psi(f(p))$, we have a map $\psi \circ f \circ \varphi^{-1} : \mathbb{C}^m \to \mathbb{C}^n$. If each function w^ν $(1 \leq \nu \leq n)$ is a holomorphic function of z^μ, f is called a **holomorphic map**. This definition is independent of the special coordinates chosen. In fact, let (U', φ') be another chart such that $U \cap U' \neq \emptyset$ and $z'^\mu = x'^\lambda + iy'^\lambda$ be the coordinates. Take a point $p \in U \cap U'$. If $w^\nu = u^\nu + iv^\nu$ is a holomorphic function with respect to z, then

$$\frac{\partial u^\nu}{\partial x'^\lambda} = \frac{\partial u^\nu}{\partial x^\mu}\frac{\partial x^\mu}{\partial x'^\lambda} + \frac{\partial u^\nu}{\partial y^\mu}\frac{\partial y^\mu}{\partial y'^\lambda} = \frac{\partial v^\nu}{\partial y^\mu}\frac{\partial y^\mu}{\partial y'^\lambda} + \frac{\partial v^\nu}{\partial x^\mu}\frac{\partial x^\mu}{\partial y'^\lambda} = \frac{\partial v^\nu}{\partial y'^\lambda}.$$

We also find $\partial u^\nu/\partial y'^\lambda = -\partial v^\nu/\partial x'^\lambda$. Thus, w^ν is holomorphic with respect to z' too. It can be shown that the holomorphic property is also independent of the choice of chart in N.

Let M and N be complex manifolds. We say M is **biholomorphic** to N if there exists a diffeomorphism $f : M \to N$ which is also holomorphic (then $f^{-1} : N \to M$ is automatically holomorphic). The map f is called a **biholomorphism**.

A **holomorphic function** is a holomorphic map $f : M \to \mathbb{C}$. There is a striking theorem; any holomorphic function on a *compact* complex manifold is *constant*. This is a generalization of the maximum principle of elementary complex analysis, see Wells (1980). The set of holomorphic functions on M is denoted by $\mathcal{O}(M)$. Similarly, $\mathcal{O}(U)$ is the set of holomorphic functions on $U \subset M$.

8.2.2   Complexifications

Let M be a differentiable manifold with $\dim_\mathbb{R} M = m$. If $f : M \to \mathbb{C}$ is decomposed as $f = g + ih$ where $g, h \in \mathcal{F}(M)$, then f is a complex-valued smooth function. The set of complex-valued smooth functions on M is called the **complexification** of $\mathcal{F}(M)$, denoted by $\mathcal{F}(M)^\mathbb{C}$. A complexified function does not satisfy the Cauchy–Riemann relation in general. For $f = g + ih \in \mathcal{F}(M)^\mathbb{C}$, the complex conjugate of f is $\overline{f} \equiv g - ih$. f is real if and only if $f = \overline{f}$.

Before we consider the complexification of T_pM, we define the complexification $V^\mathbb{C}$ of a general vector space V with $\dim_\mathbb{R} V = m$. An element of $V^\mathbb{C}$ takes the form $X + iY$ where $X, Y \in V$. The vector space $V^\mathbb{C}$ becomes a complex vector space of complex dimension m if the addition and the scalar multiplication by a complex number $a + ib$ are defined by

$$(X_1 + iY_1) + (X_2 + iY_2) = (X_1 + X_2) + i(Y_1 + Y_2)$$

$$(a + ib)(X + iY) = (aX - bY) + i(bX + aY)$$

V is a vector subspace of $V^\mathbb{C}$ since $X \in V$ and $X + i0 \in V^\mathbb{C}$ may be identified. Vectors in V are said to be **real**. The complex conjugate of $Z = X + iY$ is $\overline{Z} = X - iY$. A vector Z is real if $Z = \overline{Z}$.

A linear operator A on V is *extended* to act on $V^\mathbb{C}$ as

$$A(X + iY) = A(X) + iA(Y). \tag{8.8}$$

If $A \to \mathbb{R}$ is a linear function ($A \in V^*$), its extension is a complex-valued linear function on $V^{\mathbb{C}}$, $A : V^{\mathbb{C}} \to \mathbb{C}$. In general, any tensor defined on V and V^* is extended so that it is defined on $V^{\mathbb{C}}$ and $(V^*)^{\mathbb{C}}$. An extended tensor is complexified as $t = t_1 + it_2$, where t_1 and t_2 are tensors of the same type. The conjugate of t is $\bar{t} \equiv t_1 - it_2$. If $t = \bar{t}$, the tensor is said to be real. For example $A : V^{\mathbb{C}} \to \mathbb{C}$ is real if $\overline{A(X + iY)} = A(X - iY)$.

Let $\{e_k\}$ be a basis of V. If the basis vectors are regarded as complex vectors, the *same* basis $\{e_k\}$ becomes a basis of $V^{\mathbb{C}}$. To see this, let $X = X^k e_k$, $Y = Y^k e_k \in V$. Then $Z = X + iY$ is *uniquely* expressed as $(X^k + iY^k)e_k$. We find $\dim_{\mathbb{R}} V = \dim_{\mathbb{C}} V^{\mathbb{C}}$.

Now we are ready to complexify the tangent space $T_p M$. If V is replaced by $T_p M$, we have the complexification $T_p M^{\mathbb{C}}$ of $T_p M$, whose element is expressed as $Z = X + iY$ ($X, Y \in T_p M$). The vector Z acts on a function $f = f_1 + if_2 \in \mathcal{F}(M)^{\mathbb{C}}$ as

$$
\begin{aligned}
Z[f] &= X[f_1 + if_2] + iY[f_1 + if_2] \\
&= X[f_1] - Y[f_2] + i\{X[f_2] + Y[f_1]\}.
\end{aligned} \tag{8.9}
$$

The dual vector space $T_p^* M$ is complexified if $\omega, \eta \in T_p^* M$ are combined as $\zeta = \omega + i\eta$. The set of complexified dual vectors is denoted by $(T_p^* M)^{\mathbb{C}}$. Any tensor t is extended so that it is defined on $T_p M^{\mathbb{C}}$ and $(T_p^* M)^{\mathbb{C}}$ and then complexified.

Exercise 8.1. Show that $(T_p^* M)^{\mathbb{C}} = (T_p M^{\mathbb{C}})^*$. From now on, we denote the complexified dual vector space simply by $T_p^* M^{\mathbb{C}}$.

Given smooth vector fields $X, Y \in \mathfrak{X}(M)$, we define a complex vector field $Z = X + iY$. Clearly $Z|_p \in T_p M^{\mathbb{C}}$. The set of complex vector fields is the complexification of $\mathfrak{X}(M)$ and is denoted by $\mathfrak{X}(M)^{\mathbb{C}}$. The conjugate vector field of $Z = X + iY$ is $\bar{Z} = X - iY$. $Z = \bar{Z}$ if $Z \in \mathfrak{X}(M)$, hence $\mathfrak{X}(M)^{\mathbb{C}} \supset \mathfrak{X}(M)$. The Lie bracket of $Z = X + iY$, $W = U + iV \in \mathfrak{X}(M)^{\mathbb{C}}$ is

$$
[X + iY, U + iV] = \{[X, U] - [Y, V]\} + i\{[X, V] + [Y, U]\}. \tag{8.10}
$$

The complexification of a tensor field of type (p, q) is defined in an obvious manner. If $\omega, \eta \in \Omega^1(M)$, $\xi \equiv \omega + i\eta \in \Omega^1(M)^{\mathbb{C}}$ is a complexified one-form.

8.2.3 Almost complex structure

Since a complex manifold is also a differentiable manifold, we may use the framework developed in chapter 5. We then put appropriate *constraints* on the results. Let us look at the tangent space of a complex manifold M with $\dim_{\mathbb{C}} M = m$. The tangent space $T_p M$ is spanned by $2m$ vectors

$$
\left\{ \frac{\partial}{\partial x^1}, \ldots, \frac{\partial}{\partial x^m}; \frac{\partial}{\partial y^1}, \ldots, \frac{\partial}{\partial y^m} \right\} \tag{8.11}
$$

where $z^\mu = x^\mu + iy^\mu$ are the coordinates of p in a chart (U, φ). With the same coordinates, $T_p^* M$ is spanned by

$$\left\{ dx^1, \ldots, dx^m; dy^1, \ldots, dy^m \right\}. \tag{8.12}$$

Let us define $2m$ vectors

$$\frac{\partial}{\partial z^\mu} \equiv \frac{1}{2} \left\{ \frac{\partial}{\partial x^\mu} - i \frac{\partial}{\partial y^\mu} \right\} \tag{8.13a}$$

$$\frac{\partial}{\partial \bar{z}^\mu} \equiv \frac{1}{2} \left\{ \frac{\partial}{\partial x^\mu} + i \frac{\partial}{\partial y^\mu} \right\} \tag{8.13b}$$

where $1 \le \mu \le m$. Clearly they form a basis of the $2m$-dimensional (complex) vector space $T_p M^{\mathbb{C}}$. Note that $\overline{\partial/\partial z^\mu} = \partial/\partial \bar{z}^\mu$. Correspondingly, $2m$ one-forms

$$dz^\mu \equiv dx^\mu + i\, dy^\mu \qquad d\bar{z}^\mu \equiv dx^\mu - i\, dy^\mu \tag{8.14}$$

form the basis of $T_p^* M^{\mathbb{C}}$. They are dual to (8.13),

$$\langle dz^\mu, \partial/\partial \bar{z}^\nu \rangle = \langle d\bar{z}^\mu, \partial/\partial z^\nu \rangle = 0 \tag{8.15a}$$

$$\langle dz^\mu, \partial/\partial z^\nu \rangle = \langle d\bar{z}^\mu, \partial/\partial \bar{z}^\nu \rangle = \delta^\mu{}_\nu. \tag{8.15b}$$

Let M be a complex manifold and define a linear map $J_p : T_p M \to T_p M$ by

$$J_p \left(\frac{\partial}{\partial x^\mu} \right) = \frac{\partial}{\partial y^\mu} \qquad J_p \left(\frac{\partial}{\partial y^\mu} \right) = -\frac{\partial}{\partial x^\mu} \tag{8.16}$$

J_p is a *real* tensor of type $(1, 1)$. Note that

$$J_p^2 = -\mathrm{id}_{T_p M}. \tag{8.17}$$

Roughly speaking, J_p corresponds to the multiplication by $\pm i$. The action of J_p is independent of the chart. In fact, let (U, φ) and (V, ψ) be overlapping charts with $\varphi(p) = z^\mu = x^\mu + iy^\mu$ and $\psi(p) = w^\mu = u^\mu + iv^\mu$. On $U \cap V$, the functions $z^\mu = z^\mu(w)$ satisfy the Cauchy–Riemann relations. Then we find

$$J_p \left(\frac{\partial}{\partial u^\mu} \right) = J_p \left(\frac{\partial x^\nu}{\partial u^\mu} \frac{\partial}{\partial x^\nu} + \frac{\partial y^\nu}{\partial u^\mu} \frac{\partial}{\partial y^\nu} \right) = \frac{\partial y^\nu}{\partial v^\mu} \frac{\partial}{\partial y^\nu} + \frac{\partial x^\nu}{\partial v^\mu} \frac{\partial}{\partial x^\nu} = \frac{\partial}{\partial v^\mu}.$$

We also find that $J_p \partial/\partial v^\mu = -\partial/\partial u^\mu$. Accordingly, J_p takes the form

$$J_p = \begin{pmatrix} 0 & -I_m \\ I_m & 0 \end{pmatrix} \tag{8.18}$$

with respect to the basis (8.11), where I_m is the $m \times m$ unit matrix. Since all the components of J_p are constant at any point, we may define a smooth tensor field J whose components at p are (8.18). The tensor field J is called the **almost**

complex structure of a complex manifold M. Note that any $2m$-dimensional manifold *locally* admits a tensor field J which squares to $-I_{2m}$. However, J may be patched across charts and defined *globally* only on a complex manifold. The tensor J completely specifies the complex structure.

The almost complex structure J_p is extended so that it may be defined on $T_pM^{\mathbb{C}}$,

$$J_p(X + iY) \equiv J_pX + iJ_pY. \qquad (8.19)$$

It follows from (8.16) that

$$J_p\partial/\partial z^{\mu} = i\partial/\partial z^{\mu} \qquad J_p\partial/\partial\overline{z}^{\mu} = -i\partial/\partial\overline{z}^{\mu}. \qquad (8.20)$$

Thus, we have an expression for J_p in (anti-)holomorphic bases,

$$J_p = i\,dz^{\mu} \otimes \frac{\partial}{\partial z^{\mu}} - i\,d\overline{z}^{\mu} \otimes \frac{\partial}{\partial\overline{z}^{\mu}} \qquad (8.21)$$

whose components are given by

$$J_p = \begin{pmatrix} iI_m & 0 \\ 0 & -iI_m \end{pmatrix}. \qquad (8.22)$$

Let $Z \in T_pM^{\mathbb{C}}$ be a vector of the form $Z = Z^{\mu}\partial/\partial z^{\mu}$. Then Z is an eigenvector of J_p; $J_pZ = iZ$. Similarly, $Z = Z^{\mu}\partial/\partial\overline{z}^{\mu}$ satisfies $J_pZ = -iZ$. In this way $T_pM^{\mathbb{C}}$ of a complex manifold is separated into two *disjoint* vector spaces,

$$T_pM^{\mathbb{C}} = T_pM^+ \oplus T_pM^- \qquad (8.23)$$

where

$$T_pM^{\pm} = \{Z \in T_pM^{\mathbb{C}}|J_pZ = \pm iZ\}. \qquad (8.24)$$

We define the projection operators $\mathcal{P}^{\pm} : T_pM^{\mathbb{C}} \to T_pM^{\pm}$ by

$$\mathcal{P}^{\pm} \equiv \tfrac{1}{2}(I_{2m} \mp iJ_p). \qquad (8.25)$$

In fact, $J_p\mathcal{P}^{\pm}Z = \tfrac{1}{2}(J_p \mp iJ_p^2)Z = \pm i\mathcal{P}^{\pm}Z$ for any $Z \in T_pM^{\mathbb{C}}$. Hence,

$$Z^{\pm} \equiv \mathcal{P}^{\pm}Z \in T_pM^{\pm}. \qquad (8.26)$$

Now $Z \in T_pM^{\mathbb{C}}$ is uniquely decomposed as $Z = Z^+ + Z^-$ ($Z^{\pm} \in T_pM^{\pm}$). T_pM^+ is spanned by $\{\partial/\partial z^{\mu}\}$ and T_pM^- by $\{\partial/\partial\overline{z}^{\mu}\}$. $Z \in T_pM^+$ is called a **holomorphic vector** while $Z \in T_pM^-$ is called an **anti-holomorphic vector**. We readily verify that

$$T_pM^- = \overline{T_pM^+} = \{\overline{Z}|Z \in T_pM^+\}. \qquad (8.27)$$

Note that

$$\dim_{\mathbb{C}} T_pM^+ = \dim_{\mathbb{C}} T_pM^- = \tfrac{1}{2}\dim_{\mathbb{C}} T_pM^{\mathbb{C}} = \tfrac{1}{2}\dim_{\mathbb{C}} M.$$

Exercise 8.2. Let (U, φ) and (V, ψ) be overlapping charts on a complex manifold M and let $z^\mu = \varphi(p)$ and $w^\mu = \psi(p)$. Verify that $X = X^\mu \partial/\partial z^\mu$, expressed in the coordinates w^μ, contains a holomorphic basis $\{\partial/\partial w^\mu\}$ only. Thus, the separation of $T_p M^{\mathbb{C}}$ into $T_p M^{\pm}$ is independent of charts (note that J is defined independently of charts).

Given a complexified vector field $Z \in \mathfrak{X}(M)^{\mathbb{C}}$, we obtain a new vector field $JZ \in \mathfrak{X}(M)^{\mathbb{C}}$ defined at each point of M by $JZ|_p = J_p \cdot Z|_p$. The vector field Z is naturally separated as

$$Z = Z^+ + Z^- \qquad Z^\pm = \mathcal{P}^\pm Z \qquad (8.28)$$

where $Z^\pm = \mathcal{P}^\pm Z$. The vector field Z^+ (Z^-) is called a **holomorphic (anti-holomorphic) vector field**. Accordingly, once J is given, $\mathfrak{X}(M)^{\mathbb{C}}$ is decomposed uniquely as

$$\mathfrak{X}(M)^{\mathbb{C}} = \mathfrak{X}(M)^+ \oplus \mathfrak{X}(M)^-. \qquad (8.29)$$

$Z = Z^+ + Z^- \in \mathfrak{X}(M)^{\mathbb{C}}$ is real if and only if $Z^+ = \overline{Z^-}$.

Exercise 8.3. Let $X, Y \in \mathfrak{X}(M)^+$. Show that $[X, Y] \in \mathfrak{X}(M)^+$. [If $X, Y \in \mathfrak{X}(M)^-$, then $[X, Y] \in \mathfrak{X}(M)^-$.]

8.3 Complex differential forms

On a complex manifold, we define complex differential forms by which we will discuss such topological properties as cohomology groups.

8.3.1 Complexification of real differential forms

Let M be a differentiable manifold with $\dim_{\mathbb{R}} M = m$. Take two q-forms $\omega, \eta \in \Omega_p^q(M)$ at p and define a **complex q-form** $\zeta = \omega + i\eta$. We denote the vector space of complex q-forms at p by $\Omega_p^q(M)^{\mathbb{C}}$. Clearly $\Omega_p^q(M) \subset \Omega_p^q(M)^{\mathbb{C}}$. The conjugate of ζ is $\overline{\zeta} = \omega - i\eta$. A complex q-form ζ is real if $\zeta = \overline{\zeta}$.

Exercise 8.4. Let $\omega \in \Omega_p^q(M)^{\mathbb{C}}$. Show that

$$\overline{\omega}(V_1, \ldots, V_q) = \overline{\omega(\overline{V}_1, \ldots, \overline{V}_q)} \qquad V_i \in T_p M^{\mathbb{C}}. \qquad (8.30)$$

Show also that $\overline{\omega + \eta} = \overline{\omega} + \overline{\eta}$, $\overline{\lambda \omega} = \overline{\lambda} \overline{\omega}$ and $\overline{\overline{\omega}} = \omega$, where $\omega, \eta \in \Omega_p^q(M)^{\mathbb{C}}$ and $\lambda \in \mathbb{C}$.

A complex q-form α defined on a differentiable manifold M is a smooth assignment of an element of $\Omega_p^q(M)^{\mathbb{C}}$. The set of complex q-forms is denoted by $\Omega^q(M)^{\mathbb{C}}$. A complex q-form ζ is uniquely decomposed as $\zeta = \omega + i\eta$, where $\omega, \eta \in \Omega^q(M)$.

The exterior product of $\zeta = \omega + i\eta$ and $\xi = \varphi + i\psi$ is defined by

$$\zeta \wedge \xi = (\omega + i\eta) \wedge (\varphi + i\psi)$$
$$= (\omega \wedge \varphi - \eta \wedge \psi) + i(\omega \wedge \psi + \eta \wedge \varphi). \qquad (8.31)$$

The exterior derivative d acts on $\zeta = \omega + i\eta$ as

$$d\zeta = d\omega + i\, d\eta. \qquad (8.32)$$

d is a real operator: $\overline{d\zeta} = d\omega - i\, d\eta = d\overline{\zeta}$.

Exercise 8.5. Let $\omega \in \Omega^q(M)^{\mathbb{C}}$ and $\xi \in \Omega^r(M)^{\mathbb{C}}$. Show that

$$\omega \wedge \xi = (-1)^{qr}\xi \wedge \omega \qquad (8.33)$$
$$d(\omega \wedge \xi) = d\omega \wedge \xi + (-1)^q \omega \wedge d\xi. \qquad (8.34)$$

8.3.2 Differential forms on complex manifolds

Now we restrict ourselves to complex manifolds in which we have the decompositions $T_p M^{\mathbb{C}} = T_p M^+ \oplus T_p M^-$ and $\mathfrak{X}(M)^{\mathbb{C}} = \mathfrak{X}(M)^+ \oplus \mathfrak{X}(M)^-$.

Definition 8.2. Let M be a complex manifold with $\dim_{\mathbb{C}} M = m$. Let $\omega \in \Omega_p^q(M)^{\mathbb{C}}$ ($q \le 2m$) and r, s be positive integers such that $r + s = q$. Let $V_i \in T_p M^{\mathbb{C}}$ ($1 \le i \le q$) be vectors in either $T_p M^+$ or $T_p M^-$. If $\omega(V_1, \ldots, V_q) = 0$ unless r of the V_i are in $T_p M^+$ and s of the V_i are in $T_p M^-$, ω is said to be of **bidegree** (r, s) or simply an (r, s)-form. The set of (r, s)-forms at p is denoted by $\Omega_p^{r,s}(M)$. If an (r, s)-form is assigned smoothly at each point of M, we have an (r, s)-form defined over M. The set of (r, s)-forms over M is denoted by $\Omega^{r,s}(M)$.

Take a chart (U, φ) with the complex coordinates $\varphi(p) = z^\mu$. We take the bases (8.13) for the tangent spaces $T_p M^\pm$. The dual bases are given by (8.14). Note that dz^μ is of bidegree $(1, 0)$ since $\langle dz^\mu, \partial/\partial\bar{z}^\nu \rangle = 0$ and $d\bar{z}^\mu$ is of bidegree $(0, 1)$. With these bases, a form ω of bidegree (r, s) is written as

$$\omega = \frac{1}{r!\, s!}\omega_{\mu_1 \ldots \mu_r \nu_1 \ldots \nu_s}\, dz^{\mu_1} \wedge \ldots \wedge dz^{\mu_r} \wedge d\bar{z}^{\nu_1} \wedge \ldots \wedge d\bar{z}^{\nu_s}. \qquad (8.35)$$

The set $\{dz^{\mu_1} \wedge \ldots \wedge dz^{\mu_r} \wedge d\bar{z}^{\nu_1} \wedge \ldots \wedge d\bar{z}^{\nu_s}\}$ is the basis of $\Omega_p^{r,s}(M)$. The components are totally anti-symmetric in the μ and ν separately. Let z^μ and w^μ be two overlapping coordinates. The reader should verify that an (r, s)-form in the z^μ coordinate system is also an (r, s)-form in the w^ν system.

Proposition 8.1. Let M be a complex manifold of $\dim_{\mathbb{C}} M = m$ and ω and ξ be complex differential forms on M.

(a) If $\omega \in \Omega^{q,r}(M)$ then $\bar{\omega} \in \Omega^{r,q}(M)$.
(b) If $\omega \in \Omega^{q,r}(M)$ and $\xi \in \Omega^{q',r'}(M)$, then $\omega \wedge \xi \in \Omega^{q+q',r+r'}(M)$.

(c) A complex q-form ω is uniquely written as

$$\omega = \sum_{r+s=q} \omega^{(r,s)} \tag{8.36a}$$

where $\omega^{(r,s)} \in \Omega^{r,s}(M)$. Thus, we have the decomposition

$$\Omega^q(M)^{\mathbb{C}} = \bigoplus_{r+s=q} \Omega^{r,s}(M). \tag{8.36b}$$

The proof is easy and is left to the reader. Now any q-form ω is decomposed as

$$\begin{aligned}
\omega &= \sum_{r+s=q} \omega^{(r,s)} \\
&= \sum_{r+s=q} \frac{1}{r!s!} \omega_{\mu_1 \dots \mu_r \bar{\nu}_1 \dots \bar{\nu}_s}\, dz^{\mu_1} \wedge \dots \wedge dz^{\mu_r} \wedge d\bar{z}^{\nu_1} \wedge \dots \wedge d\bar{z}^{\nu_s}
\end{aligned}$$

$$\tag{8.37}$$

where

$$\omega_{\mu_1 \dots \mu_r \bar{\nu}_1 \dots \bar{\nu}_s} = \omega \left(\frac{\partial}{\partial z^{\mu_1}}, \dots, \frac{\partial}{\partial z^{\mu_r}}, \frac{\partial}{\partial \bar{z}^{\nu_1}}, \dots, \frac{\partial}{\partial \bar{z}^{\nu_s}} \right). \tag{8.38}$$

Exercise 8.6. Let $\dim_{\mathbb{C}} M = m$. Verify that

$$\dim_{\mathbb{R}} \Omega_p^{r,s}(M) = \begin{cases} \dbinom{m}{r}\dbinom{m}{s} & \text{if } 0 \le r, s \le m \\ 0 & \text{otherwise.} \end{cases}$$

Show also that $\dim_{\mathbb{R}} \Omega_p^q(M)^{\mathbb{C}} = \sum_{r+s=q} \dim_{\mathbb{R}} \Omega_p^{r,s}(M) = \binom{2m}{q}$.

8.3.3 Dolbeault operators

Let us compute the exterior derivative of an (r, s)-form ω. From (8.35), we find

$$\begin{aligned}
d\omega = \frac{1}{r!s!} \left(\frac{\partial}{\partial z^{\lambda}} \omega_{\mu_1 \dots \mu_r \bar{\nu}_1 \dots \bar{\nu}_s}\, dz^{\lambda} + \frac{\partial}{\partial \bar{z}^{\lambda}} \omega_{\mu_1 \dots \mu_r \bar{\nu}_1 \dots \bar{\nu}_s}\, d\bar{z}^{\lambda} \right) \\
\times dz^{\mu_1} \wedge \dots \wedge dz^{\mu_r} \wedge d\bar{z}^{\nu_1} \wedge \dots \wedge d\bar{z}^{\nu_s}.
\end{aligned} \tag{8.39}$$

$d\omega$ is a mixture of an $(r + 1, s)$-form and an $(r, s + 1)$-form. We separate the action of d according to its destinations,

$$d = \partial + \bar{\partial} \tag{8.40}$$

where $\partial : \Omega^{r,s}(M) \to \Omega^{r+1,s}(M)$ and $\bar{\partial} : \Omega^{r,s}(M) \to \Omega^{r,s+1}(M)$. For example, if $\omega = \omega_{\mu\bar{\nu}}dz^{\mu} \wedge d\bar{z}^{\nu}$, its exterior derivatives are

$$\partial\omega = \frac{\partial\omega_{\mu\bar{\nu}}}{\partial z^{\lambda}}dz^{\lambda} \wedge dz^{\mu} \wedge d\bar{z}^{\nu}$$

$$\bar{\partial}\omega = \frac{\partial\omega_{\mu\bar{\nu}}}{\partial\bar{z}^{\lambda}}d\bar{z}^{\lambda} \wedge dz^{\mu} \wedge d\bar{z}^{\nu} = -\frac{\partial\omega_{\mu\bar{\nu}}}{\partial\bar{z}^{\lambda}}dz^{\mu} \wedge d\bar{z}^{\lambda} \wedge d\bar{z}^{\nu}.$$

The operators ∂ and $\bar{\partial}$ are called the **Dolbeault operators**.

If ω is a general q-form given by (8.37), the actions of ∂ and $\bar{\partial}$ on ω are defined by

$$\partial\omega = \sum_{r+s=q} \partial\omega^{(r,s)} \qquad \bar{\partial}\omega = \sum_{r+s=q} \bar{\partial}\omega^{(r,s)}. \tag{8.41}$$

Theorem 8.1. Let M be a complex manifold and let $\omega \in \Omega^{q}(M)^{\mathbb{C}}$ and $\xi \in \Omega^{p}(M)^{\mathbb{C}}$. Then

$$\partial\partial\omega = (\partial\bar{\partial} + \bar{\partial}\partial)\omega = \bar{\partial}\bar{\partial}\omega = 0 \tag{8.42a}$$

$$\partial\bar{\omega} = \overline{\bar{\partial}\omega}, \ \bar{\partial}\bar{\omega} = \overline{\partial\omega} \tag{8.42b}$$

$$\partial(\omega \wedge \xi) = \partial\omega \wedge \xi + (-1)^{q}\omega \wedge \partial\xi \tag{8.42c}$$

$$\bar{\partial}(\omega \wedge \xi) = \bar{\partial}\omega \wedge \xi + (-1)^{q}\omega \wedge \bar{\partial}\xi. \tag{8.42d}$$

Proof. It is sufficient to prove them when ω is of bidegree (r, s).

(a) Since $d = \partial + \bar{\partial}$, we have

$$0 = d^{2}\omega = (\partial + \bar{\partial})(\partial + \bar{\partial})\omega = \partial\partial\omega + (\partial\bar{\partial} + \bar{\partial}\partial)\omega + \bar{\partial}\bar{\partial}\omega.$$

The three terms of the RHS are of bidegrees $(r + 2, s)$, $(r + 1, s + 1)$ and $(r, s + 2)$ respectively. From proposition 8.1(c), each term must vanish separately.

(b) Since $d\bar{\omega} = \overline{d\omega}$, we have

$$\partial\bar{\omega} + \bar{\partial}\bar{\omega} = d\bar{\omega} = \overline{(\partial + \bar{\partial})\omega} = \overline{\partial\omega} + \overline{\bar{\partial}\omega}.$$

Noting that $\partial\bar{\omega}$ and $\overline{\bar{\partial}\omega}$ are of bidegree $(s + 1, r)$ and $\bar{\partial}\bar{\omega}$ and $\overline{\partial\omega}$ are of $(s, r + 1)$, we conclude that $\partial\bar{\omega} = \overline{\bar{\partial}\omega}$ and $\bar{\partial}\bar{\omega} = \overline{\partial\omega}$.

(c) We assume ω is of bidegree (r, s) and ξ of (r', s'). Equation (8.42c) is proved by separating $d(\omega \wedge \xi) = d\omega \wedge \xi + (-1)^{q}\omega \wedge d\xi$, into forms of bidegrees $(r + r' + 1, s + s')$ and $(r + r', s + s' + 1)$. $\square$

Definition 8.3. Let M be a complex manifold. If $\omega \in \Omega^{r,0}(M)$ satisfies $\bar{\partial}\omega = 0$, the r-form ω is called a **holomorphic r-form**.

Let us look at a holomorphic 0-form $f \in \mathcal{F}(U)^{\mathbb{C}}$ on a chart (U, φ). The condition $\bar{\partial} f = 0$ becomes

$$\frac{\partial f}{\partial \bar{z}^{\lambda}} = 0 \qquad 1 \leq \lambda \leq m = \dim_{\mathbb{C}} M. \tag{8.43}$$

A holomorphic 0-form is just a holomorphic function, $f \in \mathcal{F}(U)^{\mathbb{C}}$. Let $\omega \in \Omega^{r,0}(M)$, where $1 \leq r \leq m = \dim_{\mathbb{C}} M$. On a chart (U, φ), we have

$$\omega = \frac{1}{r!} \omega_{\mu_1 \ldots \mu_r} dz^{\mu_1} \wedge \ldots \wedge dz^{\mu_r}. \tag{8.44}$$

Then $\bar{\partial}\omega = 0$ if and only if

$$\frac{\partial}{\partial \bar{z}^{\lambda}} \omega_{\mu_1 \ldots \mu_r} = 0$$

namely if $\omega_{\mu_1 \ldots \mu_r}$ are holomorphic functions on U.

Let $\dim_{\mathbb{C}} M = m$. The sequence of $\mathbb{C}$-linear maps

$$\Omega^{r,0}(M) \xrightarrow{\bar{\partial}} \Omega^{r,1}(M) \xrightarrow{\bar{\partial}} \cdots$$
$$\cdots \xrightarrow{\bar{\partial}} \Omega^{r,m-1}(M) \xrightarrow{\bar{\partial}} \Omega^{r,m}(M) \tag{8.45}$$

is called the **Dolbeault complex**. Note that $\bar{\partial}^2 = 0$. The set of $\bar{\partial}$-closed (r, s)-forms (those $\omega \in \Omega^{r,s}(M)$ such that $\bar{\partial}\omega = 0$) is called the (r, s)-**cocycle** and is denoted by $Z_{\bar{\partial}}^{r,s}(M)$. The set of $\bar{\partial}$-exact (r, s)-forms (those $\omega \in \Omega^{r,s}(M)$ such that $\omega = \bar{\partial}\eta$ for some $\eta \in \Omega^{r,s-1}(M)$) is called the (r, s)-**coboundary** and is denoted by $B_{\bar{\partial}}^{r,s}(M)$. The complex vector space

$$H_{\bar{\partial}}^{r,s}(M) \equiv Z_{\bar{\partial}}^{r,s}(M)/B_{\bar{\partial}}^{r,s}(M) \tag{8.46}$$

is called the (r, s)th $\bar{\partial}$-**cohomology group**, see section 8.6.

8.4 Hermitian manifolds and Hermitian differential geometry

Let M be a complex manifold with $\dim_{\mathbb{C}} M = m$ and let g be a Riemannian metric of M as a differentiable manifold. Take $Z = X+iY$, $W = U+iV \in T_p M^{\mathbb{C}}$ and extend g so that

$$g_p(Z, W) = g_p(X, U) - g_p(Y, V) + i[g_p(X, V) + g_p(Y, U)]. \tag{8.47}$$

The components of g with respect to the bases (8.13) are

$$g_{\mu\nu}(p) = g_p(\partial/\partial z^{\mu}, \partial/\partial z^{\nu}) \tag{8.48a}$$
$$g_{\mu\bar{\nu}}(p) = g_p(\partial/\partial z^{\mu}, \partial/\partial \bar{z}^{\nu}) \tag{8.48b}$$
$$g_{\bar{\mu}\nu}(p) = g_p(\partial/\partial \bar{z}^{\mu}, \partial/\partial z^{\nu}) \tag{8.48c}$$
$$g_{\bar{\mu}\bar{\nu}}(p) = g_p(\partial/\partial \bar{z}^{\mu}, \partial/\partial \bar{z}^{\nu}). \tag{8.48d}$$

We easily verify that

$$g_{\mu\nu} = g_{\nu\mu}, \qquad g_{\bar{\mu}\bar{\nu}} = g_{\bar{\nu}\bar{\mu}}, \qquad g_{\bar{\mu}\nu} = g_{\nu\bar{\mu}}, \qquad \overline{g_{\mu\bar{\nu}}} = g_{\bar{\mu}\nu}, \qquad \overline{g_{\mu\nu}} = g_{\bar{\mu}\bar{\nu}}. \tag{8.49}$$

8.4.1 The Hermitian metric

If a Riemannian metric g of a complex manifold M satisfies

$$g_p(J_pX, J_pY) = g_p(X, Y) \tag{8.50}$$

at each point $p \in M$ and for any $X, Y \in T_pM$, g is said to be a **Hermitian metric**. The pair (M, g) is called a **Hermitian manifold**. The vector J_pX is orthogonal to X with respect to a Hermitian metric,

$$g_p(J_pX, X) = g_p(J_p^2X, J_pX) = -g_p(J_pX, X) = 0. \tag{8.51}$$

Theorem 8.2. A complex manifold always admits a Hermitian metric.

Proof. Let g be any Riemannian metric of a complex manifold M. Define a new metric $\hat{g}$ by

$$\hat{g}_p(X, Y) \equiv \tfrac{1}{2}[g_p(X, Y) + g_p(J_pX, J_pY)]. \tag{8.52}$$

Clearly $\hat{g}_p(J_pX, J_pY) = \hat{g}_p(X, Y)$. Moreover, $\hat{g}$ is positive definite provided that g is. Hence, $\hat{g}$ is a Hermitian metric on M. $\qquad\square$

Let g be a Hermitian metric on a complex manifold M. From (8.50), we find that

$$g_{\mu\nu} = g\left(\frac{\partial}{\partial z^\mu}, \frac{\partial}{\partial z^\nu}\right) = g\left(J\frac{\partial}{\partial z^\mu}, J\frac{\partial}{\partial z^\nu}\right) = -g\left(\frac{\partial}{\partial z^\mu}, \frac{\partial}{\partial z^\nu}\right) = -g_{\mu\nu}$$

hence $g_{\mu\nu} = 0$. We also find that $g_{\bar{\mu}\bar{\nu}} = 0$. Thus, the Hermitian metric g takes the form

$$g = g_{\mu\bar{\nu}}dz^\mu \otimes d\bar{z}^\nu + g_{\bar{\mu}\nu}d\bar{z}^\mu \otimes dz^\nu. \tag{8.53}$$

[*Remark:* Take $X, Y \in T_pM^+$. Define an inner product h_p in T_pM^+ by

$$h_p(X, Y) \equiv g_p(X, \overline{Y}). \tag{8.54}$$

It is easy to see that h_p is a positive-definite Hermitian form in T_pM^+. In fact,

$$\overline{h(X, Y)} = \overline{g(X, \overline{Y})} = g(\overline{X}, Y) = h(Y, X)$$

and $h(X, X) = g(X, \overline{X}) = g(X_1, X_1) + g(X_2, X_2) \geq 0$ for $X = X_1 + iX_2$. This is why a metric g satisfying (8.50) is called *Hermitian*.]

8.4.2 Kähler form

Let (M, g) be a Hermitian manifold. Define a tensor field Ω whose action on $X, Y \in T_pM$ is

$$\Omega_p(X, Y) = g_p(J_pX, Y) \qquad X, Y \in T_pM. \tag{8.55}$$

Note that Ω is anti-symmetric, $\Omega(X, Y) = g(JX, Y) = g(J^2X, JY) = -g(JY, X) = -\Omega(Y, X)$. Hence, Ω defines a two-form called the **Kähler form** of a Hermitian metric g. Observe that Ω is invariant under the action of J,

$$\Omega(JX, JY) = g(J^2X, JY) = g(J^3X, J^2Y) = \Omega(X, Y). \tag{8.56}$$

If the domain is extended from T_pM to $T_pM^{\mathbb{C}}$, Ω is a two-form of bidegree $(1, 1)$. Indeed, for the metric (8.53), it is found that

$$\Omega\left(\frac{\partial}{\partial z^\mu}, \frac{\partial}{\partial z^\nu}\right) = g\left(J\frac{\partial}{\partial z^\mu}, \frac{\partial}{\partial z^\nu}\right) = ig_{\mu\nu} = 0.$$

We also have

$$\Omega\left(\frac{\partial}{\partial \bar{z}^\mu}, \frac{\partial}{\partial \bar{z}^\nu}\right) = 0, \qquad \Omega\left(\frac{\partial}{\partial z^\mu}, \frac{\partial}{\partial \bar{z}^\nu}\right) = ig_{\mu\bar{\nu}} = -\Omega\left(\frac{\partial}{\partial \bar{z}^\nu}, \frac{\partial}{\partial z^\mu}\right).$$

Thus, the components of Ω are

$$\Omega_{\mu\nu} = \Omega_{\overline{\mu\nu}} = 0 \qquad \Omega_{\mu\bar{\nu}} = -\Omega_{\bar{\nu}\mu} = ig_{\mu\bar{\nu}}. \tag{8.57}$$

We may write

$$\Omega = ig_{\mu\bar{\nu}}\, dz^\mu \otimes d\bar{z}^\nu - ig_{\bar{\nu}\mu}\, d\bar{z}^\nu \otimes dz^\mu = ig_{\mu\bar{\nu}}\, dz^\mu \wedge d\bar{z}^\nu. \tag{8.58}$$

Ω is also written as

$$\Omega = -J_{\mu\bar{\nu}}\, dz^\mu \wedge d\bar{z}^\nu \tag{8.59}$$

where $J_{\mu\bar{\nu}} = g_{\mu\bar{\lambda}}J^{\bar{\lambda}}_{\ \bar{\nu}} = -ig_{\mu\bar{\nu}}$. Ω is a real form;

$$\overline{\Omega} = -i\overline{g_{\mu\bar{\nu}}}\, d\bar{z}^\mu \wedge dz^\nu = ig_{\nu\bar{\mu}}\, dz^\nu \wedge d\bar{z}^\mu = \Omega. \tag{8.60}$$

Making use of the Kähler form, we show that any Hermitian manifold, and hence any complex manifold, is orientable. We first note that we may choose an orthonormal basis $\{\hat{e}_1, J\hat{e}_1, \ldots, \hat{e}_m, J\hat{e}_m\}$. In fact, if $g(\hat{e}_1, \hat{e}_1) = 1$, it follows that $g(J\hat{e}_1, J\hat{e}_1) = g(\hat{e}_1, \hat{e}_1) = 1$ and $g(\hat{e}_1, J\hat{e}_1) = -g(J\hat{e}_1, \hat{e}_1) = 0$. Thus $\hat{e}_1$ and $J\hat{e}_1$ form an orthonormal basis of a two-dimensional subspace. Now take $\hat{e}_2$ which is orthonormal to $\hat{e}_1$ and $J\hat{e}_1$ and form the subspace $\{\hat{e}_2, J\hat{e}_2\}$. Repeating this procedure we obtain an orthonormal basis $\{\hat{e}_1, J\hat{e}_1, \ldots, \hat{e}_m, J\hat{e}_m\}$.

Lemma 8.1. Let Ω be the Kähler form of a Hermitian manifold with $\dim_{\mathbb{C}} M = m$. Then

$$\underbrace{\Omega \wedge \ldots \wedge \Omega}_{m}$$

is a nowhere vanishing $2m$-form.

Proof. For the previous orthonormal basis, we have

$$\Omega(\hat{e}_i, J\hat{e}_j) = g(J\hat{e}_i, J\hat{e}_j) = \delta_{ij} \qquad \Omega(\hat{e}_i, \hat{e}_j) = \Omega(J\hat{e}_i, J\hat{e}_j) = 0.$$

Then it follows that

$$\underbrace{\Omega \wedge \ldots \wedge \Omega}_{m}(\hat{e}_1, J\hat{e}_1, \ldots, \hat{e}_m, J\hat{e}_m)$$

$$= \sum_P \Omega(\hat{e}_{P(1)}, J\hat{e}_{P(1)}) \ldots \Omega(\hat{e}_{P(m)}, J\hat{e}_{P(m)})$$

$$= m!\Omega(\hat{e}_1, J\hat{e}_1) \ldots \Omega(\hat{e}_m, J\hat{e}_m) = m!$$

where P is an element of the permutation group of m objects. This shows that $\Omega \wedge \ldots \wedge \Omega$ cannot vanish at any point. □

Since the *real* $2m$-form $\Omega \wedge \ldots \wedge \Omega$ vanishes nowhere, it serves as a volume element. Thus, we obtain the following theorem.

Theorem 8.3. A complex manifold is orientable.

8.4.3 Covariant derivatives

Let (M, g) be a Hermitian manifold. We define a connection which is compatible with the complex structure. It is natural to assume that a holomorphic vector $V \in T_p M^+$ parallel transported to another point q is, again, a holomorphic vector $\tilde{V}(q) \in T_q M^+$. We show later that the almost complex structure is covariantly conserved under this requirement. Let $\{z^\mu\}$ and $\{z^\mu + \Delta z^\mu\}$ be the coordinates of p and q, respectively, and let $V = V^\mu \partial/\partial z^\mu|_p$ and $\tilde{V}(q) = \tilde{V}^\mu(z + \Delta z)\partial/\partial z^\mu|_q$. We assume that (cf (7.9))

$$\tilde{V}^\mu(z + \Delta z) = V^\mu(z) - V^\lambda(z)\Gamma^\mu{}_{\nu\lambda}(z)\Delta z^\nu. \tag{8.61}$$

Then the basis vectors satisfy (cf (7.14))

$$\nabla_\mu \frac{\partial}{\partial z^\nu} = \Gamma^\lambda{}_{\mu\nu}(z) \frac{\partial}{\partial z^\lambda}. \tag{8.62a}$$

Since $\partial/\partial\bar{z}^\mu$ is a conjugate vector field of $\partial/\partial z^\mu$, we have

$$\nabla_{\bar{\mu}} \frac{\partial}{\partial\bar{z}^\nu} = \Gamma^{\bar{\lambda}}{}_{\bar{\mu}\bar{\nu}} \frac{\partial}{\partial\bar{z}^\lambda} \tag{8.62b}$$

where $\Gamma^{\bar{\lambda}}{}_{\bar{\mu}\bar{\nu}} = \overline{\Gamma^\lambda{}_{\mu\nu}}$. $\Gamma^\lambda{}_{\mu\nu}$ and $\Gamma^{\bar{\lambda}}{}_{\bar{\mu}\bar{\nu}}$ are the only non-vanishing components of the connection coefficients. Note that $\nabla_\mu \partial/\partial\bar{z}^\nu = \nabla_{\bar{\mu}}\partial/\partial z^\nu = 0$. For the dual basis, non-vanishing covariant derivatives are

$$\nabla_\mu \, dz^\nu = -\Gamma^\nu{}_{\mu\lambda} \, dz^\lambda \qquad \nabla_{\bar{\mu}} \, d\bar{z}^\nu = -\Gamma^{\bar{\nu}}{}_{\bar{\mu}\bar{\lambda}} \bar{z}^\lambda. \tag{8.63}$$

The covariant derivative of $X^+ = X^\mu \partial/\partial z^\mu \in \mathfrak{X}(M)^+$ is

$$\nabla_\mu X^+ = (\partial_\mu X^\lambda + X^\nu \Gamma^\lambda{}_{\mu\nu}) \frac{\partial}{\partial z^\lambda} \qquad (8.64)$$

where $\partial_\mu \equiv \partial/\partial z^\mu$. For $X^- = X^{\bar\mu} \partial/\partial \bar{z}^\mu \in \mathfrak{X}(M)^-$, we have

$$\nabla_\mu X^- = \partial_\mu X^{\bar\lambda} \frac{\partial}{\partial \bar{z}^\lambda} \qquad (8.65)$$

since $\Gamma^{\bar\lambda}{}_{\mu\nu} = \Gamma^{\bar\lambda}{}_{\mu\bar\nu} = 0$. As far as anti-holomorphic vectors are concerned, ∇_μ works as the ordinary derivative ∂_μ. Similarly, we have

$$\nabla_{\bar\mu} X^+ = \partial_{\bar\mu} X^\lambda \frac{\partial}{\partial z^\lambda} \qquad (8.66)$$

$$\nabla_{\bar\mu} X^- = (\partial_{\bar\mu} X^{\bar\lambda} + X^{\bar\nu} \Gamma^{\bar\lambda}{}_{\bar\mu\bar\nu}) \frac{\partial}{\partial \bar{z}^\lambda}. \qquad (8.67)$$

It is easy to generalize this to an arbitrary tensor field. For example, if $t = t_{\mu\nu}{}^{\bar\lambda} dz^\mu \otimes dx^\nu \otimes \partial/\partial \bar{z}^\lambda$, we have

$$(\nabla_\kappa t)_{\mu\nu}{}^{\bar\lambda} = \partial_\kappa t_{\mu\nu}{}^{\bar\lambda} - t_{\xi\nu}{}^{\bar\lambda} \Gamma^\xi{}_{\kappa\mu} - t_{\mu\xi}{}^{\bar\lambda} \Gamma^\xi{}_{\kappa\nu}$$

$$(\nabla_{\bar\kappa} t)_{\mu\nu}{}^{\bar\lambda} = \partial_{\bar\kappa} t_{\mu\nu}{}^{\bar\lambda} + t_{\mu\nu}{}^{\bar\xi} \Gamma^{\bar\lambda}{}_{\bar\kappa\bar\xi}.$$

We require the **metric compatibility** as in section 7.2. We demand that $\nabla_\kappa g_{\mu\bar\nu} = \nabla_{\bar\kappa} g_{\mu\bar\nu} = 0$. In components, we have

$$\partial_\kappa g_{\mu\bar\nu} - g_{\lambda\bar\nu} \Gamma^\lambda{}_{\kappa\mu} = 0 \qquad \partial_{\bar\kappa} g_{\mu\bar\nu} - g_{\mu\bar\lambda} \Gamma^{\bar\lambda}{}_{\bar\kappa\bar\mu} = 0. \qquad (8.68)$$

The connection coefficients are easily read off:

$$\Gamma^\lambda{}_{\kappa\mu} = g^{\bar\nu\lambda} \partial_\kappa g_{\mu\bar\nu} \qquad \Gamma^{\bar\lambda}{}_{\bar\kappa\bar\nu} = g^{\bar\lambda\mu} \partial_{\bar\kappa} g_{\mu\bar\nu} \qquad (8.69)$$

where $\{g^{\bar\nu\lambda}\}$ is the inverse matrix of $g_{\mu\bar\nu}$; $g_{\mu\bar\lambda} g^{\bar\lambda\nu} = \delta_\mu{}^\nu$, $g^{\bar\nu\lambda} g_{\lambda\bar\mu} = \delta^{\bar\nu}{}_{\bar\mu}$. A metric-compatible connection for which $\Gamma(\text{mixed indices}) = 0$ is called the **Hermitian connection**. By construction, this is unique and given by (8.69).

Theorem 8.4. The almost complex structure J is covariantly constant with respect to the Hermitian connection,

$$(\nabla_\kappa J)_v{}^\mu = (\nabla_{\bar\kappa} J)_v{}^\mu = (\nabla_\kappa J)_{\bar v}{}^{\bar\mu} = (\nabla_{\bar\kappa} J)_{\bar v}{}^{\bar\mu} = 0. \qquad (8.70)$$

Proof. We prove the first equality. From (8.22), we find

$$(\nabla_\kappa J)_v{}^\mu = \partial_\kappa i \delta_v{}^\mu - i \delta_\xi{}^\mu \Gamma^\xi{}_{\kappa v} + i \delta_v{}^\xi \Gamma^\mu{}_{\kappa\xi} = 0.$$

Other equalities follow from similar calculations. □

8.4.4 Torsion and curvature

The torsion tensor T and the Riemann curvature tensor R are defined by

$$T(X, Y) = \nabla_X Y - \nabla_Y X - [X, Y] \tag{8.71}$$

$$R(X, Y)Z = \nabla_X \nabla_Y Z - \nabla_Y \nabla_X Z - \nabla_{[X,Y]} Z. \tag{8.72}$$

We find that

$$T\left(\frac{\partial}{\partial z^\mu}, \frac{\partial}{\partial z^\nu}\right) = (\Gamma^\lambda{}_{\mu\nu} - \Gamma^\lambda{}_{\nu\mu})\frac{\partial}{\partial z^\lambda}$$

$$T\left(\frac{\partial}{\partial z^\mu}, \frac{\partial}{\partial \bar{z}^\nu}\right) = T\left(\frac{\partial}{\partial \bar{z}^\mu}, \frac{\partial}{\partial z^\nu}\right) = 0$$

$$T\left(\frac{\partial}{\partial \bar{z}^\mu}, \frac{\partial}{\partial \bar{z}^\nu}\right) = (\Gamma^{\bar\lambda}{}_{\bar\mu\bar\nu} - \Gamma^{\bar\lambda}{}_{\bar\nu\bar\mu})\frac{\partial}{\partial \bar{z}^\lambda}.$$

The non-vanishing components are

$$T^\lambda{}_{\mu\nu} = \Gamma^\lambda{}_{\mu\nu} - \Gamma^\lambda{}_{\nu\mu} = g^{\bar\xi\lambda}(\partial_\mu g_{\nu\bar\xi} - \partial_\nu g_{\mu\bar\xi}) \tag{8.73a}$$

$$T^{\bar\lambda}{}_{\bar\mu\bar\nu} = \Gamma^{\bar\lambda}{}_{\bar\mu\bar\nu} - \Gamma^{\bar\lambda}{}_{\bar\nu\bar\mu} = g^{\bar\lambda\xi}\left(\partial_{\bar\mu} g_{\bar\nu\xi} - \partial_{\bar\nu} g_{\bar\mu\xi}\right). \tag{8.73b}$$

As for the Riemann tensor, we find, for example, that

$$R^\kappa{}_{\lambda\mu\nu} = \partial_\mu \Gamma^\kappa{}_{\nu\lambda} - \partial_\nu \Gamma^\kappa{}_{\mu\lambda} + \Gamma^\eta{}_{\nu\lambda}\Gamma^\kappa{}_{\mu\eta} - \Gamma^\eta{}_{\mu\lambda}\Gamma^\kappa{}_{\nu\eta}.$$

If (8.69) is substituted, we find that

$$R^\kappa{}_{\lambda\mu\nu} = \partial_\mu g^{\bar\xi\kappa}\partial_\nu g_{\lambda\bar\xi} + g^{\bar\xi\kappa}\partial_\mu\partial_\nu g_{\lambda\bar\xi} - \partial_\nu g^{\bar\xi\kappa}\partial_\mu g_{\lambda\bar\xi} - g^{\bar\xi\kappa}\partial_\mu\partial_\nu g_{\lambda\bar\xi}$$
$$+ g^{\bar\xi\eta}\partial_\nu g_{\lambda\bar\xi}g^{\bar\zeta\kappa}\partial_\mu g_{\eta\bar\zeta} - g^{\bar\xi\eta}\partial_\mu g_{\lambda\bar\xi}g^{\bar\zeta\kappa}\partial_\nu g_{\eta\bar\zeta} = 0$$

where use has been made of the identity $g^{\bar\zeta\kappa}\partial_\mu g_{\eta\bar\zeta} = -g_{\eta\bar\zeta}\partial_\mu g^{\bar\zeta\kappa}$ etc. In general, we find that

$$R^\kappa{}_{\bar\lambda AB} = R^{\bar\kappa}{}_{\lambda AB} = R^A{}_{B\kappa\lambda} = R^A{}_{B\bar\kappa\bar\lambda} = 0 \tag{8.74}$$

where A and B are any (holomorphic or anti-holomorphic) indices. As a result, we are left only with the components $R^\kappa{}_{\lambda\bar\mu\nu}$, $R^\kappa{}_{\lambda\mu\bar\nu}$, $R^{\bar\kappa}{}_{\bar\lambda\bar\mu\nu}$ and $R^{\bar\kappa}{}_{\bar\lambda\mu\bar\nu}$. Note that we have a trivial symmetry $R^\kappa{}_{\lambda\bar\mu\nu} = -R^\kappa{}_{\lambda\nu\bar\mu}$. So the independent components are reduced to $R^\kappa{}_{\lambda\bar\mu\nu}$ and $R^{\bar\kappa}{}_{\bar\lambda\mu\bar\nu} = \overline{R^\kappa{}_{\lambda\bar\mu\nu}}$. We find that

$$R^\kappa{}_{\lambda\bar\mu\nu} = \partial_{\bar\mu}\Gamma^\kappa{}_{\nu\lambda} = \partial_{\bar\mu}(g^{\bar\xi\kappa}\partial_\nu g_{\lambda\bar\xi}) \tag{8.75a}$$

$$R^{\bar\kappa}{}_{\bar\lambda\mu\bar\nu} = \partial_\mu\Gamma^{\bar\kappa}{}_{\bar\nu\bar\lambda} = \partial_\mu(g^{\kappa\bar\xi}\partial_{\bar\nu}g_{\xi\bar\lambda}). \tag{8.75b}$$

Exercise 8.7. Show that

$$R_{\overline{\kappa}\lambda\overline{\mu}\nu} \equiv g_{\overline{\kappa}\xi}R^{\xi}{}_{\lambda\overline{\mu}\nu} = \partial_{\overline{\mu}}\partial_{\nu}g_{\lambda\overline{\kappa}} - g^{\overline{\eta}\xi}\partial_{\overline{\mu}}g_{\overline{\kappa}\xi}\partial_{\nu}g_{\lambda\overline{\eta}} \tag{8.76a}$$

$$R_{\kappa\overline{\lambda}\mu\overline{\nu}} \equiv g_{\kappa\overline{\xi}}R^{\overline{\xi}}{}_{\overline{\lambda}\mu\overline{\nu}} = \partial_{\mu}\partial_{\overline{\nu}}g_{\overline{\lambda}\kappa} - g^{\eta\overline{\xi}}\partial_{\mu}g_{\kappa\overline{\xi}}\partial_{\overline{\nu}}g_{\overline{\lambda}\eta} \tag{8.76b}$$

$$R_{\overline{\kappa}\lambda\mu\overline{\nu}} \equiv g_{\overline{\kappa}\xi}R^{\xi}{}_{\lambda\mu\overline{\nu}} = -R_{\overline{\kappa}\lambda\overline{\nu}\mu} \tag{8.76c}$$

$$R_{\kappa\overline{\lambda}\overline{\mu}\nu} \equiv g_{\kappa\overline{\xi}}R^{\xi}{}_{\overline{\lambda}\overline{\mu}\nu} = -R_{\kappa\overline{\lambda}\nu\overline{\mu}}. \tag{8.76d}$$

Verify the symmetries

$$R_{\overline{\kappa}\lambda\overline{\mu}\nu} = -R_{\lambda\overline{\kappa}\overline{\mu}\nu} \qquad R_{\kappa\overline{\lambda}\mu\overline{\nu}} = -R_{\overline{\lambda}\kappa\mu\overline{\nu}}. \tag{8.77}$$

Let us contract the indices of the Riemann tensor as

$$\mathfrak{R}_{\mu\overline{\nu}} \equiv R^{\kappa}{}_{\kappa\mu\overline{\nu}} = -\partial_{\overline{\nu}}(g^{\kappa\overline{\xi}}\partial_{\mu}g_{\kappa\overline{\xi}}) = -\partial_{\overline{\nu}}\partial_{\mu}\log G \tag{8.78}$$

where $G \equiv \det(g_{\mu\overline{\nu}}) = \sqrt{g}$. To obtain the last equality, we used an identity $\delta G = G g^{\mu\overline{\nu}}\delta g_{\mu\overline{\nu}}$; see (7.204). We define the **Ricci form** by

$$\mathfrak{R} \equiv i\mathfrak{R}_{\mu\overline{\nu}}\,dz^{\mu} \wedge d\overline{z}^{\nu} = i\partial\overline{\partial}\log G. \tag{8.79}$$

$\mathfrak{R}$ is a *real* form; $\overline{\mathfrak{R}} = -i\overline{\partial}\overline{\partial}\log G = -i\partial\overline{\partial}\log G = \mathfrak{R}$. From the identity $\partial\overline{\partial} = -\frac{1}{2}d(\partial - \overline{\partial})$, we find $\mathfrak{R}$ is closed; $d\mathfrak{R} \propto d^2(\partial - \overline{\partial})\log G = 0$. However, this does not imply that $\mathfrak{R}$ is exact. In fact, G is not a scalar and $(\partial - \overline{\partial})\log G$ is not defined globally. $\mathfrak{R}$ defines a non-trivial element $c_1(M) \equiv [\mathfrak{R}/2\pi] \in H^2(M; \mathbb{R})$ called the **first Chern class**. We discuss this further in section 11.2.

Proposition 8.2. The first Chern class $c_1(M)$ is invariant under a smooth change of the metric $g \to g + \delta g$.

Proof. It follows from (7.204) that $\delta\log G = g^{\mu\overline{\nu}}\delta g_{\mu\overline{\nu}}$. Then

$$\delta\mathfrak{R} = \delta i\partial\overline{\partial}\log G = i\partial\overline{\partial}g^{\mu\overline{\nu}}\delta g_{\mu\overline{\nu}} = -\frac{1}{2}d(\partial - \overline{\partial})ig^{\mu\overline{\nu}}\delta g_{\mu\overline{\nu}}.$$

Since $g^{\mu\overline{\nu}}\delta g_{\mu\overline{\nu}}$ is a scalar, $\omega \equiv -\frac{1}{2}(\partial - \overline{\partial})g^{\mu\overline{\nu}}\delta g_{\mu\overline{\nu}}$ is a well-defined one-form on M. Thus, $\delta\mathfrak{R} = d\omega$ is an exact two-form and $[\mathfrak{R}] = [\mathfrak{R} + \delta\mathfrak{R}]$, namely $c_1(M)$ is left invariant under $g \to g + \delta g$. $\square$

8.5 Kähler manifolds and Kähler differential geometry

8.5.1 Definitions

Definition 8.4. A **Kähler manifold** is a Hermitian manifold (M, g) whose Kähler form Ω is closed: $d\Omega = 0$. The metric g is called the **Kähler metric** of M. [*Warning:* Not all complex manifolds admit Kähler metrics.]

Theorem 8.5. A Hermitian manifold (M, g) is a Kähler manifold if and only if the almost complex structure J satisfies

$$\nabla_\mu J = 0 \tag{8.80}$$

where ∇_μ is the Levi-Civita connection associated with g.

Proof. We first note that for any r-form ω, $d\omega$ is written as

$$d\omega = \nabla\omega \equiv \frac{1}{r!}\nabla_\mu \omega_{\nu_1\ldots\nu_r}\,dx^\mu \wedge dx^{\nu_1} \wedge \ldots \wedge dx^{\nu_r}. \tag{8.81}$$

[For example,

$$
\begin{aligned}
\nabla\Omega &= \tfrac{1}{2}\nabla_\lambda \Omega_{\mu\nu}dx^\lambda \wedge dx^\mu \wedge dx^\nu \\
&= \tfrac{1}{2}(\partial_\lambda \Omega_{\mu\nu} - \Gamma^\kappa{}_{\lambda\mu}\Omega_{\kappa\nu} - \Gamma^\kappa{}_{\lambda\nu}\Omega_{\mu\kappa})\,dx^\lambda \wedge dx^\mu \wedge dx^\nu \\
&= \tfrac{1}{2}\partial_\lambda \Omega_{\mu\nu}dx^\lambda \wedge dx^\mu \wedge dx^\nu = d\Omega
\end{aligned}
$$

since Γ is symmetric.] Now we prove that $\nabla_\mu J = 0$ if and only if $\nabla_\mu\Omega = 0$. We verify the following equalities:

$$
\begin{aligned}
(\nabla_Z\Omega)(X, Y) &= \nabla_Z[\Omega(X, Y)] - \Omega(\nabla_Z X, Y) - \Omega(X, \nabla_Z Y) \\
&= \nabla_Z[g(JX, Y)] - g(J\nabla_Z X, Y) - g(JX, \nabla_Z Y) \\
&= (\nabla_Z g)(JX, Y) + g(\nabla_Z JX, Y) - g(J\nabla_Z X, Y) \\
&= g(\nabla_Z JX - J\nabla_Z X, Y) = g((\nabla_Z J)X, Y)
\end{aligned}
$$

where $\nabla_Z g = 0$ has been used. Since this is true for any X, Y, Z, it follows that $\nabla_Z\Omega = 0$ if and only if $\nabla_Z J = 0$. $\qquad\square$

Theorems 8.4 and 8.5 show that the Riemann structure is compatible with the Hermitian structure in the Kähler manifold.

Let g be a Kähler metric. Since $d\Omega = 0$, we have

$$
\begin{aligned}
(\partial &+ \bar{\partial})ig_{\mu\bar{\nu}}\,dz^\mu \wedge d\bar{z}^\nu \\
&= i\partial_\lambda g_{\mu\bar{\nu}}\,dz^\lambda \wedge dz^\mu \wedge d\bar{z}^\nu + i\partial_{\bar{\lambda}}g_{\mu\bar{\nu}}\,d\bar{z}^\lambda \wedge dz^\mu \wedge d\bar{z}^\nu \\
&= \tfrac{1}{2}i(\partial_\lambda g_{\mu\bar{\nu}} - \partial_\mu g_{\lambda\bar{\nu}})\,dz^\lambda \wedge dz^\mu \wedge d\bar{z}^\nu \\
&\quad + \tfrac{1}{2}i(\partial_{\bar{\lambda}}g_{\mu\bar{\nu}} - \partial_{\bar{\nu}}g_{\mu\bar{\lambda}})\,d\bar{z}^\lambda \wedge dz^\mu \wedge d\bar{z}^\nu = 0
\end{aligned}
$$

from which we find

$$\frac{\partial g_{\mu\bar{\nu}}}{\partial z^\lambda} = \frac{\partial g_{\lambda\bar{\nu}}}{\partial z^\mu} \qquad \frac{\partial g_{\mu\bar{\nu}}}{\partial \bar{z}^\lambda} = \frac{\partial g_{\mu\bar{\lambda}}}{\partial \bar{z}^\nu}. \tag{8.82}$$

Suppose that a Hermitian metric g is given on a chart U_i by

$$g_{\mu\bar{\nu}} = \partial_\mu \partial_{\bar{\nu}} \mathcal{K}_i \qquad (8.83)$$

where $\mathcal{K}_i \in \mathcal{F}(U_i)$. Clearly this metric satisfies the condition (8.82), hence it is Kähler. Conversely, it can be shown that any Kähler metric is *locally* expressed as (8.83). The function $\mathcal{K}_i$ is called the **Kähler potential** of a Kähler metric. It follows that $\Omega = i\partial\bar{\partial}\mathcal{K}_i$ on U_i.

Let (U_i, φ_i) and (U_j, φ_j) be overlapping charts. On $U_i \cap U_j$, we have

$$\frac{\partial}{\partial z^\mu} \frac{\partial}{\partial \bar{z}^\nu} \mathcal{K}_i \, dz^\mu \otimes d\bar{z}^\nu = \frac{\partial}{\partial w^\alpha} \frac{\partial}{\partial \bar{w}^\beta} \mathcal{K}_j \, dw^\alpha \otimes d\bar{w}^\beta$$

where $z = \varphi_i(p)$ and $w = \varphi_j(p)$. It then follows that

$$\frac{\partial w^\alpha}{\partial z^\mu} \frac{\partial \bar{w}^\beta}{\partial \bar{z}^\nu} \frac{\partial}{\partial w^\alpha} \frac{\partial}{\partial \bar{w}^\beta} \mathcal{K}_j = \frac{\partial}{\partial z^\mu} \frac{\partial}{\partial \bar{z}^\nu} \mathcal{K}_i. \qquad (8.84)$$

This is satisfied if and only if $\mathcal{K}_j(w, \bar{w}) = \mathcal{K}_i(z, \bar{z}) + \phi_{ij}(z) + \psi_{ij}(\bar{z})$ where ϕ_{ij} (ψ_{ij}) is holomorphic (anti-holomorphic) in z.

Exercise 8.8. Let M be a compact Kähler manifold without a boundary. Show that

$$\Omega^m \equiv \underbrace{\Omega \wedge \ldots \wedge \Omega}_{m}$$

is closed but not exact where $m = \dim_{\mathbb{C}} M$ [*Hint*: Use Stokes' theorem.] Thus, the $2m$th Betti number cannot vanish, $b^{2m} \geq 1$. We will see later that $b^{2p} \geq 1$ for $1 \leq p \leq m$.

Example 8.6. Let $M = \mathbb{C}^m = \{(z^1, \ldots, z^m)\}$. $\mathbb{C}^m$ is identified with $\mathbb{R}^{2m}$ by the identification $z^\mu \to x^\mu + iy^\mu$. Let δ be the Euclidean metric of $\mathbb{R}^{2m}$,

$$\delta\left(\frac{\partial}{\partial x^\mu}, \frac{\partial}{\partial x^\nu}\right) = \delta\left(\frac{\partial}{\partial y^\mu}, \frac{\partial}{\partial y^\nu}\right) = \delta_{\mu\nu}$$

$$\delta\left(\frac{\partial}{\partial x^\mu}, \frac{\partial}{\partial y^\nu}\right) = 0. \qquad (8.85a)$$

Noting that $J\partial/\partial x^\mu = \partial/\partial y^\mu$ and $J\partial/\partial y^\mu = -\partial/\partial x^\mu$, we find that δ is a Hermitian metric. In complex coordinates, we have

$$\delta\left(\frac{\partial}{\partial z^\mu}, \frac{\partial}{\partial z^\nu}\right) = \delta\left(\frac{\partial}{\partial \bar{z}^\mu}, \frac{\partial}{\partial \bar{z}^\nu}\right) = 0$$

$$\delta\left(\frac{\partial}{\partial z^\mu}, \frac{\partial}{\partial \bar{z}^\nu}\right) = \delta\left(\frac{\partial}{\partial \bar{z}^\mu}, \frac{\partial}{\partial z^\nu}\right) = \frac{1}{2}\delta_{\mu\nu}. \qquad (8.85b)$$

The Kähler form is given by

$$\Omega = \frac{i}{2} \sum_{\mu=1}^{m} dz^{\mu} \wedge d\bar{z}^{\mu} = \frac{i}{2} \sum_{\mu=1}^{m} dx^{\mu} \wedge dy^{\mu}. \qquad (8.86)$$

Clearly, $d\Omega = 0$ and we find that the Euclidean metric δ of $\mathbb{R}^{2m}$ is a Kähler metric of $\mathbb{C}^{m}$. The Kähler potential is

$$\mathcal{K} = \frac{1}{2} \sum z^{\mu} \bar{z}^{\mu}. \qquad (8.87)$$

The Kähler manifold $\mathbb{C}^{m}$ is called the **complex Euclid space**.

Example 8.7. Any orientable complex manifold M with $\dim_{\mathbb{C}} M = 1$ is Kähler. Take a Hermitian metric g whose Kähler form is Ω. Since Ω is a real two-form, a three-form $d\Omega$ has to vanish on M. One-dimensional compact orientable complex manifolds are known as **Riemann surfaces**.

Example 8.8. A complex projective space $\mathbb{C}P^{m}$ is a Kähler manifold. Let $(U_{\alpha}, \varphi_{\alpha})$ be a chart whose inhomogeneous coordinates are $\varphi_{\alpha}(p) = \xi_{(\alpha)}^{\nu}$, $\nu \neq \alpha$ (see example 8.3). It is convenient to introduce a tidier notation $\{\zeta^{\nu}{}_{(\alpha)} | 1 \leq \nu \leq m\}$ by

$$\xi^{\nu}{}_{(\alpha)} = \zeta^{\nu}{}_{(\alpha)} \quad (\nu \leq \alpha - 1) \qquad \xi^{\nu+1}{}_{(\alpha)} = \zeta^{\nu}{}_{(\alpha)} \quad (\nu \geq \alpha). \qquad (8.88)$$

$\{\zeta^{\nu}{}_{(\alpha)}\}$ is just a renaming of $\{\xi^{\nu}{}_{(\alpha)}\}$. Define a positive-definite function

$$\mathcal{K}_{\alpha}(p) \equiv \sum_{\nu=1}^{m} |\zeta^{\nu}{}_{(\alpha)}(p)|^{2} + 1 = \sum_{\nu=1}^{m+1} \left| \frac{z^{\nu}}{z^{\alpha}} \right|^{2}. \qquad (8.89)$$

At a point $p \in U_{\alpha} \cap U_{\beta}$, $\mathcal{K}_{\alpha}(p)$ and $\mathcal{K}_{\beta}(p)$ are related as

$$\mathcal{K}_{\alpha}(p) = \left| \frac{z^{\beta}}{z^{\alpha}} \right|^{2} \mathcal{K}_{\beta}(p). \qquad (8.90)$$

Then it follows that

$$\log \mathcal{K}_{\alpha} = \log \mathcal{K}_{\beta} + \log \frac{z^{\beta}}{z^{\alpha}} + \log \overline{\frac{z^{\beta}}{z^{\alpha}}}. \qquad (8.91)$$

Since z^{β}/z^{α} is a holomorphic function, we have $\bar{\partial} \log z^{\beta}/z^{\alpha} = 0$. Also

$$\overline{\partial \log z^{\beta}/z^{\alpha}} = \overline{\bar{\partial} \log z^{\beta}/z^{\alpha}} = 0.$$

Then it follows that

$$\partial \bar{\partial} \log \mathcal{K}_{\alpha} = \partial \bar{\partial} \log \mathcal{K}_{\beta}. \qquad (8.92)$$

A closed two-form Ω is locally defined by

$$\Omega \equiv i\partial\bar{\partial}\log\mathcal{K}_\alpha. \tag{8.93}$$

There exists a Hermitian metric whose Kähler form is Ω. Take $X, Y \in T_p\mathbb{C}P^n$ and define $g : T_p\mathbb{C}P^n \otimes T_p\mathbb{C}P^n \to \mathbb{R}$ by $g(X, Y) = \Omega(X, JY)$. To prove that g is a Hermitian metric, we have to show that g satisfies (8.50) and is positive definite. The Hermiticity is obvious since $g(JX, JY) = -\Omega(JX, Y) = \Omega(Y, JX) = g(X, Y)$. Next, we show that g is positive definite. On a chart $(U_\alpha, \varphi_\alpha)$, we obtain

$$\Omega = i\frac{\partial^2\log\mathcal{K}}{\partial\zeta^\mu\partial\bar{\zeta}^\nu}\,\mathrm{d}\zeta^\mu \wedge \mathrm{d}\bar{\zeta}^\nu \tag{8.94}$$

where we have dropped the subscript (α) to simplify the notation. If we substitute the expression (8.89) for $\mathcal{K}$ on U_α, we have

$$\Omega = i\sum_{\mu,\nu}\frac{\delta_{\mu\nu}(\sum|\zeta^\lambda|^2 + 1) - \zeta^\mu\bar{\zeta}^\nu}{(\sum|\zeta^\lambda|^2 + 1)^2}\,\mathrm{d}\zeta^\mu \wedge \mathrm{d}\bar{\zeta}^\nu. \tag{8.95}$$

Let X be a real vector, $X = X^\mu\partial/\partial\zeta^\mu + \overline{X}^\mu\partial/\partial\bar{\zeta}^\mu$ and $JX = iX^\mu\partial/\partial\zeta^\mu - i\overline{X}^\mu\partial/\partial\bar{\zeta}^\mu$. Then

$$g(X, X) = \Omega(X, JX) = 2\sum_{\mu,\nu}\frac{\delta_{\mu\nu}(\sum|\zeta^\lambda|^2 + 1) - \zeta^\mu\bar{\zeta}^\nu}{(\sum|\zeta^\lambda|^2 + 1)^2}X^\mu\overline{X}^\nu$$

$$= 2\left[\sum_\mu|X^\mu|^2\left(\sum_\lambda|\zeta^\lambda|^2 + 1\right) - \left|\sum_\mu X^\mu\zeta^\mu\right|^2\right]\left(\sum_\lambda|\zeta^\lambda|^2 + 1\right)^{-2}.$$

From the Schwarz inequality $\sum_\mu|X^\mu|^2 \cdot \sum_\lambda|\zeta^\lambda|^2 \geq \sum_\mu|X^\mu\zeta^\mu|^2$, we find the metric g is positive definite. This metric is called the **Fubini–Study metric** of $\mathbb{C}P^n$.

A few useful facts are:

(a) S^2 is the only sphere which admits a complex structure. Since $S^2 \simeq \mathbb{C}P^1$, it is a Kähler manifold.
(b) A product of two odd-dimensional spheres $S^{2m+1} \times S^{2n+1}$ always admits a complex structure. This complex structure does not admit a Kähler metric.
(c) Any complex submanifold of a Kähler manifold is Kähler.

8.5.2 Kähler geometry

A Kähler metric g is characterized by (8.82):

$$\frac{\partial g_{\mu\bar{\nu}}}{\partial z^\lambda} = \frac{\partial g_{\lambda\bar{\nu}}}{\partial z^\mu} \qquad \frac{\partial g_{\mu\bar{\nu}}}{\partial\bar{z}^\lambda} = \frac{\partial g_{\mu\bar{\lambda}}}{\partial\bar{z}^\nu}.$$

This ensures that the Kähler metric is *torsion free*:

$$T^\lambda{}_{\mu\nu} = g^{\bar\xi\lambda}(\partial_\mu g_{\nu\bar\xi} - \partial_\nu g_{\mu\bar\xi}) = 0 \tag{8.96a}$$

$$T^{\bar\lambda}{}_{\bar\mu\bar\nu} = g^{\lambda\bar\xi}(\partial_{\bar\mu} g_{\bar\nu\xi} - \partial_{\bar\nu} g_{\bar\mu\xi}) = 0. \tag{8.96b}$$

In this sense, the Kähler metric defines a connection which is very similar to the Levi-Civita connection. Now the Riemann tensor has an extra symmetry

$$R^\kappa{}_{\lambda\mu\bar\nu} = -\partial_{\bar\nu}(g^{\bar\xi\kappa}\partial_\mu g_{\lambda\bar\xi}) = -\partial_{\bar\nu}(g^{\bar\xi\kappa}\partial_\lambda g_{\mu\bar\xi}) = R^\kappa{}_{\mu\lambda\bar\nu} \tag{8.97}$$

as well as those obtained from (8.97) by known symmetry operations,

$$R^{\bar\kappa}{}_{\bar\lambda\bar\mu\nu} = R^{\bar\kappa}{}_{\bar\mu\bar\lambda\nu}, \qquad R^\kappa{}_{\lambda\bar\mu\nu} = R^\kappa{}_{\nu\bar\mu\lambda}, \qquad R^{\bar\kappa}{}_{\bar\lambda\mu\bar\nu} = R^{\bar\kappa}{}_{\bar\nu\mu\bar\lambda}. \tag{8.98}$$

The Ricci form $\mathfrak{R}$ is defined as before,

$$\mathfrak{R} = -\mathrm{i}\partial_{\bar\nu}\partial_\mu \log G \, \mathrm{d}z^\mu \wedge \mathrm{d}\bar z^\nu.$$

Because of (8.97), the components of the Ricci form agree with $Ric_{\mu\bar\nu}$; $\mathfrak{R}_{\mu\bar\nu} \equiv R^\kappa{}_{\kappa\mu\bar\nu} = R^\kappa{}_{\mu\kappa\bar\nu} = Ric_{\mu\bar\nu}$. If $Ric = \mathfrak{R} = 0$, the Kähler metric is said to be **Ricci flat**.

Theorem 8.6. Let (M, g) be a Kähler manifold. If M admits a Ricci flat metric h, then its first Chern class must vanish.

Proof. By assumption, $\mathfrak{R} = 0$ for the metric h. As was shown in the previous section, $\mathfrak{R}(g) - \mathfrak{R}(h) = \mathfrak{R}(g) = \mathrm{d}\omega$. Hence, $c_1(M)$ computed from g agrees with that computed from h and hence vanishes. $\qquad\square$

A compact Kähler manifold with vanishing first Chern class is called a **Calabi–Yau manifold**. Calabi (1957) conjectured that if $c_1(M) = 0$, the Kähler manifold M admits a Ricci-flat metric. This is proved by Yau (1977). Calabi–Yau manifolds with $\dim_\mathbb{C} M = 3$ have been proposed as candidates for superstring compactification (see Horowitz (1986) and Candelas (1988)).

8.5.3 The holonomy group of Kähler manifolds

Before we close this section, we briefly look at the holonomy groups of Kähler manifolds. Let (M, g) be a Hermitian manifold with $\dim_\mathbb{C} M = m$. Take a vector $X \in T_p M^+$ and parallel transport it along a loop c at p. Then we end up with a vector $X' \in T_p M^+$ where $X'^\mu = X^\mu h_\nu{}^\mu$. Note that ∇ does not mix the holomorphic indices with anti-holomorphic indices, hence X' has no components in $T_p M^-$. Moreover, ∇ preserves the length of a vector. These facts tell us that $(h_\mu{}^\nu(c))$ is contained in $U(m) \subset O(2m)$.

Theorem 8.7. If g is the Ricci-flat metric of an m-dimensional Calabi–Yau maifold M, the holonomy group is contained in $SU(m)$.

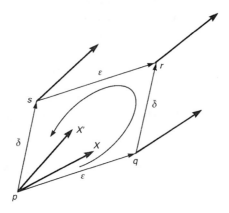

Figure 8.5. $X \in T_p M^+$ is parallel transported along $pqrs$ and comes back as a vector $X' \in T_p M^+$.

Proof. Our proof is sketchy. If $X = X^\mu \partial/\partial z^\mu \in T_p M^+$ is parallel transported along the small parallelogram in figure 8.5 back to p, we have $X' \in T_p M^+$ whose components are (cf (7.44))

$$X'^\mu = X^\mu + X^\nu R^\mu{}_{\nu\kappa\bar\lambda}\varepsilon^\kappa\bar\delta^\lambda \tag{8.99}$$

from which we find

$$h_\mu{}^\nu = \delta_\mu{}^\nu + R^\nu{}_{\mu\kappa\bar\lambda}\varepsilon^\kappa\bar\delta^\lambda. \tag{8.100}$$

$U(m)$ is decomposed as $U(m) = SU(m) \times U(1)$ in the vicinity of the unit element. In particular, the Lie algebra $\mathfrak{u}(m) = T_e(U(m))$ is separated into

$$\mathfrak{u}(m) = \mathfrak{su}(m) \oplus \mathfrak{u}(1). \tag{8.101}$$

$\mathfrak{su}(m)$ is the traceless part of $\mathfrak{u}(m)$ while $\mathfrak{u}(1)$ contains the trace. Since the present metric is Ricci flat, the $\mathfrak{u}(1)$ part vanishes,

$$R^\kappa{}_{\kappa\mu\bar\nu}\varepsilon^\mu\bar\delta^\nu = \mathfrak{R}_{\mu\bar\nu}\varepsilon^\mu\bar\delta^\nu = 0.$$

This shows that the holonomy group is contained in $SU(m)$. [*Remark:* Strictly speaking, we have only shown that the restricted holonomy group is contained in $SU(m)$. This statement remains true even when M is multiply connected.]   □

8.6 Harmonic forms and $\bar\partial$-cohomology groups

The (r, s)th $\bar\partial$-cohomology group is defined by

$$H^{r,s}_{\bar\partial}(M) \equiv Z^{r,s}_{\bar\partial}(M)/B^{r,s}_{\bar\partial}(M). \tag{8.102}$$

An element $[\omega] \in H_{\bar{\partial}}^{r,s}(M)$ is an equivalence class of $\bar{\partial}$-closed forms of bidegree (r, s) which differ from ω by a $\bar{\partial}$-exact form,

$$[\omega] = \{\eta \in \Omega^{r,s}(M) | \bar{\partial}\eta = 0, \omega - \eta = \bar{\partial}\psi, \psi \in \Omega^{r,s-1}(M)\}. \qquad (8.103)$$

Clearly $H_{\bar{\partial}}^{r,s}(M)$ is a complex vector space. Similarly to the de Rham cohomology groups, the $\bar{\partial}$-cohomology groups of $\mathbb{C}^m$ are trivial, that is, all the closed (r, s)-forms are exact. The $\bar{\partial}$-cohomology groups measure the topological non-triviality of a complex manifold M.

8.6.1 The adjoint operators $\partial^{\dagger}$ and $\bar{\partial}^{\dagger}$

Let M be a Hermitian manifold with $\dim_{\mathbb{C}} M = m$. Define the inner product between $\alpha, \beta \in \Omega^{r,s}(M)$ $(0 \leq r, s \leq m)$ by

$$(\alpha, \beta) \equiv \int_M \alpha \wedge \bar{*}\beta \qquad (8.104)$$

where $\bar{*} : \Omega^{r,s}(M) \to \Omega^{m-r,m-s}(M)$ is the **Hodge** $*$ defined by

$$\bar{*}\beta \equiv \overline{*\beta} = *\bar{\beta} \qquad (8.105)$$

where $*\beta$ is computed according to (7.173) extended to $\Omega^{r+s}(M)^{\mathbb{C}}$. [*Remark*: $*$ maps an (r, s)-form to an $(m - s, m - r)$-form since it acts on a basis of $\Omega^{r,s}(M)$, up to an irrelevant factor, as

$$* dz^{\mu_1} \wedge \ldots \wedge dz^{\mu_r} \wedge d\bar{z}^{\nu_1} \wedge \ldots \wedge d\bar{z}^{\nu_s} \sim \varepsilon^{\mu_1 \ldots \mu_r}{}_{\bar{\mu}_{r+1} \ldots \bar{\mu}_m} \varepsilon^{\bar{\nu}_1 \ldots \bar{\nu}_s}{}_{\nu_{s+1} \ldots \nu_m}$$
$$\times d\bar{z}^{\mu_{r+1}} \wedge \ldots \wedge d\bar{z}^{\mu_m} \wedge dz^{\nu_{s+1}} \wedge \ldots \wedge dz^{\nu_m}.$$

Note that the above ε-symbols are the only non-vanishing components in a Hermitian manifold. Now it follows that $\bar{*} : \Omega^{r,s}(M) \to \Omega^{m-r,m-s}(M)$.]

We define the adjoint operators $\partial^{\dagger}$ and $\bar{\partial}^{\dagger}$ of ∂ and $\bar{\partial}$ by

$$(\alpha, \partial\beta) = (\partial^{\dagger}\alpha, \beta) \qquad (\alpha, \bar{\partial}\beta) = (\bar{\partial}^{\dagger}\alpha, \beta). \qquad (8.106)$$

The operators $\partial^{\dagger}$ and $\bar{\partial}^{\dagger}$ change the bidegrees as $\partial^{\dagger} : \Omega^{r,s}(M) \to \Omega^{r-1,s}(M)$ and $\bar{\partial}^{\dagger} : \Omega^{r,s}(M) \to \Omega^{r,s-1}(M)$. Clearly $d^{\dagger} = \partial^{\dagger} + \bar{\partial}^{\dagger}$. Noting that a complex manifold M is even dimensional as a differentiable manifold, we have (see (7.184a))

$$d^{\dagger} = - * d *. \qquad (8.107)$$

Proposition 8.3.

$$\partial^{\dagger} = - * \bar{\partial}*, \qquad \bar{\partial}^{\dagger} = - * \partial *. \qquad (8.108)$$

Proof. Let $\omega \in \Omega^{r-1,s}(M)$ and $\psi \in \Omega^{r,s}(M)$. If we note that $\omega \wedge \bar{*}\psi \in \Omega^{m-1,m}(M)$ and hence $\bar{\partial}(\omega \wedge \bar{*}\psi) = 0$, we find that

$$
\begin{aligned}
d(\omega \wedge \bar{*}\psi) = \partial(\omega \wedge \bar{*}\psi) &= \partial\omega \wedge \bar{*}\psi + (-1)^{r+s-1}\omega \wedge \partial(\bar{*}\psi) \\
&= \partial\omega \wedge \bar{*}\psi + (-1)^{r+s-1}\omega \wedge (-1)^{r+s+1}\bar{*}\,\bar{*}\partial(\bar{*}\psi) \\
&= \partial\omega \wedge \bar{*}\psi + \omega \wedge \bar{*}\,\bar{*}\partial\bar{*}\psi
\end{aligned}
\tag{8.109}
$$

where use has been made of the facts $\partial\bar{*}\psi \in \Omega^{2m-r-s-1}(M)$, $\bar{*}\bar{*}\beta = *\,*\,\beta$ and (7.176a). If (8.109) is integrated over a compact complex manifold M with no boundary, we have

$$
0 = (\partial\omega, \psi) + (\omega, \bar{*}\partial\bar{*}\psi).
$$

The second term is

$$
(\omega, \bar{*}\partial\bar{*}\psi) = (\omega, *\bar{\partial}*\bar{\psi}) = (\omega, *\bar{\partial}*\psi).
$$

We finally find $0 = (\partial\omega, \psi) + (\omega, *\bar{\partial}*\psi)$, namely $\partial^{\dagger} = -*\bar{\partial}*$. The other formula $\bar{\partial}^{\dagger} = -*\partial*$ follows similarly. □

As a corollary of proposition 8.3, we have

$$
(\partial^{\dagger})^2 = (\bar{\partial}^{\dagger})^2 = 0.
\tag{8.110}
$$

8.6.2 Laplacians and the Hodge theorem

Besides the usual Laplacian $\Delta = (dd^{\dagger} + d^{\dagger}d)$, we define other Laplacians Δ_{∂} and $\Delta_{\bar{\partial}}$ on a Hermitian manifold,

$$
\Delta_{\partial} \equiv (\partial + \partial^{\dagger})^2 = \partial\partial^{\dagger} + \partial^{\dagger}\partial
\tag{8.111a}
$$

$$
\Delta_{\bar{\partial}} \equiv (\bar{\partial} + \bar{\partial}^{\dagger})^2 = \bar{\partial}\bar{\partial}^{\dagger} + \bar{\partial}^{\dagger}\bar{\partial}.
\tag{8.111b}
$$

An (r, s)-form ω which satisfies $\Delta_{\partial}\omega = 0$ $(\Delta_{\bar{\partial}}\omega = 0)$ is said to be **∂-harmonic** (**$\bar{\partial}$-harmonic**). If $\Delta_{\partial}\omega = 0$ $(\Delta_{\bar{\partial}}\omega = 0)$, ω satisfies $\partial\omega = \partial^{\dagger}\omega = 0$ $(\bar{\partial}\omega = \bar{\partial}^{\dagger}\omega = 0)$.

We have the complex version of the Hodge decomposition. Let $\text{Harm}_{\bar{\partial}}^{r,s}(M)$ be the set of $\bar{\partial}$-harmonic (r, s)-forms,

$$
\text{Harm}_{\bar{\partial}}^{r,s}(M) \equiv \{\omega \in \Omega^{r,s}(M) | \Delta_{\bar{\partial}}\omega = 0\}.
\tag{8.112}
$$

Theorem 8.8. (**Hodge's theorem**) $\Omega^{r,s}(M)$ has a unique orthogonal decomposition:

$$
\Omega^{r,s}(M) = \bar{\partial}\Omega^{r,s-1}(M) \oplus \bar{\partial}^{\dagger}\Omega^{r,s+1}(M) \oplus \text{Harm}_{\bar{\partial}}^{r,s}(M)
\tag{8.113a}
$$

namely an (r, s)-form ω is uniquely expressed as

$$\omega = \bar{\partial}\alpha + \bar{\partial}^{\dagger}\beta + \gamma \qquad (8.113b)$$

where $\alpha \in \Omega^{r,s-1}(M)$, $\beta \in \Omega^{r,s+1}(M)$ and $\gamma \in \mathrm{Harm}_{\bar{\partial}}^{r,s}(M)$.

The proof is found in lecture 22, Schwartz (1986), for example. If ω is $\bar{\partial}$-closed, we have $\bar{\partial}\omega = \bar{\partial}\bar{\partial}\bar{\partial}^{\dagger}\beta = 0$. Then $0 = \langle \beta, \bar{\partial}\bar{\partial}^{\dagger}\beta \rangle = \langle \bar{\partial}^{\dagger}\beta, \bar{\partial}^{\dagger}\beta \rangle \geq 0$ implies $\bar{\partial}^{\dagger}\beta = 0$. Thus, any closed (r, s)-form ω is written as $\omega = \gamma + \bar{\partial}\alpha$, $\alpha \in \Omega^{r,s-1}(M)$. This shows that $H_{\bar{\partial}}^{r,s}(M) \subset \mathrm{Harm}_{\bar{\partial}}^{r,s}(M)$. Note also that $\mathrm{Harm}_{\bar{\partial}}^{r,s}(M) \subset Z_{\bar{\partial}}^{r,s}(M)$ since $\bar{\partial}\gamma = 0$ for $\gamma \in \mathrm{Harm}_{\bar{\partial}}^{r,s}(M)$. Moreover, $\mathrm{Harm}_{\bar{\partial}}^{r,s}(M) \cap B_{\bar{\partial}}^{r,s}(M) = \emptyset$ since $B_{\bar{\partial}}^{r,s}(M) = \bar{\partial}\Omega^{r,s-1}(M)$ is orthogonal to $\mathrm{Harm}_{\bar{\partial}}^{r,s}(M)$. Then it follows that $\mathrm{Harm}_{\bar{\partial}}^{r,s}(M) \cong H_{\bar{\partial}}^{r,s}(M)$. If $P : \Omega^{r,s}(M) \to \mathrm{Harm}_{\bar{\partial}}^{r,s}(M)$ denotes the projection operator to a harmonic (r, s)-form, $[\omega] \in H_{\bar{\partial}}^{r,s}(M)$ has a unique harmonic representative $P\omega \in \mathrm{Harm}_{\bar{\partial}}^{r,s}(M)$.

8.6.3 Laplacians on a Kähler manifold

In a general Hermitian manifold, there exist no particular relationships among the Laplacians Δ, Δ_{∂} and $\Delta_{\bar{\partial}}$. However, if M is a Kähler manifold, they are essentially the *same*. [Note that the Levi-Civita connection is compatible with the Hermitian connection in a Kähler manifold.]

Theorem 8.9. Let M be a Kähler manifold. Then

$$\Delta = 2\Delta_{\partial} = 2\Delta_{\bar{\partial}}. \qquad (8.114)$$

The proof requires some technicalities and we simply refer to Schwartz (1986) and Goldberg (1962). This theorem puts constraints on the cohomology groups of a Kähler manifold M. A form ω which satisfies $\bar{\partial}\omega = \bar{\partial}^{\dagger}\omega = 0$ also satisfies $\partial\omega = \partial^{\dagger}\omega = 0$. Let ω be a holomorphic p-form; $\bar{\partial}\omega = 0$. Since ω contains no $\mathrm{d}\bar{z}^{\mu}$ in its expansion, we have $\bar{\partial}^{\dagger}\omega = 0$, hence $\Delta_{\bar{\partial}}\omega = (\bar{\partial}\bar{\partial}^{\dagger} + \bar{\partial}^{\dagger}\bar{\partial})\omega = 0$. According to theorem 8.9, we then have $\Delta\omega = 0$, that is *any holomorphic form is automatically harmonic* with respect to the Kähler metric. Conversely $\Delta\omega = 0$ implies $\bar{\partial}\omega = 0$, hence every harmonic form of bidegree $(p, 0)$ is holomorphic.

8.6.4 The Hodge numbers of Kähler manifolds

The complex dimension of $H_{\bar{\partial}}^{r,s}(M)$ is called the **Hodge number** $b^{r,s}$. The cohomology groups of a complex manifold are summarized by the **Hodge**

diamond,

$$\begin{pmatrix} & & & b^{m,m} & & & \\ & & b^{m,m-1} & & b^{m-1,m} & & \\ & & & \cdots & & & \\ b^{m,0} & b^{m-1,1} & & \cdots & & b^{1,m-1} & b^{0,m} \\ & & & \cdots & & & \\ & & b^{1,0} & & b^{0,1} & & \\ & & & b^{0,0} & & & \end{pmatrix}. \qquad (8.115)$$

These $(m+1)^2$ Hodge numbers are far from independent as we shall see later.

Theorem 8.10. Let M be a Kähler manifold with $\dim_{\mathbb{C}} M = m$. Then the Hodge numbers satisfy

$$\text{(a)} \qquad b^{r,s} = b^{s,r} \qquad\qquad (8.116)$$

$$\text{(b)} \qquad b^{r,s} = b^{m-r,m-s}. \qquad\qquad (8.117)$$

Proof. (a) If $\omega \in \Omega^{r,s}(M)$ is harmonic, it satisfies $\Delta_{\bar{\partial}}\omega = \Delta_{\partial}\omega = 0$. Then the (s,r)-form $\bar{\omega}$ is also harmonic, $\Delta_{\bar{\partial}}\bar{\omega} = 0$ since $\Delta_{\bar{\partial}}\bar{\omega} = \overline{\Delta_{\partial}\omega} = \overline{\Delta_{\bar{\partial}}\omega} = 0$ (note that $\Delta_{\partial} = \Delta_{\bar{\partial}}$). Thus, for any harmonic form of bidegree (r,s), there exists a harmonic form of bidegree (s,r) and *vice versa*. Thus, it follows that $b^{r,s} = b^{s,r}$. (b) Let $\omega \in \Omega^{r,s}(M)$ and $\psi \in H_{\bar{\partial}}^{m-r,m-s}(M)$. Then $\omega \wedge \psi$ is a volume element and it can be shown (Schwartz 1986) that $\int_M \omega \wedge \psi$ defines a *non-singular* map $H_{\bar{\partial}}^{r,s}(M) \times H_{\bar{\partial}}^{m-r,m-s}(M) \to \mathbb{C}$, hence the duality between $H_{\bar{\partial}}^{r,s}(M)$ and $H_{\bar{\partial}}^{m-r,m-s}(M)$. This shows that $H_{\bar{\partial}}^{r,s}(M)$ is isomorphic to $H_{\bar{\partial}}^{m-r,m-s}(M)$ as a vector space and it follows that $\dim_{\mathbb{C}} H_{\bar{\partial}}^{r,s}(M) = \dim_{\mathbb{C}} H_{\bar{\partial}}^{m-r,m-s}(M)$ hence $b^{r,s} = b^{m-r,m-s}$. $\qquad\square$

Accordingly, the Hodge diamond of a Kähler manifold is symmetric about the vertical and horizontal lines. These symmetries reduce the number of independent Hodge numbers to $(\frac{1}{2}m + 1)^2$ if m is even and $\frac{1}{4}(m+1)(m+3)$ if m is odd.

In a general Hermitian manifold, there are no direct relations between the Betti numbers and the Hodge numbers. If M is a Kähler manifold, however, theorem 8.11 establishes close relationships between them.

Theorem 8.11. Let M be a Kähler manifold with $\dim_{\mathbb{C}} M = m$ and $\partial M = \emptyset$. Then the Betti numbers b^p $(1 \le p \le 2m)$ satisfy the following conditions;

$$\text{(a)} \qquad b^p = \sum_{r+s=p} b^{r,s} \qquad\qquad (8.118)$$

$$\text{(b)} \qquad b^{2p-1} \text{ is even} \qquad (1 \le p \le m) \qquad (8.119)$$

$$\text{(c)} \qquad b^{2p} \ge 1 \qquad\qquad (1 \le p \le m) \qquad (8.120)$$

Proof. (a) $H_{\bar{\partial}}^{r,s}(M)$ is a complex vector space spanned by $\Delta_{\bar{\partial}}$-harmonic (r, s)-forms, $H_{\bar{\partial}}^{r,s}(M) = \{[\omega]|\omega \in \Omega^{r,s}(M), \Delta_{\bar{\partial}}\omega = 0\}$. Note also that, $H^p(M)$ is a real vector space spanned by Δ-harmonic p-forms, $H^p(M) = \{[\omega]|\omega \in \Omega^p(M), \Delta\omega = 0\}$. Then the complexification of $H^p(M)$ is $H^p(M)^{\mathbb{C}} = \{[\omega]|\omega \in \Omega^p(M)^{\mathbb{C}}, \Delta\omega = 0\}$. Since M is Kähler, any form ω which satisfies $\Delta_{\bar{\partial}}\omega = 0$ also satsifies $\Delta\omega = 0$ and *vice versa*. Since

$$\Omega^p(M)^{\mathbb{C}} = \oplus_{r+s=p}\Omega^{r,s}(M)$$

we find that

$$H^p(M)^{\mathbb{C}} = \oplus_{r+s=p}H^{r,s}(M).$$

Noting that $\dim_{\mathbb{R}} H^p(M) = \dim_{\mathbb{C}} H^p(M)^{\mathbb{C}}$, we obtain $b^p = \sum_{r+s=p} b^{r,s}$.

(b) From (a) and (8.116), it follows that

$$b^{2p-1} = \sum_{r+s=2p-1} b^{r,s} = 2 \sum_{\substack{r+s=2p-1 \\ r>s}} b^{r,s}.$$

Thus, b^{2p-1} must be even.

(c) The crucial observation is that the Kähler form Ω is a closed *real* two-form, $d\Omega = 0$, and the real $2p$-form

$$\Omega^p = \underbrace{\Omega \wedge \ldots \wedge \Omega}_{p}$$

is also closed, $d\Omega^p = 0$. We show that Ω^p is not exact. Suppose $\Omega^p = d\eta$ for some $\eta \in \Omega^{2p-1}(M)$. Then $\Omega^m = \Omega^{m-p} \wedge \Omega^p = d(\Omega^{m-p} \wedge \eta)$. It follows from Stokes' theorem that

$$\int_M \Omega^m = \int_M d(\Omega^{m-p} \wedge \eta) = \int_{\partial M} \Omega^{m-p} \wedge \eta = 0.$$

Since the LHS is the volume of M, this is in contradiction. Thus, there is at least one non-trivial element of $H^{2p}(M)$ and we have proved that $b^{2p} \geq 1$. $\square$

If a Kähler manifold is Ricci flat, there exists an extra relationship among the Hodge numbers, which further reduces the independent Hodge numbers, see Horowitz (1986) and Candelas (1988).

8.7 Almost complex manifolds

This and the next sections deal with spaces which are closely related to complex manifolds. These are somewhat specialized topics and may be omitted on a first reading.

8.7.1 Definitions

There are some differentiable manifolds which carry a similar structure to
complex manifolds. To study these manifolds, we somewhat relax the condition
(8.16) and require a weaker condition here.

Definition 8.5. Let M be a differentiable manifold. The pair (M, J), or simply
M, is called an **almost complex manifold** if there exists a tensor field J of type
$(1, 1)$ such that at each point p of M, $J_p^2 = -\mathrm{id}_{T_pM}$. The tensor field J is also
called the **almost complex structure**.

Since $J_p^2 = -\mathrm{id}_{T_pM}$, J_p has eigenvalues $\pm \mathrm{i}$. If there are $m + \mathrm{i}$, then there
must be an equal number of $-\mathrm{i}$, hence J_p is a $2m \times 2m$ matrix and $J_p^2 = -I_{2m}$.
Thus, M is an even-dimensional manifold. Note that not all even-dimensional
manifolds are almost complex manifolds. For example, S^4 is not an almost
complex manifold (Steenrod 1951). Note also that we now require a weaker
condition $J_p^2 = -I_{2m}$. Of course, the tensor J_p defined by (8.16) satisfies
$J_p^2 = -I_{2m}$, hence a complex manifold is an almost complex manifold. There
are almost complex manifolds which are not complex manifolds. For example, it
is known that S^6 admits an almost complex structure, although it is *not* a complex
manifold (Fröhlicher 1955).

Let us complexify a tangent space of an almost complex manifold (M, J).
Given a linear transformation J_p at T_pM such that $J_p^2 = -I_{2m}$, we extend J_p to a
$\mathbb{C}$-linear map defined on $T_pM^{\mathbb{C}}$. J_p defined on $T_pM^{\mathbb{C}}$ also satisfies $J_p^2 = -I_{2m}$,

$$J_p^2(X + \mathrm{i}Y) = J_p^2 X + \mathrm{i}J_p^2 Y = -X + \mathrm{i}(-Y) = -(X + \mathrm{i}Y)$$

where $X, Y \in T_pM$. Let us divide $T_pM^{\mathbb{C}}$ into two disjoint vector subspaces,
according to the eigenvalue of J_p,

$$T_pM^{\mathbb{C}} = T_pM^+ \oplus T_pM^- \tag{8.121}$$

where

$$T_pM^{\pm} = \{Z \in T_pM^{\mathbb{C}} | J_pZ = \pm \mathrm{i}Z\}. \tag{8.122}$$

Any vector $V \in T_pM^{\mathbb{C}}$ is written as $V = W_1 + \overline{W}_2$, where $W_1, W_2 \in T_pM^+$.
Note that $J_pV = \mathrm{i}W_1 - \mathrm{i}\overline{W}_2$. At this stage the reader might have noticed that we
can follow the classification scheme of vectors and vector fields developed for the
complex manifolds in section 8.2. In fact, the only difference is that on a complex
manifold the almost complex structure is explicitly given by (8.18), while on
an almost complex manifold, it is required to satisfy the less strict condition
$J_p^2 = -I_{2m}$. To classify the complexified tangent spaces and complexified vector
spaces, we only need the latter condition. Accordingly, we separate $T_pM^{\mathbb{C}}$ into
$T_pM^{\pm}$ and $\mathfrak{X}(M)^{\mathbb{C}}$ into $\mathfrak{X}(M)^{\pm}$, although there does not necessarily exist a basis

of $T_p M^+$ of the form $\{\partial/\partial z^\mu\}$. For example, we may still define the projection operators

$$\mathcal{P}^\pm \equiv \tfrac{1}{2}(\mathrm{id}_{T_p M} \mp iJ_p) : T_p M^\mathbb{C} \to T_p M^\pm. \tag{8.123}$$

We call a vector in $T_p M^+ (T_p M^-)$ a holomorphic (anti-holomorphic) vector and a vector field in $\mathfrak{X}(M)^+ (\mathfrak{X}(M)^-)$ a holomorphic (anti-holomorphic) vector field.

Definition 8.6. Let (M, J) be an almost complex manifold. If the Lie bracket of any holomorphic vector fields $X, Y \in \mathfrak{X}^+(M)$ is again a holomorphic vector field, $[X, Y] \in \mathfrak{X}^+(M)$, the almost complex structure J is said to be **integrable**.

Let (M, J) be an almost complex manifold. Define the **Nijenhuis tensor field** $N : \mathfrak{X}(M) \times \mathfrak{X}(M) \to \mathfrak{X}(M)$ by

$$N(X, Y) \equiv [X, Y] + J[JX, Y] + J[X, JY] - [JX, JY]. \tag{8.124}$$

Given a basis $\{e^\mu = \partial/\partial x^\mu\}$ and the dual basis $\{dx^\mu\}$, the almost complex structure is expressed as $J = J_\mu{}^\nu dx^\mu \otimes \partial/\partial x^\nu$. The component expression of N is

$$
\begin{aligned}
N(X, Y) &= (X^\nu \partial_\nu Y^\mu - Y^\nu \partial_\nu X^\mu) e_\mu \\
&\quad + J_\lambda{}^\mu \{J_\kappa{}^\nu X^\kappa \partial_\nu Y^\lambda - Y^\nu \partial_\nu (J_\kappa{}^\lambda X^\kappa)\} e_\mu \\
&\quad + J_\lambda{}^\mu \{X^\nu \partial_\nu (J_\kappa{}^\lambda Y^\kappa) - J_\kappa{}^\nu Y^\kappa \partial_\nu X^\lambda\} e_\mu \\
&\quad - \{J_\kappa{}^\nu X^\kappa \partial_\nu (J_\lambda{}^\mu Y^\lambda) - J_\kappa{}^\nu Y^\kappa \partial_\nu (J_\lambda{}^\mu X^\lambda)\} e_\mu \\
&= X^\kappa Y^\nu [-J_\lambda{}^\mu (\partial_\nu J_\kappa{}^\lambda) + J_\lambda{}^\mu (\partial_\kappa J_\nu{}^\lambda) \\
&\quad - J_\kappa{}^\lambda (\partial_\lambda J_\nu{}^\mu) + J_\nu{}^\lambda (\partial_\lambda J_\kappa{}^\mu)] e_\mu. \tag{8.125}
\end{aligned}
$$

Thus, N is indeed linear in X and Y and hence a tensor. If J is a complex structure, J is given by (8.18) and the Nijenhuis tensor field trivially vanishes.

Theorem 8.12. An almost complex structure J on a manifold M is integrable if and only if $N(A, B) = 0$ for any $A, B \in \mathfrak{X}(M)$.

Proof. Let $Z = X + iY, W = U + iV \in \mathfrak{X}(M)^\mathbb{C}$. We extend the Nijenhuis tensor field so that its action on vector fields in $\mathfrak{X}(M)^\mathbb{C}$ is given by

$$
\begin{aligned}
N(Z, W) &= [Z, W] + J[JZ, W] + J[Z, JW] - [JZ, JW] \\
&= \{N(X, U) - N(Y, V)\} + i\{N(X, V) + N(Y, U)\}. \tag{8.126}
\end{aligned}
$$

Suppose that $N(A, B) = 0$ for any $A, B \in \mathfrak{X}(M)$. From (8.126), it turns out that $N(Z, W) = 0$ for $Z, W \in \mathfrak{X}^\mathbb{C}(M)$. Let $Z, W \in \mathfrak{X}^+(M) \subset \mathfrak{X}(M)^\mathbb{C}$. Since $JZ = iZ$ and $JW = iW$, we have $N(Z, W) = 2\{[Z, W] + iJ[Z, W]\}$. By assumption, $N(Z, W) = 0$ and we find $[Z, W] = -iJ[Z, W]$ or $J[Z, W] =$

i$[Z, W]$, that is, $[Z, W] \in \mathfrak{X}^+(M)$. Thus, the almost complex structure is integrable.

Conversely, suppose that J is integrable. Since $\mathfrak{X}^{\mathbb{C}}(M)$ is a direct sum of $\mathfrak{X}^+(M)$ and $\mathfrak{X}^-(M)$, we can separate $Z, W \in \mathfrak{X}^{\mathbb{C}}(M)$ as $Z = Z^+ + Z^-$ and $W = W^+ + W^-$. Then

$$N(Z, W) = N(Z^+, W^+) + N(Z^+, W^-) + N(Z^-, W^+) + N(Z^-, W^-).$$

Since $JZ^\pm = \pm iZ^\pm$ and $JW^\pm = \pm iW^\pm$, it is easy to see that $N(Z^+, W^-) = N(Z^-, W^+) = 0$. We also have

$$\begin{aligned}
N(Z^+, W^+) &= [Z^+, W^+] + J[iZ^+, W^+] + J[Z^+, iW^+] - [iZ^+, iW^+] \\
&= 2[Z^+, W^+] - 2[Z^+, W^+] = 0
\end{aligned}$$

since $J[Z^+, W^+] = i[Z^+, W^+]$. Similarly, $N(Z^-, W^-)$ vanishes and we have shown that $N(Z, W) = 0$ for any $Z, W \in \mathfrak{X}^{\mathbb{C}}(M)$. In particular, it should vanish for $Z, W \in \mathfrak{X}(M)$. $\qquad\square$

If M is a complex manifold, the complex structure J is a constant tensor field and the Nijenhuis tensor field vanishes. What about the converse? We now state an important (and difficult to prove) theorem.

Theorem 8.13. (Newlander and Nirenberg 1957) Let (M, J) be a $2m$-dimensional almost complex manifold. If J is integrable, the manifold M is a complex manifold with the almost complex structure J.

In summary we have:

$$\frac{\text{Integrable almost}}{\text{complex structure}} = \frac{\text{Vanishing Nijenhuis}}{\text{tensor field}} = \text{Complex manifold.}$$

8.8 Orbifolds

Let M be a manifold and let G be a *discrete* group which acts on M. Then the quotient space $\Gamma \equiv M/G$ is called an **orbifold**. As we will see later there are fixed points in M, which do not transform under the action of G. These points are singular and the orbifold is not a manifold in general. Thus, even though we start with a simple manifold M, the orbifold M/G may have quite a complicated topology.

8.8.1 One-dimensional examples

To obtain a concrete idea, let us consider a simple example. Take $M = \mathbb{R}^2$ which is to be identified with the complex plane $\mathbb{C}$. Let us take $G = \mathbb{Z}_3$ and identify the points z, $e^{2\pi i/3}z$ and $e^{4\pi i/3}z$. The orbifold M/G consists of a third of the

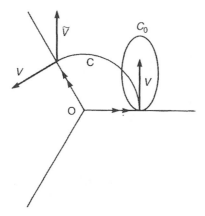

Figure 8.6. The orbifold $\mathbb{C}/\mathbb{Z}_3$ is a third of the complex plane. The edges of the orbifold are identified as shown in the figure. V becomes a vector $\tilde{V}$ after parallel transportation along C. The angle between V and $\tilde{V}$ is $2\pi/3$.

complex plane and after the identification of the edges we end up with a cone, see figure 8.6. It is interesting to see what the holonomy group of this orbifold is. We use the flat connection induced by the Euclidean metric of $\mathbb{C}$. Then, after the parallel transport of a vector V along the loop C (this is indeed a loop!), we obtain a vector $\tilde{V}$ which is different from V after the identification. Observe that the angle between V and $\tilde{V}$ is $2\pi/3$. It is easy to verify that the holomony group is $\mathbb{Z}_3$. Since the holonomy is trivial for the loop C_0 which does not encircle the origin, we find that the curvature is singular at the origin (recall that the curvature measures the non-triviality of the holonomy, see section 7.3). In general the fixed points (the origin in the present case) are singular points of the curvature. Note, however, that $\mathbb{C}/\mathbb{Z}_3$ *is* a manifold since it has an open covering homeomorphic to $\mathbb{R}^2$.

A less trivial example is obtained by taking the torus as the manifold. We identify the points z and $z + m + n e^{i\pi/3}$ ($m, n \in \mathbb{Z}$) in the complex plane; see figure 8.7(a). If we identify the edges of the parallelogram OPQR, we have the torus T^2. Let $\mathbb{Z}_3$ act on T^2 as $\alpha : z \mapsto e^{2\pi i/3}z$. We find that there are three inequivalent fixed points $z = (n/\sqrt{3})e^{\pi i/6}$ where $n = 0, 1$ and 2. This orbifold $\Gamma = \mathbb{C}/\mathbb{Z}_3$ consists of two triangles surrounding a hollow; see figure 8.7(b). If the flat connection induced by the flat metric of the torus is employed to define the parallel transport of vectors, we find that the holonomy around each fixed point is $\mathbb{Z}_3$.

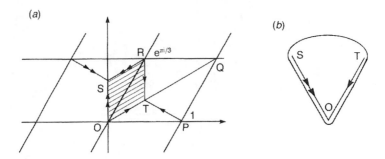

Figure 8.7. Under the action of $\mathbb{Z}_3$, points of the torus T^2 are identified. The shaded area is the orbifold $\Gamma = T^2/\mathbb{Z}_3$. If the edges of the orbifold are identified, we end up with the object in figure 8.7(b), which is homeomorphic to the sphere S^2.

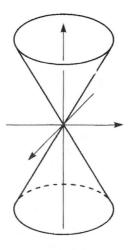

Figure 8.8. The conical singularity. The origin does not look like $\mathbb{R}^n$ or $\mathbb{C}^n$.

8.8.2 Three-dimensional examples

Orbifolds with three complex dimensions have been proposed as candidates for superstring compactification. The detailed treatment of this subject is outside the scope of this book and the reader should consult Dixson *et al* (1985, 1986) and Green *et al* (1987).

Let $T = \mathbb{C}^3/L$ be a three-dimensional complex torus, where L is a lattice in $\mathbb{C}^3$. For definiteness, let (z_1, z_2, z_3) be the coordinates of $\mathbb{C}^3$ and identify z_i and $z_i + m + ne^{\pi i/3}$. Under this identification, T is identified with a product of three tori, $T = T_1 \times T_2 \times T_3$. T admits, as before, the action of $\mathbb{Z}_3$ defined by $\alpha : z_i \mapsto e^{2\pi i/3} z_i$. If each z_i takes one of the values $0, (1/\sqrt{3})e^{i\pi/6}$, $(2/\sqrt{3})e^{\pi i/6}$, the action of α leaves the point (z_i) invariant. Thus, there are $3^3 = 27$ fixed points in the orbifold. In the present case, the fixed point is a conical singularity (figure 8.8) and the orbifold cannot be a manifold. [*Remarks*: The appearance of the conical singularity can be understood more easily from a simpler example. Let $(x, y) \in \mathbb{C}^2$ and let $\mathbb{Z}_2$ act on $\mathbb{C}^2$ as $(x, y) \mapsto \pm(x, y)$. Then the orbifold $\Gamma = \mathbb{C}^2/\mathbb{Z}_2$ has a conical singularity at the origin. In fact, let $[(x, y)] \to (x^2, xy, y^2) \equiv (X, Y, Z)$ be an embedding of Γ in $\mathbb{C}^3$. Note that X, Y and Z satisfy a relation $Y^2 = XZ$. If X, Y and Z are thought of as real variables, this is simply the equation of a cone.]

9

FIBRE BUNDLES

A manifold is a topological space which looks locally like $\mathbb{R}^m$, but not necessarily so globally. By introducing a chart, we give a local Euclidean structure to a manifold, which enables us to use the conventional calculus of several variables. A fibre bundle is, so to speak, a topological space which looks locally like a direct product of two topological spaces. Many theories in physics, such as general relativity and gauge theories, are described naturally in terms of fibre bundles.

Relevant references are Choquet-Bruhat *et al* (1982), Eguchi *et al* (1980) and Nash and Sen (1983). A complete analysis is found in Kobayashi and Nomizu (1963, 1969) and Steenrod (1951).

9.1 Tangent bundles

For clarification, we begin our exposition with a motivating example. A **tangent bundle** TM over an m-dimensional manifold M is a collection of all the tangent spaces of M:

$$TM \equiv \bigcup_{p \in M} T_p M. \tag{9.1}$$

The manifold M over which TM is defined is called the **base space**. Let $\{U_i\}$ be an open covering of M. If $x^\mu = \varphi_i(p)$ is the coordinate on U_i, an element of

$$TU_i \equiv \bigcup_{p \in U_i} T_p M$$

is specified by a point $p \in M$ and a vector $V = V^\mu(p)(\partial/\partial x^\mu)|_p \in T_p M$. Noting that U_i is homeomorphic to an open subset $\varphi(U_i)$ of $\mathbb{R}^m$ and each $T_p M$ is homeomorphic to $\mathbb{R}^m$, we find that TU_i is identified with a direct product $\mathbb{R}^m \times \mathbb{R}^m$ (figure 9.1). If $(p, V) \in TU_i$, the identification is given by $(p, V) \mapsto (x^\mu(p), V^\mu(p))$. TU_i is a $2m$-dimensional differentiable manifold. What is more, TU_i is decomposed into a direct product $U_i \times \mathbb{R}^m$. If we pick up a point u of TU_i, we can systematically decompose the information u contains into a point $p \in M$ and a vector $V \in T_p M$. Thus, we are naturally led to the concept of **projection** $\pi : TU_i \to U_i$ (figure 9.1). For any point $u \in TU_i$, $\pi(u)$ is a point $p \in U_i$ at which the vector is defined. The information about the vector

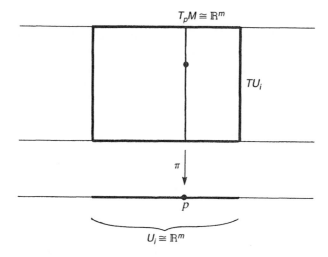

Figure 9.1. A local piece $TU_i \simeq \mathbb{R}^m \times \mathbb{R}^m$ of a tangent bundle TM. The projection π projects a vector $V \in T_pM$ to p.

is completely lost under the projection. Observe that $\pi^{-1}(p) = T_pM$. In the context of the theory of fibre bundles, T_pM is called the **fibre** at p.

It is obvious by construction that if $M = \mathbb{R}^m$, the tangent bundle itself is expressed as a direct product $\mathbb{R}^m \times \mathbb{R}^m$. However, this is not always the case and the non-trivial structure of the tangent bundle measures the topological non-triviality of M. To see this, we have to look not only at a single chart U_i but also at other charts. Let U_j be a chart such that $U_i \cap U_j \neq \emptyset$ and let $y^\mu = \psi(p)$ be the coordinates on U_j. Take a vector $V \in T_pM$ where $p \in U_i \cap U_j$. V has two coordinate presentations,

$$V = V^\mu \frac{\partial}{\partial x^\mu}\bigg|_p = \tilde{V}^\mu \frac{\partial}{\partial y^\mu}\bigg|_p. \tag{9.2}$$

It is easy to see that they are related as

$$\tilde{V}^\nu = \frac{\partial y^\nu}{\partial x^\mu}(p)V^\mu. \tag{9.3}$$

For $\{x^\mu\}$ and $\{y^\nu\}$ to be good coordinate systems, the matrix $(G_\mu^\nu) \equiv (\partial y^\nu/\partial x^\mu)$ must be non-singular: $(G_\mu^\nu) \in GL(m, \mathbb{R})$. Thus, fibre coordinates are rotated by an element of $GL(m, \mathbb{R})$ whenever we change the coordinates. The *group* $GL(m, \mathbb{R})$ is called the **structure group** of TM. In this way fibres are interwoven together to form a tangent bundle, which consequently may have quite a complicated topological structure.

We note *en passant* that the projection π can be defined globally on M. It is obvious that $\pi(u) = p$ does not depend on a special coordinate chosen. Thus, $\pi : TM \to M$ is defined globally with no reference to local charts.

Let $X \in \mathfrak{X}(M)$ be a vector field on M. X assigns a vector $X|_p \in T_pM$ at each point $p \in M$. From our viewpoint, X is looked upon as a smooth map $M \to TM$. This map is not utterly arbitrary since a point p must be mapped to a point $u \in TM$ such that $\pi(u) = p$. We define a **section** (or a **cross section**) of TM as a smooth map $s : M \to TM$ such that $\pi \circ s = \mathrm{id}_M$. If a section $s_i : U_i \to TU_i$ is defined only on a chart U_i, it is called a **local section**.

9.2   Fibre bundles

The tangent bundle in the previous section is an example of a more general framework called a fibre bundle. Definitions are now in order.

9.2.1   Definitions

Definition 9.1. A (differentiable) fibre bundle (E, π, M, F, G) consists of the following elements:

(i) A differentiable manifold E called the **total space**.
(ii) A differentiable manifold M called the **base space**.
(iii) A differentiable manifold F called the **fibre** (or **typical fibre**).
(iv) A surjection $\pi : E \to M$ called the **projection**. The inverse image $\pi^{-1}(p) = F_p \cong F$ is called the fibre at p.
(v) A Lie group G called the **structure group**, which acts on F on the left.
(vi) A set of open covering $\{U_i\}$ of M with a diffeomorphism $\phi_i : U_i \times F \to \pi^{-1}(U_i)$ such that $\pi \circ \phi_i(p, f) = p$. The map ϕ_i is called the **local trivialization** since ϕ_i^{-1} maps $\pi^{-1}(U_i)$ *onto* the direct product $U_i \times F$.
(vii) If we write $\phi_i(p, f) = \phi_{i,p}(f)$, the map $\phi_{i,p} : F \to F_p$ is a diffeomorphism. On $U_i \cap U_j \neq \emptyset$, we require that $t_{ij}(p) \equiv \phi_{i,p}^{-1} \circ \phi_{j,p} : F \to F$ be an element of G. Then ϕ_i and ϕ_j are related by a smooth map $t_{ij} : U_i \cap U_j \to G$ as (figure 9.2)

$$\phi_j(p, f) = \phi_i(p, t_{ij}(p)f). \tag{9.4}$$

The maps t_{ij} are called the **transition functions**.

[*Remarks*: We often use a shorthand notation $E \xrightarrow{\pi} M$ or simply E to denote a fibre bundle (E, π, M, F, G).

Strictly speaking, the definition of a fibre bundle should be independent of the special covering $\{U_i\}$ of M. In the mathematical literature, this definition is employed to define a **coordinate bundle** $(E, \pi, M, F, G, \{U_i\}, \{\phi_i\})$. Two coordinate bundles $(E, \pi, M, F, G, \{U_i\}, \{\phi_i\})$ and $(E, \pi, M, F, G, \{V_i\}, \{\psi_i\})$ are said to be equivalent if $(E, \pi, M, F, G, \{U_i\} \cup \{V_j\}, \{\phi_i\} \cup \{\psi_j\})$ is again a coordinate bundle. A fibre bundle is defined as an equivalence class of coordinate bundles. In practical applications in physics, however, we always employ a certain

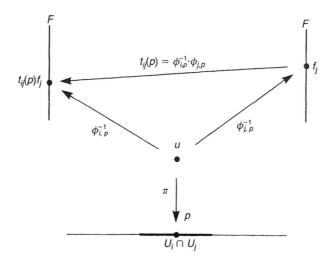

Figure 9.2. On the overlap $U_i \cap U_j$, two elements f_i, $f_j \in F$ are assigned to $u \in \pi^{-1}(p)$, $p \in U_i \cap U_j$. They are related by $t_{ij}(p)$ as $f_i = t_{ij}(p)f_j$.

definite covering and make no distinction between a coordinate bundle and a fibre bundle.]

We need to clarify several points. Let us take a chart U_i of the base space M. $\pi^{-1}(U_i)$ is a direct product diffeomorphic to $U_i \times F$, $\phi_i^{-1} : \pi^{-1}(U_i) \to U_i \times F$ being the diffeomorphism. If $U_i \cap U_j \ne \varnothing$, we have two maps ϕ_i and ϕ_j on $U_i \cap U_j$. Let us take a point u such that $\pi(u) = p \in U_i \cap U_j$. We then assign two elements of F, one by ϕ_i^{-1} and the other by ϕ_j^{-1},

$$\phi_i^{-1}(u) = (p, f_i), \qquad \phi_j^{-1}(u) = (p, f_j) \tag{9.5}$$

see figure 9.2. There exists a map $t_{ij} : U_i \cap U_j \to G$ which relates f_i and f_j as $f_i = t_{ij}(p)f_j$. This is also written as (9.4).

We require that the transition functions satisfy the following consistency conditions:

$$t_{ii}(p) = \text{identity map} \qquad (p \in U_i) \tag{9.6a}$$

$$t_{ij}(p) = t_{ji}(p)^{-1} \qquad (p \in U_i \cap U_j) \tag{9.6b}$$

$$t_{ij}(p) \cdot t_{jk}(p) = t_{ik}(p) \qquad (p \in U_i \cap U_j \cap U_k). \tag{9.6c}$$

Unless these conditions are satisfied, local pieces of a fibre bundle cannot be glued together consistently. If all the transition functions can be taken to be identity maps, the fibre bundle is called a **trivial bundle**. A trivial bundle is a direct product $M \times F$.

Given a fibre bundle $E \xrightarrow{\pi} M$, the possible set of transition functions is obviously far from unique. Let $\{U_i\}$ be a covering of M and $\{\phi_i\}$ and $\{\tilde{\phi}_i\}$ be two sets of local trivializations giving rise to the same fibre bundle. The transition functions of respective local trivializations are

$$t_{ij}(p) = \phi_{i,p}^{-1} \circ \phi_{j,p} \tag{9.7a}$$

$$\tilde{t}_{ij}(p) = \tilde{\phi}_{i,p}^{-1} \circ \tilde{\phi}_{j,p}. \tag{9.7b}$$

Define a map $g_i(p) : F \to F$ at each point $p \in M$ by

$$g_i(p) \equiv \phi_{i,p}^{-1} \circ \tilde{\phi}_{i,p}. \tag{9.8}$$

We require that $g_i(p)$ be a homeomorphism which belongs to G. This requirement must certainly be fulfilled if $\{\phi_i\}$ and $\{\tilde{\phi}_i\}$ describe the same fibre bundle. It is easily seen from (9.7) and (9.8) that

$$\tilde{t}_{ij}(p) = g_i(p)^{-1} \circ t_{ij}(p) \circ g_j(p). \tag{9.9}$$

In the practical situations which we shall encounter later, t_{ij} are the gauge transformations required for pasting local charts together, while g_i corresponds to the gauge degrees of freedom within a chart U_i. If the bundle is trivial, we may put all the transition functions to be identity maps. Then the most general form of the transition functions is

$$t_{ij}(p) = g_i(p)^{-1} g_j(p). \tag{9.10}$$

Let $E \xrightarrow{\pi} M$ be a fibre bundle. A **section** (or a **cross section**) $s : M \to E$ is a smooth map which satisfies $\pi \circ s = \mathrm{id}_M$. Clearly, $s(p) = s|_p$ is an element of $F_p = \pi^{-1}(p)$. The set of sections on M is denoted by $\Gamma(M, F)$. If $U \subset M$, we may talk of a **local section** which is defined only on U. $\Gamma(U, F)$ denotes the set of local sections on U. For example, $\Gamma(M, TM)$ is identified with the set of vector fields $\mathfrak{X}(M)$. It should be noted that not all fibre bundles admit global sections.

Example 9.1. Let E be a fibre bundle $E \xrightarrow{\pi} S^1$ with a typical fibre $F = [-1, 1]$. Let $U_1 = (0, 2\pi)$ and $U_2 = (-\pi, \pi)$ be an open covering of S^1 and let $A = (0, \pi)$ and $B = (\pi, 2\pi)$ be the intersection $U_1 \cap U_2$, see figure 9.3. The local trivializations ϕ_1 and ϕ_2 are given by

$$\phi_1^{-1}(u) = (\theta, t), \qquad \phi_2^{-1}(u) = (\theta, t)$$

for $\theta \in A$ and $t \in F$. The transition function $t_{12}(\theta)$, $\theta \in A$, is the identity map $t_{12}(\theta) : t \mapsto t$. We have two choices on B;

(I) $\phi_1^{-1}(u) = (\theta, t), \phi_2^{-1}(u) = (\theta, t)$

(II) $\phi_1^{-1}(u) = (\theta, t), \phi_2^{-1}(u) = (\theta, -t)$

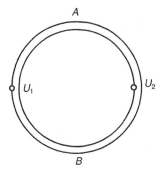

Figure 9.3. The base space S^1 and two charts U_1 and U_2 over which the fibre bundle is trivial.

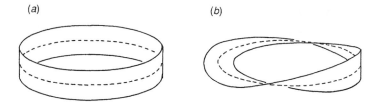

Figure 9.4. Two fibre bundles over S^1: (a) is the cylinder which is a trivial bundle $S^1 \times I$; (b) is the Möbius strip.

For case (I), we find that $t_{12}(\theta)$ is the identity map and two pieces of the local bundles are glued together to form a cylinder (figure 9.4(a)). For case (II), we have $t_{12}(\theta) : t \mapsto -t$, $\theta \in B$, and obtain the Möbius strip (figure 9.4(b)). Thus, a cylinder has the trivial structure group $G = \{e\}$ where e is the identity map of F onto F while the Möbius strip has $G = \{e, g\}$ where $g : t \mapsto -t$. Since $g^2 = e$, we find $G \cong \mathbb{Z}_2$. A cylinder is a trivial bundle $S^1 \times F$, while the Möbius strip is not. [*Remark*: The group $\mathbb{Z}_2$ is not a Lie group. This is the only occasion we use a discrete group for the structure group.]

9.2.2 Reconstruction of fibre bundles

What is the minimal information required to construct a fibre bundle? We now show that for given M, $\{U_i\}$, $t_{ij}(p)$, F and G, we can reconstruct the fibre bundle (E, π, M, F, G). This amounts to finding a unique π, E and ϕ_i from given data. Let us define

$$X \equiv \bigcup_i U_i \times F. \tag{9.11}$$

Introduce an equivalence relation $\sim$ between $(p, f) \in U_i \times F$ and $(q, f') \in U_j \times F$ by $(p, f) \sim (q, f')$ if and only if $p = q$ and $f' = t_{ij}(p)f$. A fibre

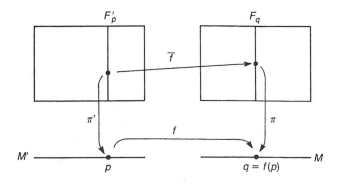

Figure 9.5. A bundle map $\bar{f} : E' \to E$ induces a map $f : M' \to M$.

bundle E is then defined as

$$E = X/\sim . \tag{9.12}$$

Denote an element of E by $[(p, f)]$. The projection is given by

$$\pi : [(p, f)] \mapsto p. \tag{9.13}$$

The local trivialization $\phi_i : U_i \times F \to \pi^{-1}(U_i)$ is given by

$$\phi_i : (p, f) \mapsto [(p, f)]. \tag{9.14}$$

The reader should verify that E, π and $\{\phi_i\}$ thus defined satisfy all the axioms of fibre bundles. Thus, the given data reconstruct a fibre bundle E uniquely.

This procedure may be employed to construct a new fibre bundle from an old one. Let (E, π, M, F, G) be a fibre bundle. Associated with this bundle is a new bundle whose base space is M, transition function $t_{ij}(p)$, structure group G and fibre F' on which G acts. Examples of associated bundles will be given later.

9.2.3 Bundle maps

Let $E \xrightarrow{\pi} M$ and $E' \xrightarrow{\pi'} M'$ be fibre bundles. A *smooth* map $\bar{f} : E' \to E$ is called a **bundle map** if it maps each fibre F'_p of E' onto F_q of E. Then $\bar{f}$ naturally induces a smooth map $f : M' \to M$ such that $f(p) = q$ (figure 9.5). Observe that the diagram

$$
\begin{array}{ccc}
E' & \xrightarrow{\bar{f}} & E \\
\pi' \downarrow & & \downarrow \pi \\
M' & \xrightarrow{f} & M
\end{array}
\qquad
\left(
\begin{array}{ccc}
u & \xrightarrow{\bar{f}} & \bar{f}(u) \\
\pi' \downarrow & & \downarrow \pi \\
p & \xrightarrow{f} & q
\end{array}
\right)
\tag{9.15}
$$

commutes. [*Caution*: A smooth map $\bar{f} : E' \to E$ is not necessarily a bundle map. It may map $u, v \in F'_p$ of E' to $\bar{f}(u)$ and $\bar{f}(v)$ on different fibres of E so that $\pi(\bar{f}(u)) \neq \pi(\bar{f}(v))$.]

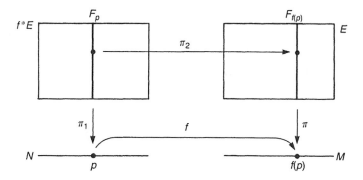

Figure 9.6. Given a fibre bundle $E \xrightarrow{\pi} M$, a map $f : N \to M$ defines a pullback bundle f^*E over N.

9.2.4 Equivalent bundles

Two bundles $E' \xrightarrow{\pi'} M$ and $E \xrightarrow{\pi} M$ are equivalent if there exists a bundle map $\bar{f} : E' \to E$ such that $f : M \to M$ is the identity map and $\bar{f}$ is a diffeomorphism:

$$\begin{array}{ccc} E' & \xrightarrow{\bar{f}} & E \\ \pi' \downarrow & & \downarrow \pi \\ M & \xrightarrow{\mathrm{id}_M} & M. \end{array} \qquad (9.16)$$

This definition of equivalent bundles is in harmony with that given in the remarks following definition 9.1.

9.2.5 Pullback bundles

Let $E \xrightarrow{\pi} M$ be a fibre bundle with typical fibre F. If a map $f : N \to M$ is given, the pair (E, f) defines a new fibre bundle over N with the same fibre F (figure 9.6). Let f^*E be a subspace of $N \times E$, which consists of points (p, u) such that $f(p) = \pi(u)$. $f^*E \equiv \{(p, u) \in N \times E | f(p) = \pi(u)\}$ is called the **pullback** of E by f. The fibre F_p of f^*E is just a copy of the fibre $F_{f(p)}$ of E. If we define $f^*E \xrightarrow{\pi_1} N$ by $\pi_1 : (p, u) \mapsto p$ and $f^*E \xrightarrow{\pi_2} E$ by $(p, u) \mapsto u$, the pullback f^*E may be endowed with the structure of a fibre bundle and we obtain the following bundle map,

$$\begin{array}{ccc} f^*E & \xrightarrow{\pi_2} & E \\ \pi_1 \downarrow & & \downarrow \pi \\ N & \xrightarrow{f} & M \end{array} \qquad \left(\begin{array}{ccc} (p, u) & \xrightarrow{\pi_2} & u \\ \pi_1 \downarrow & & \downarrow \pi \\ p & \xrightarrow{f} & f(p) \end{array} \right). \qquad (9.17)$$

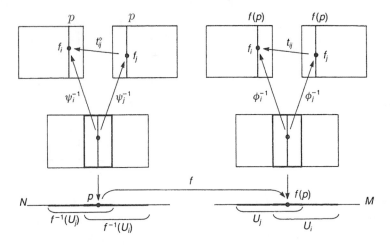

Figure 9.7. The transition function t_{ij}^* of the pullback bundle f^*E is a pullback of the transition function t_{ij} of E.

The commutativity of the diagram follows since $\pi(\pi_2(p, u)) = \pi(u) = f(p) = f(\pi_1(p, u))$ for $(p, u) \in f^*E$. In particular, if $N = M$ and $f = \mathrm{id}_M$, then two fibre bundles f^*E and E are equivalent.

Let $\{U_i\}$ be a covering of M and $\{\phi_i\}$ be local trivializations. $\{f^{-1}(U_i)\}$ defines a covering of N such that f^*E is locally trivial. Take $u \in E$ such that $\pi(u) = f(p) \in U_i$ for some $p \in N$. If $\phi_i^{-1}(u) = (f(p), f_i)$ we find $\psi_i^{-1}(p, u) = (p, f_i)$ where ψ_i is the local trivialization of f^*E. The transition function t_{ij} at $f(p) \in U_i \cap U_j$ maps f_j to $f_i = t_{ij}(f(p))f_j$. The corresponding transition function t_{ij}^* of f^*E at $p \in f^{-1}(U_i) \cap f^{-1}(U_j)$ also maps f_j to f_i; see figure 9.7. This shows that

$$t_{ij}^*(p) = t_{ij}(f(p)). \tag{9.18}$$

Example 9.2. Let M and N be differentiable manifolds with $\dim M = \dim N = m$. Let $f : N \to M$ be a smooth map. The map f induces a map $\pi_2 : TN \to TM$ such that the following diagram commutes:

$$
\begin{array}{ccc}
TN & \xrightarrow{\pi_2} & TM \\
{\scriptstyle \pi_1}\downarrow & & \downarrow{\scriptstyle \pi} \\
N & \xrightarrow{\ f\ } & M.
\end{array}
\tag{9.19}
$$

Let $W = W^\nu \partial/\partial y^\nu$ be a vector of T_pN and $V = V^\mu \partial/\partial x^\mu$ be the corresponding vector of $T_{f(p)}M$. If TN is a pullback bundle $f^*(TM)$, π_2 maps T_pN to $T_{f(p)}M$ diffeomorphically. This is possible if and only if π_2 has the maximal rank m at each point of TN. Let $\varphi(f(p)) = (f^1(y), \ldots, f^m(y))$ be the coordinates of $f(p)$ in a chart (U, φ) of M, where $y = \varphi(p)$ are the coordinates of p in a chart (V, ψ) of N. The maximal rank condition is given by $\det(\partial f^\mu(y)/\partial y^\nu) \neq 0$ for any $p \in N$.

9.2.6 Homotopy axiom

Let f and g be maps from M' to M. They are said to be **homotopic** if there exists a smooth map $F : M' \times [0, 1] \to M$ such that $F(p, 0) = f(p)$ and $F(p, 1) = g(p)$ for any $p \in M'$, see section 4.2.

Theorem 9.1. Let $E \xrightarrow{\pi} M$ be a fibre bundle with fibre F and let f and g be homotopic maps from N to M. Then f^*E and g^*E are equivalent bundles over N.

The proof is found in Steenrod (1951). Let M be a manifold which is contractible to a point. Then there exists a homotopy $F : M \times I \to M$ such that

$$F(p, 0) = p \qquad F(p, 1) = p_0$$

where $p_0 \in M$ is a fixed point. Let $E \xrightarrow{\pi} M$ be a fibre bundle over M and consider pullback bundles h_0^*E and h_1^*E, where $h_t(p) \equiv F(p, t)$. The fibre bundle h_1^*E is a pullback of a fibre bundle $\{p_0\} \times F$ and hence is a trivial bundle: $h_1^*E \simeq M \times F$. However, $h_0^*E = E$ since h_0 is the identity map. According to theorem 9.1, $h_0^*E = E$ is equivalent to $h_1^*E = M \times F$, hence E is a trivial bundle. For example, the tangent bundle $T\mathbb{R}^m$ is trivial. We have obtained the following corollary.

Corollary 9.1. Let $E \xrightarrow{\pi} M$ be a fibre bundle. E is trivial if M is contractible to a point.

9.3 Vector bundles

9.3.1 Definitions and examples

A **vector bundle** $E \xrightarrow{\pi} M$ is a fibre bundle whose fibre is a vector space. Let F be $\mathbb{R}^k$ and M be an m-dimensional manifold. It is common to call k the **fibre dimension** and denote it by $\dim E$, although the total space E is $m + k$ dimensional. The transition functions belong to $GL(k, \mathbb{R})$, since it maps a vector space onto another vector space of the same dimension isomorphically. If F is a complex vector space $\mathbb{C}^k$, the structure group is $GL(k, \mathbb{C})$.

Example 9.3. A tangent bundle TM over an m-dimensional manifold M is a vector bundle whose typical fibre is $\mathbb{R}^m$, see section 9.1. Let u be a point in TM such that $\pi(u) = p \in U_i \cap U_j$, where $\{U_i\}$ covers M. Let $x^\mu = \varphi_i(p)$

$(y^\mu = \varphi_j(p))$ be the coordinate system of U_i (U_j). The vector V corresponding to u is expressed as $V = V^\mu \partial/\partial x^\mu|_p = \tilde{V}^\mu \partial/\partial y^\mu|_p$. The local trivializations are

$$\phi_i^{-1}(u) = (p, \{V^\mu\}) \qquad \phi_j^{-1}(u) = (p, \{\tilde{V}^\mu\}). \tag{9.20}$$

The fibre coordinates $\{V^\mu\}$ and $\{\tilde{V}^\mu\}$ are related as

$$V^\mu = G^\mu{}_\nu(p)\tilde{V}^\nu \tag{9.21}$$

where $\{G^\mu{}_\nu(p)\} = \{(\partial x^\mu/\partial y^\nu)_p\} \in \mathrm{GL}(m, \mathbb{R})$. Hence, a tangent bundle is $(TM, \pi, M, \mathbb{R}^m, \mathrm{GL}(m, \mathbb{R}))$. Sections of TM are the vector fields on M; $\mathfrak{X}(M) = \Gamma(M, TM)$.

For concreteness let us work out TS^2. Let the pair $U_N \equiv S^2 - \{$South Pole$\}$ and $U_S \equiv S^2 - \{$North Pole$\}$ be an open covering of S^2. Let (X, Y) and (U, V) be the respective stereographic coordinates (example 8.1). They are related as

$$U = X/(X^2 + Y^2) \qquad V = -Y/(X^2 + Y^2). \tag{9.22}$$

Take $u \in TS^2$ such that $\pi(u) = p \in U_N \cap U_S$. Let ϕ_N and ϕ_S be the respective local trivializations such that $\phi_N^{-1}(u) = (p, V_N^\mu)$ and $\phi_S^{-1}(u) = (p, V_S^\mu)$. The transition function is

$$t_{SN}(p) = \frac{\partial(U, V)}{\partial(X, Y)} = \frac{1}{r^2}\begin{pmatrix} -\cos 2\theta & -\sin 2\theta \\ \sin 2\theta & -\cos 2\theta \end{pmatrix} \tag{9.23}$$

where we have put $X = r\cos\theta$ and $Y = r\sin\theta$. The transition of the components of the tangent vectors consists of a rotation of $\{V_i^\mu\}$ by an angle 2θ followed by a rescaling. The reader should verify that $t_{NS}(p) = t_{SN}(p)^{-1}$.

Example 9.4. Let M be an m-dimensional manifold embedded in $\mathbb{R}^{m+k}$. Let N_pM be the vector space which is normal to T_pM in $\mathbb{R}^{m+k}$, that is, $U \cdot V = 0$ with respect to the Euclidean metric in $\mathbb{R}^{m+k}$ for any $U \in N_pM$ and $V \in T_pM$. The vector space N_pM is isomorphic to $\mathbb{R}^k$. The **normal bundle**

$$NM \equiv \bigcup_{p \in M} N_pM$$

is a vector bundle with the typical fibre $\mathbb{R}^k$.

Consider the sphere S^2 embedded in $\mathbb{R}^3$. The normal bundle NS^2 is imagined as S^2 whose surface is pierced perpendicularly by straight lines. NS^2 is a trivial bundle $S^2 \times \mathbb{R}$.

A vector bundle whose fibre is one-dimensional $(F = \mathbb{R}$ or $\mathbb{C})$ is called a **line bundle**. A cylinder $S^1 \times \mathbb{R}$ is a trivial $\mathbb{R}$-line bundle. A Möbius strip is also a real line bundle. The structure group $\mathrm{GL}(1, \mathbb{R}) = \mathbb{R} - \{0\}$ or $\mathrm{GL}(1, \mathbb{C}) = \mathbb{C} - \{0\}$ is Abelian.

In the following, we often consider the **canonical line bundle** L. Recall that an element p of $\mathbb{C}P^n$ is a complex line in $\mathbb{C}^{n+1}$ through the origin (example 8.3). The fibre $\pi^{-1}(p)$ of L is defined to be the line in $\mathbb{C}^{n+1}$ which belongs to p. More formally, let $I^{n+1} \equiv \mathbb{C}P^n \times \mathbb{C}^{n+1}$ be a trivial bundle over $\mathbb{C}P^n$. If we write an element of I^{n+1} as (p, v), $p \in \mathbb{C}P^n$, $v \in \mathbb{C}^{n+1}$, L is defined by

$$L \equiv \{(p, v) \in I^{n+1} | v = ap, a \in \mathbb{C}\}.$$

The projection is $(p, v) \overset{\pi}{\to} p$.

Example 9.5. The (trivial) complex line bundle $L = \mathbb{R}^3 \times \mathbb{C}$ is associated with the non-relativistic quantum mechanics defined on $\mathbb{R}^3$. The wavefunction $\psi(x)$ is simply a section of L.

Let us consider a wavefunction $\psi(x)$ in the field of a magnetic monopole studied in section 1.9. When a monopole is at the origin, $\psi(x)$ is defined on $\mathbb{R}^3 - \{0\}$ and we have a complex line bundle over $\mathbb{R}^3 - \{0\}$. If we are interested only in the wavefunction on S^2 surrounding the monopole, we have a complex line bundle over S^2. Note that S^2 is a deformation retract of $\mathbb{R}^3 - \{0\}$.

9.3.2 Frames

On a tangent bundle TM, each fibre has a natural basis $\{\partial/\partial x^\mu\}$ given by the coordinate system x^μ on a chart U_i. We may also employ the orthonormal basis $\{\hat{e}_\alpha\}$ if M is endowed with a metric. $\partial/\partial x^\mu$ or $\{\hat{e}_\alpha\}$ is a vector field on U_i and the set $\{\partial/\partial x^\mu\}$ or $\{\hat{e}_\alpha\}$ forms linearly independent vector fields over U_i. It is always possible to choose m linearly independent tangent vectors over U_i but it is not necessarily the case throughout M. By definition, the components of the basis vectors are

$$\partial/\partial x^\mu = (0, \quad \ldots, \quad 0, \quad \underset{\mu}{1}, \quad 0, \quad \ldots, \quad 0)$$

or

$$\hat{e}_\alpha = (0, \quad \ldots, \quad 0, \quad \underset{\alpha}{1}, \quad 0, \quad \ldots, \quad 0).$$

These vectors define a (local) **frame** over U_i, see later.

Let $E \overset{\pi}{\to} M$ be a vector bundle whose fibre is $\mathbb{R}^k$ (or $\mathbb{C}^k$). On a chart U_i, the piece $\pi^{-1}(U_i)$ is trivial, $\pi^{-1}(U_i) \cong U_i \times \mathbb{R}^k$, and we may choose k linearly independent sections $\{e_1(p), \ldots, e_k(p)\}$ over U_i. These sections are said to define a **frame** over U_i. Given a frame over U_i, we have a natural map $F_p \to F$ ($= \mathbb{R}^k$ or $\mathbb{C}^k$) given by

$$V = V^\alpha e_\alpha(p) \longmapsto \{V^\alpha\} \in F. \tag{9.24}$$

The local trivialization is

$$\phi_i^{-1}(V) = (p, \{V^\alpha(p)\}). \tag{9.25}$$

By definition, we have

$$\phi_i \; (p, \{0, \quad \ldots, \quad 0, \quad \underset{\alpha}{1}, \quad 0, \quad \ldots, \quad 0\}) \; = e_\alpha(p). \tag{9.26}$$

Let $U_i \cap U_j \neq \emptyset$ and consider the change of frames. We have a frame $\{e_1(p), \ldots, e_k(p)\}$ on U_i and $\{\tilde{e}_1(p), \ldots, \tilde{e}_k(p)\}$ on U_j, where $p \in U_i \cap U_j$. A vector $\tilde{e}_\beta(p)$ is expressed as

$$\tilde{e}_\beta(p) = e_\alpha(p) G(p)^\alpha{}_\beta \tag{9.27}$$

where $G(p)^\alpha{}_\beta \in \mathrm{GL}(k, \mathbb{R})$ or $\mathrm{GL}(k, \mathbb{C})$. Any vector $V \in \pi^{-1}(p)$ is expressed as

$$V = V^\alpha e_\alpha(p) = \tilde{V}^\alpha \tilde{e}_\alpha(p). \tag{9.28}$$

From (9.27) and (9.28) we find that

$$\tilde{V}^\beta = G^{-1}(p)^\beta{}_\alpha V^\alpha \tag{9.29}$$

where $G^{-1}(p)^\beta{}_\alpha G(p)^\alpha{}_\gamma = G(p)^\beta{}_\alpha G^{-1}(p)^\alpha{}_\gamma = \delta^\beta{}_\gamma$. Thus, we find that the transition function $t_{ji}(p)$ is given by a matrix $G^{-1}(p)$.

9.3.3 Cotangent bundles and dual bundles

The **cotangent bundle** $T^*M \equiv \bigcup_{p \in M} T_p^* M$ is defined similarly to the tangent bundle. On a chart U_i whose coordinates are x^μ, the basis of $T_p^* M$ is taken to be $\{dx^1, \ldots, dx^m\}$, which is dual to $\{\partial/\partial x^\mu\}$. Let y^μ be the coordinates of U_j such that $U_i \cap U_j \neq \emptyset$. For $p \in U_i \cap U_j$, we have the transformation,

$$dy^\mu = dx^\nu \left(\frac{\partial y^\mu}{\partial x^\nu} \right)_p . \tag{9.30}$$

A one-form ω is expressed, in both coordinate systems, as

$$\omega = \omega_\mu \, dx^\mu = \tilde{\omega}_\mu \, dy^\mu$$

from which we find that

$$\tilde{\omega}_\mu = G_\mu{}^\nu(p) \omega_\nu \tag{9.31}$$

where $G_\mu{}^\nu(p) \equiv (\partial x^\nu / \partial y^\mu)_p$ corresponds to the transition function $t_{ji}(p)$. Note that $\Gamma(M, T^*M) = \Omega^1(M)$.

This cotangent bundle is easily extended to more general cases. Given a vector bundle $E \xrightarrow{\pi} M$ with the fibre F, we may define its **dual bundle** $E^* \xrightarrow{\pi} M$. The fibre F^* of E^* is the set of linear maps of F to $\mathbb{R}$ (or $\mathbb{C}$). Given a general basis $\{e_\alpha(p)\}$ of F_p, we define the dual basis $\{\theta^\alpha(p)\}$ of F_p^* by $\langle \theta^\alpha(p), e_\beta(p) \rangle = \delta^\alpha{}_\beta$.

9.3.4   Sections of vector bundles

Let s and s' be sections of a vector bundle $E \xrightarrow{\pi} M$. The vector addition and the scalar multiplication are pointwisely defined as

$$(s + s')(p) = s(p) + s'(p) \tag{9.32a}$$

$$(fs)(p) = f(p)s(p) \tag{9.32b}$$

where $p \in M$ and $f \in \mathcal{F}(M)$. The null vector 0 of each fibre is left invariant under $GL(k, \mathbb{R})$ (or $GL(k, \mathbb{C})$) and plays a distinguished role. Any vector bundle E admits a global section called the **null section** $s_0 \in \Gamma(M, E)$ such that $\phi_i^{-1}(s_0(p)) = (p, 0)$ in any local trivialization.

For example, let us consider sections of the canonical line bundle L over $\mathbb{C}P^n$. Let $\xi^\nu{}_{(\mu)}$ be the inhomogeneous coordinates and $\{z^\nu\}$ be the homogeneous coordinates on U_μ. The local section s_μ over U_μ is of the form

$$s_\mu = \{\xi^0{}_{(\mu)}, \dots, 1, \dots, \xi^n{}_{(\mu)}\} \in \mathbb{C}^{n+1}.$$

The transition from one coordinate system to the other is carried out by a scalar multiplication: $s_\nu = (z^\mu/z^\nu)s_\mu$. Let L^* be the dual bundle of L. Corresponding to s_μ, we may choose a dual section s_μ^* such that $s_\mu^*(s_\mu) = 1$. From this, we find that the transition function of s_μ^* is a multiplication by z^ν/z^μ, $s_\nu^* = (z^\nu/z^\mu)s_\mu^*$.

A fibre metric $h_{\mu\nu}(p)$ is also defined pointwisely. Let s and s' be sections over U_i. The inner product between s and s' at p is defined by

$$(s, s')_p = h_{\mu\nu}(p)s^\mu(p)s'^\nu(p) \tag{9.33a}$$

if the fibre is $\mathbb{R}^k$. If the fibre is $\mathbb{C}^k$ we define

$$(s, s')_p = h_{\mu\nu}(p)\overline{s^\mu(p)}s'^\nu(p). \tag{9.33b}$$

We have more about this subject in section 10.4.

9.3.5   The product bundle and Whitney sum bundle

Let $E \xrightarrow{\pi} M$ and $E' \xrightarrow{\pi'} M'$ be vector bundles with fibres F and F' respectively. The **product bundle**

$$E \times E' \xrightarrow{\pi \times \pi'} M \times M' \tag{9.34}$$

is a fibre bundle whose typical fibre is $F \oplus F'$. [A vector in $F \oplus F'$ is written as

$$\begin{pmatrix} V \\ W \end{pmatrix} \qquad \text{where } V \in F \text{ and } W \in F'.$$

Vector addition and scalar multiplication are defined by

$$\begin{pmatrix} V \\ W \end{pmatrix} + \begin{pmatrix} V' \\ W' \end{pmatrix} = \begin{pmatrix} V + V' \\ W + W' \end{pmatrix}$$

and

$$\lambda \begin{pmatrix} V \\ W \end{pmatrix} = \begin{pmatrix} \lambda V \\ \lambda W \end{pmatrix}.$$

Let $\{e_\alpha\}$ and $\{f_\beta\}$ be bases of F and F' respectively. Then $\{e_\alpha\} \cup \{f_\beta\}$ is a basis of $F \oplus F'$ and we find that $\dim(F \oplus F') = \dim F + \dim F'$.] If $\pi(u) = p$ and $\pi'(u') = p'$ the projection $\pi \times \pi'$ acts on $(u, u') \in E \times E'$ as

$$\pi \times \pi'(u, u') = (p, p'). \tag{9.35}$$

The fibre at (p, p') is $F_p \oplus F'_{p'}$. For example, if $M = M_1 \times M_2$, we have $TM = TM_1 \times TM_2$.

Let $E \overset{\pi}{\to} M$ and $E' \overset{\pi'}{\to} M$ be vector bundles with fibres F and F' respectively. The **Whitney sum bundle** $E \oplus E'$ is a pullback bundle of $E \times E'$ by $f : M \to M \times M$ defined by $f(p) = (p, p)$,

$$\begin{array}{ccc} E \oplus E' & \overset{\pi_2}{\longrightarrow} & E \times E' \\ \pi_1 \downarrow & & \downarrow \pi \times \pi' \\ M & \overset{f}{\longrightarrow} & M \times M. \end{array} \tag{9.36}$$

Thus, $E \oplus E' = \{(u, u') \in E \times E' | \pi \times \pi'(u, u') = (p, p)\}$. The fibre of a Whitney sum bundle is $F \oplus F'$. $(\pi \times \pi')^{-1}(p)$ is isomorphic to $\pi^{-1}(p) \oplus \pi'^{-1}(p) = F_p \oplus F'_p$. In short, $E \oplus E'$ is a bundle over M whose fibre at p is $F_p \oplus F'_p$. Let $\{U_i\}$ be an open covering of M and $\{t_{ij}^E\}$ and $\{t_{ij}^{E'}\}$ be the transition functions of E and E' respectively. Then the transition function T_{ij} of $E \oplus E'$ is a $(\dim F + \dim F') \times (\dim F + \dim F')$ matrix

$$T_{ij}(p) = \begin{pmatrix} t_{ij}^E(p) & 0 \\ 0 & t_{ij}^{E'}(p) \end{pmatrix} \tag{9.37}$$

which acts on $F \oplus F'$ on the left.

Example 9.6. Let $E = TS^2$ and $E' = NS^2$ defined in $\mathbb{R}^3$. Take $u \in TS^2$ and $v \in NS^2$ whose local trivializations are $\phi_i^{-1}(u) = (p, V)$ and $\psi_j^{-1}(v) = (q, W)$, respectively, where $p, q \in S^2$, $V \in \mathbb{R}^2$ and $W \in \mathbb{R}$. If (u, v) is a point of the product bundle $E \times E'$, we have a trivialization $\Phi_{i,j} = \phi_i \times \psi_j$ such that

$$\Phi_{i,j}^{-1}(u, v) = (p, q; V, W). \tag{9.38a}$$

If, however, $(u, v) \in E \oplus E'$, u and v satisfy the stronger condition $\pi(u) = \pi'(v)$ $(=p$, say$)$. Thus, we have

$$\Phi_i^{-1}(u, v) = (p; V, W). \tag{9.38b}$$

The Whitney sum $TS^2 \oplus NS^2$, S^2 being embedded in $\mathbb{R}^3$, is a trivial bundle over S^2, whose fibre is isomorphic to $\mathbb{R}^3$.

9.3.6 Tensor product bundles

Let $E \xrightarrow{\pi} M$ and $E' \xrightarrow{\pi'} M$ be vector bundles over M. The **tensor product bundle** $E \otimes E'$ is obtained by assigning the tensor product of fibres $F_p \otimes F'_p$ to each point $p \in M$. If $\{e_\alpha\}$ and $\{f_\beta\}$ are bases of F and F', $F \otimes F'$ is spanned by $\{e_\alpha \otimes f_\beta\}$ and, hence, $\dim(E \otimes E') = \dim E \times \dim E'$.

Let $\bigotimes^r E \equiv E \otimes \cdots \otimes E$ be the tensor product bundle of r E. If $\{e_\alpha\}$ is the basis of the fibre F of E, the fibre of $\bigotimes^r E$ is spanned by $\{e_{\alpha_1} \otimes \cdots \otimes e_{\alpha_r}\}$. If we define $\wedge$ by

$$e_\alpha \wedge e_\beta \equiv e_\alpha \otimes e_\beta - e_\beta \otimes e_\alpha \tag{9.39}$$

we have a bundle $\wedge^r(E)$ of totally anti-symmetric tensors spanned by $\{e_{\alpha_1} \wedge \ldots \wedge e_{\alpha_r}\}$. In particular, $\Omega^r(M)$, the space of r-forms on M, is identified with $\Gamma(M, \wedge^r(T^*M))$.

Exercise 9.1. Let E_1, E_2 and E_3 be vector bundles over M. Show that $\otimes$ is distributive:

$$E_1 \otimes (E_2 \oplus E_3) = (E_1 \otimes E_2) \oplus (E_1 \otimes E_3). \tag{9.40}$$

Express the transition functions of $E_1 \otimes (E_2 \oplus F_3)$ in terms of those of E_1, E_2 and E_3.

9.4 Principal bundles

9.4.1 Definitions

A principal bundle has a fibre F which is identical to the structure group G. A principal bundle $P \xrightarrow{\pi} M$ is also denoted by $P(M, G)$ and is often called a G **bundle** over M.

The transition function acts on the fibre on the left as before. In addition, we may also define the action of G on F *on the right*. Let $\phi_i : U_i \times G \to \pi^{-1}(U_i)$ be the local trivialization given by $\phi_i^{-1}(u) = (p, g_i)$, where $u \in \pi^{-1}(U_i)$ and $p = \pi(u)$. The right action of G on $\pi^{-1}(U_i)$ is defined by $\phi_i^{-1}(ua) = (p, g_i a)$, that is (figure 9.8),

$$ua = \phi_i(p, g_i a) \tag{9.41}$$

for any $a \in G$ and $u \in \pi^{-1}(p)$. Since the right action commutes with the left action, this definition is independent of the local trivializations. In fact, if $p \in U_i \cap U_j$,

$$ua = \phi_j(p, g_j a) = \phi_j(p, t_{ji}(p)g_i a) = \phi_i(p, g_i a).$$

Thus, the right multiplication is defined without reference to the local trivializations. This is denoted by $P \times G \to P$ or $(u, a) \mapsto ua$. Note that $\pi(ua) = \pi(u)$. The right action of G on $\pi^{-1}(p)$ is *transitive* since G acts on G transitively on the right and $F_p = \pi^{-1}(p)$ is diffeomorphic to G. Thus, for any

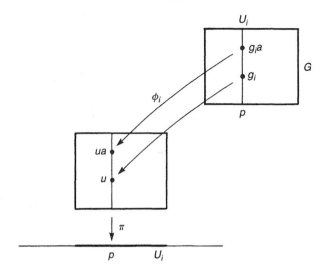

Figure 9.8. The right action of G on P.

$u_1, u_2 \in \pi^{-1}(p)$ there exists an element a of G such that $u_1 = u_2 a$. Then, if $\pi(u) = p$, we can construct the whole fibre as $\pi^{-1}(p) = \{ua | a \in G\}$. The action is also *free*; if $ua = u$ for some $u \in P$, a must be the unit element e of G. In fact, if $u = \phi_i(p, g_i)$, we have $\phi_i(p, g_i a) = \phi_i(p, g_i)a = ua = u = \phi_i(p, g_i)$. Since ϕ_i is bijective, we must have $g_i a = g_i$, that is, $a = e$.

Given a section $s_1(p)$ over U_i, we define a preferred local trivialization $\phi_i : U_i \times G \to \pi^{-1}(U_i)$ as follows. For $u \in \pi^{-1}(p)$, $p \in U_i$, there is a *unique* element $g_u \in G$ such that $u = s_i(p)g_u$. Then we define ϕ_i by $\phi_i^{-1}(u) = (p, g_u)$. In this local trivialization, the section $s_i(p)$ is expressed as

$$s_i(p) = \phi_i(p, e). \qquad (9.42)$$

This local trivialization is called the **canonical local trivialization**. By definition $\phi_i(p, g) = \phi_i(p, e)g = s_i(p)g$. If $p \in U_i \cap U_j$, two sections $s_i(p)$ and $s_j(p)$ are related by the transition function $t_{ij}(p)$ as follows

$$s_i(p) = \phi_i(p, e) = \phi_j(p, t_{ji}(p)e) = \phi_j(p, t_{ji}(p))$$
$$= \phi_j(p, e)t_{ji}(p) = s_j(p)t_{ji}(p). \qquad (9.43)$$

Example 9.7. Let P be a principal bundle with fibre $U(1) = S^1$ and the base space S^2. This principal bundle represents the topological setting of the **magnetic monopole** (section 1.9). Let $\{U_N, U_S\}$ be an open covering of S^2, U_N (U_S) being the northern (southern) hemisphere. If we parametrize S^2 by the usual polar angles, we have

$$U_N = \{(\theta, \phi) | 0 \le \theta \le \pi/2 + \varepsilon, 0 \le \phi < 2\pi\}$$
$$U_S = \{(\theta, \phi) | \pi/2 - \varepsilon \le \theta \le \pi, 0 \le \phi < 2\pi\}.$$

The intersection $U_N \cap U_S$ is a strip which is essentially the equator. Let ϕ_N and ϕ_S be the local trivializations such that

$$\phi_N^{-1}(u) = (p, e^{i\alpha_N}) \qquad \phi_S^{-1}(u) = (p, e^{i\alpha_S}) \qquad (9.44)$$

where $p = \pi(u)$. Take a transition function $t_{NS}(p)$ of the form $e^{in\phi}$ where n must be an integer so that $t_{NS}(p)$ may be uniquely defined on the equator. Since t_{NS} maps the equator S^1 to $U(1)$, this integer characterizes the homotopy group $\pi_1(U(1)) = \mathbb{Z}$. The fibre coordinates α_N and α_S are related on the equator as

$$e^{i\alpha_N} = e^{in\phi}e^{i\alpha_S}. \qquad (9.45)$$

If $n = 0$, the transition function is the unit element of $U(1)$ and we have a trivial bundle $P_0 = S^2 \times S^1$. If $n \neq 0$, the $U(1)$-bundle P_n is twisted. It is remarkable that the topological structure of a fibre bundle is characterized by an integer. The integer characterizes how two local sections are pasted together at the equator. Accordingly, the integer corresponds to the element of the homotopy group $\pi_1(U(1)) = \mathbb{Z}$.

Since $U(1)$ is Abelian, the right action and the left action are equivalent. Under the right action $g = e^{i\Lambda}$, we have

$$\phi_N^{-1}(ug) = (p, e^{i(\alpha_N + \Lambda)}) \qquad (9.46a)$$

$$\phi_S^{-1}(ug) = (p, e^{i(\alpha_S + \Lambda)}). \qquad (9.46b)$$

The right action corresponds to the $U(1)$-gauge transformation.

Example 9.8. If we identify all the infinite points of the Euclidean space $\mathbb{R}^m$, the one-point compactification $S^m = \mathbb{R}^m \cup \{\infty\}$ is obtained. If a trivial G bundle is defined over $\mathbb{R}^m$ we shall have a new G bundle over S^m after compactification, which is not necessarily trivial. Let P be an $SU(2)$ bundle over S^4 obtained from $\mathbb{R}^4$ by one-point compactification. This principal bundle represents an $SU(2)$ instanton (section 1.10). Introduce an open covering $\{U_N, U_S\}$ of S^4,

$$U_N = \{(x, y, z, t) | x^2 + y^2 + z^2 + t^2 \le R^2 + \varepsilon\}$$
$$U_S = \{(x, y, z, t) | R^2 - \varepsilon \le x^2 + y^2 + z^2 + t^2\}$$

where R is a positive constant and ε is an infinitesimal positive number. The thin intersection $U_N \cap U_S$ is essentially S^3. Let $t_{NS}(p)$ be the transition function defined at $p \in U_N \cap U_S$. Since t_{NS} maps S^3 to $SU(2)$, it is classified by $\pi_3(SU(2)) = \mathbb{Z}$. The integer characterizing the bundle is called the **instanton number**. If $t_{NS}(p)$ is taken to be the unit element $e \in SU(2)$, we have a trivial bundle $P_0 = S^3 \times SU(2)$, which corresponds to the homotopy class 0. Non-trivial bundles are obtained as follows. We first note that $SU(2) \cong S^3$ (example 4.12). An element $A \in SU(2)$ is written as

$$A = \begin{pmatrix} u & v \\ -\bar{v} & \bar{u} \end{pmatrix}$$

where $|u|^2 + |v|^2 = 1$. Separating u and v as $u = t + \mathrm{i}z$ and $v = y + \mathrm{i}x$, we find $t^2 + x^2 + y^2 + z^2 = 1$. Thus SU(2) is regarded as the unit sphere S^3 and $\pi_3(\mathrm{SU}(2)) \cong \pi_3(S^3) \cong \mathbb{Z}$ classifies maps from S^3 to $\mathrm{SU}(2) \cong S^3$. The *identity map* $f : S^3 \to S^3 \cong \mathrm{SU}(2)$ is

$$f(x, y, z, t) \mapsto \begin{pmatrix} t + \mathrm{i}z & y + \mathrm{i}x \\ -y + \mathrm{i}x & t - \mathrm{i}z \end{pmatrix}$$
$$= tI_2 + \mathrm{i}(x\sigma_x + y\sigma_y + z\sigma_z) \tag{9.47}$$

where I_2 is the 2×2 unit matrix and the σ_μ are the Pauli matrices. Let us take a point $p = (x, y, z, t) \in U_{\mathrm{N}} \cap U_{\mathrm{S}}$. If $R = (x^2 + y^2 + z^2 + t^2)^{1/2}$ denotes the radial distance of p, the vector $(x/R, y/R, z/R, t/R)$ has unit length. We assign an element of SU(2) to the point p as

$$t_{\mathrm{NS}}(p) = \frac{1}{R}\left(tI_2 + \mathrm{i}\sum_i x^i \sigma_i\right). \tag{9.48}$$

Let ϕ_{N} and ϕ_{S} be the local trivializations,

$$\phi_{\mathrm{N}}^{-1}(u) = (p, g_{\mathrm{N}}) \qquad \phi_{\mathrm{S}}^{-1}(u) = (p, g_{\mathrm{S}}) \tag{9.49}$$

where $p = \pi(u)$ and $g_{\mathrm{N}}, g_{\mathrm{S}} \in \mathrm{SU}(2)$. On $U_{\mathrm{N}} \cap U_{\mathrm{S}}$, we have

$$g_{\mathrm{N}} = \frac{1}{R}\left(tI_2 + \mathrm{i}\sum_i x^i \sigma_i\right)g_{\mathrm{S}}. \tag{9.50}$$

While (t, x) scans S^3 once, $t_{\mathrm{NS}}(p)$ sweeps SU(2) once, hence this bundle corresponds to the homotopy class 1 of $\pi_3(\mathrm{SU}(2))$. It is not difficult to see that the transition function corresponding to the homotopy class n is given by

$$t_{\mathrm{NS}}(p) = \frac{1}{R^n}\left(t\mathbf{1} + \mathrm{i}\sum_i x^i \sigma_i\right)^n. \tag{9.51}$$

To continue our study of monopoles and instantons, we have to introduce connections (the *gauge potentials*) on the fibre bundle. We will come back to these topics in the next chapter.

Example 9.9. Hopf has shown that S^3 is a U(1) bundle over S^2. The unit three-sphere embedded in $\mathbb{R}^4$ is expressed as

$$(x^1)^2 + (x^2)^2 + (x^3)^2 + (x^4)^2 = 1.$$

If we introduce $z^0 = x^1 + \mathrm{i}x^2$ and $z^1 = x^3 + \mathrm{i}x^4$, this becomes

$$|z^0|^2 + |z^1|^2 = 1. \tag{9.52}$$

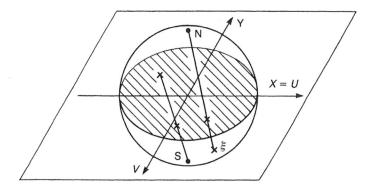

Figure 9.9. Stereographic coordinates of the sphere S^2. (X, Y) is defined with respect to the projection from the North Pole while (U, V) with respect to the projection from the South Pole.

Let us parametrize S^2 as

$$(\xi^1)^2 + (\xi^2)^2 + (\xi^3)^2 = 1.$$

The **Hopf map** $\pi : S^3 \to S^2$ is defined by

$$\xi^1 = 2(x^1 x^3 + x^2 x^4) \tag{9.53a}$$

$$\xi^2 = 2(x^2 x^3 - x^1 x^4) \tag{9.53b}$$

$$\xi^3 = (x^1)^2 + (x^2)^2 - (x^3)^2 - (x^4)^2. \tag{9.53c}$$

It is easily verified that π maps S^3 to S^2 since

$$(\xi^1)^2 + (\xi^2)^2 + (\xi^3)^2 = [(x^1)^2 + (x^2)^2 + (x^3)^2 + (x^4)^2]^2 = 1.$$

Let (X, Y) be the stereographic projection coordinates of a point in the southern hemisphere U_S of S^2 from the North Pole. If we take a complex plane which contains the equator of S^2, $Z = X + iY$ is within the circle of unit radius. We found in example 8.1 that (figure 9.9)

$$Z = \frac{\xi^1 + i\xi^2}{1 - \xi^3} = \frac{x^1 + ix^2}{x^3 + ix^4} = \frac{z^0}{z^1} \qquad (\xi \in U_S). \tag{9.54a}$$

Observe that Z is invariant under

$$(z^0, z^1) \mapsto (\lambda z^0, \lambda z^1)$$

where $\lambda \in U(1)$. Since $|\lambda| = 1$, the point $(\lambda z^0, \lambda z^1)$ is also in S^3. The stereographic coordinates (U, V) of the northern hemisphere U_N projected from the South Pole are given by

$$W = U + iV = \frac{\xi^1 - i\xi^2}{1 + \xi^3} = \frac{x^3 + ix^4}{x^1 + ix^2} = \frac{z^1}{z^0} \qquad (\xi \in U_N). \tag{9.54b}$$

Note that $Z = 1/W$ on the equator $U_N \cap U_S$.

The fibre bundle structure is given as follows. We first define the local trivializations, $\phi_S^{-1} : \pi^{-1}(U_S) \to U_S \times U(1)$ by

$$(z^0, z^1) \mapsto (z^0/z^1, z^1/|z^1|) \tag{9.55a}$$

and $\phi_N^{-1} : \pi^{-1}(U_N) \to U_N \times U(1)$ by

$$(z^0, z^1) \mapsto (z^1/z^0, z^0/|z^0|). \tag{9.55b}$$

Observe that these local trivializations are well defined on each chart. For example, $z^0 \neq 0$ on U_N, hence both $z^1/z^0 = U + iV$ and $z^0/|z^0|$ are non-singular. On the equator, $\xi^3 = 0$, we have $|z^0| = |z^1| = 1/\sqrt{2}$. Accordingly, the local trivializations on the equator are

$$\phi_S^{-1} : (z^0, z^1) \mapsto (z^0/z^1, \sqrt{2}z^1) \tag{9.56a}$$

and

$$\phi_N^{-1} : (z^0, z^1) \mapsto (z^1/z^0, \sqrt{2}z^0). \tag{9.56b}$$

The transition function on the equator is

$$t_{NS}(\xi) = \frac{\sqrt{2}z^0}{\sqrt{2}z^1} = \xi^1 + i\xi^2 \in U(1). \tag{9.57}$$

If we circumnavigate the equator, $t_{NS}(\xi)$ traverses the unit circle in the complex plane once, hence the U(1) bundle $S^3 \xrightarrow{\pi} S^2$ is characterized by the homotopy class 1 of $\pi_1(U(1)) = \mathbb{Z}$. Trautman (1977), Minami (1979) and Ryder (1980) have pointed out that a magnetic monopole of unit strength is described by the Hopf map $S^3 \xrightarrow{\pi} S^2$.

The Hopf map can be understood from a slightly different point of view. We regard S^3 as a complex one-sphere

$$S_{\mathbb{C}}^1 = \{(z^0, z^1) \in \mathbb{C}^2 \,||z^0|^2 + |z^1|^2 = 1\}.$$

Define a map $\pi : S_{\mathbb{C}}^1 \to \mathbb{C}P^1$ by

$$(z^0, z^1) \mapsto [(z^0, z^1)] = \{\lambda(z^0, z^1)|\lambda \in \mathbb{C} - \{0\}\}. \tag{9.58}$$

Under this map, points of S^3 of the form $\lambda(z^0, z^1)$, $|\lambda| = 1$ are mapped to a single point of $\mathbb{C}P^1 = S^2$. This is the Hopf map $\pi : S^3 \to S^2$ obtained earlier. This is easily generalized to the case of the quaternion $\mathbb{H}$. The quaternion algebra is defined by the product table,

$$i^2 = j^2 = k^2 = -1 \qquad ij = -ji = k$$
$$jk = -kj = i \qquad ki = -ik = j.$$

An arbitrary element of $\mathbb{H}$ is written as

$$q = t + ix + jy + kz.$$

Clearly the unit quaternion $|q| = (t^2 + x^2 + y^2 + z^2)^{1/2} = 1$ represents $S^3 \cong$ SU(2). The quaternion one-sphere is given by

$$S_{\mathbb{H}}^1 = \{(q^0, q^1) \in \mathbb{H}^2 \mid |q^0|^2 + |q^1|^2 = 1\} \tag{9.59}$$

which represents S^7. The Hopf map, in this case, takes the form

$$\pi : S_{\mathbb{H}}^1 \to \mathbb{H}P^1 \tag{9.60}$$

where $\mathbb{H}P^1$ is the quaternion projective space whose element is

$$[(q^0, q^1)] = \{\eta(q^0, q^1) \in \mathbb{H}^2 \mid \eta \in \mathbb{H} - \{0\}\}. \tag{9.61}$$

Points of S^7 with $|\eta| = 1$ are mapped under this map to a single point of $\mathbb{H}P^1 = S^4$ and we have the Hopf map

$$\pi : S^7 \to S^4. \tag{9.62}$$

The fibre is the unit quaternion $S^3 = $ SU(2). The transition function defined by the Hopf map belongs to the class 1 of $\pi_3(\mathrm{SU}(2)) \cong \mathbb{Z}$. An instanton of unit strength is described in terms of this Hopf map.

Octonions define a Hopf map $\pi : S^{15} \to S^8$. This differs from other Hopf maps in that the fibre S^7 is not really a group. So far we have not found an application of this map in physics.[1]

Example 9.10. Let H be a closed Lie subgroup of a Lie group G. We show that G is a principal bundle with fibre H and base space $M = G/H$. Define the right action of H on G by $g \mapsto ga$, $g \in G$, $a \in H$. The right action is differentiable since G is a Lie group. Define the projection $\pi : G \to M = G/H$ by the map $\pi : g \mapsto [g] = \{gh \mid h \in H\}$. Clearly, $g, ga \in G$ are mapped to the same point $[g]$ hence $\pi(g) = \pi(ga) (=[g])$. To define local trivializations, we need to define a map $f_i : G \to H$ on each chart U_i. Let s be a local section over U_i and $g \in \pi^{-1}([g])$. Define f_i by $f_i(g) = s([g])^{-1}g$. Since $s([g])$ is a section at $[g]$, it is expressed as ga for some $a \in H$ and accordingly, $s([g])^{-1}g = a^{-1}g^{-1}g = a^{-1} \in H$. Then we define the local trivialization $\phi_i : U_i \times H \to G$ by

$$\phi_i^{-1}(g) = ([g], f_i(g)). \tag{9.63}$$

It is easy to see that $f_i(ga) = f_i(g)a$ $(a \in H)$ hence $\phi_i^{-1}(ga) = (p, f_i(g)a)$ is satisfied. Useful examples are (see example 5.18)

$$\mathrm{O}(n)/\mathrm{O}(n-1) = \mathrm{SO}(n)/\mathrm{SO}(n-1) = S^{n-1} \tag{9.64}$$

$$\mathrm{U}(n)/\mathrm{U}(n-1) = \mathrm{SU}(n)/\mathrm{SU}(n-1) = S^{2n-1}. \tag{9.65}$$

---

[1] Octonions are also known as Cayley numbers. The set of octonions is a vector space over $\mathbb{R}$ but not a field. The product is neither commutative nor associative. See John C Baez, *The Octonions* math.RA/0105155 for a recent review.

9.4.2 Associated bundles

Given a principal fibre bundle $P(M, G)$, we may construct an **associated fibre bundle** as follows. Let G act on a manifold F on the left. Define an action of $g \in G$ on $P \times F$ by

$$(u, f) \to (ug, g^{-1}f) \tag{9.66}$$

where $u \in P$ and $f \in F$. Then the associated fibre bundle (E, π, M, G, F, P) is an equivalence class $P \times F/G$ in which two points (u, f) and $(ug, g^{-1}f)$ are identified.

Let us consider the case in which F is a k-dimensional vector space V. Let ρ be the k-dimensional representation of G. The **associated vector bundle** $P \times_\rho V$ is defined by identifying the points (u, v) and $(ug, \rho(g)^{-1}v)$ of $P \times V$, where $u \in P, g \in G$ and $v \in V$. For example, associated with $P(M, \mathrm{GL}(k, \mathbb{R}))$ is a vector bundle over M with fibre $\mathbb{R}^k$. The fibre bundle structure of an associated vector bundle $E = P \times_\rho V$ is given as follows. The projection $\pi_E : E \to M$ is defined by $\pi_E(u, v) = \pi(u)$. This projection is well defined since $\pi(u) = \pi(ug)$ implies $\pi_E(ug, \rho(g)^{-1}v) = \pi(ug) = \pi_E(u, v)$. The local trivialization is given by $\psi_i : U_i \times V \to \pi_E^{-1}(U_i)$. The transition function of E is given by $\rho(t_{ij}(p))$ where $t_{ij}(p)$ is that of P.

Conversely a vector bundle naturally induces a principal bundle associated with it. Let $E \xrightarrow{\pi} M$ be a vector bundle with $\dim E = k$ (i.e. the fibre is $\mathbb{R}^k$ or $\mathbb{C}^k$). Then E induces a principal bundle $P(E) \equiv P(M, G)$ over M by employing the same transition functions. The structure group G is either $\mathrm{GL}(k, \mathbb{R})$ or $\mathrm{GL}(k, \mathbb{C})$. Explicit construction of $P(E)$ is carried out following the reconstruction process described in section 9.1.

Example 9.11. Associated with a tangent bundle TM over an m-dimensional manifold M is a principal bundle called the **frame bundle** $LM \equiv \bigcup_{p \in M} L_p M$ where $L_p M$ is the set of frames at p. We introduce coordinates x^μ on a chart U_i. The bundle $T_p M$ has a natural basis $\{\partial/\partial x^\mu\}$ on U_i. A frame $u = \{X_1, \ldots, X_m\}$ at p is expressed as

$$X_\alpha = X^\mu{}_\alpha \partial/\partial x^\mu|_p \qquad 1 \le \alpha \le m \tag{9.67}$$

where $(X^\mu{}_\alpha)$ is an element $\mathrm{GL}(m, \mathbb{R})$ so that $\{X_\alpha\}$ are linearly independent. We define the local trivialization $\phi_i : U_i \times \mathrm{GL}(m, \mathbb{R}) \to \pi^{-1}(U_i)$ by $\phi_i^{-1}(u) = (p, (X^\mu{}_\alpha))$. The bundle structure of LM is defined as follows.

(i) If $u = \{X_1, \ldots, X_m\}$ is a frame at p, we define $\pi_L : LM \to M$ by $\pi_L(u) = p$.

(ii) The action of $a = (a^i{}_j) \in \mathrm{GL}(m, \mathbb{R})$ on the frame $u = \{X_1, \ldots, X_m\}$ is given by $(u, a) \mapsto ua$, where ua is a new frame at p, defined by

$$Y_\beta = X_\alpha a^\alpha{}_\beta. \tag{9.68}$$

Conversely, given any frames $\{X_\alpha\}$ and $\{Y_\beta\}$ there exists an element of $GL(m, \mathbb{R})$ such that (9.68) is satisfied. Thus, $GL(m, \mathbb{R})$ acts on LM transitively.

(iii) Let U_i and U_j be overlapping charts with the coordinates x^μ and y^μ, respectively. For $p \in U_i \cap U_j$, we have

$$X_\alpha = X^\mu{}_\alpha \partial/\partial x^\mu|_p = \tilde{X}^\mu{}_\alpha \partial/\partial y^\mu|_p \qquad (9.69)$$

where $(X^\mu{}_\alpha)$, $(\tilde{X}^\mu_\alpha) \in GL(m, \mathbb{R})$. Since $X^\mu{}_\alpha = (\partial x^\mu/\partial y^\nu)_p \tilde{X}^\mu{}_\alpha$, we find the transition function $t^L_{ij}(p)$ to be

$$t^L_{ij}(p) = ((\partial x^\mu/\partial y^\nu)_p) \in GL(m, \mathbb{R}). \qquad (9.70)$$

Accordingly, given TM, we have constructed a frame bundle LM with the same transition functions.

In general relativity, the right action corresponds to the local Lorentz transformation while the left action corresponds to the general coordinate transformation. It turns out that the frame bundle is the most natural framework in which to incorporate these transformations. If $\{X_\alpha\}$ is normalized by introducing a metric, the matrix $(X^\mu{}_\alpha)$ becomes the vierbein and the structure group reduces to $O(m)$; see section 7.8.

Example 9.12. A spinor field on M is a section of a **spin bundle** which we now define. Since $GL(k, \mathbb{R})$ has no spinor representation, we need to introduce an orthonormal frame bundle whose structure group is $SO(k)$. As we mentioned in example 4.12, $SPIN(k)$ is the universal covering group of $SO(k)$. [To define a spin bundle, we have to check whether the $SO(k)$ bundle lifts to a $SPIN(k)$ bundle over M. The obstruction to this lifting is discussed in section 11.6.]

To be specific, let us consider a spin bundle associated with the four-dimensional Lorentz frame bundle LM, where M is a four-dimensional Lorentz manifold. We are interested in a frame with a definite spacetime orientation as well as a time orientation. The structure group is then reduced to

$$O^+_\uparrow(3, 1) \equiv \{\Lambda \in O(3, 1)| \det \Lambda = +1, \Lambda_0{}^0 > 0\}. \qquad (9.71)$$

The universal covering group of $O^+_\uparrow(3, 1)$ is $SL(2, \mathbb{C})$, see example 5.16(*c*). The homomorphism $\varphi : SL(2, \mathbb{C}) \to O^+_\uparrow(3, 1)$ is a $2 : 1$ map with $\ker\varphi = \{I_2, -I_2\}$. The Weyl spinor is a section of the fibre bundle $(W, \pi, M, \mathbb{C}^2, SL(2, \mathbb{C}))$. The Dirac spinor is a section of

$$(D, \pi, M, \mathbb{C}^4, SL(2, \mathbb{C}) \oplus \overline{SL(2, \mathbb{C})}). \qquad (9.72)$$

A section of W is a $(1/2, 0)$ representation of $O^+_\uparrow(3, 1)$ and a section of $(\bar{W}, \pi, M, \mathbb{C}^2, \overline{SL(2, \mathbb{C})})$ is a $(0, 1/2)$ representation, see Ramond (1989) for example. A Dirac spinor belongs to $(1/2, 0) \oplus (0, 1/2)$.

The general structure of the spin bundle will be worked out in section 11.6.

9.4.3   Triviality of bundles

A fibre bundle is trivial if it is expressed as a direct product of the base space and the fibre. The following theorem gives the condition under which a fibre bundle is trivial.

Theorem 9.2. A principal bundle is trivial if and only if it admits a global section.

Proof. Let (P, π, M, G) be a principal bundle over M and let $s \in \Gamma(M, P)$ be a global section. This section may be used to show that there exists a homeomorphism between P and $M \times G$. If a is an element of G, the product $s(p)a$ belongs to the fibre at p. Since the right action is transitive and free, any element $u \in P$ is uniquely written as $s(p)a$ for some $p \in M$ and $a \in G$. Define a map $\Phi : P \to M \times G$ by

$$\Phi : s(p)a \mapsto (p, a). \tag{9.73}$$

It is easily verified that Φ is indeed a homeomorphism and we have shown that P is a trivial bundle $M \times G$.

Conversely, suppose $P \cong M \times G$. Let $\phi : M \times G \to P$ be a trivialization. Take a fixed element $g \in G$. Then $s_g : M \to P$ defined by $s_g(p) = \phi(p, g)$ is a global section. $\square$

Is there a corresponding theorem for vector bundles? We know that any vector bundle admits a global null section. Thus, we cannot simply replace P by E in theorem 9.2. Let us consider the associated principal bundle $P(E)$ of E. By definition, E and $P(E)$ share the same set of transition functions. Since the twisting of a bundle is described purely by the transition functions, we obtain the following corollary.

Corollary 9.2. A vector bundle E is trivial if and only if its associated principal bundle $P(E)$ admits a global section.

Problems

9.1 Let L be the real line bundle over S^1 (i.e. L is either the cylinder $S^1 \times \mathbb{R}$ or the Möbius strip). Show that the Whitney sum $L \oplus L$ is a trivial bundle. Sketch $L \oplus L$ to confirm the result.

9.2 Let Ω_n be the volume element of S^n normalized as $\int_{S^n} \Omega_n = 1$. Let $f : S^{2n-1} \to S^n$ be a smooth map and consider the pullback $f^*\Omega_n$.

(a) Show that $f^*\Omega_n$ is closed and written as $d\omega_{n-1}$, where ω_{n-1} is an $(n-1)$-form on S^{2n-1}.

(b) Show that the **Hopf invariant**

$$H(f) \equiv \int_{S^{2n-1}} \omega_{n-1} \wedge d\omega_{n-1}$$

is independent of the choice of ω_{n-1}.

(c) Show that if f is homotopic to g, then $H(f) = H(g)$.

(d) Show that $H(f) = 0$ if n is odd. [*Hint*: Use $\omega_{n-1} \wedge d\omega_{n-1} = \frac{1}{2}d(\omega_{n-1} \wedge \omega_{n-1})$.]

(e) Compute the Hopf invariant of the map $\pi : S^3 \to S^2$ defined in example 9.9.

10

CONNECTIONS ON FIBRE BUNDLES

In chapter 7 we introduced connections in Riemannian manifolds which enable us to compare vectors in different tangent spaces. In the present chapter connections on fibre bundles are defined in an abstract though geometrical way.

We first define a connection on a principal bundle. Our abstract definition is realized concretely by introducing the connection one-form whose local form is well known to physicists as a gauge potential. The Yang–Mills field strength is defined as the curvature associated with the connection. A connection on a principal bundle naturally defines a covariant derivative in the associated vector bundle. We reproduce the results obtained in chapter 7, applying our approach to tangent bundles. We conclude this chapter with a few applications of connections to physics: to gauge field theories and Berry's phase. We follow the line of Choquet-Bruhat *et al* (1982), Kobayashi (1984) and Nomizu (1981). Details will be found in the classic books by Kobayashi and Nomizu (1963, 1969). See also Daniel and Viallet (1980) for a quick review.

10.1 Connections on principal bundles

There are several equivalent definitions of a connection on a principal bundle. Our approach is based on the *separation* of tangent space $T_u P$ into 'vertical' and 'horizontal' subspaces. Although this approach seems to be abstract, it is advantageous compared with other approaches in that it clarifies the geometrical pictures involved and is defined independently of special local trivializations. Connections are also defined as $\mathfrak{g}$-valued one-forms which satisfy certain axioms. These definitions are shown to be equivalent.

We briefly summarize the basic facts on Lie groups and Lie algebras, since we shall make extensive use of these (see section 5.6 for details). Let G be a Lie group. The left action L_g and the right action R_g are defined by $L_g h = gh$ and $R_g h = hg$ for $g, h \in G$. L_g induces a map $L_{g*} : T_h(G) \to T_{gh}(G)$. A left-invariant vector field X satisfies $L_{g*} X|_h = X|_{gh}$. Left-invariant vector fields form a Lie algebra of G, denoted by $\mathfrak{g}$. Since $X \in \mathfrak{g}$ is specified by its value at the unit element e, and *vice versa*, there exists a vector space isomorphism $\mathfrak{g} \cong T_e G$. The Lie algebra $\mathfrak{g}$ is closed under the Lie bracket, $[T_\alpha, T_\beta] = f_{\alpha\beta}{}^\gamma T_\gamma$ where $\{T_\alpha\}$ is the set of generators of $\mathfrak{g}$. $f_{\alpha\beta}{}^\gamma$ are called the **structure constants**. The adjoint action ad $: G \to G$ is defined by $\mathrm{ad}_g h \equiv ghg^{-1}$. The tangent map of ad_g is called the adjoint map and is denoted by $Ad_g : T_h(G) \to T_{ghg^{-1}}(G)$. If restricted to $T_e(G) \simeq \mathfrak{g}$, Ad_g maps $\mathfrak{g}$ onto itself; $Ad_g : \mathfrak{g} \to \mathfrak{g}$ as $A \mapsto gAg^{-1}$, $A \in \mathfrak{g}$.

10.1.1 Definitions

Let u be an element of a principal bundle $P(M, G)$ and let G_p be the fibre at $p = \pi(u)$. The **vertical subspace** $V_u P$ is a subspace of $T_u P$ which is tangent to G_p at u. [*Warning*: $T_u P$ is the tangent space of P and should not be confused with the tangent space $T_p M$ of M.] Let us see how $V_u P$ is constructed. Take an element A of $\mathfrak{g}$. By the right action

$$R_{\exp(tA)}u = u\exp(tA)$$

a curve through u is defined in P. Since $\pi(u) = \pi(u\exp(tA)) = p$, this curve lies within G_p. Define a vector $A^{\#} \in T_u P$ by

$$A^{\#} f(u) = \frac{\mathrm{d}}{\mathrm{d}t} f(u\exp(tA))|_{t=0} \tag{10.1}$$

where $f : P \to \mathbb{R}$ is an arbitrary smooth function. The vector $A^{\#}$ is tangent to P at u, hence $A^{\#} \in V_u P$. In this way we define a vector $A^{\#}$ at each point of P and construct a vector field $A^{\#}$, called the **fundamental vector field** generated by A. There is a vector space isomorphism $\sharp : \mathfrak{g} \to V_u P$ given by $A \mapsto A^{\#}$. The **horizontal subspace** $H_u P$ is a complement of $V_u P$ in $T_u P$ and is uniquely specified if a connection is defined in P.

Exercise 10.1.

   (a) Show that $\pi_* X = 0$ for $X \in V_u P$.
   (b) Show that $\sharp$ preserves the Lie algebra structure:

$$[A^{\#}, B^{\#}] = [A, B]^{\#}. \tag{10.2}$$

Definition 10.1. Let $P(M, G)$ be a principal bundle. A **connection** on P is a unique separation of the tangent space $T_u P$ into the vertical subspace $V_u P$ and the horizontal subspace $H_u P$ such that

   (i) $T_u P = H_u P \oplus V_u P$.
   (ii) A smooth vector field X on P is separated into smooth vector fields $X^H \in H_u P$ and $X^V \in V_u P$ as $X = X^H + X^V$.
   (iii) $H_{ug} P = R_{g*} H_u P$ for arbitrary $u \in P$ and $g \in G$; see figure 10.1.

   The condition (iii) states that horizontal subspaces $H_u P$ and $H_{ug} P$ on the same fibre are related by a linear map R_{g*} induced by the right action. Accordingly, a subspace $H_u P$ at u generates all the horizontal subspaces on the same fibre. This condition ensures that if a point u is parallel transported, so is its constant multiple ug, $g \in G$; see later. At this point, the reader might feel rather

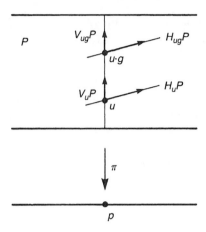

Figure 10.1. The horizontal subspace $H_{ug} P$ is obtained from $H_u P$ by the right action.

uneasy about our definition of a connection. At first sight, this definition seems to have nothing to do with the gauge potential or the field strength. We clarify these points after we introduce the connection one-form on P. We again stress that our definition, which is based on the separation $T_u P = V_u P \oplus H_u P$, is purely geometrical and is defined independently of any extra information. Although the connection becomes more tractable in the following, the geometrical picture and its intrinsic nature are generally obscured.

10.1.2 The connection one-form

In practical computations, we need to separate $T_u P$ into $V_u P$ and $H_u P$ in a systematic way. This can be achieved by introducing a Lie-algebra-valued one-form $\omega \in \mathfrak{g} \otimes T^* P$ called the **connection one-form**.

Definition 10.2. A connection one-form $\omega \in \mathfrak{g} \otimes T^* P$ is a *projection* of $T_u P$ onto the vertical component $V_u P \simeq \mathfrak{g}$. The projection property is summarized by the following requirements,

$$\text{(i)} \qquad \omega(A^{\#}) = A \qquad A \in \mathfrak{g} \tag{10.3a}$$

$$\text{(ii)} \qquad R_g^* \omega = \mathrm{Ad}_{g^{-1}} \omega \tag{10.3b}$$

that is, for $X \in T_u P$,

$$R_g^* \omega_{ug}(X) = \omega_{ug}(R_{g*} X) = g^{-1} \omega_u(X) g. \tag{10.3b'}$$

Define the horizontal subspace $H_u P$ by the kernel of ω,

$$H_u P \equiv \{ X \in T_u P | \omega(X) = 0 \}. \tag{10.4}$$

To show that this definition is consistent with definition 10.1, we prove the following proposition.

Proposition 10.1. The horizontal subspaces (10.4) satisfy

$$R_{g*}H_u P = H_{ug}P. \tag{10.5}$$

Proof. Fix a point $u \in P$ and define $H_u P$ by (10.4). Take $X \in H_u P$ and construct $R_{g*}X \in T_{ug}P$. We find

$$\omega(R_{g*}X) = R_g^*\omega(X) = g^{-1}\omega(X)g = 0$$

since $\omega(X) = 0$. Accordingly, $R_{g*}X \in H_{ug}P$. We note that R_{g*} is an invertible linear map. Hence, any vector $Y \in H_{ug}P$ is expressed as $Y = R_{g*}X$ for some $X \in H_u P$. This proves (10.5). $\qquad\square$

We have shown that the definition of the connection one-form ω is equivalent to that of the connection, since ω separates $T_u P$ into $H_u P \oplus V_u P$ in harmony with the axioms of definition 10.1. The connection one-form ω defined here is known as the **Ehresmann connection** in the literature.

10.1.3   The local connection form and gauge potential

Let $\{U_i\}$ be an open covering of M and let σ_i be a local section defined on each U_i. It is convenient to introduce a Lie-algebra-valued one-form $\mathcal{A}_i$ on U_i, by

$$\mathcal{A}_i \equiv \sigma_i^*\omega \in \mathfrak{g} \otimes \Omega^1(U_i). \tag{10.6}$$

Conversely, given a Lie-algebra-valued one-form $\mathcal{A}_i$, on U_i, we can reconstruct a connection one-form ω whose pullback by σ_i^* is $\mathcal{A}_i$.

Theorem 10.1. Given a $\mathfrak{g}$-valued one-form $\mathcal{A}_i$ on U_i and a local section $\sigma_i : U_i \rightarrow \pi^{-1}(U_i)$, there exists a connection one-form ω such that $\mathcal{A}_i = \sigma_i^*\omega$.

Proof. Let us define a $\mathfrak{g}$-valued one-form ω on P by

$$\omega_i \equiv g_i^{-1}\pi^*\mathcal{A}_i g_i + g_i^{-1}\,\mathrm{d}_P g_i \tag{10.7}$$

where d_P is the exterior derivative on P and g_i is the **canonical local trivialization** defined by $\phi_i^{-1}(u) = (p, g_i)$ for $u = \sigma_i(p)g_i$. We first show that $\sigma_i^*\omega_i = \mathcal{A}_i$. For $X \in T_p M$, we have

$$\sigma_i^*\omega_i(X) = \omega_i(\sigma_{i*}X) = \pi^*\mathcal{A}_i(\sigma_{i*}X) + \mathrm{d}_P g_i(\sigma_{i*}X)$$
$$= \mathcal{A}_i(\pi_*\sigma_{i*}X) + \mathrm{d}_P g_i(\sigma_{i*}X)$$

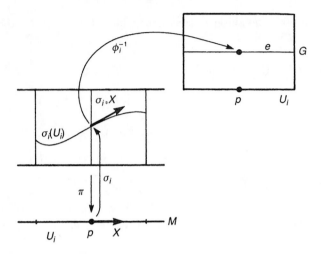

Figure 10.2. The canonical local trivialization defined by the local section σ_i over U_i.

where we have noted that $\sigma_{i*}X \in T_{\sigma_i}P$ and $g_i = e$ at σ_i, see figure 10.2. We further note that $\pi_*\sigma_{i*} = \mathrm{id}_{T_p(M)}$ and $\mathrm{d}_P g_i(\sigma_{i*}X) = 0$ since $g \equiv e$ along $\sigma_{i*}X$. Thus, we have obtained $\sigma_i^*\omega_i(X) = \mathcal{A}_i(X)$.

Next we show that ω_i satisfies the axioms of a connection one-form given in definition 10.2.

(i) Let $X = A^\# \in V_uP, A \in \mathfrak{g}$. It follows from exercise 10.1(a) that $\pi_*X = 0$. Now we have

$$\omega_i(A^\#) = g_i^{-1}\,\mathrm{d}_P g_i\,(A^\#) = g_i(u)^{-1}\,\frac{\mathrm{d}g\,(u\exp(tA))}{\mathrm{d}t}\bigg|_{t=0}$$

$$= g_i(u)^{-1}g_i(u)\,\frac{\mathrm{d}\exp(tA)}{\mathrm{d}t}\bigg|_{t=0} = A.$$

(ii) Take $X \in T_uP$ and $h \in G$. We have

$$R_h^*\omega_i(X) = \omega_i(R_{h*}X) = g_{iuh}^{-1}\mathcal{A}_i(\pi_*R_{h*}X)g_{iuh} + g_{iuh}^{-1}\,\mathrm{d}_P g_{iuh}\,(R_{h*}X).$$

Since $g_{iuh} = g_{iu}h$ and $\pi_*R_{h*}X = \pi_*X$ (note that $\pi R_h = \pi$), we have

$$R_h^*\omega_i(X) = h^{-1}g_{iu}^{-1}\mathcal{A}_i(\pi_*X)g_{iu}h + h^{-1}g_{iu}^{-1}\,\mathrm{d}_P g_{iu}\,(X)h$$

$$= h^{-1}\omega_i(X)h$$

where we have noted that

$$g_{iuh}^{-1}\,\mathrm{d}_P g_{iuh}\,(R_{h*}X) = g_{iuh}^{-1}\,\frac{\mathrm{d}}{\mathrm{d}t}g_{i\gamma(t)h}\bigg|_{t=0}$$

$$= h^{-1}g_{iu}^{-1}\,\frac{\mathrm{d}}{\mathrm{d}t}g_{i\gamma(t)}\bigg|_{t=0}h = h^{-1}g_{iu}^{-1}\,\mathrm{d}_P g_{iu}\,(X)h.$$

Here $\gamma(t)$ is a curve through $u = \gamma(0)$, whose tangent vector at u is X.

Hence, the $\mathfrak{g}$-valued one-form ω_i defined by (10.7) indeed satisfies $\mathcal{A}_i = \sigma_i^*\omega_i$ and the axioms of a connection one-form. $\square$

For ω to be defined *uniquely* on P, i.e. for the separation $T_u P = H_u P \oplus V_u P$ to be unique, we must have $\omega_i = \omega_j$ on $U_i \cap U_j$. A unique one-form ω is then defined throughout P by $\omega|_{U_i} = \omega_i$. To fulfil this condition, the local forms $\mathcal{A}_i$ have to satisfy a peculiar transformation property similar to that of the Christoffel symbols. We first prove a technical lemma.

Lemma 10.1. Let $P(M, G)$ be a principal bundle and σ_i (σ_j) be a local section over U_i (U_j) such that $U_i \cap U_j \neq \emptyset$. For $X \in T_p M$ ($p \in U_i \cap U_j$), $\sigma_{i*}X$ and $\sigma_{j*}X$ satisfy

$$\sigma_{j*}X = R_{t_{ij}*}(\sigma_{i*}X) + (t_{ij}^{-1} \, \mathrm{d}t_{ij}\,(X))^{\#} \tag{10.8}$$

where $t_{ij} : U_i \cap U_j \to G$ is the transition function.

Proof. Take a curve $\gamma : [0, 1] \to M$ such that $\gamma(0) = p$ and $\dot{\gamma}(0) = X$. Since $\sigma_i(p)$ and $\sigma_j(p)$ are related by the transition function as $\sigma_j(p) = \sigma_i(p)t_{ij}(p)$ (see (9.43)), we have

$$\sigma_{j*}X = \left.\frac{\mathrm{d}}{\mathrm{d}t}\sigma_j(\gamma(t))\right|_{t=0} = \left.\frac{\mathrm{d}}{\mathrm{d}t}\{\sigma_i(t)t_{ij}(t)\}\right|_{t=0}$$

$$= \left.\frac{\mathrm{d}}{\mathrm{d}t}\sigma_i(t) \cdot t_{ij}(p) + \sigma_i(p) \cdot \frac{\mathrm{d}}{\mathrm{d}t}t_{ij}(t)\right|_{t=0}$$

$$= R_{t_{ij}*}(\sigma_{i*}X) + \left.\sigma_j(p)t_{ij}(p)^{-1}\frac{\mathrm{d}}{\mathrm{d}t}t_{ij}(t)\right|_{t=0}$$

where $\sigma_i(t)$ stands for $\sigma_i(\gamma(t))$ and we have assumed that G is a matrix group for which $R_{g*}X = Xg$. We note that

$$t_{ij}(p)^{-1} \, \mathrm{d}t_{ij}\,(X) = \left.t_{ij}(p)^{-1}\frac{\mathrm{d}}{\mathrm{d}t}t_{ij}(t)\right|_{t=0}$$

$$= \left.\frac{\mathrm{d}}{\mathrm{d}t}[t_{ij}(p)^{-1}t_{ij}(t)]\right|_{t=0} \in T_e(G) \cong \mathfrak{g}.$$

[Note that $t_{ij}(p)^{-1}t_{ij}(\gamma(t)) = e$ at $t = 0$.] This shows that the second term of $\sigma_{j*}X$ represents the vector field $(t_{ij}^{-1}\mathrm{d}t_{ij}(X))^{\#}$ at $\sigma_j(p)$. $\square$

The compatibility condition is easily obtained by applying the connection one-form ω on (10.8). We find that

$$\sigma_j^*\omega(X) = R_{t_{ij}}^*\omega(\sigma_{i*}X) + t_{ij}^{-1} \, \mathrm{d}t_{ij}\,(X)$$

$$= t_{ij}^{-1}\omega(\sigma_{i*}X)t_{ij} + t_{ij}^{-1} \, \mathrm{d}t_{ij}\,(X)$$

where the axioms of definition 10.2 have been used. Since this is true for any $X \in T_p M$, this equation reduces to

$$\mathcal{A}_j = t_{ij}^{-1} \mathcal{A}_i t_{ij} + t_{ij}^{-1} \, \mathrm{d}t_{ij}. \tag{10.9}$$

This is the **compatibility condition** we have been seeking.

Conversely, given an open covering $\{U_i\}$, the local sections $\{\sigma_i\}$ and the local forms $\{\mathcal{A}_i\}$ which satisfy (10.9), we may construct the $\mathfrak{g}$-valued one-form ω over P. Since a non-trivial principal bundle does not admit a global section, the pullback $\mathcal{A}_i = \sigma_i^* \omega$ exists locally but not necessarily globally. In gauge theories, $\mathcal{A}_i$ is identified with the **gauge potential (Yang–Mills potential)**. As we have seen in the monopole case, the monopole field $\boldsymbol{B} = g\boldsymbol{r}/r^3$ does not admit a single gauge potential and we require at least two $\mathcal{A}_i$ to describe this U(1) bundle over S^2.

Exercise 10.2. Let $P(M, G)$ be a principal bundle over M and let U be a chart of M. Take local sections σ_1 and σ_2 over U such that $\sigma_2(p) = \sigma_1(p)g(p)$. Show that the corresponding local forms $\mathcal{A}_1$ and $\mathcal{A}_2$ are related as

$$\mathcal{A}_2 = g^{-1} \mathcal{A}_1 g + g^{-1} \, \mathrm{d}g. \tag{10.10a}$$

In components, this becomes

$$\mathcal{A}_{2\mu} = g^{-1}(p)\mathcal{A}_{1\mu}(p)g(p) + g^{-1}(p)\partial_\mu g(p) \tag{10.10b}$$

which is simply the **gauge transformation** defined in section 1.8.

Example 10.1. Let P be a U(1) bundle over M. Take overlapping charts U_i and U_j. Let $\mathcal{A}_i$ ($\mathcal{A}_j$) be a local connection form on U_i (U_j). The transition function $t_{ij} : U_i \cap U_j \to$ U(1) is given by

$$t_{ij}(p) = \exp[i\Lambda(p)] \qquad \Lambda(p) \in \mathbb{R}. \tag{10.11}$$

$\mathcal{A}_i$ and $\mathcal{A}_j$ are related as

$$\begin{aligned} \mathcal{A}_j(p) &= t_{ij}(p)^{-1} \mathcal{A}_i(p) t_{ij}(p) + t_{ij}(p)^{-1} \, \mathrm{d}t_{ij}(p) \\ &= \mathcal{A}_i(p) + i\mathrm{d}\Lambda(p). \end{aligned} \tag{10.12a}$$

In components, we have the familiar expression

$$\mathcal{A}_{j\mu} = \mathcal{A}_{i\mu} + i\partial_\mu \Lambda. \tag{10.12b}$$

Our connection $\mathcal{A}_\mu$ differs from the standard vector potential A_μ by the Lie algebra factor: $\mathcal{A}_\mu = iA_\mu$.

Here we note again that ω is defined globally over the bundle $P(M, G)$. Although there are many connection one-forms on $P(M, G)$, they share the same global information about the bundle. In contrast, an individual local piece (gauge potential) $\mathcal{A}_i$ is associated with the *trivial* bundle $\pi^{-1}(U_i)$ and cannot have any global information on P. It is ω or, equivalently, the *total* of $\{\mathcal{A}_i\}$ satisfying the compatibility condition (10.9), which carries the global information about the bundle.

10.1.4 Horizontal lift and parallel transport

Parallel transport of a vector has been defined in chapter 7 as transport *without change*. Parallel transport of an element of a principal bundle along a curve in M is provided by the 'horizontal lift' of the curve.

Definition 10.3. Let $P(M, G)$ be a G bundle and let $\gamma : [0, 1] \to M$ be a curve in M. A curve $\tilde{\gamma} : [0, 1] \to P$ is said to be a **horizontal lift** of γ if $\pi \circ \tilde{\gamma} = \gamma$ and the tangent vector to $\tilde{\gamma}(t)$ always belongs to $H_{\tilde{\gamma}(t)}P$.

Let $\tilde{X}$ be a tangent vector to $\tilde{\gamma}$. Then it satisfies $\omega(\tilde{X}) = 0$ by definition. This condition is an ordinary differential equation (ODE) and the fundamental theorem of ODEs guarantees the local existence and uniqueness of the horizontal lift.

Theorem 10.2. Let $\gamma : [0, 1] \to M$ be a curve in M and let $u_0 \in \pi^{-1}(\gamma(0))$. Then there exists a unique horizontal lift $\tilde{\gamma}(t)$ in P such that $\tilde{\gamma}(0) = u_0$.

Let us construct such a curve $\tilde{\gamma}$. Let U_i be a chart which contains γ and take a section σ_i over U_i. If there exists a horizontal lift $\tilde{\gamma}$, it may be expressed as $\tilde{\gamma}(t) = \sigma_i(\gamma(t))g_i(t)$, where $g_i(t)$ stands for $g_i(\gamma(t)) \in G$. Without loss of generality, we may take a section such that $\sigma_i(\gamma(0)) = \tilde{\gamma}(0)$, that is $g_i(0) = e$. Let X be a tangent vector to $\gamma(t)$ at $\gamma(0)$. Then $\tilde{X} = \tilde{\gamma}_* X$ is tangent to $\tilde{\gamma}$ at $u_0 = \tilde{\gamma}(0)$. Since the tangent vector $\tilde{X}$ is horizontal, it satisfies $\omega(\tilde{X}) = 0$. A slight modification of lemma 10.1 yields

$$\tilde{X} = g_i(t)^{-1}\sigma_{i*}Xg_i(t) + [g_i(t)^{-1}\, \mathrm{d}g_i\,(X)]^{\#}.$$

By applying ω on this equation, we find

$$0 = \omega(\tilde{X}) = g_i(t)^{-1}\omega(\sigma_{i*}X)g_i(t) + g_i(t)^{-1}\frac{\mathrm{d}g_i(t)}{\mathrm{d}t}.$$

Multiplying on the left by $g_i(t)$, we have

$$\frac{\mathrm{d}g_i(t)}{\mathrm{d}t} = -\,\omega(\sigma_{i*}X)g_i(t). \tag{10.13a}$$

The fundamental theorem of ODEs guarantees the existence and uniqueness of the solution of (10.13a).

Since $\omega(\sigma_{i*}X) = \sigma_i^*\omega(X) = \mathcal{A}_i(X)$, (10.13a) is expressed in a local form as

$$\frac{\mathrm{d}g_i(t)}{\mathrm{d}t} = -\mathcal{A}_i(X)g_i(t) \tag{10.13b}$$

whose formal solution with $g_i(0) = e$ is

$$g_i(\gamma(t)) = \mathcal{P}\exp\left(-\int_0^t \mathcal{A}_{i\mu}\frac{\mathrm{d}x^\mu}{\mathrm{d}t}\,\mathrm{d}t\right)$$

$$= \mathcal{P}\exp\left(-\int_{\gamma(0)}^{\gamma(t)} \mathcal{A}_{i\mu}(\gamma(t))\,\mathrm{d}x^\mu\right) \tag{10.14}$$

where $\mathcal{P}$ is a path-ordering operator along $\gamma(t)$.[1] The horizontal lift is expressed as $\tilde{\gamma}(t) = \sigma_i(\gamma(t))g_i(\gamma(t))$.

Corollary 10.1. Let $\tilde{\gamma}'$ be another horizontal lift of γ, such that $\tilde{\gamma}'(0) = \gamma(0)g$. Then $\tilde{\gamma}'(t) = \tilde{\gamma}(t)g$ for all $t \in [0, 1]$.

Proof. We first note that the horizontal subspace is right invariant, $R_{g*}H_uP = H_{ug}P$. Let $\tilde{\gamma}$ be a horizontal lift of γ. Then $\tilde{\gamma}_g : t \mapsto \tilde{\gamma}(t)g$ is also a horizontal lift of $\gamma(t)$ since its tangent vector belongs to $H_{\tilde{\gamma}g}P$. From theorem 10.2 we find $\tilde{\gamma}'$ is the unique horizontal lift which starts at $\tilde{\gamma}(0)g$.   □

Example 10.2. Let us consider the bundle $P(M, \mathbb{R}) \cong M \times \mathbb{R}$ where $M = \mathbb{R}^2 - \{0\}$. Let $\phi : ((x, y), f) \mapsto u \in P$ be a local trivialization, where (x, y) are the coordinates of M while f is that of the additive group $\mathbb{R}$. Let

$$\omega = \frac{y\mathrm{d}x - x\mathrm{d}y}{x^2 + y^2} + \mathrm{d}f$$

be a connection one-form. It is easily verified that ω satisfies the axioms of the connection one-form. In fact, for $A^\# = A\partial/\partial f$, $A \in \mathbb{R}$ being an element of the Lie algebra of additive group, we have $\omega(A^\#) = A$. Furthermore, $R_{g*}\omega = \omega = g^{-1}\omega g$, since $\mathbb{R}$ is Abelian. Let $\gamma : [0, 1] \to M$ be a curve $t \mapsto (\cos 2\pi t, \sin 2\pi t)$. Let us work out a horizontal lift which starts at $((1, 0), 0)$. Let

$$X = \frac{\mathrm{d}}{\mathrm{d}t} \equiv \frac{\mathrm{d}x}{\mathrm{d}t}\frac{\partial}{\partial x} + \frac{\mathrm{d}y}{\mathrm{d}t}\frac{\partial}{\partial y} + \frac{\mathrm{d}f}{\mathrm{d}t}\frac{\partial}{\partial f}$$

be tangent to $\tilde{\gamma}(t)$. For X to be horizontal, it must satisfy

$$0 = \omega(X) = \frac{\mathrm{d}x}{\mathrm{d}t}\frac{y}{r^2} - \frac{\mathrm{d}y}{\mathrm{d}t}\frac{x}{r^2} + \frac{\mathrm{d}f}{\mathrm{d}t} = -2\pi + \frac{\mathrm{d}f}{\mathrm{d}t}.$$

The solution is easily found to be $f = 2\pi t + \text{constant}$. We finally find the horizontal lift $\tilde{\gamma}$ passing through $((1, 0), 0)$,

$$\tilde{\gamma}(t) = ((\cos 2\pi t, \sin 2\pi t), 2\pi t) \tag{10.15}$$

which is a helix over the unit circle.

Under the group action (right or left does not matter), f translates to $f + g$, $g \in \mathbb{R}$. The shifted horizontal lift is

$$\tilde{\gamma}_g(t) = ((\cos 2\pi t, \sin 2\pi t), 2\pi t + g). \tag{10.16}$$

---

[1] $\mathcal{A}_{i\mu}(\gamma(t))$ and $\mathcal{A}_{i\nu}(\gamma(s))$ do not commute in general and the exponential in (10.14) is not well defined as it is. Let $A(t)$ and $B(t)$ be t-dependent matrices. Then the action of $\mathcal{P}$ is

$$\mathcal{P}[A(t)B(s)] = \begin{cases} A(t)B(s) & (t > s) \\ B(s)A(t) & (s > t). \end{cases}$$

Generalization to products of more matrices should be obvious.

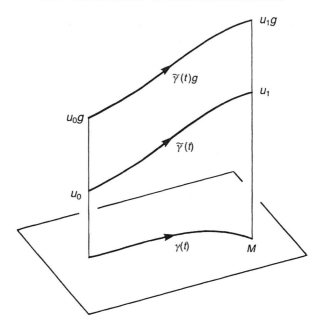

Figure 10.3. A curve $\gamma(t)$ in M and its horizontal lifts $\tilde{\gamma}(t)$ and $\tilde{\gamma}(t)g$.

Let $\gamma : [0, 1] \to M$ be a curve. Take a point $u_0 \in \pi^{-1}(\gamma(0))$. There is a unique horizontal lift $\tilde{\gamma}(t)$ of $\gamma(t)$ through u_0, and hence a unique point $u_1 = \tilde{\gamma}(1) \in \pi^{-1}(\gamma(1))$, see figure 10.3. The point u_1 is called the **parallel transport** of u_0 along the curve γ. This defines a map $\Gamma(\tilde{\gamma}) : \pi^{-1}(\gamma(0)) \to \pi^{-1}(\gamma(1))$ such that $u_0 \mapsto u_1$. If the local form (10.14) is employed, we have

$$u_1 = \sigma_i(1)\mathcal{P}\exp\left(-\int_0^1 \mathcal{A}_{i\mu}\frac{\mathrm{d}x^\mu(\gamma(t))}{\mathrm{d}t}\,\mathrm{d}t\right). \tag{10.17}$$

Corollary 10.1 ensures that $\Gamma(\tilde{\gamma})$ commutes with the right action R_g. First note that $R_g\Gamma(\tilde{\gamma})(u_0) = u_1 g$ and $\Gamma(\tilde{\gamma})R_g(u_0) = \Gamma(\tilde{\gamma})(u_0 g)$. Observe that $\tilde{\gamma}(t)g$ is a horizontal lift through $u_0 g$ *and* $u_1 g$. From the uniqueness of the horizontal lift through $u_0 g$, we have $u_1 g = \Gamma(\tilde{\gamma})(u_0 g)$, that is $R_g\Gamma(\tilde{\gamma})(u_0) = \Gamma(\tilde{\gamma})R_g(u_0)$. Since this is true for any $u_0 \in \pi^{-1}(\gamma(0))$, we have

$$R_g\Gamma(\tilde{\gamma}) = \Gamma(\tilde{\gamma})R_g. \tag{10.18}$$

Exercise 10.3. Let $\tilde{\gamma}$ be a horizontal lift of $\gamma : [0, 1] \to M$. Consider a map $\Gamma(\tilde{\gamma}^{-1}) : \pi^{-1}(\gamma(1)) \to \pi^{-1}(\gamma(0))$ where $\tilde{\gamma}^{-1}(t) = \tilde{\gamma}(1 - t)$. Show that

$$\Gamma(\tilde{\gamma}^{-1}) = \Gamma(\tilde{\gamma})^{-1}. \tag{10.19}$$

Consider two curves $\alpha : [0, 1] \to M$ and $\beta : [0, 1] \to M$ such that $\alpha(1) = \beta(0)$. Define the product $\alpha * \beta$ by

$$\alpha * \beta = \begin{cases} \alpha(2t) & 0 \le t \le \frac{1}{2} \\ \beta(2t - 1) & \frac{1}{2} \le t \le 1. \end{cases}$$

Let $\Gamma(\tilde{\alpha}) : \pi^{-1}(\alpha(0)) \to \pi^{-1}(\alpha(1))$ and $\Gamma(\tilde{\beta}) : \pi^{-1}(\beta(0)) \to \pi^{-1}(\beta(1))$. Show that

$$\Gamma(\widetilde{\alpha * \beta}) = \Gamma(\tilde{\beta}) \circ \Gamma(\tilde{\alpha}). \tag{10.20}$$

Exercise 10.4. Let us write $u \sim v$, if $u, v \in P$ are on the same horizontal lift. Show that $\sim$ is an equivalence relation.

10.2 Holonomy

10.2.1 Definitions

Let $P(M, G)$ be a principal bundle and let $\gamma : [0, 1] \to M$ be a curve whose horizontal lift through $u_0 \in \pi^{-1}(\gamma(0))$ is $\tilde{\gamma}$. In the last section, we defined a map $\Gamma(\tilde{\gamma}) : \pi^{-1}(\gamma(0)) \to \pi^{-1}(\gamma(1))$ which maps a point $u_0 = \tilde{\gamma}(0)$ to $u_1 = \tilde{\gamma}(1)$. Let us consider two curves $\alpha, \beta : [0, 1] \to M$ with $\alpha(0) = \beta(0) = p_0$ and $\alpha(1) = \beta(1) = p_1$. Take horizontal lifts $\tilde{\alpha}$ and $\tilde{\beta}$ of α and β such that $\tilde{\alpha}(0) = \tilde{\beta}(0) = u_0$. Then $\tilde{\alpha}(1)$ is not necessarily equal to $\tilde{\beta}(1)$. This shows that if we consider a *loop* $\gamma : [0, 1] \to M$ at $p = \gamma(0) = \gamma(1)$, we have $\tilde{\gamma}(0) \ne \tilde{\gamma}(1)$ in general. A loop γ defines a *transformation* $\tau_\gamma : \pi^{-1}(p) \to \pi^{-1}(p)$ on the fibre. This transformation is compatible with the right action of the group,

$$\tau_\gamma(ug) = \tau_\gamma(u)g \tag{10.21}$$

which follows immediately from (10.18). We note that τ_γ depends not only on the loop γ but also on the connection.

Example 10.3. Consider an $\mathbb{R}$-bundle over $M = \mathbb{R}^2 - \{0\}$. The connection one-form ω and the loop γ in example 10.2 define a map $\tau_\gamma : \pi^{-1}((1, 0)) \to \pi^{-1}((1, 0))$ given by $g \mapsto g + 2\pi, g \in \mathbb{R}$.

Take a point $u \in P$ with $\pi(u) = p$ and consider the set of loops $C_p(M)$ at p; $C_p(M) \equiv \{\gamma : [0, 1] \to M | \gamma(0) = \gamma(1) = p\}$. The set of elements

$$\Phi_u \equiv \{g \in G | \tau_\gamma(u) = ug, \gamma \in C_p(M)\} \tag{10.22}$$

is a subgroup of the structure group G and is called the **holonomy group** at u. The group property of Φ_u is easily derived from exercise 10.3. If α, β and $\gamma = \alpha * \beta$ are loops at p, we have $\tau_\gamma = \tau_\beta \circ \tau_\alpha$, hence

$$\tau_\gamma(u) = \tau_\beta \circ \tau_\alpha(u) = \tau_\beta(ug_\alpha) = \tau_\beta(u)g_\alpha = ug_\beta g_\alpha$$

where $\tau_\alpha(u) = ug_\alpha$ etc. This shows that

$$g_\gamma = g_\beta g_\alpha. \tag{10.23}$$

The constant loop $c : [0, 1] \mapsto p$ defines the identity transformation $\tau_c : u \mapsto u$. The inverse loop γ^{-1} of γ induces the inverse transformation $\tau_{\gamma^{-1}} = \tau_\gamma^{-1}$, hence $g_{\gamma^{-1}} = g_\gamma^{-1}$.

Exercise 10.5. (a) Let $\tau_\alpha(u) = ug_\alpha$. Show that

$$\tau_\alpha(ug) = ug(\mathrm{ad}_g \, g_\alpha) = ug(g^{-1}g_\alpha g). \tag{10.24}$$

Verify that

$$\Phi_{ua} \cong a^{-1}\Phi_u a. \tag{10.25}$$

(b) Let $u, u' \in P$ be points on the same horizontal lift $\tilde\gamma$. Show that $\Phi_u \cong \Phi_{u'}$.

(c) Suppose that M is connected. Show that all Φ_u are isomorphic to each other.

Exercise 10.6. Let $\mathcal{A}_i = \mathcal{A}_{i\mu} \, dx^\mu$ be a gauge potential over U_i and γ a loop in U_i. Let $\tau_\gamma(u) = ug_\gamma$, $u \in P$, $g_\gamma \in G$. Use (10.14) to show that

$$g_\gamma = \mathcal{P}\exp\left(-\oint_\gamma \mathcal{A}_{i\mu} \, dx^\mu\right). \tag{10.26}$$

Let $C_p^0(M)$ denote the set of loops at p, which are homotopic to the constant loop at p. The group

$$\Phi_u^0 \equiv \{g \in G | \tau_\gamma(u) = ug, \gamma \in C_p^0(M)\} \tag{10.27}$$

is called the **restricted holonomy group**.

10.3   Curvature

10.3.1   Covariant derivatives in principal bundles

We defined the exterior derivative $\mathrm{d} : \Omega^r(M) \to \Omega^{r+1}(M)$ in chapter 5. An r-form η is a real-valued form acting on vectors,

$$\eta : TM \wedge \ldots \wedge TM \to \mathbb{R}.$$

We will generalize this operation so that we can differentiate a vector-valued r-form $\phi \in \Omega^r(P) \otimes V$,

$$\phi : TP \wedge \ldots \wedge TP \to V$$

where V is a vector space of dimension k. The most general form of ϕ is $\phi = \sum_{\alpha=1}^k \phi^\alpha \otimes e_\alpha$, $\{e_\alpha\}$ being a basis of V and $\phi^\alpha \in \Omega^r(P)$.

A connection ω on a principal bundle $P(M, G)$ separates $T_u P$ into $H_u P \oplus V_u P$. Accordingly, a vector $X \in T_u P$ is decomposed as $X = X^H + X^V$ where $X^H \in H_u P$ and $X^V \in V_u P$.

Definition 10.4. Let $\phi \in \Omega^r(P) \otimes V$ and $X_1, \ldots, X_{r+1} \in T_u P$. The **covariant derivative** of ϕ is defined by

$$D\phi(X_1, \ldots, X_{r+1}) \equiv d_P \, \phi(X_1^H, \ldots, X_{r+1}^H) \tag{10.28}$$

where $d_P \, \phi \equiv d_P \, \phi^\alpha \otimes e_\alpha$.

10.3.2 Curvature

Definition 10.5. The **curvature two-form** Ω is the covariant derivative of the connection one-form ω,

$$\Omega \equiv D\omega \in \Omega^2(P) \otimes \mathfrak{g}. \tag{10.29}$$

Proposition 10.2. The curvature two-form satisfies (cf (10.3b))

$$R_a^* \Omega = a^{-1} \Omega a \qquad a \in G. \tag{10.30}$$

Proof. We first note that $(R_{a*}X)^H = R_{a*}(X^H)$ (R_{a*} preserves the horizontal subspaces) and $d_P R_a^* = R_a^* d_P$, see (5.75). By definition we find

$$
\begin{aligned}
R_a^* \Omega(X, Y) &= \Omega(R_{a*}X, R_{a*}Y) = d_P \omega((R_{a*}X)^H, (R_{a*}Y)^H) \\
&= d_P \omega(R_{a*}X^H, R_{a*}Y^H) = R_a^* \, d_P \omega(X^H, Y^H) \\
&= d_P R_a^* \omega(X^H, Y^H) \\
&= d_P (a^{-1}\omega a)(X^H, Y^H) = a^{-1} d_P \omega(X^H, Y^H) a \\
&= a^{-1} \Omega(X, Y) a
\end{aligned}
$$

where we noted that a is a constant element and hence $d_P a = 0$. $\qquad \square$

Take a $\mathfrak{g}$-valued p-form $\zeta = \zeta^\alpha \otimes T_\alpha$ and a $\mathfrak{g}$-valued q-form $\eta = \eta^\alpha \otimes T_\alpha$ where $\zeta^\alpha \in \Omega^p(P)$, $\eta^\alpha \in \Omega^q(P)$, and $\{T_\alpha\}$ is a basis of $\mathfrak{g}$. Define the commutator of ζ and η by

$$
\begin{aligned}
[\zeta, \eta] &\equiv \zeta \wedge \eta - (-1)^{pq} \eta \wedge \zeta \\
&= T_\alpha T_\beta \zeta^\alpha \wedge \eta^\beta - (-1)^{pq} T_\beta T_\alpha \eta^\beta \wedge \zeta^\alpha \\
&= [T_\alpha, T_\beta] \otimes \zeta^\alpha \wedge \eta^\beta = f_{\alpha\beta}{}^\gamma T_\gamma \otimes \zeta^\alpha \wedge \eta^\beta. \tag{10.31}
\end{aligned}
$$

If we put $\zeta = \eta$ in (10.31), when p and q are odd, we have

$$[\zeta, \zeta] = 2\zeta \wedge \zeta = f_{\alpha\beta}{}^\gamma T_\gamma \otimes \zeta^\alpha \wedge \zeta^\beta.$$

Lemma 10.2. Let $X \in H_u P$ and $Y \in V_u P$. Then $[X, Y] \in H_u P$.

Proof. Let Y be a vector field generated by $g(t)$, then

$$\mathcal{L}_Y X = [Y, X] = \lim_{t \to 0} t^{-1}(R_{g(t)*}X - X).$$

Since a connection satisfies $R_{g*}H_u P = H_{ug} P$, the vector $R_{g(t)*}X$ is horizontal and so is $[Y, X]$. $\qquad\square$

Theorem 10.3. Let $X, Y \in T_u P$. Then Ω and ω satisfy **Cartan's structure equation**

$$\Omega(X, Y) = d_P\omega(X, Y) + [\omega(X), \omega(Y)] \tag{10.32a}$$

which is also written as

$$\Omega = d_P\omega + \omega \wedge \omega. \tag{10.32b}$$

Proof. We consider the following three cases separately:

(i) Let $X, Y \in H_u P$. Then $\omega(X) = \omega(Y) = 0$ by definition. From definition 10.5, we have $\Omega(X, Y) = d_P\omega(X^H, Y^H) = d_P\omega(X, Y)$, since $X = X^H$ and $= Y^H$.

(ii) Let $X \in H_u P$ and $Y \in V_u P$. Since $Y^H = 0$, we have $\Omega(X, Y) = 0$. We also have $\omega(X) = 0$. Thus, we need to prove $d_P\omega(X, Y) = 0$. From (5.70), we obtain

$$d_P\omega(X, Y) = X\omega(Y) - Y\omega(X) - \omega([X, Y]) = X\omega(Y) - \omega([X, Y]).$$

Since $Y \in V_u P$, there is an element $V \in \mathfrak{g}$ such that $Y = V^\#$. Then $\omega(Y) = V$ is constant, hence $X\omega(Y) = X \cdot V = 0$. From lemma 10.2, we have $[X, Y] \in H_u P$ so that $\omega([X, Y]) = 0$ and we find $d_P\omega(X, Y) = 0$.

(iii) For $X, Y \in V_u P$, we have $\Omega(X, Y) = 0$. We find that, in this case,

$$d_P\omega(X, Y) = X\omega(Y) - Y\omega(X) - \omega([X, Y]) = -\omega([X, Y]).$$

We note that X and Y are closed under the Lie bracket, $[X, Y] \in V_u P$, see exercise 10.1(b). Then there exists $A \in \mathfrak{g}$ such that

$$\omega([X, Y]) = A$$

where $A^\# = [X, Y]$. Let $B^\# = X$ and $C^\# = Y$. Then $[\omega(X), \omega(Y)] = [B, C] = A$ since $[B, C]^\# = [B^\#, C^\#]$. Thus, we have shown that

$$0 = d_P\omega(X, Y) + \omega([X, Y]) = d_P\omega(X, Y) + [\omega(X), \omega(Y)].$$

Since Ω is linear and skew symmetric, these three cases are sufficient to show that (10.32) is true for any vectors.

To derive (10.32b) from (10.32a), we note that

$$\begin{aligned}
[\omega, \omega](X, Y) &= [T_\alpha, T_\beta]\omega^\alpha \wedge \omega^\beta(X, Y) \\
&= [T_\alpha, T_\beta][\omega^\alpha(X)\omega^\beta(Y) - \omega^\beta(X)\omega^\alpha(Y)] \\
&= [\omega(X), \omega(Y)] - [\omega(Y), \omega(X)] = 2[\omega(X), \omega(Y)].
\end{aligned}$$

Hence, $\Omega(X, Y) = (d_P\omega + \frac{1}{2}[\omega, \omega])(X, Y) = (d_P\omega + \omega \wedge \omega)(X, Y)$. $\qquad\square$

10.3.3 Geometrical meaning of the curvature and the Ambrose–Singer theorem

We have shown in chapter 7 that the Riemann curvature tensor expresses the non-commutativity of the parallel transport of vectors. There is a similar interpretation of curvature on principal bundles. We first show that $\Omega(X, Y)$ yields the vertical component of the Lie bracket $[X, Y]$ of horizontal vectors $X, Y \in H_u P$. It follows from $\omega(X) = \omega(Y) = 0$ that

$$\mathrm{d}_P \omega(X, Y) = X\omega(Y) - Y\omega(X) - \omega([X, Y]) = -\omega([X, Y]).$$

Since $X^H = X$, $Y^H = Y$, we have

$$\Omega(X, Y) = \mathrm{d}_P \omega(X, Y) = -\omega([X, Y]). \tag{10.33}$$

Let us consider a coordinate system $\{x^\mu\}$ on a chart U. Let $V = \partial/\partial x^1$ and $W = \partial/\partial x^2$. Take an infinitesimal parallelogram γ whose corners are $\mathrm{O} = \{0, 0, \ldots, 0\}$, $\mathrm{P} = \{\varepsilon, 0, \ldots, 0\}$, $\mathrm{Q} = \{\varepsilon, \delta, 0, \ldots, 0\}$ and $\mathrm{R} = \{0, \delta, 0, \ldots, 0\}$. Consider the horizontal lift $\tilde{\gamma}$ of γ. Let $X, Y \in H_u P$ such that $\pi_* X = \varepsilon V$ and $\pi_* Y = \delta W$. Then

$$\pi_*([X, Y]^H) = \epsilon\delta[V, W] = \epsilon\delta\left[\frac{\partial}{\partial x^1}, \frac{\partial}{\partial x^2}\right] = 0 \tag{10.34}$$

that is $[X, Y]$ is *vertical*. This consideration shows that the horizontal lift $\tilde{\gamma}$ of a loop γ fails to close. This failure is proportional to the vertical vector $[X, Y]$ connecting the initial point and the final point on the same fibre. The curvature measures this distance,

$$\Omega(X, Y) = -\omega([X, Y]) = A \tag{10.35}$$

where A is an element of $\mathfrak{g}$ such that $[X, Y] = A^\#$.

Since the discrepancy between the initial and final points of the horizontal lift of a closed curve is simply the holonomy, we expect that the holonomy group is expressed in terms of the curvature.

Theorem 10.4. (**Ambrose–Singer theorem**) Let $P(M, G)$ be a G bundle over a connected manifold M. The Lie algebra $\mathfrak{h}$ of the holonomy group Φ_{u_0} of a point $u_0 \in P$ agrees with the subalgebra of $\mathfrak{g}$ spanned by the elements of the form

$$\Omega_u(X, Y) \qquad X, Y \in H_u P \tag{10.36}$$

where $a \in P$ is a point on the same horizontal lift as u_0. [See Choquet-Bruhat *et al* (1982) for the proof.]

10.3.4 Local form of the curvature

The local form $\mathcal{F}$ of the curvature Ω is defined by

$$\mathcal{F} \equiv \sigma^* \Omega \tag{10.37}$$

where σ is a local section defined on a chart U of M (cf $\mathcal{A} = \sigma^* \omega$). $\mathcal{F}$ is expressed in terms of the gauge potential $\mathcal{A}$ as

$$\mathcal{F} = d\mathcal{A} + \mathcal{A} \wedge \mathcal{A} \tag{10.38a}$$

where d is the exterior derivative on M. The action of $\mathcal{F}$ on the vectors of TM is given by

$$\mathcal{F}(X, Y) = d\mathcal{A}(X, Y) + [\mathcal{A}(X), \mathcal{A}(Y)]. \tag{10.38b}$$

To prove (10.38a) we note that $\mathcal{A} = \sigma^* \omega$, $\sigma^* d_P \omega = d\sigma^* \omega$ and $\sigma^*(\zeta \wedge \eta) = \sigma^* \zeta \wedge \sigma^* \eta$. From Cartan's structure equation, we find

$$\mathcal{F} = \sigma^*(d_P \omega + \omega \wedge \omega) = d\sigma^* \omega + \sigma^* \omega \wedge \sigma^* \omega = d\mathcal{A} + \mathcal{A} \wedge \mathcal{A}.$$

Next, we find the component expression of $\mathcal{F}$ on a chart U whose coordinates are $x^\mu = \varphi(p)$. Let $\mathcal{A} = \mathcal{A}_\mu dx^\mu$ be the gauge potential. If we write $\mathcal{F} = \frac{1}{2} \mathcal{F}_{\mu\nu} dx^\mu \wedge dx^\nu$, a direct computation yields

$$\mathcal{F}_{\mu\nu} = \partial_\mu \mathcal{A}_\nu - \partial_\nu \mathcal{A}_\mu + [\mathcal{A}_\mu, \mathcal{A}_\nu]. \tag{10.39}$$

$\mathcal{F}$ is also called the curvature two-form and is identified with the (**Yang–Mills**) **field strength**. To avoid confusion, we call Ω the curvature and $\mathcal{F}$ the (Yang–Mills) field strength. Since $\mathcal{A}_\mu$ and $\mathcal{F}_{\mu\nu}$ are $\mathfrak{g}$-valued functions, they can be expanded in terms of the basis $\{T_\alpha\}$ of $\mathfrak{g}$ as

$$\mathcal{A}_\mu = A_\mu{}^\alpha T_\alpha \qquad \mathcal{F}_{\mu\nu} = F_{\mu\nu}{}^\alpha T_\alpha. \tag{10.40}$$

The basis vectors satisfy the usual commutation relations $[T_\alpha, T_\beta] = f_{\alpha\beta}{}^\gamma T_\gamma$. We then obtain the well-known expression

$$F_{\mu\nu}{}^\alpha = \partial_\mu A_\nu{}^\alpha - \partial_\nu A_\mu{}^\alpha + f_{\beta\gamma}{}^\alpha A_\mu{}^\beta A_\nu{}^\gamma. \tag{10.41}$$

Theorem 10.5. Let U_i and U_j be overlapping charts of M and let $\mathcal{F}_i$ and $\mathcal{F}_j$ be field strengths on the respective charts. On $U_i \cap U_j$, they satisfy the compatibility condition,

$$\mathcal{F}_j = \mathrm{Ad}_{t_{ij}^{-1}} \mathcal{F}_i = t_{ij}^{-1} \mathcal{F}_i t_{ij} \tag{10.42}$$

where t_{ij} is the transition function on $U_i \cap U_j$.

Proof. Introduce the corresponding gauge potentials $\mathcal{A}_i$ and $\mathcal{A}_j$,

$$\mathcal{F}_i = d\mathcal{A}_i + \mathcal{A}_i \wedge \mathcal{A}_i \qquad \mathcal{F}_j = d\mathcal{A}_j + \mathcal{A}_j \wedge \mathcal{A}_j.$$

Substituting $\mathcal{A}_j = t_{ij}^{-1}\mathcal{A}_i t_{ij} + t_{ij}^{-1}dt_{ij}$ into $\mathcal{F}_j$, we verify that

$$
\begin{aligned}
\mathcal{F}_j &= d\,(t_{ij}^{-1}\mathcal{A}_i t_{ij} + t_{ij}^{-1}\,dt_{ij}) \\
&\quad + (t_{ij}^{-1}\mathcal{A}_i t_{ij} + t_{ij}^{-1}\,dt_{ij}) \wedge (t_{ij}^{-1}\mathcal{A}_i t_{ij} + t_{ij}^{-1}\,dt_{ij}) \\
&= [-t_{ij}^{-1}\,dt_{ij} \wedge t_{ij}^{-1}\mathcal{A}_i t_{ij} + t_{ij}^{-1}\,d\mathcal{A}_i t_{ij} \\
&\quad - t_{ij}^{-1}\mathcal{A}_i \wedge dt_{ij} - t_{ij}^{-1}\,dt_{ij}\,t_{ij}^{-1} \wedge dt_{ij}] \\
&\quad + [t_{ij}^{-1}\mathcal{A}_i \wedge \mathcal{A}_i t_{ij} + t_{ij}^{-1}\mathcal{A}_i \wedge dt_{ij} \\
&\quad + t_{ij}^{-1}\,dt_{ij}\,t_{ij}^{-1} \wedge \mathcal{A}_i t_{ij} + t_{ij}^{-1}\,dt_{ij} \wedge t_{ij}^{-1}\,dt_{ij}] \\
&= t_{ij}^{-1}(d\mathcal{A}_i + \mathcal{A}_i \wedge \mathcal{A}_i)t_{ij} = t_{ij}^{-1}\mathcal{F}_i t_{ij}
\end{aligned}
$$

where use has been made of the identity $dt^{-1} = -t^{-1}\,dt\,t^{-1}$.   □

Exercise 10.7. The gauge potential $\mathcal{A}$ is called a **pure gauge** if $\mathcal{A}$ is written locally as $\mathcal{A} = g^{-1}\,dg$. Show that the field strength $\mathcal{F}$ vanishes for a pure gauge $\mathcal{A}$. [It can be shown that the converse is also true. If $\mathcal{F} = 0$ on a chart U, the gauge potential may be expressed *locally* as $\mathcal{A} = g^{-1}\,dg$.]

10.3.5 The Bianchi identity

Since ω and Ω are $\mathfrak{g}$-valued, we expand them in terms of the basis $\{T_\alpha\}$ of $\mathfrak{g}$ as $\omega = \omega^\alpha T_\alpha$, $\Omega = \Omega^\alpha T_\alpha$. Then (10.32b) becomes

$$\Omega^\alpha = d_P\omega^\alpha + f_{\beta\gamma}{}^\alpha \omega^\beta \wedge \omega^\gamma. \tag{10.43}$$

Exterior differentiation of (10.43) yields

$$d_P\Omega^\alpha = f_{\beta\gamma}{}^\alpha\,d_P\omega^\beta \wedge \omega^\gamma + f_{\beta\gamma}{}^\alpha \omega^\beta \wedge d_P\omega^\gamma. \tag{10.44}$$

If we note that $\omega(X) = 0$ for a horizontal vector X, we find

$$D\Omega(X, Y, Z) = d_P\Omega\,(X^H, Y^H, Z^H) = 0$$

where $X, Y, Z \in T_u P$. Thus, we have proved the **Bianchi identity**

$$D\Omega = 0. \tag{10.45}$$

Let us find the local form of the Bianchi identity. Operating with σ^* on (10.44), we find that $\sigma^*\,d_P\Omega = d \cdot \sigma^*\Omega = d\mathcal{F}$ for the LHS and

$$
\begin{aligned}
\sigma^*(d_P\omega \wedge \omega - \omega \wedge d_P\omega) &= d\sigma^*\omega \wedge \sigma^*\omega - \sigma^*\omega \wedge d\sigma^*\omega \\
&= d\mathcal{A} \wedge \mathcal{A} - \mathcal{A} \wedge d\mathcal{A} = \mathcal{F} \wedge \mathcal{A} - \mathcal{A} \wedge \mathcal{F}
\end{aligned}
$$

for the RHS. Thus, we have obtained that

$$\mathcal{D}\mathcal{F} = d\mathcal{F} + \mathcal{A} \wedge \mathcal{F} - \mathcal{F} \wedge \mathcal{A} = d\mathcal{F} + [\mathcal{A}, \mathcal{F}] = 0 \qquad (10.46)$$

where the action of $\mathcal{D}$ on a $\mathfrak{g}$-valued p-form η on M is defined by

$$\mathcal{D}\eta \equiv d\eta + [\mathcal{A}, \eta]. \qquad (10.47)$$

Note that $\mathcal{D}\mathcal{F} = d\mathcal{F}$ for $G = U(1)$.

10.4 The covariant derivative on associated vector bundles

A connection one-form ω on a principal bundle $P(M, G)$ enables us to define the covariant derivative in associated bundles of P in a natural way.

10.4.1 The covariant derivative on associated bundles

In physics, we often need to differentiate sections of a vector bundle which is associated with a certain principal bundle. For example, a charged scalar field in QED is regarded as a section of a complex line bundle associated with a $U(1)$ bundle $P(M, U(1))$. Differentiating sections covariantly is very important in constructing gauge-invariant actions.

Let $P(M, G)$ be a G bundle with the projection π_P. Let us take a chart U_i of M and a section σ_i over U_i. We take the canonical trivialization $\phi_i(p, e) = \sigma_i(p)$. Let $\tilde{\gamma}$ be a horizontal lift of a curve $\gamma : [0, 1] \rightarrow U_i$. We denote $\gamma(0) = p_0$ and $\tilde{\gamma}(0) = u_0$. Associated with P is a vector bundle $E = P \times_\rho V$ with the projection π_E, see section 9.4. Let $X \in T_pM$ be a tangent vector to $\gamma(t)$ at p_0. Let $s \in \Gamma(M, E)$ be a section, or a vector field, on M. Write an element of E as $[(u, v)] = \{(ug, \rho(g)^{-1}v | u \in P, v \in V, g \in G\}$. Taking a representative of the equivalence class amounts to fixing the gauge. We choose the following form,

$$s(p) = [(\sigma_i(p), \xi(p))] \qquad (10.48)$$

as a representative.

Now we define the parallel transport of a vector in E along a curve γ in M. Of course, a naive guess 'ξ is parallel transported if $\xi(\gamma(t))$ is constant along $\gamma(t)$' does not make sense since this statement depends on the choice of the section $\sigma_i(p)$. We define a vector to be parallel transported if it is constant with respect to a *horizontal lift* $\tilde{\gamma}$ of γ in P. In other words, a section $s(\gamma(t)) = [(\tilde{\gamma}(t), \eta(\gamma(t)))]$ is parallel transported if η is constant along $\gamma(t)$. This definition is intrinsic since if $\tilde{\gamma}'(t)$ is another horizontal lift of γ, then it can be written as $\tilde{\gamma}'(t) = \tilde{\gamma}(t)a$, $a \in G$ and we have (we omit ρ to simplify the notation)

$$[(\tilde{\gamma}(t), \eta(t))] = [(\tilde{\gamma}'(t)a^{-1}, \eta(t))] = [(\tilde{\gamma}'(t), a^{-1}\eta(t))]$$

where $\eta(t)$ stands for $\eta(\gamma(t))$. Hence, if $\eta(t)$ is constant along $\gamma(t)$, so is its constant multiple $a^{-1}\eta(t)$.

Now the definition of covariant derivative is in order. Let $s(p)$ be a section of E. Along a curve $\gamma : [0, 1] \to M$ we have $s(t) = [(\tilde{\gamma}(t), \eta(t))]$, where $\tilde{\gamma}(t)$ is an arbitrary horizontal lift of $\gamma(t)$. The covariant derivative of $s(t)$ along $\gamma(t)$ at $p_0 = \gamma(0)$ is defined by

$$\nabla_X s \equiv \left[\left(\tilde{\gamma}(0), \left. \frac{\mathrm{d}}{\mathrm{d}t} \eta(\gamma(t)) \right|_{t=0} \right) \right] \tag{10.49}$$

where X is the tangent vector to $\gamma(t)$ at p_0. For the covariant derivative to be really intrinsic, it should not depend on the *extra* information, that is the special horizontal lift. Let $\tilde{\gamma}'(t) = \tilde{\gamma}(t)a$ $(a \in G)$ be another horizontal lift of γ. If $\tilde{\gamma}'(t)$ is chosen to be *the* horizontal lift, we have a representative $[(\tilde{\gamma}'(t), a^{-1}\eta(t))]$. The covariant derivative is now given by

$$\left[\left(\tilde{\gamma}'(0), \left. \frac{\mathrm{d}}{\mathrm{d}t} \{a^{-1}\eta(t)\} \right|_{t=0} \right) \right] = \left[\left(\tilde{\gamma}'(0)a^{-1}, \left. \frac{\mathrm{d}}{\mathrm{d}t} \eta(t) \right|_{t=0} \right) \right]$$

which agrees with (10.49). Hence, $\nabla_X s$ depends only on the tangent vector X and the sections $s \in \Gamma(M, E)$ and not on the horizontal lift $\tilde{\gamma}(t)$. Our definition depends only on a curve γ and a connection and not on local trivializations. The local form of the covariant derivative is useful in practical computations and will be given later.

So far we have defined the covariant derivative at a point $p_0 = \gamma(0)$. It is clear that if X is a vector field, ∇_X maps a section s to a new section $\nabla_X s$, hence ∇_X is regarded as a map $\Gamma(M, E) \to \Gamma(M, E)$. To be more precise, take $X \in \mathfrak{X}(M)$ whose value at p is $X_p \in T_p M$. There is a curve $\gamma(t)$ such that $\gamma(0) = p$ and its tangent at p is X_p. Then any horizontal lift $\tilde{\gamma}(t)$ of γ enables us to compute the covariant derivative $\nabla_X s|_p \equiv \nabla_{X_p} s$. We also define a map $\nabla : \Gamma(M, E) \to \Gamma(M, E) \otimes \Omega^1(M)$ by

$$\nabla s(X) \equiv \nabla_X s \qquad X \in \mathfrak{X}(M) \qquad s \in \Gamma(M, E). \tag{10.50}$$

Exercise 10.8. Show that

$$\nabla_X(a_1 s_1 + a_2 s_2) = a_1 \nabla_X s_1 + a_2 \nabla_X s_2 \tag{10.51a}$$

$$\nabla(a_1 s_1 + a_2 s_2) = a_1 \nabla s_1 + a_2 \nabla s_2 \tag{10.51b}$$

$$\nabla_{(a_1 X_1 + a_2 X_2)} s = a_1 \nabla_{X_1} s + a_2 \nabla_{X_2} s \tag{10.51c}$$

$$\nabla_X(fs) = X[f]s + f \nabla_X s \tag{10.51d}$$

$$\nabla(fs) = (\mathrm{d}f)s + f \nabla s \tag{10.51e}$$

$$\nabla_{fX} s = f \nabla_X s \tag{10.51f}$$

where $a_i \in \mathbb{R}$, $s, s' \in \Gamma(M, E)$ and $f \in \mathcal{F}(M)$.

10.4.2 A local expression for the covariant derivative

In practical computations it is convenient to have a local coordinate representation of the covariant derivative. Let $P(M, G)$ be a G bundle and $E = P \times_\rho G$ be an associate vector bundle. Take a local section $\sigma_i \in \Gamma(U_i, P)$ and employ the canonical trivialization $\sigma_i(p) = \phi_i(p, e)$. Let $\gamma : [0, 1] \to M$ be a curve in U_i and $\tilde{\gamma}$ its horizontal lift, which is written as

$$\tilde{\gamma}(t) = \sigma_i(t)g_i(t) \tag{10.52}$$

where $g_i(t) \equiv g_i(\gamma(t)) \in G$. Take a section $e_\alpha(p) \equiv [(\sigma_i(p), e_\alpha{}^0)]$ of E, where $e_\alpha{}^0$ is the αth basis vector of V; $(e_\alpha{}^0)^\beta = (\delta_\alpha)^\beta$. We have

$$e_\alpha(t) = [(\tilde{\gamma}(t)g_i(t)^{-1}, e_\alpha{}^0)] = [(\tilde{\gamma}(t), g_i(t)^{-1}e_\alpha{}^0)]. \tag{10.53}$$

Note that $g_i(t)^{-1}$ acts on $e_\alpha{}^0$ to compensate for the change of basis along γ. The covariant derivative of e_α is then given by

$$\begin{aligned}
\nabla_X e_\alpha &= \left[\left(\tilde{\gamma}(0), \frac{\mathrm{d}}{\mathrm{d}t}\{g_i(t)^{-1}e_\alpha{}^0\}\Big|_{t=0} \right) \right] \\
&= \left[\left(\tilde{\gamma}(0), -g_i(t)^{-1}\left\{\frac{\mathrm{d}}{\mathrm{d}t}g_i(t)\right\}g_i(t)^{-1}e_\alpha{}^0\Big|_{t=0} \right) \right] \\
&= [(\tilde{\gamma}(0)g_i(0)^{-1}, \mathcal{A}_i(X)e_\alpha{}^0)]
\end{aligned} \tag{10.54}$$

where (10.13b) has been used. From (10.54) we find the local expression,

$$\nabla_X e_\alpha = [(\sigma_i(0), \mathcal{A}_i(X)e_\alpha{}^0)]. \tag{10.55}$$

Let $\mathcal{A}_i = \mathcal{A}_{i\mu}\, \mathrm{d}x^\mu = \mathcal{A}_{i\mu}{}^\alpha{}_\beta\, \mathrm{d}x^\mu$ where $\mathcal{A}_{i\mu}{}^\alpha{}_\beta \equiv \mathcal{A}_{i\mu}{}^\gamma(T_\gamma)^\alpha{}_\beta$. The second entry of (10.55) is

$$\mathcal{A}_i(X)e_\alpha{}^0 = \frac{\mathrm{d}x^\mu}{\mathrm{d}t}e_\beta{}^0\mathcal{A}_{i\mu}{}^\beta{}_\gamma\delta_\alpha{}^\gamma = \frac{\mathrm{d}x^\mu}{\mathrm{d}t}\mathcal{A}_{i\mu}{}^\beta{}_\alpha e_\beta{}^0.$$

Substituting this into (10.55), we finally have

$$\nabla_X e_\alpha = \left[\left(\sigma_i(0), \frac{\mathrm{d}x^\mu}{\mathrm{d}t}\mathcal{A}_{i\mu}{}^\beta{}_\alpha e_\beta{}^0 \right) \right] = \frac{\mathrm{d}x^\mu}{\mathrm{d}t}\mathcal{A}_{i\mu}{}^\beta{}_\alpha e_\beta \tag{10.56a}$$

or

$$\nabla e_\alpha = \mathcal{A}_i{}^\beta{}_\alpha e_\beta. \tag{10.56b}$$

In particular, for a coordinate curve x^μ, we have

$$\nabla_{\partial/\partial x^\mu} e_\alpha = \mathcal{A}_{i\mu}{}^\beta{}_\alpha e_\beta. \tag{10.57}$$

It is remarkable that a connection $\mathcal{A}$ on a principal bundle P completely specifies the covariant derivative on an associated bundle E (modulo representations).

Exercise 10.9. Let $s(p) = [(\sigma_i(p), \xi_i(p))] = \xi_i^\alpha(p)e_\alpha$ be a general section of E, where $\xi_i(p) = \xi_i^\alpha(p)e_\alpha{}^0$. Use the results of exercise 10.8 to verify that

$$\nabla_X s = \left[\left(\sigma_i(0), \left.\frac{d\xi_i}{dt} + \mathcal{A}_i(X)\xi_i\right|_{t=0}\right)\right] = \frac{dx^\mu}{dt}\left\{\frac{\partial\xi_i^\alpha}{\partial x^\mu} + \mathcal{A}_{i\mu}{}^\alpha{}_\beta\xi_i^\beta\right\}e_\alpha.$$

$$(10.58)$$

By construction, the covariant derivative is independent of the local trivialization. This is also observed from the local form of $\nabla_X s$. Let $\sigma_i(p)$ and $\sigma_j(p)$ be local sections on overlapping charts U_i and U_j. On $U_i \cap U_j$, we have $\sigma_j(p) = \sigma_i(p)t_{ij}(p)$. In the i-trivialization, the covariant derivative is

$$\begin{aligned}
\nabla_X s &= \left[\left(\sigma_i(0), \left.\frac{d\xi_i}{dt} + \mathcal{A}_i(X)\xi_i\right|_{t=0}\right)\right] \\
&= \left[\left(\sigma_j(0) \cdot t_{ij}^{-1}, \left.\frac{d}{dt}(t_{ij}\xi_j) + \mathcal{A}_i(X)t_{ij}\xi_j\right|_{t=0}\right)\right] \\
&= \left[\left(\sigma_j(0), \left.\frac{d\xi_j}{dt} + \mathcal{A}_j(X)\xi_j\right|_{t=0}\right)\right]
\end{aligned}$$

$$(10.59)$$

where use has been made of the condition (10.9). The last line of (10.59) is $\nabla_X s$ expressed in the j-trivialization.

We have found that the covariant derivative defined by (10.49) is independent of the horizontal lift as well as the local section. The gauge potential $\mathcal{A}_i$ transforms under the change of local trivialization so that $\nabla_X s$ is a well-defined section of E. In this sense, ∇_X is the most natural derivative on an associated vector bundle, which is compatible with the connection on the principal bundle P.

Example 10.4. Let us recover the results obtained in section 7.2. Let FM be a frame bundle over M and let TM be its associated bundle. We note $FM = P(M, \mathrm{GL}(m, \mathbb{R}))$ and $TM = FM \times_\rho \mathbb{R}^m$, where $m = \dim M$ and ρ is the $m \times m$ matrix representation of $\mathrm{GL}(m, \mathbb{R})$. Elements of $\mathfrak{gl}(m, \mathbb{R})$ are $m \times m$ matrices. Let us rewrite the local connection form $\mathcal{A}_i$ as $\Gamma^\alpha{}_{\mu\beta}\,dx^\mu$. We then find that

$$\nabla_{\partial/\partial x^\mu}e_\alpha = [(\sigma_i(0), \Gamma_\mu e_\alpha{}^0)] = \Gamma^\beta{}_{\mu\alpha}e_\beta \qquad (10.60)$$

which should be compared with (7.14). For a general section (vector field), $s(p) = [(\sigma_i(p), X_i(p))] = X_i^\alpha(p)e_\alpha$, we find

$$\nabla_{\partial/\partial x^\mu}s = \left(\frac{\partial}{\partial x^\mu}X_i^\alpha + \Gamma^\alpha{}_{\mu\beta}X^\beta\right)e_\alpha \qquad (10.61)$$

which reproduces the result of section 7.2. It is evident that the roles played by the indices α, β and μ in $\Gamma^\alpha{}_{\mu\beta}$ are very different in their characters; μ is the $\Omega^1(M)$ index while α and β are the $\mathfrak{gl}(m, \mathbb{R})$ indices.

Example 10.5. Let us consider the U(1) gauge field coupled to a complex scalar field ϕ. The relevant fibre bundles are the U(1) bundle $P(M, U(1))$ and the associated bundle $E = P \times_\rho \mathbb{C}$ where ρ is the natural identification of an element of U(1) with a complex number. The local expression for ω is $\mathcal{A}_i = \mathcal{A}_{i\mu} \, dx^\mu$, where $\mathcal{A}_{i\mu} = \mathcal{A}_i(\partial/\partial x^\mu)$ is the vector potential of Maxwell's theory. Let γ be a curve in M with tangent vector X at $\gamma(0)$. Take a local section σ_i and express a horizontal lift $\tilde{\gamma}$ of γ as $\tilde{\gamma}(t) = \sigma_i(t)e^{i\varphi(t)}$. If $1 \in \mathbb{C}$ is taken to be the basis vector, the basis section is

$$e = [(\sigma_i(p), 1)].$$

Let $\phi(p) = [(\sigma_i(p), \Phi(p))] = \Phi(p)e$ ($\Phi : M \to \mathbb{C}$) be a section of E, which is identified with a complex scalar field. With respect to $\tilde{\gamma}(t)$, the section is given by

$$\phi(t) = \Phi(t)[(\tilde{\gamma}(t), U(t)^{-1})] \tag{10.62}$$

where $U(t) = e^{i\varphi(t)}$. The covariant derivative of ϕ along γ is

$$\nabla_X \phi = \frac{d\Phi}{dt}[(\tilde{\gamma}(0), U(0)^{-1})] + \Phi(0)[(\tilde{\gamma}(0), U(0)^{-1}\mathcal{A}_i(X) \cdot 1)]$$
$$= \left(\frac{d\Phi}{dt} + \mathcal{A}_{i\mu}\Phi\frac{dx^\mu}{dt}\right)e = X^\mu\left(\frac{\partial\Phi}{\partial x^\mu} + \mathcal{A}_{i\mu}\Phi\right)e. \tag{10.63}$$

Example 10.6. Let us consider the SU(2) Yang–Mills theory on M. The relevant bundles are the SU(2) bundle $P(M, SU(2))$ and its associated bundle $E = P \times_\rho \mathbb{C}^2$, where we have taken the two-dimensional representation. The gauge potential on a chart U_i is

$$\mathcal{A}_i = \mathcal{A}_{i\mu} \, dx^\mu = A_{i\mu}{}^\alpha \left(\frac{\sigma_\alpha}{2i}\right) dx^\mu \tag{10.64}$$

where $\sigma_\alpha/2i$ are generators of SU(2), σ_α being the Pauli matrices. Let $e_\alpha{}^0$ ($\alpha = 1, 2$) be basis vectors of $\mathbb{C}^2$ and consider sections

$$e_\alpha(p) \equiv [(\sigma_i(p), e_\alpha{}^0)] \tag{10.65}$$

where $\sigma_i(p)$ defines a canonical trivialization of P over U_i. Let $\phi(p) = [(\sigma_i(p), \Phi^\alpha(p)e_\alpha{}^0)]$ be a section of E over M. Along a horizontal lift $\tilde{\gamma}(t) = \sigma_i(p)U(t)$, $U(t) \in SU(2)$, we have

$$\phi(t) = [(\tilde{\gamma}(t), U(t)^{-1}\Phi^\alpha(t)e_\alpha{}^0)]. \tag{10.66}$$

The covariant derivative of ϕ along $X = d/dt$ is

$$\nabla_X \phi = \left[\left(\tilde{\gamma}(0), U(0)^{-1}\frac{d\Phi^\alpha(0)}{dt}e_\alpha{}^0\right)\right]$$
$$+ [(\tilde{\gamma}(0), U(0)^{-1}\mathcal{A}_i(X)^\alpha{}_\beta\Phi^\beta(0)e_\alpha{}^0)]$$
$$= X^\mu\left(\frac{\partial\Phi^\alpha}{\partial x^\mu} + \mathcal{A}_{i\mu}{}^\alpha{}_\beta\Phi^\beta\right)e_\alpha \tag{10.67}$$

where (10.13b) has been used to obtain the last equality.

Exercise 10.10. Let us consider an associated adjoint bundle $E_{\mathfrak{g}} = P \times_{\mathrm{Ad}} \mathfrak{g}$ where the action of G on $\mathfrak{g}$ is the adjoint action $V \to \mathrm{Ad}_g V = g^{-1} V g$, $V \in \mathfrak{g}$ and $g \in G$. Take a local section $\sigma_i \in \Gamma(U_i, P)$ such that $\tilde{\gamma}(t) = \sigma_i(t) g(t)$. Take a section $s(p) = [(\sigma_i(p), V(p))]$ on $E_{\mathfrak{g}}$, where $V(p) = V^\alpha(p) T_\alpha$, $\{T_\alpha\}$ being the basis of $\mathfrak{g}$. Define the covariant derivative $\mathcal{D}_X s$ by

$$\mathcal{D}_X s \equiv \left[\left(\tilde{\gamma}(0), \frac{\mathrm{d}}{\mathrm{d}t} \{\mathrm{Ad}_{g(t)^{-1}} V(t)\} \Big|_{t=0} \right) \right]. \tag{10.68a}$$

Show that

$$\mathcal{D}_X s = \left[\left(\sigma_i(0), \frac{\mathrm{d}V(t)}{\mathrm{d}t} + [\mathcal{A}_i(X), V(t)] \Big|_{t=0} \right) \right]$$

$$= X^\mu \left(\frac{\partial V^\alpha}{\partial x^\mu} + f_{\beta\gamma}{}^\alpha \mathcal{A}_{i\mu}{}^\beta V^\gamma \right) [(\sigma_i(0), T_\alpha)]. \tag{10.68b}$$

10.4.3   Curvature rederived

The covariant derivative $\nabla_X s$ defines an operator $\nabla : \Gamma(M, E) \to \Gamma(M, E \otimes \Omega^1(M))$ by (10.50). More generally, the action of ∇ on a vector-valued p-form $s \otimes \eta$, $\eta \in \Omega^p(M)$, is defined by

$$\nabla(s \otimes \eta) \equiv (\nabla s) \wedge \eta + s \otimes \mathrm{d}\eta. \tag{10.69}$$

Let U_i be a chart of M and σ_i a section of P over U_i. We take the canonical local trivialization over U_i. We now prove

$$\nabla \nabla e_\alpha = e_\beta \otimes \mathcal{F}_i{}^\beta{}_\alpha \tag{10.70}$$

where $e_\alpha = [(\sigma_i, e_\alpha{}^0)] \in \Gamma(U_i, E)$. In fact, by straightforward computation, we find

$$\nabla \nabla e_\alpha = \nabla(e_\beta \otimes \mathcal{A}_i{}^\beta{}_\alpha) = \nabla e_\beta \wedge \mathcal{A}_i{}^\beta{}_\alpha + e_\beta \otimes \mathrm{d}\mathcal{A}_i{}^\beta{}_\alpha$$

$$= e_\beta \otimes (\mathrm{d}\mathcal{A}_i{}^\beta{}_\alpha + \mathcal{A}_i{}^\beta{}_\gamma \wedge \mathcal{A}_i{}^\gamma{}_\alpha) = e_\beta \otimes \mathcal{F}_i{}^\beta{}_\alpha.$$

Exercise 10.11. Let $s(p) = \xi^\alpha(p) e_\alpha(p)$ be a section of E. Show that

$$\nabla \nabla s = e_\alpha \otimes \mathcal{F}_i{}^\alpha{}_\beta \xi^\beta. \tag{10.71}$$

10.4.4   A connection which preserves the inner product

Let $E \xrightarrow{\pi} M$ be a vector bundle with a positive-definite symmetric inner product whose action is defined at each point $p \in M$ by

$$g_p : \pi^{-1}(p) \otimes \pi^{-1}(p) \to \mathbb{R}. \tag{10.72}$$

Then g is said to define a **Riemannian structure** on E. A connection ∇ is called a **metric connection** if it preserves the inner product,

$$\mathrm{d}\,[g(s, s')] = g(\nabla s, s') + g(s, \nabla s'). \tag{10.73}$$

In particular, if we take $s = e_\alpha$, $s' = e_\beta$ and set $g(e_\alpha, e_\beta) = g_{\alpha\beta}$, we find

$$\mathrm{d}g_{\alpha\beta} = \mathcal{A}_i{}^\gamma{}_\alpha g_{\gamma\beta} + \mathcal{A}_i{}^\gamma{}_\beta g_{\alpha\gamma}. \tag{10.74}$$

This should be compared with (7.30b). If $E = TM$ and, moreover, the torsion-free condition is imposed, our connection reduces to the Levi-Civita connection of the Riemannian geometry.

Given an inner product, we may take an **orthonormal frame** $\{\hat{e}_\alpha\}$ such that $g(\hat{e}_\alpha, \hat{e}_\beta) = \delta_{\alpha\beta}$. The structure group G is taken to be $O(k)$, k being the dimension of the fibre. The Lie algebra $\mathfrak{o}(k)$ is a vector space of skew symmetric matrices and the connection one-form ω satisfies

$$\omega^\alpha{}_\beta = -\omega^\beta{}_\alpha. \tag{10.75}$$

Theorem 10.6. Let E be a vector bundle with inner product g and let ∇ be the covariant derivative associated with the *orthonormal* frame. Then ∇ is a metric connection.

Proof. Since g is bilinear, it suffices to show that

$$\mathrm{d}[g(s, s')] = g(\nabla s, s') + g(s, \nabla s')$$

for $s = f\hat{e}_\alpha$ and $s' = f'\hat{e}_\beta$ where $f, f' \in \mathcal{F}(M)$. In fact, the LHS is $\mathrm{d}[g(f\hat{e}_\alpha, f'\hat{e}_\beta)] = \mathrm{d}[ff'\delta_{\alpha\beta}] = \mathrm{d}\,(ff')\delta_{\alpha\beta}$ while the RHS is

$$\begin{aligned}
g(\nabla f\hat{e}_\alpha, &f'\hat{e}_\beta) + g(f\hat{e}_\alpha, \nabla f'\hat{e}_\beta) \\
&= g(\mathrm{d}f\,\hat{e}_\alpha + f\hat{e}_\gamma\omega^\gamma{}_\alpha, f'\hat{e}_\beta) + g(f\hat{e}_\alpha, \mathrm{d}f'\,\hat{e}_\beta + f'\hat{e}_\gamma\omega^\gamma{}_\beta) \\
&= \mathrm{d}ff'\,\delta_{\alpha\beta} + ff'\omega^\gamma{}_\alpha\delta_{\gamma\beta} + f\mathrm{d}f'\,\delta_{\alpha\beta} + ff'\omega^\gamma{}_\beta\delta_{\alpha\gamma} \\
&= \mathrm{d}(ff')\,\delta_{\alpha\beta}
\end{aligned}$$

where (10.75) has been used to obtain the final equality. $\qquad\qquad\square$

10.4.5 Holomorphic vector bundles and Hermitian inner products

Definition 10.6. Let E and M be complex manifolds and $\pi : E \to M$ a holomorphic surjection. The manifold E is a **holomorphic vector bundle** if the following axioms are fulfilled.

(i) The typical fibre is $\mathbb{C}^k$ and the structure group is $GL(k, \mathbb{C})$.
(ii) The local trivialization $\phi_i : U_i \times \mathbb{C}^k \to \pi^{-1}(U_i)$ is a biholomorphism.

(iii) The transition function $t_{ij} : U_i \cap U_j \to G = \mathrm{GL}(k, \mathbb{C})$ is a holomorphic map.

For example, let M be a complex manifold with $\dim_{\mathbb{C}} M = m$. The **holomorphic tangent bundle** $TM^+ \equiv \bigcup_{p \in M} T_p M^+$ is a holomorphic vector bundle. The typical fibre is $\mathbb{C}^m$ and the local basis is $\{\partial/\partial z^{\mu}\}$.

Let h be an inner product on a holomorphic vector bundle whose action at $p \in M$ is $h_p : \pi^{-1}(p) \times \pi^{-1}(p) \to \mathbb{C}$. The most natural inner product is a **Hermitian structure** which satisfies:

(i) $h_p(u, av + bw) = a h_p(u, v) + b h_p(u, w)$, for $u, v, w \in \pi^{-1}(p), a, b \in \mathbb{C}$;
(ii) $h_p(u, v) = \overline{h_p(v, u)}, u, v \in \pi^{-1}(p)$;
(iii) $h_p(u, u) \geq 0$, $h_p(u, u) = 0$ if and only if $u = \phi_i(p, 0)$; and
(iv) $h(s_1, s_2) \in \mathcal{F}(M)^{\mathbb{C}}$ for $s_1, s_2 \in \Gamma(M, E)$.

A set of sections $\{\hat{e}_1, \ldots, \hat{e}_k\}$ is a **unitary frame** if

$$h(\hat{e}_i, \hat{e}_j) = \delta_{ij}. \tag{10.76}$$

The unitary frame bundle LM is not a holomorphic vector bundle since the structure group $\mathrm{U}(m)$ is not a complex manifold.

Given a Hermitian structure h, we define a connection which is compatible with h. The **Hermitian connection** ∇ is a linear map $\Gamma(M, E) \to \Gamma(M, E \otimes T^*M^{\mathbb{C}})$ which satisfies:

(i) $\nabla(fs) = (\mathrm{d}f)s + f\nabla s, f \in \mathcal{F}(M)^{\mathbb{C}}, s \in \Gamma(M, E)$;
(ii) $\mathrm{d}[h(s_1, s_2)] = h(\nabla s_1, s_2) + h(s_1, \nabla s_2)$; and
(iii) according to the destination, we separate the action of ∇ as $\nabla s = Ds + \bar{D}s$, Ds ($\bar{D}s$) being a $(1, 0)$-form ($(0, 1)$-form) valued section. We demand that $\bar{D} = \bar{\partial}$.

It can be shown that given E and a Hermitian metric h, there exists a *unique* Hermitian connection ∇. The curvature is defined from the Hermitian connection. Let $\{\hat{e}_1, \ldots, \hat{e}_k\}$ be a unitary frame and define the local connection form $A^{\beta}{}_{\alpha}$ by

$$\nabla \hat{e}_{\alpha} = \hat{e}_{\beta} A^{\beta}{}_{\alpha}. \tag{10.77}$$

The field strength is defined by

$$\mathcal{F} \equiv \mathrm{d}A + A \wedge A. \tag{10.78}$$

We verify that

$$\nabla \nabla \hat{e}_{\alpha} = \nabla(\hat{e}_{\beta} A^{\beta}{}_{\alpha}) = \hat{e}_{\beta} \mathcal{F}^{\beta}{}_{\alpha}. \tag{10.79}$$

We prove that both A and $\mathcal{F}$ are skew Hermitian:

$$\bar{A}^{\beta}{}_{\alpha} + A^{\alpha}{}_{\beta} = h(\nabla \hat{e}_{\alpha}, \hat{e}_{\beta}) + h(\hat{e}_{\alpha}, \nabla \hat{e}_{\beta}) = \mathrm{d}h(\hat{e}_{\alpha}, \hat{e}_{\beta}) = \mathrm{d}\delta_{\alpha\beta} = 0$$

$$\mathcal{F}^{\beta}{}_{\alpha} + \bar{\mathcal{F}}^{\alpha}{}_{\beta} = \mathrm{d}A^{\beta}{}_{\alpha} + A^{\beta}{}_{\gamma} \wedge A^{\gamma}{}_{\alpha} + \mathrm{d}\bar{A}^{\alpha}{}_{\beta} + \bar{A}^{\alpha}{}_{\gamma} \wedge \bar{A}^{\gamma}{}_{\alpha}$$

$$= \mathrm{d}(A^{\beta}{}_{\alpha} - A^{\beta}{}_{\alpha}) + A^{\beta}{}_{\gamma} \wedge A^{\gamma}{}_{\alpha} + A^{\gamma}{}_{\alpha} \wedge A^{\alpha}{}_{\gamma} = 0.$$

Thus, we have shown that

$$\mathcal{A}^\alpha{}_\beta = -\bar{\mathcal{A}}^\beta{}_\alpha \qquad \mathcal{F}^\beta{}_\alpha = -\bar{\mathcal{F}}^\alpha{}_\beta. \tag{10.80}$$

Next we show that $\mathcal{F}$ is a $(1, 1)$-form. Let $\{\hat{e}_\alpha\}$ be a unitary frame. $\mathcal{F}$ cannot have a component of bidegree-$(0, 2)$ since

$$\hat{e}_\beta \mathcal{F}^\beta{}_\alpha = \nabla\nabla\hat{e}_\alpha = (D + \bar{\partial})(D + \bar{\partial})\hat{e}_\alpha = DD\hat{e}_\alpha + (D\bar{\partial} + \bar{\partial}D)\hat{e}_\alpha.$$

It follows from $\mathcal{F}^\beta{}_\alpha = -\bar{\mathcal{F}}^\alpha{}_\beta$ that $\bar{\mathcal{F}}$ has no component of bidegree-$(0, 2)$, and, hence, $\mathcal{F}$ has no component of bidegree-$(2, 0)$ either. Thus $\mathcal{F}^\beta{}_\alpha$ is a two-form of bidegree-$(1, 1)$.

10.5 Gauge theories

As we have remarked several times, a gauge potential can be regarded as a local expression for a connection in a principal bundle. The Yang–Mills field strength is then identified with the local form of the curvature associated with the connection. We summarize here the relevant aspects of gauge theories from the geometrical viewpoint.

10.5.1 U(1) gauge theory

Maxwell's theory of electromagnetism is described by the U(1) gauge group. U(1) is Abelian and one dimensional, hence we omit all the group indices $\alpha, \beta, \ldots$ and put the structure constants $f_{\alpha\beta}{}^\gamma = 0$. Suppose the base space M is a four-dimensional Minkowski spacetime. From corollary 9.1, we find that the U(1) bundle P is trivial, namely $P = \mathbb{R}^4 \times U(1)$ and a single local trivialization over M is required. The gauge potential is simply

$$\mathcal{A} = \mathcal{A}_\mu \, dx^\mu. \tag{10.81}$$

Our gauge potential $\mathcal{A}$ differs from the usual vector potential A by the Lie algebra factor i: $\mathcal{A}_\mu = iA_\mu$. The field strength is

$$\mathcal{F} = d\mathcal{A}. \tag{10.82a}$$

In components, we have

$$\mathcal{F}_{\mu\nu} = \partial A_\nu / \partial x^\mu - \partial A_\mu / \partial x^\nu. \tag{10.82b}$$

$\mathcal{F}$ satisfies the Bianchi identity,

$$d\mathcal{F} = \mathcal{F} \wedge \mathcal{A} - \mathcal{A} \wedge \mathcal{F} = 0. \tag{10.83a}$$

This should be expected from the outset since $\mathcal{F}$ is exact, $\mathcal{F} = d\mathcal{A}$; and hence closed, $d\mathcal{F} = d^2\mathcal{A} = 0$. In components, we have

$$\partial_\lambda \mathcal{F}_{\mu\nu} + \partial_\nu \mathcal{F}_{\lambda\mu} + \partial_\mu \mathcal{F}_{\nu\lambda} = 0. \tag{10.83b}$$

If we identify the components $\mathcal{F}_{\mu\nu} \equiv iF_{\mu\nu}$ with the electric field E and the magnetic field B as

$$E_i = F_{i0}, B_i = \tfrac{1}{2}\epsilon_{ijk}F_{jk} \qquad (i, j, k = 1, 2, 3) \qquad (10.84)$$

(10.83b) reduces to two of Maxwell's equations,

$$\nabla \times E + \frac{\partial B}{\partial t} = 0 \qquad \nabla \cdot B = 0. \qquad (10.83c)$$

These equations are *geometrical* rather than *dynamical*. To find the dynamics, we have to specify the action. The **Maxwell action** $\mathcal{S}_M[\mathcal{A}]$ is a functional of $\mathcal{A}$ and is given by

$$\mathcal{S}_M[\mathcal{A}] \equiv \tfrac{1}{4}\int_{\mathbb{R}^4} \mathcal{F}_{\mu\nu}\mathcal{F}^{\mu\nu}\,\mathrm{d}^4x = -\tfrac{1}{4}\int_{\mathbb{R}^4} F_{\mu\nu}F^{\mu\nu}\,\mathrm{d}^4x. \qquad (10.85a)$$

Exercise 10.12.   (a) Let $*\mathcal{F}_{\mu\nu} \equiv \tfrac{1}{2}\mathcal{F}^{\kappa\lambda}\varepsilon_{\kappa\lambda\mu\nu}$ be the dual of $\mathcal{F}_{\mu\nu}$. Show that

$$\mathcal{S}_M[\mathcal{A}] = -\tfrac{1}{4}\int_{\mathbb{R}^4} \mathcal{F} \wedge *\mathcal{F}. \qquad (10.85b)$$

(b) Use (10.84) to show that

$$-\tfrac{1}{4}F_{\mu\nu}F^{\mu\nu} = \tfrac{1}{2}(E^2 - B^2). \qquad (10.86)$$

Show also that

$$F_{\mu\nu} * F^{\mu\nu} = B \cdot E. \qquad (10.87)$$

By the variation of $\mathcal{S}_M[\mathcal{A}]$ with respect to $\mathcal{A}_\mu$, we obtain the equation of motion,

$$\partial_\mu \mathcal{F}^{\mu\nu} = 0. \qquad (10.88a)$$

We find this equation is reduced to the second set of Maxwell's equations (in the vacuum):

$$\nabla \cdot E = 0 \qquad \nabla \times B - \frac{\partial E}{\partial t} = 0. \qquad (10.88b)$$

10.5.2   The Dirac magnetic monopole

We have studied Maxwell's theory of electromagnetism defined on $\mathbb{R}^4$. The triviality of the base space makes the U(1) bundle trivial. Poincaré's lemma ensures that the field strength $\mathcal{F}$ is globally exact: $\mathcal{F} = \mathrm{d}\mathcal{A}$. It is interesting to extend our analysis to U(1) bundles over a non-trivial base space. We assume everything is independent of time for simplicity.

The Dirac monopole is defined in $\mathbb{R}^3$ with the origin O removed. $\mathbb{R}^3 - \{0\}$ and S^2 are of the same homotopy type and the relevant bundle is a U(1) bundle $P(S^2, \mathrm{U}(1))$. S^2 is covered by two charts

$$U_N \equiv \{(\theta, \phi) | 0 \le \theta \le \tfrac{1}{2}\pi + \epsilon\} \qquad U_S \equiv \{(\theta, \phi) | \tfrac{1}{2}\pi - \epsilon \le \theta \le \pi\}$$

where θ and ϕ are polar coordinates. Let ω be an Ehresmann connection on P. Take a local section σ_N (σ_S) on U_N (U_S) and define the local gauge potentials

$$\mathcal{A}_N = \sigma_N^* \omega \qquad \mathcal{A}_S = \sigma_S^* \omega.$$

We take $\mathcal{A}_N$ and $\mathcal{A}_S$ to be of the Wu–Yang form (section 1.9),

$$\mathcal{A}_N = ig(1 - \cos\theta)\,d\phi \qquad \mathcal{A}_S = -ig(1 + \cos\theta)\,d\phi \qquad (10.89)$$

where g is the strength of the monopole.

Let t_{NS} be the transition function defined on the equator $U_N \cap U_S$. t_{NS} defines a map from S^1 (equator) to U(1) (structure group), which is classified by $\pi_1(\mathrm{U}(1)) = \mathbb{Z}$, see example 9.7. Let us write

$$t_{NS}(\phi) = \exp[i\varphi(\phi)] \qquad (\varphi : S^1 \to \mathbb{R}). \qquad (10.90)$$

The gauge potentials $\mathcal{A}_N$ and $\mathcal{A}_S$ are related on $U_N \cap U_S$ by

$$\mathcal{A}_N = t_{NS}^{-1} \mathcal{A}_S t_{NS} + t_{NS}^{-1}\,dt_{NS} = \mathcal{A}_S + i\,d\varphi. \qquad (10.91)$$

For the gauge potentials (10.89), we find

$$d\varphi = -i(\mathcal{A}_N - \mathcal{A}_S) = 2g\,d\phi.$$

While ϕ runs from 0 to 2π around the equator, $\varphi(\phi)$ takes the range

$$\Delta\varphi \equiv \int d\varphi = \int_0^{2\pi} 2g\,d\phi = 4\pi g. \qquad (10.92)$$

For t_{NS} to be defined uniquely, $\Delta\varphi$ must be a multiple of 2π,

$$\Delta\varphi/2\pi = 2g \in \mathbb{Z} \qquad (10.93)$$

which is the quantization condition of the magnetic monopole. The integer $2g$ represents the homotopy class to which this bundle belongs. This number is also obtained by considering $F_N = d\mathcal{A}_N$ and $F_S = d\mathcal{A}_S$ ($\mathcal{F}_N = iF_N$ etc). The total flux Φ is

$$\Phi = \int_{S^2} \mathbf{B} \cdot d\mathbf{S} = \int_{U_N} d\mathcal{A}_N + \int_{U_S} d\mathcal{A}_S$$

$$= \int_{S^1} \mathcal{A}_N - \int_{S^1} \mathcal{A}_S = 2g \int_0^{2\pi} d\phi = 4\pi g. \qquad (10.94)$$

Thus, the curvature, that is the pair of the field strengths $d\mathcal{A}_N$ and $d\mathcal{A}_S$, characterizes the twisting of the bundle. We discuss this further in chapter 11.

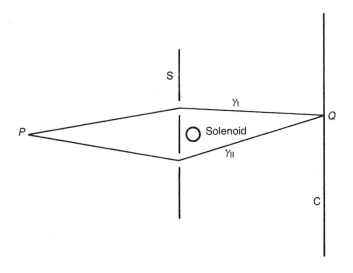

Figure 10.4. The Aharonov–Bohm experiment. $B = 0$ outside the solenoid.

10.5.3   The Aharonov–Bohm effect

In the elementary study of electromagnetism, the electric and magnetic fields (that is $F_{\mu\nu}$) are of central interest. The vector potential A and the scalar potential $\phi = A_0$ are considered to be of secondary importance. In quantum mechanics, however, there are a variety of situations in which $F_{\mu\nu}$ are not sufficient to describe the phenomena and the use of $A_\mu = (A, A_0)$ is essential. One of these examples is the **Aharonov–Bohm effect**.

The Aharonov–Bohm (AB) experiment is schematically described in figure 10.4. A beam of electrons with charge e is incoming from the far left and forms an interference pattern on the screen C. A solenoid of infinite length is placed in the middle of the beam. A shield S prevents electrons from penetrating into the solenoid. Accordingly, the electrons do not feel the magnetic field at all. What about the gauge field A_μ?

For simplicity, we make the radius of the solenoid infinitesimally small, keeping the total flux $\Phi = \int_S B \cdot dS$ fixed. It is easy to verify that

$$A(r) = \left(-\frac{y\Phi}{2\pi r^2}, \frac{x\Phi}{2\pi r^2}, 0 \right) \qquad A_0 = 0 \qquad (10.95)$$

satisfies $\int (\nabla \times A) \cdot dS = \Phi$ and $\nabla \times A = 0$ for $r \neq 0$. The vector potential does not vanish outside the solenoid. Classically, the solenoid cannot have any influence on electrons since the Lorentz force $e(v \times B)$ vanishes on the path of the beam.

In quantum mechanics, the Hamiltonian H of this system is

$$\mathcal{H} = -\frac{1}{2m}\left(\frac{\partial}{\partial x^\mu} - ieA_\mu\right)^2 + V(r) \tag{10.96}$$

where $V(r)$ represents the effect of the experimental apparatus. Semiclassically, we can distinguish between the paths γ_{I} and γ_{II} in figure 10.4. We write the wavefunction corresponding to γ_{I} (γ_{II}) as ψ_{I} (ψ_{II}) when $A = 0$. If $A \neq 0$, the wavefunction is given by the gauge-transformed form,

$$\psi_i^A(r) \equiv \exp\left(ie\int_P^r A(r')\cdot dr'\right)\psi_i(r) \qquad (i = \mathrm{I}, \mathrm{II}) \tag{10.97}$$

where P is a reference point far from the apparatus. Let us consider a superposition $\psi_{\mathrm{I}}^A + \psi_{\mathrm{II}}^A$ of wavefunctions ψ_{I}^A and ψ_{II}^A such that $\psi_{\mathrm{I}}^A(P) = \psi_{\mathrm{II}}^A(P)$. Its amplitude at a point Q on the screen is

$$
\begin{aligned}
\psi_{\mathrm{I}}^A(Q) + \psi_{\mathrm{II}}^A(Q) &= \exp\left(ie\int_{\gamma_{\mathrm{I}}} A(r')\cdot dr'\right)\psi_{\mathrm{I}}(Q) \\
&\quad + \exp\left(ie\int_{\gamma_{\mathrm{II}}} A(r')\cdot dr'\right)\psi_{\mathrm{II}}(Q) \\
&= \exp\left(ie\int_{\gamma_{\mathrm{II}}} A\cdot dr'\right)\left[\exp\left(ie\oint_\gamma A\cdot dr'\right)\psi_{\mathrm{I}}(Q) + \psi_{\mathrm{II}}(Q)\right]
\end{aligned} \tag{10.98}
$$

where $\gamma \equiv \gamma_{\mathrm{I}} - \gamma_{\mathrm{II}}$. It is evident that even though $B = 0$ at the points in space through which the electrons travel, the wavefunction depends on the vector potential A. From Stokes' theorem, we find that

$$\oint_\gamma A\cdot dr' = \int_S (\nabla\times A)\cdot dS = \int_S B\cdot dS = \Phi \tag{10.99}$$

where S is a surface bounded by γ. From this and (10.98), we find the interference pattern should be the same for two values of the fluxes Φ_a and Φ_b if

$$e(\Phi_a - \Phi_b) = 2\pi n \qquad n \in \mathbb{Z}. \tag{10.100}$$

What is the geometry underlying the Aharonov–Bohm effect? Since the problem is essentially two dimensional, we consider a region $M = \mathbb{R}^2 - \{0\}$, where the solenoid is assumed to be at the origin. The relevant bundles are the principal bundle $P(M, \mathrm{U}(1))$ and its associated bundle $E = P \times_\rho \mathbb{C}$, where $\mathrm{U}(1)$ acts on $\mathbb{C}$ in an obvious way. The bundle E is a complex line bundle over M, whose section is a wavefunction ψ.

Let us define a Lie-algebra-valued one-form $\mathcal{A} = iA = iA_\mu \, dx^\mu$. The covariant derivative associated with this local connection is $\mathcal{D} = d + \mathcal{A}$, where

$\mathcal{A}$ is given by (10.95). Since $d\mathcal{A} = \mathcal{F} = 0$, this connection is locally flat. Let us consider the unit circle S^1 which encloses the solenoid at the origin. We parametrize S^1 as $e^{i\theta}$ ($0 \le \theta \le 2\pi$) and write the connection on S^1 as

$$\mathcal{A} = i\frac{\Phi}{2\pi}d\theta. \tag{10.101}$$

This is obtained from (10.95) by putting $r = 1$. We require that the wavefunction ψ be parallel transported along S^1 with respect to this local connection, namely

$$\mathcal{D}\psi(\theta) = \left(d + i\frac{\Phi}{2\pi}d\theta\right)\psi(\theta) = 0. \tag{10.102}$$

The solution of (10.102) is easily found to be

$$\psi(\theta) = e^{-i\Phi\theta/2\pi}. \tag{10.103}$$

Taking this section ψ amounts to neglecting the velocity of the electrons. The holonomy $\Gamma : \pi^{-1}(\theta = 0) \to \pi^{-1}(\theta = 2\pi) = \pi^{-1}(\theta = 0)$ is found to be

$$\Gamma : \psi(0) \longmapsto e^{-i\Phi}\psi(0). \tag{10.104}$$

In an experiment, a toroidal permalloy (20% Fe and 80% Ni) has been used to eliminate the edge effects (Tonomura *et al* 1983). The dimensions of the permalloy are several microns and it is coated with gold to prevent electrons from penetrating into the magnetic field.

10.5.4 Yang–Mills theory

Let us consider SU(2) gauge theory defined on $\mathbb{R}^4$. The bundle which describes this gauge theory is $P(\mathbb{R}^4, SU(2))$. Since $\mathbb{R}^4$ is contractible, there is just a single gauge potential

$$\mathcal{A} = A_\mu{}^\alpha T_\alpha \, dx^\mu \tag{10.105}$$

where $T_\alpha \equiv \sigma_\alpha/2i$ generate the algebra $\mathfrak{su}(2)$,

$$[T_\alpha, T_\beta] = \epsilon_{\alpha\beta\gamma} T_\gamma.$$

The field strength is

$$\mathcal{F} \equiv d\mathcal{A} + \mathcal{A} \wedge \mathcal{A} = \tfrac{1}{2}\mathcal{F}_{\mu\nu} \, dx^\mu \wedge dx^\nu \tag{10.106a}$$

where

$$\mathcal{F}_{\mu\nu} = \partial_\mu A_\nu - \partial_\nu A_\mu + [A_\mu, A_\nu] = F_{\mu\nu}{}^\alpha T_\alpha \tag{10.106b}$$

$$F_{\mu\nu}{}^\alpha = \partial_\mu A_{\nu\alpha} - \partial_\nu A_{\mu\alpha} + \epsilon_{\alpha\beta\gamma} A_{\mu\beta} A_{\nu\gamma}. \tag{10.106c}$$

The Bianchi identity is

$$\mathcal{D}\mathcal{F} = d\mathcal{F} + [\mathcal{A}, \mathcal{F}] = 0. \tag{10.107}$$

The Yang–Mills action is

$$S_{YM}[\mathcal{A}] \equiv -\frac{1}{4} \int_M \text{tr}(\mathcal{F}_{\mu\nu}\mathcal{F}^{\mu\nu}) = \frac{1}{2} \int_M \text{tr}(\mathcal{F} \wedge *\mathcal{F}). \qquad (10.108)$$

The variation with respect to $\mathcal{A}_\mu$ yields

$$\mathcal{D}_\mu \mathcal{F}^{\mu\nu} = 0 \qquad \text{or} \qquad \mathcal{D} * \mathcal{F} = 0. \qquad (10.109)$$

10.5.5 Instantons

A path integral is well defined only on a space with a Euclidean metric. To evaluate this integral, it is important to find the local minima of the *Euclidean* action and compute the quantum fluctuations around them. Let us consider the SU(2) gauge theory on a four-dimensional Euclidean space $\mathbb{R}^4$. The local minima of this theory are known as **instantons** (or *pseudoparticles*, Belavin *et al* (1975)), see section 1.10. It is easy to verify that the Euclidean action is

$$S_{YM}^{E}[\mathcal{A}] = \frac{1}{4} \int_M \text{tr}(\mathcal{F}_{\mu\nu}\mathcal{F}^{\mu\nu}) = -\frac{1}{2} \int_M \text{tr}(\mathcal{F} \wedge *\mathcal{F}) \qquad (10.110)$$

where the Hodge $*$ is taken with respect to the Euclidean metric. As has been shown in section 1.10 the field strength corresponding to instantons is self-dual (anti-self-dual),

$$\mathcal{F}_{\mu\nu} = \pm * \mathcal{F}_{\mu\nu}. \qquad (10.111)$$

The action of a self-dual (anti-self-dual) field configuration is

$$S_{YM}^{E}[\mathcal{A}] = -\frac{1}{2} \int_M \text{tr}(\mathcal{F} \wedge *\mathcal{F}) = \mp\frac{1}{2} \int_M \text{tr}(\mathcal{F} \wedge \mathcal{F}). \qquad (10.112)$$

Let us consider the topological properties of an instanton. We require that

$$\mathcal{A}_\mu(x) \to g(x)^{-1}\partial_\mu g(x) \qquad \text{as } |x| \to L \qquad (10.113)$$

for the action to be finite, where L is an arbitrary positive number. Since $|x| = L$ is the sphere S^3, (10.113) defines a map $g : S^3 \to \text{SU}(2)$ which is classified by $\pi_3(\text{SU}(2)) \cong \mathbb{Z}$. How is this reflected upon the transition function? We compactify $\mathbb{R}^4$ by adding the infinity. We suppose the South Pole of S^4 represents the points at infinity and the North Pole the origin. Under this compactification, we separate $\mathbb{R}^4$ into two pieces and identify them with the southern hemisphere U_S and the northern hemisphere U_N of S^4 as

$$U_N = \{x \in \mathbb{R}^4 \,|\, |x| \leq L + \varepsilon\} \qquad (10.114a)$$

$$U_S = \{x \in \mathbb{R}^4 \,|\, |x| \geq L - \varepsilon\} \qquad (10.114b)$$

see figure 10.5. We assume there is no 'twist' of the gauge potential on U_S and choose

$$\mathcal{A}_S(x) \equiv 0 \qquad x \in U_S. \qquad (10.115)$$

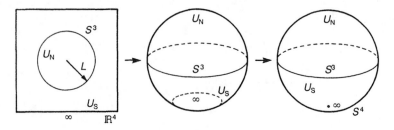

Figure 10.5. One-point compactification of $\mathbb{R}^4$ to S^4.

Then all the topological information about the bundle is contained in $\mathcal{A}_N(x)$ or the transition function $t_{NS}(x)$ on the 'equator' S^3 ($=U_N \cap U_S$). Since $\mathcal{A}_S = 0$, we have, for $x \in U_N \cap U_S$,

$$\mathcal{A}_N = t_{NS}^{-1}\mathcal{A}_S t_{NS} + t_{NS}^{-1}\,\mathrm{d}t_{NS} = t_{NS}^{-1}\,\mathrm{d}t_{NS}. \tag{10.116}$$

Thus, $g(x)$ in (10.113) is identified with the transition function $t_{NS}(x)$ and classifying the maps $g : S^3 \to \mathrm{SU}(2)$ amounts to classifying the transition functions according to $\pi_3(\mathrm{SU}(2)) = \mathbb{Z}$; see example 9.11.

We now compute the degree of a map $g : S^3 \to \mathrm{SU}(2)$ following Coleman (1979). First note that $\mathrm{SU}(2) \simeq S^3$ since

$$t^4 I_2 + t^i \sigma_i \in \mathrm{SU}(2) \leftrightarrow t^2 + (t^4)^2 = 1.$$

Thus, maps $g : S^3 \to \mathrm{SU}(2)$ are classified according to $\pi_3(\mathrm{SU}(2)) \cong \pi_3(S^3) \cong \mathbb{Z}$. We easily find the following.

(a) The constant map

$$g_0 : x \in S^3 \mapsto e \in \mathrm{SU}(2) \tag{10.117a}$$

  belongs to the class 0 (i.e. no winding) of $\pi_3(\mathrm{SU}(2))$.

(b) The *identity* map (this is, in fact, the identity map $S^3 \to S^3$)

$$g_1 : x \mapsto \frac{1}{r}[x^4 I_2 + x^i \sigma_i], r^2 = x^2 + (x^4)^2 \tag{10.117b}$$

  defines the class 1 of $\pi_3(\mathrm{SU}(2))$. The explicit form of the gauge potential corresponding to this homotopy class is given in section 1.10.

(c) The map

$$g_n \equiv (g_1)^n : x \mapsto r^{-n}[x^4 I_2 + x^i \sigma_i]^n \tag{10.117c}$$

  defines the class n of $\pi_3(\mathrm{SU}(2))$.

We recall that the strength (charge) of a magnetic monopole is given by the integral of the field strength $\mathcal{F} = \mathrm{d}\mathcal{A}$ over the sphere S^2. We expect that a similar relation exists for the instanton number. Since instantons are defined over S^4, we have to find a four-form to be integrated over S^4. A natural four-form is $\mathcal{F} \wedge \mathcal{F}$. In the following, we shall omit the exterior product symbol when this does not cause confusion ($\mathcal{F}^2$ stands for $\mathcal{F} \wedge \mathcal{F}$). Observe that $\operatorname{tr} \mathcal{F}^2$ is closed,

$$\begin{aligned}
\mathrm{d} \operatorname{tr} \mathcal{F}^2 &= \operatorname{tr}[\mathrm{d}\mathcal{F}\mathcal{F} + \mathcal{F}\,\mathrm{d}\mathcal{F}] \\
&= \operatorname{tr}\{-[\mathcal{A}, \mathcal{F}]\mathcal{F} - \mathcal{F}[\mathcal{A}, \mathcal{F}]\} = 0
\end{aligned} \tag{10.118}$$

where use has been made of the Bianchi identity $\mathrm{d}\mathcal{F} + [\mathcal{A}, \mathcal{F}] = 0$. [*Remarks:* In the present case, (10.118) seems to be trivial since any four-form on S^4 is closed. Note, however, that (10.118) remains true even on higher-dimensional manifolds.] By Poincaré's lemma, the closed form $\operatorname{tr} \mathcal{F}^2$ is *locally* exact,

$$\operatorname{tr} \mathcal{F}^2 = \mathrm{d}K \tag{10.119}$$

where K is a local three-form. Thus, $\operatorname{tr} \mathcal{F}^2$ is an element of the de Rham cohomology group $H^4(S^4)$. Later $\operatorname{tr} \mathcal{F}^2$ is identified with the second Chern character and K its Chern–Simons form, see chapter 11.

Lemma 10.3. The three-form K in (10.119) is given by

$$K = \operatorname{tr}[\mathcal{A}\,\mathrm{d}\mathcal{A} + \tfrac{2}{3}\mathcal{A}^3]. \tag{10.120}$$

Proof. A straightforward computation yields

$$\begin{aligned}
\mathrm{d}K &= \operatorname{tr}[(\mathrm{d}\mathcal{A})^2 + \tfrac{2}{3}(\mathrm{d}\mathcal{A}\,\mathcal{A}^2 - \mathcal{A}\,\mathrm{d}\mathcal{A}\mathcal{A} + \mathcal{A}^2\,\mathrm{d}\mathcal{A})] \\
&= \operatorname{tr}[(\mathcal{F} - \mathcal{A}^2)(\mathcal{F} - \mathcal{A}^2) \\
&\quad + \tfrac{2}{3}\{(\mathcal{F} - \mathcal{A}^2)\mathcal{A}^2 - \mathcal{A}(\mathcal{F} - \mathcal{A}^2)\mathcal{A} + \mathcal{A}^2(\mathcal{F} - \mathcal{A}^2)\}] \\
&= \operatorname{tr}[\mathcal{F}^2 - \mathcal{A}^2\mathcal{F} - \mathcal{F}\mathcal{A}^2 + \mathcal{A}^4 + \tfrac{2}{3}(\mathcal{F}\mathcal{A}^2 - \mathcal{A}\mathcal{F}\mathcal{A} + \mathcal{A}^2\mathcal{F} - \mathcal{A}^4)]
\end{aligned}$$

where use has been made of the identity $\mathrm{d}\mathcal{A} = \mathcal{F} - \mathcal{A}^2$. Now we note that

$$\operatorname{tr} \mathcal{A}^4 = 0 \qquad \operatorname{tr} \mathcal{A}\mathcal{F}\mathcal{A} = -\operatorname{tr} \mathcal{A}^2\mathcal{F} = -\operatorname{tr} \mathcal{F}\mathcal{A}^2.$$

For example, we have

$$\begin{aligned}
\operatorname{tr} \mathcal{A}\mathcal{F}\mathcal{A} &= \tfrac{1}{2}\operatorname{tr} \mathcal{A}_\kappa \mathcal{F}_{\lambda\mu}\mathcal{A}_\nu\,\mathrm{d}x^\kappa \wedge \mathrm{d}x^\lambda \wedge \mathrm{d}x^\mu \wedge \mathrm{d}x^\nu \\
&= -\tfrac{1}{2}\operatorname{tr} \mathcal{A}_\nu \mathcal{A}_\kappa \mathcal{F}_{\lambda\mu}\,\mathrm{d}x^\nu \wedge \mathrm{d}x^\kappa \wedge \mathrm{d}x^\lambda \wedge \mathrm{d}x^\mu = -\operatorname{tr} \mathcal{A}^2\mathcal{F}
\end{aligned}$$

where the cyclicity of the trace and the anti-commutativity of $\mathrm{d}x^\mu$ have been used. Then $\mathrm{d}K$ becomes

$$\begin{aligned}
\mathrm{d}K &= \operatorname{tr}[\mathcal{F}^2 - \mathcal{A}^2\mathcal{F} - \mathcal{F}\mathcal{A}^2 + \tfrac{2}{3}\{\mathcal{F}\mathcal{A}^2 + \tfrac{1}{2}(\mathcal{F}\mathcal{A}^2 + \mathcal{A}^2\mathcal{F}) + \mathcal{A}^2\mathcal{F}\}] \\
&= \operatorname{tr} \mathcal{F}^2
\end{aligned}$$

as has been claimed. $\qquad\qquad\qquad\qquad\qquad\qquad\qquad\qquad\qquad\qquad\qquad\qquad\square$

Lemma 10.4. Let $\mathcal{A}$ be the gauge potential of an instanton. Then it follows that

$$\int_{S^4} \operatorname{tr} \mathcal{F}^2 = -\tfrac{1}{3}\int_{S^3} \operatorname{tr} \mathcal{A}^3. \tag{10.121}$$

Proof. From Stokes' theorem, we find that

$$\int_{U_N} \mathrm{tr}\, \mathcal{F}^2 = \int_{U_N} \mathrm{d}K = \int_{S^3} K$$

where U_N is defined by (10.114) and $S^3 = \partial U_N$. Since $\mathcal{F} = 0$ on S^3, we obtain

$$K = \mathrm{tr}[A\mathrm{d}A + \tfrac{2}{3}A^3] = \mathrm{tr}[A(\mathcal{F} - A^2) + \tfrac{2}{3}A^3] = -\tfrac{1}{3}\,\mathrm{tr}\, A^3$$

on S^3, from which we find that

$$\int_{U_N} \mathrm{tr}\, \mathcal{F}^2 = \int_{S^4} \mathrm{tr}\, \mathcal{F}^2 = -\tfrac{1}{3}\int_{S^3} \mathrm{tr}\, A^3$$

where we have added $\int_{U_S} \mathrm{tr}\, \mathcal{F}^2 = 0$ since $A_S \equiv 0$.   □

Note that $\mathrm{tr}\, \mathcal{F}^2$ is invariant under the gauge transformation,

$$\mathrm{tr}\, \mathcal{F}^2 \to \mathrm{tr}[g^{-1}\mathcal{F}^2 g] = \mathrm{tr}\, \mathcal{F}^2.$$

Thus, it is reasonable to assume that $\mathrm{tr}\, \mathcal{F}^2$ indeed contains a certain amount of topological information about the bundle, which is independent of particular connections. Let us consider the gauge fields (10.117a−c) given before. We find:

(a) For $g_0(x) \equiv e$, we have $A = 0$ on S^3. Since the bundle is trivial we may take $A = 0$ throughout S^4. Then $\mathcal{F} = 0$, hence

$$\int_{S^4} \mathrm{tr}\, \mathcal{F}^2 = -\tfrac{1}{3}\int_{S^3} \mathrm{tr}\, A^3 = 0. \tag{10.122}$$

Note that this relation is true for any gauge potential which is obtained from $A = 0$ by smooth gauge transformations, that is for any gauge potential of the form $A(x) = g(x)^{-1}\,\mathrm{d}g\,(x)$, $x \in S^4$.

(b) Next consider a gauge potential whose value on S^3 is given by (10.117b) as

$$A = \frac{1}{r}(x^4 - ix^k\sigma_k)\,\mathrm{d}\left(\frac{1}{r}(x^4 + ix^l\sigma_l)\right). \tag{10.123}$$

A considerable simplification is achieved if we note that the integrand $\mathrm{tr}\, A^3$ should not depend on the point on S^3 at which it is evaluated since g_1 maps S^3 onto SU(2) $\cong S^3$ in a uniform way. So we may evaluate it at the North Pole ($x^4 = 1$, $\boldsymbol{x} = \boldsymbol{0}$) of the unit sphere. We then find $A = i\sigma_k\,\mathrm{d}x^k$ and

$$\mathrm{tr}\, A^3 = i^3\, \mathrm{tr}[\sigma_i\sigma_j\sigma_k]\,\mathrm{d}x^i \wedge \mathrm{d}x^j \wedge \mathrm{d}x^k$$
$$= 2\varepsilon_{ijk}\,\mathrm{d}x^i \wedge \mathrm{d}x^j \wedge \mathrm{d}x^k = 12\,\mathrm{d}x^1 \wedge \mathrm{d}x^2 \wedge \mathrm{d}x^3. \tag{10.124}$$

Next we note that (x^1, x^2, x^3) is a good coordinate system on *each* hemisphere of S^3 and $\omega \equiv dx^1 \wedge dx^2 \wedge dx^3$ is a volume element at the North Pole. We find

$$\int_{S^3} \text{tr} \, \mathcal{A}^3 = 12 \int_{S^3} \omega = 12(2\pi^2) = 24\pi^2$$

where $2\pi^2$ is the area of the unit sphere S^3. We finally obtain

$$-\frac{1}{8\pi^2} \int_{S^4} \text{tr} \, \mathcal{F}^2 = \frac{1}{24\pi^2} \int_{S^3} \text{tr} \, \mathcal{A}^3 = 1. \qquad (10.125)$$

(c) Next we consider the map $g_n : S^3 \to SU(2)$ given by (10.117c). We show that $g_2 = g_1 g_1$ has a winding number 2. We divide S^3 into the northern hemisphere $U_N^{(3)}$ and the southern hemisphere $U_S^{(3)}$. Given a map $g_1 : S^3 \to SU(2)$, it is always possible to transform g_1 smoothly to g_{1N} which has the winding number one and $g_{1N}(x) = e$ for $x \in U_S^{(3)}$. All the variation takes place on $U_N^{(3)}$. Similarly, g_1 may be deformed to g_{1S} with the same winding number and $g_{1S}(x) = e$ for $x \in U_N^{(3)}$. Under this deformation, g_2 becomes

$$g_2(x) \to g_2'(x) = \begin{cases} g_{1N}(x) & x \in U_N^{(3)} \\ g_{1S}(x) & x \in U_S^{(3)}. \end{cases}$$

For $\mathcal{A}(x) = g_2'(x)^{-1} \, dg_2'(x) \ (x \in S^3)$, we have

$$\frac{1}{24\pi^3} \int_{S^3} \text{tr} \, \mathcal{A}^3 = \frac{1}{24\pi^2} \left(\int_{U_N^{(3)}} \text{tr}(g_{1N}^{-1} \, dg_{1N})^3 + \int_{U_S^{(3)}} \text{tr}(g_{1S}^{-1} \, dg_{1S})^3 \right)$$

$$= 1 + 1 = 2. \qquad (10.126)$$

Repeating the same procedure we find for $\mathcal{A}(x) = g_n^{-1} \, dg_n$ that

$$-\frac{1}{8\pi^2} \int_{S^4} \text{tr} \, \mathcal{F}^2 = \frac{1}{24\pi^2} \int_{S^3} \text{tr} \, \mathcal{A}^3 = n. \qquad (10.127)$$

Collecting these results we establish the following theorem.

Theorem 10.7. The degree of mapping $g : S^3 \to SU(2)$ is given by

$$n = \frac{1}{24\pi^2} \int_{S^3} \text{tr}(g^{-1} \, dg)^3 = \frac{1}{2} \int_{S^4} \text{tr} \left(\frac{i\mathcal{F}}{2\pi} \right)^2. \qquad (10.128)$$

10.6 Berry's phase

In quantum mechanics, we define a wavefunction up to the phase. In most cases, the phase is neglected as an irrelevant factor. Berry (1984) pointed out that if the system undergoes an adiabatic change, the phase may have observable consequences.

10.6.1   Derivation of Berry's phase

Let $H(\boldsymbol{R})$ be a Hamiltonian which depends on some parameters collectively written as $\boldsymbol{R}$. Suppose $\boldsymbol{R}$ changes adiabatically as a function of time, $\boldsymbol{R} = \boldsymbol{R}(t)$. The Schrödinger equation is

$$H(\boldsymbol{R}(t))|\psi(t)\rangle = \mathrm{i}\frac{\mathrm{d}}{\mathrm{d}t}|\psi(t)\rangle. \tag{10.129}$$

We assume the system at $t = 0$ is in the nth eigenstate, $|\psi(0)\rangle = |n, \boldsymbol{R}(0)\rangle$ where

$$H(\boldsymbol{R}(0))|n, \boldsymbol{R}(0)\rangle = E_n(\boldsymbol{R}(0))|n, \boldsymbol{R}(0)\rangle. \tag{10.130}$$

What about the state $|\psi(t)\rangle$ at later time $t > 0$? We assume the system is always in the nth state, i.e. no level crossing takes place (adiabatic assumption).

Exercise 10.13. A naive guess of $|\psi(t)\rangle$ is

$$|\psi(t)\rangle = \exp\left[-\mathrm{i}\int_0^t \mathrm{d}s\, E_n(\boldsymbol{R}(s))\right]|n, \boldsymbol{R}(t)\rangle \tag{10.131}$$

where the normalized state $|n, \boldsymbol{R}(t)\rangle$ satisfies

$$H(\boldsymbol{R}(t))|n, \boldsymbol{R}(t)\rangle = E_n(\boldsymbol{R}(t))|n, \boldsymbol{R}(t)\rangle. \tag{10.132}$$

Show that (10.131) is *not* a solution of (10.129).

Since (10.131) does not satisfy the Schrödinger equation, we have to try other possibilities. Let us introduce an extra-phase $\eta_n(t)$ in the wavefunction:

$$|\psi(t)\rangle = \exp\left[\mathrm{i}\eta(t) - \mathrm{i}\int_0^t E_n(\boldsymbol{R}(s))\,\mathrm{d}s\right]|n, \boldsymbol{R}(t)\rangle. \tag{10.133}$$

Inserting (10.133) into the Schrödinger equation (10.129), we find

$$H(\boldsymbol{R}(t))|\psi(t)\rangle = E_n(\boldsymbol{R}(t))|\psi(t)\rangle$$

for the LHS (see (10.132)) and

$$\begin{aligned}
\mathrm{i}\frac{\mathrm{d}}{\mathrm{d}t}|\psi(t)\rangle = {}& \left[-\frac{\mathrm{d}\eta_n(t)}{\mathrm{d}t} + E_n(\boldsymbol{R}(t))\right]|\psi(t)\rangle \\
& + \exp\left[\mathrm{i}\eta_n(t) - \mathrm{i}\int E_n(\boldsymbol{R}(s))\,\mathrm{d}s\right]\mathrm{i}\frac{\mathrm{d}}{\mathrm{d}t}|n, \boldsymbol{R}(t)\rangle
\end{aligned}$$

for the RHS. Equating these, it is found that $\eta_n(t)$ satisifes

$$\frac{\mathrm{d}\eta_n(t)}{\mathrm{d}t} = \mathrm{i}\langle n, \boldsymbol{R}(t)|\frac{\mathrm{d}}{\mathrm{d}t}|n, \boldsymbol{R}(t)\rangle. \tag{10.134}$$

By integrating (10.134), we obtain

$$\eta_n(t) = i \int_0^t \langle n, R(s)| \frac{d}{ds} |n, R(s)\rangle ds$$

$$= i \int_{R(0)}^{R(t)} \langle n, R|\nabla_R|n, R\rangle dR \qquad (10.135)$$

where ∇_R stands for the gradient in R-space. Note that $\eta_n(t)$ is real since

$$2 \, \text{Re}\langle n, R(s)| \frac{d}{ds} |n, R(s)\rangle$$

$$= \langle n, R(s)| \frac{d}{ds} |n, R(s)\rangle + \left(\frac{d}{ds} \langle n, R(s)| \right) |n, R(s)\rangle$$

$$= \frac{d}{ds} \langle n, R(s)|n, R(s)\rangle = 0.$$

Suppose the system executes a closed loop in R-space; $R(0) = R(T)$ for some $T > 0$. We then have

$$\eta_n(T) = i \int_0^T \langle n, R(s)| \frac{d}{ds} |n, R(s)\rangle ds$$

$$= i \int_{R(0)}^{R(T)} \langle n, R|\nabla_R|n, R\rangle dR. \qquad (10.136)$$

Since $R(T) = R(0)$, the last expression seems to vanish. However, the integrand is not necessarily a total derivative and $\eta_n(T)$ may fail to vanish. The phase $\eta_n(T)$ is called **Berry's phase** (Berry 1984).

It was Simon (1983) who first recognized the deep geometrical meaning underlying Berry's phase. He observed that the origin of Berry's phase is attributed to the holonomy in the parameter space. We shall work out this point of view following Berry (1984), Simon (1983), Aitchison (1987) and Zumino (1987).

10.6.2 Berry's phase, Berry's connection and Berry's curvature

Let M be a manifold describing the parameter space and let $R = (R_1, \ldots, R_k)$ be the local coordinate. At each point R of M, we consider the normalized nth eigenstate of the Hamiltonian $H(R)$. Since a quantum state $|n; R\rangle$ cannot be distinguished from $e^{i\phi}|n; R\rangle$, a physical state is expressed by an equivalence class

$$[|R\rangle] \equiv \{g|R\rangle|g \in U(1)\} \qquad (10.137)$$

where we omit the index n since we are interested only in the nth eigenvector (figure 10.6). At each point R of M, we have a U(1) degree of freedom and we have a U(1) bundle $P(M, U(1))$ over the parameter space M. The projection is given by $\pi(g|R\rangle) = R$.

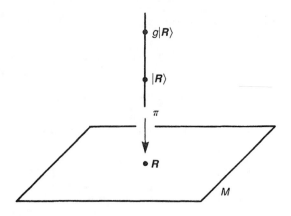

Figure 10.6. The fibre of a quantum mechanical system which depends on adiabatic parameters $\boldsymbol{R}$.

Fixing the phase of $|\boldsymbol{R}\rangle$ at each point $\boldsymbol{R} \in M$ amounts to choosing a section. Let $\sigma(\boldsymbol{R}) = |\boldsymbol{R}\rangle$ be a local section over a chart U of M. The canonical local trivialization is given by

$$\phi^{-1}(|\boldsymbol{R}\rangle) = (\boldsymbol{R}, e). \tag{10.138}$$

The 'right' action yields

$$\phi^{-1}(|\boldsymbol{R}\rangle \cdot g) = (\boldsymbol{R}, e)g = (\boldsymbol{R}, g). \tag{10.139}$$

Now that the bundle structure is defined, we provide it with a connection. Let us define **Berry's connection** by

$$\mathcal{A} = \mathcal{A}_\mu \, \mathrm{d}R^\mu \equiv \langle \boldsymbol{R}|(\mathrm{d}\,|\boldsymbol{R}\rangle) = -(\mathrm{d}\langle \boldsymbol{R}|)|\boldsymbol{R}\rangle \tag{10.140}$$

where $\mathrm{d} = (\partial/\partial R^\mu)\mathrm{d}R^\mu$ is the exterior derivative in $\boldsymbol{R}$-space. Note that $\mathcal{A}$ is anti-Hermitian since

$$0 = \mathrm{d}(\langle \boldsymbol{R}|\boldsymbol{R}\rangle) = (\mathrm{d}\langle \boldsymbol{R}|)|\boldsymbol{R}\rangle + \langle \boldsymbol{R}|\mathrm{d}|\boldsymbol{R}\rangle = \langle \boldsymbol{R}|\mathrm{d}|\boldsymbol{R}\rangle^* + \langle \boldsymbol{R}|\mathrm{d}|\boldsymbol{R}\rangle.$$

To see (10.140) is indeed a local form of a connection, we have to check the compatibility condition. Let U_i and U_j be overlapping charts of M and let $\sigma_i(\boldsymbol{R}) = |\boldsymbol{R}\rangle_i$ and $\sigma_j(\boldsymbol{R}) = |\boldsymbol{R}\rangle_j$ be the respective local sections. They are related by the transition function as $|\boldsymbol{R}\rangle_j = |\boldsymbol{R}\rangle_i t_{ij}(\boldsymbol{R})$. We then find that

$$\begin{aligned} \mathcal{A}_j(\boldsymbol{R}) &= {}_j\langle \boldsymbol{R}|\mathrm{d}|\boldsymbol{R}\rangle_j = t_{ij}(\boldsymbol{R})^{-1}{}_i\langle \boldsymbol{R}|[\mathrm{d}|\boldsymbol{R}\rangle_i t_{ij}(\boldsymbol{R}) + |\boldsymbol{R}\rangle \mathrm{d}t_{ij}(\boldsymbol{R})] \\ &= \mathcal{A}_i(\boldsymbol{R}) + t_{ij}(\boldsymbol{R})^{-1}\mathrm{d}t_{ij}(\boldsymbol{R}). \end{aligned} \tag{10.141}$$

The set of one-forms $\{\mathcal{A}_i\}$ satisfying (10.141) defines an Ehresmann connection on $P(M, \mathrm{U}(1))$.

The field strength $\mathcal{F}$ of $\mathcal{A}$ is called **Berry's curvature** and is given by

$$\mathcal{F} = d\mathcal{A} = (d\langle R|) \wedge (d|R\rangle) = \left(\frac{\partial \langle R|}{\partial R^\mu}\right)\left(\frac{\partial |R\rangle}{\partial R^\nu}\right) dR^\mu \wedge dR^\nu. \qquad (10.142)$$

After an example from atomic physics, we shall clarify how this geometrical structure is reflected in Berry's phase.

Example 10.7. Let us consider a quantum mechanical system which contains 'fast' degrees of freedom r and 'slow' degrees of freedom R. For example, we may imagine an electron moving under the potential of slowly vibrating ions. Suppose the Hamiltonian is given by

$$H = \frac{p^2}{2m} + \frac{P^2}{2M} + V(r; R) \qquad (10.143)$$

where $p(P)$ is the momentum canonical conjugate to $r(R)$. As a first approximation, we may consider the slow degrees of freedom are 'frozen' at some value R and consider an instantaneous sub-Hamiltonian

$$h(R) = \frac{p^2}{2m} + V(r; R) \qquad (10.144)$$

and the eigenvalue problem

$$h(R)|R\rangle = \epsilon_n(R)|R\rangle \qquad (10.145)$$

where $|R\rangle$ stands for the nth eigenvector $|n; R\rangle$ of the 'fast' degrees of freedom. We assume that the eigenvalue is isolated and non-degenerate. Berry's connection is $\mathcal{A}(R) = \langle R|d|R\rangle$, while the curvature is $\mathcal{F} = (d\langle R|) \wedge (d|R\rangle)$.

It is interesting to see how the fast degrees of freedom affect the slow degrees of freedom. We assume the total wavefunction is written in the form

$$\Psi(r; R) = \Phi(R)|R\rangle \qquad (10.146)$$

and find the 'effective' Schrödinger equation which $\Phi(R)$, the wavefunction of the 'slow' degrees of freedom, satisfies. The eigenvalue problem of the Hamiltonian (10.143) is

$$\begin{aligned}
H\Psi(r; R) &= -\frac{1}{2M}[\nabla_R^2 \Phi(R)|R\rangle + 2\nabla_R \Phi(R) \cdot \nabla_R|R\rangle + \Phi(R)\nabla_R^2|R\rangle] \\
&\quad - \Phi(R)\frac{1}{2m}\nabla_r^2|R\rangle + \Phi(R)V(r; R)|R\rangle \\
&= E_n(R)\Phi(R)|R\rangle.
\end{aligned}$$

If we multiply $\langle R|$ on the left and use the Schrödinger equation (10.145), this equation becomes

$$\begin{aligned}
-\frac{1}{2M}[\nabla_R^2 \Phi(R) &+ 2\nabla_R \Phi(R) \cdot \langle R|\nabla_R|R\rangle + \Phi(R)(\langle R|\nabla_R|R\rangle)^2] \\
&+ \epsilon_n(R)\Phi(R) = E_n(R)\Phi(R) \qquad (10.147)
\end{aligned}$$

where we have employed the Born–Oppenheimer approximation, in which all the matrix elements except the diagonal ones are neglected,

$$\langle n; \boldsymbol{R}|\nabla_{\boldsymbol{R}}|n'; \boldsymbol{R}\rangle = 0 \qquad n' \neq n. \tag{10.148}$$

Now the effective Hamiltonian for $|\Phi(\boldsymbol{R})\rangle$ is given by

$$H_{\text{eff}}(n) \equiv -\frac{1}{2M}\left(\frac{\partial}{\partial R^{\mu}} + \mathcal{A}_{\mu}(\boldsymbol{R})\right)^{2} + \varepsilon_{n}(\boldsymbol{R}) \tag{10.149}$$

where $\mathcal{A}_{\mu}$ is a component of Berry's connection,

$$\mathcal{A}_{\mu}(\boldsymbol{R}) = \langle \boldsymbol{R}|\frac{\partial}{\partial R^{\mu}}|\boldsymbol{R}\rangle. \tag{10.150}$$

It is remarkable that the fast degrees of freedom have induced a *vector potential* coupled to the slow degrees of freedom. Note also that the eigenvalue $\varepsilon_{n}(\boldsymbol{R})$ behaves as a potential energy in H_{eff}. This 'spontaneous creation' of the gauge symmetry reflects the phase degree of freedom of the wavefunction $|\boldsymbol{R}\rangle$.

The Schrödinger equation describing the adiabatic change is

$$H(\boldsymbol{R}(t))|\boldsymbol{R}(t), t\rangle = \mathrm{i}\frac{\mathrm{d}}{\mathrm{d}t}|\boldsymbol{R}(t), t\rangle \tag{10.151a}$$

where we note that $|\boldsymbol{R}(t), t\rangle$ has an explicit t-dependence as well as an implicit one through $\boldsymbol{R}(t)$. Berry assumes that

$$|\boldsymbol{R}(t), t\rangle = \exp\left(-\mathrm{i}\int_{0}^{t} E_{n}(t)\,\mathrm{d}t\right)\mathrm{e}^{\mathrm{i}\eta(t)}|\boldsymbol{R}(t)\rangle \tag{10.152a}$$

where $|\boldsymbol{R}\rangle$ is an instantaneous *normalized* eigenstate of $H(\boldsymbol{R})$,

$$\mathcal{H}(\boldsymbol{R})|\boldsymbol{R}\rangle = E_{n}(\boldsymbol{R})|\boldsymbol{R}\rangle \qquad \langle \boldsymbol{R}|\boldsymbol{R}\rangle = 1. \tag{10.153}$$

The first exponential is the ordinary dynamical phase while the second one is Berry's phase. It is convenient for our purpose to define an operator

$$\mathcal{H}(\boldsymbol{R}) \equiv H(\boldsymbol{R}) - E_{n}(\boldsymbol{R}) \tag{10.154}$$

to dispose of the dynamical phase. The state $|\boldsymbol{R}\rangle$ is the zero-energy eigenstate of $\mathcal{H}(\boldsymbol{R})$: $\mathcal{H}(\boldsymbol{R})|\boldsymbol{R}\rangle = 0$. The solution of the modified Schrödinger equation,

$$\mathcal{H}(\boldsymbol{R})|\boldsymbol{R}(t), t\rangle = \mathrm{i}\frac{\mathrm{d}}{\mathrm{d}t}|\boldsymbol{R}(t), t\rangle \tag{10.151b}$$

is then given by

$$|\boldsymbol{R}(t), t\rangle = \mathrm{e}^{\mathrm{i}\eta(t)}|\boldsymbol{R}(t)\rangle. \tag{10.152b}$$

We found in (10.136) that η is given by

$$\eta(t) = i \int_0^t ds \frac{dR^\mu}{ds} \langle R(s)| \frac{\partial}{\partial R^\mu} |R(s)\rangle = i \int_{R(0)}^{R(t)} \langle R|d|R\rangle. \qquad (10.155)$$

We show that Berry's phase is a holonomy associated with the connection (10.140) on $P(M, U(1))$. Take a section $\sigma(R) = |R\rangle$ over a chart U of M. Let $R : [0, 1] \to M$ be a loop in U.[2] We write a horizontal lift of $R(t)$ with respect to the connection (10.140) as

$$\tilde{R}(t) = \sigma(R(t))g(R(t)) \qquad (10.156)$$

where $g(R(0))$ is taken to be the unit element of U(1). The group element $g(t)$ satisfies (10.13b),

$$\frac{dg(t)}{dt} g(t)^{-1} = -\mathcal{A}\left(\frac{d}{dt}\right) = -\langle R(t)| \frac{d}{dt} |R(t)\rangle \qquad (10.157)$$

where $g(t)$ stands for $g(R(t))$. From $g(t) = \exp(i\eta(t))$, we obtain

$$i\frac{d\eta(t)}{dt} = -\langle R(t)| \frac{d}{dt} |R(t)\rangle$$

which is easily integrated to yield

$$\eta(1) = i \int_0^1 \langle R(s)| \frac{d}{ds} |R(s)\rangle \, ds = i \oint \langle R|d|R\rangle. \qquad (10.158)$$

Let us note that $R(0) = R(1)$, hence $|R(0)\rangle = |R(1)\rangle$. Then $\exp[i\eta(1)]$ is regarded as a holonomy (figure 10.7)

$$\tilde{R}(1) = \exp\left(-\oint \langle R|d|R\rangle\right) \cdot |R(0)\rangle. \qquad (10.159a)$$

Exercise 10.14. Let S be a surface in M, which is bounded by the loop $R(t)$. Show that

$$\tilde{R}(1) = \exp\left(-\oint_S \mathcal{F}\right) \cdot |R(0)\rangle \qquad (10.159b)$$

where $\mathcal{F}$ is given by (10.142).

Example 10.8. Let us consider a spin-$\frac{1}{2}$ particle in a magnetic field with the Hamiltonian

$$H(R) = R \cdot \sigma = \begin{pmatrix} R_3 & R_1 - iR_2 \\ R_1 + iR_2 & -R_3 \end{pmatrix}. \qquad (10.160)$$

---

[2] We shall be a little sloppy in our notation.

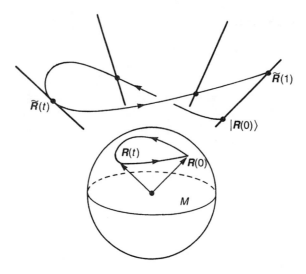

Figure 10.7. If the parameter changes adiabatically along a loop $R(t)$, the state with initial condition $|R(0)\rangle$ becomes $|\tilde{R}(1)\rangle$ which is different from $|R(0)\rangle$ in general. The difference is the holonomy and is identified with Berry's phase.

The parameter R corresponds to the applied magnetic field. This is a two-level system taking eigenvalues $\pm|R|$. Let us consider the eigenvalue $R = +|R|$. According to the prescription just described, we introduce a Hamiltonian $\mathcal{H}(R) \equiv H(R) - |R|$ and consider the zero-energy eigenstate of $\mathcal{H}(R)$ given by

$$|R\rangle_N = [2R(R + R_3)]^{-1/2} \begin{pmatrix} R + R_3 \\ R_1 + iR_2 \end{pmatrix}. \tag{10.161}$$

The gauge potential is obtained after a straightforward but tedious calculation as

$$\mathcal{A}_N = {}_N\langle R|d|R\rangle_N = -i\frac{R_2\,dR_1 - R_1\,dR_2}{2R(R + R_3)}. \tag{10.162}$$

The field strength is

$$\mathcal{F} = d\mathcal{A} = \frac{i}{2}\frac{R_1\,dR_2 \wedge dR_3 + R_2\,dR_3 \wedge dR_1 + R_3\,dR_1 \wedge dR_2}{R^3}. \tag{10.163}$$

So far we have assumed that the state $|R\rangle$ is isolated. However, this assumption breaks down if $R = 0$, in which case two eigenstates are degenerate. Surprisingly, this singularity behaves like a *magnetic monopole* in R-space. To see this, we introduce polar coordinates θ and ϕ in R-space,

$$R_1 = R\sin\theta\cos\phi \qquad R_2 = R\sin\theta\sin\phi \qquad R_3 = R\cos\theta.$$

The state (10.161) is expressed as

$$|R\rangle_N = \begin{pmatrix} \cos(\theta/2) \\ e^{i\phi} \sin(\theta/2) \end{pmatrix}. \qquad (10.164)$$

This state is singular at $\theta = \pi$, reflecting that $|R\rangle_N$ is not defined for $R_3 = -R$. Consider another eigenvector

$$|R\rangle_S \equiv e^{-i\phi}|R\rangle_N = \begin{pmatrix} e^{-i\phi} \cos(\theta/2) \\ \sin(\theta/2) \end{pmatrix}$$

$$= [2R(R - R_3)]^{-1/2} \begin{pmatrix} R_1 - iR_2 \\ R - R_3 \end{pmatrix} \qquad (10.165)$$

with the same eigenvalue. This eigenvector is singular at $\theta = 0$, that is at $R_3 = R$. Corresponding to these vectors, we have Berry's gauge potentials in polar coordinates,

$$\mathcal{A}_N = \tfrac{1}{2}i(1 - \cos\theta)\,d\phi \qquad \theta \neq \pi \qquad (10.166a)$$

$$\mathcal{A}_S = -\tfrac{1}{2}i(1 + \cos\theta)\,d\phi \qquad \theta \neq 0. \qquad (10.166b)$$

They are related by the gauge transformation,

$$\mathcal{A}_S = \mathcal{A}_N - i\,d\phi = \mathcal{A}_N + e^{i\phi}\,de^{-i\phi} \qquad (10.167)$$

where $g(\pi/2, \phi) = \exp(-i\phi)$ is identified with the transition function t_{NS}. Equation (10.166) is simply the vector potential of the Wu–Yang monopole of strength $-\tfrac{1}{2}$, see sections 1.9 and 10.5. The total flux of the monopole is $\Phi = 4\pi(-\tfrac{1}{2}) = -2\pi$.

The analogy between the present problem and the magnetic monopole is evident by now. If we fix the amplitude R of the magnetic field, the restricted parameter space is S^2. At each point R of S^2, the state has a phase degree of freedom. Thus, we are dealing with a U(1) bundle $P(S^2, \mathrm{U}(1))$, which also describes a magnetic monopole. For each choice of the parameters R, we have a fibre corresponding to the nth eigenstate $|n; R\rangle$. The fibre at R consists of the equivalence class $[|R\rangle]$ defined by (10.137). The projection π maps a state to the parameter on which it is defined: $\pi : e^{i\alpha}|R\rangle \to R \in S^2$. As we have seen, this bundle is non-trivial since it cannot be described by a single connection. The non-triviality of the bundle implies the existence of a monopole at the origin. Note that $R = 0$ (that is, $B = 0$) is a singular point at which all the eigenvalues are degenerate.

Next we turn to the problem of holonomy. Take a standard point $R(0)$ on S^2 and choose a vector $|R(0)\rangle$. We choose a loop $R(t)$ on S^2 and execute a parallel transportation of $|R(0)\rangle$ along $R(t)$, after which it comes back as a vector $\exp[i\eta(1)]|R(0)\rangle$. The additional phase η represents the holonomy

$\pi^{-1}(R) \rightarrow \pi^{-1}(R)$ and corresponds to Berry's phase. From (10.158), $\eta(1)$ is given by

$$\eta(1) = i \oint_R \mathcal{A} = i \int_S \mathcal{F} \qquad (10.168)$$

where $\mathcal{F} = d\mathcal{A}$ is the field strength and S is the surface bounded by the loop $R(t)$. It follows from (10.168) that Berry's phase $\eta(1)$ represents the 'magnetic flux' through the area S.

Exercise 10.15. Use (10.165) to show that

$$\mathcal{A}_S = \frac{i}{2} \frac{R_2 \, dR_1 - R_1 \, dR_2}{R(R - R_3)}. \qquad (10.169)$$

Show also that

$$d\phi = -\frac{R_2 \, dR_1 - R_1 \, dR_2}{(R + R_3)(R - R_3)}. \qquad (10.170)$$

Observe that $d\phi$ is singular at $R_3 = \pm R$.

Problems

10.1 Consider a two-dimensional plane M with coordinate R and a wavefunction ψ which depends on R adiabatically as $\psi = \psi(r, R)$. Let $R : [0, 1] \rightarrow M$ be a loop in M and suppose $\psi(r, R(1)) = -\psi(r, R(0))$, that is the phase of ψ changes by π after an adiabatic change along the loop. Show that there is a point within the loop at which the adiabatic assumption breaks down. See Longuet-Higgins (1975).

11

CHARACTERISTIC CLASSES

Given a fibre F, a structure group G and a base space M, we may construct many fibre bundles over M, depending on the choice of the transition functions. Natural questions we may ask ourselves are how many bundles there are over M with given F and G, and how much they differ from a trivial bundle $M \times F$. For example, we observed in section 10.5 that an SU(2) bundle over S^4 is classified by the homotopy group $\pi_3(\mathrm{SU}(2)) \cong \mathbb{Z}$. The number $n \in \mathbb{Z}$ tells us how the transition functions twist the local pieces of the bundle when glued together. We have also observed that this homotopy group is evaluated by integrating $\mathrm{tr}\, \mathcal{F}^2 \in H^4(S^4)$ over S^4, see theorem 10.7.

Characteristic classes are subsets of the cohomology classes of the base space and measure the *non-triviality* or *twisting* of a bundle. In this sense, they are *obstructions* which prevent a bundle from being a trivial bundle. Most of the characteristic classes are given by the de Rham cohomology classes. Besides their importance in classifications of fibre bundles, characteristic classes play central roles in index theorems.

Here we follow Alvalez-Gaumé and Ginsparg (1984), Eguchi *et al* (1980), Gilkey (1995) and Wells (1980). See Bott and Tu (1982), Milnor and Stasheff (1974) for more mathematical expositions.

11.1 Invariant polynomials and the Chern–Weil homomorphism

We give here a brief summary of the de Rham cohomology group (see chapter 6 for details). Let M be an m-dimensional manifold. An r-form $\omega \in \Omega^r(M)$ is *closed* if $\mathrm{d}\omega = 0$ and *exact* if $\omega = \mathrm{d}\eta$ for some $\eta \in \Omega^{r-1}(M)$. The set of closed r-forms is denoted by $Z^r(M)$ and the set of exact r-forms by $B^r(M)$. Since $\mathrm{d}^2 = 0$, it follows that $Z^r(M) \supset B^r(M)$. We define the rth de Rham cohomology group $H^r(M)$ by

$$H^r(M) \equiv Z^r(M)/B^r(M).$$

In $H^r(M)$, two closed r-forms ω_1 and ω_2 are identified if $\omega_1 - \omega_2 = \mathrm{d}\eta$ for some $\eta \in \Omega^{r-1}(M)$. Let M be an m-dimensional manifold. The formal sum

$$H^*(M) \equiv H^0(M) \oplus H^1(M) \oplus \cdots \oplus H^m(M)$$

is the cohomology ring with the product $\wedge : H^*(M) \times H^*(M) \to H^*(M)$ induced by $\wedge : H^p(M) \times H^q(M) \to H^{p+q}(M)$. Let $f : M \to N$ be a smooth map. The pullback $f^* : \Omega^r(N) \to \Omega^r(M)$ naturally induces a linear map $f^* : H^r(N) \to H^r(M)$ since f^* commutes with the exterior derivative: $f^*\,\mathrm{d}\omega = \mathrm{d}f^*\omega$. The pullback f^* preserves the algebraic structure of the cohomology ring since $f^*(\omega \wedge \eta) = f^*\omega \wedge f^*\eta$.

11.1.1 Invariant polynomials

Let $M(k, \mathbb{C})$ be the set of complex $k \times k$ matrices. Let $S^r(M(k, \mathbb{C}))$ denote the vector space of symmetric r-linear $\mathbb{C}$-valued functions on $M(k, \mathbb{C})$. In other words, a map

$$\tilde{P} : \overset{r}{\otimes} M(k, \mathbb{C}) \to \mathbb{C}$$

is an element of $S^r(M(k, \mathbb{C}))$ if it satisfies, in addition to linearity in each entry, the symmetry

$$\tilde{P}(a_1, \ldots, a_i, \ldots, a_j, \ldots, a_r)$$
$$= \tilde{P}(a_1, \ldots, a_j, \ldots, a_i, \ldots, a_r) \qquad 1 \le i, j \le r \qquad (11.1)$$

where $a_p \in \mathrm{GL}(k, \mathbb{C})$. Let

$$S^*(M(k, \mathbb{C})) \equiv \overset{\infty}{\underset{r=0}{\oplus}} S^r(M(k, \mathbb{C}))$$

denote the formal sum of symmetric multilinear $\mathbb{C}$-valued functions. We define a product of $\tilde{P} \in S^p(M(k, \mathbb{C}))$ and $\tilde{Q} \in S^q(M(k, \mathbb{C}))$ by

$$\tilde{P}\tilde{Q}(X_1, \ldots, X_{p+q})$$
$$= \frac{1}{(p+q)!} \sum_P \tilde{P}(X_{P(1)}, \ldots, X_{P(p)})\tilde{Q}(X_{P(p+1)}, \ldots, X_{P(p+q)}) \quad (11.2)$$

where P is the permutation of $(1, \ldots, p + q)$. $S^*(M(k, \mathbb{C}))$ is an algebra with this multiplication.

Let G be a matrix group and $\mathfrak{g}$ its Lie algebra. In practice, we take $G = \mathrm{GL}(k, \mathbb{C}), \mathrm{U}(k)$ or $\mathrm{SU}(k)$. The Lie algebra $\mathfrak{g}$ is a subspace of $M(k, \mathbb{C})$ and we may consider the restrictions $S^r(\mathfrak{g})$ and $S^*(\mathfrak{g}) \equiv \bigoplus_{r \ge 0} S^r(\mathfrak{g})$. $\tilde{P} \in S^r(\mathfrak{g})$ is said to be invariant if, for any $g \in G$ and $A_i \in \mathfrak{g}$, $\tilde{P}$ satisfies

$$\tilde{P}(\mathrm{Ad}_g A_1, \ldots, \mathrm{Ad}_g A_r) = \tilde{P}(A_1, \ldots, A_r) \qquad (11.3)$$

where $\mathrm{Ad}_g A_i = g^{-1} A_i g$. For example,

$$\tilde{P}(A_1, A_2, \ldots, A_r) = \mathrm{str}(A_1, A_2, \ldots, A_r)$$
$$\equiv \frac{1}{r!} \sum_P \mathrm{tr}(A_{P(1)}, A_{P(2)}, \ldots, A_{P(r)}) \qquad (11.4)$$

is symmetric, r-linear and invariant, where 'str' stands for the **symmetrized trace** and is defined by the last equality. The set of G-invariant members of $S^r(\mathfrak{g})$ is denoted by $I^r(G)$. Note that $\mathfrak{g}_1 = \mathfrak{g}_2$ does not necessarily imply $I^r(G_1) = I^r(G_2)$. The product defined by (11.2) naturally induces a multiplication

$$I^p(G) \otimes I^q(G) \to I^{p+q}(G). \tag{11.5}$$

The sum $I^*(G) \equiv \bigotimes_{r \geq 0} I^r(G)$ is an algebra with this product.

Take $\tilde{P} \in I^r(G)$. The shorthand notation for the diagonal combination is

$$P(A) \equiv \tilde{P}(\underbrace{A, A, \ldots, A}_{r}) \qquad A \in \mathfrak{g}. \tag{11.6}$$

Clearly, P is a polynomial of degree r and called an **invariant polynomial**. P is also Ad G-invariant,

$$P(\mathrm{Ad}_g A) = P(g^{-1} A g) = P(A) \qquad A \in \mathfrak{g}, g \in G. \tag{11.7}$$

For example, $\mathrm{tr}(A^r)$ is an invariant polynomial obtained from (11.4). In general, an invariant polynomial may be written in terms of a sum of products of $P_r \equiv \mathrm{tr}(A^r)$.

Conversely, any invariant polynomial P defines an invariant and symmetric r-linear form $\tilde{P}$ by expanding $P(t_1 A_1 + \cdots + t_r A_r)$ as a polynomial in t_i. Then $1/r!$ times the coefficient of $t_1 t_2 \cdots t_r$ is invariant and symmetric by construction and is called the **polarization** of P. Take $P(A) \equiv \mathrm{tr}(A^3)$, for example. Following the previous prescription, we expand $\mathrm{tr}(t_1 A_1 + t_2 A_2 + t_3 A_3)^3$ in powers of t_1, t_2 and t_3. The coefficient of $t_1 t_2 t_3$ is

$$\mathrm{tr}(A_1 A_2 A_3 + A_1 A_3 A_2 + A_2 A_1 A_3 + A_2 A_3 A_1 + A_3 A_1 A_2 + A_3 A_2 A_1)$$
$$= 3 \, \mathrm{tr}(A_1 A_2 A_3 + A_2 A_1 A_3)$$

where the cyclicity of the trace has been used. The polarization is

$$\tilde{P}(A_1, A_2, A_3) = \tfrac{1}{2} \mathrm{tr}(A_1 A_2 A_3 + A_2 A_1 A_3) = \mathrm{str}(A_1, A_2, A_3).$$

In the previous chapter, we introduced the local gauge potential $\mathcal{A} = A_\mu \, dx^\mu$ and the field strength $\mathcal{F} = \tfrac{1}{2} \mathcal{F}_{\mu\nu} \, dx^\mu \wedge dx^\nu$ on a principal bundle. We have shown that these geometrical objects describe the associated vector bundles as well. Since the set of connections $\{A_i\}$ describes the twisting of a fibre bundle, the non-triviality of a principal bundle is equally shared by its associated bundle. In fact, if (10.57) is employed as a definition of the local connection in a vector bundle, it can be defined even without reference to the principal bundle with which it is originally associated. Later, we encounter situations in which use of vector bundles is essential (the Whitney sum bundle, the splitting principle and so on).

Let $P(M, \mathbb{C})$ be a principal bundle. We extend the domain of invariant polynomials from $\mathfrak{g}$ to $\mathfrak{g}$-valued p-forms on M. For $A_i \eta_i$ ($A_i \in \mathfrak{g}, \eta \in \Omega^{p_i}(M); 1 \le i \le r$), we define

$$\tilde{P}(A_1 \eta_1, \ldots, A_r \eta_r) \equiv \eta_1 \wedge \ldots \wedge \eta_r \tilde{P}(A_1, \ldots, A_r). \tag{11.8}$$

For example, corresponding to (11.4), we have

$$\text{str}(A_1 \eta_1, \ldots, A_r \eta_r) = \eta_1 \wedge \ldots \wedge \eta_r \, \text{str}(A_1, \ldots, A_r).$$

The diagonal combination is

$$P(A\eta) \equiv \underbrace{\eta \wedge \ldots \wedge \eta}_{r} P(A). \tag{11.9}$$

The action $\tilde{P}$ or P on general elements is given by the r-linearity. In particular, we are interested in the invariant polynomial of the form $P(\mathcal{F})$ in the following. The importance of invariant polynomials resides in the following fundamental theorem.

Theorem 11.1. (**Chern–Weil theorem**) Let P be an invariant polynomial. Then $P(\mathcal{F})$ satisfies

(a) $dP(\mathcal{F}) = 0$.
(b) Let $\mathcal{F}$ and $\mathcal{F}'$ be curvature two-forms corresponding to different connections A and A'. Then the difference $P(\mathcal{F}') - P(\mathcal{F})$ is exact.

Proof. (a) It is sufficient to prove that $dP(\mathcal{F}) = 0$ for an invariant polynomial $P_r(\mathcal{F})$ which is homogeneous of degree r, since any invariant polynomial can be decomposed into homogeneous polynomials. First consider the identity,

$$\tilde{P}_r(g_t^{-1} X_1 g_t, \ldots, g_t^{-1} X_r g_t) = \tilde{P}_r(X_1, \ldots, X_r)$$

where $g_t \equiv \exp(tX)$ and $X, X_i \in \mathfrak{g}$. By putting $t = 0$ after differentiation with respect to t, we obtain

$$\sum_{i=1}^{r} \tilde{P}_r(X_1, \ldots, [X_i, X], \ldots, X_r) = 0. \tag{11.10}$$

Next, let A be a $\mathfrak{g}$-valued p-form and Ω_i be a $\mathfrak{g}$-valued p_i-form ($1 \le i \le r$). Without loss of generality, we may take $A = X\eta$ and $\Omega_i = X_i \eta_i$ where $X, X_i \in \mathfrak{g}$ and η (η_i) is a p-form (p_i-form). Define

$$[\Omega_i, A] \equiv \eta_i \wedge \eta[X_i, X]$$
$$= X_i X(\eta_i \wedge \eta) - (-1)^{p p_i} X X_i (\eta \wedge \eta_i). \tag{11.11}$$

Let us note that

$$\tilde{P}_r(\Omega_1, \ldots, [\Omega_i, A], \ldots, \Omega_r)$$
$$= \eta_1 \wedge \ldots \wedge \eta_i \wedge \eta \wedge \ldots \wedge \eta_r \, \tilde{P}_r(X_1, \ldots, X_i X, \ldots, X_r)$$
$$- (-1)^{p \cdot p_i} \eta_1 \wedge \ldots \wedge \eta \wedge \eta_i \wedge \ldots$$
$$\ldots \wedge \eta_r \, \tilde{P}_r(X_1, \ldots, X X_i, \ldots, X_r)$$
$$= \eta \wedge \eta_1 \wedge \ldots \wedge \eta_r (-1)^{p(p_1 + \cdots + p_i)}$$
$$\times \tilde{P}_r(X_1, \ldots, [X_i, X], \ldots, X_r).$$

From this and (11.10), we find

$$\sum_{i=1}^{r} (-1)^{p(p_1 + \cdots + p_i)} \tilde{P}_r(\Omega_1, \ldots, [\Omega_i, A], \ldots, \Omega_r) = 0. \qquad (11.12)$$

Next, consider the derivative,

$$d\tilde{P}_r(\Omega_1, \ldots, \Omega_r) = d(\eta_1 \wedge \ldots \wedge \eta_r) \tilde{P}_r(X_1, \ldots, X_r)$$
$$= \sum_{i=1}^{r} (-1)^{(p_1 + \cdots + p_{i-1})} (\eta_1 \wedge \ldots \wedge d\eta_i \wedge \ldots \wedge \eta_r)$$
$$\times \tilde{P}_r(X_1, \ldots, X_i, \ldots, X_r)$$
$$= \sum_{i=1}^{r} (-1)^{(p_1 + \cdots + p_{i-1})} \tilde{P}_r(\Omega_1, \ldots, d\Omega_i, \ldots, \Omega_r). \qquad (11.13)$$

Let $A = \mathcal{A}$ and $\Omega_i = \mathcal{F}$ in (11.12) and (11.13) for which $p = 1$ and $p_i = 2$. By adding 0 of the form (11.12) to (11.13) we have

$$d\tilde{P}_r(\mathcal{F}, \ldots, \mathcal{F})$$
$$= \sum_{i=1}^{r} [\tilde{P}_r(\mathcal{F}, \ldots, d\mathcal{F}, \ldots, \mathcal{F}) + \tilde{P}_r(\mathcal{F}, \ldots, [\mathcal{A}, \mathcal{F}], \ldots, \mathcal{F})]$$
$$= \sum_{i=1}^{r} \tilde{P}_r(\mathcal{F}, \ldots, \mathcal{D}\mathcal{F}, \ldots, \mathcal{F}) = 0 \qquad (11.14)$$

since $\mathcal{D}\mathcal{F} = d\mathcal{F} + [\mathcal{A}, \mathcal{F}] = 0$ (the Bianchi identity). We have proved

$$dP_r(\mathcal{F}) = d\tilde{P}_r(\mathcal{F}, \ldots, \mathcal{F}) = 0.$$

(b) Let $\mathcal{A}$ and $\mathcal{A}'$ be two connections on E and let $\mathcal{F}$ and $\mathcal{F}'$ be the respective field strengths. Define an interpolating gauge potential $\mathcal{A}_t$, by

$$\mathcal{A}_t \equiv \mathcal{A} + t\theta \qquad \theta \equiv (\mathcal{A}' - \mathcal{A}) \qquad 0 \le t \le 1 \qquad (11.15)$$

so that $\mathcal{A}_0 = \mathcal{A}$ and $\mathcal{A}_1 = \mathcal{A}'$. The corresponding field strength is

$$\mathcal{F}_t \equiv \mathrm{d}\mathcal{A}_t + \mathcal{A}_t \wedge \mathcal{A}_t = \mathcal{F} + t\mathcal{D}\theta + t^2\theta^2 \tag{11.16}$$

where $\mathcal{D}\theta = \mathrm{d}\theta + [\mathcal{A}, \theta] = \mathrm{d}\theta + \mathcal{A} \wedge \theta + \theta \wedge \mathcal{A}$. We first note that

$$P_r(\mathcal{F}') - P_r(\mathcal{F}) = P_r(\mathcal{F}_1) - P_r(\mathcal{F}_0) = \int_0^1 \mathrm{d}t \frac{\mathrm{d}}{\mathrm{d}t} P_r(\mathcal{F}_t)$$

$$= r \int_0^1 \mathrm{d}t \, \tilde{P}_r \left(\frac{\mathrm{d}}{\mathrm{d}t} \mathcal{F}_t, \mathcal{F}_t, \ldots, \mathcal{F}_t \right). \tag{11.17}$$

From (11.16), we find that

$$\frac{\mathrm{d}}{\mathrm{d}t} P_r(\mathcal{F}_t) = r \tilde{P}_r(\mathcal{D}\theta + 2t\theta^2, \mathcal{F}_t, \ldots, \mathcal{F}_t)$$

$$= r \tilde{P}_r(\mathcal{D}\theta, \mathcal{F}_t, \ldots, \mathcal{F}_t) + 2rt \tilde{P}_r(\theta^2, \mathcal{F}_t, \ldots, \mathcal{F}_t). \tag{11.18}$$

Note also that

$$\mathcal{D}\mathcal{F}_t = \mathrm{d}\mathcal{F}_t + [\mathcal{A}, \mathcal{F}_t] = -[\mathcal{A}_t, \mathcal{F}_t] + [\mathcal{A}, \mathcal{F}_t] = t[\mathcal{F}_t, \theta]$$

where use has been made of the Bianchi identity $\mathcal{D}_t\mathcal{F}_t = \mathrm{d}\mathcal{F}_t + [\mathcal{A}_t, \mathcal{F}_t] = 0$. [$\mathcal{D}$ is the covariant derivative with respect to $\mathcal{A}$ while $\mathcal{D}_t$ is that with respect to $\mathcal{A}_t$.] It then follows that

$$\mathrm{d}[\tilde{P}_r(\theta, \mathcal{F}_t, \ldots, \mathcal{F}_t)]$$
$$= \tilde{P}_r(\mathrm{d}\theta, \mathcal{F}_t, \ldots, \mathcal{F}_t) - (r - 1)\tilde{P}_r(\theta, \mathrm{d}\mathcal{F}_t, \ldots, \mathcal{F}_t)$$
$$= \tilde{P}_r(\mathcal{D}\theta, \mathcal{F}_t, \ldots, \mathcal{F}_t) - (r - 1)\tilde{P}_r(\theta, \mathcal{D}\mathcal{F}_t, \ldots, \mathcal{F}_t)$$
$$= \tilde{P}_r(\mathcal{D}\theta, \mathcal{F}_t, \ldots, \mathcal{F}_t) - (r - 1)t\tilde{P}_r(\theta, [\mathcal{F}_t, \theta], \mathcal{F}_t, \ldots, \mathcal{F}_t) \tag{11.19}$$

where we have added a 0 of the form (11.12) to change d to $\mathcal{D}$. If we take $\Omega_1 = \mathcal{A} = \theta, \Omega_2 = \cdots = \Omega_m = \mathcal{F}_t$ in (11.12), we have

$$2\tilde{P}_r(\theta^2, \mathcal{F}_t, \ldots, \mathcal{F}_t) + (r - 1)\tilde{P}_r(\theta, [\mathcal{F}_t, \theta], \mathcal{F}_t, \ldots, \mathcal{F}_t) = 0.$$

From (11.18), (11.19) and the previous identity, we obtain

$$\frac{\mathrm{d}}{\mathrm{d}t} P_r(\mathcal{F}_t) = r\mathrm{d}[\tilde{P}_r(\theta, \mathcal{F}_t, \ldots, \mathcal{F}_t)].$$

We finally find that

$$P_r(\mathcal{F}') - P_r(\mathcal{F}) = \mathrm{d}\left[r \int_0^1 \tilde{P}_r(\mathcal{A}' - \mathcal{A}, \mathcal{F}_t, \ldots, \mathcal{F}_t) \, \mathrm{d}t \right]. \tag{11.20}$$

This shows that $P_r(\mathcal{F}')$ differs from $P_r(\mathcal{F})$ by an exact form. $\square$

We define the **transgression** $TP_r(\mathcal{A}', \mathcal{A})$ of P_r by

$$TP_r(\mathcal{A}', \mathcal{A}) \equiv r \int_0^1 dt\, \tilde{P}_r\, (\mathcal{A}' - \mathcal{A}, \mathcal{F}_t, \ldots, \mathcal{F}_t) \tag{11.21}$$

where $\tilde{P}_r$ is the polarization of P_r. Transgressions will play an important role when we discuss Chern–Simons forms in section 11.5. Let $\dim M = m$. Since $P_m(\mathcal{F}')$ differs from $P_m(\mathcal{F})$ by an exact form, their integrals over a manifold M without a boundary should be the same:

$$\int_M P_m(\mathcal{F}') - \int_M P_m(\mathcal{F}) = \int_M dT P_m(\mathcal{A}', \mathcal{A}) = \int_{\partial M} P_m(\mathcal{A}', \mathcal{A}) = 0. \tag{11.22}$$

As has been proved, an invariant polynomial is closed and, in general, non-trivial. Accordingly, it defines a cohomology class of M. Theorem 11.1(b) ensures that this cohomology class is independent of the gauge potential chosen. The cohomology class thus defined is called the **characteristic class**. The characteristic class defined by an invariant polynomial P is denoted by $\chi_E(P)$ where E is a fibre bundle on which connections and curvatures are defined. [*Remark:* Since a principal bundle and its associated bundles share the same gauge potentials and field strengths, the Chern–Weil theorem applies equally to both bundles. Accordingly, E can be either a principal bundle or a vector bundle.]

Theorem 11.2. Let P be an invariant polynomial in $I^*(G)$ and E be a fibre bundle over M with structure group G.

(a) The map

$$\chi_E : I^*(G) \to H^*(M) \tag{11.23}$$

defined by $P \to \chi_E(P)$ is a homomorphism (**Weil homomorphism**).
(b) Let $f : N \to M$ be a differentiable map. For the pullback bundle f^*E of E, we have the so-called **naturality**

$$\chi_{f^*E} = f^*\chi_E. \tag{11.24}$$

Proof. (a) Take $P_r \in I^r(G)$ and $P_s \in I^s(G)$. If we write $\mathcal{F} = \mathcal{F}^\alpha T_\alpha$, we have

$$(P_r P_s)(\mathcal{F}) = \mathcal{F}^{\alpha_1} \wedge \ldots \wedge \mathcal{F}^{\alpha_r} \wedge \mathcal{F}^{\beta_1} \wedge \ldots \wedge \mathcal{F}^{\beta_s}$$
$$\times \frac{1}{(r+s)!} \tilde{P}_r(T_{\alpha_1}, \ldots, T_{\alpha_r}) \tilde{P}_n(T_{\beta_1}, \ldots, T_{\beta_s})$$
$$= P_r(\mathcal{F}) \wedge P_s(\mathcal{F}).$$

Then (a) follows since $P_r(\mathcal{F})$, $P_s(\mathcal{F}) \in H^*(M)$.

(b) Let $\mathcal{A}$ be a gauge potential of E and $\mathcal{F} = d\mathcal{A} + \mathcal{A} \wedge \mathcal{A}$. It is easy to verify that the pullback $f^*\mathcal{A}$ is a connection in f^*E. In fact, let $\mathcal{A}_i$ and $\mathcal{A}_j$ be local connections in overlapping charts U_i and U_j of M. If t_{ij} is a transition function on $U_i \cap U_j$, the transition function on f^*E is given by $f^*t_{ij} = t_{ij} \circ f$. The pullback $f^*\mathcal{A}_i$ and $f^*\mathcal{A}_j$ are related as

$$f^*\mathcal{A}_j = f^*(t_{ij}^{-1}\mathcal{A}_i t_{ij} + t_{ij}^{-1}\, \mathrm{d}t_{ij})$$
$$= (f^*t_{ij}^{-1})(f^*\mathcal{A}_i)(f^*t_{ij}) + (f^*t_{ij}^{-1})(\mathrm{d}f^*t_{ij}).$$

This shows that $f^*\mathcal{A}$ is, indeed, a local connection on f^*E. The corresponding field strength on f^*E is

$$\mathrm{d}(f^*\mathcal{A}_i) + f^*\mathcal{A}_i \wedge f^*\mathcal{A}_i = f^*[\mathrm{d}\mathcal{A}_i + \mathcal{A}_i \wedge \mathcal{A}_i] = f^*\mathcal{F}_i.$$

Hence, $f^*P(\mathcal{F}_i) = P(f^*\mathcal{F}_i)$, that is $f^*\chi_E(P) = \chi_{f^*E}(P)$. □

Corollary 11.1. Characteristic classes of a trivial bundle are trivial.

Proof. Let $E \xrightarrow{\pi} M$ be a trivial bundle. Since E is trivial, there exists a map $f : M \to \{p\}$ such that $E = f^*E_0$ where $E_0 \longrightarrow \{p\}$ is a bundle over a point p. All the de Rham cohomology groups of a point are trivial and so are the characteristic classes. Theorem 11.2(b) ensures that the characteristic classes χ_E ($= f^*\chi_{E_0}$) of E are also trivial. □

11.2 Chern classes

11.2.1 Definitions

Let $E \xrightarrow{\pi} M$ be a complex vector bundle whose fibre is $\mathbb{C}^k$. The structure group G is a subgroup of $\mathrm{GL}(k, \mathbb{C})$, and the gauge potential $\mathcal{A}$ and the field strength $\mathcal{F}$ take their values in $\mathfrak{g}$. Define the **total Chern class** by

$$c(\mathcal{F}) \equiv \det\left(I + \frac{\mathrm{i}\mathcal{F}}{2\pi}\right). \tag{11.25}$$

Since $\mathcal{F}$ is a two-form, $c(\mathcal{F})$ is a direct sum of forms of even degrees,

$$c(\mathcal{F}) = 1 + c_1(\mathcal{F}) + c_2(\mathcal{F}) + \cdots \tag{11.26}$$

where $c_j(\mathcal{F}) \in \Omega^{2j}(M)$ is called the jth **Chern class**. In an m-dimensional manifold M, the Chern class $c_j(\mathcal{F})$ with $2j > m$ vanishes trivially. Irrespective of dim M, the series terminates at $c_k(\mathcal{F}) = \det(\mathrm{i}\mathcal{F}/2\pi)$ and $c_j(\mathcal{F}) = 0$ for $j > k$. Since $c_j(\mathcal{F})$ is closed, it defines an element $[c_j(\mathcal{F})]$ of $H^{2j}(M)$.

Example 11.1. Let F be a complex vector bundle with fibre $\mathbb{C}^2$ over M, where $G = \mathrm{SU}(2)$ and dim $M = 4$. If we write the field $\mathcal{F} = \mathcal{F}^\alpha(\sigma_\alpha/2\mathrm{i})$, $\mathcal{F}^\alpha = \frac{1}{2}\mathcal{F}^\alpha{}_{\mu\nu}\, \mathrm{d}x^\mu \wedge \mathrm{d}x^\nu$, we have

$$c(\mathcal{F}) = \det\left(I + \frac{\mathrm{i}}{2\pi}\mathcal{F}^\alpha(\sigma_\alpha/2\mathrm{i})\right)$$

$$= \det \begin{pmatrix} 1 + (i/2\pi)(\mathcal{F}^3/2i) & (i/2\pi)(\mathcal{F}^1 - i\mathcal{F}^2)/2i \\ (i/2\pi)(\mathcal{F}^1 + i\mathcal{F}^2)/2i & 1 - (i/2\pi)(\mathcal{F}^3/2i) \end{pmatrix}$$

$$= 1 + \frac{1}{4}\left(\frac{i}{2\pi}\right)^2 \left(\mathcal{F}^3 \wedge \mathcal{F}^3 + \mathcal{F}^1 \wedge \mathcal{F}^1 + \mathcal{F}^2 \wedge \mathcal{F}^2\right). \tag{11.27}$$

Individual Chern classes are

$$c_0(\mathcal{F}) = 1$$

$$c_1(\mathcal{F}) = 0$$

$$\tag{11.28}$$

$$c_2(\mathcal{F}) = \left(\frac{i}{2\pi}\right)^2 \sum \frac{\mathcal{F}^\alpha \wedge \mathcal{F}^\alpha}{4} = \det\left(\frac{i\mathcal{F}}{2\pi}\right).$$

Higher Chern classes vanish identically.

For general fibre bundles, it is rather cumbersome to compute the Chern classes by expanding the determinant and it is desirable to find a formula which yields them more easily. This is done by diagonalizing the curvature form. The matrix form $\mathcal{F}$ is diagonalized by an appropriate matrix $g \in \mathrm{GL}(k, \mathbb{C})$ as $g^{-1}(i\mathcal{F}/2\pi)g = \mathrm{diag}(x_1, \ldots, x_k)$, where x_i is a two-form. This diagonal matrix will be denoted by A. For example, if $G = \mathrm{SU}(k)$, the generators are chosen to be anti-Hermitian and a Hermitian matrix $i\mathcal{F}/2\pi$ can be diagonalized by $g \in \mathrm{SU}(k)$. We have

$$\det(I + A) = \det[\mathrm{diag}(1 + x_1, 1 + x_2, \ldots, 1 + x_k)]$$

$$= \prod_{j=1}^{k}(1 + x_j)$$

$$= 1 + (x_1 + \cdots + x_k) + (x_1 x_2 + \cdots + x_{k-1} x_k)$$

$$+ \cdots + (x_1 x_2 + \cdots + x_k)$$

$$= 1 + \mathrm{tr}\, A + \tfrac{1}{2}\{(\mathrm{tr}\, A)^2 - \mathrm{tr}\, A^2\} + \cdots + \det A. \tag{11.29}$$

Observe that each term of (11.29) is an elementary symmetric function of $\{x_j\}$,

$$S_0(x_j) \equiv 1$$

$$S_1(x_j) \equiv \sum_{j=1}^{k} x_j$$

$$S_2(x_j) \equiv \sum_{i<j} x_i x_j \tag{11.30}$$

$$\vdots$$

$$S_k(x_j) \equiv x_1 x_2 \ldots x_k.$$

Since $\det(I + A)$ is an invariant polynomial, we have $P(\mathcal{F}) = P(g\mathcal{F}g^{-1}) = P(2\pi A/i)$, see (11.7). Accordingly, we have, for general $\mathcal{F}$,

$$c_0(\mathcal{F}) = 1$$

$$c_1(\mathcal{F}) = \operatorname{tr} A = \operatorname{tr}\left(g\frac{i\mathcal{F}}{2\pi}g^{-1}\right) = \frac{i}{2\pi}\operatorname{tr}\mathcal{F}$$

$$c_2(\mathcal{F}) = \tfrac{1}{2}[(\operatorname{tr}A)^2 - \operatorname{tr}A^2] = \tfrac{1}{2}(i/2\pi)^2[\operatorname{tr}\mathcal{F}\wedge\operatorname{tr}\mathcal{F} - \operatorname{tr}(\mathcal{F}\wedge\mathcal{F})] \qquad (11.31)$$

$$\vdots$$

$$c_k(\mathcal{F}) = \det A = (i/2\pi)^k \det \mathcal{F}.$$

Example 11.1 is easily verified from (11.31). [Note that the Pauli matrices (in general, any element of the Lie algebra $\mathfrak{su}(n)$ of $SU(n)$) are traceless, $\operatorname{tr}\sigma_\alpha = 0$.]

11.2.2 Properties of Chern classes

We will deal with several vector bundles in the following. We often denote the Chern class of a vector bundle E by $c(E)$. If the specification of the curvature is required, we write $c(\mathcal{F}_E)$.

Theorem 11.3. Let $E \xrightarrow{\pi} M$ be a vector bundle with $G = GL(k, \mathbb{C})$ and $F = \mathbb{C}^k$.

(a) (Naturality) Let $f : N \to M$ be a smooth map. Then

$$c(f^*E) = f^*c(E). \qquad (11.32)$$

(b) Let $F \xrightarrow{\pi'} M$ be another vector bundle with $F = \mathbb{C}^l$ and $G = GL(l, \mathbb{C})$. The total Chern class of a Whitney sum bundle $E \oplus F$ is

$$c(E \oplus F) = c(E) \wedge c(F). \qquad (11.33)$$

Proof.

(a) The naturality follows directly from theorem 11.2(a). Since the curvature of f^*E is $\mathcal{F}_{f^*E} = f^*\mathcal{F}_E$, the total Chern class of f^*E is

$$c(f^*E) = \det\left(I + \frac{i}{2\pi}\mathcal{F}_{f^*E}\right) = \det\left(I + \frac{i}{2\pi}f^*\mathcal{F}_E\right)$$

$$= f^*\det\left(I + \frac{i}{2\pi}\mathcal{F}_E\right) = f^*c(E).$$

(b) Let us consider the Chern polynomial of a matrix

$$A = \begin{pmatrix} B & 0 \\ 0 & C \end{pmatrix}.$$

[Note that the curvature of a Whitney sum bundle is block diagonal: $\mathcal{F}_{E\oplus F} = \text{diag}(\mathcal{F}_E, \mathcal{F}_F)$.] We find that

$$\det\left(I + \frac{iA}{2\pi}\right) = \det\left(\begin{array}{cc} I + \frac{iB}{2\pi} & 0 \\ 0 & I + \frac{iC}{2\pi} \end{array}\right)$$

$$= \det\left(I + \frac{iB}{2\pi}\right)\det\left(I + \frac{iC}{2\pi}\right) = c(B)c(C).$$

This relation remains true when B and C are replaced by $\mathcal{F}_E$ and $\mathcal{F}_F$, namely

$$c(\mathcal{F}_{E\oplus F}) = c(\mathcal{F}_E) \wedge c(\mathcal{F}_F)$$

which proves (11.33).   □

Exercise 11.1. (a) Let E be a trivial bundle. Use corollary 11.1 to show that

$$c(E) = 1. \tag{11.34}$$

(b) Let E be a vector bundle such that $E = E_1 \oplus E_2$ where E_1 is a vector bundle of dimension k_1 and E_2 is a trivial vector bundle of dimension k_2. Show that

$$c_i(E) = 0 \qquad k_1 + 1 \le i \le k_1 + k_2. \tag{11.35}$$

11.2.3 Splitting principle

Let E be a Whitney sum of n complex line bundles,

$$E = L_1 \oplus L_2 \oplus \cdots \oplus L_n. \tag{11.36}$$

From (11.33), we have

$$c(E) = c(L_1)c(L_2)\ldots c(L_n) \tag{11.37}$$

where the product is the exterior product of differential forms. Since $c_r(L) = 0$ for $r \ge 2$, we write

$$c(L_i) = 1 + c_1(L_i) \equiv 1 + x_i. \tag{11.38}$$

Then (11.37) becomes

$$c(E) = \prod_{i=1}^{n}(1 + x_i). \tag{11.39}$$

Comparing this with (11.29), we find that the Chern class of an n-dimensional vector bundle E is identical with that of the Whitney sum of n complex line bundles. Although E is not a Whitney sum of complex line bundles in general, as far as the Chern classes are concerned, we may pretend that this is the case. This is called the **splitting principle** and we accept this fact without proof. The general proof is found in Shanahan (1978) and Hirzebruch (1966), for example.

Intuitively speaking, if the curvature $\mathcal{F}$ is diagonalized, the complex vector space on which g acts splits into k independent pieces: $\mathbb{C}^k \to \mathbb{C} \oplus \cdots \oplus \mathbb{C}$. An eigenvalue x_i is a curvature in each complex line bundle. Since diagonalizable matrices are dense in $M(n, \mathbb{C})$, any matrix may be approximated by a diagonal one as closely as we wish. Hence, the splitting principle applies to any matrix. As an exercise, the reader may prove (11.33) using the splitting principle.

11.2.4 Universal bundles and classifying spaces

By now the reader must have some acquaintance with characteristic classes. Before we close this section, we examine these from a slightly different point of view emphasizing their role in the classification of fibre bundles. Let $E \xrightarrow{\pi} M$ be a vector bundle with fibre $\mathbb{C}^k$. It is known that we can always find a bundle $\bar{E} \xrightarrow{\pi'} M$ such that

$$E \oplus \bar{E} \cong M \times \mathbb{C}^n \qquad (11.40)$$

for some $n \geq k$. The fibre F_p of E at $p \in M$ is a k-plane lying in $\mathbb{C}^n$. Let $G_{k,n}(\mathbb{C})$ be the Grassmann manifold defined in example 8.4. The manifold $G_{k,n}(\mathbb{C})$ is the set of k-planes in $\mathbb{C}^n$. Similarly to the canonical line bundle, we define the canonical k-plane bundle $L_{k,n}(\mathbb{C})$ over $G_{k,n}(\mathbb{C})$ with the fibre $\mathbb{C}^k$. Consider a map $f : M \to G_{k,n}(\mathbb{C})$ which maps a point p to the k-plane F_p in $\mathbb{C}^n$.

Theorem 11.4. Let M be a manifold with $\dim M = m$ and let $E \xrightarrow{\pi} M$ be a complex vector bundle with the fibre $\mathbb{C}^k$. Then there exists a natural number N such that for $n > N$,

(a) there exists a map $f : M \to G_{k,n}(\mathbb{C})$ such that

$$E \cong f^* L_{k,n}(\mathbb{C}) \qquad (11.41)$$

(b) $f^* L_{k,n}(\mathbb{C}) \cong g^* L_{k,n}(\mathbb{C})$ if and only if $f, g : M \to G_{k,n}(\mathbb{C})$ are homotopic.

The proof is found in Chern (1979). For example, if $E \xrightarrow{\pi} M$ is a complex line bundle, then there exists a bundle $\bar{E} \xrightarrow{\pi'} M$ such that $E \oplus \bar{E} \cong M \times \mathbb{C}^n$ and a map $f : M \to G_{1,n}(\mathbb{C}) \cong \mathbb{C}P^{n-1}$ such that $E = f^* L$, L being the canonical line bundle over $\mathbb{C}P^{n-1}$. Moreover, if $f \sim g$, then $f^* L$ is equivalent to $g^* L$. Theorem 11.4 shows that the classification of vector bundles reduces to that of the homotopy classes of the maps $M \to G_{k,n}(\mathbb{C})$.

It is convenient to define the **classifying space** $G_k(\mathbb{C})$. Regarding a k-plane in $\mathbb{C}^n$ as that in $\mathbb{C}^{n+1}$, we have natural inclusions.

$$G_{k,k}(\mathbb{C}) \hookrightarrow G_{k,k+1}(\mathbb{C}) \hookrightarrow \cdots \hookrightarrow G_k(\mathbb{C}) \qquad (11.42)$$

where

$$G_k(\mathbb{C}) \equiv \bigcup_{n=k}^{\infty} G_{k,n}(\mathbb{C}). \qquad (11.43)$$

Correspondingly, we have the **universal bundle** $L_k \to G_k(\mathbb{C})$ whose fibre is $\mathbb{C}^k$. For any complex vector bundle $E \xrightarrow{\pi} M$ with fibre $\mathbb{C}^k$, there exists a map $f : M \to G_k(\mathbb{C})$ such that $E = f^* L_k(\mathbb{C})$.

Let $E \xrightarrow{\pi} M$ be a vector bundle. A characteristic class χ is defined as a map $\chi : E \to \chi(E) \in H^*(M)$ such that

$$\chi(f^* E) = f^* \chi(E) \qquad \text{(naturality)} \qquad (11.44a)$$
$$\chi(E) = \chi(E') \qquad \text{if } E \text{ is equivalent to } E'. \qquad (11.44b)$$

The map f^* on the LHS of (11.44a) is a pullback of the bundle while f^* on the RHS is that of the cohomology class. Since the homotopy class $[f]$ of $f : M \to G_k(\mathbb{C})$ uniquely defines the pullback

$$f^* : H^*(G_k) \to H^*(M) \qquad (11.45)$$

an element $\chi(E) = f^* \chi(G_k)$ proves to be useful in classifying complex vector bundles over M with $\dim E = k$. For each choice of $\chi(G_k)$, there exists a characteristic class in E.

The Chern class $c(E)$ is also defined axiomatically by

(i) $\quad c(f^* E) = f^* c(E) \qquad \text{(naturality)} \qquad (11.46a)$

(ii) $\quad c(E) = c_0(E) \oplus c_1(E) \oplus \cdots \oplus c_k(E)$

$\qquad c_i(E) \in H^{2i}(M); \; c_i(E) = 0 \qquad i > k \qquad (11.46b)$

(iii) $\quad c(E \oplus F) = c(E)c(E) \qquad \text{(Whitney sum)} \qquad (11.46c)$

(iv) $\quad c(L) = 1 + x \qquad \text{(normalization)} \qquad (11.46d)$

L being the canonical line bundle over $\mathbb{C}P^n$. It can be shown that these axioms uniquely define the Chern class as (11.25).

11.3 Chern characters

11.3.1 Definitions

Among the characteristic classes, the Chern characters are of special importance due to their appearance in the Atiyah–Singer index theorem. The **total Chern character** is defined by

$$\mathrm{ch}(\mathcal{F}) \equiv \mathrm{tr} \exp\left(\frac{i\mathcal{F}}{2\pi}\right) = \sum_{j=1} \frac{1}{j!} \mathrm{tr}\left(\frac{i\mathcal{F}}{2\pi}\right)^j. \qquad (11.47)$$

The jth **Chern character** $ch_j(\mathcal{F})$ is

$$ch_j(\mathcal{F}) \equiv \frac{1}{j!} \, \text{tr} \left(\frac{i\mathcal{F}}{2\pi} \right)^j. \tag{11.48}$$

If $2j > m = \dim M$, $ch_j(\mathcal{F})$ vanishes, hence $ch(\mathcal{F})$ is a polynomial of finite order.
   Let us diagonalize $\mathcal{F}$ as

$$\frac{i\mathcal{F}}{2\pi} \to g^{-1} \left(\frac{i\mathcal{F}}{2\pi} \right) g = A \equiv \text{diag}(x_1, \ldots, x_k) \qquad g \in \text{GL}(k, \mathbb{C}).$$

The total Chern character is expressed as

$$\text{tr}[\exp(A)] = \sum_{j=1}^{k} \exp(x_j). \tag{11.49}$$

In terms of the elementary symmetric functions $S_r(x_j)$, the total Chern character becomes

$$\sum_{j=1}^{k} \exp(x_j) = \sum_{j=1}^{k} \left(1 + x_j + \frac{1}{2!}x_j^2 + \frac{1}{3!}x_j^3 + \cdots \right)$$

$$= k + S_1(x_j) + \frac{1}{2!}[S_1(x_j)^2 - 2S_2(x_j)] + \cdots. \tag{11.50}$$

Accordingly, each Chern character is expressed in terms of the Chern classes as

$$ch_0(\mathcal{F}) = k \tag{11.51a}$$

$$ch_1(\mathcal{F}) = c_1(\mathcal{F}) \tag{11.51b}$$

$$ch_2(\mathcal{F}) = \tfrac{1}{2}[c_1(\mathcal{F})^2 - 2c_2(\mathcal{F})] \tag{11.51c}$$

$$\vdots$$

where k is the fibre dimension of the bundle.

Example 11.2. Let P be a U(1) bundle over S^2. If $\mathcal{A}_N$ and $\mathcal{A}_S$ are the local connections on U_N and U_S defined in section 10.5, the field strength is given by $\mathcal{F}_i = d\mathcal{A}_i$ ($i = N, S$). We have

$$ch(\mathcal{F}) = 1 + \frac{i\mathcal{F}}{2\pi} \tag{11.52}$$

where we have noted that $\mathcal{F}^n = 0$ ($n \geq 2$) on S^2. This bundle describes the magnetic monopole. The magnetic charge $2g$ given by (10.94) is an integer expressed in terms of the Chern character as

$$N = \frac{i}{2\pi} \int_{S^2} \mathcal{F} = \int_{S^2} ch_1(\mathcal{F}). \tag{11.53}$$

Let P be an SU(2) bundle over S^4. The total Chern class of P is given by (11.27). The total Chern character is

$$\mathrm{ch}(\mathcal{F}) = 2 + \mathrm{tr}\left(\frac{i\mathcal{F}}{2\pi}\right) + \frac{1}{2}\,\mathrm{tr}\left(\frac{i\mathcal{F}}{2\pi}\right)^2. \tag{11.54}$$

Ch$(\mathcal{F})$ terminates at $\mathrm{ch}_2(\mathcal{F})$ since $\mathcal{F}^n = 0$ for $n \geq 3$. Moreover, $\mathrm{tr}\,\mathcal{F} = 0$ for $G = \mathrm{SU}(2)$, $n \geq 2$. As we found in section 10.5, the instanton number is given by

$$\frac{1}{2}\int_{S^4}\mathrm{tr}\left(\frac{i\mathcal{F}}{2\pi}\right)^2 = \int_{S^4}\mathrm{ch}_2(\mathcal{F}). \tag{11.55}$$

In both cases, ch_j measures how the bundle is twisted when local pieces are patched together.

Example 11.3. Let P be a U(1) bundle over a $2m$-dimensional manifold M. The mth Chern character is

$$\frac{1}{m!}\mathrm{tr}\left(\frac{i\mathcal{F}}{2\pi}\right)^m = \frac{1}{m!}\left(\frac{i}{2\pi}\right)^m\left[\frac{1}{2}\mathcal{F}_{\mu\nu}\,dx^\mu \wedge dx^\nu\right]^m$$

$$= \frac{1}{m!}\left(\frac{i}{4\pi}\right)^m \mathcal{F}_{\mu_1\nu_1}\ldots\mathcal{F}_{\mu_m\nu_m}\,dx^{\mu_1}\wedge dx^{\nu_1}\wedge\ldots\wedge dx^{\mu_m}\wedge dx^{\nu_m}$$

$$= \left(\frac{i}{4\pi}\right)^m \epsilon^{\mu_1\nu_1\ldots\mu_m\nu_m}\mathcal{F}_{\mu_1\nu_1}\ldots\mathcal{F}_{\mu_m\nu_m}\,dx^1\wedge\ldots\wedge dx^{2m}$$

which describes the U(1) anomaly in $2m$-dimensional space, see chapter 13.

Example 11.4. Let L be a complex line bundle. It then follows that

$$\mathrm{ch}(L) = \mathrm{tr}\exp\left(\frac{i\mathcal{F}}{2\pi}\right) = e^x = 1 + x \qquad x \equiv \frac{i\mathcal{F}}{2\pi}. \tag{11.56}$$

For example, let $L \xrightarrow{\pi} \mathbb{C}P^1$ be the canonical line bundle over $\mathbb{C}P^1 = S^2$. The Fubini–Study metric yields the curvature

$$\mathcal{F} = -\partial\bar{\partial}\ln(1 + |z|^2) = -\frac{dz\wedge d\bar{z}}{(1 + z\bar{z})^2} \tag{11.57}$$

see example 8.8. In real coordinates $z = x + iy = r\exp(i\theta)$, we have

$$\mathcal{F} = 2i\frac{dx\wedge dy}{(1 + x^2 + y^2)^2} = 2i\frac{r\,dr\wedge d\theta}{(1 + r^2)^2}. \tag{11.58}$$

From $\mathrm{ch}(\mathcal{F}) = 1 + \mathrm{tr}(i\mathcal{F}/2\pi)$, we have

$$\mathrm{ch}_1(\mathcal{F}) = -\frac{1}{\pi}\frac{r\,dr\wedge d\theta}{(1 + r^2)^2}. \tag{11.59}$$

Ch$_1(L)$, the integral of $\mathrm{ch}_1(\mathcal{F})$ over S^2 is an integer,

$$\mathrm{Ch}_1(L) = -\frac{1}{\pi}\int\frac{r\,dr d\theta}{(1 + r^2)^2} = -\int_1^\infty t^{-2}\,dt = -1. \tag{11.60}$$

11.3.2 Properties of the Chern characters

Theorem 11.5.   (a) (Naturality) Let $E \xrightarrow{\pi} M$ be a vector bundle with $F = \mathbb{C}^k$. Let $f : N \to M$ be a smooth map. Then

$$\mathrm{ch}(f^*E) = f^*\mathrm{ch}(E). \tag{11.61}$$

(b) Let E and F be vector bundles over a manifold M. The Chern characters of $E \otimes F$ and $E \oplus F$ are given by

$$\mathrm{ch}(E \otimes F) = \mathrm{ch}(E) \wedge \mathrm{ch}(F) \tag{11.62a}$$

$$\mathrm{ch}(E \oplus F) = \mathrm{ch}(E) \oplus \mathrm{ch}(F). \tag{11.62b}$$

Proof. (a) follows from theorem 11.2(a).

(b) These results are immediate from the definition of the ch-polynomial. Let

$$\mathrm{ch}(A) = \sum \frac{1}{j!} \mathrm{tr} \left(\frac{\mathrm{i}A}{2\pi} \right)^j$$

be a polynomial of a matrix A. Suppose A is a tensor product of B and C, $A = B \otimes C = B \otimes I + I \otimes C$ (note that $\mathcal{F}_{E \otimes F} = \mathcal{F}_E \otimes I + I \otimes \mathcal{F}_F$). Then we find that

$$\mathrm{ch}(B \otimes C) = \sum_j \frac{1}{j!} \left(\frac{\mathrm{i}}{2\pi} \right)^j \mathrm{tr}(B \otimes I + I \otimes C)^j$$

$$= \sum_j \frac{1}{j!} \left(\frac{\mathrm{i}}{2\pi} \right)^j \sum_{m=1}^j \binom{j}{m} \mathrm{tr}(B^m)\, \mathrm{tr}(C^{j-m})$$

$$= \sum_m \frac{1}{m!} \mathrm{tr} \left(\frac{\mathrm{i}B}{2\pi} \right)^m \sum_n \frac{1}{n!} \mathrm{tr} \left(\frac{\mathrm{i}C}{2\pi} \right)^n = \mathrm{ch}(B)\mathrm{ch}(C).$$

Equation (11.62a) is proved if B is replaced by $\mathcal{F}_E$ and C by $\mathcal{F}_F$.

If A is block diagonal,

$$A = \begin{pmatrix} B & 0 \\ 0 & C \end{pmatrix} = B \oplus C$$

we have

$$\mathrm{ch}(B \oplus C) = \sum \frac{1}{j!} \left(\frac{\mathrm{i}}{2\pi} \right)^j \mathrm{tr}(B \oplus C)^j$$

$$= \sum \frac{1}{j!} \left(\frac{1}{2\pi} \right)^j [\mathrm{tr}(B^j) + \mathrm{tr}(C^j)] = \mathrm{ch}(B) + \mathrm{ch}(C).$$

This relation remains true when A, B and C are replaced by $\mathcal{F}_{E \oplus F}$, $\mathcal{F}_E$ and $\mathcal{F}_F$ respectively.   □

Let us see how the splitting principle works in this case. Let L_j $(1 \leq j \leq k)$ be complex line bundles. From (11.62b) we have, for $E = L_1 \oplus L_2 \oplus \cdots \oplus L_k$,

$$\text{ch}(E) = \text{ch}(L_1) \oplus \text{ch}(L_2) \oplus \cdots \oplus \text{ch}(L_k). \tag{11.63}$$

Since $\text{ch}(L_i) = \exp(x_i)$, we find

$$\text{ch}(E) = \prod_{j=1}^{k} \exp(x_j) \tag{11.64}$$

which is simply (11.50). Hence, the Chern character of a general vector bundle E is given by that of a Whitney sum of k complex line bundles. The characteristic classes themselves cannot differentiate between two vector bundles of the same base space and the same fibre dimension. What is important is their *integral* over the base space.

11.3.3 Todd classes

Another useful characteristic class associated with a complex vector bundle is the **Todd class** defined by

$$\text{Td}(\mathcal{F}) = \prod_{j} \frac{x_j}{1 - e^{-x_j}} \tag{11.65}$$

where the splitting principle is understood. If expanded in powers of x_j, $\text{Td}(\mathcal{F})$ becomes

$$\text{Td}(\mathcal{F}) = \prod_{j} \left(1 + \frac{1}{2}x_j + \sum_{k \geq 1}(-1)^{k-1} \frac{B_k}{(2k)!} x_j^{2k} \right)$$
$$= 1 + \frac{1}{2}\sum_{j} x_j + \frac{1}{12}\sum_{j} x_j^2 + \frac{1}{4}\sum_{j<k} x_j x_k + \cdots$$
$$= 1 + \frac{1}{2}c_1(\mathcal{F}) + \frac{1}{12}[c_1(\mathcal{F})^2 + c_2(\mathcal{F})] + \cdots \tag{11.66}$$

where the B_k are the **Bernoulli numbers**

$$B_1 = \tfrac{1}{6} \qquad B_2 = \tfrac{1}{30} \qquad B_3 = \tfrac{1}{42} \qquad B_4 = \tfrac{1}{30} \qquad B_5 = \tfrac{5}{66} \qquad \ldots.$$

The first few terms of (11.66) are:

$$\text{Td}_0(\mathcal{F}) = 1 \tag{11.67a}$$
$$\text{Td}_1(\mathcal{F}) = \tfrac{1}{2}c_1 \tag{11.67b}$$
$$\text{Td}_2(\mathcal{F}) = \tfrac{1}{12}(c_1^2 + c_2) \tag{11.67c}$$
$$\text{Td}_3(\mathcal{F}) = \tfrac{1}{24}c_1 c_2 \tag{11.67d}$$
$$\text{Td}_4(\mathcal{F}) = \tfrac{1}{720}(-c_1^4 + 4c_1^2 c_2 + 3c_2^2 + c_1 c_3 - c_4) \tag{11.67e}$$
$$\text{Td}_5(\mathcal{F}) = \tfrac{1}{1440}(-c_1^3 c_2 + 3c_1 c_2^2 + c_1^2 c_3 - c_1 c_4) \tag{11.67f}$$

where c_i stands for $c_i(\mathcal{F})$.

Exercise 11.2. Let E and F be complex vector bundles over M. Show that

$$\mathrm{Td}(E \oplus F) = \mathrm{Td}(E) \wedge \mathrm{Td}(F). \tag{11.68}$$

11.4 Pontrjagin and Euler classes

In the present section we will be concerned with the characteristic classes associated with a real vector bundle.

11.4.1 Pontrjagin classes

Let E be a real vector bundle over an m-dimensional manifold M with $\dim_{\mathbb{R}} E = k$. If E is endowed with the fibre metric, we may introduce orthonormal frames at each fibre. The structure group may be reduced to $O(k)$ from $GL(k, \mathbb{R})$. Since the generators of $o(k)$ are skew symmetric, the field strength $\mathcal{F}$ of E is also skew symmetric. A skew-symmetric matrix A is not diagonalizable by an element of a subgroup of $GL(k, \mathbb{R})$. It is, however, reducible to block diagonal form as

$$A \to \begin{pmatrix} 0 & \lambda_1 & & & & 0 \\ -\lambda_1 & 0 & & & & \\ & & 0 & \lambda_2 & & \\ & & -\lambda_2 & 0 & & \\ 0 & & & & \ddots & \end{pmatrix}$$

$$\to \begin{pmatrix} i\lambda_1 & & & & & \\ & -i\lambda_1 & & & 0 & \\ & & i\lambda_2 & & & \\ & & & -i\lambda_2 & & \\ 0 & & & & \ddots & \end{pmatrix} \tag{11.69}$$

where the second diagonalization is achieved only by an element of $GL(k, \mathbb{C})$. If k is odd, the last diagonal element is set to zero. For example, the generator of $o(3) = \mathfrak{so}(3)$ generating rotations around the z-axis is

$$T_z = \begin{pmatrix} 0 & 1 & 0 \\ -1 & 0 & 0 \\ 0 & 0 & 0 \end{pmatrix}.$$

The **total Pontrjagin class** is defined by

$$p(\mathcal{F}) \equiv \det\left(I + \frac{\mathcal{F}}{2\pi}\right). \tag{11.70}$$

From the skew symmetry $\mathcal{F}^t = -\mathcal{F}$, it follows that

$$\det\left(I + \frac{\mathcal{F}}{2\pi}\right) = \det\left(I + \frac{\mathcal{F}^t}{2\pi}\right) = \det\left(I - \frac{\mathcal{F}}{2\pi}\right).$$

Therefore, $p(\mathcal{F})$ is an *even* function in $\mathcal{F}$. The expansion of $p(\mathcal{F})$ is

$$p(\mathcal{F}) = 1 + p_1(\mathcal{F}) + p_2(\mathcal{F}) + \cdots \tag{11.71}$$

where $p_j(\mathcal{F})$ is a polynomial of order $2j$ and is an element of $H^{4j}(M; \mathbb{R})$. We note that $p_j(\mathcal{F}) = 0$ for either $2j > k = \dim E$ or $4j > \dim M$.[1]

Let us diagonalize $\mathcal{F}/2\pi$ as

$$\frac{\mathcal{F}}{2\pi} \rightarrow A \equiv \begin{pmatrix} -ix_1 & & & & \\ & ix_1 & & 0 & \\ & & -ix_2 & & \\ & 0 & & ix_2 & \\ & & & & \ddots \end{pmatrix} \tag{11.72}$$

where $x_k \equiv -\lambda_k/2\pi$, λ_k being the eigenvalues of $\mathcal{F}$. The sign has been chosen to simplify the Euler class defined here. The generating function of $p(\mathcal{F})$ is given by

$$p(\mathcal{F}) = \det(I + A) = \prod_{i=1}^{[k/2]}(1 + x_i^2) \tag{11.73}$$

where

$$[k/2] =\longrightarrow \begin{cases} k/2 & \text{if } k \text{ is even} \\ (k-1)/2 & \text{if } k \text{ is odd.} \end{cases}$$

In (11.73) only *even* powers appear, reflecting the skew symmetry. Each **Pontrjagin class** is computed from (11.73) as

$$p_j(\mathcal{F}) = \sum_{i_1 < i_2 < \ldots < i_j}^{[k/2]} x_{i_1}^2 x_{i_2}^2 \ldots x_{i_j}^2. \tag{11.74}$$

To write $p_j(\mathcal{F})$ in terms of the curvature two-form $\mathcal{F}/2\pi$, we first note that

$$\text{tr}\left(\frac{\mathcal{F}}{2\pi}\right)^{2j} = \text{tr } A^{2j} = 2(-1)^j \sum_{i=1}^{[k/2]} x_i^{2j}.$$

---

[1] Although $p_m(\mathcal{F}) = 0$, $p_m(B)$ need not vanish for a matrix B. p_m will be used to define the Euler class later.

It then follows that

$$p_1(\mathcal{F}) = \sum_i x_i^2 = -\frac{1}{2}\left(\frac{1}{2\pi}\right)^2 \mathrm{tr}\,\mathcal{F}^2 \tag{11.75a}$$

$$p_2(\mathcal{F}) = \sum_{i<j} x_i^2 x_j^2 = \frac{1}{2}\left[\left(\sum_i x_i^2\right)^2 - \sum_i x_i^4\right]$$

$$= \frac{1}{8}\left(\frac{1}{2\pi}\right)^4 [(\mathrm{tr}\,\mathcal{F}^2)^2 - 2\,\mathrm{tr}\,\mathcal{F}^4] \tag{11.75b}$$

$$p_3(\mathcal{F}) = \sum_{i<j<k} x_i^2 x_j^2 x_k^2$$

$$= \frac{1}{48}\left(\frac{1}{2\pi}\right)^6 [-(\mathrm{tr}\,\mathcal{F}^2)^3 + 6\,\mathrm{tr}\,\mathcal{F}^2\,\mathrm{tr}\,\mathcal{F}^4 - 8\,\mathrm{tr}\,\mathcal{F}^6] \tag{11.75c}$$

$$p_4(\mathcal{F}) = \sum_{i<j<k<l} x_i^2 x_j^2 x_k^2 x_l^2$$

$$= \frac{1}{384}\left(\frac{1}{2\pi}\right)^8 [(\mathrm{tr}\,\mathcal{F}^2)^4 - 12(\mathrm{tr}\,\mathcal{F}^2)^2\,\mathrm{tr}\,\mathcal{F}^4 + 32\,\mathrm{tr}\,\mathcal{F}^2\,\mathrm{tr}\,\mathcal{F}^6$$

$$+ 12(\mathrm{tr}\,\mathcal{F}^4)^2 - 48\,\mathrm{tr}\,\mathcal{F}^8] \tag{11.75d}$$

$$\vdots$$

$$p_{[k/2]}(\mathcal{F}) = x_1^2 x_2^2 \ldots x_{[k/2]}^2 = \left(\frac{1}{2\pi}\right)^k \det\mathcal{F}. \tag{11.75e}$$

The reader should verify that

$$p(E \oplus F) = p(E) \wedge p(F). \tag{11.76}$$

It is easy to guess that the Pontrjagin classes are written in terms of Chern classes. Since Chern classes are defined only for complex vector bundles, we must complexify the fibre of E so that complex numbers make sense. The resulting vector bundle is denoted by $E^{\mathbb{C}}$. Let A be a skew-symmetric real matrix. We find that

$$\det(I + iA) = \det\begin{pmatrix} 1+x_1 & & & & 0 \\ & 1-x_1 & & & \\ & & 1+x_2 & & \\ 0 & & & 1-x_2 & \\ & & & & \ddots \end{pmatrix}$$

$$= \prod_{i=1}^{[k/2]}(1 - x_i^2) = 1 - p_1(A) + p_2(A) - \cdots$$

from which it follows that

$$p_j(E) = (-1)^j c_{2j}(E^{\mathbb{C}}). \tag{11.77}$$

Example 11.5. Let M be a four-dimensional Riemannian manifold. When the orthonormal frame $\{\hat{e}_\alpha\}$ is employed, the structure group of the tangent bundle TM may be reduced to O(4). Let $\mathcal{R} = \frac{1}{2}\mathcal{R}_{\alpha\beta}\theta^\alpha \wedge \theta^\beta$ be the curvature two-form ($\mathcal{R}$ should not be confused with the scalar curvature). For the tangent bundle, it is common to write $p(M)$ instead of $p(\mathcal{R})$. We have

$$\det\left(I + \frac{\mathcal{R}}{2\pi}\right) = 1 - \frac{1}{8\pi^2}\operatorname{tr}\mathcal{R}^2 + \frac{1}{128\pi^4}[(\operatorname{tr}\mathcal{R}^2)^2 - 2\operatorname{tr}\mathcal{R}^4]. \tag{11.78}$$

Each **Pontrjagin class** is given by

$$p_0(M) = 1 \tag{11.79a}$$

$$p_1(M) = -\frac{1}{8\pi^2}\operatorname{tr}\mathcal{R}^2 = -\frac{1}{8\pi^2}\mathcal{R}_{\alpha\beta}\mathcal{R}_{\beta\alpha} \tag{11.79b}$$

$$p_2(M) = \frac{1}{128\pi^4}[(\operatorname{tr}\mathcal{R}^2)^2 - 2\operatorname{tr}\mathcal{R}^4] = \left(\frac{1}{2\pi}\right)^4 \det\mathcal{R}. \tag{11.79c}$$

Although $p_2(M)$ vanishes as a differential form, we need it in the next subsection to compute the Euler class.

11.4.2 Euler classes

Let M be a $2l$-dimensional orientable Riemannian manifold and let TM be the tangent bundle of M. We denote the curvature by $\mathcal{R}$. It is always possible to reduce the structure group of TM down to SO($2l$) by employing an orthonormal frame. The **Euler class** e of M is defined by the square root of the $4l$-form p_l,

$$e(A)e(A) = p_l(A). \tag{11.80}$$

Both sides should be understood as functions of a $2l \times 2l$ matrix A and not of the curvature $\mathcal{R}$, since $p_l(\mathcal{R})$ vanishes identically. However, $e(M) \equiv e(\mathcal{R})$ thus defined is a $2l$-form and, indeed, gives a volume element of M. If M is an odd-dimensional manifold we define $e(M) = 0$, see later.

Example 11.6. Let $M = S^2$ and consider the tangent bundle TS^2. From example 7.14, we find the curvature two-form,

$$\mathcal{R}_{\theta\phi} = -\mathcal{R}_{\phi\theta} = \sin^2\theta \frac{d\theta \wedge d\phi}{\sin\theta} = \sin\theta\, d\theta \wedge d\phi$$

where we have noted that $g_{\theta\theta} = \sin^2\theta$. Although $p_1(S^2) = 0$ as a differential form, we compute it to find the Euler form. We have

$$p_1(S^2) = -\frac{1}{8\pi^2}\operatorname{tr}\mathcal{R}^2 = -\frac{1}{8\pi^2}[\mathcal{R}_{\theta\phi}\mathcal{R}_{\phi\theta} + \mathcal{R}_{\phi\theta}\mathcal{R}_{\theta\phi}]$$

$$= \left(\frac{1}{2\pi}\sin\theta\, d\theta \wedge d\phi\right)^2$$

from which we read off

$$e(S^2) = \frac{1}{2\pi} \sin\theta \, d\theta \wedge d\phi. \tag{11.81}$$

It is interesting to note that

$$\int_{S^2} e(S^2) = \frac{1}{2\pi} \int_0^{2\pi} d\phi \int_0^{\pi} d\theta \sin\theta = 2 \tag{11.82}$$

which is the Euler characteristic of S^2, see section 2.4. This is not just a coincidence. Let us take another convincing example, a torus T^2. Since T^2 admits a flat connection, the curvature vanishes identically. It then follows that $e(T^2) \equiv 0$ and $\chi(T^2) = 0$. These are special cases of the **Gauss–Bonnet theorem**,

$$\int_M e(M) = \chi(M) \tag{11.83}$$

for a compact orientable manifold M. If M is odd dimensional both e and χ vanish, see (6.39).

In general, the determinant of a $2l \times 2l$ skew-symmetric matrix A is a square of a polynomial called the **Pfaffian** $\mathrm{Pf}(A)$, [2]

$$\det A = \mathrm{Pf}(A)^2. \tag{11.84}$$

We show that the Pfaffian is given by

$$\mathrm{Pf}(A) = \frac{(-1)^l}{2^l l!} \sum_P \mathrm{sgn}(P) A_{P(1)P(2)} A_{P(3)P(4)} \cdots A_{P(2l-1)P(2l)} \tag{11.85}$$

where the phase has been chosen for later convenience. We first note that a skew-symmetric matrix A can be block diagonalized by an element of $O(2l)$ as

$$S^t A S = \Lambda = \begin{pmatrix} \begin{matrix} 0 & \lambda_1 \\ -\lambda_1 & 0 \end{matrix} & & & \\ & \begin{matrix} 0 & \lambda_2 \\ -\lambda_2 & 0 \end{matrix} & & \\ & & \ddots & \\ & & & \begin{matrix} 0 & \lambda_l \\ -\lambda_l & 0 \end{matrix} \end{pmatrix}. \tag{11.86}$$

It is easy to see that

$$\det A = \det \Lambda = \prod_{i=1}^l \lambda_i^2.$$

[2] See proposition 1.3. The definition here differs in phase from that in section 1.5. It turns out to be convenient to choose the present phase convention in the definition of the Euler class.

To compute $\mathrm{Pf}(\Lambda)$, we note that the non-vanishing terms in (11.85) are of the form $A_{12}A_{34}\ldots A_{2l-1,2l}$. Moreover, there are 2^l ways of changing the suffices as $A_{ij} \to A_{ji}$, such as

$$A_{12}A_{34}\ldots A_{2l-1,2l} \to A_{21}A_{34}\ldots A_{2l-1,2l}$$

and $l!$ permutations of the pairs of indices, for example,

$$A_{12}A_{34}\ldots A_{2l-1,2l} \to A_{34}A_{12}\ldots A_{2l-1,2l}.$$

Hence, we have

$$\mathrm{Pf}(\Lambda) = (-1)^l A_{12}A_{34}\ldots A_{2l-1,2l} = (-1)^l \prod_{i=1}^{l} \lambda_i.$$

Thus, we conclude that a block diagonal matrix Λ satisfies

$$\det \Lambda = \mathrm{Pf}(\Lambda)^2.$$

To show that (11.84) is true for any skew-symmetric matrices (not necessarily block diagonal) we use the following lemma,[3]

$$\mathrm{Pf}(X^{t}AX) = \mathrm{Pf}(A)\det X. \tag{11.87}$$

If $S^{t}AS = \Lambda$ for $S \in O(2l)$, we have $A = S\Lambda S^{t}$, hence

$$\mathrm{Pf}(S\Lambda S^{t}) = \mathrm{Pf}(\Lambda)\det S = (-1)^l \prod_{i=1}^{l} \lambda_i \det S.$$

We finally find $\det A = \mathrm{Pf}(A)^2$ for a skew-symmetric matrix A.

Note that $\mathrm{Pf}(A)$ is $SO(2l)$ invariant but changes sign under an improper rotation S ($\det S = -1$) of $O(2l)$.

Exercise 11.3. Show that the determinant of an odd-dimensional skew-symmetric matrix vanishes. This is why we put $e(M) = 0$ for an odd-dimensional manifold.

The **Euler class** is defined in terms of the curvature $\mathcal{R}$ as

$$e(M) = \mathrm{Pf}(\mathcal{R}/2\pi)$$
$$= \frac{(-1)^l}{(4\pi)^l l!} \sum_{P} \mathrm{sgn}(P)\mathcal{R}_{P(1)P(2)} \ldots \mathcal{R}_{P(2l-1)P(2l)}. \tag{11.88}$$

---

[3] Since $\det(X^{t}AX) = (\det X)^2 \det A$, we have $\mathrm{Pf}(X^{t}AX) = \pm\mathrm{Pf}(A)\det X$. Here the plus sign should be chosen since $\mathrm{Pf}(I^{t}AI) = \mathrm{Pf}(A)$.

The generating function is obtained by taking $x_j = -\lambda_i/2\pi$,

$$e(x) = x_1 x_2 \ldots x_l = \prod_{i=1}^{l} x_i. \qquad (11.89)$$

The phase $(-1)^l$ has been chosen to simplify the RHS.

Example 11.7. Let M be a four-dimensional orientable manifold. The structure group of TM is $SO(4)$, see example 11.5. The Euler class is obtained from (11.88) as

$$e(M) = \frac{1}{2(4\pi)^2} \epsilon^{ijkl} \mathcal{R}_{ij} \wedge \mathcal{R}_{kl}. \qquad (11.90)$$

This is in agreement with the result of example 11.5. The relevant Pontrjagin class is

$$p_2(M) = \frac{1}{128\pi^4}[(\operatorname{tr} \mathcal{R}^2)^2 - 2\operatorname{tr} \mathcal{R}^4] = x_1^2 x_2^2.$$

Since $e(M) = x_1 x_2$, we have $p_2(M) = e(M) \wedge e(M)$. This is written as a matrix identity,

$$\frac{1}{128\pi^4}[(\operatorname{tr} A^2)^2 - 2\operatorname{tr} A^4] = \left(\frac{1}{2(4\pi)^4} \epsilon^{ijkl} A_{ij} A_{kl}\right)^2.$$

11.4.3 Hirzebruch L-polynomial and $\hat{A}$-genus

The **Hirzebruch L-polynomial** is defined by

$$L(x) = \prod_{j=1}^{k} \frac{x_j}{\tanh x_j}$$

$$= \prod_{j=1}^{k} \left(1 + \sum_{n \geq 1}(-1)^{n-1} \frac{2^{2n}}{(2n)!} B_n x_j^{2n}\right) \qquad (11.91)$$

where the B_n are Bernoulli numbers, see (11.66). The function $L(x)$ is even in x_j and can be written in terms of the Pontrjagin classes,

$$L(\mathcal{F}) = 1 + \tfrac{1}{3}p_1 + \tfrac{1}{45}(-p_1^2 + 7p_2) + \tfrac{1}{945}(2p_1^3 - 13p_1 p_2 + 62 p_3) + \cdots \quad (11.92)$$

where p_j stands for $p_j(\mathcal{F})$. From the splitting principle, we find that

$$L(E \oplus F) = L(E) \wedge L(F). \qquad (11.93)$$

The $\hat{A}$ (**A-roof**) **genus** $\hat{A}(\mathcal{F})$ is defined by

$$\hat{A}(\mathcal{F}) = \prod_{j=1}^{k} \frac{x_j/2}{\sinh(x_j/2)}$$

$$= \prod_{j=1}^{k} \left(1 + \sum_{n \geq 1}(-1)^n \frac{(2^{2n} - 2)}{(2n)!} B_n x_j^{2n}\right). \qquad (11.94)$$

This is an even function of x_j and can be expanded in p_j. $\hat{A}$ is also called the **Dirac genus** by physicists. It satisfies

$$\hat{A}(E \oplus F) = \hat{A}(E) \wedge \hat{A}(F). \tag{11.95}$$

$\hat{A}$ is written in terms of the Pontrjagin classes as

$$\hat{A}(\mathcal{F}) = 1 - \tfrac{1}{24}p_1 + \tfrac{1}{5760}(7p_1^2 - 4p_2)$$
$$+ \tfrac{1}{967\,680}(-31p_1^3 + 44p_1p_2 - 16p_3) + \cdots . \tag{11.96}$$

Example 11.8. Let M be a compact connected and orientable four-dimensional manifold. Let us consider the symmetric bilinear form $\sigma : H^2(M; \mathbb{R}) \times H^2(M; \mathbb{R}) \to \mathbb{R}$ defined by

$$\sigma([\alpha], [\beta]) = \int_M \alpha \wedge \beta. \tag{11.97}$$

σ is a $b^2 \times b^2$ symmetric matrix where $b^2 = \dim H^2(M; \mathbb{R})$ is the Betti number. Clearly σ is non-degenerate since $\sigma([\alpha], [\beta]) = 0$ for any $[\alpha] \in H^2(M; \mathbb{R})$ implies $[\beta] = 0$. Let p (q) be the number of positive (negative) eigenvalues of σ. The **Hirzebruch signature** of M is

$$\tau(M) \equiv p - q. \tag{11.98}$$

According to the **Hirzebruch signature theorem** (see section 12.5), this number is also given in terms of the L-polynomial as

$$\tau(M) = \int_M L_1(M) = \tfrac{1}{3} \int_M p_1(M). \tag{11.99}$$

11.5 Chern–Simons forms

11.5.1 Definition

Let $P_j(\mathcal{F})$ be an arbitrary $2j$-form characteristic class. Since $P_j(\mathcal{F})$ is closed, it can be written locally as an exact form by Poincaré's lemma. Let us write

$$P_j(\mathcal{F}) = \mathrm{d}Q_{2j-1}(\mathcal{A}, \mathcal{F}) \tag{11.100}$$

where $Q_{2j-1}(\mathcal{A}, \mathcal{F}) \in \mathfrak{g} \otimes \Omega^{2j-1}(M)$. [*Warning:* This cannot be true globally. If $P_j = \mathrm{d}Q_{2j-1}$ globally on a manifold M without boundary, we would have

$$\int_M P_{m/2} = \int_M \mathrm{d}Q_{m-1} = \int_{\partial M} Q_{m-1} = 0$$

where $m = \dim M$.] The $2j - 1$ from $Q_{2j-1}(\mathcal{A}, \mathcal{F})$ is called the **Chern–Simons form** of $P_j(\mathcal{F})$. From the proof of theorem 11.2(b), we find that Q is given by the transgression of P_j,

$$Q_{2j-1}(\mathcal{A}, \mathcal{F}) = T P_j(\mathcal{A}, 0) = j \int_0^1 \tilde{P}_j(\mathcal{A}, \mathcal{F}_t, \ldots, \mathcal{F}_t) \, \mathrm{d}t \tag{11.101}$$

where $\tilde{P}_j$ is the polarization of P_j, $\mathcal{F} = \mathrm{d}A + A^2$ and we set $A' = \mathcal{F}' = 0$. Since Q_{2j-1} depends on $\mathcal{F}$ and A, we explicitly quote the A-dependence. Of course, A' can be put equal to zero only on a local chart over which the bundle is trivial.

Suppose M is an even-dimensional manifold ($\dim M = m = 2l$) such that $\partial M \neq \emptyset$. Then it follows from Stokes' theorem that

$$\int_M P_l(\mathcal{F}) = \int_M \mathrm{d}Q_{m-1}(A, \mathcal{F}) = \int_{\partial M} Q_{m-1}(A, \mathcal{F}). \tag{11.102}$$

The LHS takes its value in integers, and so does the RHS. Thus Q_{m-1} is a characteristic class in its own right and it describes the topology of the boundary ∂M.

11.5.2 The Chern–Simons form of the Chern character

As an example, let us work out the Chern–Simons form of a Chern character $\mathrm{ch}_j(\mathcal{F})$. The connection A_t which interpolates between 0 and A is

$$A_t = tA \tag{11.103}$$

the corresponding curvature being

$$\mathcal{F}_t = t\,\mathrm{d}A + t^2 A^2 = t\mathcal{F} + (t^2 - t)A^2. \tag{11.104}$$

We find from (11.21) that

$$Q_{2j-1}(A, \mathcal{F}) = \frac{1}{(j-1)!} \left(\frac{\mathrm{i}}{2\pi}\right)^j \int_0^1 \mathrm{d}t\ \mathrm{str}(A, \mathcal{F}_t^{j-1}). \tag{11.105}$$

For example,

$$Q_1(A, \mathcal{F}) = \frac{\mathrm{i}}{2\pi} \int_0^1 \mathrm{d}t\ \mathrm{tr}\,A = \frac{\mathrm{i}}{2\pi}\,\mathrm{tr}\,A \tag{11.106a}$$

$$Q_3(A, \mathcal{F}) = \left(\frac{\mathrm{i}}{2\pi}\right)^2 \int_0^1 \mathrm{d}t\ \mathrm{str}(A, t\mathrm{d}A + t^2 A^2)$$

$$= \frac{1}{2}\left(\frac{\mathrm{i}}{2\pi}\right)^2 \mathrm{tr}\left(A\mathrm{d}A + \frac{2}{3}A^3\right). \tag{11.106b}$$

$$Q_5(A, \mathcal{F}) = \frac{1}{2}\left(\frac{\mathrm{i}}{2\pi}\right)^3 \int_0^1 \mathrm{d}t\ \mathrm{str}[A, (t\mathrm{d}A + t^2 A^2)^2]$$

$$= \frac{1}{6}\left(\frac{\mathrm{i}}{2\pi}\right)^3 \mathrm{tr}\left[A(\mathrm{d}A)^2 + \frac{3}{2}A^3\mathrm{d}A + \frac{3}{5}A^5\right]. \tag{11.106c}$$

Exercise 11.4. Let $\mathcal{F}$ be the field strength of the SU(2) gauge theory. Write down the component expression of the identity $\mathrm{ch}_2(\mathcal{F}) = \mathrm{d}Q_3(A, \mathcal{F})$ to verify that (cf lemma 10.3)

$$\mathrm{tr}[\epsilon^{\kappa\lambda\mu\nu}\mathcal{F}_{\kappa\lambda}\mathcal{F}_{\mu\nu}] = \partial_\kappa[2\epsilon^{\kappa\lambda\mu\nu}\,\mathrm{tr}(A_\lambda\partial_\mu A_\nu + \tfrac{2}{3}A_\lambda A_\mu A_\nu)]. \tag{11.107}$$

11.5.3 Cartan's homotopy operator and applications

For later purposes, we define Cartan's homotopy formula following Zumino (1985) and Alvarez-Gaumé and Ginsparg (1985). Let

$$\mathcal{A}_t = \mathcal{A}_0 + t(\mathcal{A}_1 - \mathcal{A}_0) \qquad \mathcal{F}_t = d\mathcal{A}_t + \mathcal{A}_t^2 \tag{11.108}$$

as before. Define an operator l_t by

$$l_t \mathcal{A}_t = 0 \qquad l_t \mathcal{F}_t = \delta t (\mathcal{A}_1 - \mathcal{A}_0). \tag{11.109}$$

We require that l_t be an anti-derivative,

$$l_t (\eta_p \omega_q) = (l_t \eta_p) \omega_q + (-1)^p \eta_p (l_t \omega_q) \tag{11.110}$$

for $\eta_p \in \Omega^p(M)$ and $\omega_q \in \Omega^q(M)$. We verify that

$$(dl_t + l_t d)\mathcal{A}_t = l_t(\mathcal{F}_t - \mathcal{A}_t^2) = \delta t (\mathcal{A}_1 - \mathcal{A}_0) = \delta t \frac{\partial \mathcal{A}_t}{\partial t}$$

and

$$\begin{aligned}
(dl_t + l_t d)\mathcal{F}_t &= d[\delta t (\mathcal{A}_1 - \mathcal{A}_0)] + l_t[\mathcal{D}_t \mathcal{F}_t - \mathcal{A}_t \mathcal{F}_t + \mathcal{F}_t \mathcal{A}_t] \\
&= \delta t[d(\mathcal{A}_1 - \mathcal{A}_0) + \mathcal{A}_t(\mathcal{A}_1 - \mathcal{A}_0) + (\mathcal{A}_1 - \mathcal{A}_0)\mathcal{A}_t] \\
&= \delta t \mathcal{D}_t (\mathcal{A}_1 - \mathcal{A}_0) = \delta t \frac{\partial \mathcal{F}_t}{\partial t}
\end{aligned}$$

where we have used the Bianchi identity $\mathcal{D}_t \mathcal{F}_t = 0$. This shows that for any polynomial $S(\mathcal{A}, \mathcal{F})$ of $\mathcal{A}$ and $\mathcal{F}$, we obtain

$$(dl_t + l_t d)S(\mathcal{A}_t, \mathcal{F}_t) = \delta t \frac{\partial}{\partial t} S(\mathcal{A}_t, \mathcal{F}_t). \tag{11.111}$$

On the RHS, S should be a polynomial of $\mathcal{A}$ and $\mathcal{F}$ *only* and not of $d\mathcal{A}$ or $d\mathcal{F}$: if S does contain them, $d\mathcal{A}$ should be replaced by $\mathcal{F} - \mathcal{A}^2$ and $d\mathcal{F}$ by $\mathcal{D}\mathcal{F} - [\mathcal{A}, \mathcal{F}] = -[\mathcal{A}, \mathcal{F}]$. Integrating (11.111) over [0, 1], we obtain **Cartan's homotopy formula**

$$S(\mathcal{A}_1, \mathcal{F}_1) - S(\mathcal{A}_0, \mathcal{F}_0) = (dk_{01} + k_{01}d)S(\mathcal{A}_t, \mathcal{F}_t) \tag{11.112}$$

where the **homotopy operator** k_{01} is defined by

$$k_{01}S(\mathcal{A}_t, \mathcal{F}_t) \equiv \int_0^1 \delta t \, l_t S(\mathcal{A}_t, \mathcal{F}_t). \tag{11.113}$$

To operate k_{01} on $S(\mathcal{A}, \mathcal{F})$, we first replace $\mathcal{A}$ and $\mathcal{F}$ by $\mathcal{A}_t$ and $\mathcal{F}_t$, respectively, then operate l_t on $S(\mathcal{A}_t, \mathcal{F}_t)$ and integrate over t.

Example 11.9. Let us compute the Chern–Simons form of the Chern character using the homotopy formula. Let $S(\mathcal{A}, \mathcal{F}) = \mathrm{ch}_{j+1}(\mathcal{F})$ and $\mathcal{A}_1 = \mathcal{A}$, $\mathcal{A}_0 = 0$. Since $\mathrm{d\,ch}_{j+1}(\mathcal{F}) = 0$, we have

$$\mathrm{ch}_{j+1}(\mathcal{F}) = (\mathrm{d}k_{01} + k_{01}\mathrm{d})\mathrm{ch}_{j+1}(\mathcal{F}_t) = \mathrm{d}[k_{01}\mathrm{ch}_{j+1}(\mathcal{F}_t)].$$

Thus, $k_{01}\mathrm{ch}_{j+1}(\mathcal{F})$ is identified with the Chern–Simons form $Q_{2j+1}(\mathcal{A}, \mathcal{F})$. We find that

$$
\begin{aligned}
k_{01}\mathrm{ch}_{j+1}(\mathcal{F}_t) &= \frac{1}{(j+1)!}k_{01}\,\mathrm{tr}\left(\frac{i\mathcal{F}}{2\pi}\right)^{j+1} \\
&= \frac{1}{(j+1)!}\left(\frac{i}{2\pi}\right)^{j+1}\int_0^1 \delta t\, l_t\, \mathrm{tr}(\mathcal{F}_t^{j+1}) \\
&= \frac{1}{j!}\left(\frac{i}{2\pi}\right)^{j+1}\int_0^1 \delta t\, \mathrm{str}(\mathcal{A}, \mathcal{F}_t^j) \qquad (11.114)
\end{aligned}
$$

in agreement with (11.105).

Although a characteristic class is gauge invariant, the Chern–Simons form need not be so. As an application of Cartan's homotopy formula, we compute the change in $Q_{2j+1}(\mathcal{A}, \mathcal{F})$ under $\mathcal{A} \to \mathcal{A}^g = g^{-1}(\mathcal{A} + \mathrm{d})g$, $\mathcal{F} \to \mathcal{F}^g = g^{-1}\mathcal{F}g$. Consider the interpolating families $\mathcal{A}_t^g$ and $\mathcal{F}_t^g$ defined by

$$\mathcal{A}_t^g \equiv tg^{-1}\mathcal{A}g + g^{-1}\mathrm{d}g \qquad (11.115a)$$

$$\mathcal{F}_t^g \equiv \mathrm{d}\mathcal{A}_t^g + (\mathcal{A}_t^g)^2 = g^{-1}\mathcal{F}_t g \qquad (11.115b)$$

where $\mathcal{F}_t \equiv t\mathcal{F} + (t^2 - t)\mathcal{A}^2$. Note that $\mathcal{A}_0^g = g^{-1}\mathrm{d}g$, $\mathcal{A}_1^g = \mathcal{A}^g$, $\mathcal{F}_0^g = 0$ and $\mathcal{F}_1^g = \mathcal{F}^g$. Equation (11.112) yields

$$Q_{2j+1}(\mathcal{A}^g, \mathcal{F}^g) - Q_{2j+1}(g^{-1}\mathrm{d}g, 0) = (\mathrm{d}k_{01} + k_{01}\mathrm{d})Q_{2j+1}(\mathcal{A}_t^g, \mathcal{F}_t^g). \quad (11.116)$$

For example, let Q_{2j+1} be the Chern–Simons form of the Chern character $\mathrm{ch}_{j+1}(\mathcal{F})$. Since $\mathrm{d}Q_{2j+1}(\mathcal{A}_t^g, \mathcal{F}_t^g) = \mathrm{ch}_{j+1}(\mathcal{F}_t^g) = \mathrm{ch}_{j+1}(\mathcal{F}_t)$, we have

$$
\begin{aligned}
k_{01}\,\mathrm{d}Q_{2j+1}(\mathcal{A}_t^g, \mathcal{F}_t^g) &= k_{01}\mathrm{ch}_{j+1}(\mathcal{F}_t^g) \\
&= k_{01}\mathrm{ch}_{j+1}(\mathcal{F}_t) = Q_{2j+1}(\mathcal{A}, \mathcal{F}) \qquad (11.117)
\end{aligned}
$$

where the result of example 11.9 has been used to obtain the final equality. Collecting these results, we write (11.116) as

$$Q_{2j+1}(\mathcal{A}^g, \mathcal{F}^g) - Q_{2j+1}(\mathcal{A}, \mathcal{F}) = Q_{2j+1}(g^{-1}\mathrm{d}g, 0) + \mathrm{d}\alpha_{2j} \qquad (11.118)$$

where α_{2j} is a $2j$-form defined by

$$
\begin{aligned}
\alpha_{2j}(\mathcal{A}, \mathcal{F}, v) &\equiv k_{01}Q_{2j+1}(\mathcal{A}_t^g, \mathcal{F}_t^g) \\
&= k_{01}Q_{2j+1}(\mathcal{A}_t + v, \mathcal{F}_t) \qquad (11.119)
\end{aligned}
$$

where $v \equiv \mathrm{d}g \cdot g^{-1}$. [Note that $Q_{2j+1}(\mathcal{A}, \mathcal{F}) = Q_{2j+1}(g\mathcal{A}g^{-1}, g\mathcal{F}g^{-1})$.] The first term on the RHS of (11.118) is

$$
\begin{aligned}
Q_{2j+1}(g^{-1}\mathrm{d}g, 0) &= \frac{1}{j!}\left(\frac{\mathrm{i}}{2\pi}\right)^{j+1}\int_0^1 \delta t\, \mathrm{tr}[g^{-1}\mathrm{d}g\{(t^2-t)(g^{-1}\mathrm{d}g)^2\}^j] \\
&= \frac{1}{j!}\left(\frac{\mathrm{i}}{2\pi}\right)^{j+1}\mathrm{tr}[(g^{-1}\mathrm{d}g)^{2j+1}]\int_0^1 \delta t\,(t^2-t)^j \\
&= (-1)^j\frac{j!}{(2j+1)!}\left(\frac{\mathrm{i}}{2\pi}\right)^{j+1}\mathrm{tr}[(g^{-1}\mathrm{d}g)^{2j+1}] \quad (11.120)
\end{aligned}
$$

where we have noted that $\mathcal{F}_t = (t^2-t)(g^{-1}\mathrm{d}g)^2$ and

$$
\int_0^1 \delta t\,(t^2-t)^j = (-1)^j B(j+1, j+1) = (-1)^j\frac{(j!)^2}{(2j+1)!}
$$

B being the beta function. The $2j+1$ form $Q_{2j+1}(g\mathrm{d}g, 0)$ is closed and, hence, locally exact: $\mathrm{d}Q_{2j+1}(g^{-1}\mathrm{d}g, 0) = \mathrm{ch}_{j+1}(0) = 0$.

As for α_{2j} we have, for example,

$$
\begin{aligned}
\alpha_2 &= \frac{1}{2}\left(\frac{\mathrm{i}}{2\pi}\right)^2\int_0^1 l_t\, \mathrm{tr}[(\mathcal{A}_t + v)\mathcal{F}_t - \tfrac{1}{3}(\mathcal{A}_t + v)^3] \\
&= \frac{1}{2}\left(\frac{\mathrm{i}}{2\pi}\right)^2\int_0^1 \delta t\, \mathrm{tr}(-t\mathcal{A}^2 - v\mathcal{A}) \\
&= -\frac{1}{2}\left(\frac{\mathrm{i}}{2\pi}\right)^2\mathrm{tr}(v\mathcal{A}) \quad (11.121)
\end{aligned}
$$

where we have noted that

$$
\mathrm{tr}\,\mathcal{A}^2 = \mathrm{d}x^\mu \wedge \mathrm{d}x^\nu\, \mathrm{tr}(\mathcal{A}_\mu\mathcal{A}_\nu) = -\mathrm{d}x^\nu \wedge \mathrm{d}x^\mu\, \mathrm{tr}(\mathcal{A}_\nu\mathcal{A}_\mu) = 0.
$$

Example 11.10. In three-dimensional spacetime, a gauge theory may have a gauge-invariant mass term given by the Chern–Simons three-form (Jackiw and Templeton 1981, Deser *et al* 1982a, b). Since the Chern–Simons form changes by a locally exact form under a gauge transformation, the action remains invariant. We restrict ourselves to the U(1) gauge theory for simplicity. Consider the Lagrangian (we put $\mathcal{A} = \mathrm{i}A, \mathcal{F} = \mathrm{i}F$)

$$
\mathcal{L} = -\tfrac{1}{4}F_{\mu\nu}F^{\mu\nu} + \tfrac{1}{4}m\epsilon^{\lambda\mu\nu}F_{\lambda\mu}A_\nu \quad (11.122)
$$

where $F_{\mu\nu} = \partial_\mu A_\nu - \partial_\nu A_\mu$. Note that the second term is the Chern–Simons form of the second Chern character F^2 (modulo a constant factor) of the U(1) bundle. The field equation is

$$
\partial_\mu F^{\mu\nu} + m * F^\nu = 0 \quad (11.123)
$$

where

$$*F^\mu = \tfrac{1}{2}\epsilon^{\mu\kappa\lambda}F_{\kappa\lambda} \qquad F^{\mu\nu} = \epsilon^{\mu\nu\lambda}*F_\lambda.$$

The Bianchi identity

$$\partial_\mu * F^\mu = 0 \tag{11.124}$$

follows from (11.123) as a consequence of the skew symmetry of $F^{\mu\nu}$. It is easy to verify that the field equation is invariant under a gauge transformation,

$$A_\mu \to A_\mu + \partial_\mu\theta \tag{11.125}$$

while the Lagrangian changes by a total derivative,

$$\mathcal{L} \to -\tfrac{1}{4}F^{\mu\nu}F_{\mu\nu} + \tfrac{1}{4}m\epsilon^{\lambda\mu\nu}F_{\lambda\mu}(A_\nu + \partial_\nu\theta) = \mathcal{L} + \tfrac{1}{2}m\partial_\nu(*F^\nu\theta). \tag{11.126}$$

Equation (11.106b) shows that the last term on the RHS is identified with

$$Q_3(A^\theta, F^\theta) - Q_3(A, F) \sim (A + \mathrm{d}\theta)\,\mathrm{d}A - A\,\mathrm{d}A \sim \mathrm{d}(\theta\mathrm{d}A).$$

If we assume that F falls off at large spacetime distances, this term does not contribute to the action:

$$\int \mathrm{d}^3x\mathcal{L} \to \int \mathrm{d}^3x\mathcal{L} + \frac{m}{2} \int \mathrm{d}^3x\partial_\nu(*F^\nu\theta) = \int \mathrm{d}^3x\mathcal{L}. \tag{11.127}$$

Let us show that (11.122) describes a *massive* field. We first write (11.123) as

$$\epsilon^{\mu\nu\alpha}\partial_\mu * F_\alpha = -m * F^\nu.$$

Multiplying $\varepsilon_{\kappa\lambda\nu}$ on both sides, we have

$$\partial_\lambda * F_\kappa - \partial_\kappa * F_\lambda = -mF_{\kappa\lambda}.$$

Taking the ∂^λ-derivative and using (11.124), we find that

$$(\partial^\lambda\partial_\lambda + m^2) * F_\kappa = 0 \tag{11.128}$$

which shows that $*F_\kappa$ is a massive vector field of mass m.

11.6   Stiefel–Whitney classes

The last example of the characteristic classes is the Stiefel–Whitney class. In contrast to the rest of the characteristic classes, the Stiefel–Whitney class cannot be expressed in terms of the curvature of the bundle. The Stiefel–Whitney class is important in physics since it tells us whether a manifold admits a spin or not. Let us start with a brief review of a spin bundle.

11.6.1 Spin bundles

Let $TM \xrightarrow{\pi} M$ be a tangent bundle with $\dim M = m$. The bundle TM is assumed to have a fibre metric and the structure group G is taken to be $O(m)$. If, furthermore, M is orientable, G can be reduced down to $SO(m)$. Let LM be the frame bundle associated with TM. Let t_{ij} be the transition function of LM which satisfies the consistency condition (9.6)

$$t_{ij} t_{jk} t_{ki} = I \qquad t_{ii} = I.$$

A spin structure on M is defined by the transition function $\tilde{t}_{ij} \in \mathrm{SPIN}(m)$ such that

$$\varphi(\tilde{t}_{ij}) = t_{ij} \qquad \tilde{t}_{ij} \tilde{t}_{jk} \tilde{t}_{ki} = I \qquad \tilde{t}_{ii} = I \qquad (11.129)$$

where φ is the double covering $\mathrm{SPIN}(m) \to SO(m)$. The set of $\tilde{t}_{ij}$ defines a **spin bundle** $PS(M)$ over M and M is said to admit a **spin structure** (of course, M may admit many spin structures depending on the choice of $\tilde{t}_{ij}$).

It is interesting to note that not all manifolds admit spin structures. Non-admittance of spin structures is measured by the second Stiefel–Whitney class which takes values in the Čech cohomology group $H^2(M; \mathbb{Z}_2)$.

11.6.2 Čech cohomology groups

Let $\mathbb{Z}_2$ be the *multiplicative* group $\{-1, +1\}$. A **Čech r-cochain** is a function $f(i_0, i_1, \ldots, i_r) \in \mathbb{Z}_2$, defined on $U_{i_0} \cap U_{i_1} \cap \ldots \cap U_{i_r} \neq \emptyset$, which is totally symmetric under an arbitrary permutation P,

$$f(i_{P(0)}, \ldots, i_{P(r)}) = f(i_0, \ldots, i_r).$$

Let $C^r(M, \mathbb{Z}_2)$ be the multiplicative group of Čech r-cochains. We define the coboundary operator $\delta : C^r(M; \mathbb{Z}_2) \to C^{r+1}(M; \mathbb{Z}_2)$ by

$$(\delta f)(i_0, \ldots, i_{r+1}) = \prod_{j=0}^{r+1} f(i_0, \ldots, \hat{i}_j, \ldots, i_{r+1}) \qquad (11.130)$$

where the variable below the $\hat{}$ is omitted. For example,

$$(\delta f_0)(i_0, i_1) = f_0(i_1) f_0(i_0) \qquad f_0 \in C^0(M; \mathbb{Z}_2)$$

$$(\delta f_1)(i_0, i_1, i_2) = f_1(i_1, i_2) f_1(i_0, i_2) f_1(i_0, i_1) \qquad f_1 \in C^1(M; \mathbb{Z}_2).$$

Since we employ the multiplicative notation, the unit element of $C^r(M; \mathbb{Z}_2)$ is denoted by 1. We verify that δ is nilpotent:

$$(\delta^2 f)(i_0, \ldots, i_{r+2}) = \prod_{j,k=1}^{r+1} f(i_0, \ldots, \hat{i}_j, \ldots, \hat{i}_k, \ldots, i_{r+2}) = 1$$

since -1 always appears an even number of times in the middle expression (for example if $f(i_0, \ldots, \hat{i}_j, \ldots, \hat{i}_k, \ldots, i_{r+2}) = -1$, we have $f(i_0, \ldots, \hat{i}_k, \ldots, \hat{i}_j, \ldots, i_{r+2}) = -1$ from the symmetry of f). Thus, we have proved, for any Čech r-cochain f, that

$$\delta^2 f = 1. \tag{11.131}$$

The **cocycle group** $Z^r(M; \mathbb{Z}_2)$ and the **coboundary group** $B^r(M; \mathbb{Z}_2)$ are defined by

$$Z^r(M; \mathbb{Z}_2) = \{ f \in C^r(M; \mathbb{Z}_2) | \delta f = 1 \} \tag{11.132}$$

$$B^r(M; \mathbb{Z}_2) = \{ f \in C^r(M; \mathbb{Z}_2) | f = \delta f', \, f' \in C^{r-1}(M; \mathbb{Z}_2). \tag{11.133}$$

Now the rth **Čech cohomology group** $H^r(M; \mathbb{Z}_2)$ is defined by

$$H^r(M; \mathbb{Z}_2) = \ker \delta_r / \mathrm{im} \delta_{r-1} = Z^r(M; \mathbb{Z}_2) / B^r(M; \mathbb{Z}_2). \tag{11.134}$$

11.6.3 Stiefel–Whitney classes

The **Stiefel–Whitney class** w_r is a characteristic class which takes its values in $H^r(M; \mathbb{Z}_2)$. Let $TM \xrightarrow{\pi} M$ be a tangent bundle with a Riemannian metric. The structure group is $O(m)$, $m = \dim M$. We assume $\{U_i\}$ is a *simple* open covering of M, which means that the intersection of any number of charts is either empty or contractible. Let $\{e_{i\alpha}\}$ $(1 \le \alpha \le m)$ be a local orthonormal frame of TM over U_i. We have $e_{i\alpha} = t_{ij} e_{j\alpha}$ where $t_{ij} : U_i \cap U_j \to O(m)$ is the transition function. Define the Čech 1-cochain $f(i, j)$ by

$$f(i, j) \equiv \det(t_{ij}) = \pm 1. \tag{11.135}$$

This is, indeed, an element of $C^1(M; \mathbb{Z}_2)$ since $f(i, j) = f(j, i)$. From the cocycle condition $t_{ij} t_{jk} t_{ki} = I$, we verify that

$$\begin{aligned} \delta f(i, j, k) &= \det(t_{ij}) \det(t_{jk}) \det(t_{ki}) \\ &= \det(t_{ij} t_{jk} t_{ki}) = 1. \end{aligned} \tag{11.136}$$

Hence, $f \in Z^1(M, \mathbb{Z}_2)$ and it defines an element $[f]$ of $H^1(M; \mathbb{Z}_2)$. Now we show that this element is independent of the local frame chosen. Let $\{\bar{e}_{i\alpha}\}$ be another frame over U_i such that $\bar{e}_{i\alpha} = h_i e_{i\alpha}$, $h_i \in O(m)$. From $\bar{e}_{i\alpha} = \bar{t}_{ij} \bar{e}_{j\alpha}$, we find $\bar{t}_{ij} = h_i t_{ij} h_j^{-1}$. If we define the 0-cochain f_0 by $f_0(i) \equiv \det h_i$, we find that

$$\begin{aligned} \tilde{f}(i, j) &= \det(h_i t_{ij} h_j^{-1}) = \det(h_i) \det(h_j) \det(t_{ij}) \\ &= \delta f_0(i, j) f(i, j) \end{aligned}$$

where use has been made of the identity $\det h_j^{-1} = \det h_j$ for $h_j \in O(m)$. Thus, f changes by an exact amount and still defines the same cohomology class $[f]$.[4]

[4] Note that the multiplicative notation is being used.

This special element $w_1(M) \equiv [f] \in H^1(M; \mathbb{Z}_2)$ is called the **first Stiefel–Whitney class**.

Theorem 11.6. Let $TM \xrightarrow{\pi} M$ be a tangent bundle with fibre metric. M is orientable if and only if $w_1(M)$ is trivial.

Proof. If M is orientable, the structure group may be reduced to $SO(m)$ and $f(i, j) = \det(t_{ij}) = 1$, and hence $w_1(M) = 1$, the unit element of $\mathbb{Z}_2$. Conversely, if $w_1(M)$ is trivial, f is a coboundary; $f = \delta f_0$. Since $f_0(i) = \pm 1$, we can always choose $h_i \in O(m)$ such that $\det(h_i) = f_0(i)$ for each i. If we define the new frame $\bar{e}_{i\alpha} = h_i e_{i\alpha}$, we have transition functions $\tilde{t}_{ij}$ such that $\det(\tilde{t}_{ij}) = 1$ for any overlapping pair (i, j) and M is orientable. [Suppose $f(i, j) = \det t_{ij} = -1$ for some pair (i, j). Then we may take $f_0(i) = -1$ and $f_0(j) = +1$, hence $\det \tilde{t}_{ij} = -\det t_{ij} = +1$.] $\quad\square$

Theorem 11.6 shows that the first Stiefel–Whitney class is an obstruction to the orientability. Next we define the second Stiefel–Whitney class. Suppose M is an m-dimensional orientable manifold and TM is its tangent bundle. For the transition function $t_{ij} \in SO(m)$, we consider a 'lifting' $\tilde{t}_{ij} \in SPIN(m)$ such that

$$\varphi(\tilde{t}_{ij}) = t_{ij} \qquad \tilde{t}_{ji} = \tilde{t}_{ij}^{-1} \tag{11.137}$$

where $\varphi : SPIN(m) \to SO(m)$ is the $2 : 1$ homomorphism (note that we have an option $t_{ij} \leftrightarrow \tilde{t}_{ij}$ or $-\tilde{t}_{ij}$). This lifting always exists locally. Since

$$\varphi(\tilde{t}_{ij}\tilde{t}_{jk}\tilde{t}_{ki}) = t_{ij}t_{jk}t_{ki} = I$$

we have $\tilde{t}_{ij}\tilde{t}_{jk}\tilde{t}_{ki} \in \ker\varphi = \{\pm I\}$. For $\tilde{t}_{ij}$ to define a spin bundle over M, they must satisfy the cocycle condition,

$$\tilde{t}_{ij}\tilde{t}_{jk}\tilde{t}_{ki} = I. \tag{11.138}$$

Define the Čech 2-cochain $f : U_i \cap U_j \cap U_k \to \mathbb{Z}_2$ by

$$\tilde{t}_{ij}\tilde{t}_{jk}\tilde{t}_{ki} = f(i, j, k)I. \tag{11.139}$$

It is easy to see that f is symmetric and closed. Thus, f defines an element $w_2(M) \in H^2(M, \mathbb{Z}_2)$ called the **second Stiefel–Whitney class**. It can be shown that $w_2(M)$ is independent of the local frame chosen.

Exercise 11.5. Suppose we take another lift $-\tilde{t}_{ij}$ of t_{ij}. Show that f changes by an exact amount under this change. Accordingly, $[f]$ is independent of the lift. [*Hint:* Show that $f(i, j, k) \to f(i, j.k)\delta f_1(i, j, k)$ where $f_1(i, j)$ denotes the sign of $\pm\tilde{t}_{ij}$.]

Theorem 11.7. Let TM be the tangent bundle over an orientable manifold M. There exists a spin bundle over M if and only if $w_2(M)$ is trivial.

Proof. Suppose there exists a spin bundle over M. Then we define a set of transition functions $\tilde{t}_{ij}$ such that $\tilde{t}_{ij}\tilde{t}_{jk}\tilde{t}_{ki} = I$ over any overlapping charts U_i, U_j and U_k, hence $w_2(M)$ is trivial. Conversely, suppose $w_2(M)$ is trivial, namely

$$f(i, j, k) = \delta f_1(i, j, k) = f_1(j, k)f_1(i, k)f_1(k, i)$$

f_1 being a 1-cochain. We consider the 1-cochain $f_1(i, j)$ defined in exercise 11.5. If we choose new transition functions $\tilde{t}'_{ij} \equiv \tilde{t}_{ij}f_1(i, j)$, we have

$$\tilde{t}'_{ij}\tilde{t}'_{jk}\tilde{t}'_{ki} = [\delta f_1(i, j, k)]^2 = I$$

and, hence, $\{\tilde{t}'_{ij}\}$ defines a spin bundle over M. $\square$

We outline some useful results:

(a)

$$w_1(\mathbb{C}P^m) = 1 \qquad w_2(\mathbb{C}P^m) = \begin{cases} 1 & m \text{ odd} \\ x & m \text{ even} \end{cases} \qquad (11.140)$$

x being the generator of $H^2(\mathbb{C}P^m; \mathbb{Z}_2)$.

(b)

$$w_1(S^m) = w_2(S^m) = 1 \qquad (11.141)$$

(c)

$$w_1(\Sigma_g) = w_2(\Sigma_g) = 1 \qquad (11.142)$$

Σ_g being the Riemann surface of genus g.

12

INDEX THEOREMS

In physics, we often consider a differential operator defined on a manifold M. Typical examples will be the Laplacian, the d'Alembertian and the Dirac operator. From the mathematical point of view, these operators are regarded as maps of sections

$$D : \Gamma(M, E) \to \Gamma(M, F)$$

where E and F are vector bundles over M. For example, the Dirac operator is a map $F(M, E) \to F(M, E)$, E being a spin bundle over M. If inner products are defined on E and F, it is possible to define the adjoint of D,

$$D^\dagger : \Gamma(M, F) \to \Gamma(M, E).$$

Since it is a differential operator, D carries analytic information on the spectrum and its degeneracy. In what follows, we are interested in the zero eigenvectors of D and $D^\dagger$,

$$\ker D \equiv \{s \in \Gamma(M, E) | Ds = 0\}$$
$$\ker D^\dagger \equiv \{s \in \Gamma(M, F) | D^\dagger s = 0\}.$$

The **analytical index** is defined by

$$\text{ind } D = \dim \ker D - \dim \ker D^\dagger.$$

Surprisingly, this analytic quantity is a topological invariant expressed in terms of an integral of an appropriate characteristic class over M, which provides purely topological information on M. This interplay between analysis and topology is the main ingredient of the index theorem.

Our exposition follows Eguchi *et al* (1980), Gilkey (1984), Shanahan (1978), Kulkarni (1975) and Booss and Bleecker (1985). The reader should consult these references for details. Alvarez (1985) contains a brief summary of this subject along with applications to anomalies and strings.

12.1 Elliptic operators and Fredholm operators

In the following, we will be concerned with differential operators defined on vector bundles over a compact manifold M without a boundary. We exclusively deal with a nice class of differential operators called the Fredholm operators.

12.1.1   Elliptic operators

Let E and F be complex vector bundles over a manifold M. A differential operator D is a linear map

$$D : \Gamma(M, E) \to \Gamma(M, F). \qquad (12.1)$$

Take a chart U of M over which E and F are trivial. We denote the local coordinates of U as x^μ. We introduce the following multi-index notation,

$$M \equiv (\mu_1, \mu_2, \dots, \mu_m) \qquad \mu_j \in \mathbb{Z}, \mu_j \geq 0$$
$$|M| \equiv \mu_1 + \mu_2 + \cdots + \mu_m$$
$$D_M = \frac{\partial^{|M|}}{\partial x^M} \equiv \frac{\partial^{\mu_1 + \cdots + \mu_m}}{\partial (x^1)^{\mu_1} \dots \partial (x^m)^{\mu_m}}.$$

If $\dim E = k$ and $\dim F = k'$, the most general form of D is

$$[Ds(x)]^\alpha = \sum_{\substack{|M| \leq N \\ 1 \leq a \leq k}} A^{M\alpha}{}_a(x) D_M s^a(x) \qquad 1 \leq \alpha \leq k' \qquad (12.2)$$

where $s(x)$ is a section of E. Note that x denotes a point whose coordinates are x^μ. This slight abuse simplifies the notation. $A^M \equiv (A^M)^\alpha{}_a$ is a $k \times k'$ matrix which may depend on the position x. The positive integer N in (12.2) is called the **order** of D. We are interested in the case in which $N = 1$ (the Dirac operator) and $N = 2$ (the Laplacian). For example, if F is a spin bundle over M, the Dirac operator $D \equiv i\gamma^\mu \partial_\mu + m : \Gamma(M, E) \to \Gamma(M, E)$ acts on a section $\psi(x)$ of E as

$$[D\psi(x)]^\alpha = i(\gamma^\mu)^\alpha{}_\beta \partial_\mu \psi^\beta(x) + m\psi^\alpha(x).$$

The **symbol** of D is a $k \times k'$ matrix

$$\sigma(D, \xi) \equiv \sum_{|M|=N} A^{M\alpha}{}_a(x)\xi_M \qquad (12.3)$$

where ξ is a real m-tuple $\xi = (\xi_1, \dots, \xi_m)$. The symbol is also defined independently of the coordinates as follows. Let $E \xrightarrow{\pi} M$ be a vector bundle and let $p \in M, \xi \in T_p^* M$ and $s \in \pi_E^{-1}(p)$. Take a section $\tilde{s} \in \Gamma(M, E)$ such that $\tilde{s}(p) = s$ and a function $f \in \mathcal{F}(M)$ such that $f(p) = 0$ and $df(p) = \xi \in T_p^* M$. Then the symbol may be defined by

$$\sigma(D, \xi)s = \frac{1}{N!} D(f^N \tilde{s})|_p. \qquad (12.4)$$

The factor f^N automatically picks up the Nth-order term due to the condition $f(p) = 0$. Equation (12.4) yields the same symbol as (12.3).

If the matrix $\sigma(D, \xi)$ is invertible for each $x \in M$ and each $\xi \in \mathbb{R}^m - \{0\}$, the operator D is said to be **elliptic**. Clearly this definition makes sense only when $k = k'$. It should be noted that the symbol for a composite operator $D = D_1 \circ D_2$ is a composite of the symbols, namely $\sigma(D, \xi) = \sigma(D_1, \xi)\sigma(D_2, \xi)$. This shows that composites of elliptic operators are also elliptic. In general, powers and roots of elliptic operators are elliptic.

Example 12.1. Let x^μ be the natural coordinates in $\mathbb{R}^m$. If E and F are real line bundles over $\mathbb{R}^m$, the Laplacian $\Delta : \Gamma(\mathbb{R}^m, E) \to \Gamma(\mathbb{R}^m, F)$ is defined by

$$\Delta \equiv \frac{\partial^2}{\partial(x^1)^2} + \cdots + \frac{\partial^2}{\partial(x^m)^2}. \tag{12.5}$$

According to (12.3), the symbol is

$$\sigma(\Delta, \xi) = \sum_\mu (\xi_\mu)^2.$$

This is in agreement with the result obtained from (12.4),

$$\sigma(\Delta, \xi)s = \frac{1}{2}\Delta(f^2\tilde{s})|_p = \frac{1}{2}\sum \frac{\partial^2}{\partial(x^\mu)^2}(f^2\tilde{s})|_p$$

$$= \frac{1}{2}\left(f^2\Delta\tilde{s} + 2f\Delta f\tilde{s} + 2f\sum \frac{\partial f}{\partial x^\mu}\frac{\partial \tilde{s}}{\partial x^\mu} + 2\sum \frac{\partial f}{\partial x^\mu}\frac{\partial f}{\partial x^\mu}\tilde{s}\right)\Bigg|_p$$

$$= \sum (\xi_\mu)^2 s.$$

This symbol is clearly invertible for $\xi \neq 0$, and hence Δ is elliptic.
However, the d'Alembertian

$$\Box \equiv \frac{\partial^2}{\partial(x^1)^2} + \cdots + \frac{\partial^2}{\partial(x^{m-1})^2} - \frac{\partial^2}{\partial(x^m)^2} \tag{12.6}$$

is not elliptic since the symbol

$$\sigma(\Box, \xi) = (\xi^1)^2 + \cdots + (\xi^{m-1})^2 - (\xi^m)^2$$

vanishes everywhere on the light cone,

$$(\xi^m)^2 = (\xi^1)^2 + \cdots + (\xi^{m-1})^2.$$

Exercise 12.1. Let $M = \mathbb{R}^2$ and consider a differential operator D of order two. The symbol of D is of the form

$$\sigma(D, \xi) = A_{11}\xi^1\xi^1 + 2A_{12}\xi^1\xi^2 + A_{22}\xi^2\xi^2.$$

Show that D is elliptic if and only if $\sigma(D, \xi) = 1$ is an ellipse in ξ-space.

12.1.2 Fredholm operators

Let $D : \Gamma(M, E) \to \Gamma(M, F)$ be an elliptic operator. The **kernel** of D is the set of null eigenvectors

$$\ker D \equiv \{s \in \Gamma(M, E) | Ds = 0\}. \tag{12.7}$$

Suppose E and F are endowed with fibre metrics, which will be denoted $\langle\ ,\ \rangle_E$ and $\langle\ ,\ \rangle_F$, respectively. The **adjoint** $D^\dagger : \Gamma(M, F) \to \Gamma(M, E)$ of D is defined by

$$\langle s', Ds \rangle_F \equiv \langle D^\dagger s', s \rangle_E \tag{12.8}$$

where $s \in \Gamma(M, E)$ and $s' \in \Gamma(M, F)$. We define the **cokernel** of D by

$$\mathrm{coker}\, D \equiv \Gamma(M, F)/\mathrm{im} D. \tag{12.9}$$

Among elliptic operators we are interested in a class of operators whose kernels and cokernels are finite dimensional. An elliptic operator D which satisfies this condition is called a **Fredholm operator**. The **analytical index**

$$\mathrm{ind}\, D \equiv \dim \ker D - \dim \mathrm{coker}\, D \tag{12.10}$$

is well defined for a Fredholm operator. Henceforth, we will be concerned only with Fredholm operators. It is known from the general theory of operators that elliptic operators on a *compact* manifold are Fredholm operators. Theorem 12.1 shows that $\mathrm{ind}\, D$ is also expressed as

$$\mathrm{ind}\, D = \dim \ker D - \dim \ker D^\dagger. \tag{12.11}$$

Theorem 12.1. Let $D : \Gamma(M, E) \to \Gamma(M, F)$ be a Fredholm operator. Then

$$\mathrm{coker}\, D \cong \ker D^\dagger \equiv \{s \in \Gamma(M, F) | D^\dagger s = 0\}. \tag{12.12}$$

Proof. Let $[s] \in \mathrm{coker}\, D$ be given by

$$[s] = \{s' \in \Gamma(M, F) | s' = s + Du, u \in \Gamma(M, E)\}.$$

We show that there is a surjection $\ker D^\dagger \to \mathrm{coker}\, D$, namely any $[s] \in \mathrm{coker}\, D$ has a representative $s_0 \in \ker D^\dagger$. Define s_0 by

$$s_0 \equiv s - D \frac{1}{D^\dagger D} D^\dagger s. \tag{12.13}$$

We find $s_0 \in \ker D^\dagger$ since $D^\dagger s_0 = D^\dagger s - D^\dagger D (D^\dagger D)^{-1} D^\dagger s = D^\dagger s - D^\dagger s = 0$. Next, let $s_0, s_0' \in \ker D^\dagger$ and $s_0 \neq s_0'$. We show that $[s_0] \neq [s_0']$ in $\Gamma(M, F)/\mathrm{im}\, D$. If $[s_0] = [s_0']$, there is an element $u \in \Gamma(M, E)$ such that $s_0 - s_0' = Du$. Then $0 = \langle u, D^\dagger(s_0 - s_0') \rangle_E = \langle u, D^\dagger Du \rangle_E = \langle Du, Du \rangle_F \geq 0$, hence $Du = 0$, which contradicts our assumption $s_0 \neq s_0'$. Thus, the map $s_0 \mapsto [s]$ is a bijection and we have established that $\mathrm{coker}\, D \cong \ker D^\dagger$. $\square$

12.1.3 Elliptic complexes

Consider a sequence of Fredholm operators,

$$\cdots \to \Gamma(M, E_{i-1}) \xrightarrow{D_{i-1}} \Gamma(M, E_i) \xrightarrow{D_i} \Gamma(M, E_{i+1}) \xrightarrow{D_{i+1}} \cdots \qquad (12.14)$$

where $\{E_i\}$ is a sequence of vector bundles over a compact manifold M. The sequence (E_i, D_i) is called an **elliptic complex** if D_i is *nilpotent* (that is $D_i \circ D_{i-1} = 0$) for any i. The reader may refer to $\Gamma(M, E_i) = \Omega_i(M)$ and $D_i = \mathrm{d}$ (exterior derivative) for example. The adjoint of $D_i : \Gamma(M, E_i) \to \Gamma(M, E_{i+1})$ is denoted by

$$D_i^\dagger : \Gamma(M, E_{i+1}) \to \Gamma(M, E_i).$$

The **Laplacian** $\Delta_i : \Gamma(M, E_i) \to \Gamma(M, E_i)$ is

$$\Delta_i \equiv D_{i-1} D^\dagger{}_{i-1} + D_i^\dagger D_i. \qquad (12.15)$$

The Hodge decomposition also applies to the present case,

$$s_i = D_{i-1} s_{i-1} + D^\dagger{}_i s_{i+1} + h_i \qquad (12.16)$$

where $s_{i\pm1} \in \Gamma(M, E_{i\pm1})$ and h_i is in the kernel of Δ_i, $\Delta_i h_i = 0$.

Analogously to the de Rham cohomology groups, we define

$$H^i(E, D) \equiv \ker D_i / \mathrm{im} D_{i-1}. \qquad (12.17)$$

As in the case of the de Rham theory, it can be shown that $H^i(E, D)$ is isomorphic to the kernel of Δ_i. Accordingly, we have

$$\dim H^i(E, D) = \dim \mathrm{Harm}^i(E, D) \qquad (12.18)$$

where $\mathrm{Harm}^i(E, D)$ is a vector space spanned by $\{h_i\}$. The **index** of this elliptic complex is defined by

$$\mathrm{ind}\, D \equiv \sum_{i=0}^{m}(-1)^i \dim H^i(E, D) = \sum_{i=0}^{m}(-1)^i \dim \ker \Delta_i. \qquad (12.19)$$

The index thus defined generalizes the Euler characteristic, see example 12.2.

How is this related to (12.10)? Consider the complex $\Gamma(M, E) \xrightarrow{D} \Gamma(M, F)$. We may formally add zero on both sides,

$$0 \xhookrightarrow{i} \Gamma(M, E) \xrightarrow{D} \Gamma(M, F) \xrightarrow{\varphi} 0 \qquad (12.20)$$

where i is the inclusion. The index according to (12.19) is

$$\dim \ker D - \{\dim \Gamma(M, F) - \dim \mathrm{im} D\} = \dim \ker D - \dim \mathrm{coker} D$$

where we have noted that $\dim \operatorname{im} i = 0$, $\ker \varphi = \Gamma(M, F)$ and $\operatorname{coker} D = \ker \varphi / \operatorname{im} D$. Thus, (12.19) yields the same index as (12.10).

It is often convenient to work with a two-term elliptic complex which has the same index as the original elliptic complex (E, D). This *rolling up* is carried out by defining

$$E_+ \equiv \bigoplus_r E_{2r}, \quad E_- \equiv \bigoplus_r E_{2r+1} \tag{12.21}$$

which are called the **even bundle** and the **odd bundle**, respectively. Correspondingly we consider the operators

$$A \equiv \bigoplus_r (D_{2r} + D^{\dagger}{}_{2r-1}), A^{\dagger} \equiv \bigoplus_r (D_{2r+1} + D^{\dagger}{}_{2r}). \tag{12.22}$$

We readily verify that $A : \Gamma(M, E_+) \to \Gamma(M, E_-)$ and $A^{\dagger} : \Gamma(M, E_-) \to \Gamma(M, E_+)$. From A and $A^{\dagger}$, we construct the two Laplacians

$$\Delta_+ \equiv A^{\dagger} A = \bigoplus_{r,s} (D_{2r+1} + D^{\dagger}{}_{2r})(D_{2s} + D^{\dagger}{}_{2s-1})$$

$$= \bigoplus_r (D_{2r-1} D^{\dagger}{}_{2r-1} + D^{\dagger}{}_{2r} D_{2r}) = \bigoplus_r \Delta_{2r} \tag{12.23a}$$

$$\Delta_- \equiv A A^{\dagger} = \bigoplus_r \Delta_{2r+1}. \tag{12.23b}$$

Then we have

$$\operatorname{ind}(E_{\pm}, A) = \dim \ker \Delta_+ - \dim \ker \Delta_-$$

$$= \sum (-1)^r \dim \ker \Delta_r = \operatorname{ind}(E, D). \tag{12.24}$$

Example 12.2. Let us consider the de Rham complex $\Omega(M)$ over a compact manifold M without a boundary,

$$0 \xrightarrow{i} \Omega^0(M) \xrightarrow{d} \Omega^1(M) \xrightarrow{d} \cdots \xrightarrow{d} \Omega^m(M) \xrightarrow{d} 0 \tag{12.25}$$

where $m = \dim M$ and d stands for $d_r : \Omega^r(M) \to \Omega^{r+1}(M)$. $H^r(E, D)$ defined by (12.25) agrees with the de Rham cohomology group $H_r(M, \mathbb{R})$. The index is identified with the Euler characteristic,

$$\operatorname{ind}(\Omega^*(M), d) = \sum_{r=0}^m (-1)^r \dim H^r(M; \mathbb{R}) = \chi(M). \tag{12.26}$$

We found in chapter 7 that $b^r \equiv \dim H^r(M.\mathbb{R})$ agrees with the number of linearly independent harmonic r-forms: $\dim H^r(M, \mathbb{R}) = \dim \operatorname{Harm}^r(M) = \dim \ker \Delta_r$, where Δ_r is the Laplacian

$$\Delta_r = (d + d^{\dagger})^2 = d_{r-1} d^{\dagger}{}_{r-1} + d^{\dagger}{}_r d_r \tag{12.27}$$

$d_r^\dagger : \Omega^{r+1}(M) \to \Omega^r(M)$ being the adjoint of d_r. Now we find that

$$\chi(M) = \sum_{r=0}^{m}(-1)^r \dim \ker \Delta_r. \tag{12.28}$$

This relation is very interesting since the LHS is a purely topological quantity which can be computed by triangulating M, for example, while the RHS is given by the solution of an analytic equation $\Delta_r u = 0$. We noted in example 11.6 that $\chi(M)$ is given by integrating the Euler class over M: $\chi(M) = \int_M e(TM)$. Now (12.28) reads

$$\sum_{r=1}^{m}(-1)^r \dim \ker \Delta_r = \int_M e(TM). \tag{12.29}$$

This is a typical form of the index theorem. The RHS is an analytic index while the LHS is a topological index given by the integral of certain characteristic classes. In section 12.3, we derive (12.29) from the Atiyah–Singer index theorem.

The two-term complex is given by

$$\Omega^+(M) \equiv \bigoplus_r \Omega^{2r}(M) \qquad \Omega^-(M) \equiv \bigoplus_r \Omega^{2r+1}(M). \tag{12.30}$$

The corresponding operators are

$$A \equiv \bigoplus_r (d_{2r} + d^\dagger_{2r-1}) \qquad A^\dagger \equiv \bigoplus_r (d_{2r-1} + d^\dagger_{2r}). \tag{12.31}$$

It is left as an exercise to the reader to show that

$$\mathrm{ind}(\Omega^\pm(M), A) = \dim \ker A_+ - \dim \ker A_- = \chi(M). \tag{12.32}$$

12.2 The Atiyah–Singer index theorem

12.2.1 Statement of the theorem

Theorem 12.2. (**Atiyah–Singer index theorem**) Let (E, D) be an elliptic complex over an m-dimensional compact manifold M without a boundary. The index of this complex is given by

$$\mathrm{ind}(E, D) = (-1)^{m(m+1)/2} \int_M \mathrm{ch}\left(\bigoplus_r (-1)^r E_r\right) \frac{\mathrm{Td}(TM^{\mathbb{C}})}{e(TM)}\bigg|_{\mathrm{vol}}. \tag{12.33}$$

In the integrand of the RHS, only m-forms are picked up, so that the integration makes sense. [*Remarks*: The division by $e(TM)$ can really be carried out at the formal level. If m is an odd integer, the index vanishes identically, see below. Original references are Atiyah and Singer (1968a, b), Atiyah and Segal (1968).]

The proof of theorem 12.2 is found in Shanahan (1978), Palais (1965) and Gilkey (1984). The proof found there is based on either K-theory or the heat kernel formalism. In section 13.2, we give a proof of the simplest version of the Atiyah–Singer (AS) index theorem for a spin complex. Recently physicists have found another proof of the theorem making use of supersymmetry. This proof is outlined in sections 12.9 and 12.10. Interested readers should consult Alvarez-Gaumé (1983) and Friedan and Windey (1984, 1985) for further details.

The following corollary is a direct consequence of theorem 12.2.

Corollary 12.1. Let $\Gamma(M, E) \overset{D}{\to} \Gamma(M, F)$ be a two-term elliptic complex. The index of D is given by

$$\text{ind } D = \dim \ker D - \dim \ker D^\dagger$$

$$= (-1)^{m(m+1)/2} \int_M (\text{ch}E - \text{ch}F) \frac{\text{Td}(TM^{\mathbb{C}})}{e(TM)}\bigg|_{\text{vol}}. \qquad (12.34)$$

12.3 The de Rham complex

Let M be an m-dimensional compact orientable manifold with no boundary. By now we are familiar with the de Rham complex,

$$\cdots \overset{d}{\to} \Omega^{r-1}(M)^{\mathbb{C}} \overset{d}{\to} \Omega^r(M)^{\mathbb{C}} \overset{d}{\to} \Omega^{r+1}(M)^{\mathbb{C}} \overset{d}{\to} \cdots \qquad (12.35)$$

where $\Omega^r(M)^{\mathbb{C}} = \Gamma(M, \wedge^r T^*M^{\mathbb{C}})$. We complexified the forms so that we may apply the AS index theorem. The exterior derivative satisfies $d^2 = 0$. To show that (12.35) is an elliptic complex, we have to show that d is elliptic. To find the symbol for d, we note that

$$\sigma(d, \xi)\omega = d(f\tilde{s})|_p = df \wedge \tilde{s} + f d\tilde{s}|_p = \xi \wedge \omega$$

where $p \in M, \omega \in \Omega^r_p(M)^{\mathbb{C}}, f(p) = 0, df(p) = \xi, \tilde{s} \in \Omega^r(M)^{\mathbb{C}}$ and $\tilde{s}(p) = \omega$; see (12.4). We find

$$\sigma(d, \xi) = \xi \wedge. \qquad (12.36)$$

This defines a map $\Omega^r(M)^{\mathbb{C}} \to \Omega^{r+1}(M)^{\mathbb{C}}$ and is non-singular if $\xi \neq 0$. Thus, we have proved that d $: \Omega^r(M)^{\mathbb{C}} \to \Omega^{r+1}(M)^{\mathbb{C}}$ is elliptic and, hence, (12.35) is an elliptic complex. Note, however, that the operator d $: \Omega^k(M) \to \Omega^{k+1}(M)$ is not Fredholm since $\ker d$ is infinite dimensional. To apply the index theorem to this complex, we have to consider the de Rham cohomology group $H^r(M)$ instead. The operator d is certainly Fredholm on this space.

Let us find the index theorem for this complex. We note that $\dim_{\mathbb{C}} H^r(M; \mathbb{C}) = \dim_{\mathbb{R}} H^r(M; \mathbb{R})$. Hence, the analytical index is

$$\text{ind } d = \sum_{r=0}^m (-1)^r \dim_{\mathbb{C}} H^r(M; \mathbb{C})$$

$$= \sum (-1)^r \dim_{\mathbb{R}} H^r(M; \mathbb{R}) = \chi(M) \qquad (12.37)$$

where $\chi(M)$ is the Euler characteristic of M. Suppose M is even dimensional, $m = 2l$. The RHS of (12.33) gives the topological index

$$(-1)^{l(2l+1)} \int_M \mathrm{ch}\left(\bigoplus_{r=0}^{m} (-1)^r \wedge^r T^*M^{\mathbb{C}} \right) \frac{\mathrm{Td}(TM^{\mathbb{C}})}{e(TM)} \bigg|_{\mathrm{vol}}. \qquad (12.38)$$

The splitting principle yields

$$\mathrm{ch}\left(\bigoplus_{r=0}^{m} (-1)^r \wedge^r T^*M^{\mathbb{C}} \right)$$

$$= 1 - \mathrm{ch}(T^*M^{\mathbb{C}}) + \mathrm{ch}(\wedge^2 T^*M^{\mathbb{C}}) + \cdots + (-1)^m \mathrm{ch}(\wedge^m T^*M^{\mathbb{C}})$$

$$= 1 - \sum_{i=1}^{m} e^{-x_i}(TM^{\mathbb{C}}) + \sum_{i<j} e^{-x_i} e^{-x_j}(TM^{\mathbb{C}}) + \cdots$$

$$+ (-1)^m e^{-x_1} e^{-x_2} \ldots e^{-x_m}(TM^{\mathbb{C}})$$

$$= \prod_{i=1}^{m} (1 - e^{-x_i})(TM^{\mathbb{C}})$$

where we have noted that $x_i(T^*M^{\mathbb{C}}) = -x_i(TM^{\mathbb{C}})$. [Let L be a complex line bundle and L^* be its dual bundle. $L \otimes L^*$ is a bundle whose section is a map $\mathbb{C} \to \mathbb{C}$ at each fibre of L. $L \otimes L^*$ has a global section which vanishes nowhere (the identity map, for example) from which we can show $L \otimes L^*$ is a trivial bundle. We have $c_1(L \otimes L^*) = c_1(L) + c_1(L^*) = 0$, hence $x(L^*) = -x(L)$. The splitting principle yields $x_i(T^*M^{\mathbb{C}}) = -x_i(TM^{\mathbb{C}})$.] We also have

$$\mathrm{Td}(TM^{\mathbb{C}}) = \prod_{i=1}^{m} \frac{x_i}{1 - e^{-x_i}}(TM^{\mathbb{C}})$$

$$e(TM) = \prod_{i=1}^{l} x_i(TM^{\mathbb{C}}).$$

Substituting these in (12.38), we have

$$\mathrm{ind}\, \mathrm{d} = \int_M (-1)^{l(2l+1)}(-1)^l \left(\prod_{i=1}^{l} x_i(TM^{\mathbb{C}}) \right) = \int_M e(TM). \qquad (12.39)$$

If m is odd, it can be shown that (Shanahan (1978), p22)

$$\mathrm{ind}\, \mathrm{d} = 0 \qquad (12.40)$$

which is in harmony with the fact that $e(TM) = 0$ if $\dim M$ is odd. In any case, the index theorem for the de Rham complex is

$$\chi(M) = \int_M e(TM). \qquad (12.41)$$

Example 12.3. Let M be a two-dimensional orientable manifold without boundary. Equation (12.41) reads

$$\chi(M) = \frac{1}{4\pi} \int_M \epsilon^{\alpha\beta} \mathcal{R}_{\alpha\beta} = \frac{1}{2\pi} \int_M \mathcal{R}_{12} \qquad (12.42a)$$

which is the celebrated Gauss–Bonnet theorem. For dim $M = 4$, it reads as

$$\chi(M) = \frac{1}{32\pi^2} \int_M \epsilon^{\alpha\beta\gamma\delta} \mathcal{R}_{\alpha\beta} \wedge \mathcal{R}_{\gamma\delta}. \qquad (12.42b)$$

12.4 The Dolbeault complex

We recall some elementary facts about complex manifolds (see chapter 8 for details). Let M be a compact complex manifold of complex dimension m without a boundary. Let $z^\mu = x^\mu + \mathrm{i}y^\mu$ be the local coordinates and $\bar{z}^\mu = x^\mu - \mathrm{i}y^\mu$ their complex conjugates. TM^+ denotes the tangent bundle spanned by $\{\partial/\partial z^\mu\}$ and $TM^- = \overline{TM^+}$ the complex conjugate bundle spanned by $\{\partial/\partial\bar{z}^\mu\}$. The dual of TM^+ is denoted by T^*M^+ and spanned by $\{\mathrm{d}z^\mu\}$ while that of TM^- is $T^*M^- = \overline{T^*M^+}$ spanned by $\{\mathrm{d}\bar{z}^\mu\}$. The space $\Omega^r(M)^\mathbb{C}$ of complexified r-forms is decomposed as

$$\Omega^r(M)^\mathbb{C} = \bigoplus_{p+q=r} \Omega^{p,q}(M)$$

where $\Omega^{p,q}(M)$ is the space of the (p, q)-forms, which is spanned by a basis of the form

$$\mathrm{d}z^{\mu_1} \wedge \ldots \wedge \mathrm{d}z^{\mu_p} \wedge \mathrm{d}\bar{z}^{\nu_1} \wedge \ldots \wedge \mathrm{d}\bar{z}^{\nu_q}.$$

The exterior derivative is decomposed as $\mathrm{d} \equiv \partial + \bar{\partial}$ where

$$\partial = \mathrm{d}z^\mu \wedge \partial/\partial z^\mu \qquad \bar{\partial} = \mathrm{d}\bar{z}^\mu \wedge \partial/\partial\bar{z}^\mu.$$

They satisfy $\partial\bar{\partial} + \bar{\partial}\partial = \partial^2 = \bar{\partial}^2 = 0$. We have the sequences

$$\cdots \xrightarrow{\bar{\partial}} \Omega^{p,q}(M) \xrightarrow{\bar{\partial}} \Omega^{p,q+1}(M) \xrightarrow{\bar{\partial}} \cdots \qquad (12.43a)$$

$$\cdots \xrightarrow{\partial} \Omega^{p,q}(M) \xrightarrow{\partial} \Omega^{p+1,q}(M) \xrightarrow{\partial} \cdots. \qquad (12.43b)$$

We are interested in the first sequence with $p = 0$,

$$\cdots \xrightarrow{\bar{\partial}} \Omega^{0,q}(M) \xrightarrow{\bar{\partial}} \Omega^{0,q+1}(M) \xrightarrow{\bar{\partial}} \cdots. \qquad (12.44)$$

This sequence is called the **Dolbeault complex**.

To show that (12.44) is an elliptic complex, we compute the symbol for $\bar{\partial}$. Let $\xi = \xi^{0,1} + \xi^{1,0}$ be a *real* one-form at $p \in M$, where $\xi^{0,1} \in \Omega_p^{0,1}(M)$ and

$$\xi^{1,0} = \overline{\xi^{0,1}} \in \Omega_p^{1,0}(M).$$

Take an anti-holomorphic r-form $\omega \in \Omega^{0,r}(M)$. We find

$$\sigma(\bar{\partial}, \xi)\omega = \bar{\partial}(f\tilde{s}) = \bar{\partial}f \wedge \tilde{s} + f\bar{\partial}\tilde{s}|_p = \xi^{0,1} \wedge \omega$$

where $f(p) = 0$, $\bar{\partial}f(p) = \xi^{0,1}$, $\tilde{s} \in \Omega^{0,r}(M)$ and $\tilde{s}(p) = \omega$. We have

$$\sigma(\bar{\partial}, \xi) = \xi^{0,1} \wedge . \tag{12.45}$$

From a similar argument to that given in the previous section, it follows that the symbol (12.45) is elliptic. Thus, the Dolbeault complex (12.44) is an elliptic complex.

The AS index theorem takes the form

$$\text{ind}\,\bar{\partial} = \int_M \text{ch}\left(\sum_r (-1)^r \wedge^r T^*M^-\right) \frac{\text{Td}(TM^{\mathbb{C}})}{e(TM)}\bigg|_{\text{vol}}. \tag{12.46}$$

The LHS is computed as follows. We first note that

$$\ker \bar{\partial}_r / \text{im}\bar{\partial}_{r-1} = H^{0,r}(M)$$

where $H^{0,r}(M)$ is the $\bar{\partial}$-cohomology group. Then the LHS is

$$\text{ind}\,\bar{\partial} = \sum_{r=0}^{n}(-1)^r b^{0,r} \tag{12.47}$$

where $b^{0,r} \equiv \dim_{\mathbb{C}} H^{0,r}(M)$ is the Hodge number. This index is called the **arithmetic genus** of M.

Simplification of the topological index can be carried out as in the case of the de Rham complex. We refer the reader to Shanahan (1978) for the technical details. We have

$$\sum_{r=1}^{n}(-1)^r b^{0,r} = \int_M \text{Td}(TM^+) \tag{12.48}$$

where $\text{Td}(TM^+)$ is the Todd class of TM^+.

12.4.1 The twisted Dolbeault complex and the Hirzebruch–Riemann–Roch theorem

In the Dolbeault complex, we may replace $\Omega^{0,r}(M)$ by the tensor product bundles $\Omega^{0,r}(M) \otimes V$, where V is a holomorphic vector bundle over M,

$$\cdots \xrightarrow{\bar{\partial}_V} \Omega^{0,r-1}(M) \otimes V \xrightarrow{\bar{\partial}_V} \Omega^{0,r}(M) \otimes V \xrightarrow{\bar{\partial}_V} \cdots . \tag{12.49}$$

The AS index theorem of this complex reduces to the **Hirzebruch–Riemann–Roch theorem**,

$$\text{ind}\,\bar{\partial}_V = \int_M \text{Td}(TM^+)\text{ch}(V). \tag{12.50}$$

For example, if $m = \dim_\mathbb{C} M = 1$, we have

$$
\begin{aligned}
\text{ind } \bar{\partial}_V &= \tfrac{1}{2} \dim V \int_M c_1(TM^+) + \int_M c_1(V) \\
&= (2 - g) \dim V + \int_M \frac{i\mathcal{F}}{2\pi}
\end{aligned}
\tag{12.51}
$$

since it can be shown that

$$
\int_M c_1(TM^+) = \int_M e(TM) = 2 - g
$$

g being the genus of M.

12.5 The signature complex

12.5.1 The Hirzebruch signature

Let M be a compact orientable manifold of even dimension, $m = 2l$. Let $[\omega]$ and $[\eta]$ be the elements of the 'middle' cohomology group $H^l(M; \mathbb{R})$. We consider a bilinear form $H^l(M; \mathbb{R}) \times H^l(M; \mathbb{R}) \to \mathbb{R}$ defined by

$$
\sigma([\omega], [\eta]) \equiv \int_M \omega \wedge \eta
\tag{12.52}
$$

cf example 11.8. This definition is independent of the representatives of $[\omega]$ and $[\eta]$. The form σ is symmetric if l is even ($m \equiv 0 \bmod 4$) and anti-symmetric if l is odd ($m \equiv 2 \bmod 4$). Poincaré duality shows that the bilinear form σ has the maximal rank $b^l = \dim H^l(M; \mathbb{R})$ and is, hence, non-degenerate. If $l = 2k$ is even, the symmetric form σ has real eigenvalues, b^+ of which are positive and b^- of which are negative ($b^+ + b^- = b^l$). The **Hirzebruch signature** is defined by

$$
\tau(M) \equiv b^+ - b^-.
\tag{12.53}
$$

If l is odd, $\tau(M)$ is defined to vanish (an anti-symmetric form has pure imaginary eigenvalues). In the following, we set $l = 2k$.

The Hodge $*$ satisfies $*^2 = 1$ when acting on a $2k$-form in a $4k$-dimensional manifold M and hence $*$ has eigenvalues ± 1. Let $\text{Harm}^{2k}(M)$ be the set of harmonic $2k$-forms on M. We note that $\text{Harm}^{2k}(M) \cong H^{2k}(M; \mathbb{R})$ and each element of $H^{2k}(M; \mathbb{R})$ has a unique harmonic representative. $\text{Harm}^{2k}(M)$ is separated into disjoint subspaces,

$$
\text{Harm}^{2k}(M) = \text{Harm}^{2k}_+(M) \oplus \text{Harm}^{2k}_-(M)
\tag{12.54}
$$

according to the eigenvalue of $*$. This separation block diagonalizes the bilinear form σ. In fact, for $\omega^\pm \in \text{Harm}^{2k}_\pm(M)$,

$$
\sigma(\omega^+, \omega^+) = \int_M \omega^+ \wedge \omega^+ = \int_M \omega^+ \wedge *\omega^+ = (\omega^+, \omega^+) > 0
$$

where (ω^+, ω^+) is the standard positive-definite inner product defined by (7.181). We also find

$$\sigma(\omega^-, \omega^-) = -\int_M \omega^- \wedge *\omega^- = -(\omega^-, \omega^-) < 0$$

$$\sigma(\omega^+, \omega^-) = -\int_M \omega^+ \wedge *\omega^- = -\int_M \omega^- \wedge *\omega^+ = -\sigma(\omega^+, \omega^-) = 0$$

where we have noted that $\alpha \wedge *\beta = \beta \wedge *\alpha$ for any forms α and β. Hence, σ is block diagonal with respect to $\mathrm{Harm}^{2k}_+(M) \oplus \mathrm{Harm}^{2k}_-(M)$ and, moreover, $b^\pm = \dim_{\mathbb{R}} \mathrm{Harm}^{2k}_\pm(M)$. Now $\tau(M)$ is expressed as

$$\tau(M) = \dim \mathrm{Harm}^{2k}_+(M) - \dim \mathrm{Harm}^{2k}_-(M). \tag{12.55}$$

Exercise 12.2. Let dim $M = 4k$. Show that

$$\tau(M) = \chi(M) \bmod 2. \tag{12.56}$$

[*Hint*: Use the Poincaré duality to show that $\chi(M) = b^{2k} \bmod 2$.]

12.5.2 The signature complex and the Hirzebruch signature theorem

Let M be an m-dimensional compact Riemannian manifold without a boundary and let g be the given metric. Consider an operator

$$\mathfrak{D} \equiv \mathrm{d} + \mathrm{d}^\dagger. \tag{12.57}$$

$\mathfrak{D}$ is a square root of the Laplacian: $\mathfrak{D}^2 = \mathrm{dd}^\dagger + \mathrm{d}^\dagger\mathrm{d} = \Delta$. To show that $\mathfrak{D}$ is elliptic, it suffices to verify that Δ is elliptic since the symbol of a product of operators is the product of symbols. Let us compute the symbol of Δ. As for d, we have $\sigma(\mathrm{d}, \xi)\omega = \xi \wedge \omega$. As for $\mathrm{d}^\dagger$, it can be shown that (Palais 1965, pp77–8)

$$\sigma(\mathrm{d}^\dagger, \xi) = -i_\xi. \tag{12.58}$$

Here $i_\xi : \Omega^r_p(M) \to \Omega^{r-1}_p(M)$ is an interior product defined by (cf. (5.79))

$$i_\xi(\mathrm{d}x^{\mu_1} \wedge \ldots \wedge \mathrm{d}x^{\mu_r})$$

$$\equiv \sum_{j=1}^r (-1)^{j+1} g^{\mu_j \mu} \xi_\mu \, \mathrm{d}x^{\mu_1} \wedge \ldots \wedge \mathrm{d}\hat{x}^{\mu_j} \wedge \ldots \wedge \mathrm{d}x^{\mu_r}$$

where the one-form under $\hat{}$ is omitted and we put $\xi = \xi_\mu \, \mathrm{d}x^\mu$. Now the symbol of the Laplacian is obtained from (12.58) as

$$\sigma(\Delta, \xi)\omega = \sigma(\mathrm{dd}^\dagger + \mathrm{d}^\dagger\mathrm{d}, \xi)\omega = -[\xi \wedge i_\xi(\omega) + i_\xi(\xi \wedge \omega)]$$

$$= -i_\xi(\xi) \wedge \omega = -\|\xi\|^2 \omega$$

where ω is an arbitrary r-form and the norm $\| \ \|$ is taken with respect to the given Riemannian metric. Finally, we obtain

$$\sigma(\Delta, \xi) = -\|\xi\|^2. \tag{12.59}$$

Thus, the Laplacian Δ is elliptic and so is $\mathfrak{D} = d + d^\dagger$.

Since the Laplacian $\Delta = \mathfrak{D}^2$ is self-dual on $\Omega^*(M)$, the index of Δ vanishes trivially. It is also observed that $\mathfrak{D} = \mathfrak{D}^\dagger$ on $\Omega^*(M)$ and, hence, ind $\mathfrak{D} = 0$. To construct a non-trivial index theorem, we have to find a complex on which $\mathfrak{D} \neq \mathfrak{D}^\dagger$.

Exercise 12.3. Consider the restriction $\mathfrak{D}^e$ of $\mathfrak{D}$ to even forms, $\mathfrak{D}^e : \Omega^e(M)^{\mathbb{C}} \to \Omega^o(M)^{\mathbb{C}}$ where $\Omega^e(M)^{\mathbb{C}} \equiv \oplus\Omega^{2i}(M)^{\mathbb{C}}$ and $\Omega^o(M)^{\mathbb{C}} \equiv \oplus\Omega^{2i+1}(M)^{\mathbb{C}}$. The adjoint of $\mathfrak{D}^e$ is $\mathfrak{D}^o \equiv \mathfrak{D}^{e\dagger} : \Omega^o(M)^{\mathbb{C}} \to \Omega^e(M)^{\mathbb{C}}$. Show that

$$\text{ind } \mathfrak{D}^e = \dim \ker \mathfrak{D}^e - \dim \ker \mathfrak{D}^o = \chi(M).$$

[*Hint*: Prove $\ker \mathfrak{D}^e = \oplus \text{Harm}^{2i}(M)$ and $\ker \mathfrak{D}^o = \oplus \text{Harm}^{2i+1}(M)$. This complex, although non-trivial, does not yield anything new.]

If $\dim M = m = 2l$, we have $* * \eta = (-1)^r \eta$ for $\eta \in \Omega^r(M)^{\mathbb{C}}$. We define an operator $\pi : \Omega^r(M)^{\mathbb{C}} \to \Omega^{m-r}(M)^{\mathbb{C}}$ by

$$\pi \equiv i^{r(r-1)+l} *. \tag{12.60}$$

Observe that π is a 'square root' of $(-1)^r * * = 1$. In fact, for $\omega \in \Omega^r(M)^{\mathbb{C}}$,

$$\pi^2 \omega = i^{r(r-1)+l} \pi(*\omega) = i^{r(r-1)+l+(2l-r)(2l-r-1)+l} * *\omega$$
$$= i^{2r^2} * *\omega = (-1)^r * *\omega = \omega \tag{12.61}$$

where we have noted that $r \equiv r^2 \mod 2$. We easily verify (exercise) that

$$\{\pi, \mathfrak{D}\} = \pi\mathfrak{D} + \mathfrak{D}\pi = 0. \tag{12.62}$$

Let π act on $\Omega^*(M)^{\mathbb{C}} = \oplus\Omega^r(M)^{\mathbb{C}}$. Since $\pi^2 = 1$, the eigenvalues of π are ± 1. Then we have a decomposition of $\Omega^*(M)^{\mathbb{C}}$ into the ± 1 eigenspaces $\Omega^\pm(M)$ of π as

$$\Omega^*(M)^{\mathbb{C}} = \Omega^+(M) \oplus \Omega^-(M). \tag{12.63}$$

Since $\mathfrak{D}$ anti-commutes with π, the restriction of $\mathfrak{D}$ to $\Omega^+(M)$ defines an elliptic complex called the **signature complex**,

$$\mathfrak{D}_+ : \Omega^+(M) \to \Omega^-(M) \tag{12.64}$$

where $\mathfrak{D}_+ \equiv \mathfrak{D}|_{\Omega^+(M)}$. The index of the signature complex is

$$\text{ind } \mathfrak{D}_+ = \dim \ker \mathfrak{D}_+ - \dim \ker \mathfrak{D}_-$$
$$= \dim \text{Harm}(M)^+ - \dim \text{Harm}(M)^- \tag{12.65}$$

where $\mathfrak{D}_- \equiv \mathfrak{D}_+^\dagger : \Omega^-(M) \to \Omega^+(M)$ and $\mathrm{Harm}(M)^\pm \equiv \{\omega \in \Omega^\pm(M) | \mathfrak{D}_{\pm}\omega = 0\}$. On the RHS of (12.65), all the contributions except those from the harmonic l-forms cancel out. To see this, we separate $\ker \mathfrak{D}_+$ and $\ker \mathfrak{D}_-$ as

$$\ker \mathfrak{D}_\pm = \mathrm{Harm}^l(M)^\pm \oplus \sum_{0 \le r < l} [\mathrm{Harm}^r(M)^\pm \oplus \mathrm{Harm}^{m-r}(M)^\pm]$$

where $\mathrm{Harm}^r(M)^\pm \equiv \mathrm{Harm}(M)^\pm \cap \Omega^r(M)$. If $\omega \in \mathrm{Harm}^r(M)$, we have $\omega \pm \pi\omega \in \mathrm{Harm}^r(M)^\pm \oplus \mathrm{Harm}^{m-r}(M)^\pm$. Then a map $\omega + \pi\omega \to \omega - \pi\omega$ defines an isomorphism between $\mathrm{Harm}^r(M)^+ \oplus \mathrm{Harm}^{m-r}(M)^+$ and $\mathrm{Harm}^r(M)^- \oplus \mathrm{Harm}^{m-r}(M)^-$. Now the index simplifies as

$$\mathrm{ind}\, \mathfrak{D}_+ = \dim \mathrm{Harm}^{2k}(M)^+ - \dim \mathrm{Harm}^{2k}(M)^- \tag{12.66}$$

where we put $l = 2k$ as before (the index vanishes if l is odd). It is important to note that $\mathrm{Harm}^{2k}(M)^\pm = \mathrm{Harm}^{2k}_\pm(M)$ since $\pi = *$ in $\mathrm{Harm}^{2k}(M)$, see (12.54). Now the index (12.66) reduces to the **Hirzebruch signature**,

$$\mathrm{ind}\, \mathfrak{D}_+ = \tau(M). \tag{12.67}$$

The derivation of the topological index is rather technical and we simply quote the result from Shanahan (1978). Let $\wedge^\pm T^*M^\mathbb{C}$ be the subspace of $\wedge T^*M^\mathbb{C}$ such that $\Omega^\pm(M) = \Gamma(M, \wedge^\pm T^*M^\mathbb{C})$. Then we have

$$\text{topological index} = (-1)^l \int_M \mathrm{ch}(\wedge^+ T^*M^\mathbb{C} - \wedge^- T^*M^\mathbb{C}) \left.\frac{\mathrm{Td}(TM^\mathbb{C})}{e(TM)}\right|_{\mathrm{vol}}$$

$$= 2^l \int_M \prod_{i=1}^l \left.\frac{x_i/2}{\tanh x_i/2}\right|_{\mathrm{vol}} = \int_M \prod_{i=1}^l \left.\frac{x_i}{\tanh x_i}\right|_{\mathrm{vol}}$$

where the last equality is true only for the $2l$-forms in the expansion and $x_i = x_i(TM^\mathbb{C})$. Now we have obtained the **Hirzebruch signature theorem**

$$\tau(M) = \int_M L(TM)|_{\mathrm{vol}} \tag{12.68}$$

where L is the Hirzebruch L-polynomial defined by (11.91). Since L is even in x_i, $\tau(M)$ vanishes if $m = 2 \bmod 4$. For example, $\tau(M) = 0$ for $m = 2$. If $m = 4$, we have

$$\tau(M) = \int_M \frac{1}{3} p_1(TM) = -\frac{1}{24\pi^2} \int \mathrm{tr}\, \mathcal{R}^2. \tag{12.69}$$

As in the case of the Dolbeault complex, we may twist the signature complex, see Eguchi *et al* (1980), for example.

12.6 Spin complexes

The final example of classical complexes is the spin complex. This complex is very important in physics since it describes Dirac fields interacting with gauge fields and/or gravitational fields.

12.6.1 Dirac operator

Let us consider a spin bundle $S(M)$ over an m-dimensional orientable manifold M. We shall denote the set of sections of this bundle by $\Delta(M) = \Gamma(M, S(M))$. We assume that $m = 2l$ is an even integer. The spin group SPIN(m) is generated by m Dirac matrices $\{\gamma^\alpha\}$, which satisfy

$$\gamma^{\alpha\dagger} = \gamma^\alpha \tag{12.70a}$$

$$\{\gamma^\alpha, \gamma^\beta\} = 2\delta^{\alpha\beta}. \tag{12.70b}$$

Throughout this chapter we assume that the metric has the Euclidean signature. The Clifford algebra is generated by

$$1; \gamma^\alpha; \gamma^{\alpha_1}\gamma^{\alpha_2} \ (\alpha_1 < \alpha_2); \ldots;$$

$$\gamma^{\alpha_1} \ldots \gamma^{\alpha_k} \ (\alpha_1 < \ldots < \alpha_k); \ldots; \gamma^1 \ldots \gamma^{2l}.$$

The last generator is of particular importance and we define

$$\gamma^{m+1} \equiv i^l \gamma^1 \ldots \gamma^m. \tag{12.71}$$

Our convention is such that $(\gamma^{m+1})^2 = I$ and $(\gamma^{m+1})^\dagger = \gamma^{m+1}$. It can be shown from the general theory of the Clifford algebra that the γ^x are represented by $2^l \times 2^l$ matrices with complex entries. It is convenient to take a representation of $\{\gamma^x\}$ such that γ^{m+1} is diagonal,

$$\gamma^{m+1} = \begin{pmatrix} \mathbf{1} & 0 \\ 0 & -\mathbf{1} \end{pmatrix} \tag{12.72}$$

where $\mathbf{1}$ here is the $2^{l-1} \times 2^{l-1}$ unit matrix.

Example 12.4. For $m = 2$, we take

$$\gamma^0 = \sigma_2 \qquad \gamma^1 = \sigma_1 \qquad \gamma^3 = i\gamma^0\gamma^1 = \sigma_3$$

σ_α being the Pauli matrices,

$$\sigma_1 = \begin{pmatrix} 0 & 1 \\ 1 & 0 \end{pmatrix} \qquad \sigma_2 = \begin{pmatrix} 0 & -i \\ i & 0 \end{pmatrix} \qquad \sigma_3 = \begin{pmatrix} 1 & 0 \\ 0 & -1 \end{pmatrix}.$$

For $m = 4$, we may take

$$\gamma^\beta = \begin{pmatrix} 0 & i\alpha^\beta \\ -i\bar{\alpha}^\beta & 0 \end{pmatrix} \qquad \alpha^\beta = (I_2, -i\sigma), \bar{\alpha}^\beta = (I_2, i\sigma)$$

$$\gamma^5 = -\gamma^0\gamma^1\gamma^2\gamma^3 = \begin{pmatrix} I_2 & 0 \\ 0 & -I_2 \end{pmatrix}.$$

A Dirac spinor $\psi \in \Delta(M)$ is an irreducible representation of the Clifford algebra but *not* that of SPIN($2l$). Irreducible representations of SPIN($2l$) are obtained by separating $\Delta(M)$ according to the eigenvalues of γ^{m+1}. Since $(\gamma^{m+1})^2 = I$, the eigenvalues of γ^{m+1}, called the **chirality**, must be ± 1. Then $\Delta(M)$ is separated into two eigenspaces

$$\Delta(M) = \Delta^+(M) \oplus \Delta^-(M) \tag{12.73}$$

where $\gamma^{m+1}\psi^\pm = \pm \psi^\pm$ for $\psi^\pm \in \Delta^\pm(M)$. The projection operators $\mathcal{P}^\pm$ onto $\Delta^\pm$ are given by

$$\mathcal{P}^+ \equiv \frac{1}{2}(I + \gamma^{m+1}) = \begin{pmatrix} 1 & 0 \\ 0 & 0 \end{pmatrix} \tag{12.74a}$$

$$\mathcal{P}^- \equiv \frac{1}{2}(I - \gamma^{m+1}) = \begin{pmatrix} 0 & 0 \\ 0 & 1 \end{pmatrix}. \tag{12.74b}$$

Thus, we may write[1]

$$\psi^+ = \begin{pmatrix} \psi^+ \\ 0 \end{pmatrix} \in \Delta^+(M), \qquad \psi^- = \begin{pmatrix} 0 \\ \psi^- \end{pmatrix} \in \Delta^-(M). \tag{12.75}$$

The reader should verify that $\mathcal{P}^+ + \mathcal{P}^- = \mathbf{1}, (\mathcal{P}^\pm)^2 = \mathcal{P}^\pm, \mathcal{P}^+\mathcal{P}^- = 0, \mathcal{P}^\pm\psi^\pm = \psi^\pm$ and $\mathcal{P}^\pm\psi^\mp = 0$.

The **Dirac operator** in a curved space is given by (section 7.10)

$$i\slashed{\nabla}\psi \equiv i\gamma^\mu \nabla_{\partial/\partial x^\mu}\psi = i\gamma^\mu(\partial_\mu + \omega_\mu)\psi \tag{12.76}$$

where $\omega_\mu = \frac{1}{2}i\omega_\mu{}^{\alpha\beta}\Sigma_{\alpha\beta}$ is the spin connection and $\gamma^\mu = \gamma^\alpha e_\alpha{}^\mu$. Let us prove that $i\slashed{\nabla}$ is elliptic. Let f be a function defined near $p \in M$ such that $f(p) = 0$ and $i\gamma^\mu \partial_\mu f(p) = i\gamma^\mu \xi_\mu \equiv i\slashed{\xi}$.[2] Take a section $\tilde{\psi} \in \Delta(M)$ such that $\tilde{\psi}(p) = \psi$. From (12.4), we have

$$\sigma(i\slashed{\nabla}, \xi)\psi = i\slashed{\nabla}(f\tilde{\psi})|_p = (i\slashed{\nabla}f)\tilde{\psi}|_p = i\slashed{\xi}\psi$$

which shows that

$$\sigma(i\slashed{\nabla}, \slashed{\xi}) = i\slashed{\xi}. \tag{12.77}$$

If we note that $\slashed{\xi}\slashed{\xi} = \xi_\alpha \xi_\beta \gamma^\alpha \gamma^\beta = \xi^\mu \xi_\mu$, we find that (12.77) is invertible for $i\slashed{\xi} \neq 0$, hence $i\slashed{\nabla}$ is an elliptic operator.

It can be shown that $\{\gamma^\alpha\}$ is taken in the form

$$\gamma^\beta = \begin{pmatrix} 0 & i\alpha_\beta \\ -i\bar{\alpha}_\beta & 0 \end{pmatrix} \qquad \alpha^\dagger{}_\beta = \bar{\alpha}_\beta \tag{12.78}$$

---

[1] Note the minor abuse of the notation.
[2] For a vector $A = A^\mu e_\mu$, $\slashed{A}$ denotes $\gamma^\mu A_\mu$.

see example 12.4 for $m = 2$ and 4. Then (12.76) becomes

$$i\slashed{\nabla} = \begin{pmatrix} 0 & D^\dagger \\ D & 0 \end{pmatrix} \tag{12.79}$$

where

$$D \equiv \bar{\alpha}^\beta e_\beta{}^\mu (\partial_\mu + \omega_\mu) \qquad D^\dagger \equiv -\alpha^\beta e_\beta{}^\mu (\partial_\mu + \omega_\mu). \tag{12.80}$$

Hence, $D^\dagger$ is, indeed, the adjoint of D (note that $\partial_\mu + \omega_\mu$ is anti-Hermitian). For

$$\begin{pmatrix} \psi^+ \\ 0 \end{pmatrix} \in \Delta^+(M)$$

we have

$$i\slashed{\nabla} \begin{pmatrix} \psi^+ \\ 0 \end{pmatrix} = \begin{pmatrix} 0 & D^\dagger \\ D & 0 \end{pmatrix} \begin{pmatrix} \psi^+ \\ 0 \end{pmatrix} = \begin{pmatrix} 0 \\ D\psi^+ \end{pmatrix}$$

while for

$$\begin{pmatrix} 0 \\ \psi^- \end{pmatrix} \in \Delta^-(M)$$

we have

$$i\slashed{\nabla} \begin{pmatrix} 0 \\ \psi^- \end{pmatrix} = \begin{pmatrix} D^\dagger \psi^- \\ 0 \end{pmatrix}.$$

Hence, $D = i\slashed{\nabla}P^+ : \Delta^+(M) \to \Delta^-(M)$ and $D^\dagger = i\slashed{\nabla}P^- : \Delta^-(M) \to \Delta^+(M)$. Now we have a two-term complex

$$\Delta^+(M) \underset{D^\dagger}{\overset{D}{\rightleftarrows}} \Delta^-(M) \tag{12.81}$$

called the **spin complex**. The analytical index of this complex is

$$\operatorname{ind} D = \dim \ker D - \dim \ker D^\dagger = \nu_+ - \nu_- \tag{12.82}$$

where ν_+ (ν_-) is the number of zero-energy modes of chirality $+ (-)$.

Let us apply the AS index theorem to this case. Without getting into the details of the Clifford algebra and the spin complex, we simply write down the result. The AS index theorem for the spin complex (12.81) is

$$\nu_+ - \nu_- = \int_M \operatorname{ch}(\Delta^+(M) - \Delta^-(M)) \left. \frac{\operatorname{Td}(TM^\mathbb{C})}{e(TM)} \right|_{\text{vol}}$$

$$= \int_M \hat{A}(TM)|_{\text{vol}} \tag{12.83}$$

where $\hat{A}$ is the **Dirac genus** defined by (11.94). Since $\hat{A}$ contains only $4j$-forms, $\nu_+ - \nu_-$ vanishes unless $m = 0 \bmod 4$. Of course, this does not necessarily imply $\nu_+ = \nu_- = 0$. The proof of (12.83) will be given later in sections 12.9 and 12.10.

12.6.2 Twisted spin complexes

In physics, a spinor field may belong to a representation of a group G. For example, the quark field in QCD belongs to the $\mathbf{3}$ of SU(3). A spinor which belongs to a representation of G is a section of the product bundle $S(M) \otimes E$, where E is an associated vector bundle of $P(M, G)$ in an appropriate representation. The Dirac operator $D_E : \Delta^+(M) \otimes E \to \Delta(M)^- \otimes E$ in this case is

$$D_E = i\gamma^\alpha e_\alpha{}^\mu (\partial_\mu + \omega_\mu + \mathcal{A}_\mu) \mathcal{D}_+ \tag{12.84}$$

where $\mathcal{A}_\mu$ is the gauge potential on E. The AS index theorem for this twisted spin complex is

$$\nu_+ - \nu_- = \int_M \hat{A}(TM) \mathrm{ch}(E)|_{\mathrm{vol}}. \tag{12.85}$$

For dim $M = 2$, we have

$$\nu_+ - \nu_- = \int_M \mathrm{ch}_1(E) = \frac{i}{2\pi} \int_M \mathrm{tr}\, \mathcal{F} \tag{12.86}$$

while for dim $M = 4$,

$$\nu_+ - \nu_- = \int_M [\mathrm{ch}_2(E) + \hat{A}_1(TM)\mathrm{ch}_0(E)]$$
$$= \frac{-1}{8\pi^2} \int_M \mathrm{tr}\, \mathcal{F}^2 + \frac{\dim E}{192\pi^2} \int_M \mathrm{tr}\, \mathcal{R}^2. \tag{12.87}$$

Example 12.5. Let
$$M = T^{2l} = \underbrace{S^1 \times \cdots \times S^1}_{2l \text{ times}}.$$

Then we find

$$\hat{A}(TM) = \hat{A}\left(\overset{2l}{\underset{1}{\oplus}} TS^1\right) = \prod_1^{2l} \hat{A}(TS^1) = 1.$$

We also have $\hat{A}(TS^{2l}) = 1$. Accordingly, the index of these bundles is

$$\nu_+ - \nu_- = \int_M \mathrm{ch}(E)|_{\mathrm{vol}}. \tag{12.88}$$

Example 12.6. Let us consider the monopole bundle $P(S^2, U(1))$. If $\mathcal{A}$ is the local gauge potential, the field strength is $\mathcal{F} = d\mathcal{A}$. The index theorem is

$$\nu_+ - \nu_- = \frac{i}{2\pi} \int_{S^2} \mathcal{F} = -\frac{1}{2\pi} \int_{S^2} F \tag{12.89}$$

where $\mathcal{F} = iF$. As was shown in section 10.5, the RHS represents the winding number $\pi_1(U(1)) = \mathbb{Z}$ and analytical information (the LHS) is now expressed in a topological way (the RHS).

Let $P(S^4, SU(2))$ be the instanton bundle. Expression (12.88) reads as

$$\nu_+ - \nu_- = \int_{S^4} \mathrm{ch}_2(\mathcal{F}) = \frac{-1}{8\pi^2} \int_{S^4} \mathrm{tr}\, \mathcal{F}^2. \tag{12.90}$$

The RHS represents the instanton number $k \in \pi_3(SU(2)) = \mathbb{Z}$. Note that $k > 0$ if $\mathcal{F} = *\mathcal{F}$ while $k < 0$ if $\mathcal{F} = -*\mathcal{F}$. It can be shown that $\nu_- = 0$ $(\nu_+ = 0)$ if $k > 0$ $(k < 0)$, see Jackiw and Rebbi (1977). For example, let $\mathcal{F}$ be self-dual. Suppose $\psi^- \in \ker D^\dagger = \ker DD^\dagger$. From (12.80), we find that

$$DD^\dagger \psi^- = [(\partial_\mu + \mathcal{A}_\mu)^2 + 2\mathrm{i}\bar{\sigma}_{\mu\nu} \mathcal{F}^{\mu\nu}]\psi^- = 0$$

where $\bar{\sigma}_{\mu\nu} \equiv (1/4\mathrm{i})(\alpha^\mu \bar{\alpha}^\nu - \alpha^\nu \bar{\sigma}^\mu)$. It is easily verified that $\bar{\sigma}^{\mu\nu}$ is anti-self-dual $(\bar{\sigma}^{\mu\nu} = -*\bar{\sigma}^{\nu\mu})$ and hence $\bar{\sigma}_{\mu\nu} \mathcal{F}^{\mu\nu} = 0$. Since $(\partial_\mu + \mathcal{A}_\mu)^2$ is a positive-definite operator, it has no normalizable bound states. This verifies that $\ker D^\dagger = \emptyset$.

12.7 The heat kernel and generalized ζ-functions

As we mentioned in section 12.2, there are several methods of proving the AS index theorem. The heat kernel is relatively accessible to physicists and it also has many applications to other problems in physics. The generalized ζ-function is related to the heat kernel and also has relevance in physics.

12.7.1 The heat kernel and index theorem

Let E be a complex vector bundle over an m-dimensional compact manifold M. Let $\Delta : \Gamma(M, E) \to \Gamma(M, E)$ be an elliptic operator with eigenvectors $|n\rangle$ such that

$$\Delta |n\rangle = \lambda_n |n\rangle. \tag{12.91}$$

We denote the set of eigenvalues of Δ by $\mathrm{Spec}\, \Delta$. We assume that Δ is non-negative, i.e. all the eigenvalues are non-negative. Suppose there are n_0 modes $|0, i\rangle, 1 \le i \le n_0$ with vanishing eigenvalue. In other words,

$$\dim \ker \Delta = n_0. \tag{12.92}$$

These modes are called the **zero modes**. Define the **heat kernel** $h(t)$ by

$$h(t) \equiv \mathrm{e}^{-t\Delta}. \tag{12.93}$$

It is convenient to represent $h(t)$ in the coordinate basis as

$$h(x, y; t) \equiv \langle x|h(t)|y\rangle = \langle x| \sum_n \mathrm{e}^{-t\Delta} |n\rangle \langle n|y\rangle$$

$$= \sum_n \mathrm{e}^{-t\lambda_n} \langle x|n\rangle \langle n|y\rangle. \tag{12.94}$$

Multiple eigenstates should be counted as many times as they appear. We assume $\langle x|n \rangle$ is orthonormal: $\int \langle n|x \rangle \langle x|m \rangle dx = \delta_{mn}$. The convergence of (12.93) for $t > 0$ is guaranteed since Δ is non-negative. Taking the limit $t \to \infty$, we have

$$\lim_{t \to \infty} h(x, y; t) = \sum_{i=1}^{n_0} \langle x|0, i \rangle \langle 0, i|y \rangle \tag{12.95}$$

where the summation is over the zero modes $|0, i \rangle$ only. Thus, $h = e^{-t\Delta}$ tends to be the projection operator onto the space of zero modes as

$$e^{-t\Delta} \overset{t \to \infty}{\longrightarrow} \sum_{i=1}^{n_0} |0, i \rangle \langle 0, i|. \tag{12.96}$$

Define

$$\tilde{h}(t) \equiv \int h(x, x; t)\, dx = \sum_n e^{-t\lambda_n}. \tag{12.97}$$

Then it follows from (12.95) that

$$n_0 = \lim_{t \to \infty} \tilde{h}(t). \tag{12.98}$$

It is easy to verify that h satisfies the **heat equation**,

$$\left(\frac{\partial}{\partial t} + \Delta_x \right) h(x, y; t) = 0. \tag{12.99}$$

If Δ is the conventional Laplacian, (12.99) reduces to the ordinary heat equation. The initial condition is

$$h(x, y; 0) = \sum_n \langle x|n \rangle \langle n|y \rangle = \delta(x - y) \tag{12.100}$$

where the last equality follows from the completeness of the eigenvectors.

Exercise 12.4. Let $u(x, t)$ be a solution of (12.99) such that $u(x, 0) = u(x)$. Show that

$$u(x, t) = \int h(x, y; t)u(y)\, dy. \tag{12.101}$$

[*Hint:* First verify that (12.101) satisfies the initial condition, next that it is a solution of the heat equation.]

It is known that the solution of (12.99) has an asymptotic expansion for $t \to \varepsilon$ given by

$$h(x, x; \varepsilon) = \sum_i a_i(x)\varepsilon^i \tag{12.102}$$

see Gilkey (1984). Similarly, $h(t)$ has an expansion

$$\tilde{h}(\epsilon) \equiv \sum_i a_i \varepsilon^i \tag{12.103}$$

where $a_i = \int a_i(x)dx$.

Let E and F be complex vector bundles over M and $D : \Gamma(M, E) \to \Gamma(M, F)$ be an elliptic operator. We define two Laplacians

$$\Delta_E \equiv D^\dagger D : \Gamma(M, E) \to \Gamma(M, E) \tag{12.104a}$$

$$\Delta_F \equiv DD^\dagger : \Gamma(M, F) \to \Gamma(M, F). \tag{12.104b}$$

It is important to note that they have the same non-vanishing eigenvalues including the degeneracy. To see this, let $\Delta_E|\lambda\rangle = \lambda|\lambda\rangle$. Then there is a vector $D|\lambda\rangle \in \Gamma(M, F)$ such that

$$\Delta_F(D|\lambda\rangle) = DD^\dagger D|\lambda\rangle = D\Delta_E|\lambda\rangle = \lambda(D|\lambda\rangle).$$

Note that $D|\lambda\rangle \neq 0$ since $\ker \Delta_E = \ker D$. Conversely, if $|\mu\rangle \in \Gamma(M, F)$ satisfies $\Delta_F|\mu\rangle = \mu|\mu\rangle$, then $D^\dagger|\mu\rangle \in \Gamma(M, E)$ is an eigenvector of Δ_E with the same eigenvalue μ. Thus, we have found the symmetry[3]

$$\text{Spec}' \Delta_E = \text{Spec}' \Delta_F \tag{12.105}$$

where the prime denotes that the zero eigenmodes are omitted.

Define two heat kernels h_E and h_F by

$$h_E(x, y, t) = \sum e^{-\lambda_n} \langle x|n\rangle\langle n|y\rangle \tag{12.106a}$$

$$h_F(x, y, t) = \sum e^{-\mu_m} \langle x|m\rangle(m|y). \tag{12.106b}$$

We have

$$\lim_{t\to\infty} \tilde{h}_E(t) = \dim \ker \Delta_E = \dim \ker D \tag{12.107a}$$

$$\lim_{t\to\infty} \tilde{h}_F(t) = \dim \ker \Delta_F = \dim \ker D^\dagger. \tag{12.107b}$$

What is more interesting is the index of D. Since $\ker D = \ker \Delta_E$ and $\ker D^\dagger = \ker \Delta_F$, we have

$$\text{ind} D = \dim \ker D - \dim \ker D^\dagger = \dim \ker \Delta_E - \dim \ker \Delta_F$$
$$= \lim_{t\to\infty} [\tilde{h}_E(t) - \tilde{h}_F(t)] = \tilde{h}_E(t) - \tilde{h}_F(t). \tag{12.108}$$

The final equality follows since the t-dependent part of $\tilde{h}_E(t) - \tilde{h}_F(t)$ cancels out by the symmetry (12.105). We expand $\tilde{h}_E(t)$ and $\tilde{h}_F(t)$ as

$$\tilde{h}_E(t) = \sum a_i^E t^i \qquad \tilde{h}_F(t) = \sum a_i^F t^i.$$

---

[3] This is a kind of 'supersymmetry', see section 12.10.

Picking up t-independent terms, we have

$$\text{ind } D = a_0^E - a_0^F = \int dx\, [a_0^E(x) - a_0^F(x)]\, dx \tag{12.109}$$

where $a_0^{E,F}(x)$ are defined in (12.102).

In general, $a_0^{E,F}(x)$ are local invariants written in terms of curvature two-forms. In section 13.2, we use the heat kernel to prove the index theorem

$$\text{ind } D = \nu_+ - \nu_- = \int_M \text{ch}(\mathcal{F})|_{\text{vol}}$$

for the twisted spin complex over a manifold with $\hat{A}(TM) = 1$.

Exercise 12.5. Let D, $D^\dagger$, Δ_E and Δ_F be as before. Show that

$$I(s) \equiv \text{tr}\left[\frac{s}{\Delta_E + s} - \frac{s}{\Delta_F + s}\right] \qquad \text{Re } s > 0 \tag{12.110}$$

is independent of s. Show also that $I(s) = \text{ind } D$.

12.7.2 Spectral ζ-functions

Let E and F be vector bundles over M. Define a new function

$$\zeta_E(x, y; s) \equiv \sum_n {}' \langle x|n\rangle \langle n|y\rangle \lambda_n^{-s} \qquad \text{Re } s > 0 \tag{12.111}$$

where $\Delta_E|n\rangle = \lambda_n|n\rangle$ and the prime denotes the omission of the zero modes ($\lambda_n = 0$). A function $\zeta_F(x, y; s)$ may similarly be defined for Δ_F. The functions h_E and ζ_E are related by the **Mellin transformation**. To see this, we recall the definition of the Γ-function,

$$\Gamma(s) \equiv \int_0^\infty t^{s-1} e^{-t}\, dt = \lambda^s \int_0^\infty t^{s-1} e^{-\lambda t}\, dt$$

where λ is taken to be strictly positive. From this we find

$$\Gamma(s)\zeta(x, y; s) = \sum_n {}' \int_0^\infty t^{s-1} e^{-\lambda_n t} \langle x|n\rangle \langle n|y\rangle\, dt$$

$$= \int_0^\infty t^{s-1}\left[h(x, y; t) - \sum_i \langle x|0, i\rangle \langle 0, i|y\rangle\right] dt. \tag{12.112}$$

We also note that

$$\zeta_\Delta(s) \equiv \int_M \zeta(x, x; s)\, dx = \sum_n {}' \lambda_n^{-s} \tag{12.113}$$

is the spectral ζ-function defined in (1.158).

Exercise 12.6. Verify that

$$\Delta^{-s} f(x) = \int \zeta(x, y; s) f(y) \, dy \tag{12.114}$$

where the general power of an operator may be defined in the sense of an eigenvalue, namely we put $\Delta^{-s}|n\rangle = \lambda_n^{-s}|n\rangle$. Re s is assumed to be sufficiently large so that (12.114) is well defined. [*Hint:* Use the completeness of the eigenvectors.]

Example 12.7. The following example is taken from Kulkarni (1975). Let $M = S^1 = \{e^{i\theta}\}$ and $E = F = $ a trivial line bundle over S^1 (a cylinder). Take an operator $\Delta \equiv -\partial^2/\partial\theta^2$. From the eigenvalue equation,

$$-\frac{\partial^2 e^{in\theta}}{\partial\theta^2} = n^2 e^{in\theta} \qquad n \in \mathbb{Z}$$

we find that

$$\lambda_n = n^2 \qquad \langle\theta|n\rangle = (2\pi)^{-1/2} e^{in\theta}.$$

The heat kernel is

$$h(\theta_1, \theta_2; t) = \sum e^{-n^2 t} \langle\theta_1|n\rangle\langle n|\theta_2\rangle$$

$$= \frac{1}{2\pi}\left(1 + {\sum}' e^{-n^2 t} e^{in(\theta_1 - \theta_2)}\right) \tag{12.115}$$

while

$$\zeta(\theta_1, \theta_2; s) = {\sum}' n^{-2s} \langle\theta_1|n\rangle\langle n|\theta_2\rangle$$

$$= \frac{1}{2\pi}{\sum}' n^{-2s} e^{in(\theta_1 - \theta_2)}. \tag{12.116}$$

We easily verify that $\tilde{h}(t) = 1 + {\sum}' e^{-n^2 t}$ satisfies

$$1 + 2\int_1^\infty e^{-x^2 t} \, dx < \tilde{h}(t) < 1 + 2\int_0^\infty e^{-x^2 t} \, dx.$$

We then find from these inequalities that

$$\int_{-\infty}^{+\infty} e^{-x^2 t} \, dx - 1 < \tilde{h}(t) < \int_{-\infty}^{+\infty} e^{-x^2 t} \, dx + 1$$

or by putting the value

$$\int e^{-x^2 t} \, dx = \sqrt{\pi} t^{-1/2}$$

we find

$$\sqrt{\pi} t^{-1/2} - 1 < \tilde{h}(t) < \sqrt{\pi} t^{-1/2} + 1.$$

This shows that

$$\lim_{t \to 0^+} \tilde{h}(t) \sim \sqrt{\pi} t^{-1/2}. \tag{12.117}$$

In general, the asymptotic series starts with $t^{-\dim M/2}$.

12.8 The Atiyah–Patodi–Singer index theorem

So far we have been concerned with index theorems defined on a compact manifold without a boundary. In practical situations in physics, we often need to find an index of an operator defined over a base space M with a boundary. The extensions of the AS index theorem to these cases are discussed here. Our argument is restricted to the spin bundle over M since this is the only situation we shall be concerned with in chapter 13.

12.8.1 η-invariant and spectral flow

Let $i\slashed{\nabla}$ be a Hermitian Dirac operator defined on an odd-dimensional manifold M, $\dim M = 2l + 1$. Since $i\slashed{\nabla}$ is Hermitian, the eigenvalues λ_k are real. We define the η-**invariant** of $i\slashed{\nabla}$ by the spectral asymmetry of $i\slashed{\nabla}$,

$$\eta \equiv \sum_{\lambda_k > 0} 1 - \sum_{\lambda_k < 0} 1. \tag{12.118}$$

This is not well defined and requires a proper regularization. For example, we may define η by $\lim_{s \to 0} \eta(s)$ where

$$\eta(s) \equiv \sum_k{}' \operatorname{sgn}(\lambda_k) |\lambda_k|^{-2s} \qquad \operatorname{Re} s > 0. \tag{12.119}$$

It can be shown that, under proper boundary conditions, $\eta(s)$ has no pole at $s = 0$.

Exercise 12.7. Use the Mellin transformation

$$\frac{1}{2} \Gamma\left(\frac{s+1}{2}\right) a^{-(s+1)/2} = \int_0^\infty \mathrm{d}x\, x^s \mathrm{e}^{-ax^2} \qquad a > 0$$

to verify that

$$\eta(s) = \frac{2}{\Gamma(\frac{1}{2}(s+1))} \int_0^\infty \mathrm{d}x\, x^s \operatorname{tr} i\slashed{\nabla} \mathrm{e}^{-x^2 (i\slashed{\nabla})^2}. \tag{12.120}$$

Suppose a Dirac field is interacting with an external gauge potential $\mathcal{A}_t, t \in [0, 1]$. The Dirac operator $i\slashed{\nabla}(\mathcal{A}_t)$ has a t-dependent eigenvalue problem. If an eigenvalue of $i\slashed{\nabla}(\mathcal{A}_t)$ crosses zero, the η-invariant jumps by ± 2. This jump

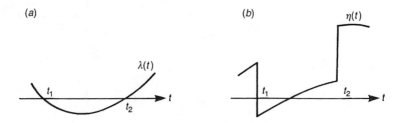

Figure 12.1. Whenever an eigenvalue λ crosses zero (a), the η-invariant jumps by ± 2 (b). The sign depends on the way in which λ crosses zero.

denotes the **spectral flow** from $\lambda \gtrless 0$ modes to $\lambda \lessgtr 0$ modes; if η jumps by $+2$ (-2), there is a flow of a state from $\lambda < 0$ to $\lambda > 0$ ($\lambda > 0$ to $\lambda < 0$), see figure 12.1. In addition to the discontinuous change associated with the spectral flow, $i\slashed{\nabla}$ also has a continuous variation η_c. We have

$$\eta(t=1) - \eta(t=0) = \int_0^1 dt \frac{d\eta_c}{dt} + 2 \times (\text{spectral flow}). \qquad (12.121)$$

12.8.2 The Atiyah–Patodi–Singer (APS) index theorem

Let us consider a $(2l+2)$-dimensional Dirac operator

$$i\hat{D}_{2l+2} = i\sigma_1 \frac{\partial}{\partial t} + \sigma_2 \otimes i\slashed{\nabla}(\mathcal{A}_t) = \begin{pmatrix} 0 & D \\ D^\dagger & 0 \end{pmatrix} \qquad (12.122a)$$

where

$$D = i\partial_t - \slashed{\nabla}(\mathcal{A}_t) \qquad D^\dagger = i\partial_t + \slashed{\nabla}(\mathcal{A}_t). \qquad (12.122b)$$

[*Remark:* The positions of D and $D^\dagger$ are reversed since

$$\gamma^{2l+3} = \begin{pmatrix} -1 & 0 \\ 0 & 1 \end{pmatrix}$$

for our choice of γ-matrices; cf (12.79).]

Theorem 12.3. (**Atiyah–Patodi–Singer theorem**) Let M be an odd-dimensional manifold and $i\slashed{\nabla}(\mathcal{A}_t)$ a Dirac operator on M interacting with an external gauge field $\mathcal{A}_t$. Then,

$$\begin{aligned}
\text{ind } D &= \dim \ker D - \dim \ker D^\dagger \\
&= \int_{M \times I} \hat{A}(\mathcal{R}) \text{ch}(\mathcal{F})|_{\text{vol}} - \tfrac{1}{2}[\eta(i\slashed{\nabla}(\mathcal{A}_1)) - \eta(i\slashed{\nabla}(\mathcal{A}_0))].
\end{aligned}$$

$$(12.123)$$

The general argument shows that the continuous part η_c of the η-invariant satisfies

$$\int_0^1 dt\, \frac{d\eta_c}{dt} = 2\int_{M\times I} \hat{A}(\mathcal{R})\mathrm{ch}(\mathcal{F})|_{\mathrm{vol}}. \tag{12.124}$$

Then the RHS of (12.123) is simply the spectral flow

$$-\frac{1}{2}[\eta(t=1) - \eta(t=0)] + \frac{1}{2}\int_0^1 dt\, \frac{d\eta_c}{dt} = -\text{spectral flow}.$$

Thus, we find another expression for the APS index theorem,

$$\mathrm{ind}\, i\hat{D}_{2l+2} = -\text{spectral flow}. \tag{12.125}$$

The proof of the APS index theorem in its most general form is found in Atiyah *et al* (1975a, b, 1976). The physicists' proof is found in Alvarez-Gaumé *et al* (1985). We use the APS index theorem to study the odd-dimensional parity anomaly in section 13.6.

Example 12.8. To see why the spectral flow appears in the index theorem, we consider an example taken from Atiyah (1985). Let $M = S^1$ and θ be its coordinate. Consider a Hermitian operator

$$i\nabla_t \equiv i\left(\frac{\partial}{\partial\theta} - it\right) = i\partial_\theta + t \qquad t \in \mathbb{R}. \tag{12.126}$$

The term $-it$ is thought of as a U(1) gauge potential. The eigenvector and the eigenvalue of $i\nabla_t$ are

$$\psi_{n,t}(\theta) = \frac{1}{\sqrt{2\pi}}e^{-in\theta}(n \in \mathbb{Z}) \qquad \lambda_n(t) = n + t.$$

Since $\mathrm{Spec}\, i\nabla_t = \mathrm{Spec}\, i\nabla_{t+1}$, the family of operators $i\nabla_t$ is periodic in t with the period 1, see figure 12.2. This periodicity manifests itself in the gauge equivalence of $i\nabla_t$ and $i\nabla_{t+1}$:

$$i\nabla_{t+1} = e^{i\theta}i\nabla_t e^{-i\theta}.$$

There is precisely unit spectral flow from $\lambda < 0$ to $\lambda > 0$ at $t = 0$ while t changes from $-\varepsilon$ to $1 - \varepsilon$, ε being a small positive number. From $i\nabla_t$, we construct a two-dimensional Dirac operator

$$i\slashed{D}_2 \equiv i\sigma_1 \otimes \frac{\partial}{\partial t} + \sigma_2 \otimes i\nabla_t = \begin{pmatrix} 0 & D \\ D^\dagger & 0 \end{pmatrix} \tag{12.127a}$$

where

$$D \equiv i\partial_t + \partial_\theta - it \qquad D^\dagger \equiv i\partial_t - \partial_\theta + it. \tag{12.127b}$$

These operators act on functions which satisfy the boundary conditions

$$\phi(\theta + 2\pi, t) = \phi(\theta, t) \qquad \phi(\theta, t + 1) = e^{i\theta}\phi(\theta, t). \tag{12.128}$$

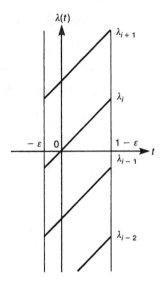

Figure 12.2. Time evolution of the eigenvalues of $i\nabla_t$. Spec $i\nabla_t$ has period 1. The ith eigenvalue crosses zero at $t = 0$ and, hence, there is a unit spectral flow.

Let $\phi_0 \in \ker D^\dagger$. We have a Fourier expansion

$$\phi_0(\theta, t) = \sum a_n(t)e^{-in\theta}.$$

It follows from $D^\dagger\phi_0 = 0$ that

$$a_n'(t) + (n + t)a_n(t) = 0$$

which is easily solved to yield

$$a_n(t) = c_n \exp\left(-\frac{(n + t)^2}{2}\right).$$

The boundary conditions (12.128) require that

$$\sum_n c_n \exp\left(-\frac{(n + t + 1)^2}{2}\right)e^{-in\theta} = \sum_n c_n \exp\left(-\frac{(n + t)^2}{2}\right)e^{-i(n-1)\theta}$$

from which we find that c_n is independent of n. Thus, $\ker D^\dagger$ is one dimensional and is spanned by the theta function,

$$\phi_0(\theta, t) = \sum \exp\left(-\frac{(n + t)^2}{2} - in\theta\right). \qquad (12.129)$$

Suppose $\tilde{\phi}_0(\theta, t) \in \ker D$. If we put $\tilde{\phi}_0(\theta, t) = \sum b_n(t) e^{-in\theta}$, $b_n(t)$ satisfies

$$b_n'(t) - (n + t)b_n(t) = 0.$$

The solution of this equation is

$$b_n(t) = b_n(0) \exp \frac{(n + t)^2}{2}$$

and, hence, $\tilde{\phi}_0$ cannot be normalized. This shows that

$$\text{ind}\, D = \dim \ker D - \dim \ker D^\dagger = -1$$

which agrees with $-$(spectral flow).

12.9 Supersymmetric quantum mechanics

We present, in the next section, the *physicists'* proof of the index theorem in its simplest setting. The proof is heavily based on path integral formulation of supersymmetric quantum mechanics (SUSYQM), which will be outlined in the present section.

We have studied the path integral quantization of bosons and fermions. If these particles are combined together, there appears a new symmetry called **supersymmetry**. We will introduce a special class of SUSYQM later, which turns out to be crucial in the proof of an index theorem.

This and the next sections may be read separately from the previous sections. The necessary tools are supplied to make these sections self-contained. Our exposition follows Alvarez (1995) and Nakahara (1998). Original references are Alvarez-Gaumé L (1983) and Friedan and Windey (1984, 1985).

12.9.1 Clifford algebra and fermions

We restrict ourselves to a particle moving in $\mathbb{R}^3$ to start with. More general settings will be studied later. Let $\{\psi_i\} = \{\psi_1, \psi_2, \psi_3\}$ be *real* Grassmann variables, where $i = 1, 2, 3$ labels the coordinate index. They satisfy the algebra

$$\{\psi_i, \psi_j\} = 0$$

Let us consider the Lagrangian

$$L = \frac{i}{2}\psi_i \dot{\psi}_i - \frac{i}{2}\epsilon_{ijk} B_i \psi_j \psi_k \tag{12.130}$$

where B_i is a real number. The canonical conjugate momentum for ψ_i is

$$\pi_i \equiv \frac{\partial L}{\partial \dot{\psi}_i} = -\frac{i}{2}\psi_i.$$

Then the Hamiltonian is

$$H = -\dot{\psi}_i \frac{i}{2}\psi_i - L = \frac{i}{2}\epsilon_{ijk} B_i \psi_j \psi_k. \tag{12.131}$$

The Poincaré one-form of this system is

$$\theta = \frac{i}{2}\psi_i \, d\psi_i. \tag{12.132}$$

The corresponding symplectic two-form is

$$\omega = d\theta = \frac{i}{2} \, d\psi_i \wedge d\psi_i \tag{12.133}$$

from which we obtain the Poisson bracket

$$[\psi_j, i\psi_k]_{\mathrm{PB}} = i\delta_{jk}. \tag{12.134}$$

Quantization of the system is achieved by replacing this Poisson bracket by the anti-commutation relation

$$\{\psi_j, \psi_k\} = \delta_{jk}. \tag{12.135}$$

This anti-commutation relation is called the **Clifford algebra** in $\mathbb{R}^3$. Let σ_i be the ith component of the Pauli matrices. It is easily verified from the observation

$$\{\sigma_j, \sigma_k\} = 2\delta_{jk}$$

that $\psi_i = \sigma_i/\sqrt{2}$ is the two-dimensional representation of the Clifford algebra. It is known that the finite-dimensional irreducible representation of the Clifford algebra is unique (modulo conjugate transformations). Thus, the Hilbert space of this system turns out to be $\mathcal{H} = \mathbb{C}^2$. The Hamiltonian is rewritten in terms of the Pauli matrices as

$$H = -\tfrac{1}{2}\boldsymbol{B} \cdot \boldsymbol{\sigma}. \tag{12.136}$$

This Hamiltonian is known as the **Pauli Hamiltonian** and describes a spin in a magnetic field.

Similarly, the Clifford algebra defined in $\mathbb{R}^{2n}$ and $\mathbb{R}^{2n+1}$ acts on the Hilbert space $\mathcal{H} = \mathbb{C}^{2^n}$.

12.9.2 Supersymmetric quantum mechanics in flat space

The Pauli Hamiltonian is made only of the spin coordinates ψ_i and is independent of the space coordinate x_k. Accordingly, it cannot describe a travelling spin. Now the Hamiltonian is modified so that the spin may move around the space. This can be realized by adding a kinetic term to the Hamiltonian. Let us consider a spin in $\mathbb{R}^d$ and put $\boldsymbol{B} = 0$ to obtain the Hamiltonian

$$L = \frac{1}{2}\dot{x}_k \dot{x}_k + \frac{i}{2}\psi_k \dot{\psi}_k. \tag{12.137}$$

The coefficients of this Lagrangian have been chosen so that the system has a supersymmetry defined later. The canonically conjugate momenta are $p_k = \dot{x}_k$ and $\pi_k = -i\psi_k/2$, from which we obtain the Poisson brackets of the system

$$[x_j, x_k]_{\text{PB}} = [p_j, p_k]_{\text{PB}} = 0 \qquad [x_j, p_k]_{\text{PB}} = [\psi_j, \psi_k]_{\text{PB}} = \delta_{jk}.$$

It is easy to derive (anti)commutation relations from these Poisson brackets. The canonical (anti)commutation relations are

$$[x_j, x_k] = [p_j, p_k] = 0 \qquad [x_j, p_k] = \{\psi_j, \psi_k\} = \delta_{jk}. \tag{12.138}$$

The Hamiltonian is

$$H = \dot{x}_j p_j - \dot{\psi}_j \frac{i}{2}\psi_j - L = \frac{1}{2}p^2 = -\frac{1}{2}\Delta \tag{12.139}$$

where $\Delta = \sum_{k=1}^{d} \partial_k^2$ is the d-dimensional Laplacian. The Hilbert space on which H acts is $L^2(\mathbb{R}^d) \otimes \mathbb{C}^{2^n}$, where $L^2(\mathbb{R}^d)$ stands for the set of square-integrable functions in $\mathbb{R}^d$ and $n \equiv [d/2]$ is the integer part of $d/2$.

Variation of the Lagrangian yields

$$\delta L = \dot{x}_j \frac{d}{dt}\delta x_j + \frac{i}{2}\delta\psi_j\dot{\psi}_j + \frac{i}{2}\psi_j\frac{d}{dt}\delta\psi_j.$$

Let us verify that the Lagrangian is invariant under the following **supersymmetry transformation**

$$\delta x_j = i\epsilon\psi_j \qquad \delta\psi_j = -\epsilon\dot{x}_j \tag{12.140}$$

where ϵ is an 'infinitesimal' real Grassmann constant. In fact,

$$\begin{aligned}
\delta L &= i\dot{x}_j\epsilon\dot{\psi}_j - \frac{i}{2}\epsilon\dot{x}_j\dot{\psi}_j - \frac{i}{2}\psi_j\epsilon\ddot{x}_j \\
&= i\dot{x}_j\epsilon\dot{\psi}_j - \frac{i}{2}\epsilon\dot{x}_j\dot{\psi}_j - \frac{i}{2}\frac{d}{dt}(\psi_j\epsilon\dot{x}_j) + \frac{i}{2}\dot{\psi}_j\epsilon\dot{x}_j \\
&= -\frac{i}{2}\frac{d}{dt}(\psi_j\epsilon\dot{x}_j) \tag{12.141}
\end{aligned}$$

and the action $S = \int L\,dt$ is left invariant. The corresponding charge (the generator) is called the **supercharge** and defined through the Noether's theorem as[4]

$$\epsilon Q \equiv i\epsilon p_j\psi_j = i\epsilon\psi_j p_j = i\epsilon\psi_j\dot{x}_j. \tag{12.142}$$

Exercise 12.8. Show that

$$\delta x_j = [x_j, \epsilon Q] \tag{12.143}$$
$$\delta\psi_j = \{\psi_j, \epsilon Q\}. \tag{12.144}$$

---

[4] Note that the mass of the particle is set to unity and hence we have $p_j = \dot{x}_j$.

These equations show that Q is the generator of SUSY transformations.

Let us take $d = 2n$ to be an even integer and quantize the system in the following. We introduce the matrix representation $\psi_j = \gamma_j/\sqrt{2}$, which is the generalization of the two-dimensional representation introduced in the previous subsection. Here γ_j are the d-dimensional Dirac matrices that satisfy the Clifford algebra

$$\{\gamma_j, \gamma_k\} = 2\delta_{ij}. \tag{12.145}$$

The Hamiltonian acts on the Hilbert space

$$\mathcal{H} = L_2(\mathbb{R}^{2n}) \otimes \mathbb{C}^{2^n}.$$

The supercharge takes the form, upon diagonalizing the coordinate,

$$Q = i\psi_j p_j = \frac{1}{\sqrt{2}}\gamma_j \frac{\partial}{\partial x_j}. \tag{12.146}$$

The operator

$$\displaystyle{\not{\partial}} \equiv \gamma_j \frac{\partial}{\partial x_j} \tag{12.147}$$

is nothing but the Dirac operator in Euclidean space $\mathbb{R}^{2n}$ and plays an important role in the proof of the index theorem.

The hypercharge Q transforms in an interesting way under an SUSY transformation (12.140)

$$\delta Q = i(\delta\psi_j)\dot{x}_j + i\psi_j \frac{d}{dt}\delta x_j = i(-\epsilon\dot{x}_j)\dot{x}_j + i\psi_j(i\epsilon\dot{\psi}_j)$$

$$= -i\epsilon\dot{x}_j\dot{x}_j + \epsilon\psi_j\dot{\psi}_j = -2i\epsilon\left(\frac{1}{2}\dot{x}_j\dot{x}_j + \frac{i}{2}\psi_j\dot{\psi}_j\right)$$

$$= -2i\epsilon L. \tag{12.148}$$

Namely, the variation of the supercharge under an infinitesimal SUSY transformation is the Lagrangian!

We next consider the relation between the supercharge and the Hamiltonian of the system. Let us consider successive SUSY transformations with Grassmann parameters ϵ_1 and ϵ_2. If a transformation with ϵ_1 is applied first and then ϵ_2 next, we obtain

$$x_j \overset{\epsilon_1}{\to} x_j + i\epsilon_1\psi_j \overset{\epsilon_2}{\to} x_j + i(\epsilon_1 + \epsilon_2)\psi_j - i\epsilon_1\epsilon_2\dot{x}_j$$

$$\psi_j \overset{\epsilon_1}{\to} \psi_j - \epsilon_1\dot{x}_j \overset{\epsilon_2}{\to} \psi_j - (\epsilon_1 + \epsilon_2)\dot{x}_j - i\epsilon_1\epsilon_2\dot{\psi}_j$$

while if the order of the SUSY transformations is reversed,

$$x_j \to x_j + i(\epsilon_1 + \epsilon_2)\psi_j - i\epsilon_2\epsilon_1\dot{x}_j$$

$$\psi_j \to \psi_j - (\epsilon_1 + \epsilon_2)\dot{x}_j - i\epsilon_2\epsilon_1\dot{\psi}_j.$$

We find, from these results, the commutation relation of the SUSY variations:

$$[\delta_{\epsilon_2}, \delta_{\epsilon_1}] = \delta_{\epsilon_2}\delta_{\epsilon_1} - \delta_{\epsilon_1}\delta_{\epsilon_2} = -2i\epsilon_1\epsilon_2\frac{\partial}{\partial t}. \tag{12.149}$$

The observation that the commutation relation of two SUSY transformations is a time derivative, i.e. the Hamiltonian, suggests that the anti-commutation relation of the supercharge, the generator of the SUSY transformation, also yields the Hamiltonian. In fact,

$$\begin{aligned}
\{Q, Q\} = 2Q^2 &= 2(ip_j\psi_j)(ip_k\psi_k) \\
&= -p_jp_k(\psi_j\psi_k + \psi_k\psi_j) = -p_jp_k\delta_{jk} \\
&= -2H.
\end{aligned}$$

After all, the SUSY algebra reduces to

$$Q^2 = -H. \tag{12.150}$$

Since Q is anti-Hermitian, the Hamiltonian is a Hermite operator with non-negative spectrum.

In summary, we proved in equations (12.148) and (12.141) that

$$\delta Q = -2i\epsilon L \qquad \delta L = \frac{1}{2}\epsilon\frac{dQ}{dt}. \tag{12.151}$$

If these equations are compared with the SUSY transformations (12.140) of the coordinates x_j and ψ_j, we readily notice that the roles played by bosonic quantities (x_j and L) and the fermionic quantities (ψ_j and Q) are interchanged. Note that the variation of the supercharge Q in (12.151) is always a time derivative of the Lagrangian L. This observation is crucial in constructing a SUSY-invariant Lagrangian out of a supercharge Q.

12.9.3 Supersymmetric quantum mechanics in a general manifold

Let M be a Riemannian manifold with dim $M = 2n$. The Riemannian metric is

$$ds^2 = g_{\mu\nu}\,dx^\mu\,dx^\nu$$

and the inner product of two vectors X and Y with respect to this metric is denoted as

$$\langle X, Y\rangle = g_{\mu\nu}X^\mu Y^\nu.$$

The vector $\psi^\mu(t)$ belongs to $TM_{x(t)}$ at each instant of time t. Therefore, $\psi^\mu(t)$ obeys the ordinary transformation rule for a vector under the coordinate transformation $x^\mu \to x'^\mu = x'^\mu(x^\nu)$:

$$\psi^\mu \to \psi'^\mu = \frac{\partial x'^\mu}{\partial x^\nu}\psi^\nu. \tag{12.152}$$

Then, under the SUSY transformation $\delta \equiv \delta_\epsilon$, the coordinates transform as

$$\delta x'^\mu = \frac{\partial x'^\mu}{\partial x^\nu} \delta x^\nu = \frac{\partial x'^\mu}{\partial x^\nu} i\epsilon \psi^\nu = i\epsilon \psi'^\mu$$

and

$$\begin{aligned}
\delta \psi'^\mu &= \frac{\partial^2 x'^\mu}{\partial x^\nu \partial x^\lambda} \delta x^\lambda \psi^\nu + \frac{\partial x'^\mu}{\partial x^\nu} \delta \psi^\nu \\
&= \frac{\partial^2 x'^\mu}{\partial x^\nu \partial x^\lambda} i\epsilon \psi^\lambda \psi^\nu + \frac{\partial x'^\mu}{\partial x^\nu}(-i\epsilon \dot{x}^\nu) = -\epsilon \dot{x}'^\mu
\end{aligned}$$

where the anti-commutativity of Grassmann numbers has been used to obtain the last equality. These transformation rules show that the SUSY transformation is covariant under the coordinate transformation $x^\mu \to x'^\mu$.

The supercharge Q introduced in the previous subsection should be generalized on the manifold M as

$$Q = i\langle \dot{x}, \psi \rangle = ig_{\mu\nu}(x)\dot{x}^\mu \psi^\nu. \tag{12.153}$$

The SUSY-invariant Lagrangian on M is constructed from the SUSY variation of this Q as

$$\begin{aligned}
\delta Q &= i\partial_\lambda g_{\mu\nu} \delta x^\lambda \dot{x}^\mu \psi^\nu + ig_{\mu\nu} \delta \dot{x}^\mu \psi^\nu + ig_{\mu\nu} \dot{x}^\mu \delta \psi^\nu \\
&= i\partial_\lambda g_{\mu\nu} i\epsilon \psi^\lambda \dot{x}^\mu \psi^\nu + ig_{\mu\nu}(i\epsilon \dot{\psi}^\mu)\psi^\nu + ig_{\mu\nu} \dot{x}^\mu(-\epsilon \dot{x}^\nu) \\
&= -2i\epsilon \left[\frac{1}{2} g_{\mu\nu} \dot{x}^\mu \dot{x}^\nu + \frac{i}{2} g_{\mu\nu} \psi^\nu \dot{\psi}^\mu \right. \\
&\qquad \left. -\frac{i}{2} \dot{x}^\mu \frac{1}{2}\left(\partial_\lambda g_{\mu\nu} - \partial_\nu g_{\mu\lambda} - \partial_\mu g_{\lambda\nu} \right) \psi^\lambda \psi^\nu \right] \\
&= -2i\epsilon \left(\frac{1}{2} g_{\mu\nu} \dot{x}^\mu \dot{x}^\nu + \frac{i}{2} g_{\mu\nu} \psi^\nu \dot{\psi}^\mu + \frac{i}{2} \dot{x}^\mu g_{\lambda\rho} \Gamma^\rho{}_{\mu\nu} \psi^\lambda \psi^\nu \right)
\end{aligned}$$

where

$$\Gamma^\nu{}_{\lambda\mu} = \tfrac{1}{2} g^{\nu\rho} \left(\partial_\lambda g_{\rho\mu} + \partial_\mu g_{\lambda\rho} - \partial_\rho g_{\lambda\nu} \right)$$

is the Christoffel symbol associated with the Levi-Civita connection. Note the symmetry $\Gamma^\lambda{}_{\mu\nu} = \Gamma^\lambda{}_{\nu\mu}$. By comparing this δQ with (12.151), we read off the Lagrangian,

$$\begin{aligned}
L &= \frac{1}{2} g_{\mu\nu}(x)\dot{x}^\mu \dot{x}^\nu + \frac{i}{2} g_{\mu\nu}(x)\psi^\mu \left(\frac{d\psi^\nu}{dt} + \dot{x}^\lambda \Gamma^\nu{}_{\lambda\kappa}(x)\psi^\kappa \right) \\
&= \frac{1}{2}\langle \dot{x}, \dot{x} \rangle + \frac{i}{2}\left\langle \psi, \frac{D\psi}{Dt} \right\rangle. \tag{12.154}
\end{aligned}$$

Here $D\psi/Dt$ is the covariant derivative of ψ along the curve $x(t)$.

Exercise 12.9. Show that the SUSY variation of the Lagrangian is proportional to the time derivative of the supercharge,

$$\delta L = \frac{1}{2}\epsilon \frac{dQ}{dt}. \tag{12.155}$$

The quantum version of the supercharge is

$$Q \sim g_{\mu\nu} p^{\mu} \gamma^{\nu} \tag{12.156}$$

that is the Dirac operator $\slashed{\partial}$ on M.

Let us define some symbols that will be employed in the next section. The connection one-form is

$$\Gamma^{\mu}_{\ \nu} = dx^{\lambda} \Gamma^{\mu}_{\ \lambda\nu} \tag{12.157}$$

while the Riemann curvature two-form is

$$\mathcal{R}^{\mu}_{\ \nu} = d\Gamma^{\mu}_{\ \nu} + \Gamma^{\mu}_{\ \sigma} \wedge \Gamma^{\sigma}_{\ \nu}. \tag{12.158}$$

The Riemann curvature two-form is expanded in terms of $dx^{\rho} \wedge dx^{\sigma}$ to yield

$$\mathcal{R}^{\mu}_{\ \nu} = \tfrac{1}{2} R^{\mu}_{\ \nu\rho\sigma} \, dx^{\rho} \wedge dx^{\sigma} \tag{12.159}$$

the component of which is the ordinary Riemann curvature tensor. This component is also written in terms of the connection ∇_{μ} as

$$R^{\kappa}_{\ \lambda\mu\nu} = \left\langle dx^{\kappa}, \nabla_{\mu}\nabla_{\nu}\frac{\partial}{\partial x^{\lambda}} - \nabla_{\nu}\nabla_{\mu}\frac{\partial}{\partial x^{\lambda}} \right\rangle$$
$$= \partial_{\mu}\Gamma^{\kappa}_{\ \nu\lambda} - \partial_{\nu}\Gamma^{\kappa}_{\ \mu\lambda} + \Gamma^{\eta}_{\ \nu\lambda}\Gamma^{\kappa}_{\ \mu\eta} - \Gamma^{\eta}_{\ \mu\lambda}\Gamma^{\kappa}_{\ \nu\eta}. \tag{12.160}$$

12.10   Supersymmetric proof of index theorem

The proof of the index theorem in its simplest setting will be given in the present section by making use of the supersymmetric quantum mechanics developed in the previous section.

12.10.1   The index

Let us consider vector bundles $E_{\pm} \xrightarrow{\pi} M$, $E = E_{+} \oplus E_{-}$ and let $\mathcal{D}$ be an elliptic differential operator acting as

$$\mathcal{D} : \Gamma(M, E^{+}) \to \Gamma(M, E^{-}).$$

It is possible, by using the fibre norm, to define the adjoint of $\mathcal{D}$ as

$$\mathcal{D}^{\dagger} : \Gamma(M, E^{-}) \to \Gamma(M, E^{+}).$$

Assuming that $\mathcal{D}$ is Fredholm, the index

$$\operatorname{Ind}\mathcal{D} = \dim\ker\mathcal{D} - \dim\ker\mathcal{D}^\dagger \qquad (12.161)$$

is well defined.

Theorem 12.4. The number ind $\mathcal{D}$ is invariant under a 'small' deformation of $\mathcal{D}$.

Proof. Note, first, that $\mathcal{D}\mathcal{D}^\dagger$ and $\mathcal{D}^\dagger\mathcal{D}$ are non-negative and, hence, it follows that

$$\ker\mathcal{D} = \ker\mathcal{D}^\dagger\mathcal{D} \qquad \ker\mathcal{D}^\dagger = \ker\mathcal{D}\mathcal{D}^\dagger.$$

Let $\{\phi_n\}$ be the orthonormal set of eigensections of $\mathcal{D}^\dagger\mathcal{D} : \Gamma(M, E^+) \to \Gamma(M, E^+)$:

$$(\mathcal{D}^\dagger\mathcal{D})\phi_n = \lambda_n\phi_n.$$

Define $\psi_n \equiv \mathcal{D}\phi_n/\sqrt{\lambda_n}$ for $\lambda_n > 0$, namely $\phi_n \in (\ker\mathcal{D})^\perp$. Then we find that ψ_n is an eigensection with the same eigenvalue λ_n, namely $\psi_n \in (\ker\mathcal{D}^\dagger)^\perp$ since

$$(\mathcal{D}\mathcal{D}^\dagger)\psi_n = \mathcal{D}(\mathcal{D}^\dagger\mathcal{D}\phi_n)/\sqrt{\lambda_n} = \lambda_n\mathcal{D}\phi_n/\sqrt{\lambda_n} = \lambda_n\psi_n.$$

Note also that $\{\psi_n\}$ is an orthonormal eigensection,

$$\langle\psi_n|\psi_m\rangle = \frac{1}{\sqrt{\lambda_n\lambda_m}}\langle\phi_n|\mathcal{D}^\dagger\mathcal{D}|\phi_m\rangle = \frac{\lambda_m}{\sqrt{\lambda_n\lambda_m}}\delta_{nm} = \delta_{nm}.$$

Thus, it follows that there is a natural isomorphism between $(\ker\mathcal{D})^\perp$ and $(\ker\mathcal{D}^\dagger)^\perp$. Note, however, that there exists no such isomorphism between $\ker\mathcal{D}$ and $\ker\mathcal{D}^\dagger$. Suppose N states in $\ker\mathcal{D}$ obtain non-vanishing eigenvalues as a result of a small perturbation of the operator $\mathcal{D}$ and $\dim\ker\mathcal{D}$ decreases by N. Then it follows from this observation that the same number of states must also leave $\ker\mathcal{D}^\dagger$. Otherwise $(\ker\mathcal{D})^\perp$ is no longer isomorphic to $(\ker\mathcal{D}^\dagger)^\perp$. Similary, if $\dim\ker\mathcal{D}$ increases by N, $\dim\ker\mathcal{D}^\dagger$ must also increase by N to keep the pairing properties of $(\ker\mathcal{D})^\perp$ and $(\ker\mathcal{D}^\dagger)^\perp$. Therefore, ind $\mathcal{D}$ is invariant under small perturbations of $\mathcal{D}$. $\qquad\square$

Theorem 12.5. Let $\mathcal{D}$ be a Fredholm differential operator. Then its index is given by

$$\operatorname{ind}\mathcal{D} = \operatorname{Tr}e^{-\beta\mathcal{D}^\dagger\mathcal{D}} - \operatorname{Tr}e^{-\beta\mathcal{D}\mathcal{D}^\dagger} \qquad (12.162)$$

where $\beta > 0$ is a real constant. In fact, the index is independent of β.

Proof. The traces in (12.162) are over $\{\phi_n\}$ and $\{\psi_n\}$, respectively. Let $\{\phi_i^0\}$ and $\{\psi_i^0\}$ be orthonormal eigensections of $\ker \mathcal{D}$ and $\ker \mathcal{D}^\dagger$, respectively, and $1 \leq i \leq \dim \ker \mathcal{D}$ and $1 \leq j \leq \dim \ker \mathcal{D}^\dagger$. Then it follows that

$$
\begin{aligned}
\mathrm{Tr}\, e^{-\beta \mathcal{D}^\dagger \mathcal{D}} &- \mathrm{Tr}\, e^{-\beta \mathcal{D} \mathcal{D}^\dagger} \\
&= \sum_{\lambda_n \neq 0} \langle \phi_n | e^{-\beta \mathcal{D}^\dagger \mathcal{D}} | \phi_n \rangle - \sum_{\lambda_n \neq 0} \langle \psi_n | e^{-\beta \mathcal{D} \mathcal{D}^\dagger} | \psi_n \rangle \\
&\quad + \sum_i \langle \phi_i^0 | \phi_i^0 \rangle - \sum_j \langle \psi_j^0 | \psi_j^0 \rangle \\
&= \sum_{\lambda_n \neq 0} e^{-\beta \lambda_n} \left(\langle \phi_n | \phi_n \rangle - \langle \psi_n | \psi_n \rangle \right) + \sum_i 1 - \sum_j 1 \\
&= \dim \ker \mathcal{D} - \dim \ker \mathcal{D}^\dagger \\
&= \mathrm{ind}\, \mathcal{D}.
\end{aligned}
$$

Since the summations over i and j are independent of β, $\mathrm{ind}\, \mathcal{D}$ thus defined is independent of β. □

The trace that appears in theorem 12.5 is identified with the heat kernel. Let $E = E_+ \oplus E_-$ and define a differential operator acting on E by[5] (cf equation (12.79))

$$
\mathrm{i} Q \equiv \begin{pmatrix} 0 & \mathcal{D}^\dagger \\ \mathcal{D} & 0 \end{pmatrix} : E \to E. \tag{12.163}
$$

Moreover, define a 'Hamiltonian' and a matrix Γ by

$$
H = (\mathrm{i} Q)^2 = \begin{pmatrix} \mathcal{D}^\dagger \mathcal{D} & 0 \\ 0 & \mathcal{D} \mathcal{D}^\dagger \end{pmatrix} \qquad \Gamma = \begin{pmatrix} 1 & 0 \\ 0 & -1 \end{pmatrix}. \tag{12.164}
$$

Since Q thus defined is anti-Hermitian, the operator H is Hermite and non-negative. The index of $\mathcal{D}$ is rewritten in a compact form by making use of Γ as

$$
\mathrm{ind}\, \mathcal{D} = \mathrm{Tr}\, \Gamma e^{-\beta H}. \tag{12.165}
$$

Let M be a spin manifold, for which the second Stiefel–Whitney class $w_2(M)$ is trivial. Accordingly, the $\mathrm{SO}(k)$ principal bundle over M may be lifted to the $\mathrm{SPIN}(k)$ principal bundle as

$$
\begin{array}{ccc}
\mathrm{SO}(k) & \to & \mathrm{SPIN}(k). \\
\downarrow \pi & & \\
M & &
\end{array}
$$

Let $E = \Delta(M)$ be this spin bundle. Then, associated with $\Delta(M)$ is a Clifford algebra $\{\gamma^\mu, \gamma^\nu\} = 2\delta^{\mu\nu}$. Let us define the **chirality operator**

$$
\gamma_{2n+1} \equiv \mathrm{i}^n \gamma_1 \gamma_2 \dots \gamma_{2n}. \tag{12.166}
$$

---

[5] The operator Q will be identified with the supercharge later.

It follows from $\gamma_{2n+1}^2 = 1$ that the eigenvalues of γ_{2n+1} are restricted to be ± 1, which we call **chirality**.

Exercise 12.10. Use the Clifford algebra to show that

$$\gamma_{2n+1}^2 = 1 \qquad \{\gamma_\mu, \gamma_{2n+1}\} = 0.$$

The set of sections $\Gamma(M, \Delta)$ for an even k is not an irreducible representation of SPIN(k) but can be decomposed into two subspaces according to the chirality as

$$\Gamma(M, \Delta) = \Gamma(M, \Delta^+) \oplus \Gamma(M, \Delta^-) \qquad (12.167)$$

where $\psi_\pm \in \Gamma(M, \Delta^\pm)$ satisfy $\gamma_{2n+1}\psi_\pm = \pm\psi_\pm$. We assign the **fermion number** $F = 0$ to sections in $\Gamma(M, \Delta^+)$ while $F = 1$ for those in $\Gamma(M, \Delta^-)$. Then the Γ defined in (12.164) can be written as

$$\Gamma = (-1)^F. \qquad (12.168)$$

It is clear that the operator Q flips the chirality and hence $\{Q, \Gamma\} = 0$.

Let Q be the Dirac operator on M and let $\Gamma = \gamma_{2n+1}$. In fact, it follows from exercise 12.11 that $\{Q, \gamma_{2n+1}\} = 0$ and γ_{2n+1} is identified with $(-1)^F$. When Γ is diagonalized as in (12.164), the chirality eigensections are expressed as[6]

$$\psi_+ = \begin{pmatrix} \psi_+ \\ 0 \end{pmatrix} \qquad \psi_- = \begin{pmatrix} 0 \\ \psi_- \end{pmatrix}. \qquad (12.169)$$

It should be then clear that $\mathcal{D} : \Gamma(M, \Delta^+) \to \Gamma(M, \Delta^-)$ and $\mathcal{D}^\dagger : \Gamma(M, \Delta^-) \to \Gamma(M, \Delta^+)$ are identified with D and $D^\dagger$, respectively, in (12.79). Accordingly, the index of the Dirac operator is defined as

$$\text{ind } Q = \dim \ker D - \dim \ker D^\dagger. \qquad (12.170)$$

Physicists often call the sections in $\ker D$ and $\ker D^\dagger$ **zero modes**. Then, the index of the Dirac operator is the difference between the number of positive and negative chirality zero modes. This index has a path integral expression as we see in the next subsection.

12.10.2   Path integral and index theorem

Let us consider a Dirac operator Q on a $2n$-dimensional spin manifold M. We employ Euclidean time $(t \to -it)$ from now on.

Let $H = (\mathrm{i}Q)^2 = \frac{1}{2}g_{\mu\nu}p^\mu p^\nu$ be the Hamiltonian corresonding to Q. Then the index of the Dirac operator has a path integral expression

$$\text{ind } Q = \text{Tr } \Gamma e^{-\beta H} = \text{Tr}(-1)^F e^{-\beta H}$$

$$= \int_{\text{PBC}} \mathcal{D}x \, \mathcal{D}\psi \, e^{-\int_0^\beta \mathrm{d}t \, L} \qquad (12.171)$$

---

[6] Note the slight abuse of notations. The symbols $\psi_\pm$ have been used to denote sections in $\Gamma(M, S)$ as well as those in $\Gamma(M, \Delta^\pm)$.

where the Lagrangian L has been introduced in (12.154),

$$L = \frac{1}{2}g_{\mu\nu}(x)\dot{x}^\mu\dot{x}^\nu + \frac{1}{2}g_{\mu\nu}(x)\psi^\mu\frac{D\psi^\nu}{Dt} \qquad (12.172)$$

and PBC stands for the boundary condition in which the path integral is over functions satisfying a periodic boundary condition over $[0, \beta]$. The factor $(-1)^F$ disappears if the anti-periodic boundary condition for the fermionic variables is changed into a periodic one. This can be seen from the following observation. In the path integral formalism, the trace with $(-1)^F$ is (see section 1.5)

$$\text{tr}(-1)^F e^{-\beta H} = \sum_n \langle n|(-1)^F e^{-\beta H}|n\rangle$$

$$= \int d\theta^* \, d\theta \, \langle -\theta|(-1)^F e^{-\beta H}|\theta\rangle e^{-\theta^*\theta} \qquad (12.173)$$

where $F = c^\dagger c$ is the Fermion number operator. By noting that

$$|\theta\rangle = |0\rangle + |1\rangle\theta \qquad (-1)^F|\theta\rangle = |0\rangle - |1\rangle\theta = |-\theta\rangle$$

this integral is cast into the form

$$\int d\theta^* \, d\theta \, \langle\theta|e^{-\beta H}|\theta\rangle e^{-\theta^*\theta}. \qquad (12.174)$$

Thus, by eliminating $(-1)^F$, we have to change the boundary condition to a periodic one.

This path integral is evaluated in the rest of this section to show that it reduces to a topological index obtained from the Dirac $\hat{A}$-genus.

The SUSY transformation in Euclidean time is obtained by the replacement $t \to -it$ in (12.140) as

$$\delta x^\mu = i\epsilon\psi^\mu \qquad \delta\psi^\mu = -i\epsilon\dot{x}^\mu.$$

As was shown in the previous subsection, the index is independent of β and, hence, we may consider the limit $\beta \downarrow 0$ in computing the trace. By rescaling the time parameter as $t = \beta s$, we cast the action into the form

$$\int_0^\beta dt \left[\frac{1}{2}g_{\mu\nu}(x)\dot{x}^\mu\dot{x}^\nu + \frac{1}{2}g_{\mu\nu}(x)\psi^\mu\frac{D\psi^\nu}{Dt} \right]$$

$$= \int_0^1 ds \left[\frac{1}{\beta}\frac{1}{2}g_{\mu\nu}(x)\frac{dx^\mu}{ds}\frac{dx^\nu}{ds} + \frac{1}{2}g_{\mu\nu}(x)\psi^\mu\frac{D\psi^\nu}{Ds} \right]. \qquad (12.175)$$

Thus, any path with $\dot{x} \neq 0$ has an exponentially small contribution to the path integral in the limit $\beta \downarrow 0$. Accordingly, the contributions to the path integral come only from paths $x(t) = $ constant in this limit. Clearly, these paths satisfy the periodic boundary condition.

The periodic boundary condition forces us to take the set of loops in M, which we will denote as $L(M)$, as the configuration space of the bosonic coordinates. To apply the saddle point method to the evaluation of the path integral, we have to find the set $\mathcal{M}$ of the extrema of the action, namely the solutions of the classical Euler–Lagrange equations

$$-g_{\lambda\mu}(x)\frac{D\dot{x}^{\mu}}{Dt} + \frac{1}{2}R_{\mu\nu\lambda\rho}\psi^{\mu}\psi^{\nu}\dot{x}^{\rho} = 0 \tag{12.176}$$

$$\frac{D\psi^{\mu}}{Dt} = \frac{d\psi^{\mu}}{dt} + \dot{x}^{\lambda}\Gamma^{\mu}_{\lambda\nu}\psi^{\nu} = 0. \tag{12.177}$$

It is instructive to outline the derivation of these equations since the anti-commutativity of Grassmann numbers and the symmetries of the Riemann tensor are fully utilized. The Euler–Lagrange equation for ψ^{μ} is

$$0 = \frac{\partial L}{\partial \psi^{\rho}} - \frac{d}{dt}\left(\frac{\partial L}{\partial \dot{\psi}^{\rho}}\right)$$
$$= \frac{1}{2}g_{\rho\nu}\frac{D\psi^{\nu}}{Dt} - \frac{1}{2}g_{\kappa\nu}\psi^{\kappa}\dot{x}^{\lambda}\Gamma^{\nu}_{\lambda\rho} + \frac{1}{2}\frac{d}{dt}\left(g_{\rho\nu}\psi^{\nu}\right)$$
$$= \frac{1}{2}\left[g_{\rho\nu}\frac{D\psi^{\nu}}{Dt} - g_{\kappa\nu}\dot{x}^{\lambda}\Gamma^{\nu}_{\lambda\rho}\psi^{\kappa} + \left(\partial_{\lambda}g_{\rho\nu}\right)\dot{x}^{\lambda}\psi^{\nu} + g_{\rho\nu}\dot{\psi}^{\nu}\right].$$

By multiplying both sides by $g^{\mu\rho}$ and summing over ρ, we have

$$0 = \frac{D\psi^{\mu}}{Dt} - g^{\mu\rho}g_{\kappa\nu}\dot{x}^{\lambda}\Gamma^{\nu}_{\lambda\rho}\psi^{\kappa} + g^{\mu\rho}\left(\partial_{\lambda}g_{\rho\nu}\right)\dot{x}^{\lambda}\psi^{\nu} + \dot{\psi}^{\mu}$$
$$= \frac{D\psi^{\mu}}{Dt} + \dot{\psi}^{\mu} + \dot{x}^{\lambda}\left[g^{\mu\rho}\left(\partial_{\lambda}g_{\rho\nu}\right) - g^{\mu\rho}g_{\nu\kappa}\Gamma^{\kappa}_{\lambda\rho}\right]\psi^{\nu} = 2\frac{D\psi^{\mu}}{Dt}$$

which proves (12.177). Here, use has been made of the identity

$$g^{\mu\rho}[(\partial_{\lambda}g_{\rho\nu}) - \tfrac{1}{2}(\partial_{\lambda}g_{\nu\rho} + \partial_{\rho}g_{\nu\lambda} - \partial_{\nu}g_{\lambda\rho})]$$
$$= g^{\mu\rho}\tfrac{1}{2}\left(\partial_{\lambda}g_{\rho\nu} + \partial_{\nu}g_{\lambda\rho} - \partial_{\rho}g_{\nu\lambda}\right) = \Gamma^{\mu}_{\nu\lambda}$$

in the square brackets in the second line above.

Let us prove the equation of motion for x^{μ} next. We find

$$\frac{\partial L}{\partial x^{\mu}} - \frac{d}{dt}\left(\frac{\partial L}{\partial \dot{x}^{\mu}}\right)$$
$$= \frac{1}{2}(\partial_{\mu}g_{\alpha\beta})\dot{x}^{\alpha}\dot{x}^{\beta} + \frac{1}{2}(\partial_{\mu}g_{\alpha\beta})\psi^{\alpha}\frac{D\psi^{\beta}}{Dt} + \frac{1}{2}g_{\alpha\beta}\psi^{\alpha}\dot{x}^{\lambda}\partial_{\mu}\Gamma^{\beta}_{\lambda\kappa}\psi^{\kappa}$$
$$\quad - \frac{d}{dt}\left(g_{\mu\nu}\dot{x}^{\nu} + \frac{1}{2}g_{\alpha\beta}\psi^{\alpha}\Gamma^{\beta}_{\mu\kappa}\psi^{\kappa}\right)$$
$$= -[g_{\mu\nu}\ddot{x}^{\nu} + \tfrac{1}{2}(\partial_{\lambda}g_{\mu\nu} + \partial_{\nu}g_{\mu\lambda} - \partial_{\mu}g_{\nu\lambda})\dot{x}^{\nu}\dot{x}^{\lambda}]$$

$$+ \tfrac{1}{2}[g_{\alpha\beta}\partial_\mu\Gamma^\beta_{\lambda\kappa} - \partial_\lambda g_{\alpha\beta}\Gamma^\beta_{\mu\kappa} - g_{\alpha\beta}\partial_\lambda\Gamma^\beta_{\mu\kappa}]\psi^\alpha\psi^\kappa\dot{x}^\lambda$$

$$+ \tfrac{1}{2}g_{\alpha\beta}\dot{x}^\lambda\Gamma^\alpha_{\lambda\gamma}\psi^\gamma\Gamma^\beta_{\mu\kappa}\psi^\kappa + \tfrac{1}{2}g_{\alpha\beta}\psi^\alpha\Gamma^\beta_{\mu\kappa}\dot{x}^\lambda\Gamma^\kappa_{\lambda\nu}\psi^\nu$$

$$= -g_{\mu\nu}\frac{D\dot{x}^\nu}{Dt} + \frac{1}{2}[g_{\alpha\beta}\partial_\mu\Gamma^\beta_{\lambda\kappa} - g_{\alpha\beta}\partial_\lambda\Gamma^\beta_{\mu\kappa} - \partial_\lambda g_{\alpha\beta}\Gamma^\beta_{\mu\kappa}$$

$$+ g_{\gamma\beta}\Gamma^\gamma_{\lambda\alpha}\Gamma^\beta_{\mu\kappa} + g_{\alpha\beta}\Gamma^\beta_{\mu\gamma}\Gamma^\gamma_{\lambda\kappa}]\psi^\alpha\psi^\kappa\dot{x}^\lambda$$

$$= -g_{\mu\nu}\frac{D\dot{x}^\nu}{Dt} + \frac{1}{2}(\partial_\mu\Gamma^\beta_{\lambda\kappa} - \partial_\lambda\Gamma^\beta_{\mu\kappa} + \Gamma^\beta_{\mu\kappa}\Gamma^\gamma_{\lambda\kappa})\psi^\alpha\psi^\kappa\dot{x}^\lambda$$

$$+ \tfrac{1}{2}(g_{\gamma\beta}\Gamma^\gamma_{\lambda\alpha} - \partial_\lambda g_{\alpha\beta})\Gamma^\beta_{\mu\kappa}\psi^\alpha\psi^\kappa\dot{x}^\lambda.$$

The last term of the last line of this equation is written as

$$[g_{\gamma\beta}\tfrac{1}{2}g^{\gamma\nu}(\partial_\lambda g_{\nu\alpha} + \partial_\alpha g_{\nu\lambda} - \partial_\nu g_{\lambda\alpha}) - \partial_\lambda g_{\alpha\beta}]\Gamma^\beta_{\mu\nu}\psi^\alpha\psi^\kappa\dot{x}^\lambda$$

$$= -\tfrac{1}{2}(\partial_\lambda g_{\alpha\beta} + \partial_\beta g_{\lambda\alpha} - \partial_\alpha g_{\lambda\beta})\Gamma^\beta_{\mu\nu}\psi^\alpha\psi^\kappa\dot{x}^\lambda$$

$$= -\Gamma_{\alpha\lambda\beta}\Gamma^\beta_{\mu\kappa}\psi^\alpha\psi^\kappa\dot{x}^\lambda$$

$$= -g_{\alpha\beta}\Gamma^\beta_{\lambda\beta}\Gamma^\beta_{\mu\kappa}\psi^\alpha\psi^\kappa\dot{x}^\lambda$$

from which we obtain

$$0 = -g_{\mu\nu}\frac{D\dot{x}^\nu}{Dt} + \frac{1}{2}(\partial_\mu\Gamma^\beta_{\lambda\kappa} - \partial_\lambda\Gamma^\beta_{\mu\kappa} + \Gamma^\beta_{\mu\gamma}\Gamma^\gamma_{\lambda\kappa} - \Gamma^\beta_{\lambda\gamma}\Gamma^\gamma_{\mu\kappa})\psi^\alpha\psi^\kappa\dot{x}^\lambda$$

$$= -g_{\mu\nu}\frac{D\dot{x}^\nu}{Dt} + \frac{1}{2}R_{\alpha\kappa\mu\lambda}\psi^\alpha\psi^\kappa\dot{x}^\lambda.$$

Equation (12.176) follows by renaming dummy indices.

Let us come back to the study of the solutions of the equations of motion (12.176) and (12.177). Clearly, the pair $x =$ constant and $\psi =$ constant is one of solutions. Therefore, $x_p : t \mapsto p \in M$ is always contained in the solutions, which may be written as $M \subset \mathcal{M}$. Equation (12.176) reduces to the geodesic equation when $\psi = 0$ but not necessarily so in general. When the fundamental group $\pi_1(M)$ is non-trivial, there exist non-contractible geodesics in general. Their contributions to the path integral, however, vanish exponentially as $\exp(-c/\beta)$ as $\beta \downarrow 0$ and, hence, are negligible.

Before we proceed to the proof of the index theorem, we need to explain the **saddle point method**. Let us start with a simple example. Consider the integral

$$Z = \int_{-\infty}^{\infty} \frac{dx}{\sqrt{2\pi\hbar}} e^{-f(x)/\hbar}.$$

The function $f(x)$ is assumed to have only one minimum at $x = x_0$ and that $f(x) \to \infty$ as $x \to \pm\infty$. Let us consider the asymmptotic expansion of the integral Z when the limit $\hbar \to 0$ is taken. Put $x = x_0 + \sqrt{\hbar}y$ and expand $f(x)$ at x_0. Taking $f'(x_0) = 0$ into account, we obtain the expansion

$$f(x) = f(x_0) + \frac{1}{2!}\hbar y^2 f''(x_0) + \frac{1}{3!}\hbar^{3/2}y^3 f^{(3)}(x_0) + \frac{1}{4!}\hbar^2 y^4 f^{(4)}(x_0) + \cdots.$$

If this expansion is substituted into Z, we have

$$
Z = e^{-f(x_0)/\hbar} \int_{-\infty}^{\infty} \frac{dy}{\sqrt{2\pi}}
$$
$$
\times \exp\left[-\frac{1}{2}y^2 f''(x_0) - \left(\frac{1}{3!}\hbar^{1/2}y^3 f^{(3)}(x_0) + \frac{1}{4!}\hbar y^4 f^{(4)}(x_0) + \cdots \right) \right].
$$

Let us define the moment of y by

$$
\langle y^n \rangle = \frac{\displaystyle\int \frac{dy}{\sqrt{2\pi}} \, y^n e^{-y^2 f''(x_0)/2}}{\displaystyle\int \frac{dy}{\sqrt{2\pi}} \, e^{-y^2 f''(x_0)/2}}.
$$

Then we finally obtain the expansion of Z as

$$
Z = \frac{e^{-f(x_0)/\hbar}}{\sqrt{f''(x_0)}} \left\langle \exp\left[-\frac{1}{3!}\hbar^{1/2}y^3 f^{(3)}(x_0) - \frac{1}{4!}\hbar y^4 f^{(4)}(x_0) \cdots \right] \right\rangle.
$$

One might think that one will get terms of order $O(\hbar^{1/2})$ if $\langle \cdots \rangle$ is expanded. However, this is not the case since $\langle y^3 \rangle = 0$ and one has $\langle \cdots \rangle = 1 + O(\hbar)$ in reality. In the proof of the following index theorem, the parameter $\hbar$ is replaced by β. The index is, however, independent of β and we conclude that terms of order $O(\beta)$ vanish and, hence, we need to take only the extrema of the action and the second-order fluctuations thereof into account.

Exercise 12.11. Use the previous expansion to prove the **Staring formula**

$$
n! \simeq \sqrt{2\pi n}e^{-n}n^n \tag{12.178}
$$

for $n \gg 1$.

Let us come back to SUSYQM. We take the second-order fluctuation around the solutions of the classical equations of motion in evaluating Z. The principal contribution to the path integral comes from the solution $x = x_0$ and $\psi = \psi_0$. We employ the **Riemann normal coordinate** based at $x = x_0$ to make our life easier. This is to take a coordinate system in which the metric tensor satisfies conditions[7]

$$
g_{\mu\nu}(x_0) = \delta_{\mu\nu} \qquad \frac{\partial}{\partial x^\lambda} g_{\mu\nu}(x_0) = 0.
$$

Thus, we have $g \equiv \det g = 1$. We define the fluctuations in this coordinate system as

$$
x^\mu(t) = x_0^\mu + \xi^\mu(t)
$$
$$
\psi^\mu(t) = \psi_0^\mu + \eta^\mu(t).
$$

[7] Of course, this choice does not imply that the Riemann tensor vanishes in general.

Note here that $dx^\mu = d\xi^\mu$, $d\psi^\mu = d\eta^\mu$. The second-order expansion of the action is now written as

$$S_2 = \int_0^\beta dt \left[\frac{1}{2} \frac{d\xi^\mu}{dt} \frac{d\xi^\mu}{dt} + \frac{1}{2} \eta^\mu \frac{d\eta^\mu}{dt} + \frac{1}{2} \tilde{\mathcal{R}}_{\mu\nu}(x_0) \xi^\mu \frac{d\xi^\nu}{dt} \right] \qquad (12.179)$$

where we have put

$$\tilde{\mathcal{R}}_{\mu\nu}(x_0) = \tfrac{1}{2} R_{\mu\nu\rho\sigma}(x_0) \psi_0^\rho \psi_0^\sigma.$$

Needless to say, the zeroth-order action $S_0 = S(x_0, \psi_0)$ vanishes identically.
     Let us evaluate the index

$$\text{ind } Q = \int \mathcal{D}\xi \mathcal{D}\eta e^{-S_2} \qquad (12.180)$$

using the second-order action S_2. Here we have taken the translational invariance of the path integral measure $\mathcal{D}x\mathcal{D}\psi = \mathcal{D}\xi\mathcal{D}\eta$. Taking the periodic boundary condition of ξ, η into account, their Fourier expansions are given by

$$\xi^\mu = \frac{1}{\sqrt{\beta}} \sum_{n=-\infty}^{\infty} \xi_n^\mu e^{2\pi i n t/\beta}$$

$$\eta^\mu = \frac{1}{\sqrt{\beta}} \sum_{n=-\infty}^{\infty} \eta_n^\mu e^{2\pi i n t/\beta}.$$

The fluctuation operator for ξ in S_2 is

$$-\delta_{\mu\nu} \frac{d^2}{dt^2} + \tilde{\mathcal{R}}_{\mu\nu} \frac{d}{dt}$$

while that for η is

$$\delta_{\mu\nu} \frac{d}{dt}.$$

We have to consider the zero modes ξ_0^μ and η_0^μ, for which $n = 0$, separately in the following Gaussian integrals.[8] Taking these into account, we write

$$\text{ind } Q = \mathcal{N} \int \prod_{\mu=1}^{d} \frac{d\xi_0^\mu}{\sqrt{2\pi}} d\eta_0^\mu \left[\text{Det}_{\text{PBC}}' \left(\delta_{\mu\nu} \frac{d}{dt} \right) \right]^{1/2}$$

$$\times \left[\text{Det}_{\text{PBC}}' \left(-\delta_{\mu\nu} \frac{d^2}{dt^2} + \tilde{\mathcal{R}}_{\mu\nu}(x_0) \frac{d}{dt} \right) \right]^{-1/2}$$

$$= \mathcal{N} \int \prod_{\mu=1}^{d} \frac{d\xi_0^\mu}{\sqrt{2\pi}} d\eta_0^\mu \left[\text{Det}_{\text{PBC}}' \left(-\delta_{\mu\nu} \frac{d}{dt} + \tilde{\mathcal{R}}_{\mu\nu}(x_0) \right) \right]^{-1/2} \qquad (12.181)$$

[8] The integrations over ξ_0 and η_0 are equivalent with those over x_0 and ψ_0.

where $\prime$ indicates that the zero modes are omitted while $\mathcal{N}$ is the normalization factor, which takes care of the ambiguities associated with the ordering of Grassmann numbers. Let us evaluate this factor now.

Since ind Q is independent of β, we put $\beta = 1$ for simplicity. We also simplify our calculation by choosing the metric to be $g_{\mu\nu} = \delta_{\mu\nu}$. Then the fermion and boson parts separate completely. The fermionic part is evaluated, by noting $H_{\text{fermion}} = 0$, to yield

$$
\begin{aligned}
\operatorname{Tr} \gamma_{2n+1} &= \int_{\text{PBC}} \mathcal{D}\psi \, e^{-\frac{1}{2} \int_0^1 \psi \cdot \dot{\psi} \, dt} \\
&= \mathcal{N}_f \operatorname{Det}'_{\text{PBC}}(\delta_{\mu\nu}\partial_t)^{1/2} \int d\psi_0^1 \cdots d\psi_0^{2n},
\end{aligned}
$$

where ψ_0^μ is the zero mode. The determinant is evaluated as follows. First, note that the argument in section 1.5 shows that the determinant is, in fact,

$$
\operatorname{Det}'_{\text{PBC}} (\partial_t + \omega) = \lim_{\varepsilon \to 0} \operatorname{Det}' ((1 - \varepsilon\omega)\partial_t + \omega)
$$

where we have introduced the harmonic oscillator frequency ω, which will be set to zero at the end of the calculation. The 'partition function' is

$$
\begin{aligned}
\operatorname{tr}(-1)^F e^{-\beta H} &= 2 \sinh(\beta\omega/2) \\
&= e^{\beta\omega/2} \operatorname{Det}'_{\text{PBC}} ((1 - \varepsilon\omega)\partial_t + \omega).
\end{aligned} \tag{12.182}
$$

Therefore, the determinant in the limit $\omega \to 0$ is

$$
\operatorname{Det}'_{\text{PBC}} (\partial_t) = \lim_{\omega \to 0} e^{-\beta\omega/2} 2 \sinh(\beta\omega/2) = 1. \tag{12.183}
$$

Thus, we finally obtained

$$
\operatorname{Tr} \gamma_{2n+1} = \mathcal{N}_f \int d\psi_0^1 \ldots d\psi_0^{2n}. \tag{12.184}
$$

We insert

$$
\gamma_{2n+1} = i^n \gamma_0^1 \cdots \gamma_0^{2n} = (2i)^n \psi_0^1 \cdots \psi_0^{2n}
$$

further in the trace. Since $\operatorname{Tr} \gamma_{2n+1}^2 = \operatorname{Tr} I = 2^n$, we obtain

$$
\operatorname{Tr} \gamma_{2n+1}^2 = 2^n = \mathcal{N}_f \int d\psi_0^1 \ldots d\psi_0^{2n} (2i)^n \psi_0^1 \cdots \psi_0^{2n} = \mathcal{N}_f(-2i)^n
$$

which leads to

$$
\mathcal{N}_f = i^n.
$$

Next, we evaluate the normalization factor $\mathcal{N}_b$ of the boson part. If we employ imaginary time in (1.101) to obtain $\langle x, 1 | x, 0 \rangle = (2\pi)^{-1/2}$, we have

$$
\int \mathcal{D}x^\mu e^{-\frac{1}{2} \int_0^1 \dot{x}^{\mu\,2}} = \mathcal{N}_b \frac{1}{\operatorname{Det}^{1/2}(-\delta_{\mu\nu}\partial_t^2)} \int \prod_{\mu=1}^{2n} \frac{dx^\mu}{\sqrt{2\pi}} = (2\pi)^{-n} \int \prod_{\mu=1}^{2n} dx^\mu.
$$

The determinant is evaluated using the ζ-function regularization as in section 1.4. The eigenvalue of $-\mathrm{d}^2/\mathrm{d}t^2$ with the periodic boundary condition is $\lambda_n = (2n\pi/\beta)^2$ and then

$$\mathrm{Det}'_{\mathrm{PBC}}\left(-\frac{\mathrm{d}^2}{\mathrm{d}t^2}\right) = \prod_{n\in\mathbb{Z},n\neq0}\left(\frac{2\pi n}{\beta}\right)^2.$$

The spectral ζ-function is

$$\zeta_{-\mathrm{d}^2/\mathrm{d}t^2}(s) = \sum_{n\in\mathbb{Z},n\neq0}^{\infty}\left[\left(\frac{2n\pi}{\beta}\right)^2\right]^{-s} = 2\left(\frac{\beta}{2\pi}\right)^{2s}\zeta(2s)$$

from which we find

$$\zeta'_{-\mathrm{d}^2/\mathrm{d}t^2}(0) = 4\log(\beta/2\pi)\mathrm{e}^{2s\,\log(\beta/2\pi)}\zeta(2s) + 4\mathrm{e}^{2s\,\log(\beta/2\pi)}\zeta'(2s)|_{s=0}$$
$$= 4[\log(\beta/2\pi)\zeta(0) + \zeta'(0)] = -2\log\beta.$$

Therefore, the determinant is

$$\mathrm{Det}'_{\mathrm{PBC}}\left(-\frac{\mathrm{d}^2}{\mathrm{d}t^2}\right) = \exp[-\zeta'_{-\mathrm{d}^2/\mathrm{d}t^2}(0)] = \beta^2. \tag{12.185}$$

By putting $\beta = 1$, we find $\mathrm{Det}'_{\mathrm{PBC}}\left(-\mathrm{d}^2/\mathrm{d}t^2\right) = 1$. Thus, we have obtained the normalization factor

$$\mathcal{N}_b = 1.$$

Putting these results together, we have shown that $\mathcal{N} = \mathcal{N}_f\mathcal{N}_b = \mathrm{i}^n$. Accordingly, the index is expressed as

$$\mathrm{ind}\,Q = \mathrm{i}^n \int \prod_{\mu=1}^{d}\frac{\mathrm{d}\xi_0^{\mu}}{\sqrt{2\pi}}\mathrm{d}\eta_0^{\mu}\left[\mathrm{Det}_{\mathrm{PBC}}'\left(-\delta_{\mu\nu}\frac{\mathrm{d}}{\mathrm{d}t} + \tilde{\mathcal{R}}_{\mu\nu}(x_0)\right)\right]^{-1/2}. \tag{12.186}$$

Let us evaluate the functional determinant in (12.186). Since the Fermi variables are contained only in $\tilde{\mathcal{R}}_{\mu\nu}(x_0)$ and this is Grassmann-even, we pretend this part is a commuting number for the time being. The anti-symmetry of the Riemann tensor implies that $\tilde{\mathcal{R}}_{\mu\nu}(x_0)$ satisfies $\tilde{\mathcal{R}}_{\mu\nu} = -\tilde{\mathcal{R}}_{\nu\mu}$. Therefore, it is possible, in an even-dimensional manifold M, to block-diagonalize $\tilde{\mathcal{R}}_{\mu\nu}$ in the form

$$\tilde{\mathcal{R}}_{\mu\nu} = \begin{pmatrix} 0 & y_1 & & & \\ -y_1 & 0 & & & \\ & & \ddots & & \\ & & & 0 & y_n \\ & & & -y_n & 0 \end{pmatrix}. \tag{12.187}$$

Let us concentrate on the first block. The operator

$$-\delta_{\mu\nu}\frac{\mathrm{d}}{\mathrm{d}t} + \tilde{\mathcal{R}}_{\mu\nu}(x_0)$$

is real and, hence, the eigenvalues are made of complex conjugate pairs. Let us express the determinant of this block in terms of the product of these complex eigenvalues. We find

$$\det' \begin{pmatrix} -\dfrac{\mathrm{d}}{\mathrm{d}t} & y_1 \\ -y_1 & -\dfrac{\mathrm{d}}{\mathrm{d}t} \end{pmatrix} = \mathrm{Det}' \left(\frac{\mathrm{d}^2}{\mathrm{d}t^2} + y_1^2 \right) = \prod_{n \neq 0} \left(y_1^2 - (2\pi n/\beta)^2 \right)$$

$$= \left[\prod_{n \geq 1} \left(\frac{2\pi n}{\beta} \right)^2 \prod_{n \geq 1} \left[1 - \left(\frac{y_1 \beta}{2\pi n} \right)^2 \right] \right]^2$$

$$= \left(\frac{\sin \beta y_1/2}{y_1/2} \right)^2. \tag{12.188}$$

Now the index is expressed as

$$\mathrm{ind}\, Q = \mathrm{i}^n \int \prod_{\mu=1}^{2n} \frac{\mathrm{d}\xi_0^\mu}{\sqrt{2\pi}} \mathrm{d}\eta_0^\mu \prod_{j=1}^{n} \frac{y_j/2}{\sin \beta y_j/2}. \tag{12.189}$$

The product with respect to j is written as

$$\frac{1}{\beta^{d/2}} \det \left(\frac{\beta \tilde{\mathcal{R}}/2}{\sin \beta \tilde{\mathcal{R}}/2} \right)^{1/2}.$$

Note that any Taylor expansion with respect to $\tilde{\mathcal{R}}$ terminates at finite order since $\tilde{\mathcal{R}}^p = 0$ for $p > d/2$.

We have evaluated the contributions of the second-order fluctuations around a particular pair x_0, ψ_0 so far. Now we need to take the contributions coming from all the solutions to the classical equations of motion into account. We have noted before that the set $\mathcal{M}$ of the solutions of the equations of motion contains the constant solution (x_0, ψ_0) as a subset and that the contributions from non-constant solutions are exponentially small as $\beta \downarrow 0$. Therefore, we neglect all periodic solutions except for constant solutions. If we note the expansion

$$x^\mu = x_0^\mu + \frac{1}{\sqrt{\beta}} \xi_0^\mu + \cdots$$

we find that the integral over x_0 is equivalent with that over $\xi_0/\sqrt{\beta}$, namely $\mathrm{d}x_0^\mu = \mathrm{d}\xi_0^\mu/\sqrt{\beta}$. This argument is also applied to the Grassmannian zero mode and we find $d\psi_0^\mu = \sqrt{\beta} d\eta_0^\mu$. In summary, the index is now written as

$$\text{ind } Q = i^n \int \prod_{\mu=1}^{2n} \frac{dx_0^\mu}{\sqrt{2\pi}} d\psi_0^\mu \frac{1}{\beta^{d/2}} \det \left(\frac{\beta\tilde{\mathcal{R}}/2}{\sin\beta\tilde{\mathcal{R}}/2} \right)^{1/2}. \qquad (12.190)$$

We make the following change of variables to erase the apparent β-dependence of the index,

$$\psi_0^\mu = \frac{\chi_0^\mu}{\sqrt{2\pi\beta}}, \qquad d\psi_0^\mu = \sqrt{2\pi\beta}\, dx_0^\mu.$$

Substituting

$$\beta\tilde{\mathcal{R}}_{\mu\nu} = \frac{1}{2\pi}\frac{1}{2}\mathcal{R}_{\mu\nu\rho\sigma}\chi_0^\rho\chi_0^\sigma$$

into the integrand, we obtain

$$\text{ind } Q = i^n \int \prod_{\mu=1}^{2n} dx_0^\mu\, dx_0^\mu\, \det \left(\frac{\frac{1}{2}\frac{1}{2\pi}\frac{1}{2}\mathcal{R}_{\mu\nu\rho\sigma}(x_0)\chi_0^\rho\chi_0^\sigma}{\sin\frac{1}{2}\frac{1}{2\pi}\frac{1}{2}\mathcal{R}_{\mu\nu\rho\sigma}(x_0)\chi_0^\rho\chi_0^\sigma} \right)^{1/2}. \qquad (12.191)$$

This is the Atiyah–Singer index theorem for the Dirac operator.

Let us rewrite the previous theorem in a more familiar form. Note that only terms of order $2n$ in χ in the integrand yield non-vanishing contributions upon integration over $\prod dx_0^\mu$. Note also that $\prod dx_0^\mu$ is just an ordinary volume element. Then define the curvature two-form

$$\mathcal{R}_{\mu\nu} = \tfrac{1}{2} R_{\mu\nu\rho\sigma}\, dx^\rho \wedge dx^\sigma. \qquad (12.192)$$

Then note that $\mathcal{R}/\sin\mathcal{R}$ is even in $\mathcal{R}$ and, hence, the integral is non-vanishing only when n is even, that is only when d is a multiple of four. If this is the case, the factor i^n takes only ± 1 and we can formally replace the integrand as

$$i^n \frac{\mathcal{R}}{\sin\mathcal{R}} \rightarrow \frac{\mathcal{R}}{\sinh\mathcal{R}}.$$

The reader should verify the first few terms. Then the index is now written in the well-known form as

$$\text{ind } Q = \int_M \det \left(\frac{\frac{1}{2}\frac{1}{2\pi}\mathcal{R}}{\sinh\frac{1}{2}\frac{1}{2\pi}\mathcal{R}} \right)^{1/2}.$$

We, moreover, define the $\hat{A}$-genus. Since $\mathcal{R}$ is anti-symmetric, it can be block-diagonalized as

$$\frac{1}{2\pi}\mathcal{R}_{\mu\nu} = \begin{pmatrix} 0 & x_1 & & & \\ -x_1 & 0 & & & \\ & & \ddots & & \\ & & & 0 & x_n \\ & & & -x_n & 0 \end{pmatrix}.$$

Then define the $\hat{A}$-genus of M by

$$\hat{A}(M) = \prod_{j=1}^{n} \frac{x_j/2}{\sinh x_j/2} \tag{12.193}$$

where the RHS is defined by its formal expansion with respect to x_j.

In summary, we have proved the Atiyah–Singer index theorem in the simplest setting (the spin complex).

Theorem 12.6. (**Index theorem for a spin complex**) The index of a Dirac operator defined in M is

$$\text{ind}\, Q = \int_M \hat{A}(M). \tag{12.194}$$

Problems

12.1 In the text, we dealt only with compact manifolds. The extension of the AS index theorem to non-compact manifolds is the Callias–Bott–Seely index theorem (Callias 1978, Bott and Seely 1978). Here we consider the simplest case studied by Hirayama (1983). Consider a pair of operators

$$L \equiv \frac{1}{i}\frac{d}{dx} - iW(x) \qquad L^\dagger \equiv \frac{1}{i}\frac{d}{dx} + iW(x)$$

where $W(+\infty) = \mu$ and $W(-\infty) = \lambda$.

(a) Show that $\text{Spec}'\, L^\dagger L = \text{Spec}'\, LL^\dagger$, where the prime indicates that the zero eigenvalues are omitted.

(b) Show that

$$J(z) \equiv \text{tr}\left(\frac{z}{L^\dagger L + z} - \frac{z}{LL^\dagger + z}\right) = \frac{1}{2}\left(\frac{\mu}{(\mu^2 + z)^{1/2}} - \frac{\lambda}{(\lambda^2 + z)^{1/2}}\right).$$

13

ANOMALIES IN GAUGE FIELD THEORIES

In particle physics, symmetry principles are some of the most important concepts in model building. Symmetries play crucial roles for the theory to be renormalizable and unitary. The Lagrangian must be chosen so that it fulfils the observed symmetry. Note, however, that the symmetry of the Lagrangian is *classical*. There is no warranty that symmetry of the Lagrangian may be elevated to a *quantum* symmetry, i.e., the symmetry of the effective action. If the classical symmetry of the Lagrangian cannot be maintained in the process of quantization, the theory is said to have an *anomaly*. There are many types of anomaly: the chiral anomaly, gauge anomaly, gravitational anomaly, supersymmetry anomaly and so on. Each adjective refers to the symmetry under consideration. In the present chapter we look at the geometrical and topological structures of the anomalies appearing in gauge theories.

We follow closely Alvarez-Gaumé (1986), Alvarez-Gaumé and Ginsparg (1985) and Sumitani (1985). See Rennie (1990) and Bartlmann (1996) for a complete analysis of the subject. Mickelsson (1989) and Nash (1991) have a section on anomalies from a more mathematical point of view.

13.1 Introduction

Before we introduce topological and geometrical methods to anomalies, we give a brief survey of the subject here. Let ψ be a massless Dirac field in four-dimensional space interacting with an external gauge field $\mathcal{A}_\mu = A_\mu{}^\alpha T_\alpha$, where $\{T_\alpha\}$ is the set of anti-Hermitian generators of the gauge group G which is compact and semisimple (SU(N), for example). The theory is described by the Lagrangian

$$\mathcal{L} = i\bar{\psi}\gamma^\mu(\partial_\mu - \mathcal{A}_\mu)\psi. \tag{13.1}$$

The Lagrangian is invariant under the usual (local) gauge transformation

$$\psi(x) \to g^{-1}\psi(x) \qquad \mathcal{A}_\mu(x) \to g^{-1}[\mathcal{A}_\mu(x) + \partial_\mu]g. \tag{13.2}$$

It also has a *global* symmetry,

$$\psi(x) \to e^{i\gamma_5\alpha}\psi(x) \qquad \bar{\psi}(x) \to \bar{\psi}(x)e^{i\gamma_5\alpha} \tag{13.3}$$

called the **chiral symmetry**. The chiral current j_5 derived from this symmetry is

$$j_5^\mu \equiv \bar{\psi}\gamma^\mu\gamma_5\psi. \tag{13.4}$$

In general, whether the symmetry of a Lagrangian is retained under quantization is not a trivial question. In fact, it has been shown that the chiral symmetry of $\mathcal{L}$ is destroyed at the quantum level. Adler (1969) and Bell and Jackiw (1969) have shown by computing the triangle diagram with an external axial current and two external vector currents that the naive conservation law $\partial_\mu j_5^\mu = 0$ is violated,

$$
\begin{aligned}
\partial_\mu j_5^\mu &= \frac{1}{16\pi^2}\epsilon^{\kappa\lambda\mu\nu}\,\mathrm{tr}\,\mathcal{F}_{\kappa\lambda}\mathcal{F}_{\mu\nu} \\
&= \frac{1}{4\pi^2}\,\mathrm{tr}\left[\epsilon^{\kappa\lambda\mu\nu}\partial_\kappa\left(\mathcal{A}_\lambda\partial_\mu\mathcal{A}_\nu + \frac{2}{3}\mathcal{A}_\lambda\mathcal{A}_\mu\mathcal{A}_\nu\right)\right]
\end{aligned} \tag{13.5}
$$

where tr is a trace over the group indices. The current j_5^μ which appears in (13.5) has no group index, and, hence, (13.5) is called the **Abelian anomaly**.

It is interesting to study the behaviour of a current which carries the group index. Consider a Weyl fermion ψ which couples with an external gauge field. The non-Abelian gauge current of the theory also satisfies an anomalous conservation law which defines the **non-Abelian anomaly**. The action is given by

$$\mathcal{L} \equiv \psi^\dagger(\mathrm{i}\slashed{\nabla})\mathcal{P}_+\psi \qquad \mathcal{P}_\pm = \tfrac{1}{2}(I \pm \gamma^5). \tag{13.6}$$

The Lagrangian has the gauge symmetry

$$\mathcal{A}_\mu \to g^{-1}(\mathcal{A}_\mu + \partial_\mu)g \qquad \psi \to g^{-1}\psi. \tag{13.7}$$

The corresponding non-Abelian current is

$$j^{\mu\alpha} \equiv \psi^\dagger\gamma^\mu T^\alpha\mathcal{P}_+\psi. \tag{13.8}$$

It has been shown by Bardeen (1969) and Gross and Jackiw (1972) that, up to the one-loop level, the current is not conserved,

$$(\mathcal{D}_\mu j_\delta^\mu)^\alpha = \frac{1}{24\pi^2}\,\mathrm{tr}\left[T^\alpha\partial_\kappa\epsilon^{\kappa\lambda\mu\nu}\left(\mathcal{A}_\lambda\partial_\mu\mathcal{A}_\nu + \frac{1}{2}\mathcal{A}_\lambda\mathcal{A}_\mu\mathcal{A}_\nu\right)\right]. \tag{13.9}$$

At first sight, the RHSs of (13.5) and (13.9) look very similar. However, the difference between the normalization and the numerical factors of $\frac{2}{3}$ and $\frac{1}{2}$ have a deep topological origin. We shall see later that the Abelian anomaly in $(2l + 2)$ dimensions and the non-Abelian anomaly in $2l$ dimensions are closely related but in an unexpected manner.

13.2 Abelian anomalies

Henceforth, we work in an even-dimensional manifold M ($\dim M = m = 2l$) with a Euclidean signature. Four-dimensional results will readily be obtained by putting $m = 4$. We assume our system is non-chiral, namely, the gauge field couples to the right and the left components in the same way. Our convention is

$$\gamma^{\mu\dagger} = \gamma^\mu \qquad \{\gamma^\mu, \gamma^\nu\} = 2\delta^{\mu\nu} \qquad \gamma^{m+1} = (i)^l \gamma^1 \dots \gamma^m$$

$$\gamma^{m+1\dagger} = \gamma^{m+1} \qquad (\gamma^{m+1})^2 = +I.$$

The Lie group generators $\{T_\alpha\}$ satisfy

$$T^\dagger{}_\alpha = -T_\alpha \qquad [T_\alpha, T_\beta] = f_{\alpha\beta}{}^\gamma T_\gamma \qquad \text{tr}(T^\alpha T^\beta) = -\tfrac{1}{2}\delta^{\alpha\beta}.$$

13.2.1 Fujikawa's method

Among several methods of deriving anomalies, Fujikawa's way (Fujikawa 1979, 1980, 1986) reveals the topological and geometrical nature of the problem most directly. This method is equivalent to the heat kernel proof of the relevant index theorem.

Let ψ be a massless Dirac field interacting with an external non-Abelian gauge field $\mathcal{A}_\mu$. The effective action $W[\mathcal{A}]$ is given by

$$e^{-W[\mathcal{A}]} = \int \mathcal{D}\psi \mathcal{D}\bar{\psi} e^{-\int dx\, \bar{\psi} i\slashed{\nabla} \psi} \tag{13.10}$$

where $i\slashed{\nabla} = i\gamma^\mu \nabla_\mu = i\gamma^\mu(\partial_\mu + \omega_\mu + \mathcal{A}_\mu)$, with $\omega_\mu = \tfrac{1}{2}\omega_{\mu\alpha\beta}\Sigma^{\alpha\beta}$ being the spin connection of the background space. We compactify the space in such a way that the geometry (the spin connection) plays no role. For example, this can be achieved by compactifying $\mathbb{R}^4$ to $S^4 = \mathbb{R}^4 \cup \{\infty\}$, for which the Dirac genus $\hat{A}(TM)$ is trivial; see example 12.5. If this is the case, the spin connection is irrelevant and may be dropped from $i\slashed{\nabla}$. The classical action $\int dx \bar{\psi} i\slashed{\nabla}\psi$ is invariant with respect to the chiral rotation,

$$\psi \to e^{i\gamma^{m+1}\alpha}\psi \quad \bar{\psi} \to \bar{\psi}e^{i\gamma^{m+1}\alpha}. \tag{13.11}$$

We expand ψ and $\bar{\psi}$ as

$$\psi = \sum_i a_i \psi_i \qquad \bar{\psi} = \sum_i \bar{b}_i \psi^\dagger{}_i \tag{13.12}$$

where a_i and $\bar{b}_i$ are anti-commuting Grassmann variables,

$$\{a_i, a_j\} = 0 \qquad \{\bar{b}_i, \bar{b}_j\} = 0 \qquad \{a_i, \bar{b}_j\} = 0$$

and ψ_i is an eigenvector of the Dirac operator

$$i\slashed{\nabla}\psi_i = \lambda_i \psi_i. \tag{13.13}$$

Since $i\slashed{\nabla}$ is Hermitian, λ_i is real. Since M is compact, ψ_i can be normalized as

$$\langle \psi_i | \psi_j \rangle = \int dx\, \psi_i^\dagger(x) \psi_j(x) = \delta_{ij}.$$

Now the path integrals over ψ and $\bar{\psi}$ are replaced by those over a_i and $\bar{b}_i$.

Consider an infinitesimal chiral transformation,

$$\psi(x) \to \psi(x) + i\alpha(x)\gamma^{m+1}\psi(x) \tag{13.14a}$$

$$\bar{\psi}(x) \to \bar{\psi}(x) + i\bar{\psi}(x)\alpha(x)\gamma^{m+1}. \tag{13.14b}$$

As usual, we take $\alpha = \alpha(x)$ to be x-dependent. Under this change, the classical action transforms as

$$
\begin{aligned}
\int dx\, \bar{\psi} i\slashed{\nabla}\psi &\to \int dx\, (\bar{\psi} + i\bar{\psi}\alpha\gamma^{m+1}) i\slashed{\nabla}(\psi + i\alpha\gamma^{m+1}\psi) \\
&= \int dx\, \bar{\psi} i\slashed{\nabla}\psi + i\int dx\, [\alpha\bar{\psi}\gamma^{m+1} i\slashed{\nabla}\psi + \bar{\psi} i\slashed{\nabla}(\alpha\gamma^{m+1}\psi)] \\
&= \int dx\, \bar{\psi} i\slashed{\nabla}\psi - \int dx\, [\alpha\bar{\psi}\gamma^{m+1}\gamma^\mu(\partial_\mu + A_\mu)\psi \\
&\quad + \bar{\psi}\gamma^\mu(\partial_\mu + A_\mu)(\alpha\gamma^{m+1}\psi)] \\
&= \int dx\, \bar{\psi} i\slashed{\nabla}\psi + \int dx\, \alpha(x)\partial_\mu j_{m+1}^\mu(x) \tag{13.15}
\end{aligned}
$$

where we have used the anti-commutation relations $\{\gamma^\mu, \gamma^{m+1}\} = 0$ and

$$j_{m+1}^\mu(x) \equiv \bar{\psi}(x)\gamma^\mu\gamma^{m+1}\psi(x) \tag{13.16}$$

is the **chiral current**. This is the higher-dimensional analogue of j_5^μ defined previously. If (13.15) were the only change caused by (13.14), naive application of the Ward–Takahashi relation would imply the conservation of the axial current $\partial_\mu j_{m+1}^\mu = 0$. In quantum theory, however, we have an additional change, namely the change of the path integral measure. Define the chiral-rotated fields by

$$\psi' = \psi + i\alpha\gamma^{m+1}\psi = \sum_i a_i'\psi_i \tag{13.17a}$$

$$\bar{\psi}' = \bar{\psi} + i\bar{\psi}\alpha\gamma^{m+1} = \sum_i \bar{b}_i'\psi_i^\dagger. \tag{13.17b}$$

Now the measure changes as

$$\int \prod_i da_i\, d\bar{b}_i \to \int \prod_i da_i'\, d\bar{b}_i'. \tag{13.18}$$

From the orthonormality of $\{\psi_i\}$, we find that

$$
\begin{aligned}
a_i' &= \langle \psi_i | \psi' \rangle = \langle \psi_i | (1 + i\alpha\gamma^{m+1})\psi \rangle \\
&= \sum_j \langle \psi_i | (1 + i\alpha\gamma^{m+1})\psi_j \rangle a_j \equiv \sum_j C_{ij} a_j \tag{13.19a}
\end{aligned}
$$

where

$$C_{ij} = \langle \psi_i | (1 + i\alpha\gamma^{m+1})\psi_j \rangle = \delta_{ij} + i\alpha \langle \psi_i | \gamma^{m+1}\psi_j \rangle. \tag{13.20}$$

The measure in terms of the new variables is

$$
\begin{aligned}
\prod da'_j &= [\det C_{ij}]^{-1} \prod da_i = \exp(-\operatorname{tr}\ln C_{ij}) \prod da_i \\
&= \exp[-\operatorname{tr}\ln(I + i\alpha\langle\psi_i|\gamma^{m+1}\psi_j\rangle)] \prod da_i \\
&\approx \exp(-\operatorname{tr}i\alpha\langle\psi_i|\gamma^{m+1}\psi_j\rangle) \prod da_i \\
&= \exp\left(-i\alpha \sum_i \langle\psi_i|\gamma^{m+1}\psi_i\rangle\right) \prod da_i
\end{aligned} \tag{13.21}
$$

where the inverse of the determinant appears since a_i and a'_i are Grassmann variables, see Berezin (1966).[1] As for $\bar{b}_i \to \bar{b}'_i$, we have

$$\bar{b}'_i = \sum_j \bar{b}_j \langle \psi_j | (1 + i\alpha\gamma^{m+1}) | \psi_i \rangle = \sum_j C_{ji} \bar{b}_j. \tag{13.19b}$$

The Jacobian for the change $\bar{b}_i \to \bar{b}'_i$ agrees with (13.21). Thus, the measure transforms under the chiral rotation (13.17) as

$$\prod_i da_i\, d\bar{b}_i \to \prod_i da'_i\, d\bar{b}'_i \exp\left(-2i \int dx\, \alpha(x) \sum_n \psi_n^\dagger(x)\gamma^{m+1}\psi_n(x)\right). \tag{13.22}$$

Now the effective action has two expressions:

$$
\begin{aligned}
e^{-W[\mathcal{A}]} &= \int \prod_i da_i\, d\bar{b}_i\, \exp\left(-\int dx\, \bar{\psi} i \slashed{\nabla} \psi\right) \\
&= \int \prod_i da'_i\, d\bar{b}'_i\, \exp\left(-\int dx\, \bar{\psi} i \slashed{\nabla} \psi - \int dx\, \alpha(x)\partial_\mu j^\mu_{m+1}(x)\right. \\
&\quad \left. - 2i \int dx\, \alpha(x) A(x)\right)
\end{aligned} \tag{13.23}
$$

where

$$A(x) \equiv \sum_i \psi^\dagger_i(x)\gamma^{m+1}\psi_i(x). \tag{13.24}$$

Since $\alpha(x)$ is arbitrary, we have

$$\partial_\mu j^\mu_{m+1}(x) = -2iA(x). \tag{13.25}$$

[1] See section 1.5. For example, we have $\int a\, da = \int ca\, d(ca) = 1$, $c \in \mathbb{R}$ and a being a real Grassmann number. This shows that $d(ca) = da/c$.

Thus, naive conservation of an axial current does not hold in quantum theory. This non-conservation of the current j^{μ}_{m+1} is called the **Abelian anomaly** (or **chiral anomaly** or **axial anomaly**).

How is this related to the topology? Let us look at the Jacobian (13.22) and assume that $\alpha(x)$ is independent of x.[2] The integral in (13.22) is not well defined and must be regularized. We introduce the Gaussian cut-off (heat kernel regularization) as

$$\int dx\, A(x) = \int dx \sum_i \psi^{\dagger}{}_i(x) \gamma^{m+1} \psi_i(x) \exp[-(\lambda_i/M)^2]|_{M \to \infty}$$

$$= \sum_i \langle \psi_i | \gamma^{m+1} \exp[-(i\nabla\!\!\!/\,/M)^2] | \psi_i \rangle |_{M \to \infty}. \tag{13.26}$$

In (13.26), $1/M^2$ corresponds to the 'time' parameter t in the previous chapter and $M \to \infty$ implies $t \to \varepsilon$. Let $|\psi_i\rangle$ be an eigenstate of $i\nabla\!\!\!/$ with *non-vanishing* eigenvalue λ_i. Among the eigenstates, there exists a state $|\psi_i\rangle^{\chi} \equiv \gamma^{m+1}|\psi\rangle$ with eigenvalue $-\lambda_i$:

$$i\nabla\!\!\!/\,|\psi_i\rangle^{\chi} = i\nabla\!\!\!/\,\gamma^{m+1}|\psi_i\rangle = -\gamma^{m+1} i\nabla\!\!\!/\,|\psi_i\rangle$$

$$= -\lambda_i \gamma^{m+1}|\psi_i\rangle = -\lambda_i |\psi_i\rangle^{\chi}$$

where use has been made of the anti-commutation relation $\{\gamma^{m+1}, i\nabla\!\!\!/\,\} = 0$. Since $i\nabla\!\!\!/$ is a Hermitian operator, eigenvectors which belong to different eigenvalues are orthogonal, hence $\langle \psi_i | \psi_i \rangle^{\chi} = \langle \psi_i | \gamma^{m+1} | \psi_i \rangle = 0$. This shows that

$$\langle \psi_i | \gamma^{m+1} \exp[-(i\nabla\!\!\!/\,/M)^2] | \psi_i \rangle = \langle \psi_i | \gamma^{m+1} | \psi_i \rangle \exp[-(\lambda_i/M)^2] = 0.$$

Thus, the contribution to the RHS of (13.26) comes only from the zero-energy modes. Let $|0, i\rangle$ be the zero-energy modes of $i\nabla\!\!\!/$, $(1 \le i \le n_0)$. They are not in an irreducible representation of the spin algebra and should be classified according to the eigenvalue of γ^{m+1}. We write

$$\gamma^{m+1}|0, i\rangle_{\pm} = \pm|0, i\rangle_{\pm}. \tag{13.27}$$

Then, (13.26) becomes

$$\int dx\, A(x) = \sum_i \langle \psi_i | \gamma^{m+1} \exp[-(i\nabla\!\!\!/\,/M)^2] | \psi_i \rangle |_{M \to \infty}$$

$$= \sum_{+\,i} \langle 0, i | 0, i \rangle_{+} - \sum_{-\,i} \langle 0, i | 0, i \rangle_{-}$$

$$= \nu_+ - \nu_- = \text{ind}\, i\nabla\!\!\!/\,_+ \tag{13.28}$$

where ν_+ (ν_-) is the number of zero-energy modes with positive (negative) chirality ($\nu_+ + \nu_- = n_0$) and $i\nabla\!\!\!/\,_+$ is defined by

$$i\nabla\!\!\!/ = \begin{pmatrix} 0 & i\nabla\!\!\!/\,_- \\ i\nabla\!\!\!/\,_+ & 0 \end{pmatrix} \qquad i\nabla\!\!\!/\,_- = (i\nabla\!\!\!/\,_+)^{\dagger}.$$

[2] We are looking at the zero-momentum Ward–Takahashi relation.

The Atiyah–Singer index theorem now comes into the problem.

To show that (13.28), indeed, represents an integral of the relevant Chern character, we first note that

$$(i\not{V})^2 = -\gamma^\mu\gamma^\nu\nabla_\mu\nabla_\nu = -\{\delta^{\mu\nu} + \tfrac{1}{2}[\gamma^\mu, \gamma^\nu]\}\tfrac{1}{2}[\{\nabla_\mu, \nabla_\nu\} + \mathcal{F}_{\mu\nu}]$$
$$= -\nabla_\mu\nabla^\mu - \tfrac{1}{4}[\gamma^\mu, \gamma^\nu]\mathcal{F}_{\mu\nu} \tag{13.29}$$

where use has been made of the relation $[\nabla_\mu, \nabla_\nu] = \mathcal{F}_{\mu\nu}$. Then

$$A(x) = \sum_i \langle \psi_i | x \rangle \langle x | \gamma^{m+1} \exp[(\nabla^2 + \tfrac{1}{4}[\gamma^\mu, \gamma^\nu]\mathcal{F}_{\mu\nu})/M^2]|\psi_i\rangle|_{M\to\infty}. \tag{13.30}$$

Let us take $m = 4$ for definiteness. We introduce the plane wave basis as

$$\langle x | \psi_i \rangle = \int \frac{d^4k}{(2\pi)^4} \langle x | k \rangle \langle k | \psi_i \rangle.$$

Then (13.30) becomes

$$A(x) = \int \frac{dk}{(2\pi)^4} \int \frac{dk'}{(2\pi)^4} \sum_i \langle \psi_i | k' \rangle \langle k' | x \rangle$$

$$\times \gamma^{m+1} \exp[(\nabla^2 + \tfrac{1}{4}[\gamma^\mu, \gamma^\nu]\mathcal{F}_{\mu\nu})/M^2]\langle x | k \rangle \langle k | \psi_i \rangle \bigg|_{\substack{M\to\infty \\ y\to x}}$$

$$= \int \frac{dk}{(2\pi)^4} \operatorname{tr} \gamma^{m+1} \exp[(-k^2 + \tfrac{1}{4}[\gamma^\mu, \gamma^\nu]\mathcal{F}_{\mu\nu})/M^2]_{M\to\infty} \tag{13.31}$$

where use has been made of the completeness property

$$\sum_i \langle k | \psi_i \rangle \langle \psi_i | k' \rangle = (2\pi)^4 \delta^4(k - k').$$

In (13.31), we have replaced ∇^2 by the symbol $-k^2$ since the residual terms containing $\mathcal{A}$ do not survive in the limit $M \to \infty$. If we put $\tilde{k}^\mu \equiv k^\mu/M$, (13.31) becomes

$$A(x) = \operatorname{tr}[\gamma^5 \exp(\tfrac{1}{4}[\gamma^\mu, \gamma^\nu]\mathcal{F}_{\mu\nu}/M^2)]M^4 \int \frac{d\tilde{k}}{(2\pi)^4} \exp(-\tilde{k}^2).$$

We expand the first exponential and use

$$\operatorname{tr}\gamma^5 = \operatorname{tr}\gamma^5\gamma^\mu\gamma^\nu = 0 \qquad \operatorname{tr}\gamma^5\gamma^\kappa\gamma^\lambda\gamma^\mu\gamma^\nu = -4\epsilon^{\kappa\lambda\mu\nu}$$

$$\int d\tilde{k} \exp(-\tilde{k}^2) = \pi^2$$

to obtain

$$A(x) = \frac{1}{2} \operatorname{tr}\left[\gamma^5 \frac{1}{4^2}\{[\gamma^\mu, \gamma^\nu]\mathcal{F}_{\mu\nu}\}^2\right]\frac{1}{16\pi^2}$$

$$= \frac{-1}{32\pi^2} \operatorname{tr}\epsilon^{\kappa\lambda\mu\nu}\mathcal{F}_{\kappa\lambda}(x)\mathcal{F}_{\mu\nu}(x). \tag{13.32}$$

Note that the higher-order terms in the expansion of the exponential vanish in the limit $M \to \infty$. The anomalous conservation law (13.25) now becomes

$$\partial_\mu j_5^\mu = \frac{1}{16\pi^2} \operatorname{tr}\epsilon^{\kappa\lambda\mu\nu}\mathcal{F}_{\kappa\lambda}\mathcal{F}_{\mu\nu}$$

$$= \frac{1}{4\pi^2} \operatorname{tr}[\epsilon^{\kappa\lambda\mu\nu}\partial_\kappa(\mathcal{A}_\lambda\partial_\mu\mathcal{A}_\nu + \tfrac{2}{3}\mathcal{A}_\lambda\mathcal{A}_\mu\mathcal{A}_\nu)]. \tag{13.33}$$

This is regarded as a local version of the AS index theorem. Let us write (13.33) in terms of the field strength $\mathcal{F} = \frac{1}{2}\mathcal{F}_{\mu\nu}\,\mathrm{d}x^\mu \wedge \mathrm{d}x^\nu$. We easily verify that

$$\nu_+ - \nu_- = \int_M \mathrm{d}x\,\partial_\mu j_{m+1}^\mu = \int_M \mathrm{ch}_2(\mathcal{F}). \tag{13.34}$$

This is the index theorem for a twisted spinor complex with trivial background geometry ($\hat{A}(TM) = 1$).

For dim $M = m = 2l$, we have the following identity:

$$\nu_+ - \nu_- = \int_M \mathrm{d}x\,\partial_\mu j_{m+1}^\mu = \int_M \mathrm{ch}_l(\mathcal{F}) = \int_M \frac{1}{l!} \operatorname{tr}\left(\frac{i\mathcal{F}}{2\pi}\right)^l. \tag{13.35}$$

13.3 Non-Abelian anomalies

In the last section we considered the chiral current which is a gauge singlet (no gauge indices). Now we turn to the study of the gauge current $j^\mu{}_\alpha$ where α is the gauge index. Here we consider a chiral theory in which the gauge field $\mathcal{A}$ couples only to the left-handed *Weyl* fermion ψ. Suppose ψ transforms in a complex representation r of the gauge group G. For example, suppose ψ belongs to a **3** of SU(3). The effective action $W_r[\mathcal{A}]$ is given by

$$e^{-W_r[\mathcal{A}]} = \int \mathcal{D}\psi\mathcal{D}\bar{\psi} \exp\left(-\int \mathrm{d}x\,\bar{\psi}i\nabla_+\psi\right) \tag{13.36}$$

where

$$i\nabla_+ = i\gamma^\mu(\partial_\mu + \mathcal{A}_\mu)\mathcal{P}_+ \qquad \mathcal{P}_\pm = \tfrac{1}{2}(1 \pm \gamma^{m+1}). \tag{13.37}$$

The gauge current is

$$j^\mu{}_\alpha = i\bar{\psi}\gamma^\mu T_\alpha\mathcal{P}_+\psi. \tag{13.38}$$

Let $v = v^\alpha T_\alpha$ be an infinitesimal gauge transformation parameter, $g = 1 - v$ under which we have

$$\mathcal{A}_\mu \to (1 + v)(\mathcal{A}_\mu + \mathrm{d})(1 - v) = \mathcal{A}_\mu - \mathcal{D}_\mu v \qquad (13.39)$$

where $\mathcal{D}_\mu v \equiv \partial_\mu v + [\mathcal{A}_\mu, v]$ is the covariant derivative for a field in the adjoint representation. The effective action transforms as

$$
\begin{aligned}
W_r[\mathcal{A}] &\to W_r[\mathcal{A} - \mathcal{D}v] \\
&= W_r[\mathcal{A}] - \int \mathrm{d}x \,\, \mathrm{tr}\left(\mathcal{D}v \frac{\delta}{\delta \mathcal{A}} W_r[\mathcal{A}]\right) \\
&= W_r[\mathcal{A}] - \int \mathrm{d}x \,\, \mathrm{tr}(\partial_\mu v^\alpha + f_{\alpha\beta\gamma} A_\mu{}^\beta v^\gamma) \frac{\delta}{\delta A_\mu{}^\alpha} W_r[\mathcal{A}] \\
&= W_r[\mathcal{A}] + \int \mathrm{d}x \,\, \mathrm{tr}\left(v^\alpha \mathcal{D} \frac{\delta}{\delta \mathcal{A}} W_r[\mathcal{A}]_\alpha\right).
\end{aligned}
\qquad (13.40)
$$

Since

$$\frac{\delta}{\delta A_\mu{}^\alpha} W_r[\mathcal{A}] = \langle \mathrm{i}\bar{\psi}\gamma^\mu T_\alpha \tfrac{1}{2}(1 + \gamma^{m+1})\psi \rangle_{\mathcal{A}} = \langle j^\mu{}_\alpha \rangle$$

we obtain

$$W_r[\mathcal{A} - \mathcal{D}v] - W_r[\mathcal{A}] = \int \mathrm{d}x \,\, \mathrm{tr}(v^\alpha \mathcal{D}_\mu \langle j^\mu \rangle_\alpha). \qquad (13.41)$$

We are naively tempted to regard (13.36) as $\det(\mathrm{i}\slashed{\nabla}) = \prod \lambda_i'$, λ_i being the 'eigenvalue' of $\mathrm{i}\slashed{\nabla}$. A subtlety arises here: $\mathrm{i}\slashed{\nabla}_+$ maps sections of $S_+ \otimes E$ to those of $S_- \otimes E$, where E is the vector bundle associated with the G bundle and $S_\pm$ are spin bundles with chirality $\pm$. Accordingly, the equation $\mathrm{i}\slashed{\nabla}_+\psi = \lambda\psi$ is meaningless. To avoid this difficulty, we formally introduce a *Dirac* spinor ψ and define

$$\mathrm{e}^{-W_r[\mathcal{A}]} = \int \mathcal{D}\psi \mathcal{D}\bar{\psi} \exp\left(-\int \mathrm{d}x \,\, \bar{\psi} \mathrm{i}\hat{D}\psi\right) \qquad (13.42)$$

where $\mathrm{i}\hat{D}$ is defined by

$$\mathrm{i}\hat{D} \equiv \mathrm{i}\gamma^\mu(\partial_\mu + \mathrm{i}A_\mu \mathcal{P}_+) = \begin{pmatrix} 0 & \mathrm{i}\slashed{\partial}_- \\ \mathrm{i}\slashed{\nabla}_+ & 0 \end{pmatrix} \qquad (13.43)$$

where we have diagonalized γ^{m+1}. In (13.43), the gauge field $\mathcal{A}$ couples only to the positive chirality field. Now the eigenvalue problem $\mathrm{i}\hat{D}\psi_i = \lambda_i \psi_i$ is well defined. Note that $\mathrm{i}\hat{D}$ is not Hermitian and λ_i is a complex number in general. Moreover, we need to introduce right and left eigenfunctions separately by

$$\mathrm{i}\hat{D}\psi_i = \lambda_i \psi_i \qquad (13.44\mathrm{a})$$

$$\chi^\dagger{}_i(\mathrm{i}\overleftarrow{\hat{D}}) = \lambda_i \chi^\dagger{}_i \qquad (\mathrm{i}\hat{D})^\dagger \chi_i = \bar{\lambda}_i \chi_i. \qquad (13.44\mathrm{b})$$

Since $\int \chi_i^\dagger \psi_j \, dx = 0$ for $i \neq j$, we may choose an orthonormal basis,

$$\int \chi^\dagger_i \psi_j \, dx = \delta_{ij}. \tag{13.45}$$

It should be noted that the eigenvalue λ_i is *not* gauge invariant. This follows from the observation that

$$g(i\hat{D}(A^g))g^{-1} = gi\gamma^\mu[\partial_\mu + g^{-1}(A_\mu + \partial_\mu)g\mathcal{P}_+]g^{-1}$$
$$= i\hat{D}(A) - i\slashed{\partial}gg^{-1} + i\slashed{\partial}gg^{-1}\mathcal{P}_+ \neq i\hat{D}(A). \tag{13.46}$$

If the equality were to hold in (13.46), $g^{-1}\psi_i$ would satisfy $i\hat{D}(A^g)g^{-1}\psi_i = \lambda_i g^{-1}\psi_i$ when $i\hat{D}(A)\psi_i = \lambda_i \psi_i$. Then $\mathrm{Spec}\, i\hat{D}(A)$ would be gauge invariant. Although individual eigenvalues are not gauge invariant, the absolute value of the product of eigenvalues of $i\hat{D}$ is gauge invariant. In fact,

$$\det(i\hat{D})\det((i\hat{D})^\dagger) = \det(i\hat{D}(i\hat{D})^\dagger)$$
$$= \det\begin{pmatrix} (i\slashed{\partial}_-)(i\slashed{\partial}_+) & 0 \\ 0 & (i\slashed{\nabla}_+)(i\slashed{\nabla}_-) \end{pmatrix}$$
$$= \det(i\slashed{\partial}_- i\slashed{\partial}_+)\det(i\slashed{\nabla}_+ i\slashed{\nabla}_-) \tag{13.47}$$

where $i\slashed{\partial}_+ = (i\slashed{\partial}_-)^\dagger$ and $i\slashed{\nabla}_- = (i\slashed{\nabla}_+)^\dagger$. This is simply the Dirac determinant (up to an irrelevant factor $\det(i\slashed{\partial}_- i\slashed{\partial}_+)$),

$$[\det(i\slashed{\nabla})]^2 = \det\begin{pmatrix} i\slashed{\nabla}_- i\slashed{\nabla}_+ & 0 \\ 0 & i\slashed{\nabla}_+ i\slashed{\nabla}_- \end{pmatrix} = [\det(i\slashed{\nabla}_+ i\slashed{\nabla}_-)]^2 \tag{13.48}$$

where $i\slashed{\nabla}$ is given by

$$i\slashed{\nabla} = \begin{pmatrix} 0 & i\slashed{\nabla}_- \\ i\slashed{\nabla}_+ & 0 \end{pmatrix}. \tag{13.49}$$

The Dirac determinant is gauge invariant, hence so is $|\det(i\hat{D})|$. It then follows that $\mathrm{Re}\, W_r[A]$ is gauge invariant since

$$\exp(-W_r[A])\exp(-\overline{W_r[A]}) = \det(i\hat{D})\det((i\hat{D})^\dagger) \propto \det(i\slashed{\nabla}_+ i\slashed{\nabla}_-)$$

is gauge invariant. Therefore, only the *imaginary part* of $W_r[A]$, that is the *phase* of $\det(i\hat{D})$, may gain an anomalous variation under gauge transformations.

The anomaly may be computed by evaluating the Jacobian as before. The functional measure is taken to be $\prod_i da_i \, d\bar{b}_i$. We consider an infinitesimal gauge transformation,

$$A \to A - \mathcal{D}v \qquad \psi \to \psi + v\psi_+ \qquad \bar{\psi} \to \bar{\psi} - \bar{\psi}_- v \tag{13.50}$$

where the gauge transformation rotates the positive chirality parts only. The Jacobian factor is

$$\int dx \, \mathrm{tr}\, v(x) \sum_n \langle n|x\rangle \gamma^{m+1} \langle x|n\rangle \tag{13.51}$$

where $\langle x|n\rangle = \psi_n(x)$ and $\langle n|x\rangle = \chi_n^\dagger(x)$ (note that $\langle n|$ is *not* the Hermitian conjugate of $|n\rangle$). This integral is ill defined and must be regularized. As before, we employ the Gaussian regulator,

$$\int dx \lim_{\substack{M\to\infty \\ x\to y}} \operatorname{tr} v(x) \sum_n \langle n|y\rangle \gamma^{m+1} \langle x|e^{-(i\hat{D})^2/M^2}|n\rangle$$

$$= \int dx \lim_{\substack{M\to\infty \\ x\to y}} \operatorname{tr} v(x)\gamma^{m+1} e^{-(i\hat{D}_x)^2/M^2}\delta(x-y) \qquad (13.52)$$

where use has been made of the completeness relation

$$\sum_n |n\rangle\langle n| = I. \qquad (13.53)$$

It follows from (13.41) and (13.52) that

$$\int dx \, v^\alpha \mathcal{D}_\mu \left(\frac{\delta}{\delta A_\mu{}^\alpha} W_r[A] \right) = \int dx \lim_{\substack{M\to\infty \\ x\to y}} \operatorname{tr}[v\gamma^{m+1} e^{-(i\hat{D}_x)^2/M^2}\delta(x-y)].$$

$$(13.54)$$

In the present case W_r really changes under (13.50). The trace may be written as

$$\operatorname{tr}[v\gamma^{m+1} e^{-(i\hat{D}_x)^2/M^2}] = \operatorname{tr}[v(\mathcal{P}_+ - \mathcal{P}_-)e^{-(i\slashed{\partial}-i\slashed{\nabla}_+)-(i\slashed{\nabla}_-i\slashed{\partial}_+)/M^2}]$$

$$= \operatorname{tr}[vP_+e^{(i\slashed{\partial}i\slashed{\nabla})/M^2}] - \operatorname{tr}[vP_-e^{(i\slashed{\nabla}i\slashed{\partial})/M^2}]. \qquad (13.55)$$

(13.55) can be evaluated in the plane wave basis, which is straightforward but tedious (see Gross and Jackiw (1972), for example). We derive the non-Abelian anomaly from a topological viewpoint in the next section. For $m = 4$, the anomalous variation is

$$W_r[A - \mathcal{D}v] - W_r[A] = \int dx \, v^\alpha \mathcal{D}_\mu \langle j^\mu \rangle_\alpha$$

$$= \frac{1}{24\pi^2} \int dx \, \operatorname{tr}\{v^\alpha T_\alpha \epsilon^{\kappa\lambda\mu\nu}\partial_\kappa [A_\lambda \partial_\mu A_\nu + \tfrac{1}{2}A_\lambda A_\mu A_\nu]\}$$

$$= \frac{1}{24\pi^2} \int \operatorname{tr}\{vd[AdA + \tfrac{1}{2}A^3]\}. \qquad (13.56)$$

The anomalous divergence of the gauge current is

$$\mathcal{D}_\mu \langle j^\mu \rangle_\alpha = \frac{1}{24\pi^2} \operatorname{tr}\{T_\alpha \epsilon^{\kappa\lambda\mu\nu}\partial_\kappa [A_\lambda \partial_\mu A_\nu + \tfrac{1}{2}A_\lambda A_\mu A_\nu]\}. \qquad (13.57)$$

This should be compared with (13.33). There are two differences between these results: the two-thirds in front of A^3 is replaced by a half and the overall factor is different.

13.4   The Wess–Zumino consistency conditions

13.4.1   The Becchi–Rouet–Stora operator and the Faddeev–Popov ghost

Let $W[A]$ be the effective action of the Weyl fermion in the complex representation r of the gauge group G.[3] In the previous section, we observed that the change of $W[A]$ under an infinitesimal gauge transformation $\delta_v A = -Dv$ is given by

$$\delta_v W[A] = -\int (D_\mu v)^\alpha \frac{\delta}{\delta A_\mu{}^\alpha} W[A] = \int v^\alpha D_\mu \langle j^\mu \rangle_\alpha. \qquad (13.58)$$

Following Stora (1984) and Zumino (1985) we introduce the BRS operator $\mathcal{S}$ and the Faddeev–Popov ghost ω. Let $\Omega^m(G)$ be the set of maps from S^m to G.[4] In addition to the ordinary exterior derivative d, we introduce another exterior derivative $\mathcal{S}$ on $\Omega^m(G)$ which we call the **Becchi–Rouet–Stora (BRS) operator**. In general, $\mathcal{S}$ is defined on an infinite-dimensional space but we may also consider the restriction of $\mathcal{S}$ to a finite-dimensional compact subspace of $\Omega^m(G)$, such as S^n, parametrized by λ^α. Then $\mathcal{S}$ may be written as $\mathcal{S} \equiv d\lambda^\alpha \partial/\partial\lambda^\alpha$. We require that d and $\mathcal{S}$ be anti-derivatives,

$$d^2 = \mathcal{S}^2 = d\mathcal{S} + \mathcal{S}d = 0. \qquad (13.59)$$

If we define $\Delta \equiv d + \mathcal{S}$, Δ is clearly nilpotent,

$$\Delta^2 = d^2 + d\mathcal{S} + \mathcal{S}d + \mathcal{S}^2 = 0. \qquad (13.60)$$

Under the action of $g = g(x, \lambda^\alpha)$, $\mathcal{A}$ transforms as

$$\mathcal{A} \to A \equiv g^{-1}(\mathcal{A} + d)g. \qquad (13.61)$$

Note that $\mathcal{A}$ is independent of λ while A depends on λ through g. Define the **Faddeev–Popov (FP) ghost** by

$$\omega \equiv g^{-1}\mathcal{S}g. \qquad (13.62)$$

The actions of $\mathcal{S}$ on A and ω are found to be

$$\begin{aligned}
\mathcal{S}A &= \mathcal{S}[g^{-1}(\mathcal{A} + d)g] = -g^{-1}\mathcal{S}gA - g^{-1}\mathcal{A}\mathcal{S}g + g^{-1}\mathcal{S}(dg) \\
&= -\omega A - (A - g^{-1}\,dg)\omega - g^{-1}\,d(\mathcal{S}g) \\
&= -\omega A - A\omega - d\omega \equiv -D_A\omega \qquad (13.63a) \\
\mathcal{S}\omega &= -g^{-1}\mathcal{S}gg^{-1}\mathcal{S}g = -\omega^2. \qquad (13.63b)
\end{aligned}$$

---

[3] We drop the representation index r to simplify the expression.
[4] The set $\Omega^m(G)$ should not be confused with $\Omega^m(M)$, the set of m-forms on M. The distinction should be clear from the context.

It is easy to verify that S is nilpotent on A and ω and, hence, on any polynomial of A and ω as it should be; see exercise 13.1. Define the field strength of A by

$$F \equiv dA + A^2 = g^{-1}\mathcal{F}g. \tag{13.64}$$

We also define

$$\mathbb{A} \equiv g^{-1}(\mathcal{A} + \Delta)g = A + g^{-1}Sg = A + \omega \tag{13.65a}$$

$$\mathbb{F} \equiv \Delta\mathbb{A} + \mathbb{A}^2 = g^{-1}\mathcal{F}g = F \tag{13.65b}$$

where (13.65b) follows since $\mathcal{F} = d\mathcal{A} + \mathcal{A}^2 = \Delta\mathcal{A} + \mathcal{A}^2$ (note that $S\mathcal{A} = 0$). It is found from theorem 10.1 that $\mathbb{A}$ is an Ehresmann connection on the principal bundle and $\mathbb{F}$ its associated curvature two-form.

The existence of a non-Abelian anomaly implies that $W[A]$ does not vanish under the action of the BRS operator S (ω roughly corresponds to v; see (13.39) and (13.63a)),

$$SW[A] = G[\omega, A]. \tag{13.66}$$

Since $W[A]$ is independent of ω, S acts through A only. Before we write down the Wess–Zumino consistency condition for the non-Abelian anomaly, we stop here and consider the physical meaning of the BRS operator and the FP ghost.

Exercise 13.1. Verify from (13.63) that the actions of S on A and ω are nilpotent,

$$S^2 A = 0 \qquad S^2 \omega = 0. \tag{13.67}$$

13.4.2 The BRS operator, FP ghost and moduli space

To find the physical meaning of S and ω, we need to examine the topology of the gauge fields (Atiyah and Jones 1978, Singer 1985, Sumitani 1985). Let $\mathfrak{A}$ be the space of all gauge potential configurations on S^m. For definiteness, we take $m = 4$ but the generalization to arbitrary m is obvious. The topology of $\mathfrak{A}$ is trivial since, for any gauge potential configurations $\mathcal{A}_1$ and $\mathcal{A}_2$, the combination $t\mathcal{A}_1 + (1 - t)\mathcal{A}_2$ $(0 \le t \le 1)$ is again a gauge potential on S^4. Note, however, that $\mathfrak{A}$ does not describe the physical configuration space of the gauge theory. We have to identify those field configurations which are connected by G-gauge transformations. Let $\mathfrak{G}$ be the space of all gauge transformations on S^4 ($\mathfrak{G} = \Omega^4(G)$ in our previous notation). Then the physical configuration space must be identified with $\mathfrak{A}/\mathfrak{G}$, called the **moduli space** of the gauge theory. We have seen in section 10.5 that the gauge field configuration on S^4 is classified by the transition function $g : S^3 \to G$, S^3 being the equator of S^4. In the present case, $\mathfrak{A}/\mathfrak{G}$ is classified by the transition function on the equator $S^3 \to G$ and, hence,

$$\mathfrak{A}/\mathfrak{G} \simeq \Omega^3(G). \tag{13.68}$$

Thus, each connected component of $\mathfrak{A}/\mathfrak{G}$ is labelled by the instanton number k. This component is denoted by $\Omega^4_k(G)$.

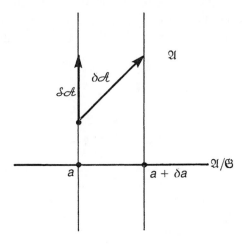

Figure 13.1. The BRS operator $\mathcal{S}$ is the restriction of δ along the fibre.

We note that the space $\mathfrak{A}$ has a natural projection $\pi : \mathfrak{A} \to \mathfrak{A}/\mathfrak{G}$ and can be made into a fibre bundle whose fibre is $\mathfrak{G}$, see figure 13.1. Let $a \in \mathfrak{A}$ be a representative of the class $[a] \in \mathfrak{A}/\mathfrak{G}$ and let

$$A(x) = g^{-1}(x)(a(x) + \mathrm{d})g(x) \tag{13.69}$$

be an element of $\mathfrak{A}$ in $[a]$. We denote the exterior derivative operator in $\mathfrak{A}$ by δ, which is a *functional* variation and should not be confused with the usual derivative d; see Leinaas and Olaussen (1982). If δ is applied on (13.69), we find that

$$\begin{aligned}
\delta A &= -g^{-1}\delta g A + g^{-1}\delta a g - g^{-1}a\delta g - g^{-1}\mathrm{d}\,(\delta g) \\
&= g^{-1}\delta a g - \mathrm{d}\,(g^{-1}\delta g) - g^{-1}\delta g A - A g^{-1}\delta g \\
&= g^{-1}\delta a g - \mathcal{D}_A(g^{-1}\delta g)
\end{aligned} \tag{13.70}$$

where $\mathcal{D}_A = \mathrm{d} + [A, \quad]$. The first term of (13.70) represents the derivative of A along $\mathfrak{A}/\mathfrak{G}$ while the second represents that along the fibre; see figure 13.1. The BRS transformation $\mathcal{S}$ is obtained by restricting the variation δ along the fibre,

$$\mathcal{S}A \equiv \delta A|_{\text{fibre}} = -\mathcal{D}_A \omega \tag{13.71a}$$

where the FP ghost ω is $g^{-1}\mathcal{S}g \equiv g^{-1}\delta g|_{\text{fibre}}$. We also find that

$$\mathcal{S}\omega = \delta\omega|_{\text{fibre}} = -g^{-1}\mathcal{S}g g^{-1}\mathcal{S}g = -\omega^2 \tag{13.71b}$$

which reproduces (13.63a).

13.4.3   The Wess–Zumino conditions

Exercise 13.1 shows that S is *nilpotent* on any polynomial f of A and ω,

$$S^2 f(\omega, A) = 0. \tag{13.72}$$

The nilpotency is required by the interpretation of S as an exterior derivative operator. In particular, we should have

$$SG[\omega, A] = S^2 W[A] = 0. \tag{13.73}$$

This condition is called the **Wess–Zumino consistency condition** (WZ condition) and can be used to determine the non-Abelian anomaly (Wess and Zumino 1971, Stora 1984, Zumino 1985, Zumino *et al* 1984). If the anomaly G is mathematically well defined, G should satisfy the WZ condition. This condition is so strong that once the first term of $G[\omega, A]$ is given, the anomaly is completely pinned down.

13.4.4   Descent equations and solutions of WZ conditions

Stora (1984) and Zumino (1985) constructed the solution of WZ conditions as follows. The *Abelian* anomaly in $(2l + 2)$-dimensional space is given by

$$ch_{l+1}(F) = \frac{1}{(l+1)!} \, \text{tr} \left(\frac{iF}{2\pi} \right)^{l+1} \tag{13.74}$$

where $F = dA + A^2$, $A = g^{-1}(A + d)g$ as before. Let $Q_{2l+1}(A, F)$ be the Chern–Simons form of $ch_{l+1}(F)$,

$$ch_{l+1}(F) = dQ_{2l+1}(A, F). \tag{13.75}$$

Since the algebraic structure of the triplet $(\triangle, \mathbb{A}, \mathbb{F})$ is exactly the same as that of (d, A, F), we also have

$$ch_{l+1}(\mathbb{F}) = \triangle Q_{2l+1}(\mathbb{A}, \mathbb{F}) = \triangle Q_{2l+1}(A + \omega, F) \tag{13.76}$$

where we have noted that $\mathbb{A} = A + \omega$ and $\mathbb{F} = F$. If we expand $Q_{2l+1}(\mathbb{A}, \mathbb{F}) = Q_{2l+1}(A + \omega, F)$ in powers of ω, we have

$$Q_{2l+1}(\mathbb{A}, \mathbb{F}) = Q_{2l+1}^0(A, F) + Q_{2l}^1(\omega, A, F) + Q_{2l-1}^2(\omega, A, F)$$
$$+ \cdots + Q_0^{2l+1}(\omega, A, F) \tag{13.77}$$

where Q_r^s is sth order in ω and $r + s = 2l + 1$.

We now note that $ch_{l+1}(\mathbb{F}) = ch_{l+1}(F)$ since $\mathbb{F} = F = g^{-1}\mathcal{F}g$. In terms of the Chern–Simons forms, this can be expressed as

$$\triangle Q_{2l+1}(\mathbb{A}, \mathbb{F}) = dQ_{2l+1}(A, F). \tag{13.78}$$

Substituting (13.77) into (13.78), we have

$$(d + S)[Q^0_{2l+1}(A, F) + Q^1_{2l}(\omega, A, F)$$
$$+ \cdots + Q^{2l+1}_0(\omega, A, F)] = dQ^0_{2l+1}(A, F). \tag{13.79}$$

If we collect terms of the same order in ω, we have the '**descent equations**'

$$SQ^0_{2l+1}(A, F) + dQ^1_{2l}(\omega, A, F) = 0 \tag{13.80a}$$
$$SQ^1_{2l}(\omega, A, F) + dQ^2_{2l-1}(\omega, A, F) = 0 \tag{13.80b}$$

$$\vdots$$

$$SQ^{2l}_1(\omega, A, F) + dQ^{2l+1}_0(\omega, A, F) = 0 \tag{13.80c}$$
$$SQ^{2l+1}_0(\omega, A, F) = 0. \tag{13.80d}$$

Note here that S increases the degree of ω by one, see (13.63). Let us look at the $2l$-form $Q^1_{2l}(\omega, A, F)$. If we put

$$G[\omega, A, F] \equiv \int_M Q^1_{2l}(\omega, A, F) \tag{13.81}$$

$G[\omega, A, F]$ satisfies the WZ condition,

$$SG[\omega, A, F] = \int_M SQ^1_{2l}(\omega, A, F) = -\int_M dQ^2_{2l-1}(\omega, A, F)$$
$$= -\int_{\partial M} Q^2_{2l-1}(\omega, A, F) = 0$$

where we have assumed that M has no boundary and use has been made of (13.80b). This shows that once $Q^1_{2l}(\omega, A, F)$ is obtained, the anomaly $G[\omega, A, F]$ is easily found.

Proposition 13.1. Q^1_{2l} defined here is given by

$$Q^1_{2l}(\omega, A, F) = \left(\frac{i}{2\pi}\right)^{l+1} \frac{1}{(l-1)!} \int_0^1 St(1-t) \, \text{str}[\omega d(A\mathcal{F}^{l-1}_t)]. \tag{13.82}$$

[*Note:* In the proof, we tentatively drop the normalization factor $(i/2\pi)^{l+1}$ to simplify the expressions. This factor will be recovered at the very end.]

Proof. We start with (11.105),

$$Q_{2l+1}(A + \omega, \mathcal{F}) = \frac{1}{l!} \int_0^1 St \, \text{tr}[(A + \omega)\hat{\mathcal{F}}^l_t]$$

where

$$\hat{\mathcal{F}}_t \equiv t\mathcal{F} + (t^2 - t)(A + \omega)^2$$
$$= \mathcal{F}_t + (t^2 - t)\{A, \omega\} + (t^2 - t)\omega^2$$
$$\mathcal{F}_t \equiv d\,(tA) + (tA)^2.$$

If we substitute $\hat{\mathcal{F}}_t$ into Q_{2l+1} and collect terms of first order in ω, we have:

$$\frac{1}{l!}\int_0^1 \delta t\ \text{tr}[\omega\mathcal{F}_t^l + (t^2 - t)(A[A, \omega]\mathcal{F}_t^{l-1} + A\mathcal{F}_t[A, \omega]\mathcal{F}_t^{l-2}$$

$$+ \cdots + A\mathcal{F}_t^{l-1}[A, \omega])]$$

$$= \frac{1}{l!}\int \delta t\ \text{str}[\omega\mathcal{F}_t^l + (t^2 - t)A(\mathcal{F}_t^{l-1}[A, \omega]]$$

$$+ \mathcal{F}_t^{l-2}[A, \omega]\mathcal{F}_t + \cdots)]$$

$$= \frac{1}{l!}\int \delta t\ \text{str}[\omega\mathcal{F}_t^l + (t^2 - t)lA[A, v]\mathcal{F}_t^{l-1}]$$

$$= \frac{1}{l!}\int \delta t\ \text{str}[\omega\mathcal{F}_t^l + l(t^2 - t)([A, A]\omega\mathcal{F}_t^{l-1} + A\omega[A, \mathcal{F}_t^{l-1}])]$$

$$= \frac{1}{l!}\int \delta t\ \text{str}[\omega\{\mathcal{F}_t^l + l(t - 1)(t[A, A]\mathcal{F}_t^{l-1} - A[A_t, \mathcal{F}_t^{l-1}])\}]$$

where str is the symmetrized trace defined by (11.8). Now we use

$$\mathcal{D}_t\mathcal{F}_t^{l-1} \equiv d\mathcal{F}_t^{l-1} + [A_t, \mathcal{F}_t^{l-1}] = 0$$

$$\frac{\partial \mathcal{F}_t}{\partial t} = dA + t[A, A]$$

to change the final line of the previous equation to

$$\frac{1}{l!}\int \delta t\ \text{str}\left[\omega\left\{\mathcal{F}_t^l + l(t - 1)\left[\left(\frac{\partial \mathcal{F}_t}{\partial t} - dA\right)\mathcal{F}_t^{l-1} + Ad\mathcal{F}_t^{l-1}\right]\right\}\right]$$

$$= \frac{1}{l!}\int \delta t\ \text{str}\left[\omega\left\{\mathcal{F}_t^l + l(1 - t)d(A\mathcal{F}_t^{l-1}) + (t - 1)\frac{\partial \mathcal{F}_t^l}{\partial t}\right\}\right].$$

Integrating by parts, we find that

$$Q_{2l}^1(\omega, A, \mathcal{F}) = \frac{1}{(l - 1)!}\int \delta t(1 - t)\ \text{str}[\omega d(A\mathcal{F}_t^{l-1})].$$

If we recover the normalization, we finally have

$$Q_{2l}^1(\omega, A, \mathcal{F}) = \left(\frac{i}{2\pi}\right)^{l+1}\frac{1}{(l - 1)!}\int_0^1 \delta t(1 - t)\ \text{str}[\omega d(A\mathcal{F}_t^{l-1})].\qquad\square$$

For $m = 2l = 2$ and $m = 4$, we have

$$Q_2^1(\omega, A, F) = \left(\frac{i}{2\pi}\right)^2 \text{tr}(\omega dA) \tag{13.83a}$$

$$Q_4^1(\omega, A, F) = \frac{1}{6}\left(\frac{i}{2\pi}\right)^3 \text{str}(\omega d(AdA + \tfrac{1}{2}A^3)). \tag{13.83b}$$

These results are also verified by direct computations. Up to the normalization factor, (13.83b) yields the non-Abelian anomaly in four-dimensional space; see (13.56).

Sumitani (1984) pointed out that the approach to the non-Abelian anomalies here is *ad hoc* and does not clarify the following points:

(1) The WZ condition (13.73) does not fix the normalization of the anomaly and, moreover, the uniqueness of the solution is far from trivial.
(2) It is not clear why we should start from the Abelian anomaly in $(m + 2)$-dimensional space.

To answer these questions we need to develop a more elaborate index theorem called the family index theorem; see Atiyah and Singer (1984), Singer (1985) and Sumitani (1984, 1985). In the next section, we outline the physicists' approach to this problem, closely following the work of Alvarez-Gaumé and Ginsparg (1984).

13.5 Abelian anomalies *versus* non-Abelian anomalies

Let us consider an m-dimensional Euclidean space ($m = 2l$) which is compactified to $S^m = \mathbb{R}^m \cup \{\infty\}$ and let G be a semisimple gauge group which is simply connected (like SU(N) for which $\pi_1(\text{SU}(N))$ is trivial). Consider a one-parameter family of gauge transformations $g(\theta, x)$ ($0 \leq \theta \leq 2\pi$) such that

$$g(0, x) = g(2\pi, x) = e. \tag{13.84}$$

Without loss of generality, we may normalize g so that $g(\theta, x_0) = e$ at a point $x_0 \in S^m$. The map $g : S^1 \times S^m \to G$ is classified according to the homotopy class $\pi_{m+1}(G)$. To see this we define the **smash product** $X \wedge Y$ of topological spaces X and Y by the direct product $X \times Y$ with $X \vee Y \equiv (x_0 \times X) \cup (X \times y_0)$ shrunk to a point. From figure 13.2, we easily find that $S^1 \wedge S^m = S^m \wedge S^1 = S^{m+1}$.[5] Repeated applications of this yield

$$S^m \wedge S^n = S^{m+n}. \tag{13.85}$$

In the case which interests us, the conditions (13.84) make the direct product $S^1 \times S^m$ look topologically like $S^1 \wedge S^m = S^{m+1}$. Thus, g is regarded as a map

[5] The readers may convince themselves by explicitly drawing $S^1 \wedge S^1 = S^2$.

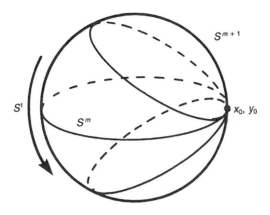

Figure 13.2. The smash product $S^1 \wedge S^m \simeq S^{m+1}$.

from S^{m+1} to G and is classified by $\pi_{m+1}(G)$. Since we have a one-parameter family in the space $\mathfrak{G} = \Omega^m(G)$, we also have $\pi_{m+1}(G) = \pi_1(\mathfrak{G})$. In practice, we take $G = \mathrm{SU}(N)$ for which we have

$$\pi_{m+1}(\mathrm{SU}(N)) = \mathbb{Z} \qquad N \geq \tfrac{1}{2}m + 1. \tag{13.86}$$

Now we take a 'reference' gauge field $\mathcal{A}$ in the zero instanton sector $\Omega_0^m(G)$ for which we may assume, without loss of generality, that the Dirac operator (13.49) has no zero modes. Consider a one-parameter family of gauge potentials

$$\mathcal{A}^{g(\theta)}(x) \equiv g^{-1}(\theta, x)(\mathcal{A}(x) + \mathrm{d})g(\theta, x) \tag{13.87}$$

where θ parametrizes S^1. In section 13.3, we observed that $|\det i\hat{D}|$ is gauge invariant (see (13.47)) and only the *phase* of $\det i\hat{D}$ may gain an anomalous variation under a gauge transformation. This, in particular, implies that $\det i\hat{D}$ does not vanish for any θ. We write

$$\exp\{-W_r[\mathcal{A}^{g(\theta)}]\} = \det i\hat{D}(\mathcal{A}^{g(\theta)}) = [\det i\mathcal{\slashed{V}}(\mathcal{A})]^{1/2} \exp[iw(\mathcal{A}, \theta)] \tag{13.88}$$

where $i\mathcal{\slashed{V}}$ is the Dirac operator (13.49) and $\exp[iw(\mathcal{A}, \theta)]$ is the anomalous phase associated with the gauge transformation (13.87). Next we consider a *two-parameter* family of gauge fields $\mathcal{A}^{t,\theta}$ ($0 \leq t \leq 1$) which interpolates between $\mathcal{A} = 0$ and $\mathcal{A}^{g(\theta)}$,

$$\mathcal{A}^{t,\theta} \equiv t\mathcal{A}^{g(\theta)} \qquad (0 \leq t \leq 1). \tag{13.89}$$

The parameter space specified by (t, θ) is considered to be a two-dimensional unit disc D^2 with polar coordinates (t, θ). On the boundary of the disc, $\partial D^2 = S^1$, the modulus of $\det i\hat{D}(\mathcal{A}^{1,\theta})$ is a non-vanishing constant. The phase $e^{iw(\mathcal{A},\theta)}$ now defines a map S^1 $(=\partial D^2) \to S^1$ $(=\mathrm{U}(1))$; see figure 13.3. As we move around

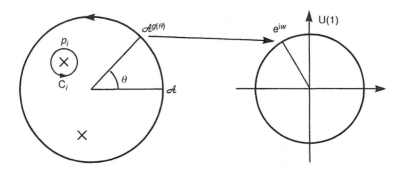

Figure 13.3. The phase of the effective action $W[\mathcal{A}^{g(\theta)}]$ defines a map $S^1 \to U(1)$ by $\theta \mapsto e^{iw(\mathcal{A},\theta)}$. On the disc, there are points $\{p_i\}$ at which $\det i\hat{D}(\mathcal{A}^{t,\theta})$ vanishes. The winding number of the map $S^1 \to U(1)$ is obtained by summing a winding number along C_i.

the boundary of the disc, the phase winds around the unit circle. The winding number of this map is an integer

$$\mathcal{N} = \frac{1}{2\pi} \int_0^{2\pi} \frac{\partial w(\mathcal{A}, \theta)}{\partial \theta} d\theta. \tag{13.90}$$

We find below that $\mathcal{N}$ is derived from the Abelian anomaly in $(m+2)$ dimensions.

Exercise 13.2. Show that

$$W[\mathcal{A}^{g(2\pi)}] - W[\mathcal{A}^{g(0)}] = -2\pi i\mathcal{N}. \tag{13.91}$$

Since $g(2\pi) = g(0)$, (13.91) may be regarded as a Berry phase.

13.5.1 *m* dimensions *versus* *m* + 2 dimensions

We recall that our reference gauge field $\mathcal{A}$ supports no zero modes of the operator $i\hat{D}(\mathcal{A})$. Since $|\det i\hat{D}(\mathcal{A}^{g(0)})| = |\det i\hat{D}(\mathcal{A})| \neq 0$, the operator $i\hat{D}(\mathcal{A}^{g(0)})$ does not admit zero modes either. Of course, $i\hat{D}(\mathcal{A}^{t,\theta})$ may have zero modes since $\mathcal{A}^{t,\theta}$ is *not* obtained from $\mathcal{A}$ by a gauge transformation in general. Suppose it has a zero mode at $p_i = (t_i, \theta_i)$. We assume they are isolated points. Since $\det i\hat{D}(\mathcal{A}^{t,\theta})$ is a regularized product of eigenvalues, it vanishes at p_i. The phase of $\det i\hat{D}(\mathcal{A}^{t,\theta})$ may be homotopically non-trivial only around these points. Moreover, the winding number at p_i is determined by the eigenvalue which vanishes at p_i. For example, if $\lambda_n(t, \theta)$ vanishes at p_i it should be of the form

$$\lambda_n(t, \theta) = f(t, \theta)e^{iw_i(t,\theta)} \tag{13.92}$$

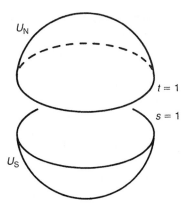

Figure 13.4.

where $f(t_i, \theta_i) = 0$. The winding number at p_i is

$$m_i = \frac{1}{2\pi} \int_{C_i} \frac{\mathrm{d}}{\mathrm{d}s} w_i(t, \theta) \, \mathrm{d}s \tag{13.93}$$

where C_i is a small contour surrounding p_i, see figure 13.3. Continuously deforming the loop $S^1 = \partial D^2$ into a sum of small circles C_i enclosing p_i, we find that the total winding number is

$$\mathcal{N} = \frac{1}{2\pi} \int_{S^1} \mathrm{d}\theta \, \frac{\partial}{\partial \theta} w(\mathcal{A}, \theta) = \sum_i m_i. \tag{13.94}$$

Now we show that the winding number $\mathcal{N}$ is related to the index theorem in $(m + 2)$-dimensional space ($m = 2l$): $\mathcal{N} = \text{ind} \, i\nabla\!\!\!\!/_{m+2}$ where $i\nabla\!\!\!\!/_{m+2}$ is the Dirac operator on $S^2 \times S^m$ defined later. Let us consider a gauge theory defined on $D^2 \times S^m$ whose coordinates are (t, θ, x). To avoid the boundary term, we add another piece, $D^2 \times S^m$, with coordinates (s, θ, x), to form a manifold $S^2 \times S^m$ without a boundary; see figure 13.4. We call the patch (t, θ) the northern hemisphere U_N and (s, θ) the southern hemisphere U_S. On the equator S^1 of S^2, we have $t = s = 1$. We choose the following local gauge potentials

$$\mathcal{A}_N(t, \theta, x) = \mathcal{A}^{t, \theta} + g^{-1} \, \mathrm{d}_\theta g \qquad (t, \theta) \in U_N \tag{13.95a}$$

$$\mathcal{A}_S(s, \theta, x) = \mathcal{A} \qquad (s, \theta) \in U_S \tag{13.95b}$$

where $\mathcal{A}$ is the reference gauge field introduced previously. To elevate $\mathcal{A}_N = \mathcal{A}_{N\mu} \, \mathrm{d}x^\mu$ and $\mathcal{A}_S = \mathcal{A}_{S\mu} \, \mathrm{d}x^\mu$ to the globally defined connection on the G bundle over $S^2 \times S^m$ we define the $(m + 2)$-dimensional gauge potentials

$$\mathbb{A}_N(t, \theta, x) = (\mathcal{A}_t, \mathcal{A}_\theta, \mathcal{A}_\mu) = (0, 0, \mathcal{A}_{N\mu}) \tag{13.96a}$$

$$\mathbb{A}_S(s, \theta, x) = (\mathcal{A}_s, \mathcal{A}_\theta, \mathcal{A}_\mu) = (0, 0, \mathcal{A}_{S\mu}). \tag{13.96b}$$

On the equator ($t = s = 1$), we have $\mathbb{A}_N = g^{-1}(\mathbb{A}_S + \Delta)g$, where $\Delta = d + d_\theta + d_t$ (note that $d_t g = 0$). Thus, $\mathbb{A} = \{\mathbb{A}_N, \mathbb{A}_S\}$ defines a global connection on $S^2 \times S^m$. Consider a *Dirac* operator $i\slashed{\nabla}_{m+2}$ which couples to $\mathbb{A}$. The index theorem for $i\slashed{\nabla}_{m+2}$ is given by

$$\text{ind}\, i\slashed{\nabla}_{m+2} = \mathcal{N}_+ - \mathcal{N}_- = \int_{S^2 \times S^m} \text{ch}_{l+1}(\mathbb{F}) \tag{13.97}$$

where $\mathbb{F} = \Delta\mathbb{A} + \mathbb{A}^2$ and $\mathcal{N}_+$ ($\mathcal{N}_-$) is the number of $+$ $(-)$ chirality zero modes of $i\slashed{\nabla}_{m+2}$ (chirality is defined in an $(m+2)$-space).

Alvarez-Gaumé and Ginsparg (1984) have shown, using an adiabatic perturbative computation, that each winding number m_i must be ± 1. Moreover, the Dirac operator $i\slashed{\nabla}_{m+2}$ has a zero mode at $p_i = (t_i, \theta_i)$ with $(m+2)$-dimensional chirality $\chi = m_i = \pm 1$. Then the total winding number $\mathcal{N} = \sum m_i$ is given by the index $\mathcal{N}_+ - \mathcal{N}_-$. Now we have

$$\text{ind}\, i\slashed{\nabla}_{m+2} = \int_{S^2 \times S^m} \text{ch}_{l+1}(\mathbb{F}) = \frac{1}{2\pi} \int_0^{2\pi} d\theta\, \frac{\partial w(\mathcal{A}, \theta)}{\partial \theta}. \tag{13.98}$$

We easily find the non-Abelian anomaly from (13.98) including the normalization. Since $\text{ch}_{l+1}(\mathbb{F}) = dQ_{m+1}(\mathbb{A}, \mathbb{F})$, we have

$$\int_{S^2 \times S^m} \text{ch}_{l+1}(\mathbb{F}) = \int_{D^2 \times S^m} \text{ch}_{l+1}(\mathbb{F}_N) + \int_{D^2 \times S^m} \text{ch}_{l+1}(\mathbb{F}_S)$$

$$= \int_{S^1 \times S^m} [Q_{m+1}(\mathbb{A}_N, \mathbb{F}_N)|_{t=1} - Q_{m+1}(\mathbb{A}_S, \mathbb{F}_S)|_{s=1}]. \tag{13.99}$$

From (11.118), we find that

$$Q_{m+1}(\mathbb{A}_N, \mathbb{F}_N)|_{t=1} - Q_{m+1}(\mathbb{A}_S, \mathbb{F}_S)|_{s=1}$$
$$= Q_{m+1}(g^{-1}\Delta g, 0) + \Delta\alpha_m$$
$$= (-1)^l \left(\frac{i}{2\pi}\right)^{l+1} \frac{l!}{(m+1)!} \text{tr}(g^{-1}\Delta g)^{m+1} + \Delta\alpha_m. \tag{13.100}$$

The index theorem is now given by

$$\text{ind}\, i\slashed{\nabla}_{m+2} = (-1)^l \left(\frac{i}{2\pi}\right)^{l+1} \frac{l!}{(m+1)!} \int_{S^1 \times S^m} \text{tr}(g^{-1}\Delta g)^{m+1}. \tag{13.101}$$

Theorem 10.7 states that $\int_{S^3} \text{tr}(g^{-1}dg)^3$ yields the winding number of the map $g : S^3 \to SU(2)$. In the same manner, (13.101) represents the winding number of the map $g : S^{m+1} \to G$ and is classified by $\pi_{m+1}(G)$ (note that $S^1 \wedge S^m = S^{m+1}$).

Finally, we show that the non-Abelian anomaly should be identified with Q_m^1. We first note that

$$\int_{S^1 \times S^m} Q_{m+1}(\mathbb{A}_S, \mathbb{F}_S) = 0$$

since the integrand is independent of $d\theta$ and, thus, cannot be a volume element of $S^1 \times S^m$. Then we have

$$\text{ind} \, i\mathbb{V}_{m+2} = \int_{S^1 \times S^m} Q_{m+1}(A^{g(\theta)} + \omega, \mathcal{F}^{g(\theta)}) \tag{13.102}$$

where $\omega = g^{-1} d_\theta g$ and $\mathcal{F}^{g(\theta)} = dA^{g(\theta)} + (A^{g(\theta)})^2 = g(\theta)^{-1} \mathcal{F} g(\theta)$. If the integrand in (13.102) is expanded in ω, only the term *linear* in $d\theta$ contributes to the integral. This term $Q_m^1(\omega, A^{g(\theta)}, \mathcal{F}^{g(\theta)})$ is proportional to $d\theta \wedge$ (volume element in S^m) and, hence, is a volume element of $S^1 \times S^m$. We now have

$$\delta_\omega W[A] = \int_{S^m} \text{tr} \, \omega \mathcal{D}_\mu \frac{\delta W[A]}{\delta A_\mu}$$

$$= \text{id}_\theta \, w(\theta, A) = 2\pi i \int_{S^m} Q_m^1(\omega, A^{g(\theta)}, \mathcal{F}^{g(\theta)}). \tag{13.103}$$

The explicit form of Q_m^1 is given by (13.82). For $m = 4$, we find that

$$\int \text{tr} \, \omega \mathcal{D}_\mu \frac{\delta W[A]}{\delta A_\mu} = 2\pi i \int_{S^4} Q_4^1(\omega, A^{g(\theta)}, \mathcal{F}^{g(\theta)})$$

$$= \frac{1}{24\pi^2} \int_{S^4} \text{tr} \, \omega \, d \left[A^{g(\theta)} \, dA^{g(\theta)} + \frac{1}{2}(A^{g(\theta)})^3 \right] \tag{13.104}$$

Putting $\theta = 0$ $(g = e)$, we reproduce the anomalous divergence

$$\mathcal{D}_\mu \langle j^\mu \rangle_\alpha = \frac{1}{24\pi^2} \text{tr} \, T_\alpha \epsilon^{\kappa\lambda\mu\nu} \partial_\kappa \left[A_\lambda \partial_\mu A_\nu + \frac{1}{2} A_\lambda A_\mu A_\nu \right] \tag{13.105}$$

which is in agreement with (13.56). The present method guarantees that the WZ condition yields the correct result. Moreover, it reproduces the anomalous divergence including the normalization which cannot be fixed by the WZ condition alone.

13.6   The parity anomaly in odd-dimensional spaces

So far, we have been working in even-dimensional spaces. One of the reasons for this is that $SO(2l + 1)$ has real or pseudo-real spinor representations but no *complex* representations, hence no gauge anomaly is expected. However, we can show that gauge theories in odd-dimensional spaces have a different kind of anomaly called the 'parity anomaly', in which the parity symmetry of the classical action is not maintained through quantization. It should be noted that the parity anomaly in $2l + 1$ dimensions is related to the Abelian anomaly in $2l + 2$ dimensions as was pointed out by Alvarez-Gaumé *et al* (1985).

13.6.1   The parity anomaly

Let M be a $(2l + 1)$-dimensional Riemannian manifold. We distinguish one dimension from the others; that is we assume that M is of the form $\mathbb{R} \times \mathfrak{M}$ or $S^1 \times \mathfrak{M}$, where $\mathfrak{M}$ is a $2l$-dimensional compact manifold without a boundary. We denote the coordinate of $\mathbb{R}$ or S^1 by t while that of $\mathfrak{M}$ is denoted by x. The index 0 denotes the component in t-space while μ denotes that in x-space. For example, the components of the γ-matrices are $\{\gamma^0, \gamma^\mu \ (1 \le \mu \le 2l)\}$.

Define the 'parity' operation P by

$$\mathcal{A}_0(t, x) \to \mathcal{A}_0^P(t, x) = -\mathcal{A}_0(-t, x)$$
$$\mathcal{A}_\mu(t, x) \to \mathcal{A}_\mu^P(t, x) = \mathcal{A}_\mu(-t, x)$$
$$\psi(t, x) \to \psi^P(t, x) = i\gamma_0\psi(-t, x)$$
$$\bar{\psi}(t, x) \to \bar{\psi}^P(t, x) = i\bar{\psi}(-t, x)\gamma_0.$$

The classical action is invariant under the parity operation,

$$\int dt\, dx\, \bar{\psi} i\slashed{\nabla}\psi \to -\int dt\, dx\, \bar{\psi}(-t, x)\gamma^0 i[\gamma^0(\partial_0 - \mathcal{A}_0(-t, x))$$
$$+ \gamma^\mu(\partial_\mu + \mathcal{A}_\mu(-t, x))]\gamma^0\psi(-t, x)$$
$$= \int dt\, dx\, \bar{\psi}(t, x) i[\gamma^0(\partial_0 + \mathcal{A}_0(t, x))$$
$$+ \gamma^\mu(\partial_\mu + \mathcal{A}_\mu(t, x))]\psi(t, x)$$

where we put $t \to -t$ in the final line. Let us see whether this invariance is observed by the effective action. The effective action is given by the regularized product of the eigenvalues of $i\slashed{\nabla}$. We employ the **Pauli–Villars regularization** to regulate the product, that is

$$\mathcal{L}_{\text{reg}} \equiv \bar{\chi} i\slashed{\nabla}\chi + iM\bar{\chi}\chi \tag{13.106}$$

is added to the original Lagrangian. The Pauli–Villars regulator χ is a spinor which obeys *bosonic* statistics and the limit $M \to \infty$ is understood. The regularized determinant is

$$e^{-W[\mathcal{A}]} = \frac{\det i\slashed{\nabla}}{\det(i\slashed{\nabla} + iM)} = \prod_i \frac{\lambda_i}{\lambda_i + iM} \tag{13.107}$$

where we noted that χ is bosonic. Here λ_i is the ith eigenvalue of $i\slashed{\nabla}$; $i\slashed{\nabla}\psi_i = \lambda_i\psi_i$. Under the parity operation, eigenvalues change sign,

$$i[\gamma^0(\partial_0 - \mathcal{A}_0(-t, x)) + \gamma^i(\partial_i + \mathcal{A}_i(-t, x))]i\gamma^0\psi_i(-t, x)$$
$$= i\gamma^0[\gamma^0(-\partial_\tau - \mathcal{A}_0(\tau, x)) - \gamma^i(\partial_i + \mathcal{A}_i(\tau, x))]i\psi(\tau, x)$$
$$= -\lambda_i i\gamma^0\psi_i(\tau, x)$$

where $\tau = -t$. This shows that the effective action $W[\mathcal{A}]$ transforms under the parity operation P as

$$W[\mathcal{A}] \to W[\mathcal{A}^{\mathrm{P}}] = -\ln \prod \frac{-\lambda_i}{-\lambda_i + iM} = \overline{W[\mathcal{A}]} \qquad (13.108)$$

where the bar denotes complex conjugation. (13.108) shows that the imaginary part of W is identified with the parity-violating part

$$W[\mathcal{A}] - W[\mathcal{A}^{\mathrm{P}}] = 2 \operatorname{Im} W[\mathcal{A}]. \qquad (13.109)$$

$\operatorname{Im} W[\mathcal{A}]$ is given by the η-invariant defined in section 12.8. In fact,

$$\operatorname{Im} W[\mathcal{A}] = \lim_{M \to \infty} \operatorname{Im} \left(-\sum_i \ln \frac{\lambda_i}{\lambda_i + iM} \right) = \lim_{M \to \infty} \sum_i \tan^{-1}(M/\lambda_i)$$

$$= \frac{\pi}{2} \left(\sum_{\lambda > 0} 1 - \sum_{\lambda < 0} 1 \right) = \frac{\pi}{2} \eta. \qquad (13.110)$$

Thus, the Pauli–Villars regulator gives a regularized form for the η-invariant. We finally have

$$\operatorname{Im} W[\mathcal{A}] = \frac{\pi}{2} \eta = \frac{\pi}{2} \lim_{s \to 0} \sum_i{}' \operatorname{sgn} \lambda_i |\lambda_i|^{-2s} \qquad (13.111)$$

where the prime indicates the omission of zero modes.

13.6.2 The dimensional ladder: 4–3–2

It is remarkable that the parity anomaly (13.110) is closely related to the chiral anomaly in a $(2l + 2)$-dimensional space (Alvarez-Gaumé et al 1985). Following Forte (1987), we look at the **dimensional ladder**,

four-dimensional Abelian anomaly

$\downarrow$

three-dimensional parity anomaly   (13.112)

$\downarrow$

two-dimensional non-Abelian anomaly.

We take $M_4 = S^2 \times S^2$ as a four-dimensional space. The Abelian anomaly is given by the index

$$\operatorname{ind} i \slashed{\mathcal{V}}_4 = \mathcal{N}_+ - \mathcal{N}_- = \int_{S^2 \times S^2} \partial_\mu j_5^\mu = \int_{S^2 \times S^2} \operatorname{ch}_2(\mathbb{F}). \qquad (13.113)$$

As before, $\mathcal{N}_+$ ($\mathcal{N}_-$) is the number of positive (negative) chirality zero modes. Let Q_3 be the Chern–Simons form of $\operatorname{ch}_2(\mathbb{F})$; $\operatorname{ch}_2(\mathbb{F}) = \mathrm{d}Q_3(\mathbb{A}, \mathbb{F})$. Then

$\mathcal{N} \equiv \mathcal{N}_+ - \mathcal{N}_-$ is given by

$$
\begin{aligned}
\mathcal{N} &= \int_{S^2 \times S^2} \mathrm{ch}_2(\mathbb{F}) = \int_{U_N \times S^2} \mathrm{d}Q_3(\mathbb{A}_N, \mathbb{F}_N) + \int_{U_S \times S^2} \mathrm{d}Q_3(\mathbb{A}_S, \mathbb{F}_S) \\
&= \int_{S^1 \times S^2} [Q_3(\mathbb{A}_N, \mathbb{F}_N) - \mathrm{d}Q_3(\mathbb{A}_S, \mathbb{F}_S)] \\
&= \frac{1}{24\pi^2} \int_{S^1 \times S^2} \mathrm{tr}(g^{-1}\mathrm{d}g)^3
\end{aligned}
\tag{13.114}
$$

where g is the gauge transformation connecting $\mathbb{A}_N$ and $\mathbb{A}_S$; $\mathbb{A}_N = g^{-1}(\mathbb{A}_S + \mathrm{d} + \mathrm{d}_\theta)g$. In the previous section, we have shown that $\mathcal{N}$ also represents the non-Abelian anomaly

$$
\mathcal{N} = \frac{1}{2\pi} \int_0^{2\pi} \mathrm{d}\theta\, \frac{\partial w(\mathcal{A}, \theta)}{\partial \theta}
\tag{13.115a}
$$

where w is defined by

$$
\det \mathrm{i}\hat{D}(\mathcal{A}^{g(\theta)}) = \mathrm{e}^{\mathrm{i}w(\mathcal{A}, \theta)} \det \mathrm{i}\hat{D}(\mathcal{A}).
\tag{13.115b}
$$

Here $\mathcal{A}$ is the reference gauge potential and

$$
\mathcal{A}^{g(\theta)} = g^{-1}(x, \theta)(\mathcal{A} + \mathrm{d})g(x, \theta)\, \mathrm{i}\hat{D} = \partial\!\!\!/ + \mathcal{A}\mathcal{P}_+.
$$

Next, we show that $\mathcal{N}$ is also related to the parity anomaly in three-dimensional space. Let $\mathrm{i}\nabla\!\!\!\!/_3$ be a three-dimensional Dirac operator and define a four-dimensional Dirac operator by

$$
\mathrm{i}D\!\!\!\!/_4[\mathcal{A}] \equiv \mathrm{i}\sigma_1 \otimes I \frac{\partial}{\partial t} + \sigma_2 \otimes \mathrm{i}\nabla\!\!\!\!/_3[\mathcal{A}_t]
\tag{13.116}
$$

where $\mathcal{A}_t$ is a one-parameter family of gauge potentials interpolating $\mathcal{A}_0 = \mathcal{A}_{t=0}$ and $\mathcal{A}_1 = \mathcal{A}_{t=1}$. The Atiyah–Patodi–Singer index theorem (section 12.8) is

$$
\mathrm{ind}\, \mathrm{i}D\!\!\!\!/_4 = -\int_{S^2 \times S^1 \times I} \mathrm{ch}_2(\mathcal{F}) + \tfrac{1}{2}[\eta(t=1) - \eta(t=0)]
\tag{13.117}
$$

where we have noted that the Dirac genus $\hat{A}$ is trivial on $S^2 \times S^1 \times I$. Suppose $\mathcal{A}_0$ and $\mathcal{A}_1$ are related by a gauge transformation,

$$
\mathcal{A}_1 = g^{-1}(\mathcal{A}_0 + \mathrm{d})g
\tag{13.118a}
$$

and consider an interpolating potential

$$
\mathcal{A}_t \equiv t\mathcal{A}_1 + (1 - t)\mathcal{A}_0.
\tag{13.118b}
$$

Since the spectrum of $i\nabla_3$ is gauge invariant, in particular $\text{Spec}\, i\nabla_3(\mathcal{A}_0) = \text{Spec}\, i\nabla_3(\mathcal{A}_1)$, the η-invariant is also gauge invariant.[6] Then $\eta(t=0) = \eta(t=1)$ and the APS index theorem (13.117) yields

$$\text{spectral flow} = \text{ind}\, i\slashed{D}_4(\mathcal{A}_t)$$

$$= \int_{S^2 \times S^2} \text{ch}_2(\mathcal{F}) = \int_{S^1 \times S^2} [Q_3(\mathcal{A}_1, \mathcal{F}_1) - Q_3(\mathcal{A}_0, \mathcal{F}_0)]$$

$$= \int_{S^1 \times S^2} Q_3(g^{-1}\, \text{d}g, 0) = \mathcal{N}. \qquad (13.119)$$

Thus, the spectral flow of the three-dimensional theory is given by the index $\mathcal{N}$.

In summary, the map $g : S^2 \times S^1 \to G$ is understood in three different ways:

(1) g is a transition function at the boundary of two patches of a G bundle over $S^2 \times S^2$. It yields the index $\mathcal{N}$ of the four-dimensional Abelian anomaly.

(2) Suppose $\mathcal{A}_0$ and $\mathcal{A}_1 = g^{-1}(\mathcal{A}_0 + \text{d})g$ are gauge potentials on $S^2 \times S^1$. The gauge transformation function g measures the spectral flow $\mathcal{N}$ between $\text{Spe}\, i\nabla_3(\mathcal{A}_0)$ and $\text{Spe}\, i\nabla_3(\mathcal{A}_1)$.

(3) $g : S^2 \times S^1 \to G$ induces a map $S^1 \to \mathfrak{G}$, the winding number $\mathcal{N}$ of which is identified with the non-Abelian anomaly in two-dimensional space.

Thus, we have obtained the '**dimensional ladder**' 4–3–2. The extension to higher dimensions is obvious.

---

[6] Note that there is no gauge anomaly in odd-dimensional spaces.

14

BOSONIC STRING THEORY

In the present chapter, we study the one-loop amplitude of bosonic string theory. Our example is the simplest one: closed, oriented bosonic strings in 26-dimensional Euclidean space.[1] The action is the Polyakov action

$$S = \frac{1}{2\pi} \int_{\Sigma_g} d^2\xi \, \sqrt{\gamma} \gamma^{\alpha\beta} \partial_\alpha X^\mu \partial_\beta X_\mu - \frac{\lambda}{4\pi} \int_{\Sigma_g} d^2\xi \, \sqrt{\gamma} \mathcal{R} \qquad (14.1)$$

where Σ_g is a Riemann surface with genus g. The second term is proportional to the Euler characteristic $\chi = 2 - 2g$ and, hence, determines the relative ratio of multi-loop amplitudes; the g-loop amplitude is proportional to $\exp(-\lambda g)$. We have not written down the possible counter terms explicitly.

In the following sections, we work out the path integral formalism of bosonic strings. We first develop the necessary mathematical tools, namely differential geometry on Riemann surfaces. Then the path integral expression for the vacuum amplitude is written down. As an example, we compute the one-loop vacuum amplitude. Our exposition is based on D'Hoker and Phong (1986), Polchinski (1986) and Moore and Nelson (1986). There are many surveys of these topics, for example, Alvarez-Gaumé and Nelson (1986), Bagger (1987), D'Hoker and Phong (1988) and Weinberg (1988).

14.1 Differential geometry on Riemann surfaces

Riemann surfaces are real two-dimensional manifolds without boundary. In our study of topology and geometry, we referred to them in various places. Here we summarize the basic facts on Riemann surfaces, which will make this chapter self-contained. We also introduce several new aspects of Riemann surfaces, which provide enough background for the study of bosonic string amplitudes.

14.1.1 Metric and complex structure

Let Σ_g be a Riemann surface of genus g. It was shown in example 7.9 that we may introduce, in any chart U, the **isothermal coordinates** (ξ^1, ξ^2) in which the metric is conformally flat:

$$g = e^{2\sigma(\xi)}(d\xi^1 \otimes d\xi^1 + d\xi^2 \otimes d\xi^2). \qquad (14.2)$$

---

[1] The reason for $D = 26$ will be clarified in section 14.2.

Introduce the complex coordinates

$$z = \xi^1 + i\xi^2 \qquad \bar{z} = \xi^1 - i\xi^2. \tag{14.3}$$

Forms and vectors are spanned by

$$dz = d\xi^1 + i\,d\xi^2 \qquad d\bar{z} = d\xi^1 - i\,d\xi^2 \tag{14.4a}$$

$$\partial_z = \frac{1}{2}\left(\frac{\partial}{\partial \xi^1} - i\frac{\partial}{\partial \xi^2}\right) \qquad \partial_{\bar{z}} = \frac{1}{2}\left(\frac{\partial}{\partial \xi^1} + i\frac{\partial}{\partial \xi^2}\right). \tag{14.4b}$$

In terms of the complex coordinates, the metric takes the form

$$g = \tfrac{1}{2}e^{2\sigma(z,\bar{z})}[dz \otimes d\bar{z} + d\bar{z} \otimes dz]. \tag{14.5}$$

The components of g are

$$g_{z\bar{z}} = g_{\bar{z}z} = \tfrac{1}{2}e^{2\sigma} \qquad g_{zz} = g_{\bar{z}\bar{z}} = 0 \tag{14.6a}$$

$$g^{z\bar{z}} = g^{\bar{z}z} = 2e^{-2\sigma} \qquad g^{zz} = g^{\bar{z}\bar{z}} = 0. \tag{14.6b}$$

Let V be another chart of Σ_g such that $U \cap V \neq \emptyset$. Let $(w, \bar{w})$ be the complex coordinates in V. The metric in V is

$$g = e^{2\sigma'(w,\bar{w})}\,dw \otimes d\bar{w}. \tag{14.7}$$

The two expressions (14.5) and (14.7) should agree on $U \cap V$,

$$e^{2\sigma(z,\bar{z})}\,dz \otimes d\bar{z} = e^{2\sigma'(w,\bar{w})}\,dw \otimes d\bar{w}.$$

Since

$$dw \otimes d\bar{w} = [(\partial w/\partial z)\,dz + (\partial w/\partial \bar{z})d\bar{z}] \otimes [(\partial \bar{w}/\partial z)\,dz + (\partial \bar{w}/\partial \bar{z})d\bar{z}]$$
$$\propto dz \otimes d\bar{z}$$

we must have $\partial w/\partial \bar{z} = \partial \bar{w}/\partial z = 0$. [Another possibility, $\partial w/\partial z = \partial \bar{w}/\partial \bar{z} = 0$ is ruled out if $(z, \bar{z})$ and $(w, \bar{w})$ define the same orientation.] Thus, it follows that

$$w = w(z) \qquad \bar{w} = \bar{w}(\bar{z}) \tag{14.8}$$

which verifies that Σ_g is a complex manifold. We also have

$$e^{2\sigma(z,\bar{z})} = e^{2\sigma'(w,\bar{w})}|\partial w/\partial z|^2. \tag{14.9}$$

14.1.2 Vectors, forms and tensors

Let $M = \Sigma_g$. The components of vector fields $V^z \partial/\partial z \in TM^+$ and $V^{\bar{z}}\partial/\partial \bar{z} \in TM^-$ transform as

$$V^w = (\partial w/\partial z)V^z \qquad V^{\bar{w}} = (\partial \bar{w}/\partial \bar{z})V^{\bar{z}}. \tag{14.10}$$

The components of differential forms $w_z \, dz \in \Omega^{1,0}(M)$ and $w_{\bar{z}} \, d\bar{z} \in \Omega^{0,1}(M)$ transform as

$$\omega_w = (\partial w/\partial z)^{-1}\omega_z \qquad \omega_{\bar{w}} = (\partial \bar{w}/\partial \bar{z})^{-1}\omega_{\bar{z}}. \tag{14.11}$$

These are identified with sections of the holomorphic (anti-holomorphic) line bundles over $M = \Sigma_g$, for which the transition functions are holomorphic (anti-holomorphic). The metric provides a natural isomorphism between TM^+ and $\Omega^{0,1}(M)$ through

$$\omega_{\bar{z}} = g_{\bar{z}z}V^z, \qquad V^z = g^{z\bar{z}}\omega_{\bar{z}}. \tag{14.12}$$

Similarly, TM^- is isomorphic to $\omega^{1,0}(M)$:

$$\omega_z = g_{z\bar{z}}V^{\bar{z}}, \qquad V^{\bar{z}} = g^{\bar{z}z}\omega_z. \tag{14.13}$$

In general, given an arbitrary tensor, the metric allows us to trade all the $\bar{z}$-indices for z-indices. It is easy to see that

$$T^{\overbrace{z...z}^{q_1}\overbrace{\bar{z}...\bar{z}}^{q_2}}_{\underbrace{z...z}_{p_1}\underbrace{\bar{z}...\bar{z}}_{p_2}} \rightarrow T^{\overbrace{z...z}^{q_1+p_2}}_{\underbrace{z...z}_{p_1+q_2}} = (g_{z\bar{z}})^{q_2}(g^{z\bar{z}})^{p_2}\, T^{\overbrace{z...z}^{q_1}\overbrace{\bar{z}...\bar{z}}^{q_2}}_{\underbrace{z...z}_{p_1}\underbrace{\bar{z}...\bar{z}}_{p_2}}. \tag{14.14}$$

This correspondence is an isomorphism. For example, observe that

$$T_{z\bar{z}}{}^{\bar{z}} \rightarrow g^{z\bar{z}}g_{z\bar{z}}T_{z\bar{z}}{}^{\bar{z}} = T_z{}^z{}_z.$$

Thus, it is only necessary to consider tensors with pure z-indices. For these tensors, we assign the helicity. Since T has z-indices only, it transforms under $z \rightarrow w$ as

$$T \rightarrow \left(\frac{\partial w}{\partial z}\right)^n T \tag{14.15}$$

where $n \in \mathbb{Z}$ is given by the number of upper z-indices minus the number of lower z-indices. For example,

$$T^{zz}{}_z \rightarrow T^{ww}{}_w = \left(\frac{\partial w}{\partial z}\right)T^{zz}{}_z.$$

All that matters is the *difference* between the number of upper indices and the number of lower indices. The tensor $T^z{}_z$ is left invariant under $z \rightarrow w$ and is regarded as a scalar. The number n is called the **helicity**. The set of helicity-n tensors is denoted by $\mathcal{T}^n$:

$$\mathcal{T}^n \equiv \{T^{\overbrace{z...z}^{q}}{}_{\underbrace{z...z}_{p}} \,|\, q - p = n\}. \tag{14.16}$$

The helicity characterizes the irreducible representation of $U(1) = SO(2)$.

So far we have assumed n is an integer. It can be shown that $n = \frac{1}{2}$ corresponds to the spinor field on Σ_g. In fact, the existence of spinors on the Riemann surfaces is guaranteed by the triviality of the second Stiefel–Whitney class of Σ_g. The set $\mathcal{T}^1$ is identified with the holomorphic line bundle K over Σ_g. Then $\mathcal{T}^{1/2}$ is the *square root* of K: $S_+^2 = K = \mathcal{T}^1$ where S_+ is the positive-chirality spin bundle. Similarly, we have $\mathcal{T}^{-1} = \bar{K} = S_-^2$ where S_- is the negative-chirality spin bundle.[2]

Example 14.1. In real indices, the helicity ± 1 vectors are given by $V^1 \pm iV^2$. This follows since

$$V^1 \frac{\partial}{\partial \xi^1} + V^2 \frac{\partial}{\partial \xi^2} = (V^1 + iV^2)\partial_z + (V^1 - iV^2)\partial_{\bar{z}}.$$

We put $V^z = V^1 + iV^2$ and $V^{\bar{z}} = V^1 - iV^2 \simeq V_z$. The helicity ± 2 tensors are $T^{11} \pm iT^{22}$, where T is a *symmetric traceless* tensor of rank two. In fact, we find

$$T^{11}\left(\frac{\partial}{\partial \xi^1} \otimes \frac{\partial}{\partial \xi^1} - \frac{\partial}{\partial \xi^2} \otimes \frac{\partial}{\partial \xi^2}\right) + T^{12}\left(\frac{\partial}{\partial \xi^1} \otimes \frac{\partial}{\partial \xi^2} + \frac{\partial}{\partial \xi^2} \otimes \frac{\partial}{\partial \xi^1}\right)$$
$$= 2(T^{11} + iT^{12})\partial_z \otimes \partial_z + 2(T^{11} - iT^{12})\partial_{\bar{z}} \otimes \partial_{\bar{z}}.$$

Clearly $T^{zz} = 2(T^1 + iT^{12})$ has helicity $+2$ and $T^{\bar{z}\bar{z}} = 2(T^{11} - iT^{12})$ has helicity -2 (note that $g_{z\bar{z}}g_{z\bar{z}}T^{\bar{z}\bar{z}} = T_{zz}$).

14.1.3 Covariant derivatives

The only non-vanishing Christoffel symbols of Σ_g are (see (8.69))

$$\Gamma^z_{zz} = g^{z\bar{z}}\partial_z g_{z\bar{z}} = 2\partial_z\sigma \qquad \Gamma^{\bar{z}}_{\bar{z}\bar{z}} = g^{\bar{z}z}\partial_{\bar{z}} g_{z\bar{z}} = 2\partial_{\bar{z}}\sigma. \tag{14.17}$$

For tensors in $\mathcal{T}^n$, we define two kinds of covariant derivative: $\nabla^z_{(n)} : \mathcal{T}^n \to \mathcal{T}^{n+1}$ and $\nabla_z^{(n)} : \mathcal{T}^n \to \mathcal{T}^{n-1}$. Let

$$T^{\overbrace{z\ldots z}^{q}}{}_{\underbrace{z\ldots z}_{p}} \in \mathcal{T}^n \qquad (q - p = n).$$

We define

$$\nabla^z_{(n)} T^{z\ldots z}{}_{z\ldots z} = g^{z\bar{z}}\nabla_{\bar{z}} T^{z\ldots z}{}_{z\ldots z}$$
$$= g^{z\bar{z}}[\partial_{\bar{z}} + (q - p)\Gamma^z_{\bar{z}z}]T^{z\ldots z}{}_{z\ldots z}$$
$$= g^{z\bar{z}}\partial_{\bar{z}} T^{z\ldots z}{}_{z\ldots z} \tag{14.18a}$$
$$\nabla_z^{(n)} T^{z\ldots z}{}_{z\ldots z} = \nabla_z T^{z\ldots z}{}_{z\ldots z}$$
$$= [\partial_z + (q - p)\Gamma^z_{zz}]T^{z\ldots z}{}_{z\ldots z}$$
$$= (\partial_z + 2n\partial_z\sigma)T^{z\ldots z}{}_{z\ldots z}. \tag{14.18b}$$

---

[2] We use $S_\pm$, instead of $\Delta_\pm$, to denote the spin bundles. The symbol $\Delta^\pm$ is reserved for Laplacians.

In (14.18b), $2n\partial_z\sigma$ acts like a gauge potential $\mathcal{A}$. We also define covariant derivatives with respect to $\bar{z}$,

$$\nabla^{\bar{z}}_{(n)} = g^{\bar{z}z}\nabla^{(n)}_z, \qquad \nabla^{(n)}_{\bar{z}} = g_{\bar{z}z}\nabla^z_{(n)}. \tag{14.19}$$

The curvature two-form of K and the scalar curvature associated with the Christoffel symbols are

$$\mathcal{F} = R^z{}_{zz\bar{z}}\,dz \wedge d\bar{z} = -\partial_{\bar{z}}(2\partial_z\sigma)\,dz \wedge d\bar{z}$$
$$= -2\partial_z\partial_{\bar{z}}\sigma\,dz \wedge d\bar{z} \tag{14.20a}$$
$$\mathcal{R} = g^{\bar{z}z}\,Ric_{\bar{z}z} + g^{z\bar{z}}\,Ric_{z\bar{z}} = -8e^{-2\sigma}\partial_z\partial_{\bar{z}}\sigma. \tag{14.20b}$$

Exercise 14.1. Verify that

$$\nabla^{\bar{z}}_{(n)} = 2e^{-2\sigma}\partial_{\bar{z}} \qquad \nabla^{(n)}_z = e^{-2n\sigma}\partial_z e^{2n\sigma} \tag{14.21a}$$
$$\nabla^{\bar{z}}_{(n)} = 2e^{-2(n+1)\sigma}\partial_z e^{2n\sigma} \qquad \nabla^{(n)}_{\bar{z}} = \partial_{\bar{z}}. \tag{14.21b}$$

$\nabla^z_{(n)}$ and $\nabla^{(n)}_z$ are mutual adjoints with respect to a properly defined inner product. Let $T, U \in \mathcal{T}^n$. We require that the inner product be invariant under a holomorphic change of the coordinate $z \to w$. Since

$$g_{z\bar{z}} \to |dw/dz|^{-2}g_{z\bar{z}} \qquad d^2z\sqrt{g} \to d^2w\sqrt{g}$$
$$\bar{T} \to (\overline{dw/dz})^n\bar{T} \qquad U \to (dw/dz)^nU.$$

We find the combination

$$(T, U) \equiv \int d^2z\sqrt{g}(g_{z\bar{z}})^n\bar{T}U \tag{14.22}$$

is invariant under holomorphic coordinate transformations. Take $T \in \mathcal{T}^n$ and $U \in \mathcal{T}^{n+1}$. We find that

$$(U, \nabla^z_{(n)}T) = \int d^2z\,e^{2\sigma}2^{-n-1}e^{2(n+1)\sigma}\bar{U}2e^{-2\sigma}\partial_{\bar{z}}T$$
$$= -2^{-n}\int d^2z\,T\partial_{\bar{z}}[e^{(2n+1)\sigma}\bar{U}] \qquad \text{(partial integration)}$$
$$= -2^{-n}\int d^2z\,Te^{(2n+1)\sigma}\overline{[\partial_zU + (2n+1)(\partial_z\sigma)U]}$$
$$= -\int d^2z\,\sqrt{g}(g_{z\bar{z}})^n[\nabla^{(n+1)}_zU]\bar{T} = (-\nabla^{(n+1)}_zU, T).$$

This shows that

$$(\nabla^z_{(n)})^{\dagger} = -\nabla^{(n+1)}_z. \tag{14.23a}$$

Exercise 14.2. Show that

$$(\nabla_z^{(n)})^\dagger = -\nabla_{(n-1)}^z. \tag{14.23b}$$

We define two kinds of Laplacian $\Delta_{(n)}^\pm : \mathcal{T}^n \to \mathcal{T}^{n\pm1} \to \mathcal{T}^n$ by

$$\Delta_{(n)}^+ \equiv -\nabla_z^{(n+1)}\nabla_{(n)}^z = -2e^{-2\sigma}[\partial_z\partial_{\bar{z}} + 2n(\partial_z\sigma)\partial_{\bar{z}}] \tag{14.24a}$$

$$\Delta_{(n)}^- \equiv -\nabla_{(n-1)}^z\nabla_z^{(n)} = -2e^{-2\sigma}[\partial_z\partial_{\bar{z}} + 2n(\partial_z\sigma)\partial_{\bar{z}} + 2n(\partial_z\partial_{\bar{z}}\sigma)]. \tag{14.24b}$$

Then it follows that

$$\Delta_{(n)}^+ - \Delta_{(n)}^- = 4ne^{-2\sigma}(\partial_z\partial_{\bar{z}}\sigma) = -\tfrac{1}{2}n\mathcal{R}. \tag{14.25}$$

This shows, in particular, that

$$\Delta_{(0)}^+ = \Delta_{(0)}^- \ (\equiv \Delta_{(0)}). \tag{14.26}$$

14.1.4   The Riemann–Roch theorem

Here we derive a version of the Riemann–Roch theorem from the Atiyah–Singer index theorem following D'Hoker and Phong (1988).

Theorem 14.1. (**Riemann–Roch theorem**) Let Σ_g be a Riemann surface of genus g. Then the index of the operator $\nabla_z^{(n)}$ is

$$\dim_{\mathbb{C}} \ker \nabla_z^{(n)} - \dim_{\mathbb{C}} \ker \nabla_{(n-1)}^z = (2n-1)(g-1). \tag{14.27}$$

Proof. We use the heat kernel to evaluate the index. We first note that $\ker \nabla_z^{(n)} = \ker \Delta_{(n)}^+$ and $\ker \nabla_{(n-1)}^z = \ker \Delta_{(n-1)}^+$ (see (14.24)). The heat kernel $\mathcal{K}_n^+$ of $\Delta_{(n)}^+$ satisfies

$$\left(\frac{\partial}{\partial t} + \Delta_{(n)}^+\right)\mathcal{K}_n^+(z, w; t) = \left(\frac{\partial}{\partial t} + \Delta - V_n\right)\mathcal{K}_n^+(z, w; t) = 0$$

where $\Delta \equiv -2\partial_z\partial_{\bar{z}}$ is the flat-space Laplacian and

$$V_n \equiv \Delta - \Delta_{(n)}^+ = (1 - e^{-2\sigma})\Delta + 4ne^{-2\sigma}\partial_z\sigma\partial_{\bar{z}}.$$

The Laplacian Δ also defines a heat kernel by

$$\left(\frac{\partial}{\partial t} + \Delta\right) K(z, w; t) = 0$$

which is easily solved to yield

$$K(z, w; t) = \frac{1}{4\pi t}e^{-|z-w|^2/2t}.$$

The perturbative computation and iteration yield

$$\mathcal{K}_n^+(z, z'; t) = K(z, z'; t)$$

$$+ \int_0^t ds \int dw\, K(z, w; t - s) V_n(w) \mathcal{K}_n^+(w, z'; s)$$

$$= K(z, z'; t) + \int ds \int dw\, K(z, w; t - s) V_n(w) K(w, z'; s)$$

$$+ \int ds \int ds' \int dv \int dw\, K(z, v; t - s) V_n(v)$$

$$\times K(v, w; s - s') V_n(w) K(w, z'; s')$$

$$+ \cdots .$$

We are particularly interested in $\mathcal{K}_n^+(z, z; t)$, t being small,

$$\mathcal{K}_n^+(z, z; t) = \frac{1}{4\pi t} + \int_0^t ds \int dw\, K(z, w; t - s) V_n(w) K(w, z; s) + \mathcal{O}(t).$$

$$(14.28)$$

If we take a coordinate system in which $\sigma = 0$ at z, we have

$$\sigma(w) \simeq 0 + \partial_z \sigma (w - z) + \partial_{\bar{z}} \sigma (\bar{w} - \bar{z})$$
$$+ \tfrac{1}{2}[\partial_z^2 \sigma (w - z)^2 + \partial_{\bar{z}}^2 \sigma (\bar{w} - \bar{z})^2 + 2\partial_z \partial_{\bar{z}} \sigma |w - z|^2] + \cdots .$$

Due to rotational symmetry in two-dimensional space, only those terms with one z-derivative and one $\bar{z}$-derivative survive in the integral in (14.28). Terms proportional to $\partial_z \sigma \partial_{\bar{z}} \sigma$ cancel between the second and third terms in the expansion and we are left with terms proportional to $\partial_z \partial_{\bar{z}} \sigma$. Now we have to evaluate

$$\int_0^t ds \int d^2w\, K(z, w; t - s)$$

$$\times [2\partial_z \partial_{\bar{z}} \sigma |\bar{w} - \bar{z}|^2 \Delta_w + 4n(\bar{w} - \bar{z})\partial_z \partial_{\bar{z}} \sigma \partial_{\bar{w}}] K(w, z; s).$$

From the identities

$$\int d^2w\, K(z, w; t - s)|w - z|^2 \Delta_w K(w, z; s)$$

$$= \frac{1}{16\pi^2 s^2 (t - s)} \int d^2w |w|^2 \exp\left(-\frac{t}{2s(t - s)}|w|^2\right)$$

$$- \frac{1}{32\pi^2 s^3 (t - s)} \int d^2w |w|^4 \exp\left(-\frac{t}{2s(t - s)}|w|^2\right)$$

$$= \frac{(t - s)(2s - t)}{2\pi t^3}$$

and

$$\int d^2w \, K(z, w; t - s)(\bar{z} - \bar{w})\partial_{\bar{w}} K(w, z; s)$$

$$= \frac{1}{32\pi^2 s^2(t-s)} \int d^2w \exp\left(-\frac{t}{2s(t-s)}|w|^2\right) = \frac{t-s}{4\pi t^2}$$

we find that

$$\mathcal{K}_n^+(z, z; t) = \frac{1}{4\pi t} + \frac{1 + 3n}{12\pi}\Delta\sigma + \mathcal{O}(t). \qquad (14.29a)$$

We also have the diagonal part of the heat kernel $\mathcal{K}_n^-$ for $\Delta_{(n)}^-$,

$$\mathcal{K}_n^-(z, z; t) = \frac{1}{4\pi t} + \frac{1 - 3n}{12\pi}\Delta\sigma + \mathcal{O}(t). \qquad (14.29b)$$

From (14.29) and (14.20b), we obtain

$$\mathrm{ind}\,\nabla_z^{(n)} = \int d^2z \left(\frac{1 - 3n}{12\pi} - \frac{1 + 3(n-1)}{12\pi}\right)\Delta\sigma = \frac{1 - 2n}{8\pi}\int d^2x\,\mathcal{R}$$

$$= -\frac{2n-1}{2}\chi(\Sigma_g) = (2n-1)(g-1)$$

where

$$\chi = \frac{1}{4\pi}\int d^2x\,\mathcal{R} = 2 - 2g$$

is the Euler characteristic of Σ_g.

14.2 Quantum theory of bosonic strings

Now we are ready to introduce Polyakov's formulation of bosonic strings, which is based on the path integral over geometries. Since the string action contains an enormous symmetry, we have to pay special attention to counting independent geometries once and only once. This is achieved by the Faddeev–Popov trick. Our argument will be restricted to the simplest case, namely closed orientable bosonic strings; the theory is defined on Riemann surfaces.

14.2.1 Vacuum amplitude of Polyakov strings

According to the general prescription of the path integral formalism, the partition function (vacuum-to-vacuum amplitude) of the string theory is given by

$$Z = \sum_{g=0}^{\infty} Z_g = \sum_{g=0}^{\infty} \int \mathcal{D}X\mathcal{D}\gamma e^{-S[X,\gamma]} \qquad (14.30)$$

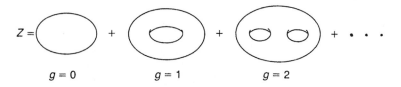

Figure 14.1. The total vacuum amplitude is given by summing over g-loop amplitudes.

see figure 14.1. To avoid confusion, we denote the genus by g and the metric by γ. The sum over genera amounts to the sum over the topologies. Z_g is the g-loop amplitude and is obtained by integrating over all metrics γ and all embeddings X. As we shall see later, the measure $\mathcal{D}X\mathcal{D}\gamma$ is not well defined and we need some modifications. The string action $S[X, \gamma]$ is taken to be

$$S[X, \gamma] \equiv \frac{1}{2} \int d^2\xi \, \sqrt{\gamma}\gamma^{\alpha\beta}\partial_\alpha X^\mu \partial_\beta X_\mu + \frac{\lambda}{4\pi} \int d^2\xi \, \sqrt{\gamma}\mathcal{R}. \tag{14.31}$$

The first term is the Polyakov action. The second term is proportional to the Euler characteristic

$$\chi = \frac{1}{4\pi} \int d^2\xi \sqrt{\gamma}\mathcal{R} = 2 - 2g$$

and serves as the string coupling constant; the amplitude of a loop with genus g is suppressed by the factor $e^{-2\lambda g}$. Since this term is a topological invariant, it does not affect the dynamics of the string. We are interested in Riemann surfaces of a fixed genus g and drop this term. The first term of the action has the following symmetries (section 7.11):

(A) Diff(Σ_g), the group of diffeomorphisms $f : \Sigma_g \to \Sigma_g$. Let $\xi^\alpha \to \xi'^\alpha(\xi)$ be the coordinate expression for f. The new metric is the pullback of the old one whose coordinate component expression is

$$\gamma_{\alpha\beta} \to f^*\gamma_{\alpha\beta} = \frac{\partial\xi^\gamma}{\partial\xi'^\alpha} \frac{\partial\xi^\delta}{\partial\xi'^\beta}\gamma_{\gamma\delta}. \tag{14.32}$$

The embedding also gets transformed as

$$X^\mu \to f^*X^\mu = X^\mu f. \tag{14.33}$$

The invariance of the classical action takes the form

$$S[X, \gamma] = S[f^*X, f^*\gamma]. \tag{14.34}$$

(B) Weyl(Σ_g), the group of two-dimensional Weyl rescalings

$$\gamma_{\alpha\beta} \to \hat{\gamma}_{\alpha\beta} \equiv e^\phi \gamma_{\alpha\beta} \tag{14.35}$$

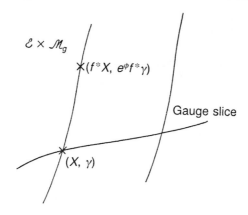

Figure 14.2. An element of $\mathcal{E} \times \mathcal{M}_g$ is obtained by the action of $\mathrm{Diff}(\Sigma_g) * \mathrm{Weyl}(\Sigma_g)$ on an element (X, γ) in the gauge slice.

where $\phi \in \mathcal{F}(\Sigma_g)$. The conformal invariance of S takes the form

$$S[X, \gamma] = S[X, \hat{\gamma}]. \tag{14.36}$$

The symmetries (A) and (B) must be preserved under quantization, otherwise the theory has anomalies.

According to the standard Faddeev–Popov formalism, the degrees of freedom corresponding to these symmetries have to be omitted when we define Z_g. For example, the string geometry specified by the pairs (X_1, γ_1) and (X_2, γ_2) should not be counted independently if they are related by an element of $\mathrm{Diff}(\Sigma_g)$. Similarly, (X, γ) and $(X, e^\phi \gamma)$ should not be counted as independent configurations. Unless special attention is paid, we would count the same configurations infinitely many times, which leads to disastrous divergences. It turns out that the space of all the geometries (X, γ) can be separated into equivalence classes (the **gauge slice**), any two points of which cannot be connected by these symmetries, see figure 14.2.

To be more mathematical, let $\mathcal{E}$ be the space of all the embeddings $X : \Sigma_g \to \mathbb{R}^D$ and let $\mathcal{M}_g$ be the space of all the metrics defined on Σ_g. Naively, the path integral is defined over $\mathcal{E} \times \mathcal{M}_g$. Because of the symmetries (A) and (B), however, the integral should be restricted to the quotient space $(\mathcal{E} \times \mathcal{M}_g)/G$ where $G = \mathrm{Diff}(\Sigma_g) * \mathrm{Weyl}(\Sigma_g)$ is the gauge group.[3] The action of (f, e^ϕ) on $(X, \gamma) \in \mathcal{E} \times \mathcal{M}_g$ is

$$(f, e^\phi)(X, \gamma) = (f^* X, e^\phi f^* \gamma). \tag{14.37}$$

The quotient $\mathcal{M}_g/G$ is called the **moduli space** of Σ_g and is denoted by $\mathrm{Mod}(\Sigma_g)$. We are also interested in the subgroup $\mathrm{Diff}_0(\Sigma_g)$ of $\mathrm{Diff}(\Sigma_g)$, which

---

[3] Here $*$ denotes the semi-direct product. Note that $\mathrm{Diff}(\Sigma_g) \cap \mathrm{Weyl}(\Sigma_g) \neq \emptyset$. We shall come back to this point later.

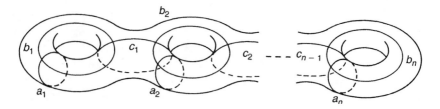

Figure 14.3. The mapping class group (MCG) is generated by Dehn twists around a_i, b_i and c_i ($1 \leq i \leq g$).

is a connected component of the identity map. The quotient space $\text{Teich}(\Sigma_g) \equiv \mathcal{M}_g/\text{Diff}_0(\Sigma_g) * \text{Weyl}(\Sigma_g)$ is called the **Teichmüller space** of Σ_g. The general theory of Riemann surfaces shows that $\text{Teich}(\Sigma_g)$ is a finite-dimensional universal covering space of $\text{Mod}(\Sigma_g)$. Explicitly, we have

$$\dim_{\mathbb{R}} \text{Teich}(\Sigma_g) = \begin{cases} 0 & g = 0 \\ 2 & g = 1 \\ 6g - 6 & g \geq 2. \end{cases} \tag{14.38}$$

The group $\text{Diff}(\Sigma_g)/\text{Diff}_0(\Sigma_g)$ is known as the **modular group** (MG) or the **mapping class group** (MCG). The MCG is generated by the **Dehn twists** defined in example 8.2. For the torus with genus g, the MCG is generated by $3g - 1$ Dehn twists around a_i, b_i and c_i in figure 14.3. Unfortunately, these $3g - 1$ Dehn twists are not the minimal set of the generators. The general form of MCG for $g \geq 2$ is not well understood.

From these arguments, the meaningful partition function turns out to be

$$Z_g \equiv \int_{\mathcal{E} \times \mathcal{M}_g} \frac{\mathcal{D}X\mathcal{D}\gamma}{V(\text{Diff} * \text{Weyl})} e^{-S[X,\gamma]} \tag{14.39}$$

where $V(\text{Diff} * \text{Weyl})$ is the (infinite) volume of the space of $\text{Diff}(\Sigma_g)*\text{Weyl}(\Sigma_g)$ and takes care of the infinite overcounting of the same geometry. The order (the number of elements) of MCG is denoted by $|\text{MCG}|$. Clearly,

$$V(\text{Diff} * \text{Weyl}) = |\text{MCG}|V(\text{Diff}_0 * \text{Weyl}). \tag{14.40}$$

14.2.2 Measures of integration

We have to define a sensible measure to carry out the integration (14.39) so that the physical degrees of freedom and the gauge degrees of freedom are separated. This separation of degrees of freedom requires the Jacobian,

$$\mathcal{D}\gamma\mathcal{D}X \rightarrow J(\mathcal{D} \text{ physical})(\mathcal{D} \text{ gauge}). \tag{14.41}$$

To find this Jacobian, we note that the Jacobian on a manifold M agrees with that on TM. To see this, let x^μ (y^μ) be a coordinate of a chart U (V) of M such that $U \cap V \neq \emptyset$. The Jacobian of the coordinate change is $J = \det(\partial y^\mu / \partial x^\nu)$. Take $V \in T_p M$. In components, we have $V = u^\mu \partial / \partial x^\mu = v^\mu \partial / \partial y^\mu$, where

$$v^\mu = u^\nu (\partial y^\mu / \partial x^\nu). \tag{14.42}$$

$\{u^\mu\}$ and $\{v^\mu\}$ are fibre coordinates of $T_p M$. The Jacobian $\hat{J}$ associated with this coordinate change is

$$\hat{J} = \det(\partial v^\mu / \partial u^\nu) = \det(\partial y^\mu / \partial x^\nu) = J. \tag{14.43}$$

This shows that the Jacobian at $p \in M$ is the same as that on $T_p M$. The Jacobian $\hat{J}$ depends on p but not on the vector itself, since J depends only on p.

Example 14.2. Let (x, y) and (r, θ) be coordinates of $\mathbb{R}^2$, where $x = r \cos \theta$ and $y = r \sin \theta$. The Jacobian of the coordinate change is

$$J = \det \frac{\partial(x, y)}{\partial(r, \theta)} = r.$$

Let us take

$$V = v_x \partial / \partial x + v_y \partial / \partial y = v_r \partial / \partial r + v_\theta \partial / \partial \theta \in T_p \mathbb{R}^2.$$

(v_x, v_y) and (v_r, v_θ) serve as fibre coordinates of $T_p \mathbb{R}^2$. Since

$$v_x = v_r \partial x / \partial r + v_\theta \partial x / \partial \theta \qquad v_y = v_r \partial y / \partial r + v_\theta \partial y / \partial \theta$$

the associated Jacobian $\hat{J}$ is easily calculated to be

$$\hat{J} = \det[\partial(v_x, v_y) / \partial(v_r, v_\theta)] = \begin{vmatrix} \partial x / \partial r & \partial x / \partial \theta \\ \partial y / \partial r & \partial y / \partial \theta \end{vmatrix} = J.$$

Let us derive this Jacobian in an indirect but suggestive way. We normalize the measure $d^2 v$ as[4]

$$1 = \int d^2 v \, \exp(-\tfrac{1}{2} \|v\|^2) = \int dv_x \, dv_y \, \exp[-\tfrac{1}{2}(v_x^2 + v_y^2)].$$

We also have $\|v^2\|^2 = v_r^2 + r^2 v_\theta^2$. Noting that the Jacobian is independent of v_r and v_θ, we have

$$1 = J \int dv_r \, dv_\theta \, \exp[-\tfrac{1}{2}(v_r^2 + r^2 v_\theta^2)] = J r^{-1}$$

[4] This normalization of the measure differs by a constant factor from the conventional one.

from which we find $J = r$. We use this procedure to find the functional measure of string theory.[5]

This analysis enables us to write

$$\mathcal{D}\delta\gamma\mathcal{D}\delta X = J\mathcal{D}\delta(\text{physical})\mathcal{D}\delta(\text{gauge}) \qquad (14.44)$$

where $\delta\gamma$ (δX) is a small variation of the metric γ (the embedding X) and is regarded as an element of $T_\gamma(\mathcal{M}_g)$ ($T_X\mathcal{E}$). The meaning of the RHS becomes clear in a moment.

Consider the diffeomorphism generated by an infinitesimal vector field δv on Σ_g. Since δv is infinitesimal, it belongs to $\text{Diff}_0(\Sigma_g)$ rather than the full group $\text{Diff}(\Sigma_g)$. The changes of the metric and the embedding under δv are (see (7.120))

$$\delta_\mathrm{D}\gamma_{\alpha\beta} = (\mathcal{L}_{\delta v}\gamma)_{\alpha\beta} = \nabla_\alpha\delta v_\beta + \nabla_\beta\delta v_\alpha \qquad \delta_\mathrm{D}X = \delta v^\alpha \partial_\alpha X. \qquad (14.45)$$

The changes of γ and X under an infinitesimal Weyl rescaling $e^{\delta\phi}$ are

$$\delta_\mathrm{W}\gamma_{\alpha\beta} = \delta\phi\gamma_{\alpha\beta} \qquad \delta_\mathrm{W}X = 0. \qquad (14.46)$$

These changes belong to unphysical (gauge) degrees of freedom. In general, a small change of metric is given by

$$\delta\gamma_{\alpha\beta} = \delta_\mathrm{W}\gamma_{\alpha\beta} + \delta_\mathrm{D}\gamma_{\alpha\beta} + (\text{physical change})$$
$$= \delta\phi\gamma_{\alpha\beta} + \nabla_\alpha\delta v_\beta + \nabla_\beta\delta v_\alpha + \delta t^i\frac{\partial}{\partial t^i}\gamma_{\alpha\beta}(t) \qquad (14.47)$$

where the last term is called the **Teichmüller deformation** of the metric, which can neither be described by a diffeomorphism nor by a Weyl rescaling. As mentioned before, $\{i\}$ is a finite set, $1 \leq i \leq n = \dim_\mathbb{R}\text{Teich}(\Sigma_g)$. It is convenient for later purposes to separate $\delta\gamma$ into a traceless part and a part with a non-zero trace. We write

$$\delta\gamma_{\alpha\beta} = \delta\bar{\phi}\gamma_{\alpha\beta} + (P_1\delta v)_{\alpha\beta} + \delta t^i T_{i\alpha\beta}(t) \qquad (14.48)$$

where $T_{i\alpha\beta}$ is the traceless part of the Teichmüller deformation,

$$T_{i\alpha\beta} \equiv \frac{\partial\gamma_{\alpha\beta}}{\partial t^i} - \frac{1}{2}\gamma_{\alpha\beta}\gamma^{\gamma\delta}\frac{\partial\gamma_{\gamma\delta}}{\partial t^i}. \qquad (14.49)$$

The operator P_1 is defined by

$$(P_1\delta v)_{\alpha\beta} \equiv \nabla_\alpha\delta v_\beta + \nabla_\beta\delta v_\alpha - \gamma_{\alpha\beta}(\nabla_\gamma\delta v_\gamma) \qquad (14.50)$$

and picks up the traceless part of $\delta_\mathrm{D}\gamma_{\alpha\beta}$ while $\delta\bar{\phi}$ is defined by

$$\delta\bar{\phi} = \delta\phi + \left(\nabla_\gamma\delta v^\gamma + \text{trace part of } \delta t\frac{\partial\gamma}{\partial t}\right) \qquad (14.51)$$

---

[5] It should be kept in mind that we introduce the tangent space only to obtain the Jacobian. The tangent space itself has no physical relevance.

where we do not need the explicit form in the parentheses.

As for the embeddings, we consider the quotient $\mathcal{E}/\mathrm{Diff}(\Sigma_g)$. An arbitrary embedding X is obtained by the action of $\mathrm{Diff}(\Sigma_g)$ on some $\tilde{X} \in \mathcal{E}/\mathrm{Diff}(\Sigma_g)$. Then a small change of the embedding is expressed as

$$\delta X = \delta v^\alpha \partial_\alpha \tilde{X} + \delta \tilde{X} \tag{14.52}$$

where the first term represents the change of X generated by δv while the second is not associated with diffeomorphisms. Now the measure should look like

$$\mathcal{D}\delta\gamma\mathcal{D}\delta X = J\, \mathrm{d}^n t\, \mathcal{D}\delta v \mathcal{D}\delta\phi \mathcal{D}\delta\tilde{X}. \tag{14.53}$$

To define the measure, we need to specify a metric on the tangent space, see example 14.2. We restrict ourselves to the so called *ultralocal* metric which is quadratic and depends on $\gamma_{\alpha\beta}$ but not on $\partial\gamma_{\alpha\beta}$. Define a metric for symmetric second-rank tensors by

$$\|\delta h\|_\gamma^2 = \int \mathrm{d}^2\xi \sqrt{\gamma}(G^{\alpha\beta\gamma\delta} + u\gamma^{\alpha\beta}\gamma^{\gamma\delta})\delta h_{\alpha\beta}\delta h_{\gamma\delta} \tag{14.54a}$$

where $u > 0$ is an arbitrary constant and

$$G^{\alpha\beta\gamma\delta} \equiv \gamma^{\alpha\gamma}\gamma^{\beta\delta} + \gamma^{\alpha\delta}\gamma^{\beta\gamma} - \gamma^{\alpha\beta}\gamma^{\gamma\delta}. \tag{14.55}$$

It is readily verified that G is the projection operator to the traceless part $(\mathrm{tr}\, G^{\alpha\beta\gamma\delta}\delta h_{\gamma\delta} = \gamma_{\alpha\beta} G^{\alpha\beta\gamma\delta}\delta h_{\gamma\delta} = 0)$ while $u\gamma^{\alpha\beta}\gamma^{\gamma\delta}$ is that to the trace part. In a finite-dimensional manifold, a metric defines a natural volume element. In the present case, however, the measure cannot be defined explicitly and we have to define it implicitly in terms of the Gaussian integral (see example 14.2),

$$\int \mathcal{D}\delta h \exp(-\tfrac{1}{2}\|\delta h\|_\gamma^2) = 1. \tag{14.56a}$$

Similarly, the metrics for a scalar $\delta\phi$, a vector δv and a map δX^μ are defined by

$$\|\delta\phi\|_\gamma^2 = \int \mathrm{d}^2\xi \sqrt{\gamma}\delta\phi^2 \tag{14.54b}$$

$$\|\delta v\|_\gamma^2 = \int \mathrm{d}^2\xi \sqrt{\gamma}\gamma_{\alpha\beta}\delta v^\alpha \delta v^\beta \tag{14.54c}$$

$$\|\delta X\|_\gamma^2 = \int \mathrm{d}^2\xi \sqrt{\gamma}\delta X^\mu \delta X_\mu. \tag{14.54d}$$

With these metrics, the measures are defined by

$$\int \mathcal{D}\delta\phi \exp(-\tfrac{1}{2}\|\delta\phi\|_\gamma^2) = 1 \tag{14.56b}$$

$$\int \mathcal{D}\delta v \exp(-\tfrac{1}{2}\|\delta v\|_\gamma^2) = 1 \tag{14.56c}$$

$$\int \mathcal{D}\delta X \exp(-\tfrac{1}{2}\|\delta X\|_\gamma^2) = 1. \tag{14.56d}$$

Exercise 14.3. Show that $\|\delta\gamma\|_\gamma^2$ and $\|\delta X\|_\gamma^2$ are invariant under $\mathrm{Diff}(\Sigma_g)$ but not under $\mathrm{Weyl}(\Sigma_g)$. This is the possible origin of conformal anomalies, see (14.84).

Before we proceed further, we need to clarify the overlap between $\mathrm{Diff}_0(\Sigma_g)$ and $\mathrm{Weyl}(\Sigma_g)$. Suppose $\delta v \in \ker P_1$, that is,

$$P_1 \delta v = \nabla_\alpha \delta v_\beta + \nabla_\beta \delta v_\alpha - \gamma_{\alpha\beta}(\nabla_\gamma \delta v^\gamma) = 0. \tag{14.57}$$

We find, for such δv, that $\delta_D \gamma_{\alpha\beta} = (\nabla_\gamma \delta v^\gamma)\gamma_{\alpha\beta}$. A vector $\delta v \in \ker P_1$ is identified with the **conformal Killing vector** (CKV), see section 7.7. It is important to note that δ_D and δ_W yield the same metric deformations if $\delta\phi$ is taken to be $\nabla_\gamma \delta v^\gamma$. Thus, the set of the CKVs is identified with the overlap between $\mathrm{Diff}_0(\Sigma_g)$ and $\mathrm{Weyl}(\Sigma_g)$. Let there be k independent CKVs on Σ_g and denote these by $\Phi_s^\alpha (1 \le s \le k)$. It is known from the theory of Riemann surfaces that

$$k = \begin{cases} 6 & g = 0 \\ 2 & g = 1 \\ 0 & g \ge 2. \end{cases} \tag{14.58}$$

We separate δv into a part generated by the CKV, and its orthogonal complement, which we write as

$$\delta v^\alpha = \delta\tilde{v}^\alpha + \delta a^s \Phi_s^\alpha. \tag{14.59}$$

The tangent vector δX is also decomposed as

$$\delta X = \delta\tilde{X} + \delta\tilde{v}^\alpha \partial_\alpha \tilde{X}^\mu + \delta a^s \Phi_s^\alpha \partial_\alpha \tilde{X}^\mu. \tag{14.60}$$

The functional measures now become

$$\mathcal{D}\delta\gamma \mathcal{D}\delta X \rightarrow J\, d^n \delta\, t \mathcal{D}\delta\phi \mathcal{D}\delta\tilde{v}\, d^k \delta\, a \mathcal{D}\delta\tilde{X} \tag{14.61}$$

where we noted that the t- and a-parameters are finite dimensional.

Let $\mathrm{Diff}_0^\perp(\Sigma_g)$ be the subspace of $\mathrm{Diff}_0(\Sigma_g)$, which is orthogonal to the CKV. We have

$$V(\mathrm{Diff}_0) = V(\mathrm{Diff}_0^\perp) \cdot V(\mathrm{CKV}) \tag{14.62}$$

$$V(\mathrm{Diff}_0 * \mathrm{Weyl}) = V(\mathrm{Diff}_0^\perp)V(\mathrm{Weyl})$$
$$= V(\mathrm{Diff}_0)V(\mathrm{Weyl})/V(\mathrm{CKV}). \tag{14.63}$$

Take a slice $\hat{\gamma}(t)$ of $\mathcal{M}_g$. The slice is parametrized by n Teichmüller parameters. Any metric $\tilde{\gamma}$ related to $\hat{\gamma}$ by $G = \mathrm{Diff}(\Sigma_g) * \mathrm{Weyl}(\Sigma_g)$ is written as

$$\tilde{\gamma} = f^*(e^\phi \hat{\gamma}) \qquad f \in \mathrm{Diff}(\Sigma_g), e^\phi \in \mathrm{Weyl}(\Sigma_g). \tag{14.64}$$

We express a small deformation $\delta\tilde{\gamma}$ at $\tilde{\gamma}$ as a pullback of a deformation $\delta\gamma$ at $\gamma \equiv e^{\delta\phi}\hat{\gamma}$: $\delta\tilde{\gamma} = f^*(\delta\gamma)$. Note that $\delta\gamma$ is a small diffeomorphism at the *origin* of $\mathrm{Diff}_0(\Sigma_g)$ and, hence, can be described by a vector field δv. As was shown in exercise 14.3, $\mathrm{Diff}(\Sigma_g)$ is the isometry of the relevant vector spaces. It then follows that

$$\|\delta\tilde{\gamma}\|_{\tilde{\gamma}}^2 = \|f^*(\delta\gamma)\|_{f^*\gamma}^2 = \|\delta\gamma\|_\gamma^2 \qquad \gamma = \mathrm{e}^\phi \hat{\gamma}. \tag{14.65}$$

At the point γ, we decompose $\delta\gamma$ as

$$\delta\gamma_{\alpha\beta} = \delta\phi\gamma_{\alpha\beta} + (P_1\delta\tilde{v})_{\alpha\beta} + \delta t^i\, T_{i\alpha\beta} \tag{14.66}$$

where $\delta\phi$ has been redefined so that it includes the trace parts of the Teichmüller deformation and $\nabla_\alpha\delta v_\beta + \nabla_\beta\delta v_\alpha$, see (14.51).

Exercise 14.4. Show that $T_{i\alpha\beta}$ at γ is related to $\hat{T}_{i\alpha\beta}$ at $\hat{\gamma}$ as

$$T_{i\alpha\beta} = \mathrm{e}^\phi \hat{T}_{i\alpha\beta}. \tag{14.67}$$

Now we are ready to give the explicit form of the measure. We first find the Jacobian associated with the change $\mathcal{D}\delta v \to \mathcal{D}\delta\tilde{v}\mathrm{d}^k\delta a$. We have

$$\begin{aligned}
1 &= \int \mathcal{D}\delta v \exp(-\tfrac{1}{2}\|\delta v\|_\gamma^2) \\
&= J \int \mathcal{D}\delta\tilde{v}\, \mathrm{d}^k\delta\, a \exp(-\tfrac{1}{2}\|\delta\tilde{v}\|_\gamma^2 - \tfrac{1}{2}\|\delta a^s\, \Phi_s\|_\gamma^2) \\
&= J[\det(\Phi_s, \Phi_r)]^{-1/2}
\end{aligned} \tag{14.68a}$$

where

$$(\Phi_s, \Phi_r) = \int \mathrm{d}^2\xi\,\sqrt{\gamma}\,\gamma_{\alpha\beta}\Phi_s^\alpha \Phi_r^\beta. \tag{14.68b}$$

[*Remark:* Although the matrix element (14.68b) is defined for $\gamma = \mathrm{e}^\phi \hat{\gamma}$, we can show that it is independent of e^ϕ. To see this, let us take a CKV $\hat{\Phi}_s^\alpha$ of the metric $\hat{\gamma}$; $\hat{\nabla}_\alpha\hat{\Phi}_{s\beta} + \hat{\nabla}_\beta\hat{\Phi}_{s\alpha} = \hat{\gamma}_{\alpha\beta}\hat{\nabla}\hat{\Phi}_s^\gamma$, where $\hat{\nabla}$ is the covariant derivative with respect to $\hat{\gamma}$ and $\hat{\Phi}_{s\alpha} \equiv \hat{\gamma}_{\alpha\beta}\hat{\Phi}_s^\beta$. A simple calculation shows that $\Phi_{s\alpha} = \gamma_{\alpha\beta}\hat{\Phi}_s^\beta = \mathrm{e}^\phi \hat{\Phi}_{s\alpha}$ satisfies

$$\begin{aligned}
\nabla_\alpha\Phi_{s\beta} + \nabla_\beta\Phi_{s\alpha} &= \mathrm{e}^\phi(\hat{\nabla}_\alpha\hat{\Phi}_{s\beta} + \hat{\nabla}_\beta\hat{\Phi}_{s\alpha} + \hat{\gamma}_{\alpha\beta}\Phi_s^\gamma\partial_\gamma\phi) \\
&= \mathrm{e}^\phi\hat{\gamma}_{\alpha\beta}(\hat{\nabla}_\gamma\Phi_s^\gamma + \Phi_s^\gamma\partial_\gamma\phi) = \gamma_{\alpha\beta}\nabla_\gamma\Phi_s^\gamma
\end{aligned}$$

∇ being the covariant derivative with respect to γ. Thus, $\Phi_s^\alpha = \hat{\Phi}_s^\alpha$ is a CKV of the metric $\gamma = \mathrm{e}^\phi\hat{\gamma}$ and the CKV are taken to be ϕ independent.] Equation (14.68a) shows that

$$\mathcal{D}\delta v = [\det(\Phi_r, \Phi_s)]^{1/2}\mathcal{D}\delta\tilde{v}\, \mathrm{d}^k\delta\, a. \tag{14.69}$$

Now the total measure is written as

$$J[\det(\Phi_r, \Phi_s)]^{1/2}\,\mathrm{d}^n t\, \mathcal{D}\delta\phi\mathcal{D}\delta\tilde{v}\, \mathrm{d}^k\delta\, a\mathcal{D}\delta\tilde{X} \tag{14.70}$$

where J takes care of the rest of the variable changes.

The Jacobian J is now obtained from (14.60), (14.66), (14.70) and the definition of the measures (14.56). We have

$$1 = \int \mathcal{D}\delta\gamma \mathcal{D}\delta X \exp\left(-\tfrac{1}{2}\|\delta\gamma\|^2_\gamma - \tfrac{1}{2}\|\delta X\|^2_\gamma\right)$$

$$= J \det^{1/2}(\Phi, \Phi) \int \mathrm{d}^n\delta\, t \mathcal{D}\delta\tilde{v}\mathcal{D}\delta\phi\, \mathrm{d}^k\delta\, a \mathcal{D}\delta\tilde{X}$$

$$\times \exp\left[-\frac{1}{2}\left\|\delta\phi\gamma_{\alpha\beta} + (P_1\delta\tilde{v})_{\alpha\beta} + \delta t^i \frac{\partial\gamma_{\alpha\beta}}{\partial t^i}\right\|^2 \right.$$

$$\left. -\frac{1}{2}\|\delta\tilde{X} + \delta\tilde{v}^\alpha\partial_\alpha\tilde{X} + \delta a^s \Phi^\alpha_s \partial_\alpha\tilde{X}\|^2\right]$$

$$= J \det^{1/2}(\Phi, \Phi) \int \mathrm{d}^n\delta\, t \mathcal{D}\delta\tilde{v}\ldots\exp(-\tfrac{1}{2}\|MV\|^2) \qquad (14.71)$$

where

$$V = \begin{pmatrix} \delta t \\ \delta\phi \\ \delta\tilde{v} \\ \delta a \\ \delta\tilde{X} \end{pmatrix} \qquad M = \left(\begin{array}{ccc|cc} \partial\gamma/\partial t & \gamma & P_1 & 0 & 0 \\ \hline 0 & 0 & \partial\tilde{X} & \Phi\cdot\partial\tilde{X} & 1 \end{array}\right) \equiv \begin{pmatrix} A & 0 \\ C & B \end{pmatrix}.$$

$$(14.72)$$

The matrix in the exponent of (14.71) is

$$M^\dagger M = \begin{pmatrix} A^\dagger & C^\dagger \\ 0 & B^\dagger \end{pmatrix}\begin{pmatrix} A & 0 \\ C & B \end{pmatrix} = \begin{pmatrix} A^\dagger A + C^\dagger C & C^\dagger B \\ B^\dagger C & B^\dagger B \end{pmatrix}$$

$$= \begin{pmatrix} I & * \\ 0 & B^\dagger B \end{pmatrix}\begin{pmatrix} A^\dagger A & 0 \\ ** & I \end{pmatrix} \qquad (14.73)$$

where $*$ and $**$ are irrelevant. The last expression has been obtained from the identity,

$$\begin{pmatrix} A & B \\ C & D \end{pmatrix} = \begin{pmatrix} I & B \\ 0 & D \end{pmatrix}\begin{pmatrix} A - BD^{-1}C & 0 \\ D^{-1}C & I \end{pmatrix}.$$

The Gaussian integrals in (14.71) are readily evaluated to yield

$$1 = J \det^{1/2}(\Phi, \Phi) \det^{-1/2}(M^\dagger M)$$

$$= J \det^{1/2}(\Phi, \Phi)[\det(A^\dagger A) \det(B^\dagger B)]^{-1/2}. \qquad (14.74)$$

To compute $\det^{1/2}(A^\dagger A)$, we need to evaluate $\|\delta\gamma\|^2_\gamma$. We have

$$\|\delta\gamma\|^2_\gamma = \int \mathrm{d}^2\xi\, \sqrt{\gamma}(G^{\alpha\beta\gamma\delta} + u\gamma^{\alpha\beta}\gamma^{\gamma\delta})$$

$$\times [\delta\phi\gamma_{\alpha\beta} + (P_1\delta\tilde{v})_{\alpha\beta} + \delta t^i T_{i\alpha\beta}][\delta\phi\gamma_{\gamma\delta} + (P_1\delta\tilde{v})_{\gamma\delta} + \delta t^j T_{j\gamma\delta}]$$

$$= 4u\|\delta\phi\|^2_\gamma + \|P_1\delta\tilde{v}\|^2 + \delta t^i \delta t^j (T_i, T_j) + 2\delta t^i (P_1\delta\tilde{v}, T_i).$$

$$(14.75)$$

In general, T_i is not orthogonal to $P_1 \delta v$. To separate T_i into parts orthogonal to $P_1 \delta v$ and parallel to $P_1 \delta v$, we need to define the adjoint $P_1^\dagger$ of P_1. P_1 is an elliptic operator which takes a vector field into a traceless symmetric tensor field. Thus, $P_1^\dagger$ maps symmetric traceless tensors to vectors. For a symmetric traceless tensor δh, we have

$$
\begin{aligned}
(P_1 \delta v, \delta h) &= \int d^2 \xi \sqrt{\gamma} G^{\alpha\beta\gamma\delta} (P_1 \delta v)_{\alpha\beta} \delta h_{\gamma\delta} \\
&= \int d^2 \xi \sqrt{\gamma} (\nabla^\alpha \delta v^\beta + \nabla^\beta \delta v^\alpha) \delta h_{\alpha\beta} \\
&= \int d^2 \xi \sqrt{\gamma} \delta v^\alpha (-2\nabla^\beta) \delta h_{\alpha\beta} \equiv (\delta v, P_1^\dagger \delta h)
\end{aligned}
$$

where the inner product in the last expression is defined by (14.54c). Thus, it follows that

$$
(P_1^\dagger \delta h)_\alpha = -2\nabla^\beta \delta h_{\alpha\beta}. \tag{14.76}
$$

Suppose δh is orthogonal to $P_1 \delta v$. From the previous discussion, we have $(P_1 \delta v, \delta h) = (\delta v, P_1^\dagger \delta h) = 0$. Since δv is arbitrary, δh must be an element of $\ker P_1^\dagger$, see figure 14.4. Now T_i may be separated as

$$
T_i = \mathcal{P}_0 T_i + \mathcal{P}_\perp T_i \tag{14.77a}
$$

where the projection operators $\mathcal{P}_0$ and $\mathcal{P}_\perp$ are defined by

$$
\mathcal{P}_0 \equiv 1 - P_1 \frac{1}{P_1^\dagger P_1} P_1^\dagger \qquad \mathcal{P}_\perp \equiv P_1 \frac{1}{P_1^\dagger P_1} P_1^\dagger. \tag{14.77b}
$$

It is easy to verify that $\mathcal{P}_0 + \mathcal{P}_1 = 1$, $\mathcal{P}_0 \mathcal{P}_\perp = 0$, $P_1^\dagger \mathcal{P}_0 = 0$, $P_1^\dagger \mathcal{P}_\perp = P_1^\dagger$, $\mathcal{P}_0 T_i = T_i$ and $\mathcal{P}_\perp T_i = 0$ for $T_i \in \ker P_1^\dagger$ etc. Thus (14.77a) is an orthogonal decomposition of T_i. We write $\mathcal{P}_\perp T_i = P_1 u_i$, where

$$
u_i = \frac{1}{P_1^\dagger P_1} P_1^\dagger T_i.
$$

Let $\{\psi_r\}$ ($1 \leq r \leq n$) be a real basis of $\ker P_1^\dagger$, which is not necessarily orthonormal. Then T_i can be expanded as (figure 14.5)

$$
T_i = \sum_r \psi_r Q_{ri} + P_1 u_i. \tag{14.78}
$$

Taking an inner product between T_i and ψ_r, we find that

$$
Q_{ri} = \sum_s [(\psi, \psi)^{-1}]_{rs} (\psi_s, T_i). \tag{14.79}
$$

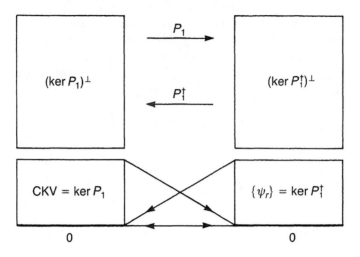

Figure 14.4. The map P_1 and its adjoint $P_1^\dagger$.

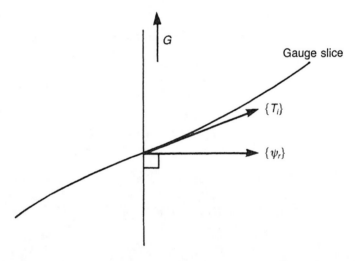

Figure 14.5. $\{T_i\}$ spans the deformation tangent to the gauge slice while $\{\psi_r\}$ spans ker $P_1^\dagger$.

Finally, $\delta\gamma$ is decomposed into mutually orthogonal pieces as

$$\delta\gamma = \delta\phi\gamma + P_1(\delta\tilde{v} + \delta t^i u_i) + \delta t^i \psi_r Q_{ri}. \qquad (14.80a)$$

Correspondingly, the space of the metric deformation $\{\delta\gamma\}$ separates into the direct sum

$$\{\delta\gamma\} = \{\text{conf}\} \oplus \{\text{im } P_1\} \oplus \{\text{ker } P_1^\dagger\}. \qquad (14.80b)$$

Substituting (14.80a) into (14.75), we obtain

$$
\begin{aligned}
||\delta\gamma||^2 = {} & 4u||\delta\phi||^2 + ||P_1\delta\bar{v}||^2 \\
& + \delta t^i \delta t^j (T_i, \psi_r)_\gamma [(\psi, \psi)_\gamma^{-1}]_{rs} (\psi_s, T_j)_\gamma
\end{aligned}
\tag{14.81}
$$

where $\delta\bar{v} \equiv \delta\tilde{v} + \delta t^i u_i$ and the inverse in the last term refers to the inverse of the matrix $(a_{rs}) = ((\psi_r, \psi_s))$. If we put $\mathcal{V}_1^t = (\delta t, \delta\phi, \delta\bar{v})$, we find that

$$
\begin{aligned}
\det{}^{-1/2}(A^\dagger A) &= \int d^n \delta\, t \mathcal{D}\delta\phi \mathcal{D}\delta\bar{v} \exp(-\tfrac{1}{2}\mathcal{V}_1^t A^\dagger A \mathcal{V}_1) \\
&= \int \mathcal{D}\delta\phi \exp(-2u||\delta\phi||^2) \int \mathcal{D}\delta\bar{v} \exp(-\tfrac{1}{2}||P_1\bar{v}||^2) \\
&\quad \times \int d^n \delta\, t \exp\{-\tfrac{1}{2}\delta t^i (T_i, \psi_r)[(\psi, \psi)^{-1}]_{rs}(\psi_s, T_j)\delta t^j\} \\
&\propto (\det P_1^\dagger P_1)^{-1/2} \left(\frac{\det(T, \psi)^2}{\det(\psi, \psi)} \right)^{-1/2} .
\end{aligned}
\tag{14.82}
$$

Collecting the results (14.71) and (14.82), we have

$$
1 = J \det{}^{1/2}(\Phi, \Phi) \det{}^{-1/2} B^\dagger B \det{}^{-1/2} P_1^\dagger P_1 \left(\frac{\det(T, \psi)^2}{\det(\psi, \psi)} \right)^{-1/2} .
$$

The g-loop partition function is then given by

$$
\begin{aligned}
Z_g = {} & \int \frac{d^n t\, \mathcal{D}\bar{v}\mathcal{D}\phi \det \tilde{X}}{V(\mathrm{Diff} * \mathrm{Weyl})} \det{}^{1/2} B^\dagger B \det{}^{-1/2}(\Phi, \Phi) \\
& \times \left(\det P_1^\dagger P_1 \frac{\det(T, \psi)^2}{\det(\psi, \psi)} \right)^{1/2} e^{-S}.
\end{aligned}
\tag{14.83}
$$

The integral over a (the CKV) has been omitted since it is already included in the ϕ-integration. Naively, the integral over $\bar{v}$ yields $V(\mathrm{Diff}_0^\perp)$ and that over ϕ yields $V(\mathrm{Weyl})$. However, as exercise 14.3 shows, the measures $\mathcal{D}X$ and $\mathcal{D}\gamma$ depend on the conformal factor. Polyakov (1981) has shown that, under the conformal transformation $\gamma \to e^{2\phi}\gamma$, the measures transform as

$$
\mathcal{D}X \to \exp\left(\frac{D}{24\pi^2} \int d^2\xi \sqrt{\gamma}(\gamma^{\alpha\beta}\partial_\alpha\phi\partial_\beta\phi + \mathcal{R}\phi) \right) \mathcal{D}X
\tag{14.84a}
$$

$$
\mathcal{D}\gamma \to \exp\left(\frac{-26}{24\pi^2} \int d^2\xi \sqrt{\gamma}(\gamma^{\alpha\beta}\partial_\alpha\phi\partial_\beta\phi + \mathcal{R}\phi) \right) \mathcal{D}\gamma.
\tag{14.84b}
$$

Thus, the measure $\mathcal{D}X\mathcal{D}\gamma$ is conformally invariant if and only if $D = 26$. This number 26 is called the **critical dimension**. Henceforth, we always assume that

$D = 26$. Now (14.83) simplifies as

$$Z_g = \frac{1}{|\text{MCG}|} \int d^n t \, \mathcal{D}\tilde{X} \, \det{}^{1/2} B^\dagger B \, \det{}^{-1/2}(\Phi, \Phi)$$

$$\times \left(\det P_1^\dagger P_1 \frac{\det(T, \psi)^2}{\det(\psi, \psi)} \right)^{1/2} e^{-S}. \tag{14.85}$$

We perform the X-integration to eliminate $\det^{1/2} B^\dagger B$. We have

$$1 = \int \mathcal{D}\delta X \exp(-\tfrac{1}{2}||\delta X||^2)$$

$$= J \int \mathcal{D}\delta \tilde{X} \, d^k \delta \, a \exp(-\tfrac{1}{2}||\delta \tilde{X} + \delta a^s \Phi_s^\alpha \partial_\alpha \tilde{X}||^2)$$

$$= J \int \mathcal{D}\delta \tilde{X} \exp(-\tfrac{1}{2}||\delta \tilde{X}||^2) \int d^k \delta \, a \exp(-\tfrac{1}{2}||\delta a^s \Phi_s^\alpha \partial_\alpha \tilde{X}||^2)$$

$$= J \det{}^{-1/2}(B^\dagger B)$$

and hence $\det^{1/2}(B^\dagger B)$ is identified with the Jacobian of the transformation $X \to (\tilde{X}, a)$. Thus, it follows that

$$\int \mathcal{D}\tilde{X} \, \det{}^{1/2} B^\dagger B e^{-S} = \int \frac{\mathcal{D}X}{V(\text{CKV})} e^{-S} \tag{14.86}$$

where $V(\text{CKV}) = \int d^k a$ is the volume of the CKV.

The integration over X is readily carried out. Let us write

$$\int_{\mathcal{E}} \mathcal{D}X e^{-S} = \int_{\mathcal{E}} \mathcal{D}X \exp[-\tfrac{1}{2}(X, \Delta X)] \tag{14.87a}$$

where

$$\Delta = -\frac{1}{\sqrt{\gamma}} \partial_\alpha \sqrt{\gamma} \gamma^{\alpha\beta} \partial_\beta \tag{14.87b}$$

is the Laplacian acting on 0-forms, see (7.188). We write down the explicit form of the path integral (14.87a). Let ψ_n be the eigenfunction of Δ,

$$\Delta \psi_n = \lambda_n \psi_n \qquad \lambda_n \in [0, \infty) \tag{14.88}$$

where ψ_n are normalized as

$$(\psi_n, \psi_m) = \int d^2\xi \, \sqrt{\gamma} \psi_n \psi_m = \delta_{nm}.$$

The eigenvalue λ is non-negative since Δ is positive definite. Let us expand X^μ in ψ_n as

$$X^\mu = \sum_{n=0}^{\infty} a_n^\mu \psi_n = X_0^\mu + X'^\mu \qquad a_n^\mu \in \mathbb{R} \tag{14.89}$$

where $X_0^\mu = a_0^\mu \psi_0$ is the zero eigenfunction of Δ and X'^μ are the remaining degrees of freedom. Correspondingly, the path integral (14.87a) is written as

$$\int \mathcal{D}X \exp[-\tfrac{1}{2}(X, \Delta X)] = \int \prod_{n,\mu} da_n^\mu \exp\left(-\tfrac{1}{2}\sum_{n,\mu}\lambda_n(a_n^\mu)^2\right)$$

$$= \int \prod_\mu da_0^\mu \int \prod_{n\neq 0}\prod_\mu da_n^\mu \exp\left(-\tfrac{1}{2}\sum_{n,\mu}\lambda_n(a_n^\mu)^2\right)$$

$$= \left(\int \prod_\mu da_0^\mu\right)(\det'\Delta)^{-13} \tag{14.90}$$

where the prime indicates that the zero mode is omitted. To integrate over the zero mode, we note that the *normalized* eigenvector ψ_0 is given by[6]

$$\psi_0 = \left(\frac{1}{\int d^2\xi\sqrt{\gamma}}\right)^{1/2}. \tag{14.91}$$

From $X_0^\mu = a_0^\mu \psi_0$, we have

$$\int \prod_\mu da_0^\mu = \int \prod_\mu dX_0^\mu (\psi_0)^{-26} = V\left(\frac{1}{\int d^2\xi\sqrt{\gamma}}\right)^{-13} \tag{14.92}$$

where $V = \int \prod dX_0^\mu$ is the spacetime volume. Collecting the results (14.90) and (14.92), we find that

$$\int \mathcal{D}X e^{-S} = \left(\frac{\det'\Delta}{\int d^2\xi\sqrt{\gamma}}\right)^{-13} \tag{14.93}$$

where we have dropped V and other irrelevant constants.

Finally, we have obtained the expression for the g-loop partition function

$$Z_g = \int_{\text{Mod}} \frac{d^n t}{V(\text{CKV})} \frac{\det(T, \psi)}{\det^{1/2}(\psi, \psi) \det^{1/2}(\Phi, \Phi)}$$

$$\times [\det' P_1^\dagger P_1]^{1/2}\left(\frac{\det'\Delta}{\int d^2\xi\sqrt{\gamma}}\right)^{-13} \tag{14.94}$$

where we have noted that

$$\frac{1}{|\text{MCG}|}\int_{\text{Teich}} d^n t = \int_{\text{Mod}} d^n t. \tag{14.95}$$

If $g \geq 2$, the Riemann surfaces have no CKV and (14.95) reduces to

$$Z_g = \int_{\text{Mod}} d^n t \, \frac{\det(T, \psi)}{\det^{1/2}(\psi, \psi)}(\det' P_1^\dagger P_1)^{1/2}\left(\frac{\det'\Delta}{\int d^2\xi\sqrt{\gamma}}\right)^{-13}. \tag{14.96}$$

---

[6] Since ψ_0 satisfies $\Delta\psi_0 = 0$, it is a harmonic function. Any harmonic function on a Riemann surface must be a *constant* by the maximum principle.

14.2.3 Complex tensor calculus and string measure

Since any Riemann surface admits complex structures, we may take advantage of this fact to compute string amplitudes. Many beautiful aspects of string theory are revealed only when these complex structures are explicitly taken into account. Here we rewrite the partition function in the language of complex differential geometry.

We first fix the gauge in $\mathcal{M}_g$ by choosing the isothermal coordinate system

$$\gamma = \tfrac{1}{2}e^{2\sigma}[dz \otimes d\bar{z} + d\bar{z} \otimes dz]$$

where $\gamma_{z\bar{z}} = \gamma_{\bar{z}z} = \tfrac{1}{2}\exp 2\sigma.$[7] Then the deformation of γ under a diffeomorphism generated by δv is (cf (14.45))

$$\delta_D \gamma_{zz} = 2\nabla_z^{(-1)}\delta v_z$$
$$\delta_D \gamma_{z\bar{z}} = \nabla_z \delta v_{\bar{z}} + \nabla_{\bar{z}}\delta v_z = \gamma_{z\bar{z}}(\nabla_z^{(1)}\delta v^z + \nabla_{(-1)}^z\delta v_z). \tag{14.97}$$

Similarly, $\delta_W \gamma$ generated by an infinitesimal conformal change is (cf (14.46))

$$\delta_W \gamma_{z\bar{z}} = \delta\phi \gamma_{z\bar{z}} \qquad \delta_W \gamma_{zz} = 0. \tag{14.98}$$

To see the action of the operator P_1 on vectors, we take $\delta v^z \in \mathcal{T}^1$ and $\delta v_z \in \mathcal{T}^{-1}$. From (14.50), we find that

$$(P_1\delta v)^{zz} = 2\nabla_{(1)}^z\delta v^z \in \mathcal{T}^2 \tag{14.99a}$$
$$(P_1\delta v)_{zz} = 2\nabla_z^{(-1)}\delta v_z \in \mathcal{T}^{-2}. \tag{14.99b}$$

This shows that P_1 is a map:

$$P_1 = \begin{pmatrix} \nabla_{(1)}^z & 0 \\ 0 & \nabla_z^{(-1)} \end{pmatrix} : \mathcal{T}^1 \oplus \mathcal{T}^{-1} \to \mathcal{T}^2 \oplus \mathcal{T}^{-2}. \tag{14.100}$$

Similarly, $P_1^\dagger$ maps traceless symmetric tensors to vectors. For $\delta h^{zz} \in \mathcal{T}^2$ and $\delta h_{zz} \in \mathcal{T}^{-2}$, we have

$$(P_1^\dagger \delta h)^z = \nabla_z^{(2)}\delta h^{zz} \in \mathcal{T}^1 \tag{14.101a}$$
$$(P_1^\dagger \delta h)_z = \nabla_{(-2)}^z\delta h_{zz} \in \mathcal{T}^{-1}. \tag{14.101b}$$

Thus, $P_1^\dagger$ is a map:

$$P_1^\dagger = \begin{pmatrix} \nabla_z^{(2)} & 0 \\ 0 & \nabla_{(-2)}^z \end{pmatrix} : \mathcal{T}^2 \oplus \mathcal{T}^{-2} \to \mathcal{T}^1 \oplus \mathcal{T}^{-1}. \tag{14.102}$$

---

[7] In fact, the gauge is not uniquely fixed with this choice. We will invoke the *uniformization theorem* later to fix the gauge completely.

The product $P_1^\dagger P_1$ is

$$P_1^\dagger P_1 = \begin{pmatrix} \nabla_z^{(2)} \nabla_{(1)}^z & 0 \\ 0 & \nabla_{(-2)}^z \nabla_z^{(-1)} \end{pmatrix} : \mathcal{T}^1 \oplus \mathcal{T}^{-1} \to \mathcal{T}^1 \oplus \mathcal{T}^{-1}. \quad (14.103)$$

Accordingly, the determinant in (14.96) becomes

$$(\det' P_1^\dagger P_1)^{1/2} = (\det' \nabla_z^{(2)} \nabla_{(1)}^z \det' \nabla_{(-2)}^z \nabla_z^{(-1)})^{1/2}$$
$$= (\det' \Delta_{(1)}^+ \Delta_{(-1)}^-)^{1/2} \quad (14.104)$$

where $\Delta_{(n)}^\pm$ are the Laplacians. We show that the spectrum of $\Delta_{(1)}^+$ is the same as that of $\Delta_{(-1)}^-$. Take an eigenfunction δv^z of $\Delta_{(1)}^+$,

$$\Delta_{(1)}^+ \delta v^z = -2e^{-4\sigma} \partial_z e^{2\sigma} \partial_{\bar{z}} \delta v^z = \lambda \delta z^z \quad (14.105)$$

where (14.21a) has been used. The eigenvalue λ is a non-negative real number (note $\Delta_{(n)}^\pm$ are positive-definite Hermitian operators). Then we find

$$\Delta_{(-1)}^- (\gamma_{z\bar{z}} \overline{\delta v^z}) = -e^{-2\sigma} \partial_{\bar{z}} e^{2\sigma} \partial_z \overline{\delta v^z} = -e^{-2\sigma} \overline{\partial_z e^{2\sigma} \partial_{\bar{z}} \delta v^z}$$
$$= -\gamma_{z\bar{z}} 2e^{-4\sigma} \overline{\partial_z e^{2\sigma} \partial_{\bar{z}} \delta v^z} = \lambda \gamma_{z\bar{z}} \overline{\delta v^z} \quad (14.106)$$

which shows that $\gamma_{z\bar{z}} \overline{\delta v^z}$ is an eigenfunction of $\Delta_{(-1)}^-$ with the same eigenvalue λ. It is easy to see that the converse is also true, see exercise 14.5. Thus, $\Delta_{(1)}^+$ and $\Delta_{(-1)}^-$ share the same eigenvalues and $\det' \Delta_{(1)}^+ = \det' \Delta_{(-1)}^-$. Now (14.104) becomes

$$(\det' P_1^\dagger P_1)^{1/2} = \det' \Delta_{(-1)}^- = \det' \Delta_{(1)}^+. \quad (14.107)$$

Exercise 14.5. Let δv_z be an eigenvector of $\Delta_{(-1)}^-$ with an eigenvalue λ. Show that $\gamma^{z\bar{z}} \overline{\delta v_z}$ is an eigenvector of $\Delta_{(1)}^+$ with the same eigenvalue.

The physical change of the metric is the Teichmüller deformation $\delta \tau^i \mu_i$, where τ^i (μ_i) is the complex counterpart of t^i (T_i). From our experience, we know that the relevant part of the Teichmüller deformation is *symmetric* and *traceless* in the real basis. In the complex basis, this amounts to $\mu_{iz\bar{z}} = \mu_{i\bar{z}z} = 0$. Accordingly, the general variation of the metric is given by

$$\delta \gamma_{zz} = \nabla_z^{(-1)} \delta \tilde{v}_z + \delta \tau^i \mu_{izz} \quad (14.108a)$$
$$\delta \gamma_{z\bar{z}} = \delta \phi \gamma_{z\bar{z}} \quad (14.108b)$$

where we have redefined $\delta \phi$ so that it includes the variation of $\delta \gamma_{z\bar{z}}$ due to δv (note that $\delta_D \gamma_{z\bar{z}} \propto \gamma_{z\bar{z}}$). In (14.108a), $\delta \tilde{v}$ does not contain the CKV, that is, $\delta \tilde{v} \in (\ker \nabla_z^{(-1)})^\perp$.

To carry out the orthogonal decomposition of $\{\delta\gamma\}$, we need to define the inner products in various spaces. The most natural choices are

$$\|\delta\gamma_{zz}\|^2 = \int d^2z \, \sqrt{\gamma} \overline{\delta\gamma_{zz}} \delta\gamma^{zz} \tag{14.109a}$$

$$\|\delta\gamma_{z\bar{z}}\|^2 = \int d^2z \, \sqrt{\gamma} \overline{\delta\gamma_{z\bar{z}}} \delta\gamma^{z\bar{z}} \tag{14.109b}$$

and

$$\|\delta v_z\|^2 = \int d^2z \, \sqrt{\gamma} \gamma_{z\bar{z}} \overline{\delta v^z} \delta v^z. \tag{14.109c}$$

Note that $\delta\gamma_{zz} dz \otimes dz$ and $\delta\gamma_{z\bar{z}} dz \otimes d\bar{z}$ are different tensors; we have to specify the inner product separately.

Following the argument in the previous subsection, we introduce the orthogonal decomposition,

$$\delta\gamma_{zz} = \nabla_z^{(-1)} \delta\tilde{v}_z + \delta\tau^i \mu_{izz} = \nabla_z^{(-1)} \delta\bar{v}_z + \delta\tau^i \phi_{izz} \tag{14.110}$$

where $\delta\bar{v} = \delta\tilde{v} + $ (projection of $\delta\tau^i \mu_{izz}$ into $\{im \, \nabla_z^{(-1)}\}$). The orthogonality of $\nabla_z^{(-1)} \delta\bar{v}_z$ and ϕ_{izz} implies

$$0 = (\nabla_z^{(-1)} \delta v_z, \phi_{izz}) = \int d^2z \, \sqrt{\gamma} \overline{\delta v_z} (-\nabla_{(-2)}^z \phi_{izz})$$

where we have noted that $\nabla_z^{(-1)\dagger} = -\nabla_{(-2)}^z$. Thus, we find that (figure 14.6)

$$\phi_{izz} \in \ker \nabla_{(-2)}^z. \tag{14.111}$$

The explicit form of $\nabla_{(-2)}^z$ shows that $\partial_{\bar{z}}\phi_{izz} = 0$, that is $\ker \nabla_{(-2)}^z$ is the set of holomorphic tensors of helicity -2. The tensor $\phi_i = \phi_{izz} dz \otimes dz$ is called the **quadratic differential** while $\mu_i = \mu_{izz} dz \otimes dz$ is the **Beltrami differential**, see figure 14.7. In practical computations, it is often convenient to specify the gauge slice by the Beltrami differential, see later. Now we have established that

$$\{\ker P_1^\dagger\} = \{\text{Quadratic differential}\} = \{\ker \nabla_{(-2)}^z\}. \tag{14.112}$$

The Riemann–Roch theorem (14.27) takes the form

$$\dim_{\mathbb{C}} \ker \nabla_z^{(-1)} - \dim_{\mathbb{C}} \ker \nabla_{(-2)}^z = 3 - 3g. \tag{14.113}$$

Now we have separated $\{\delta\gamma\}$ into mutually orthogonal pieces

$$\{\delta\gamma\} = \{\text{conf}\} \oplus \{im \, \nabla_z^{(-1)}\} \oplus \{\ker \nabla_{(-2)}^z\} + cc \tag{14.114}$$

which should be compared with (14.80b). The measure becomes

$$\mathcal{D}\delta\gamma \mathcal{D}\delta X \rightarrow J \, d^n \delta\tau \mathcal{D}\delta\bar{v} \mathcal{D}\delta\phi \mathcal{D}\delta\tilde{X} \, d^k \delta a \tag{14.115}$$

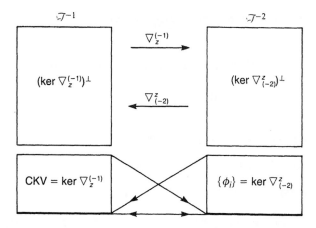

Figure 14.6. The map $\nabla_z^{(-)}$ and its adjoint $\nabla_{(-2)}^z$.

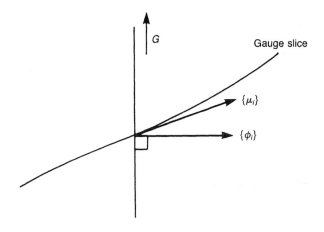

Figure 14.7. The Beltrami differential $\{\mu_i\}$ spans the deformation tangent to the gauge slice while $\{\phi_i\}$ spans $\ker \nabla_{(-2)}^z$.

where n and k are the complex dimensions of the Teichmüller space and the CKV, respectively. The Jacobian is obtained by repeating the argument in the previous subsection and we find that

$$
Z_g = \int \mathcal{D}\gamma\,\mathcal{D}X\,\frac{1}{V(\text{Diff*Weyl})}\,e^{-S}
$$
$$
= \int_{\text{Mod}} d^n\tau\,\mathcal{D}X\,\frac{\det{}' \Delta_{(1)}^+}{V(\text{CKV})}\,\frac{|\det(\mu,\phi)|^2}{\det(\phi,\phi)\det(\Phi,\Phi)}\,e^{-S}. \tag{14.116}
$$

Since we are integrating over complex variables, the power of a half in (14.96) does not appear in (14.116). The X-integration yields

$$Z_g = \int_{\text{Mod}} \frac{d^n \tau}{V(\text{CKV})} \frac{|\det(\mu, \phi)|^2}{\det(\phi, \phi) \det(\Phi, \Phi)}$$

$$\times \det' \Delta_{(1)}^+ \left(\frac{\det' \Delta}{\int d^2 z \sqrt{\gamma}} \right)^{-13}. \tag{14.117}$$

14.2.4 Moduli spaces of Riemann surfaces

The spaces $\text{Mod}(\Sigma_g)$ and $\text{Teich}(\Sigma_g)$ have been defined as

$$\text{Mod}(\Sigma_g) \equiv \mathcal{M}_g / \text{Diff}(\Sigma_g) \qquad \text{Teich}(\Sigma_g) \equiv \mathcal{M}_g / \text{Diff}_0(\Sigma_g).$$

They are related through MCG $\equiv$ $\text{Diff}(\Sigma_g)/\text{Diff}_0(\Sigma_g)$ as $\text{Mod}(\Sigma_g) = \text{Teich}(\Sigma_g)/\text{MCG}$. We look at these objects more closely here. We first note:

g	$\dim_{\mathbb{C}}$ CKV	CKV	$\dim_{\mathbb{C}}$ Teich(Σ_g)	MCG	
0	3	SL(2, C)	0	SL(2, R)	(14.118)
1	1	U(1) × U(1)	1	SL(2, Z)	
≥ 2	0	empty	$3g - 3$	?	

[*Remark*: MCG for $g \geq 2$ can be expressed by $3g - 1$ Dehn twists which are, however, not minimal.] From (14.118), we immediately conclude that $Z_0 = 0$ since the Teichmüller space is a single point and the volume of SL(2, $\mathbb{C}$) is infinite. Of course, this does not imply that the three amplitudes with vertex operators vanish. In general, $\text{Mod}(\Sigma_g)$ is topologically non-trivial although $\text{Teich}(\Sigma_g)$ is. $\text{Teich}(\Sigma_g)$ is a universal covering space of $\text{Mod}(\Sigma_g)$ and the topological non-triviality comes from MCG.

In actual computations, the uniformization theorem is very useful. In the previous subsection, we first chose the Beltrami differential μ_i, then changed the basis to $\phi_i \in \ker P_1^\dagger$. Our initial choice μ_i is motivated by the uniformization theorem.

Theorem 14.2. (**Uniformization theorem**) Let Σ_g be a torus with genus g. Then it is conformally related to the constant-curvature Riemann surface, which is given by the following:

g	Riemann surface	Metric	sign$\mathcal{R}$	
0	$\mathbb{C} \cup \{\infty\}$	$ds^2 = dz \otimes d\bar{z}/(1 + z\bar{z})^2$	+	(14.119)
1	$\mathbb{C}/L$	$ds^2 = dz \otimes d\bar{z}$	0	
≥ 2	H/G	$ds^2 = dz \otimes d\bar{z}/(\text{Im } z)^2$	−	

where L is a lattice in $\mathbb{C}$ (see example 8.2), H the upper half-plane and $G \subset$ SL(2, $\mathbb{R}$) is called the **Fuchsian group**. The metric for $g \geq 2$ is the **Poincaré metric**, see example 7.6.

The proof of this theorem is found in Farkas and Kra (1980), for example. Thanks to this theorem, we may always take constant-curvature metrics to form the gauge slice in $\mathcal{M}_g$. This corresponds to a special choice of the Beltrami differential μ_i. This slice defines the **Weil–Petersson measure**:

$$\int d^n \tau \, \frac{|\det(\mu, \phi)|^2}{\det(\phi, \phi)} = \int d(\text{Weil–Petersson}) \qquad (14.120)$$

see D'Hoker and Phong (1986).

Exercise 14.6. Compute the scalar curvature of the metrics given in (14.119). Verify that they are independent of z and $\bar{z}$.

14.3 One-loop amplitudes

As an illustration of the formalism developed in the previous section, we compute the one-loop vacuum-to-vacuum amplitude of the closed orientable bosonic string theory. Since $\dim_{\mathbb{C}} \text{Teich}(\Sigma_1) = 1$ and $\dim_{\mathbb{C}} \ker \nabla_z^{(-1)} = 1$, we have

$$Z_1 = \int_{\text{Mod}} \frac{d\tau}{V(\text{CKV})} \frac{|(\mu, \phi)|^2}{(\phi, \phi) \cdot (\Phi, \Phi)} \det' \Delta_{(1)}^+ \left(\frac{\det' \Delta}{\int d^2 \xi \sqrt{\gamma}} \right)^{-13}. \qquad (14.121)$$

To evaluate (14.121) we need to take several steps.

14.3.1 Moduli spaces, CKV, Beltrami and quadratic differentials

In example 8.2, we have shown that the complex structure, namely the conformal structure, of the torus is specified by a complex parameter τ ($\text{Im} \, \tau > 0$). Figure 8.3 shows the moduli space

$$\text{Mod}(\Sigma_g) = \mathcal{M}_1/G = \text{Teich}(\Sigma_g)/SL(2, \mathbb{Z}) = H/SL(2, \mathbb{Z})$$

where H is the upper half-plane.

Take the torus T_τ specified by the Teichmüller parameter $\tau = \tau_1 + i\tau_2$ ($\tau_2 > 0$). As a representative, we take a torus in figure 14.8. The metric in $\mathbb{C}$ naturally induces a flat metric (as guaranteed by the uniformization theorem)

$$\gamma = \tfrac{1}{2}[dz \otimes d\bar{z} + d\bar{z} \otimes dz]. \qquad (14.122)$$

The CKV are globally defined holomorphic vectors. We take $\Phi = \alpha \partial/\partial z$ as the normalized basis of the CKV. The condition $(\Phi, \Phi) = 1$ yields $\int d^2 z |\alpha|^2 = \tau_2 |\alpha|^2 = 1$, that is $\alpha = \tau_2^{-1/2}$ (we have dropped the phase). The vector Φ generates translations in the complex plane,

$$z \to z' = z + \tau_2^{-1/2}(v^1 + iv^2). \qquad (14.123)$$

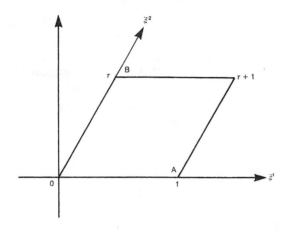

Figure 14.8. The parallelogram whose complex structure is parametrized by τ.

We must note, however, that the translation is defined modulo the lattice; $\tau_2^{-1/2}(v^1 + iv^2)$ and $\tau_2^{-1/2}(v^1 + iv^2) + (m + \tau n)$ yield the identical translation. This forces $\tau_2^{-1/2}(v_1 + iv^2)$ to lie within the parallelogram of figure 14.8. Since

$$\tau_2 = \int d^2 z = \tau_2^{-1} \int d^2 v$$

$V(\mathrm{CKV})$ is found to be

$$V(\mathrm{CKV}) = \int d^2 v = \tau_2^2. \qquad (14.124)$$

Our next task is to evaluate the Weil–Petersson measure. On the torus there is one quadratic differential ϕ. Since $\phi \in \mathcal{T}^{-2}$ is a globally defined holomorphic differential, it must be of the form,

$$\phi = a \, dz \otimes dz \qquad a \in \mathbb{C}. \qquad (14.125)$$

To find the Beltrami differential, we evaluate the change of the metric under a small variation of τ. For this purpose, it is convenient to introduce the ξ^α-coordinate system in figure 14.8. The point A corresponds to $(1, 0)$ and B to $(0, 1)$. Accordingly, we have $z = \xi^1 + \tau \xi^2$. Under a small change $\delta\tau$ of the Teichmüller parameter, we have, up to a conformal factor,

$$|dz|^2 \rightarrow |d\xi^1 + (\tau + \delta\tau)d\xi^2|^2 = |dz + \delta\tau d\xi^2|^2$$
$$= \left| dz + d\tau \frac{dz - d\bar{z}}{2i\tau_2} \right|^2 = \left| dz + \delta\tau \frac{id\bar{z}}{2\tau_2} \right|.$$

Comparing this with (14.110), we find that

$$\mu_{zz} = i/2\tau_2. \qquad (14.126)$$

Here $(\delta\tau)\mu$ is the complex conjugate of $(\delta\tau)\mu$ in (14.110). Of course, this is a reparametrization of the Teichmüller space and does not affect the results. If the reader feels awkward with this, s/he may choose $\bar{\tau}$ as the Teichmüller parameter. From (14.125) and (14.126), we have, up to irrelevant constants,

$$(\mu, \phi) = \int d^2z \, \overline{\mu^{zz}} \phi_{zz} = \frac{i}{2\tau_2} a\tau_2 \propto a$$

$$(\phi, \phi) = \int d^2z \, \overline{\phi^{zz}} \phi_{zz} = a^2\tau_2.$$

Finally, we have obtained

$$\frac{|(\mu, \phi)|^2}{(\phi, \phi)} = \tau_2^{-1}. \tag{14.127}$$

14.3.2 The evaluation of determinants

We first consider $\det' P_1^\dagger P_1 = \det' \Delta_{(1)}^+$. Since we take a flat metric, the Laplacian takes quite a simple form,

$$\Delta_{(1)}^+ = -2\partial_z \partial_{\bar{z}} = \Delta \tag{14.128}$$

where Δ is the Laplacian defined by (14.87b). Since

$$\int d^2\xi \, \sqrt{\gamma} = \int d^2z = \tau_2$$

the amplitude (14.121) reduces to

$$Z_1 = \int_{\text{Mod}} \frac{d\tau}{\tau_2^2} \frac{\det' \Delta}{\tau_2} \left(\frac{\det' \Delta}{\tau_2}\right)^{-13} \tag{14.129}$$

$$\begin{array}{ccc} \uparrow & \uparrow & \uparrow \\ V(\text{CKV}) & \text{W-P} & \int d^2z \end{array}$$

where we have used (14.124) and (14.127). We have factorized the integrand so that the modular invariance is manifest, see exercise 14.7.

Let us compute the spectrum of Δ. It is convenient to express the Laplacian in ξ^α-coordinates. From

$$\xi^1 = i(\bar{\tau}z - \tau\bar{z})/2\tau_2 \qquad \xi^2 = (z - \bar{z})/2i\tau_2 \tag{14.130}$$

we readily find that

$$\Delta = -\frac{1}{2\tau_2^2}[|\tau|^2(\partial_1)^2 - 2\tau_1\partial_1\partial_2 + (\partial_2)^2] \tag{14.131}$$

where $\partial_1 = \partial/\partial\xi^1$ etc. The eigenfunction satisfying the periodic boundary condition on the torus is

$$\psi_{m,n}(\xi) = \exp[2\pi i(n\xi^1 + m\xi^2)] \qquad (m, n) \in \mathbb{Z}^2. \tag{14.132}$$

Substituting this into (14.131), we find the eigenvalue

$$\lambda_{m,n} = \frac{2\pi^2}{\tau_2^2}(m - \tau n)(m - \bar{\tau} n).$$ (14.133)

The determinant is expressed as an infinite product:

$$\det' \Delta = \prod_{m,n}' \frac{2\pi^2}{\tau_2^2}|m + \tau n|^2$$ (14.134)

the product being taken for all integers $(m, n) \neq (0, 0)$.

Clearly $\det' \Delta$ is ill defined and needs to be regularized. Let us introduce the **Eisenstein series** (Siegel 1980, Lang 1987) defined by

$$E(\tau, s) \equiv \sum_{m,n}' \frac{\tau_2^s}{|m + \tau n|^{2s}}$$ (14.135)

the summation being taken for all integers $(m, n) \neq (0, 0)$. This series converges for Re $s > 1$ and can be analytically continued to the complex s-plane. The series $E(\tau, s)$ has a simple pole at $s = 1$ where we have a Laurent expansion,

$$E(\tau, s) = \frac{\pi}{s - 1} + 2\pi[\gamma - \ln 2 - \ln(\sqrt{\tau_2}|\eta(\tau)|^2)] + \mathcal{O}(s - 1).$$ (14.136)

This expression is known as the **Kronecker first limit formula** and is essential for our purposes. In (14.136), $\gamma = 0.57721\ldots$ is Euler's constant and $\eta(\tau)$ is the **Dedekind η-function**

$$\eta(\tau) \equiv e^{i\pi\tau/12} \prod_{n>1}(1 - e^{2i\pi n\tau}).$$ (14.137)

Neglecting constant factors, we have

$$\frac{\det' \Delta}{\tau_2} = \exp\left(-\ln\tau_2 + \sum'\ln\frac{|m + \tau n|^2}{\tau_2^2}\right)$$
$$= \exp\left(-\ln\tau_2 - \frac{\partial}{\partial s}[\tau_2^s E(\tau, s)]\Big|_{s=0}\right)$$
$$= \exp\{-\ln\tau_2[1 + E(\tau, 0)] - E'(\tau, 0)\}.$$ (14.138)

To evaluate the exponent, we note the functional equation,

$$\pi^{-s}\Gamma(s)E(\tau, s) = \pi^{-(1-s)}\Gamma(1 - s)E(\tau, 1 - s).$$ (14.139)

Taking the limit $s \to 0$ in (14.139), we have

$$sE(\tau, 1 - s) = \pi^{1-2s}\frac{\Gamma(1 + s)}{\Gamma(1 - s)}E(\tau, s)$$
$$= \pi(1 - 2s\ln\pi + \cdots)\frac{(1 - \gamma s + \cdots)}{(1 + \gamma s + \cdots)}[E(\tau, 0) + E'(\tau, 0)s + \cdots]$$
$$= \pi E(\tau, 0) + [-2(\ln\pi + \gamma)E(\tau, 0) + E'(\tau, 0)]\pi s + \cdots.$$

From (14.136), we also have

$$sE(\tau, 1 - s) = -\pi + 2\pi s[\gamma - \ln 2 - \ln(\sqrt{\tau_2}|\eta(\tau)|^2)] + \cdots.$$

Equating the coefficients of s^0 and s^1, we find that

$$E(\tau, 0) = -1 \tag{14.140a}$$

$$E'(\tau, 0) = -2[\ln 2\pi + \ln(\sqrt{\tau_2}|\eta(\tau)|^2)]. \tag{14.140b}$$

Substituting (14.140) into (14.138), we obtain

$$\frac{\det' \Delta}{\tau_2} = \exp[-E'(\tau, 0)] = \tau_2|\eta(\tau)|^4. \tag{14.141}$$

Finally, it follows from (14.129) and (14.141) that

$$Z_1 = \int_{\text{Mod}} \frac{d\tau}{\tau_2^2} \tau_2^{-12}|\eta(\tau)|^{-48}. \tag{14.142}$$

A neat form of Z_1 is obtained if we define the **discriminant**

$$\Delta(\tau) \equiv (2\pi)^{12}\eta(\tau)^{24}. \tag{14.143}$$

Up to an irrelevant constant, the one-loop amplitude is

$$Z_1 = \int_{\text{Mod}} \frac{d\tau}{\tau_2^2} \tau_2^{-12}|\Delta(\tau)|^{-2}. \tag{14.144}$$

$\Delta(\tau)$ is known as the **cusp form** of weight 12, implying

$$\Delta\left(\frac{a\tau + b}{c\tau + d}\right) = (c\tau + d)^{12}\Delta(\tau) \tag{14.145}$$

and $c(0) = 0$, where the $c(n)$ are the Fourier coefficients,

$$\Delta(\tau) = \sum_{n \geq 0} c(n)e^{2\pi n i\tau}. \tag{14.146}$$

Higher genus amplitudes are given by the cusp forms of other weights, see Belavin and Knizhnik (1986), Moore (1986), Gilbert (1986) and Morozov (1987).

Exercise 14.7. Show that

$$\eta(\tau + 1) = e^{\pi i/12}\eta(\tau) \qquad \eta(-1/\tau) = (-i\tau)^{1/2}\eta(\tau) \tag{14.147}$$

where the branch is chosen so that $\sqrt{z} > 0$ if $z > 0$. Use this result to show that $d\tau/\tau_2^2$ and $\tau_2^{-12}|\eta(\tau)|^{-48}$ are independently invariant under $\tau \to \tau + 1$ and $\tau \to -1/\tau$.

REFERENCES

Adler S L 1969 *Phys. Rev.* **177** 2426

Aitchison I J R 1987 *Acta Phys. Pol.* B **18** 207

Alvarez O 1985 Topological methods in field theory *Berkeley Preprint* UCB-PTH-85/43

Alvarez O 1995 *Geometry and Quantum Field Theory* ed D S Freed *et al* (Providence, RI: American Mathematical Society) p 271

Alvarez-Gaumé L 1983 Commun. Math. Phys. **90** 161

——1986 *Fundamental Problems of Gauge Field Theory (Erice, 1985)* ed V Gelo and A S Wightman (New York: Plenum)

Alvarez-Gaumé L and Della Pietra S 1985 *Recent Developments in Quantum Field Theory* ed J Ambjørn *et al* (Amsterdam: Elsevier) p 95

Alvarez-Gaumé L, Della Pietra S and Moore G 1985 *Ann. Phys., NY* **163** 288

Alvarez-Gaumé L and Ginsparg P 1984 *Nucl. Phys.* B **243** 449

——1985 *Ann. Phys., NY* **161** 423

Alvarez-Gaumé L and Nelson P 1986 *Supersymmetry, Supergravity, and Superstrings '86* ed B de Wit and M Grisaru (Singapore: World Scientific)

Anderson P W and Brinkman W F 1975 *The Helium Liquids* ed J G M Armitage and I E Farquhar (New York: Academic) p 315

Anderson P W and Toulouse G 1977 *Phys. Rev. Lett.* **38** 408

Armstrong M A 1983 *Basic Topology* (New York: Springer)

Atiyah M F 1985 *Arbeitstagung Bonn 1984* ed F Hirzebruch, J Schwermer and S Suter (Berlin and Heidelberg: Springer) p 251

Atiyah M F and Jones J D S 1978 *Commun. Math. Phys.* **61** 97

Atiyah M F, Patodi V and Singer I M 1975a *Math. Proc. Camb. Phil. Soc.* **77** 43

——1975b *Math. Proc. Camb. Phil. Soc.* **77** 405

——1976 *Math. Proc. Camb. Phil. Soc.* **79** 71

Atiyah M F and Segal G B 1968 *Ann. Math.* **87** 531

Atiyah M F and Singer I M 1968a *Ann. Math.* **87** 485

——1968b *Ann. Math.* **87** 546

——1984 *Proc. Natl Acad. Sci., USA* **81** 2597

Bagger J 1987 *The Santa Fe TASI-87* ed R Slansky and G West (Singapore: World Scientific)

Bailin D and Love A L 1996 *Introduction to Gauge Field Theory* revised edn (Bristol and New York: Adam Hilger)

Bardeen W A 1969 *Phys. Rev.* **184** 1848

Belavin A A and Knizhnik V G 1986 *Sov. Phys.–JETP* **64** 214

Belavin A A and Polyakov A M 1975 *JETP Lett.* **22** 245

Belavin A A, Polyakov A M, Schwartz A S and Tyupkin Yu S 1975 *Phys. Lett.* B **59** 85

Bell J and Jackiw R 1969 *Nuovo Cimento* A **60** 47

Berezin F A 1966 *The Method of Second Quantization* (New York and London: Academic)

Berry M 1984 *Proc. R. Soc.* A **392** 45

——1989 *Principles of Cosmology and Gravitation* 2nd edn (Bristol: Adam Hilger)

Bertlmann R A 1996 *Anomalies in Quantum Field Theory* (Oxford: Oxford University Press)

Booss B and Bleecker D D 1985 *Topology and Analysis: The Atiyah–Singer Index Formula and Gauge-Theoretic Physics* (New York: Springer)

Bott R and Seeley R 1978 *Commun. Math. Phys.* **62** 235

Bott R and Tu L W 1982 *Differential Forms in Algebraic Topology* (New York: Springer)

Buchholtz L J and Fetter A L 1977 *Phys. Rev.* B **15** 5225

Calabi E 1957 *Algebraic Geometry and Topology: A Symposium in Honor of S Lefschetz* (Princeton, NJ: Princeton University Press)

Callias C 1978 *Commun. Math. Phys.* **62** 213

Candelas P 1988 *Superstrings '87* (Singapore: World Scientific)

Cheng T-P and Li L-F 1984 *Gauge Theory of Elementary Particle Physics* (New York and Oxford: Oxford University Press)

Chern S S 1979 *Complex Manifolds without Potential Theory* 2nd edn (New York: Springer)

Choquet-Bruhat Y and DeWitt-Morette C with Dillard-Bleick M 1982 *Analysis, Manifolds and Physics* revised edn (Amsterdam: North-Holland)

Coleman S 1979 *The Whys of Subnuclear Physics* ed A Zichichi (New York: Plenum)

Crampin M and Pirani F A E 1986 *Applicable Differential Geometry* (Cambridge: Cambridge University Press)

Croom F H 1978 *Basic Concepts of Algebraic Topology* (New York: Springer)

Daniel M and Viallet C M 1980 *Rev. Mod. Phys.* **52** 175

Das A 1993 *Field Theory* (Singapore: World Scientific)

Deser S, Jackiw R and Templeton S 1982a *Phys. Rev. Lett.* **48** 975

——1982b *Ann. Phys., NY* **140** 372

D'Hoker E and Phong D 1986 *Nucl. Phys.* B **269** 205

——1988 *Rev. Mod. Phys.* **60** 917

Dirac P A M 1931 *Proc. R. Soc.* A **133** 60

Dixon L, Harvey J, Vafa C and Witten E 1985 *Nucl. Phys.* B **261** 678

——1986 *Nucl. Phys.* B **274** 285

Dodson C T J and Poston T 1977 *Tensor Geometry* (London: Pitman)

Donaldson S K 1983 *J. Diff. Geom.* **18** 279

Eells J and Lemaire L 1968 *Bull. London Math. Soc.* **10** 1

Eguchi T, Gilkey P B and Hanson A J 1980 *Phys. Rep.* **66** 213

Farkas H M and Kra I 1980 *Riemann Surfaces* (New York: Springer)

Federbush P 1987 *Bull. Am. Math. Soc. (N.S.)* **17** 93

Flanders H 1963 *Differential Forms with Applications to the Physical Sciences* (New York: Academic, reprint Dover)

Forte S 1987 *Nucl. Phys.* B **288** 252

Fraleigh J B 1976 *A First Course in Abstract Algebra* (Reading, MA: Addison-Wesley)

Freed D S and Uhlenbeck K 1984 *Instantons and Four-Manifolds* (New York: Springer)

Friedan D and Windey P 1984 *Nucl. Phys.* B **235** 395

——1985 *Physica* D **15** 71

Frödlicher A 1955 *Math. Ann.* **129** 50

Fujikawa K 1979 *Phys. Rev. Lett.* **42** 1195

——1980 *Phys. Rev.* D **21** 2848; *Phys. Rev.* D **22** 1499(E)

——1986 in *Superstrings, Supergravity and Unified Theories* ed G Furlan *et al* (Singapore: World Scientific) p 230

Gilbert G 1986 *Nucl. Phys.* B **277** 102

Gilkey P B 1995 *Invariance Theory, the Heat Equation and the Atiyah–Singer Index Theorem* 2nd edn (Boca Raton, FL: Chemical Rubber Company)

Goldberg S I 1962 *Curvature and Homology* (New York: Academic)

Green M B, Schwarz J H and Witten E 1987 *Superstring Theories* vols I and II (Cambridge: Cambridge University Press)

Greenberg M J and Harper J R 1981 *Algebraic Topology: A First Course* (Reading, MA: Benjamin/Cummings)

Greene R E 1987 *Differential Geometry (Lecture Notes in Mathematics 1263)* ed V L Hansen (Berlin and Heidelberg: Springer) p 228

Griffiths P and Harris J 1978 *Principles of Algebraic Geometry* (New York: Wiley)

Gross D J and Jackiw R 1972 *Phys. Rev.* D **6** 477

Gunning R C 1962 *Lectures on Modular Forms* (Princeton, NJ: Princeton University Press)

Hawking S 1977 *Commun. Math. Phys.* **55** 133

Hicks N 1965 *Notes on Differential Geometry* (Princeton, NJ: Van Nostrand)

Hirayama M 1983 *Prog. Theor. Phys.* **70** 1444

Hirzebruch F 1966 *Topological Methods in Algebraic Geometry* 3rd edn (Berlin and Heidelberg: Springer)

Horowitz G 1986 *Unified String Theories* ed M Green and D Gross (Singapore: World Scientific) p 635

Huang K 1982 *Quarks, Leptons and Gauge Fields* (Singapore: World Scientific)

Ito K (ed) 1987 *Encyclopedic Dictionary of Mathematics* 3rd edn (Cambridge, MA: MIT Press)

Jackiw R and Rebbi C 1977 *Phys. Rev.* D **16** 1052

Jackiw R and Templeton S 1981 *Phys. Rev.* D **23** 2291

Kleinert H 1990 *Path Integrals* (Singapore: World Scientific)

Kobayashi S 1984 *Introduction to the Theory of Connections* (Yokohama: Department of Mathematics, Keio University) (in Japanese)

Kobayashi S and Nomizu K 1963 *Foundations of Differential Geometry* vol I (New York: Interscience)

——1969 *Foundations of Differential Geometry* vol II (New York: Interscience)

Koblitz N 1984 *Introduction to Elliptic Curves and Modular Forms* (New York: Springer)

Kulkarni R S 1975 *Index Theorems of Atiyah–Bott–Patodi and Curvature Invariants* (Montréal: Les Presses de l'Université de Montréal)

Lang S 1987 *Elliptic Functions* 2nd edn (New York: Springer)

Leggett A J 1975 *Rev. Mod. Phys.* **47** 331

Leinaas J M and Olaussen K 1982 *Phys. Lett.* B **108** 199

Lightman A P, Press W H, Price R H and Teukolsky S A 1975 *Problem Book in Relativity and Gravitation* (Princeton, NJ: Princeton University Press)

Longuet-Higgins H C 1975 *Proc. R. Soc.* A **344** 147

Maki K and Tsuneto T 1977 *J. Low-Temp. Phys.* **27** 635

Matsushima Y 1972 *Differentiable Manifolds* (New York: Dekker)

Mermin N D 1978 in *Quantum Liquids* ed J Ruvalds and T Regge (Amsterdam: North-Holland) p 195

——1979 *Rev. Mod. Phys.* **51** 591

Mermin N D and Ho T-L 1976 *Phys. Rev. Lett.* **36** 594

Milnor J 1956 *Ann. Math.* **64** 394

Mickelsson J 1989 *Current Algebras and Groups* (New York: Plenum)

Milnor J W and Stasheff J D 1974 *Characteristic Classes* (Princeton, NJ: Princeton University Press)

Minami S 1979 *Prog. Theor. Phys.* **62** 1128

Mineev V P 1980 *Sov. Sci. Rev.* A **2** 173

Misner C W 1978 *Phys. Rev.* D **18** 4510

Misner C W, Thorne K S and Wheeler J A 1973 *Gravitation* (San Francisco, CA: Freeman)

Moore G 1986 *Phys. Lett.* B **176** 369

Moore G and Nelson P 1986 *Nucl. Phys.* B **266** 58

Morozov 1987 *Sov. J. Nucl. Phys.* **45** 181

Nakahara M 1998 *Path Integrals and Their Applications* (Tokyo: Graduate School of Mathematical Sciences, University of Tokyo)

Nambu Y 1970 *Lectures at the Copenhagen Symposium* unpublished

Nash C 1991 *Differential Topology and Quantum Field Theory* (London: Academic)

Nash C and Sen S 1983 *Topology and Geometry for Physicists* (London: Academic)

Newlander A and Nirenberg L 1957 *Ann. Math.* **65** 391

Nomizu K 1981 *Introduction to Modern Differential Geometry* (Tokyo: Shokabo) (in Japanese)

Palais R S 1965 *Seminars on the Atiyah–Singer Index Theorem* (Princeton, NJ: Princeton University Press)

Polchinski J 1986 *Commun. Math. Phys.* **104** 37

Polyakov A M 1981 *Phys. Lett.* B **103** 207

Price R H 1982 *Am. J. Phys.* **50** 300

Rabin J M 1995 *Geometry and Quantum Field Theory* ed D S Freed *et al* (Providence, RI: American Mathematical Society) p 183

Ramond P 1989 *Field Theory: A Modern Primer* 2nd edn (Reading, MA: Benjamin/Cummings)

Rennie R 1990 *Adv. Phys.* **39** 617

Ryder L H 1980 *J. Phys. A: Math. Gen.* **13** 437

——1996 *Quantum Field Theory* 2nd edn (Cambridge: Cambridge University Press)

Sakita B 1985 *Quantum Theory of Many-Variable System and Fields* (Singapore: World Scientific)

Sánchez N 1988 *Harmonic Mappings, Twistors, and a-Models* ed P Gauduchon (Singapore: World Scientific) p 270

Sattinger D H and Weaver O L 1986 *Lie Groups and Algebras with Applications to Physics, Geometry, and Mechanics* (New York: Springer)

Scherk J 1975 *Rev. Mod. Phys.* **47** 123

Schutz B F 1980 *Geometrical Methods of Mathematical Physics* (Cambridge: Cambridge University Press)

Schwartz L 1986 *Lectures on Complex Analytic Manifolds* (Berlin and Heidelberg: Springer)

Shanahan P 1978 *The Atiyah–Singer Index Theorem: An Introduction* (Berlin and Heidelberg: Springer)

Shankar R 1977 *J. Physique* **38** 1405

Siegel C L 1980 *Advanced Analytic Number Theory* (Bombay: Tata Institute of Fundamental Research)

Simon B 1983 *Phys. Rev. Lett.* **51** 2167

Singer I M 1985 *Soc. Math. de France, Astérisque* hors série 323

Steenrod N 1951 *The Topology of Fibre Bundles* (Princeton, NJ: Princeton University Press)

Stora R 1984 in *Progress in Gauge Field Theory* ed G 't Hooft *et al* (New York: Plenum) p 543

Sumitani T 1984 *J. Phys. A: Math. Gen.* **17** L811

——1985 *MSc Thesis Soryushiron-Kenkyu* **71** 65 (in Japanese)

Swanson M S 1992 *Path Integrals and Quantum Processes* (Boston, MA: Academic)

Tonomura A, Umezaki H, Matsuda T, Osakabe N, Endo J and Sugita Y 1983 *Phys. Rev. Lett.* **51** 331

Toulouse G and Kléman M 1976 *J. Physique Lett.* **37** L149

Trautman A 1977 *Int. J. Theor. Phys.* **16** 561

Tsuneto T 1982 *The Structure and Properties of Matter* ed T Matsubara (Berlin: Springer) p 101

Wald R M 1984 *General Relativity* (Chicago, IL: The University of Chicago Press)

Warner F W 1983 *Foundations of Differentiable Manifolds and Lie Groups* (New York: Springer)

Weinberg S 1972 *Gravitation and Cosmology: Principles and Applications of the General Theory of Relativity* (New York: Wiley)

——1988 *Strings and Superstrings: Jerusalem Winter School for Theoretical Physics* ed S Weinberg (Singapore: World Scientific)

Wells R O 1980 *Differential Analysis on Complex Manifolds* (New York: Springer)

Wess J and Zumino B 1971 *Phys. Lett.* B **37** 95

Whitehead G W 1978 *Elements of Homotopy Theory* (New York: Springer)

Wu T T and Yang C N 1975 *Phys. Rev.* D **12** 3845

Yang C N and Mills R L 1954 *Phys. Rev.* **96** 191

Utiyama R 1956 *Phys. Rev.* **101** 1597

Yau S-T 1977 *Proc. Natl Acad. Sci., USA* **74** 1798

Zumino B 1985 *Relativity, Groups and Topology II* vol 3, ed B S DeWitt and R Stora (Amsterdam: North-Holland) p 1291

——1987 Geometry and physics *Berkeley Preprint* UCB/pTH-87/13

Zumino B, Wu Y-S and Zee A 1984 *Nucl. Phys.* B **239** 477

INDEX